Water Science and Application 5

Ancient Floods, Modern Hazards

Principles and Applications of Paleoflood Hydrology

P. Kyle House
Robert H. Webb
Victor R. Baker
Daniel R. Levish
Editors

With accompanying CD-ROM compiled and designed by P. Kyle House and Larry Mayer

American Geophysical Union
Washington, DC

Ancient Floods, Modern Hazards: Principles and Applications of Paleoflood Hydrology
Water Science and Application 5

Library of Congress Cataloging-in-Publication Data
Ancient floods, modern hazards : principles and applications of paleoflood hydrology / P. Kyle House ... [et al.], editors ; with accompanying CD-ROM compiled and designed by Larry Mayer and P. Kyle House.
p. cm. – (Water science and application ; 5)
ISBN 0-87590-354-1
1. Paleohydrology. 2. Floods. I. House, P. Kyle. II. Series.

QE39.5.P27 A53 2001
551.48'9–dc21 2001053365

ISBN 0-87590-354-1
ISSN 1526-758X

Cover: Upper portion of the Dogleg Bar reach of the Owyhee River Canyon, southeastern Oregon. The Dogleg bar (lower right) is the lowest of three discrete river terraces, each composed predominantly of volcanic boulders deposited by extreme floods. The two higher terraces are discernible on the opposite side of the river in the right and center of the photo. Photograph by Kyle House, May 2001.

Printed in the United States of America.

CONTENTS

CONTENTS

Perspective

PREFACE

Flooding is the most pervasive and costly natural hazard regularly faced by society. Extreme floods constitute significant agents of physical change on portions of the Earth's surface and leave diagnostic imprints in the geologic record, both ancient and recent. Most large floods result from the integrated effects of complex, multi-scaled interactions in Earth's atmosphere. Flood records from gaging stations typically have insufficient length to adequately characterize the actual temporal context of hydrologic extremes like large floods. The foregoing statements have a common theme—paleoflood hydrology—an interdisciplinary scientific method of reconstructing the history of flooding on rivers and streams from the basis of physical evidence of flooding. From a societal perspective, paleoflood hydrology provides a direct record of the types of floods that are most likely to cause the greatest damage. As a result, flood plain managers and regulatory agencies should benefit greatly from the research and interpretations that paleoflood hydrology offers.

Paleoflood hydrology uses principles of geology, hydrology, and fluid dynamics to exploit the fact that floods can leave long-lived physical evidence in their wake. The primary goal in paleoflood hydrology is to extend flood chronologies over periods of time ranging from centuries to millennia. Paleoflood chronologies can form the basis for proxy chronologies of flood producing events, provide insight into changes in river channel morphology, and serve as a standard for comparison of modeled conceptions of floods and flood hazards on specific rivers and streams. Paleoflood hydrology can make its greatest contribution to society in the applied domain of flood-risk evaluation and hazard management. However, the explicit and effective assimilation of paleoflood hydrology into the realm of flood-hazard management remains to be fully realized.

This book provides a comprehensive overview of the present state of knowledge in paleoflood hydrology and offers new perspectives on its applicability to a variety of related subjects. It is not intended to serve as an explicit "how-to" manual on paleoflood hydrology as achieved by the book *Flood Geomorphology* by Baker et al. published in 1988. Instead, the present volume, which offers an in-depth perspective on specific aspects of the field and their applications, can serve as a framework for future developments and applications. The scope of information presented here will satisfy both the investigator who wishes to learn basic modalities of paleoflood research and the experienced investigator seeking conceptual and methodological insights as guidance and inspiration. Without question, the collective bibliography of each of the chapters will serve as an authoritative and comprehensive reference list for those scientists wishing to delve deeper into the subject. Overall, the book emphasizes recent, significant improvements in the methodological and conceptual basis for paleoflood studies, provides quantitative and statistical tests of the common assumptions that underlie paleoflood research, and presents the results of recent applications and variations on basic methods and concepts.

Each contributed chapter represents an outgrowth of papers and discussions at the Second International Paleoflood Conference held in Prescott, Arizona (September 27-29, 1999). This conference, dedicated to presenting the state-of-the-art in paleoflood hydrology, was organized with sponsorship from the U.S. Bureau of Reclamation, the Hydrological Sciences Program of the National Science Foundation, the USGS Western Regional Competitive Grants Program in Water Resources Research, the INQUA Global Commission on Continental Paleohydrology (GLOCOPH), the Desert Research Institute, the Nevada Bureau of Mines and Geology, and the Arizona Geological Survey. Each paper in the book has been subjected to independent peer and editorial review and, in some cases, offers only a glimpse of the breadth of presentations at the conference. A pool of 45 reviewers (27 of whom were entirely external to the monograph) donated their time and expertise to enable us to complete this process, and we gratefully acknowledge their assistance. The compilation of this book would not have been possible without the persistence and dedication of each contributing author, and we thank them all. We would also like to acknowledge the efforts expended on behalf of this book and the conference from which it was developed; in particular, Terri Garside, Jack Hursh, Kris Pizarro, and Susan Tingley, all of the Nevada Bureau of Mines and Geology, helped in innumerable ways.

One technical note about this book:

We include a CD-ROM that contains figures and data from selected papers, largely to allow the authors to provide ancillary information for their chapters, such as digital images from the chapter, supplementary images not in the text, and data tables. In some instances, the CD provides some color figures that give additional information yet were too expensive to reproduce within the text. The inclusion of the CD is not necessarily essential to the content of the monograph, but it is an important supplement that has allowed some authors to share their data and to include color images as a component of their chapter at minimal expense.

P. Kyle House
University of Nevada
Reno, Nevada

Robert H. Webb
U. S. Geological Survey
Tucson, Arizona

Victor R. Baker
University of Arizona
Tucson, Arizona

Daniel R. Levish
U. S. Bureau of Reclamation
Denver, Colorado

CD-ROM INFORMATION

The CD-ROM that accompanies this book contains figures and data from selected papers, including digital images in text, supplementary images not in text, data tables, and color figures. Please note this convention for CD-ROM figures throughout the book:

- In chapters where authors contributed all figures as printed for inclusion on the CD, the first call-out in the text will include the statement "also on CD."
- If the figure is black and white in text but has a corresponding color version on the CD, it will be called out as a "plate" followed by the statement "for color version see CD" or something similar.
- Where the CD contains figures that are supplementary (but not essential) to the text, they are called-out as "CD Figures" or "CD Plates" depending on their status as black and white or color, respectively.

The Scientific and Societal Value of Paleoflood Hydrology

Victor R. Baker

University of Arizona, Tucson, Arizona

Robert H. Webb

U.S. Geological Survey, Tucson, Arizona

P. Kyle House

Nevada Bureau of Mines and Geology, Reno, Nevada

Flood damages in the United States continue to increase despite increased spending for flood control and mitigation, suggesting that new methods for estimating and characterizing flood hazards would be valued by society. Paleoflood hydrology was developed in the 1970s as a way to understand the magnitude of extreme flooding in central Texas and has evolved considerably in the last 30 years to become an important scientific endeavor with broad scientific and social relevance. The interdisciplinary methods of paleoflood hydrology have evolved to overcome difficulties, real and perceived, involving (1) inaccuracies in estimating the ages of floods, (2) inaccuracies in reconstructing flood discharges, (3) lack of robust statistical methods for incorporating paleoflood data in flood-frequency analysis, and (4) the effects of climatic shifts and nonstationarity. Of these, only the fourth represents a real problem that is faced by both conventional flood hydrology and paleoflood hydrology. Its resolution requires further collection of flood data, both from gaging stations on unregulated rivers and paleoflood sites. The basic techniques of paleoflood hydrology lead to the use of paleoflood data in flood-hazard analysis, both in terms of traditional flood-frequency analysis and the potential use of flood history within the context of flood-hazard assessment. Because paleoflood data reduce the uncertainty in estimates of long return-period floods, use of this technique offers substantial societal benefit, particularly in the design or retrofitting of dams or other significant floodplain structures. Improving the general understanding of hydroclimatic effects on flood frequency represents one of the most important contributions that paleoflood hydrology can make in the future. With continued methodological and conceptual refinement, paleoflood hydrology has the potential to guide important changes in the traditional paradigms of flood-frequency analysis and flood-hazard assessment.

Ancient Floods, Modern Hazards: Principles
and Applications of Paleoflood Hydrology
Water Science and Application Volume 5, pages 1-19

INTRODUCTION

Paleoflood hydrology is the science of reconstructing the magnitude and frequency of large floods using geological evidence and a variety of interdisciplinary techniques. Within the last 30 years, paleoflood hydrology has emerged as an important and highly relevant component of earth science with numerous applications to the understanding of flood occurrences and the evaluation of flood hazards. Paleofloods are events generally recorded outside of gaging records, or they are historical floods that can be placed within a long-term historical or geological context. They can be extremely ancient; for example, there is plentiful evidence for catastrophic floods preserved in regions of Earth affected by glaciation in the late Pleistocene, and there is clear evidence of catastrophic floods from billions of years ago preserved on the surface of Mars [*Baker*, 2001]. Nevertheless, the paleofloods of greatest concern to humans are those associated with climatic conditions most like the present. Such floods have generally occurred in Holocene, or approximately the last 10,000 years.

The science of extreme floods has been revolutionized in recent decades by two major advances. First, computational methods have allowed the modeling of the diverse range of phenomena envisioned for extreme flooding, and second, a whole world of data on extreme floods has been opened up by the rapid development of scientific techniques of paleoflood hydrology.

The purpose of this chapter is to review the evolution of paleoflood hydrology as a distinct scientific field. We describe its basic principles (with particular reference to chapters in this monograph in addition to seminal contributions in the field) and discuss the need for paleoflood techniques and data to become more widely integrated into the realm of conventional hydrology and engineering. In this vein, we discuss the problem of persistently increasing flood damages in the United States despite unprecedented computational advances in conventional flood hydrology, how the techniques of paleoflood hydrology were developed to address the estimation of extreme floods, and how paleoflood data can be efficiently integrated into flood-frequency analysis. We stress the value of paleoflood data in a societal context, given that this type of data can improve confidence in estimates of flood hazard and potentially reduce damages from flooding.

FLOOD DAMAGES IN THE UNITED STATES

The great paradox of flood science is that, despite immense expenditures on flood-hazard prevention and advances in statistical methods for evaluating flood hazards, damages from floods continue to rise at an alarming rate in the United States [*Pielke and Downton*, 2000]. In real dollars (not adjusted for inflation), flood damages between 1903 and 1999 peaked with the floods associated with the 1993 El Niño and its effects on California, Arizona, and states along the Mississippi River. The 25-year running mean of flood damages has risen from \$0.043 to \$3.15 billion dollars per year (Figure 1a). Even adjusted for inflation to 1997 dollars, flood damages have increased dramatically (Figure 1b).

Flood damages are increased by a number of factors not related to estimation of the magnitude and frequency of flooding. Some flood damages are by design; for example, farmers may lose crops grown deliberately in floodways that are designed to be inundated infrequently during major floods. *Pielke and Downton* [2000] attribute the increases in flood damage to increased precipitation amounts, which implies climatic nonstationarity (see later sections). In many cases, however, flood damages are high because zoning to minimize flood damages is inadequate [*van der Vink et al.*, 1998].

The basic flaw of flood science is that conventional methods for estimating flood hazard, while rigorous statistically, lack understanding of the flood process and typically are based on less than 100 years of gaging data. One of the largest benefits of paleoflood hydrology is the reconstruction of flood discharges over thousands of years, which establishes the long-term context of short hydrologic records and thereby greatly increases our ability to estimate flood hazards.

A BRIEF HISTORY OF PALEOFLOOD HYDROLOGY

Less than two decades have passed since the introduction of the term "paleoflood hydrology" [*Kochel and Baker*, 1982]. Nearly all the conventional hydrological uses of paleoflood data were developed during this period [*e.g.*, *Kochel and Baker*, 1982, 1988; *Stedinger and Baker*, 1987]. Paleofloods are floods that occurred in the past but were not recorded, either by stream-gaging stations or by other indirect means of human observation or measurement. Paleofloods are closely related to, but can be distinguished from, floods that occurred historically with observed stages; these are termed "historical floods." In general, paleofloods are indicated by the effects of their passage that remain preserved long enough for later interpretation. A flood occurring in a remote area, without a gaging record or human observation, can be described as a paleoflood even though it occurred very recently because the magnitude of the flood is reconstructed using geological evidence.

One of the first descriptions of what we now interpret as slackwater flood deposits (SWD) was made by *Tarr* [1892], who worked along the Colorado River in central Texas. In 1922, J.E. Stewart recognized the importance of such sediments as paleostage indicators (PSI) in a study of the Skagit River in Washington [*Stewart and Bodhaine*,

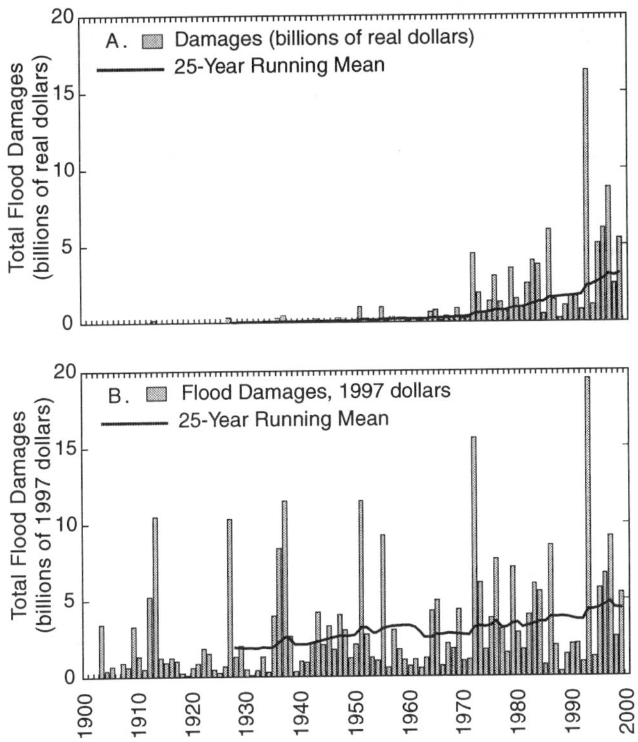

Figure 1. Flood damages in the United States, 1903-1999. A. Damages in billions of real dollars. B. Damages in billions of dollars adjusted to 1997. [*Pielke and Downton*, 2000; data source: http://www.nws.noaa.gov/oh/hic/flood_stats/Flood_loss_time_series.htm]

1961; *Costa*, 1987a]. Similar studies of historical flood deposits were conducted for the Ohio [*Mansfield*, 1938] and Connecticut Rivers [*Jahns*, 1947]. However, the cataclysmic late Pleistocene Missoula flood deposits of eastern Washington [*Bretz*, 1929; *Bretz et al.*, 1956] provided the greatest stimulus to paleoflood studies in the United States [*Costa*, 1987a; *Patton*, 1987]. Research on Pleistocene floods in the United States continues to this day [*e.g.*, *Fenton et al.*, this volume; *Knudsen et al.*, this volume].

Working in the same region studied by *Tarr* [1892], and applying experience gained by studies of the late Pleistocene Missoula floods [*Baker*, 1973], the major concepts of SWD-PSI paleoflood hydrology were mostly formulated during the 1970s. *Baker* [1974, p. 63] observed that floods in central and west Texas deposit sediment in eddies that form at the mouths of tributaries, and the elevations of the deposits provide minimum stages of past floods. *Patton et al.* [1979] and *Kochel et al.* [1982] improved and added to the techniques. The 10,000-year paleoflood record of the Pecos River in west Texas was incorporated into a flood-frequency analysis [*Patton and Baker*, 1977; *Baker et al.*, 1979], the first augmentation of a flood record with paleoflood data. By the early 1980s, a new, innovative and considerably longer viewpoint was possible in flood hydrology.

In 1976, the newly developing subdiscipline of SWD-PSI paleoflood hydrology was thought to have the following problems: (1) it was presumed impossible to determine the ages of ancient floods to a accuracy desirable for flood-frequency analysis, (2) it was unlikely that the discharge of an ancient flow event could be estimated accurately, (3) there was no appropriate statistical procedure that would incorporate paleoflood data to make accurate estimates of the frequency of large floods, and (4) past changes in climate and(or) land-use practices would invalidate any flood-frequency analysis that incorporated paleoflood data to address future flood hazard. These types of criticisms persisted into the mid- to late-1980s [*Hosking and Wallis*, 1986a, 1986b; *Yevjevich and Harmancioglu*, 1988].

These initial and, to a lesser extent, continuing criticisms of presumed problems with paleoflood hydrology are inevitable for a young and evolving discipline. In fact, early criticism of paleoflood techniques prompted further innovation and intellectual development of its concepts and applications. The intensity of criticism has waned overall as the basic principles of the discipline have been repeatedly tested and proven in a multitude of settings and contexts. Nonetheless, this brief review of the basic criticisms helps to frame the context of the development of modern paleoflood hydrology as well as to illuminate some misconceptions that persist to the present day.

Many critics of the techniques appear to be unfamiliar with basic concepts of geology as they are applied to paleoflood hydrology, and they are overconfident in the assumed accuracy/validity of more traditional methods. As an example, criticism (1) reflects misconceptions about the accuracy of modern geochronology.

Criticism (2) is curious because it is the conventional stream gages that have great difficulty in accurately recording extreme floods. Streamflow gaging stations are established and rated most reliably for common, relatively small discharges. The very rare flood usually does not occur in a gaging record, and when it does, its direct measurement is usually compromised by complex hydraulics and flood damage (*e.g.*, Figure 2), and the flood magnitude is computed using indirect methods that are typically used in paleoflood magnitude determination [*Webb and Jarrett*, this volume]. Although the high energy of large floods may damage or destroy a gaging station, their effects on the landscape and sedimentation are such that the resulting paleoflood record stands the most chance of long-term preservation. Advances in hydraulic modeling [*Webb and Jarrett*, this volume; *Denlinger et al.*, this volume] make the methods used for paleodischarge estimation equivalent to that at gaging stations.

Figure 2. Photographs showing gaging stations compromised during large floods. A. The Santa Ynez River, southern California, showing the high-water line of the January 25, 1969, flood of an estimated 2,266 m^3/s [see *Levish et al.*, 1997]. Remnants of the stilling well (arrow) for the Crawford Reach stream gage, 1.6 km downstream of Bradbury Dam, lies in the high-water line of the flood (Dan Levish). B. The Verde River, Arizona, at the Tangle Creek gaging station showing flood waters near the estimated discharge of 3511 m3/s on February 20, 1993 (USGS). The gage house is at lower left. The gaging station recorded the stage during the flood, which is unusual for an event of this size.

Criticism (3) mostly arises from statistical questions and assumptions that errors and uncertainties in paleoflood data degrade the value of gaging data in flood-frequency analysis. This is an easily refuted criticism [*Blainey et al.*, this volume], and new methods explicitly account for uncertainties in flood magnitude and timing [*O'Connell et al.*, in press].

Criticism (4) will be discussed later in this paper. The ramifications of hydroclimatically induced nonstationarity in flood-frequency bolster the need for using paleoflood and related paleoenvironmental data to more fully understand the complex, often time-variant linkages between floods and climate [*e.g.*, *Redmond et al.*, this volume].

In each instance, each category of criticism, regardless of its basis in fact, has served and continues to serve the important purpose of ultimately generating thoughtful responses that involve further testing, refinement, and improvement of techniques and concepts of paleoflood hydrology. The publication of this monograph is a testament to the strength and viability of the concepts of paleoflood hydrology and the maturity of its development as a distinct scientific field. It will also serve the purpose of eliminating lingering misconceptions about its techniques and will, ideally, enable more efficient integration of paleoflood hydrology into the realm of conventional flood science.

PALEOFLOOD TECHNIQUES

Many techniques are available for the inference of past flood-flow parameters using principles of geomorphology and related aspects of Quaternary stratigraphy, geochronology and sedimentology. The most commonly used procedures can be divided into the following categories: (1) regime-based paleoflow estimates (RBPE), (2) paleocompetence studies, (3) floodplain stratigraphy, and (4) paleostage estimates and bounds. Portions of the following discussion is generalized from *Baker* [1987] and *Baker* [2000].

Regime-Based Paleoflow Estimates

Regime-based paleoflow estimates use empirically derived relations to estimate the values of relatively high-probability flow events, such as the mean annual flood or bankfull discharge, from paleochannel dimensions, sediment types, paleochannel gradients, and other field evidence. The relations apply exclusively to alluvial channels, which adjust their width, depth, sediment transport, and slope to some representative value of streamflow discharge. These variables are related to discharge by regression expressions derived from observations of relations in modern alluvial channels. RBPE studies have been summarized by *Dury* [1976], *Ethridge and Schumm* [1978], and *Williams* [1984, 1988]. Because nearly all regime relations apply only to relatively frequent floods, they are of little use for estimating the frequency of extreme floods, but offer important insights into more moderate (average) flow conditions.

Paleocompetence Studies

Paleocompetence studies use empirical regression or theoretical expressions to relate very large sedimentary particles to hydraulic conditions necessary for their entrainment, transport, or deposition. The hydraulic variables generally include shear stress, velocity, or stream power [*Jarrett*, 1991; *O'Connor*, 1993; *Carling et al.*, this volume]. Paleohydrologic applications usually apply to limiting hydraulic conditions of maximum floods and require measurements of paleochannel dimensions, slopes, and maximum particle sizes. Where local controls on sediment availability and transport are not known, paleocompetence studies can be of very low accuracy [*Church*, 1978; *Maizels*, 1983]. However, limiting curves for probable causative conditions [*Williams*, 1983] and use of several independent calculation approaches to data [*Costa*, 1983] may allow bracketing of the uncertainty in paleodischarge measurements.

Analysis of Floodplain Stratigraphy

The use of floodplain stratigraphy for paleoflood hydrology has a rich heritage [*Costa*, 1987a]. *Costa* [1978] reviews research in this area to the late 1970s; recent reviews include *Gregory et al.* [1995], *Branson et al.* [1996], *Benito et al.* [1998], and *Knox* [2000]. *Knox and Daniels* [this volume] discuss applications of this approach to watersheds in the upper Mississippi River basin.

Paleostage-Based Flow Data

Paleostages are estimated by documenting flood-induced erosion or deposition near maximum water levels of large floods. The most accurate and best-preserved records of paleostage flow data have been found to occur in stable-boundary river reaches, typically narrow bedrock canyons, because these reaches are most conducive to the accumulation and preservation of SWD-PSI. SWD consist of sedimentary particles with high settling velocities that accumulate relatively rapidly from suspension during major floods [see *House et al.*, this volume; *Webb et al.*, this volume]. Caves [*Springer*, this volume], alcoves [*Webb et al.*, this volume], and tributary mouths [*House et al.*, this volume] are particularly good sites for deposition of SWD. PSI are other high-water marks, including silt plastered on bedrock walls and scour of regolith materials marginal to the channel [see *Jarrett and England*, this volume] and can occur in a variety of geometric and geomorphic settings.

The best reaches generally have relatively narrow, deep cross sections that induce large changes in flood stage from relatively small changes in flood discharge. Floods in these reaches produce very high values of stream power per unit area [*Baker and Costa*, 1987; *Baker and Pickup*, 1987], which results in transport of coarse sediment in suspension. Sediment that crosses the boundary separating high-velocity, energetic flow from local slackwater zones (eddies) is rapidly deposited near or at the level of the maximum flood stage [see *Jarrett and England*, this volume; *Webb et al.*, this volume; *House et al.*, this volume; *Kite et al.*, this volume].

Although large floods may be very erosive within their effective flow boundaries, removing the evidence of preceding flows, they typically add deposits to the slackwater sites. Thus, slackwater sites preserve a sedimentary sequence of deposits representing the largest floods in the river since the start of deposition. The most common type of slackwater deposit is a single, vertically accreting section, although inset deposits and more complex geometric relations also occur [see *House et al.*, this volume]. Much of SWD-PSI analysis consists of interpreting the sequence and timing of flow events from this preserved record.

A regional *SWD-PSI* paleoflood study can have up to six major phases [*Baker*, 2000]: (1) identification of key river reaches that preserve appropriate paleoflood evidence, (2) stratigraphic analysis and age-estimation of paleoflood deposits to establish a flood chronology, (3) hydraulic analysis of the study reach, (4) relation of the paleoflood evidence to the channel hydraulics to determine paleoflood discharge(s), (5) flood-frequency analysis of paleoflood data collected in phases (2) through (4) in concert with relevant gaging station data, and (6) analysis of temporal and spatial patterns in paleoflood data from multiple sites. Many of these phases can apply to any type of paleoflood study, but we emphasize their importance to SWD-PSI work here.

Field Reconnaissance. Reconnaissance for potential study sites is essential before investing the substantial time and resources necessary for SWD-PSI analysis. A considerable amount of experience has been acquired for identifying appropriate reaches [*e.g.*, *Baker*, 1987; *Baker et al.*, 1988]. Regional factors of importance include (1) confined canyons in resistant geological materials, (2) relatively high concentrations of suspended sediment, (3) appropriate sites for slackwater accumulation outside effective flow boundaries, (4) minimal channel bed instabilities over the period of the paleoflood record, and (5) stability of SWD-PSI sites over the time scale of interest. In general, arid, semiarid, and savanna climates have proven best relative to factors (2) and (5), but certain optimum sites provide useful records in humid regions as well [*Kite et al.*, this volume; *Springer*, this volume].

The emergence of high-resolution remote sensing allows improvements both in field reconnaissance and reconstruction of flow areas. Airborne geophysical methods may allow extension of paleoflood techniques to alluvial channels, where deposition on floodplains creates complicated flow fields [*Pickup et al.*, this volume]. Landsat data, which the United States began collecting in the early 1970s, may be used to assess the spatial extent of flood damages [*Mayer and Pearthree*, this volume] and identify study sites with potentially interesting flood histories. Image analysis of digital photography using automated, high-resolution mapping methods may alter the way in which field descriptions of paleoflood deposits are made [*Denlinger et al.*, this volume]. All of these approaches are most appropriate in arid regions with low vegetation cover, which allows remote sensors to detect changes in sedimentary deposits.

Stratigraphy and Geochronology. Accuracy in estimating the ages of paleofloods is a function of the accuracy and precision of the technique used. Dendrochronology, which can date floods to the nearest year, is the most accurate technique, but it is limited to the ages of trees growing next to rivers [*Sigafoos*, 1961, 1964; *Yanosky and Jarrett*, this volume]. Dendro-corrected radiocarbon dating [*Stuiver*, 1982; *Stuiver and Reimer*, 1993; *Stuiver et al.*, 1998] of organic material entrained in or associated with the flood deposits is the standard approach [*Kochel and Baker*, 1982, 1988] and was extended to the post-1952 era using bomb-enhanced ^{14}C in the atmosphere [*Baker et al.*, 1985; *Ely et al.*, 1992]. Floods in the time period 10,000 to 350 years ago are most amenable to radiocarbon dating. Uncertainties inherent in the counting techniques range from about ±40-90 years. In addition to counting errors, other errors are introduced by small sample sizes and contamination by younger carbon not transported by the flood. The effects of these errors can easily be minimized.

Association of organic material to the transporting flood may be the most important problem in ^{14}C analyses [*Ely et al.*, 1992; *Webb et al.*, this volume; *House et al.*, this volume; *Levish*, this volume]. Floods during the period A.D. 1650 to 1950 are difficult to date accurately by radiocarbon analysis because of loading of the atmosphere in CO_2 from fossil carbon during the industrial revolution [*Stuiver*, 1982]. However, numerous other dating procedures may be used for such relatively young floods, including historical documents, distinctive archaeological materials [*e.g.*, *House and Baker*, 2001], and radioisotopes derived from atmospheric nuclear testing [particularly ^{14}C and ^{137}Cs; *Baker et al.*, 1985; *Ely et al.*, 1992].

Recent developments in methods using mineral luminescence have potential for accurately dating fine-grained slackwater deposits [*Aitken*, 1998; *Forman et al.*, 2000]. The key assumption in this analysis is that sediment—typically quartz crystals—in the flow are exposed to sufficient sunlight to reset their luminescence characteristics. Certain assumptions in these methods may not be met in conditions typical of sediment transport and deposition by large floods in bedrock canyons, however, and their potential applicability to deposits associated with inundation of broad floodplains may be considerably higher. Also, typical errors in luminescence studies are higher than with radiocarbon analysis.

Surface-exposure dating using cosmogenically formed isotopes is an emerging method for dating flood terraces and similar fluvial landforms [see *Fenton et al.*, this volume]. Cosmogenic dating requires a surface that is exposed to bombardment by cosmic rays, and the most ideal sites are flat surfaces at high elevations, and as a result most cosmogenic dating has been done on glacial moraines [*Cerling and Craig*, 1994]. The canyon walls at many SWD-PSI sites shield exposed sediments considerably and reduce the method's effectiveness. At present, the lowest errors associated with cosmogenic dating typically are ±500 years, which reduces this method to mostly being useful to older surfaces where no other methods are available (*e.g.*, Pleistocene river gravels). Fluvial terraces comprised of pebbles and cobbles, which seldom contain datable carbon, often can be dated by cosmogenic isotopes, making this method ideal for flood deposits on outwash plains or isolated terraces.

Other tools have been used in relative dating and correlation. Deposits can be placed in a hierarchy by age based on the stratigraphic position of the deposit [*Costa*, 1978], on the presence of historical or archaeological materials [*Patton and Dibble*, 1982; *Ely and Baker*, 1985; *House and Baker*, 2001], and on the development of material properties that change as a function of time, such as soil color and degree of induration [*Baker et al.*, 1983]. The resolution of these methods depends upon context and could range from ±1-1,000 years or more.

Hydraulic Analysis. The hydraulics of SWD-PSI reaches pose the same problem of indirect measurement as would be the case for any large flood discharge. In conventional practice, water-surface profiles for various candidate discharges are predicted from a one-dimensional flow model fit to the channel geometry of a reach [*O'Connor and Webb*, 1988; *Webb and Jarrett*, this volume]. *Webb and Jarrett* [this volume] discuss the detailed concerns with such paleodischarge calculations, which are particularly difficult in sinuous bedrock canyons [*Webb et al.*, this volume]. The emergence of multidimensional modeling, as discussed by *Denlinger et al.* [this volume] and *Wohl* [this volume], may reduce some of the uncertainties associated with reconstructing flood discharge in complex reaches as well as answer questions as to maximum potential flood size over a finite time period [*Levish*, this volume].

Data Structuring. For most SWD-PSI studies, discharges and ages of ancient floods can be further analyzed. These data may be structured in various ways depending on the intended use of the information; the structuring scheme required by the intended use of the data will then feed back to the field program. Experience has shown that it is relatively easy to reconstruct the largest paleoflood in a given time period by SWD-PSI techniques. However, the reconstruction and completeness of record for smaller paleofloods requires greater time and resources. A balance may need to be achieved between value of added information and the cost of obtaining it.

Baker [1989] discussed the use of multiple site studies and correlation techniques in relation to the concept of event censoring levels. A censoring level is a threshold discharge (or stage) above which the paleoflood record can be assumed to be complete over a specified time interval.

Completeness means that no actual discharge (or stage) occurred in the time interval that exceeded the censoring level without leaving its evidence in the paleoflood record. In practice the largest paleofloods usually leave the most durable records, so setting the censoring level depends on the information extracted on the smaller floods. The geometry of natural SWD-PSI sites ensures that complete samples of floods immediately smaller than the maximum will be preserved [*Baker*, 1989]. However, each field situation will differ in how low the censoring level can be set with confidence.

The censoring structure is most important in regard to statistical flood-frequency analysis of the data, discussed later in this paper as well as in *Blainey et al.* [this volume]. Censoring structure also affects record completeness, and thus has a bearing on paleoenvironmental and other interpretations that may assume complete records of paleofloods [*e.g.*, *House et al.*, this volume].

Other uses of paleoflood data may require different considerations, particularly the use of nonexceedance thresholds, or paleohydrologic bounds [*Levish*, this volume]. This approach can be considered as a special case of paleostage-based flow estimation in which evidence is found to document that a given stage was not exceeded over some time period. The approach is less sensitive to record completeness than it is to the necessity that the largest floods over discrete periods of time are accountable. All the considerations of floodplain stratigraphy, geochronology, hydraulic analysis (particularly multidimensional), and data structuring apply to studies of paleohydrologic bounds. However, there are generally more opportunities for establishing that a particular stage, perhaps corresponding to an undisturbed soil profile on a river terrace, was not exceeded than there are opportunities to establish detailed records of the largest floods in a given time period. Because the potential for quantifying nonexceedance bounds exists for most rivers, nonexceedance techniques are of great interest for achieving nearly universal practical application of paleoflood hydrology for extending record length.

Spatial and Temporal Patterns. Regionalization uses spatial extension to expand the sample size of annual floods for statistical analysis. Problems with regionalization procedures include the lack of understanding of the physical factors controlling the statistical distribution of floods. This limits the rational basis for establishing regional homogeneity. Regions of relatively low flood variability, such as England or much of the eastern United States, seem to best meet criteria for effective regionalization studies, including relatively long at-site conventional records and homogeneity. Regions of high flood variability (which pose the most difficult problems in flood-hazard assessment), such as the southwestern United States, are heterogeneous and generally have short gaging records. Moreover, relatively large rivers may have complex, unique physical factors dictating their flood response. Paleoflood data can supplement conventional data in addressing these issues when the two procedures are combined [*Stedinger and Baker*, 1987; *Tasker and Stedinger*, 1987].

O'Connor et al. [this volume] describe how the world's largest floods—as reconstructed through analysis of both historical and paleoflood records—are inextricably linked with geography and geologic setting. Many of the largest floods are either from areas with the potential for tropical cyclone incursions or where natural dams—either made of ice or landslide materials—have failed. Thus, a global study of paleofloods would be a worthy effort in advancing the scientific understanding of the largest floods and their causes [*Baker*, 1995]. Progress toward this end is summarized, somewhat incompletely, in Figure 3 (compare to Figure 1 in *Kochel and Baker*, 1988). A principal pattern in this distribution is the general tendency for paleoflood investigations to be focused on arid and semiarid regions. As recent studies indicate, broader regional applicability is not only possible but increasingly necessary as the need for characterization of flood hazards and frequency on the basis of physical flood data becomes more apparent.

FLOOD-FREQUENCY ANALYSIS: PURPOSES AND PHILOSOPHIES

The practice and science of flood-frequency analysis developed for two main purposes, both of which ultimately should directly benefit society. Most flood-frequency analysis in the United States is done by engineers for the purposes of hydrologic design [*Thomas*, 1985]. The purposes of hydrologic design are as varied as designing culverts to pass certain sizes of floods to designing a dam to not fail owing to extremely high inflows to the reservoir. Because hydrologic design has the definite purpose of building a structure to not fail as a result of some hydrologic process, most hydrologic design strategies are strict procedures that are reproducible and defensible in a court of law [*Klêmes*, 1994].

Klêmes [1994], a particularly strong critic of conventional flood frequency analysis, presented the following extreme characterization of codified, uniform methods of calculating flood hazard. (1) Because users require the probabilities of occurrence for extreme floods, an industry of flood-frequency analysis has grown to fill the need. (2) Flood frequency cannot be accurately calculated because gaging data typically are of insufficient length to quantitatively define extreme flood hazard for most watersheds. (3) Rigorous probabilistic theories have been developed within the hydrologic sciences that may be applicable, if flood data from gaging stations can be defined as identically distributed, independent and random variables.

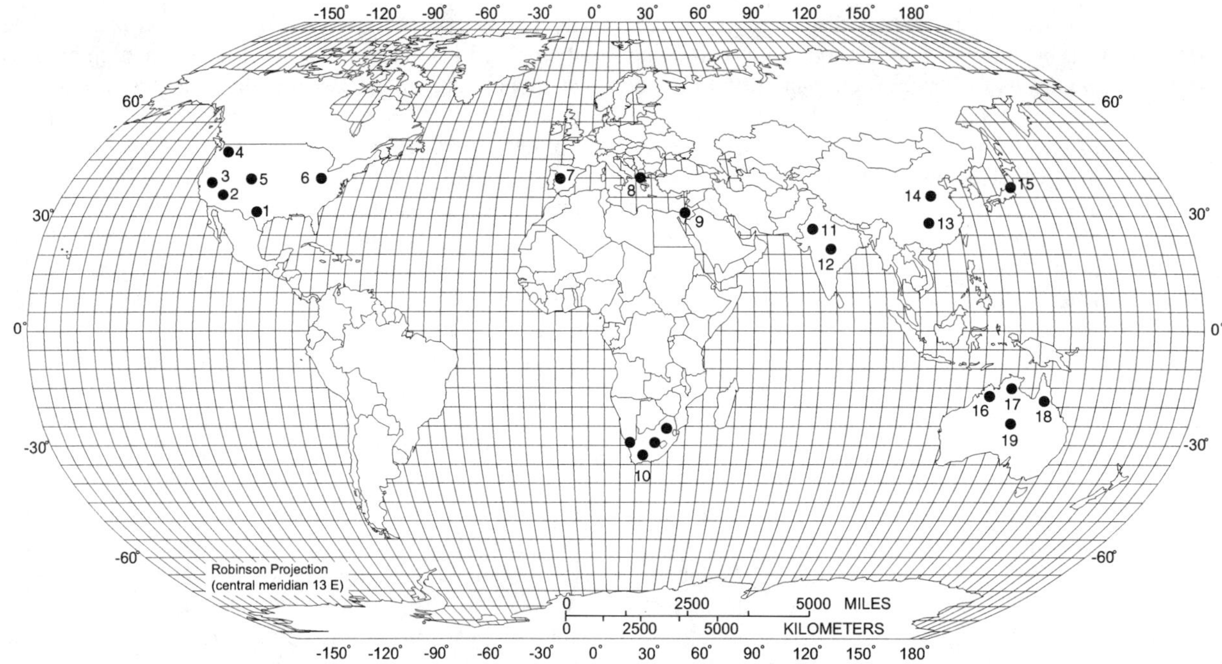

Figure 3. Map showing global distribution of representative SWD-PSI paleoflood hydrology studies. This map shows the general distribution of areas surveyed to date. Many more studies have been done in most of the areas than are indicated by the selected references. 1. West and central Texas [*Kochel et al.*, 1982]. 2. Southwestern U.S. [*Ely et al.*, 1993; *Enzel et al.*, 1993]. 3. East slope, Sierra Nevada [*Mann*, 2000; *Kellogg*, 2001]. 4. Columbia River [*Chatters and Hoover*, 1994]. 5. Colorado [*Jarrett and Tomlinson*, 2000; *Klawon et al.*, 2000]. 6. Central Appalachians [*Kite et al.*, this volume]. 7. West-central Spain [*Benito et al.*, in press]. 8. Northwestern Greece [*Lewin et al.*, 1991]. 9. Negev Desert [*Greenbaum et al.*, 2000, this volume]. 10. South Africa [*Zawada*, 1996; *Zawada and Hattingh*, 1994]. 11. Northwestern India [*Kale et al.*, 2000]. 12. Central India [*Ely et al.*, 1996]. 13. Chang Jiang [*Baker and Kochel*, 1988]. 14. Huang He [*Fucheng et al.*, 1985]. 15. East-central Japan [*Jones et al.*, 2001]. 16. Northwestern Australia [*Gillieson et al.*, 1991]. 17. North-central Australia [*Baker and Pickup*, 1987]. 18. Northeastern Australia [*Wohl et al.*, 1994]. 19. Central Australia [*Baker et al.*, 1987; *Pickup et al.*, 1988; *Pickup et al.*, this volume].

(4) Real floods, particularly the largest ones, generally do not satisfy these simplistic assumptions. (5) Nonetheless, because of the growing need to calculate the hazard, the analyst must make simplistic assumptions to calculate flood quantiles. (6) To compensate for these obvious problems, some hydrologists believe that increasing emphasis should be placed on development of theoretical models with ever increasing mathematical rigor and precision.

Klêmês' [1994] perspective contains considerable insight into institutional, legal, and professional realities that are faced by persons engaging in flood-frequency analysis and flood hazard management. More recently, *Pielke* [1999] highlighted flaws in flood frequency analysis in his enumeration of 'Nine Fallacies of Floods" that relate to a larger set of fundamental problems in dealing with and understanding flood hazards in the United States. Each of these critiques centers on the fact that existing methods contain critical weaknesses and could be improved considerably. This dilemma can be overcome in part at least by the acceptance and incorporation of paleoflood data and similar geologic information into a new paradigm of flood hazard evaluation.

Why have such codified procedures developed when their operation seems so contrary to common sense [*Klêmês*, 2000]? The answer lies in the purpose towards which one directs flood-frequency analysis. As idealized by *Pielke and Byerly* [1998], research funding drives institutionalized flood research in a highly linearized manner (Figure 4a). In flood studies [*Pielke*, 1996], society is presumed to benefit from flood-hazard regulations or designs based on statistical models, which in turn are grounded in basic research. More rigorous and precise research drives more accurate model predictions that (presumably) yield better regulations and designs—all authoritatively imposed to achieve benefit.

The flaw in this strategy is that wise human action does not immediately follow from providing the best possible technical information. The paradox of increasing flood

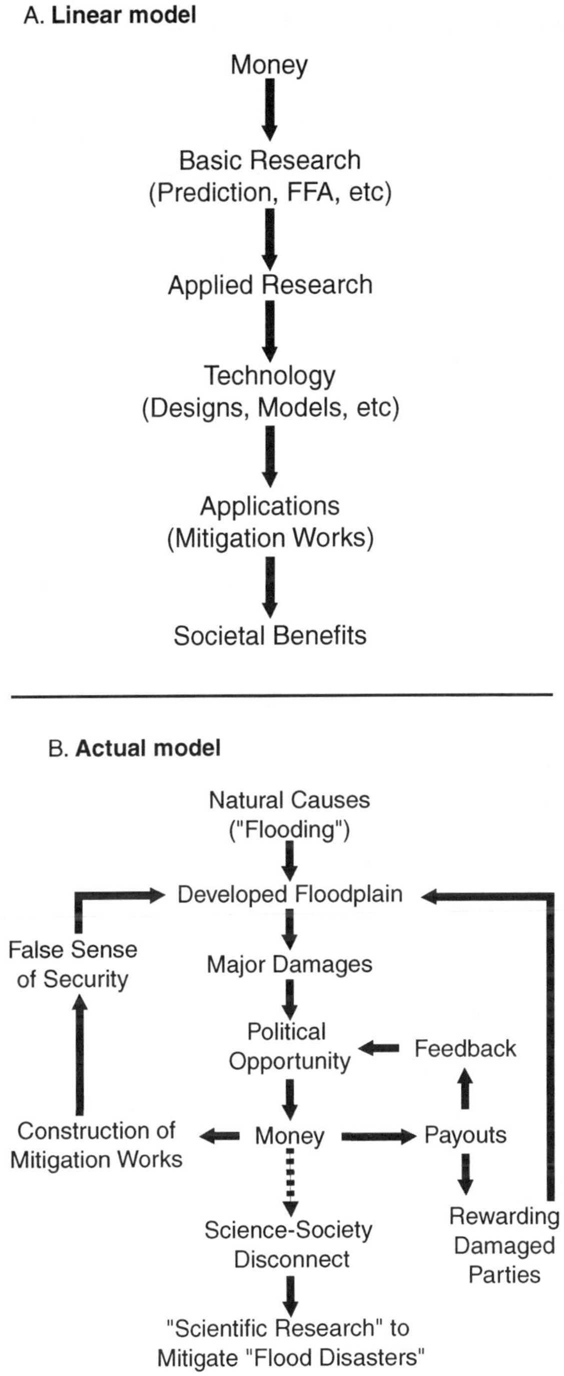

Figure 4. A. Linear model of flood-science-society relationships [after *Pielke*, 1996]. B. Actual flood-society relationships [this study].

damages (Figure. 1) paralleling increased advances in flood science and hazard mitigation reflects this defect in the linear model. What actually happens is that scientific research is disconnected from a complex of political, societal, and human perception factors (Figure 4b) that actually exacerbate flood damages, despite the best intentioned scientific and technical input. An alternative to this state of affairs is for flood science to actually influence human perception more directly, by calling attention to problems that demand political solutions [*Baker*, 1998]. The point is not for flood scientists to dictate what people are supposed to do about flood hazards, but for a science of floods that so compels peoples' perceptions about floods that they actually take the actions necessary to mitigate hazardous consequences. Instead of thinking up the best possible science, one works towards the science that yields the best possible human outcome.

In contrast to the economically driven nature of traditional flood science, a better understanding of the temporal and spatial nature of floods could help constrain the question of flood potential. Floods may be clustered in time, either in response to changes in land use, such as the occurrence of a large fire in a watershed, or hydroclimatic variability. Channel change associated episodic downcutting of alluvial channels [*Webb et al.*, 1991] significantly affects flood hazard by reducing or eliminating the short-term probability of inundation. The sizes of floods may change through time, possibly in response to changes in flood hydroclimatology [*Ely et al.*, 1993]. Because of these considerations, the scientific study of floods in the order in which they occur and not just their frequency could have large societal benefits, both in terms of new techniques for characterizing flood frequency as well as a different perception of flood hazards by people living near rivers.

These two approaches to flood science require distinctly different approaches to understanding and calculating flood recurrence. Most flood-frequency analysis is based on consideration of annual floods in the *frequency domain*, where the temporal order of flood occurrence is not considered. On the other hand, many scientists are interested in the occurrence of floods in the *time domain*, or *historical domain*, where the order in which floods occur is of paramount importance to understanding their causes and potential recurrence, and in communicating the reality of flood hazards to those needing to mitigate their effects. Analysis of floods in the historical domain is in its infancy and will not be elaborated further in this paper. The frequency domain of flooding is discussed in the following section.

PALEOFLOOD DATA AND THE FREQUENCY DOMAIN

Traditional flood-frequency analysis is based on the assumptions that (1) floods are random in space and time, (2) floods at a specific place (*e.g.*, a gaging station) arise from a single probability distribution, and (3) the first two moments of that probability distribution are time-invariant (type I stationarity; *Box and Jenkins* [1971]). The risk associated with flooding is calculated as a future probability of occurrence. The most commonly used quantile for risk analyses–the 100-year flood–is a misnomer because the real definition of this quantile is a flood of 0.01 probability that could occur next year. In that sense, flood-frequency analysis in hydrology is analogous to long-term weather forecasting in climatology [*Redmond et al.*, this volume]. *Benson* [1962] reviewed the evolution of methods for estimating flood frequency; and a recent update is provided by *Stedinger* [2000].

Methods for Fitting Probability Distributions

Flood-frequency analysis in the United States is specified by the *U.S. Water Resources Council* [1982] in Bulletin 17B to use a log-Pearson type III distribution with a specific algorithm for incorporating historical and (or) paleoflood data. Neither of these specifications are necessarily based on sound science; in fact, the debate over which probability distribution is most appropriate continues to rage [*National Research Council*, 1988; *Stedinger*, 2000] with many scientists preferring the generalized extreme-value distribution [*e.g.*, *Francis et al.*, 1994].

The methods specified in Bulletin 17B use all types of floods—both large and small—to estimate return periods for very large floods. As a result, floods that occur during drought periods are combined with those that occur during very wet periods, resulting in mixed populations being analyzed as a single population. A more rigorous and viable approach would be to censor all floods low recurrence intervals (*e.g.*, less than the 2- or 5-year flood) and only estimate design quantiles using larger floods that presumably are from a more homogeneous population [J.R. Stedinger, pers. commun., 1989].

Bulletin 17B uses the method of moments to fit annual peak floods to the log-Pearson type III distribution. Maximum-likelihood analysis offers another method for fitting data to probability distributions, but the procedure is more mathematically and computationally complex. Using a Monte Carlo analysis, *Blainey et al.* [this volume] found that maximum-likelihood analysis resulted in fits with lower root-mean-squared errors than did method of moments for gaging data alone. This suggests that the traditional method for analyzing flood frequency is also the least efficient with flood data, even when no historical or paleoflood information is included.

Another problem with traditional flood-frequency analysis is the short lengths of most gaging records. Most hydrologic design for floodplain structures is based on the 100-year flood, but few records from the United States are of that length. Therefore, any estimates of the 100-year flood (floodplain design) or higher recurrence intervals (dams, critical structures) are obtained through extrapolation of shorter records. Because most major rivers in the United States are regulated, usually with at least some flood-control component of dam operations, future gaging records will not necessarily supply additional years of useful data. Paleoflood data offer an efficient means to add record length to the flood-frequency analysis for these rivers.

Incorporation of Paleoflood Data into Flood-Frequency Analysis

Incorporation of paleoflood data into flood-frequency analysis would appear to be the ideal means for reducing the estimation error of flood quantiles. Most paleoflood records exceed 1,000 years in length, with many records in the southwestern United States being 4,000 years long [*Ely et al.*, 1993] or longer [*Patton and Dibble*, 1982]. The record length alone decreases the 95% confidence limits for design recurrence intervals, such as the 100-year flood [*Webb and Rathburn*, 1989; *Blainey et al.*, this volume], which leads to decreased statistical risk.

Benson [1950] proposed early methods for incorporating historical data into flood-frequency analysis. Subsequently, three methods have been used to incorporate historical and paleoflood data into flood-frequency analysis. The first is given in *U.S. Water Resources Council* [1982] using an algorithm developed by *Matalas and Jacobs* [1964]. This method has been described as statistically inefficient [*Cohn and Stedinger*, 1987]. In response to this problem, *Stedinger and Cohn* [1986] developed a maximum-likelihood estimator (MLE) approach that used hierarchical censoring thresholds to depict historical and paleoflood data and fit them to probability distributions.

Cohn et al. [1997] developed the third method, the expected-moments algorithm (EMA), which uses a method of moments approach to incorporate censored data into a log-Pearson type III distribution. *Cohn et al.* [2001] added confidence limits to EMA. *Hirsch and Stedinger* [1987] developed a rigorous method for plotting positions for historical and paleoflood data, fixing another problem with the procedures specified by *U.S. Water Resources Council* [1982].

Both MLEs and EMA have been widely applied in paleoflood hydrology. *Stedinger et al.* [1988] first applied the maximum-likelihood method to paleoflood data collected by *Partridge and Baker* [1985], followed by *Webb and Rathburn*'s [1989] re-analysis of paleoflood data from west Texas [*Kochel et al.*, 1982] and southern Utah [*Webb et al.*, 1988]. *Wohl et al.* [1994] applied MLEs to flood-frequency questions in Australia and, more recently, the methods have been applied to rivers in the western United States [*Mann*, 2000; *Kellogg*, 2001]. EMA has been applied in several studies in southern Utah [*Webb et al.*, this volume; *Orchard*, 2001]. Recent applications of MLEs in a Bayesian statistical framework have demonstrated that various types of paleoflood data and their related measurement uncertainties can be efficiently incorporated into flood-frequency analysis [*O'Connell et al.*, 1997; *O'Connell et al.*, in press].

Some data errors, such as uncertainties in indirectly estimated discharges and nonstationarity of flood series, affect both paleoflood and gaging data alike. As discussed by *Blainey et al.* [this volume], the effects of most errors in paleoflood data are negated by the increase in record length when compared with the errors typically associated with gaging data.

Hydroclimatology, Flood-Frequency Analysis, and Nonstationarity

In the broadest sense, floods are a hydroclimatic process, and the frequency of floods may be affected by climatic variability [*Redmond et al.*, this volume]. Indeed, *Pielke and Downton* [2000] concluded that the increase in flood damages in recent decades (Figure 1) is best explained, but not necessarily caused, by climatic variability. *Ely et al.* [1993] made broad paleoclimatic inferences from clusters of paleofloods in radiocarbon times; their analysis does not include the effect of measurement error on radiocarbon samples, preservation issues, or the statistical inference of single floods as being representative of general hydroclimatological conditions. Nevertheless, *Ely et al.* [1993] found that the largest floods in the last 4,000 years in the southwestern United States were generally clustered into distinct groups, suggesting the possibility of periods when flood frequency is increased for hydroclimatic reasons. Their results agree in general with *Hirschboeck* [1987, 1988] and *Webb and Betancourt* [1992], who worked on 20th century floods in the same region.

Several attempts have been made to incorporate hydroclimatic information into flood-frequency analysis. *Waylen and Woo* [1982] proposed dividing annual flood series into hydroclimatic groups, fitting those groups separately to probability distributions, then combining the separate distributions using mixed-population analysis. *Hirschboeck* [1985] analyzed hydroclimatically distinct floods from the partial duration series, defined as all floods above an arbitrary base discharge, in the Gila River basin. However, the El Niño – Southern Oscillation (ENSO) greatly affects flood probability on an interannual basis, further violating the assumption that annual floods are independent [*Cayan and Webb*, 1992].

In certain parts of the United States, flood frequency is strongly related to hydroclimatic processes controlled at the global scale. The Santa Cruz River in southern Arizona offers a superb example of a river with different hydroclimatic populations of floods (Figure 5A). *Webb and Betancourt* [1992] defined three hydroclimatic groups of floods affecting the Santa Cruz River, used the partial-duration series as a censoring threshold, fit separate probability distributions using maximum-likelihood analysis, and combined the distributions using the method of *Waylon and Woo* [1982] to calculate flood quantiles. Unfortunately, they found that hydroclimatic homogeneity only may have occurred in 30-year intervals in the 20th century, further reducing the amount of information available for flood-frequency analysis.

Webb and Betancourt [1992] showed graphically how hydroclimatic forcing can affect flood-quantile estimates by sequentially calculating flood frequency with addition of floods from a period of increased flood potential (Figure 5). The estimates of the 100-year flood in 2000 were outside the 95% confidence limits calculated in 1970 (Figure 5b), primarily because of increases in the sample standard deviation (Figure 5d) and radical fluctuations in the calculated skew coefficient of the annual flood series (Figure 5e). As a result, the local regulatory agency re-established its design flood to be about 10% higher than the largest flood of the record (Figure 5a) with no return period assigned.

As shown in Figure 6, flood frequency on two other rivers in Arizona is weakly related to the Pacific Decadal Oscillation (PDO), a measure of oceanic response to climatic fluctuations [*Mantua et al.*, 1997]. In this region, periods of positive PDO are periods when larger floods tend to occur; periods of negative PDO tend to be drought years [*Redmond et al.*, this volume]. If PDO could be accurately forecast, this type of hydroclimatic modeling could allow an explicit accounting for hydroclimatically induced nonstationarity in flood frequency. This example suggests that the emerging field of long-range climatic forecasting based on regional or global climatic indices may provide useful information on flood hazard.

Nearly all mathematical treatments of hydrologic processes assume that the underlying probability distribution for a random variable, such as flood recurrence, remains constant through time. Stationarity is the required assumption of most methods for estimating flood-exceedance probabilities, including the analysis of gaging data alone. The Interagency Advisory Committee on Water Data [*U.S. Water Resources Council*, 1982, p. 6], for example, requires that "... the array of flood information is a reliable and representative time sample of random homogeneous events." However, changes in climate, vegetation, and watershed conditions may dramatically alter flood responses, as documented by periods of differing flood-frequency characteristics in some very long paleoflood records [*Patton and Dibble*, 1982; *Knox*, 1985; *Webb and Baker*, 1987; *Webb et al.*, 1988].

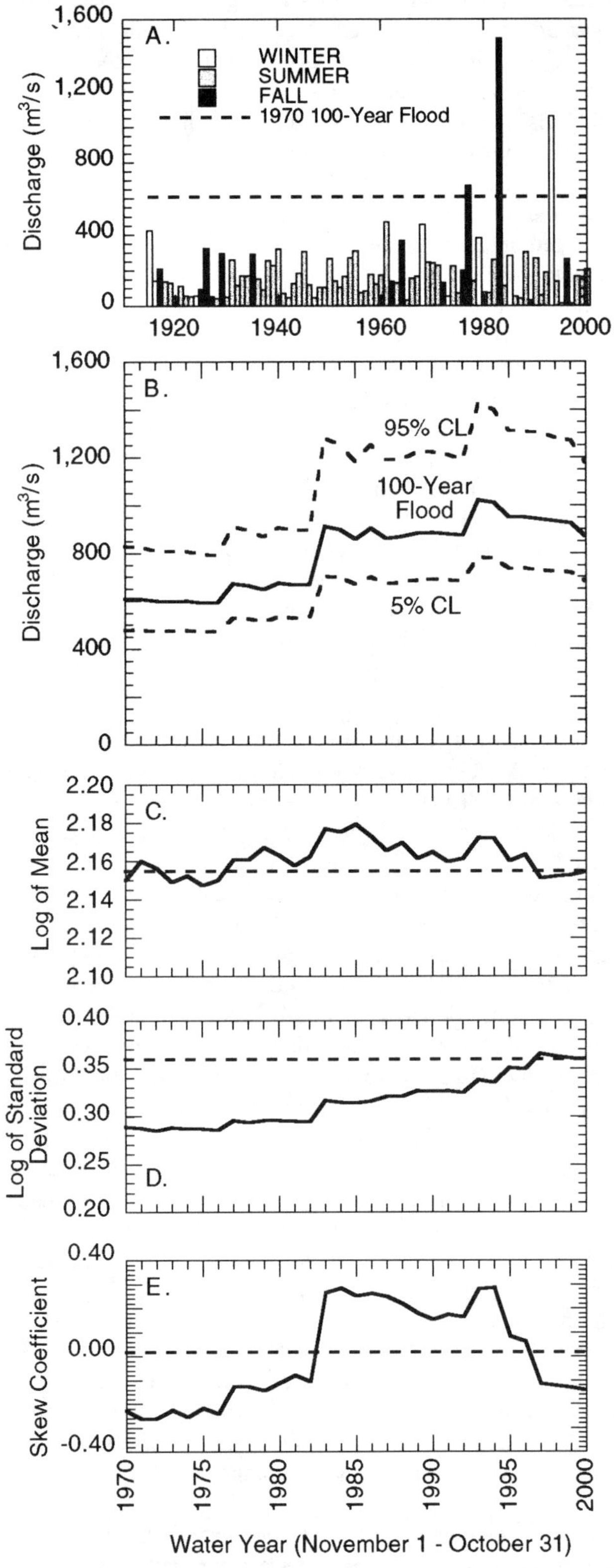

Figure 5. The annual flood series (a) and changes in the 100-year flood (B) and moments of the annual flood series (C-E) for the Santa Cruz River at Tucson, Arizona (09482500). The 100-year flood and moments were calculated sequentially beginning with the period of record as of 1970 (initial n=54 years)(updated from *Webb and Betancourt*, 1992). Large floods from dissipating tropical cyclones and severe winter frontal systems dominated flood frequency after 1970, as opposed to local summer thunderstorms in most of the preceding half century.

One of the major lessons from paleoflood hydrology is the perspective it affords on flood populations. Many rivers in the southwestern United States, for example, have a largest flood that is 2-4 times larger than the second highest event in records 50-75 years long [*Webb et al.*, this volume; *Orchard*, 2001]. This suggests that extremely large floods may arise from a different probability distribution than more common floods, in accord with the ideas that the hydroclimatic conditions associated with large floods are extraordinary in themselves [*Redmond et al.*, this volume]. One possible way to handle this statistically is to only analyze the largest floods, typically those preserved in the paleoflood record [*Wohl et al.*, 1994] or using the largest flood in the paleoflood record to anchor the tail of a single distribution defined for convenience by smaller floods [*Blainey et al.*, this volume].

For paleoflood records, the key stationarity questions center on climate variability. Deviations from purely random-event generation occur because catastrophic floods are generated in many cases by non-random occurrences of atmospheric phenomena [*Hirschboeck*, 1987, 1988]. The scientifically interesting goal is to document such occurrences by developing long-term paleoflood data sets at multiple localities [*Ely et al.*, 1993]. Without such information, it is impossible to evaluate the validity of any flood-frequency analysis, particularly regionalized flood frequency, especially those based only the temporally restricted data from gaging stations.

Ultimately, nonstationarity in flood-frequency analysis is a problem with both gaging records and the combination of paleoflood and gaging records. The ultimate impact of nonstationarity will have to be viewed in the context of hydroclimatological models for long-period climatic variations. Such models hold the promise of meshing paleoflood studies and other aspects of flood geomorphology with the relatively short time scales [*Hirschboeck*, 1987].

Enveloping Curves

Because of these problems, alternatives to parametric probability distributions and the analyses specified by *U.S. Water Resources Council* [1982] have been developed. The simplest alternative is the use of enveloping curves, first

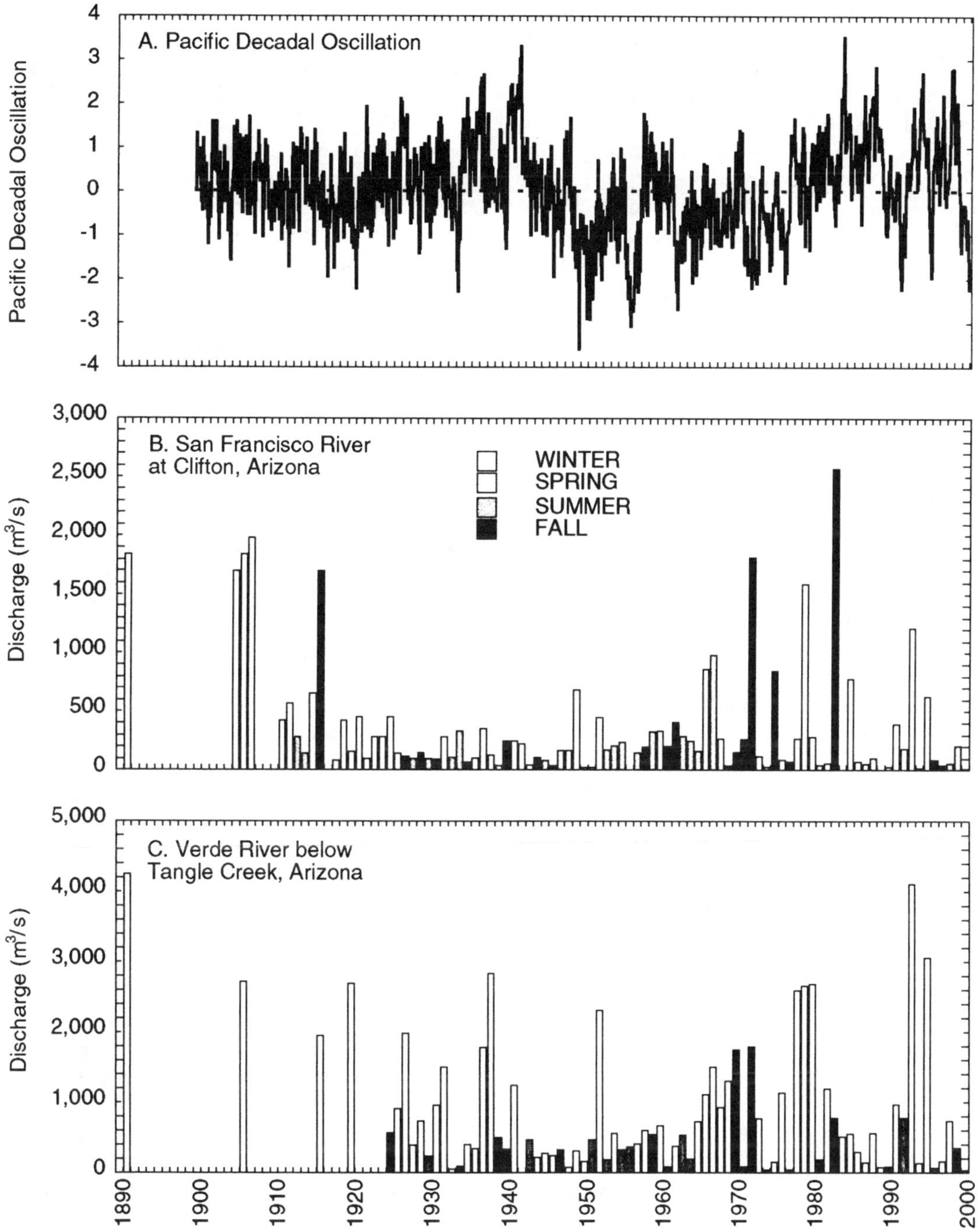

Figure 6. A. The Pacific Decadal Oscillation (PDO), a measure of factors affecting climate in the western United States [*Mantua et al.*, 1997]. B. Annual flood series for the San Francisco River near Clifton, Arizona (09444500). C. Annual flood series for the Verde River below Tangle Creek, Arizona (09508500) [for paleoflood record, see *House et al.*, this volume].

proposed by *Creager* [1939]. *Costa* [1987b] presented an enveloping curve for the largest floods in the continental United States, updating the work of *Crippen and Bue* [1977], and *Enzel et al.* [1993] developed an enveloping curve for the Colorado River drainage that incorporated paleoflood data. Enveloping curves are now used as a reality check for paleoflood discharges and some historical flood estimates [*House and Pearthree*, 1995; *Webb and*

Jarrett, this volume] and have been applied in many studies [*e.g.*, *Webb et al.*, this volume; *House and Baker*, 2001; *Orchard*, 2001].

There are problems with envelope curves, not the least of which is the fact that the sizes of extremely large floods are analyzed as a univariate function of drainage area [*Matthai*, 1990]. Partial-area contributions, which may be important in some types of rainfall-induced floods, cannot be accounted for using this method. Other variables, particularly drainage-basin shape and orientation, latitude, longitude, and the presence of topography that might steer storms, may be much more important in determining the size of floods [*Benson*, 1964]. Nonetheless, the enveloping curve based on paleoflood data implies that it represents the largest flood over a long period of time [*e.g.*, 4,000 years in the case of *Enzel et al.*, 1993] making enveloping curves heuristically analogous to probable-maximum floods, but based on long-term data.

Regionalization and Paleoflood Data

One of the best future research opportunities in flood-frequency analysis is the incorporation of paleoflood data in regionalized flood-frequency analysis. Regionalization typically is used to calculate flood quantiles with confidence limits for ungaged sites [*Tasker and Stedinger*, 1987, 1989]. The most common method applies generalized least squares to flood quantiles calculated from gaging stations to develop a method that can be applied to ungaged sites with similar characteristics. An example of this method, applied to the arid and semiarid regions of the western United States, appears in *Thomas et al.* [1997]. The error associated with regionalization techniques can be decreased with the use of paleoflood data. *Jin and Stedinger* [1989] demonstrated the value of historical (or paleoflood) information in regionalization procedures that employ generalized maximum-likelihood estimators.

THE VALUE OF PALEOFLOOD INFORMATION TO SOCIETY

Hydrologists and geologists are increasingly asked to aid in the hydrological design of structures at risk for extreme floods. Design criteria, of course, are scaled to the importance of the project. Although paleoflood data may be useful to the question of whether a nuclear power plant is outside of the 500-year floodplain or not, it has little relevance to the design of culverts, which may be designed to pass only the 25-year flood. The question of the safety of human lives and protection of vital floodplain structures is one for which collection of paleoflood data is ideally suited.

One of the most controversial topics in hydrologic design involves the retrofitting of dams to pass the Probable Maximum Flood (PMF). The PMF is based on the most severe combination of hydrologic and meteorologic conditions that are considered reasonably possible for a drainage under study [*Cudworth*, 1989], and the subsequent hypothetical flood is routed downstream using watershed models. This flood is neither probable nor maximum and is essentially a best guess at the best design discharge [see discussion in *Lave et al.*, 1990]. In some cases, the PMF appears to be unreasonably large in comparison to the paleoflood record [*Levish et al.*, 1997]. The cost of retrofitting dams to pass the PMF is extremely high [*Lave et al.*, 1990], prompting some observers to question whether dams can or should be retrofitted [*Graham*, 2000].

Returning to Figure 4, monetary costs often steer the direction of hydrologic design. Incorporation of paleoflood data into flood-frequency analysis offers a relatively inexpensive means for evaluating whether a PMF is reasonable or not, and therefore whether an extremely expensive retrofit of a dam's spillway should be authorized. However, the future of paleoflood research should continue to include explicit analyses of the errors of this technique, because, as noted previously, some paleoflood data incorporated in flood-frequency analysis may come under the scrutiny of a court of law.

Legal considerations aside, perception of flood hazards by the general public is a topic that must be addressed by all flood-frequency analysts. Conveying a topic as complicated as flood frequency, with all its uncertainty and mathematical rigor, is an extremely difficult job for a public-relations specialist, much less a scientist. Moreover, flood scientists are tasked with discussing a concept—the 100-year flood—which is vague at best and difficult to defend at worst. Returning to the example of the Santa Cruz River at Tucson, three 100-year floods, as calculated using data through 1970, occurred in the 17 years of 1977 through 1993 (Figure 5a). Public knowledge of such occurrences, which are now widespread in the United States, erodes public confidence in scientific prediction of flood frequency and decreases the credibility of flood-inundation maps.

What can we do to both increase our ability to predict flood hazard and convince an increasingly skeptical public that flood hazards are real? There are no easy answers as to the best method for predicting flood hazards, but many promising statistical methods, as discussed here and in *Stedinger* [2000], have been developed. What is lacking is a way to effectively communicate the very real flood hazard to society. Signs along rivers showing historical flood levels help (Figure 7); it is difficult to argue that a floodplain should be developed when historical inscriptions show that it has been repeatedly inundated. These inscriptions provide a public display of what we actually know about floods and their inundation levels.

Figure 7. High-water marks painted on buildings at Harper's Ferry, West Virginia (Ellen Wohl). Thcsc 20th century floods result from the combined flows of the Shenandoah and Potomac Rivers. The highest mark—in 1936—represents a discharge of 6,800 m^3/s.

Paleoflood data are essentially geologic sign posts along rivers, recording past flood stages. They are more difficult to read than ones painted by humans, but their lesson should be critical to property owners, land managers, and flood-frequency analysts. Flood scientists should focus more on providing real flood data, particularly because flood-control operation of dams has truncated the potential gaging record length for many rivers. Collection of paleoflood data is one means of correcting this problem.

A more revolutionary approach comes from the context of the historical domain of flood information. As with human history [*Carr*, 1961], the science of floods is not restricted to a shallow inventory of events and dates. The historian determines causes and uses those insights in a dual and reciprocal fashion to improve our understanding of the past in light of the present and vice versa [*Carr*, 1961, p. 14]. The purpose of this inquiry is also clear:

> The historian distills from the experience of the past...that part...amenable to rational explanation and interpretation, and from it draws conclusions which may serve as a guide to action [*Carr*, 1961, p. 136].

The history of past floods, not the historical idealization of their future possibilities, is the understandable basis upon which people may be guided to effective action. This is a social issue, not merely a matter of physical science. But if we wish this science to avoid failed consequences (Figures 1 and 4), it must be seriously engaged to, not detached from, its societal application.

Considering floods in their temporal context is essential to an understanding of the potential for nonstationarity, which affects any type of probabilistic flood forecasting. Although some regions, such as southern Arizona, may have nonstationary flood records, flood scientists have not addressed the spatial scale of the problem to determine how much of the United States, much less the rest of the planet, is affected. Flood hydroclimatology, and particularly the use of hydroclimatic models to explain past occurrences of floods, will be one of the exciting future directions of this discipline.

At a minimum, paleoflood information can help establish a tangible temporal context in which to place and evaluate seemingly anomalous individual floods or sequences of events, such as those three 100-year floods in 17 years on the Santa Cruz River (Figure 5). This is a type of perceptual context that in many ways is related to the numerical context that paleoflood data can provide by augmenting short gaging records. Because statistical extrapolations of flood probabilities can be improved by the augmentation of short records with paleoflood data, the public's general awareness of the reality of flood hazards can be greatly enhanced by presenting tangible physical evidence that indicates "floods have occurred here with some regularity during the geologically recent past." If flood recurrence intervals and related floodplain maps show some reasonable degree of consistency with the physical information that bears on the occurrence of actual floods, than public confidence and compliance will likely be improved.

The "best" method for estimating flood hazard—whether with a single probability distribution, using multiple distributions, by using a nonparametric approach, or even with a PMF that is actually probable—has yet to be determined. What is clearly apparent is that the basic task of collecting additional gaging data to increase record length will be extremely difficult owing to flow regulation. Also, the disparity between the short-term need to address flood hazards and the long-term program required to actually develop an adequate sample of floods cannot be completely solved by collecting more data at more gaging stations or developing more sophisticated statistical models. A major part of the solution is to collect paleoflood data with explicitly defined uncertainties. At this point in time, it is clear that methods for the collection, synthesis, and analysis of paleoflood information have reached an acceptably high level of scientific credibility that it can and should be integrated into the realm of traditional practice.

Acknowledgments. The authors thank Dan Levish and Ellen Wohl for providing photographs for this article. Levish, P.G. Griffiths, and S.M. Wiele critically reviewed this manuscript.

REFERENCES

Aitken, M.J., 1998, *Introduction to Optical Dating*, Oxford University Press, Oxford, UK, 262 pp.

Baker, V.R., Paleohydrology and sedimentology of Lake Missoula flooding in eastern Washington, *Geol. Soc. Amer. Spec. Pap. 144*, 79 pp., 1973.

Baker, V.R., Techniques and problems in estimating Holocene flood discharges, *Amer. Quat. Assoc., Abstracts of the 3rd Ann. Mtg*, p. 63, Univ. of Wisconsin, Madison, 1974.

Baker, V.R., Geology, determinism, and risk assessment, in *Scientific Basis of Water-Resource Management*, pp. 109-117, Natl. Acad. Press, Washington, DC, 1982.

Baker, V.R., Paleoflood hydrology and extraordinary flood events, *J. Hydrol., 96*, 79-99, 1987.

Baker, V.R., Magnitude and frequency of palaeofloods, in *Floods, their hydrological, sedimentological and geomorphological implications* edited by K. Beven and P. Carling, pp. 171-183, John Wiley, Chichester, UK, 1989.

Baker, V.R., Global paleohydrological change, *Quaestiones Geographicae, Spec. Issue 4*, 2735, 1995.

Baker, V.R., Hydrological understanding and societal action, *J. Amer. Water Resour. Assoc., 34*, 1-7, 1998.

Baker, V.R., Paleoflood hydrology and the estimation of extreme floods, in *Inland Flood Hazards*, edited by E.E. Wohl, pp. 359-377, Cambridge University Press, Cambridge UK., 2000.

Baker, V.R., Water and the Martian landscape, *Nature, 412*, 228-236, 2001.

Baker, V.R., and G. Pickup, Flood geomorphology of the Katherine Gorge, Northern Territory, Australia, *Geol. Soc. Amer. Bull., 98*, 635-646, 1987.

Baker, V.R., and J.E. Costa, Flood power, in *Catastrophic Flooding,* edited by L. Mayer and D. Nash, pp. 1-21, Allen and Unwin, Boston, 1987.

Baker, V.R., and R.C. Kochel, Flood sedimentation in bedrock fluvial systems, in *Flood Geomorphology,* edited by V.R. Baker, R.C. Kochel, and P.C. Patton, pp. 123-137, John Wiley and Sons, New York, 1988.

Baker, V.R., R.C. Kochel, and P.C. Patton, Long-term flood frequency analysis using geological data, *IAHS-AISH Publ. 128*, 3-9, 1979.

Baker, V.R., R.C. Kochel, P.C. Patton, and G. Pickup, Paleohydrologic analysis of Holocene flood slack-water sediments, in *Modern and ancient fluvial systems*, pp. 229-239, *Int. Assoc. Sed. Spec. Publ. 6*, 1983.

Baker, V.R., G. Pickup, and H.A. Polach, Radiocarbon dating of flood events, Katherine Gorge, Northern Territory, Australia, *Geology, 13*, 344-347, 1985.

Baker, V.R., G. Pickup, and R.H. Webb, Paleoflood hydrologic analysis at ungaged sites, central and northern Australia, in *Regional Flood Frequency Analysis,* edited by V.P. Singh, pp. 321-334, D. Reidel, Boston, 1987.

Baker, V.R., R.C. Kochel, and P.C. Patton, editors, *Flood Geomorphology,* 503 pp., John Wiley and Sons, New York, 1988.

Benito, G., V.R. Baker, and K.J. Gregory, editors, *Palaeo-hydrology and Environmental Change*, 353 pp., Wiley, Chichester, UK, 1998.

Benito, G., A. Sopeña, Y. Sánchez-Moya, M.J. Machado, and A. Pérez-González, Palaeoflood record of the Tagus River (central Spain) during the Pleistocene and Holocene, *Geomorph.*, in press.

Benson, M.A., Use of historical data in flood frequency analysis, *Trans. Amer. Geophys. Union, 31*, 419-424, 1950.

Benson, M.A., Evolution of methods for evaluating the occurrence of floods, *U.S. Geol. Surv. Water-Supply Pap. 1580-A*, 30 pp., 1962.

Benson, M.A., Factors affecting the occurrence of floods in the southwest, *U.S. Geol. Surv. Water-Supply Pap. 1580-D*, 70 pp., 1964.

Box, G.E.P., and G.M. Jenkins, *Time series analysis*, 553 p., Holden-Day, San Francisco, 1971.

Branson, J., A.G. Brown, and K.J. Gregory, editors, *Global Continental Changes: The Context of Palaeohydrology*, 272 pp., Geol. Soc., London, 1996.

Bretz, J.H., Valley deposits immediately east of the Channeled Scablands of Washington, *J. Geol., 37*, 393-427, 1929.

Bretz, J.H., H.T.U. Smith, and G.E. Neff, Channeled Scabland of Washington; new data and interpretations, *Geol. Soc. Amer. Bull., 67*, 957-1049, 1956.

Carr, E.H., *What is History?*, Vintage Books, NY, 209 pp., 1961.

Cayan, D.R., and Webb, R.H., 1992, El Niño/Southern Oscillation and streamflow in the western United States, pp. 29-68 in *El Niño, Historical and Paleoclimatic Aspects of the Southern Oscillation* edited by H.F. Diaz and V. Markgraf, Cambridge Univ. Press, Cambridge, England, 1992.

Cerling, T.E., and H. Craig, Geomorphology and in situ cosmogenic isotopes, *Ann. Rev. Earth Plan. Sci., 22*, 273-317, 1994.

Chatters, J.C. and K.A. Hoover, Response of the Columbia River fluvial system to Holocene climatic change, *Quat. Res., 37*, 42-59, 1994.

Church, M.A., Paleohydrological reconstructions from a Holocene valley fill, in *Fluvial Sedimentology,* edited by A.D.

Miall, pp. 743-772, Can. Soc. Petrol. Geol., Memoir 5, 1978.

Cohn, T.A., and J.R. Stedinger, Use of historical information in a maximum-likelihood framework, *J. Hydrol., 96*, 215-223, 1987.

Cohn, T.A., W.L. Lane, and W.G. Baier, An algorithm for computing moments-based flood quantile estimates when historical information is available, *Water Resour. Res., 33*, 2089-2096, 1997.

Cohn, T.A., W.L. Lane, and J.R. Stedinger, Confidence intervals for expected moments algorithm flood quantile estimates, *Water Resour. Res., 37*, 1695-1706, 2001.

Costa, J.E., Holocene stratigraphy in flood-frequency analysis, *Water Resour. Res., 14*, 626-632, 1978.

Costa, J.E., Paleohydraulic reconstruction of flash-flood peaks from boulder deposits in the Colorado Front Range, *Geol. Soc. Amer. Bull., 94*, 986-1004, 1983.

Costa, J.E., A history of paleoflood hydrology in the United States, 1800-1970, in *History of Hydrology,* edited by E.R. Landa and S. Ince, pp. 49-53, Amer. Geophys. Union, History of Geophysics, Vol. 3,Washington, DC, 1987a.

Costa, J.E., Hydraulics and basin morphometry of the largest flash floods in the conterminous United States, *J. Hydrol., 93*, 313-338, 1987b.

Creager, W.P., Possible and probable future floods, *Civ. Eng. N.Y., 9*, 668-670, 1939.

Crippen, J.R., and C.D. Bue, Maximum floodflows in the conterminous United States, *U.S. Geol. Surv. Water-Supply Paper 1887*, 52 pp., 1977.

Cudworth, A.G., *The Flood Hydrology Manual*, 243 pp., Bureau of Reclamation, Washington, D.C., 1989.

Dury, G.H., Discharge prediction, present and former, from channel dimensions, *J. Hydrol., 30*, 219-245, 1976.

Ely, L.L., Response of extreme floods in the south-western United States to climatic variations in the late Holocene, *Geomorph., 19*, 175-201, 1997.

Ely, L.L., and V.R. Baker, Reconstructing paleoflood hydrology with slackwater deposits: Verde River, Arizona, *Phys. Geog., 6*, 103-126, 1985.

Ely, L.L., R.H. Webb, and Y. Enzel, Accuracy of post-bomb ^{137}Cs and ^{14}C in dating fluvial deposits, *Quat. Res., 38*, 196-204, 1992.

Ely, L.L., Y. Enzel, V.R. Baker, and D.R., Cayan, A 5000-year record of extreme floods and climate change in the southwestern United States, *Science, 262*, 410-412, 1993.

Ely, L.L., Y. Enzel, V.R. Baker, V.S. Kale, and S. Mishra, Changes in the magnitude and frequency of late Holocene monsoon floods on the Narmada River, central India, *Geol. Soc. Amer. Bull., 108*, 1134-1148, 1996.

Enzel, Y., L.L. Ely, P.K. House, V.R. Baker, and R.H. Webb, Paleoflood evidence for a natural upper bound to flood magnitudes in the Colorado River basin, *Water Resour. Res., 29*, 2287-2297, 1993.

Ethridge, F.G., and S.A. Schumm, Reconstructing paleochannel morphologic and flow characteristics —Limitations and assessment, pp. 703-721 in *Fluvial Sedimentology*, edited by A.D. Miall, Can. Soc. Petrol. Geol., Mem. 5, 1978.

Forman, S.L., J. Pierson, and K. Leper, Luminescence geochronology, in *Quaternary Geochronology: Methods and Applications,* edited by J.S. Noller, J.M. Sowers, and W.R. Lettis, pp. 157-176, American Geophysical Union, Washington, D.C., 2000.

Francis, F., J.D. Salas, and D.C. Boes, Flood frequency analysis with systematic and historical or paleoflood data based on the two-parameter general extreme value models, *Water Resour. Res., 30*, 1653-1664, 1994.

Fucheng, S., Y. Yungjun, and H. Manhua, Investigation and verification of extraordinary large floods of the Yellow River in China, in *Proceedings of the U.S. China Bilateral Symposium on the Analysis of Extraordinary Flood Events*, Nanjing, China, 1985.

Gillieson, D., D.I. Smith, M. Greenaway, M. Ellaway, Flood history of the limestone ranges in the Kimberley region, western Australia, *Appl. Geogr., 11*, 105-123, 1991.

Graham, W.J., Should dams be modified for the probable maximum flood?, *J. Amer. Water Resour. Assoc., 36*, 953-963, 2000.

Greenbaum, N., A.P. Schick, and V.R. Baker, The paleoflood record of a hyperarid catchment, Nahal Zin, Negev Desert, Israel, *Earth Surf. Proc. Landforms, 25*, 951-971, 2000.

Gregory, K.J., L. Starkel, and V.R. Baker, editors, *Global Continental Palaeohydrology*, 334 pp., Wiley, Chichester, UK, 1995.

Hirsch, R.M., and J.R. Stedinger, Plotting positions for historical floods and their precision, *Water Resour. Res., 23*, 715-727, 1987.

Hirschboeck, K.K., *Hydroclimatology of flow events in the Gila River basin, central and southern Arizona*, 335 pp., unpubl. Ph.D. thesis, Univ. Ariz., Tucson, AZ, 1985.

Hirschboeck, K.K., Catastrophic flooding and atmospheric circulation anomalies, in *Catastrophic flooding,* edited by L. Mayer and D. Nash, pp. 25-56, Allen and Unwin, London, 1987.

Hirschboeck, K.K., Flood hydroclimatology, in *Flood Geomorphology,* edited by V.R. Baker, R.C Kochel, and P.C. Patton, pp. 27-49, John Wiley and Sons, New York, 1988.

Hosking, J.R., and J.R. Wallis, The value of historical data in flood frequency analysis, *Water Resour. Res., 22*, 1606-1612, 1986a.

Hosking, J.R., and J.R. Wallis, Paleoflood hydrology and flood frequency analysis, *Water Resour. Res., 22*, 543-550, 1986b.

House, P.K., and V.R. Baker, Paleohydrology of flash floods in small desert watersheds in western Arizona, *Water Resour. Res., 37*, 1825-1839, 2001.

House, P.K. and P.A. Pearthree, A geomorphologic and hydrologic evaluation of an extraordinary flood discharge estimate: Bronco Creek, Arizona, *Water Resour. Res., 31*, 3059-3073, 1995.

Jahns, R.H., Geologic features of the Connecticut Valley, Massachusetts, as related to recent floods, *U.S. Geol. Surv. Water Supply Pap. 996*, 158 pp., 1947.

Jarrett, R.D., Paleohydrology and its value in analyzing floods and droughts, *U.S. Geol. Surv. Water Supply Pap. 2375*, 105-116, 1991.

Jarrett, R.D., and E.M. Tomlinson, Regional interdisciplinary paleoflood approach to assess extreme flood potential, *Water Resour. Res., 36*, 2957-2984, 2000.

Jin, J., and J.R. Stedinger, Flood frequency analysis with regional and historical information, *Water Resour. Res., 25*, 925-936, 1989.

Jones, A.P., H. Shimazu, T. Oguchi, M. Okuno, and M. Tokutake, Late Holocene slackwater deposits on the Nakagawa River, Tochigi Prefecture, Japan, *Geomorph.*, *39*, 39-51, 2001.

Kale, V.S., L.L. Ely, Y. Enzel, and V.R. Baker, Geomorphic and hydrologic aspects of monsoon floods on the Narmada and Tapi Rivers in central India, *Geomorph.*, *10*, 157-168, 1994.

Kellogg, M.J., 2001, *Paleoflood hydrology of the Carson River, Nevada and California*, 128 pp., unpubl. M.S. Thesis, Univ. Nevada, Reno, NV, 2001.

Klawon, J.E., D.R. Levish, D.R.H. O'Connell, Paleofloods and paleosols on the North Platte River: Stratigraphic records of exceedance and nonexceedance, *Geol. Soc. Amer. Abs. Progs.*, *32*, A-509, 2000.

Klêmes, V., Statistics and probability: Wrong remedies for a confused hydrologic modeler, in *Statistics for the Environment 2: Water Related Issues*, edited by V. Barnett and K.F. Turkman, pp. 345-366, John Wiley and Sons, London, 1994.

Klêmes, V., *Common Sense and Other Heresies: Selected Papers on Hydrology and Water Resources Engineering*, Can. Water Resour. Assoc., Cambridge, Ontario, 378 pp., 2000.

Knox, J.C., Responses of floods to Holocene climatic change in the upper Mississippi Valley, *Quat. Res.*, *23*, 287-300, 1985.

Knox, J.C., Sensitivity of modern and Holocene floods to climate change, *Quat. Sci. Rev.*, *19*, 439-457, 2000.

Kochel, R.C., and V.R. Baker, Paleoflood hydrology, *Science*, *215*, 353-361, 1982.

Kochel, R.C., and V.R. Baker, Paleoflood analysis using slack-water deposits, in *Flood Geomorphology,* edited by V.R. Baker, R.C. Kochel, and P.C. Patton, pp. 357-376, John Wiley and Sons, New York, 1988.

Kochel, R.C., V.R. Baker, and P.C. Patton, Paleohydrology of southwest Texas, *Water Resour. Res.*, *18*, 1165-1183, 1982.

Lave, L.B., D. Resendiz-Carrillo, and F.C. McMichael, Safety goals for high-hazard dams: Are dams too safe?, *Water Resour. Res.*, 26(7), 1383-1391, 1990.

Levish, D.R., D.A. Ostenaa, and D.R.H. O'Connell, Paleoflood hydrology and dam safety, in *Waterpower '97*, pp. 2205-2214, Proc. Intl. Conf. Hydropower, Aug. 5-8, 1997, Atlanta, GA, 1997.

Lewin, J., M.G. Macklin, and J.C. Woodward, Late Quaternary fluvial sedimentation in the Voidomatis basin, Epirus, N.W. Greece, *Quat. Res.*, *35*, 103-115, 1991.

Maizels, J.K., Palaeovelocity and palaeodischarge determination for coarse gravel deposits, in *Background to Palaeohydrology,* edited by K.J. Gregory, pp. 101-139, John Wiley, Chichester, UK, 1983.

Mann, M.P., 2000, *Use of geomorphic information in extending the flood record of the West Walker River, California*, unpubl. MS Thesis, Univ. Nev., Reno, NV, 189 p.

Mansfield, G.R., Flood deposits of the Ohio River, January-February 1937, a study of sedimentation, pp. 693-736, *in* Grover, N.C., *U.S. Geol. Surv. Water Supply Pap. 838,* 1938.

Mantua, N.J., S.R. Hare, Y. Zhang, J.M. Wallace, and R.C. Francis, A Pacific interdecal climate oscillation with impacts on salmon production, *Bull. Amer. Met. Soc., 78*, 1069-1079, 1997.

Matalas, N., and B. Jacobs, A correlation procedure for augmenting hydrologic data, *U.S. Geol. Surv. Prof. Paper 434-E*, 1964.

Matthai, H.F., Floods, pp. 97-120 *in* M.G. Wolman and H.C. Riggs, *Surface-water hydrology*, vol. O-1, The Geology of North America, Geol. Soc. Amer., Boulder, CO, 1990.

National Research Council, *Estimating probabilities of extreme floods, Methods and recommended research,* 141 pp., Comm. Tech. Est. Probs. Extreme Floods, Natl. Res. Coun., Natl. Acad. Press, Washington, DC, 1988.

O'Connell, D.O., D. Levish, and D. Ostenaa, Bayesian flood frequency analysis with paleohydrologic bounds for late Holocene paleofloods, Santa Ynez River, California, in *Twenty Years Later: What Have We Learned Since the Big Thompson Flood?*, edited by E. Gruntfest, pp. 183-196, Natural Haz. Res. and Application Info. Center, Spec. Pub., 33, Univ. Colo., Boulder, CO, 1997.

O'Connell, D.R.H., D.A. Ostenaa, D.R. Levish, and R.E. Klinger, Bayesian flood frequency analysis with paleohydrologic bound data, *Water Resour. Res.*, in press.

O'Connor, J.E., Hydrology, hydraulics, and geomorphology of the Bonneville Flood, *Geol. Soc. Am. Spec. Pap. 274*, 83 pp., 1993.

O'Connor, J.E., and R.H. Webb, Hydraulic modeling for paleoflood analysis, in *Flood Geomorphology,* edited by V.R. Baker, R.C. Kochel, and P.C. Patton, pp. 383-402, John Wiley, New York, 1988.

Orchard, K.L., *Paleoflood hydrology of the San Juan River, southeastern Utah*, 137 pp., unpubl. M.S. thesis, Dept. Hydrol. Water Resour., Univ. Ariz., Tucson, AZ, 2001.

Partridge, J.B., and V.R. Baker, Paleoflood hydrology of the Salt River, Arizona, *Earth Surf. Proc. Landf.*, *12*, 109-125, 1985.

Patton, P.C., Measuring the rivers of the past: A history of fluvial paleohydrology, in *History of Hydrology,* edited by E.R. Landa and S. Ince, pp. 55-67, American Geophysical Union, History of Geophysics, No. 3, Washington, DC, 1987.

Patton, P.C., and V.R. Baker, Geomorphic response of central Texas stream channels to catastrophic rainfall and runoff, in *Geomorphology of Arid Regions, Publ. in Geomorph.*, edited by D.O. Doehring, pp. 189-217, State Univ. New York, Binghamton, NY, 1977.

Patton, P.C., and Dibble, D.S., Archaeologic and geomorphic evidence for the paleohydrologic record of the Pecos River in west Texas, *Amer. J. Science*, *282*, 97-121, 1982.

Patton, P.C., V.R. Baker, and R.C. Kochel, Slackwater deposits: A geomorphic technique for the interpretation of fluvial paleohydrology, in *Adjustments of the Fluvial System*, edited by D.D. Rhodes and G.P. Williams, pp. 225-252, Kendall-Hunt, Dubuque, IO, 1979.

Pickup, G., G. Allan, and V.R. Baker, History, palaeochannels and palaeofloods of the Finke River, central Australia, in *Fluvial geomorphology of Australia*, edited by R.F. Warner, pp. 177-200, Academic Press, Sydney, Australia, 1988.

Pielke, R.A., Jr., *Midwest flood of 1993: Weather, climate, and societal impacts*, 159 pp., Env. Soc. Impacts Group, Natl. Center Atmos. Res., Boulder, CO, 1996.

Pielke, R.A., Jr., Nine fallacies of floods, *Clim. Change*, 42, 413-438, 1999.

Pielke, R.A., Jr., and R. Byerly, Jr., Beyond basic and applied, *Phys. Today*, 51(2), 42-46, 1998.

Pielke, R.A., Jr., and M.W. Downton, Precipitation and damaging floods: Trends in the United States, 1932-97, *J. Climate*, 13, 3625-3637, 2000.

Sigafoos, R.S., Vegetation in relation to flood frequency near Washington, DC, *U.S. Geol. Surv. Prof. Pap. 424-C*, 248-249, 1961.

Sigafoos, R.S., Botanical evidence of floods and flood-plain deposition, *U.S. Geol. Surv. Prof. Pap. 485-A*, 1-35, 1964.

Stedinger, J.R., Flood frequency analysis and statistical estimation of flood risk, in *Inland Flood Hazards: Human, Riparian, and Aquatic Communities*, edited by E.E. Wohl, pp. 334-358, 2000.

Stedinger, J.R., and T.A. Cohn, Flood frequency analysis with historical and paleoflood information, *Water Resour. Res., 22*, 273-286, 1986.

Stedinger, J.R., and V.R. Baker, Surface water hydrology: Historical and paleoflood information, *Rev. Geophys., 25*, 119-124, 1987.

Stedinger, J.R., R. Therivel, and V.R. Baker, The use and value of historical and paleoflood information in flood frequency analysis, *8th Annual USCOLD Lecture*, January 11-15, 1988 Meeting, U.S. Committee on Large Dams, Phoenix, AZ, 1988.

Stewart, J.E., and G.L. Bodhaine, Floods in the Skagit River basin, Washington, *U.S. Geol. Surv. Water Supply Pap. 1527*, 66 pp., 1961.

Stuiver, M., A high-precision calibration of the A.D. radiocarbon time scale, *Radiocarbon, 24*, 1-26, 1982.

Stuiver, M., and P.J. Reimer, Extended ^{14}C database and revised CALIB radiocarbon calibration program (ver. 4.1), *Radiocarbon, 35*, 215-230, 1993.

Stuiver, M., P.J. Reimer, E. Bard, J.W. Beck, G.S. Burr, K.A. Hughen, B. Kromer, F.G. McCormac, J.v.d. Plicht, and M. Spurk, INTCAL98 Radiocarbon age calibration 24,000-0 cal BP., *Radiocarbon, 40*, 1041-1083, 1998.

Tarr, R.S., A hint with respect to the origin of terraces in glaciated regions, *Amer. J. Sci., 144*, 59-61, 1892.

Tasker, G.D., and J.R. Stedinger, Regional regression of flood characteristics employing historical information, *J. Hydrol., 96*, 255-264, 1987.

Tasker, G.D., and J.R. Stedinger, An operational GLS model for hydrologic regression, *J. Hydrol., 111*, 361-375, 1989.

Thomas, B.E., H.W. Hjalmarson, and S.D. Waltemeyer, Methods of estimating magnitude and frequency of floods in the southwestern United States, *U.S. Geol. Surv. Water-Supply Pap. 2433*, 195 pp., 1997.

Thomas, W.O., Jr., A uniform technique for flood frequency analysis, *J. Water Resour. Plan. Manage., 111*, 321-337, 1985.

U.S. Water Resources Council, *Guidelines for determining flood flow frequency*, U.S. Interagency Adv. Comm. Water Data, Hydrol. Subcomm., Bulletin 17B, Washington, DC, 1982.

van der Vink, G., R.M. Allen, J. Chapin, M. Crooks, W. Fraley, J. Krantz, A.M. Lavigne, A. LeCuyer, E.K. MacColl, W.J. Morgan, B. Ries, E. Robinson, K. Rodriquez, M. Smith, and K. Sponberg, Why the United States is becoming more vulnerable to natural disasters, *EOS, 79*, 533-537, 1998.

Waylen, P., and M.K. Woo, Prediction of annuals floods generated by mixed processes, *Water Resour. Res., 18*, 1283-1286, 1982.

Webb, R.H., and V.R. Baker, Changes in hydrologic conditions related to large floods on the Escalante River, South-central Utah, in *Regional flood frequency,* edited by V.P. Singh, pp. 309-323, D. Reidel, Boston, 1987.

Webb, R.H., and S.L. Rathburn, Paleoflood hydrologic research in the Southwestern United States, *Trans. Res. Rec., 1201*, 9-21, 1989.

Webb, R.H., and J.L. Betancourt, Climatic variability and flood frequency of the Santa Cruz River, Pima County, Arizona, *U.S. Geol. Surv. Water-Supply Pap. 2379*, 40 pp., 1992.

Webb, R. H., J.E. O'Connor, and V.R. Baker, Paleohydrologic reconstruction of flood frequency on the Escalante River, South-central Utah, in *Flood Geomorphology,* edited by V.R. Baker, R.C. Kochel, and P.C. Patton, pp. 403-418, John Wiley, New York, 1988.

Webb, R.H., S.S. Smith, and V.A.S. McCord, *Historic Channel Change of Kanab Creek, Southern Utah and Northern Arizona*, Grand Canyon Natl. Hist. Assoc. Mono. 9, 91 pp., 1991.

Williams, G.P., Paleohydrological methods and some Swedish fluvial environments:, I-Cobble and boulder deposits, *Geograf. Annaler, 65A*, 227-243, 1983.

Williams, G.P., Paleohydrologic equations for rivers, in *Developments and Applications of Geomorphology* edited by J.E. Costa and P.J. Fleisher, pp. 353-367, Springer-Verlag, Berlin, 1984.

Williams, G.P., Paleofluvial estimates from dimensions of former channels and meanders, in *Flood Geomorphology,* edited by V.R. Baker, R.C. Kochel, and P.C. Patton, pp. 321-334, John Wiley, New York, 1988.

Wohl, E.E., R.H. Webb, V.R. Baker, and G. Pickup, Sedimentary flood records in the bedrock canyons of rivers in the monsoonal region of Australia, *Water Resour. Papers, 107*, 102 pp., 1994.

Yevjevich, V., and N.B. Harmancioglu, Research needs on flood characteristics, in *Application of Frequency and Risk in Water Resources,* edited by V.P. Singh, pp. 1-21, D. Reidel, Boston, 1987.

Zawada, P.K., *Palaeoflood hydrology of selected South African rivers*, 348 pp., unpubl. Ph.D. thesis, Univ. Port Elizabeth, South Africa, 1996.

Zawada, P.K., and J. Hattingh, Studies on the palaeoflood hydrology of South African rivers, *South African J. Sci., 90*, 567-568, 1994.

V.R. Baker, Department of Hydrology and Water Resources, University of Arizona, Tucson, AZ 85721.

R.H. Webb, U.S. Geological Survey, 1675 W. Anklam Rd., Tucson, AZ 85745.

P.K. House, Nevada Bureau of Mines and Geology, University of Nevada, Reno, NV 89557.

Climate Variability and Flood Frequency at Decadal to Millennial Time Scales

Kelly T. Redmond[1], Yehouda Enzel[2], P. Kyle House[3], Franco Biondi[4]

The climate research community has generally accepted the notion of the climate system as an entity that routinely produces nonstationary time series. Because there is an intimate tie between climate and weather, and between weather and floods, it should not be surprising to encounter nonstationary flood series with some regularity. There is an increasingly large body of paleohydrology and paleoclimate data that offers significant promise in better understanding of the climate system and how it varies over time scales formerly inaccessible to close analysis by those who have concentrated on systematic records. The same information can provide insights into ways to place contemporary systematic total and peak streamflow measurements into longer time contexts spanning centuries or more. Most observations and the majority of studies to date have concentrated on arid and semi-arid regions, where confounding influences are minimized and interpretations pertaining to the use of paleoflood data are cleaner. Building from prior studies, this work furnishes further examples utilizing additional paleohydrology data from the southwest United States. El Niño/La Niña connections to climate in this region and elsewhere have received intense scrutiny. For that reason ENSO is used as one example mechanism (although there are certainly others) to illustrate how climate variability can lead to flood variability on a variety of time scales of interest to flood frequency analysts. Flood frequency analysis is essentially an exercise in forecasting, and the quality of the forecast of the future distribution is influenced considerably by the degree to which all scales, local to global, of the flood-producing system are understood. There are, however, logistical barriers and limitations in using paleohydrology data, and care is required in selecting appropriate records for the application at hand.

INTRODUCTION

Climate elements vary on all time scales. As the mean state of climate fluctuates, the associated probability distributions of particular types of events occurring at smaller time scales also fluctuate. Among such smaller time scale events is a subset we normally call "weather." With the exception of failures of artificial structures or glacial floods, it is a further subset of these weather events that, individually or in combination, produces nearly all river floods. To perform analyses of the frequency of these floods, certain assumptions must be made about the underlying probability distributions. It is one of those assumptions that we wish to

[1]Western Regional Climate Center, Desert Research Institute, Reno, Nevada
[2]Institute of Earth Sciences, Hebrew University of Jerusalem, Jerusalem, Israel
[3]Nevada Bureau of Mines and Geology, University of Nevada, Reno, Nevada
[4] Department of Geography, University of Nevada, Reno, Nevada

Ancient Floods, Modern Hazards: Principles and Applications of Paleoflood Hydrology
Water Science and Application Volume 5, pages 21-45

discuss and examine more closely in this paper, namely, that these distributions do not vary with time.

Since climate varies, and flood occurrence is tied to the behavior of climate, then flood frequencies must be expected to vary also. Proxy information from pre-historical (prefixed as "paleo-") sources plays a dual role in establishing how climate varies, and in how flood occurrence varies, at the longer time scales. In modern times, climate and flood records can be obtained independently. In the more distant past, this distinction becomes more blurry. As will be shown, the clearest and cleanest such examples are in arid regions of the earth.

What we call "climate" can be viewed in two ways. In a descriptive sense, climate consists of the entire set of statistics that completely characterize observed behavior (means, all higher moments, autocorrelations, spatial relationships, etc.) at a time scale of interest. Climate may also be thought of as the entire distribution of possibilities (sequences of events), consistent with current boundary conditions, that the system can exhibit, at the time scale of interest. The specific record we possess constitutes a particular observed realization (one of many) that has been drawn from this distribution. Climate in this case acts in essence as a kind of "enabling" factor. These distinctions follow the logic of Bryson [1997], who refers to them with a lowercase or uppercase "C" respectively: "climate" is what did happen, "Climate" is what could happen. If climate is viewed as "setting the stage" for allowable possibilities (the second definition), it exerts a kind of "control" over possible behavior and outcomes. It is in this sense that the meaning of the term "control" is intended in this paper, by causing modifications to the characteristics of the underlying probability distributions of smaller time scale events such as flood-producing episodes.

Although climate does vary on all time scales, the decision-making time scale for flood-frequency analysis is typically in the range of a few decades to a few centuries, and millennia for critical hydraulic structures. Consequently, it is on those scales where most of the emphasis here will be placed. The shorter of these time scales can be easily evaluated with a variety of systematic data types, but we must rely on proxy information such as paleoclimate and paleoflood records to address longer time frames. Such information can improve our understanding about context and limitations of approaches to evaluating flood frequency. In some cases it can be directly incorporated into an estimation procedure [e.g., Blainey et al., this volume; Stedinger and Cohn, 1986; Levish et al., 1997].

The presence of temporal variability in climate and flood frequency is clearly evident in recent high-resolution systematic data for shorter time scales in western North America. The first thorough examination of this connection, consisting of both a literature survey and an analysis of Arizona flood records, was performed by Hirschboeck [1985], and later elaborated [Hirschboeck, 1987a, 1987b, 1988]. These studies highlighted the link between climate variability and flood frequency distributions (including changes in the mixtures of these distributions), and also the associated causative mechanisms, and have served as a foundation for subsequent studies by others.

In the context of this monograph, we are particularly interested in evaluating the information content of paleoflood and paleohydrological information with respect to long-term climate variability. There are numerous instances in the literature of reasonably successful attempts to interpret paleoflood and paleohydrologic data in the context of climate variability [Ely et al., 1993, 1996; Knox, 1983, 2000; Enzel et al., 1989; Benito et al., 1996; Chatters and Hoover, 1986; Fuller et al., 1998]. The intent here is not to replicate such studies, but rather to review examples and to generally evaluate the relative merits of drawing such inferences from higher-resolution data types. At the onset, we note that there are compelling lines of evidence within each type of data—high resolution systematic data through low resolution paleo—that support the notion that climate does vary on all time scales, and similarly, so do a variety of intimately related hydrologic processes governing annual streamflow characteristics and floods.

In a recent paper Knox [2000] summarized over twenty years of research on modern floods, Holocene paleofloods, and related alluvial stratigraphy in the Upper Mississippi Valley. He concluded that the mean state of the climate exerts an influence or control on flood frequency and compared his results with other flood and Holocene paleoflood studies in the United States and elsewhere [e.g., Lamb, 1977; Wells, 1990; Said, 1993; Enzel, 1992; Ely et al., 1993, 1996; Benito et al., 1996; Ely, 1997]. Also, quite recently, Hirschboeck et al. [2000] summarized regional paleoflood records from various parts of the world and likewise concluded that there is abundant evidence for changes in large flood frequency over time scales of hundreds to thousands of years, and that these changes are inextricably linked to climate variability. These studies collectively provide persuasive support for the control of climate on the frequency of large floods across a wide range of time scales: from decadal scales during the 20th century to millennial scales during the Holocene. Knox [2000] and the others have argued that even minor climate variations at these time scales resulted in significant changes in magnitude and frequencies of floods.

CLIMATE VARIABILITY AND FLOOD FREQUENCY

The issue at hand is not an obscure or abstract scientific problem. It encompasses some of the basic concepts in flood hydrology and has tremendous impact on the way we practice flood risk analysis. In two clear regional examples, Knox [1983, 1984, 2000; upper Mississippi River] and

Webb and Betancourt [1992; Santa Cruz River, southern Arizona] have shown that the twentieth century experienced dramatic changes in flood frequency and that these changes are directly related to decadal-scale climate changes. Using almost 100 years of record they showed that the 1920s to 1950s had very different flood exceedance probabilities from the prior and later decades and that variations in synoptic-scale, flood-producing causes have been responsible for the decadal-scale variations in the probabilities of floods. The examples illustrate that if flood-risk analyses for these basins were based on data from those three decades the resulting design flood would be much lower than the design flood based either on the entire record or on the records from the other decades. This simple observation shows, when considering flood risks, the utility and need for longer flood records and for simultaneous observation of the long-term climate and its relationship to floods, in order to establish and understand historical context. As another example, a dramatic increase in the frequency of very large floods occurred on the American River (Figure 1, also on CD) in the central Sierra Nevada of California during the period 1950-1997 compared with 1907-1950 [NRC, 1999] (Figure 2, also on CD). Flood records from the earlier period were used to design Folsom Reservoir, which protects the city of Sacramento immediately downstream. Although reflecting a significant scientific question for climate research, this also has posed a very difficult practical dilemma for California's capital city in terms of how to respond to changing flood hazards. The flood increase clearly represents a real shift in the flood generating function [National Research Council, 1999], but it is not certain what mechanism or mechanisms are responsible.

Recent climate research has shown that the two to three decades from the 1920s to the 1950s appear to differ from the rest of the 20th century in certain respects for large regions. For example, the relationships between the oceans and the atmosphere seem to be different in the tropical Pacific during this interval. Furthermore, the relationships between simple oceanic and atmospheric indices of tropical Pacific behavior and large-scale continental hydrology are also different from those of earlier and later decades in parts of eastern and western North America [e.g., Tourre et al., 1999; McCabe and Dettinger, 1999], and for the region that experiences the Asian monsoon [e.g., Torrence and Webster, 1999]. Numerous hydrologic projects in the world were initiated and built during the 1950s and 1960s, with their design floods either based solely or mainly on prior data from decades with possibly different large scale climate behavior. The earlier example of the Colorado River Compact, signed in November 1922, is particularly well known [Stockton and Jacoby, 1975; Stockton, 1975]. The consequences of the associated decisions continue to reverberate every day—80 years later—in the southern half of the western United States. Better knowledge of the river's long-term history would have been extraordinarily helpful to society. Paleoflood and paleoclimate data clearly can provide an important role in evaluating the long-term context of short systematic records

In the absence of information to the contrary, the traditional approach to estimation of hydrologic risk appears reasonable. When long-term relationships between climate and floods are clearly present, however, willingness to question conventional assumptions—discussed below—is warranted.

Stationarity as an Assumption

A basic assumption contained in the United States federal guidelines for determining flood flow frequencies, emulated in much of the world, is that "...flood flows are not affected by climatic trends or cycles" and as a result the "annual maximum peak flows may be considered a sample of random and independent events" [U.S. Water Resources Council, 1981]. Flood characteristics are usually treated as "i.i.d"— independent and identically distributed (i.e., drawn from exactly the same population). These assumptions allow flood hydrologists to use any available short-term data without necessity for consideration of climate variability on time scales longer than the record length. They form the basis for operational practices employed by a large community of individuals and organizations. If these underlying assumptions of uniform climatic conditions and stationary mean and variance of the annual flood

Figure 1. Map of the southwestern United States showing the locations of river basins discussed in the text.

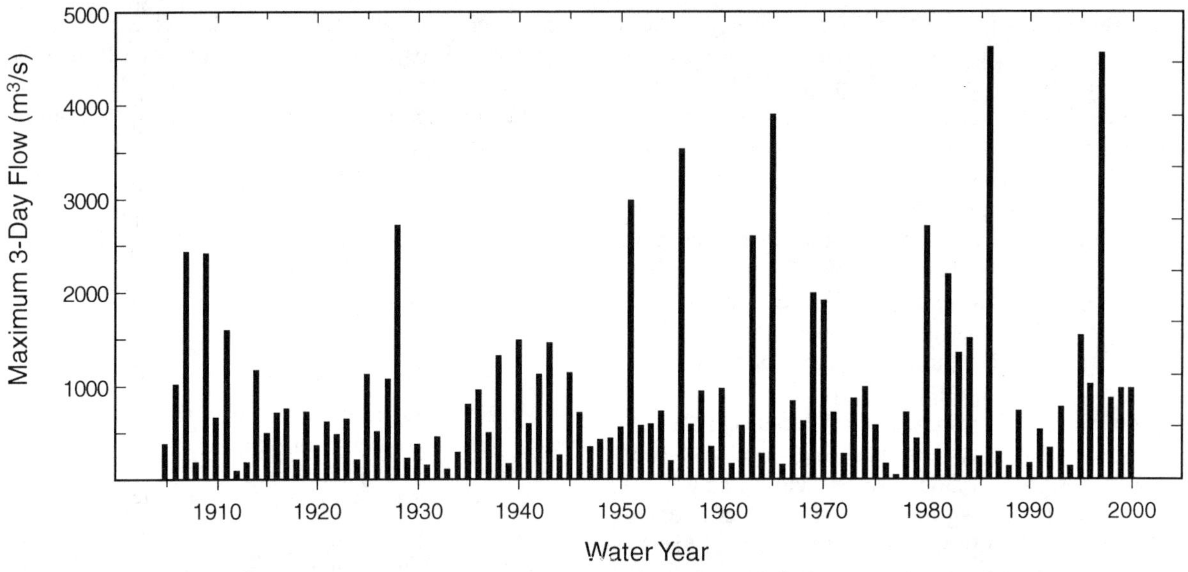

Figure 2. Annual maximum 3-day flow of the American River at Fair Oaks, California, 1905-2000. Expressed as average discharge, in cubic meters per second. Plotted in ending year.

series are not correct, modifications in traditional approaches merit consideration. It should also be noted that the operational guidelines referenced above have general provisions that do allow consideration of non-i.i.d. circumstances, but they are not often invoked.

The goal of our effort here is to use a few well-documented cases to challenge the widely held but unrealistic view that flood-frequency characteristics are invariant and independent of the potentially controlling influence of climate at longer time scales and larger space scales. The assumption that recent gaged records are free of nonstationarity has not been universally established, is at least open to question in many circumstances, and is clearly untrue in certain locations during the observed recent past. (We restrict this discussion to gages where artificial influences are clearly minimal, as determined by corroboration from other types of climate evidence.) It seems quite likely that such climate effects on flood characteristics will occur at all time scales. An associated and implicit assumption is that Climate (with a capital "C", the probability distribution from which events are drawn) changes "slowly." This is increasingly open to question [e.g., Taylor, 1999, and references therein]; the observed rates of Climate change span a wide range of magnitudes.

FLOOD FREQUENCY ANALYSES AS FORECASTS

Flood frequency analysis should most properly be viewed as a forecast problem. Since all the events of interest lie in the future, flood-risk estimation is of necessity a de facto forecast. Past statistics are utilized, of course, but the unstated assumption is that the future will be like the past. This assumption is itself a forecast. More specifically, the assumption is that hydroclimatic statistics of the future will be like the hydroclimatic statistics of the past. Even neglecting the potential effects of global climate change, our experience base for testing this assumption is typically limited to only a few decades to a century or so in most cases. There is clearly much room for doubt about the validity of such an assumption. If only systematic data are at our disposal, we are unable to resolve this question. Presumably the practitioner has a time frame of interest for the application at hand: the next decade, next 50, or next 200 years.

With respect to the use of past measurements or evidence from any era, modern or paleo, the most general question which dominates the decision-making framework is this: Which part of the climatic past is relevant to which part of the climatic future?

For the case above, what part of the past best reflects the next 10, 50, or 200 years? If we were somehow given a perfect flood data set stretching into the distant past, what part would we, or should we, choose to utilize for the practical decision at hand, and on what basis? No conclusive or encompassing answer has been found; this is part of the research agenda for the future.

Until we better understand the flood-generation system—in its entirety—well enough to predict its future characteristics, we must face the uncertainty inherent in dealing with this pervasive and largely unanswerable (for now) question. In so doing, it is important that we evaluate all relevant information available to help characterize this complex problem. A variety of paleohydrologic and paleoclimatologic data are appropriate for this.

Unfortunately, we are typically in the position of being able to state more confidently what is not the case rather than what is the case. In part because of the history of assumed stationarity, there has been little incentive to develop alternate and complementary methods to those in common use despite increasingly available paleohydrologic data and methodologies.

USES FOR PALEOHYDROLOGY DATA

Paleohydrologic information has several potential roles in flood-frequency analysis. Those roles can be further subdivided. One category involves direct applications which are specific to an individual basin:

1. Bracketing "what's possible." Even if not expressed in strictly quantitative frequency terms, the existence of tangible, physical evidence of such extremes does carry more weight with certain key audiences than theoretical considerations do, whether rightly or wrongly from the perspectives of technical experts.
2. Extreme distributions. In some circumstances such data can provide constraints on the behavior of the upper tails of frequency distributions. The result in this case is quantitative information useful to better define the distributions.

A second category involves application of a more general nature, not necessarily specific to any particular basin, but useful nonetheless:

3. Climate reconstruction and synthesis. In this case paleohydrology information is blended with paleoclimate and modern observations, obtained from all possible sources, and interpreted within a larger framework, with the help of careful studies that numerically simulate the dynamics of climate. Theoretical and observational approaches inform each other—the paleo data are not "acting alone." We thereby acquire a better knowledge of how the entire climate system operates, in all its spatial-temporal complexity, on longer time scales. The resulting benefit here is not specific to the individual basin in question; our subject is a highly interconnected system, with the parts and the whole communicating in both directions. In the long run this indirect approach will prove just as crucial to understanding the flood regime of a particular basin as the more direct approaches, but the path to that knowledge will be indirect and unpredictable.

The blending of paleohydrologic data and climatologic analysis can enhance our overall understanding of climate variability and the likely responses of hydrologic systems. It is possible to choose and compare data types that span a broad range of time scales to identify behavior not apparent from systematic and historic data alone. Paleohydrologic data are tangible records of events that have transpired within the realm of then-prevailing climate. The assumed value of paleohydrologic data is predicated on the validity of using evidence of past conditions as representative of possible future conditions.

A relatively new consideration is the prospect of anthropogenic climate change. With respect to the uncertainty in flood-frequency analysis associated with the prospect of global climate change we offer this comment: Which is more likely: a) that a multi-century to millennial record of large floods captures the variability that a river system will exhibit over (say) the next century?, or b) that the behavior of near-future climate will be so different as to lie completely outside the bounds of what has occurred over at least portions of the period of the paleohydrology or paleoclimate record (which may span several millennia)?

It should be noted that the relationship between "climate" signals and "flood" signals retained in paleo records typically shows greater clarity in arid and semi-arid locations [e.g. Baker et al., 1995]. In many such climates, a large fraction of the annual precipitation occurs during one or a few events. In such places wet regimes are almost synonymous with flood regimes. In many cases, such floods have only one or two main causes. Additionally, evidence remains better preserved, and for longer time periods, and has often not yet been significantly disturbed by human activities, even though populations are now flocking to desert landscapes.

Paleohydrologic and Paleoclimatic Data Types and Applications

Three principal types of paleohydrologic data bear on the discussion presented here: fluvial, lacustrine, and dendrochronological. Fluvial and lacustrine data include geomorphic, sedimentologic, and stratigraphic evidence that document individual extreme events (paleoflood data), or past hydrologic regimes associated with specific long-term average hydrologic conditions (paleoflow data) [e.g., Wohl and Enzel, 1995; Baker, 1987, 1989; numerous references in this volume]. The temporal resolution of typical fluvial and lacustrine data varies from decades at best to several centuries at worst. This evidence is linked to the occurrence of discrete, extreme flood events and can also provide constraints on maximum flood magnitudes over long periods of time.

By contrast, dendrochronologic records (time series of tree ring properties) have annual resolution, an attribute we wish to exploit whenever possible. These properties typically include width, but may extend to density and isotopic ratios. In many geographic and climatic regimes, tree-ring properties are related to annual precipitation and annual streamflow. In some of these settings, particularly in arid regions, tree-growth and annual streamflow are sensitive to seasonal precipitation input. In such areas it is possible to transform tree-ring chronologies into accurate proxies for annual stream flow [e.g. Stockton, 1975; Meko et al., 2001]. In certain favorable circumstances—an important caveat—single individual precipitation events provide the dominant contribution to annual precipitation and to tree growth for particular years, and generate a large flood as they do so.

When such relatively uncommon factors are present simultaneously, tree ring information can provide useful, even invaluable, input for flood studies, especially large floods. In most other cases it cannot. The most favorable circumstances occur nearly exclusively in arid environments, and apply only to the largest and rarest floods. It is also required that the drainage basin be of sufficient size that those largest floods are either exclusively or nearly always caused by large scale synoptic events such as intense rain-dominated winter storms. Only a few basins meet these criteria. Several of these can be found in our study area, the southwestern United States. Fortunately, such events often provide substantial enough soil-moisture recharge [e.g., Stamos et al., 2001] for trees to take advantage. Despite the need for caution in choosing basins and events, the potential ability to date flood events to a specific year is presently unique to tree rings, and motivates further analysis.

Techniques are emerging that enable separate construction of early-wood / late-wood series, using ring density [e.g., Meko and Baisan, 2000] to isolate the monsoon and winter precipitation components of the Arizona climate. Although reconstructions from ring-width measurements cannot fully resolve individual floods, the frequency and magnitude of paleofloods can be obtained by dendrochronological dating of scars left on riparian trees by floating debris [McCord, 1996, Gottesfeld, 1996].

Limitations and Barriers to Use of Paleo Information

The paleohydrology information that is available has probably been neglected in conventional flood-frequency analysis for several main reasons:

Many types of longer and older climate records generally consist of time-averaged quantities rather than events. It is assumed that instantaneous flood events will not be detectable in most long-term low-time-resolution paleoclimatic records. Even tree rings, with their relatively high temporal resolution (for paleoclimate data), cannot sufficiently resolve events of interest, unless very particular circumstances (discussed above) apply. The mismatch of scales becomes more problematic when we seek accurate representations of the frequency distributions of floods or storms with lifetimes of up to a few days during intervals which, for example, might stretch 365,242 days (one millennium).

The relative ease of practicing flood hydrology without consideration of fluctuations of climate has obvious attractions.

In many parts of the world the time intervals during which systematic flood records and climate records are simultaneously available are relatively short, and thus inadequate for establishing flood-climate relationships. In many cases it is also unclear which spatial scale of climate should be considered in establishing such relationships, which may further vary among flood types.

The literature has only a few persuasive examples of the climatically controlled, non-stationary nature of flood series at time scales covering the 20th century to late Holocene [e.g., Knox, 1983, 1984, 2000; Hirschboeck 1985, 1987a, 1988; Enzel, 1992; Webb and Betancourt, 1992; Ely et al., 1993].

The connection to climate may be subtle, or distant, or lagged (see ENSO, below, for example), and require special techniques and disciplinary knowledge to first suspect and then tease out. It is thus perhaps not surprising, then, that testing long historical and gaged flood series for relations to climate is not a common practice in the hydrologic community. This is in spite of the fact that the availability of hydrologic data in digital form has increased tremendously in recent years, as has the potential of computers to analyze large data sets at very high speeds.

If non-stationarity is indeed found, then what? A decision is still needed.

The relative dearth of precedents and examples, the difficulty in establishing the nature of the relationships, the need to have some probability distribution for the future period, and the inertia of the status quo may be part of the disconnect which exists between the research and operational communities on these issues.

Extending flood records beyond the historical observations is a difficult and laborious task that demands training from fields traditionally not familiar to hydrologists. Developing or analyzing a paleoflood chronology, the main type of event-specific flood proxy available, requires a combined education in both the geological and hydrological sciences. The integration of paleoflood information with climate information calls for understanding the linkages between meteorology, climatology, and associated fields like oceanography and hydrology. Such combined knowledge is difficult to obtain and research groups with such multidisciplinary experience are relatively uncommon. These combinations are even more rare in practitioners. But, as this volume attests, interest in the discipline and its applications are increasing rapidly. Nonetheless, there are presently only a few regions where the connections have been established between paleofloods or other extreme hydrologic event chronologies and climate variations.

Also rare are synthesized regional chronologies based on several basins, but these too, are increasing in abundance. Among these are basins and regions located in Northern Coastal Peru [Wells, 1990; Fontugne et al., 1999], the Upper Mississippi Valley [Knox, 1993; 1999; 2000], the Lower Colorado River [Ely et al., 1993; Ely, 1997], southern California [Enzel et al., 1989; Enzel, 1992; Enzel and Wells, 1997], the northeastern United States [Brown et al., 2000], southern Israel [Greenbaum et al., 2000], central India [Ely et al., 1996; Kale et al., 1997], and central Spain [Benito et al., 1996, 1998]. Each of these diverse studies

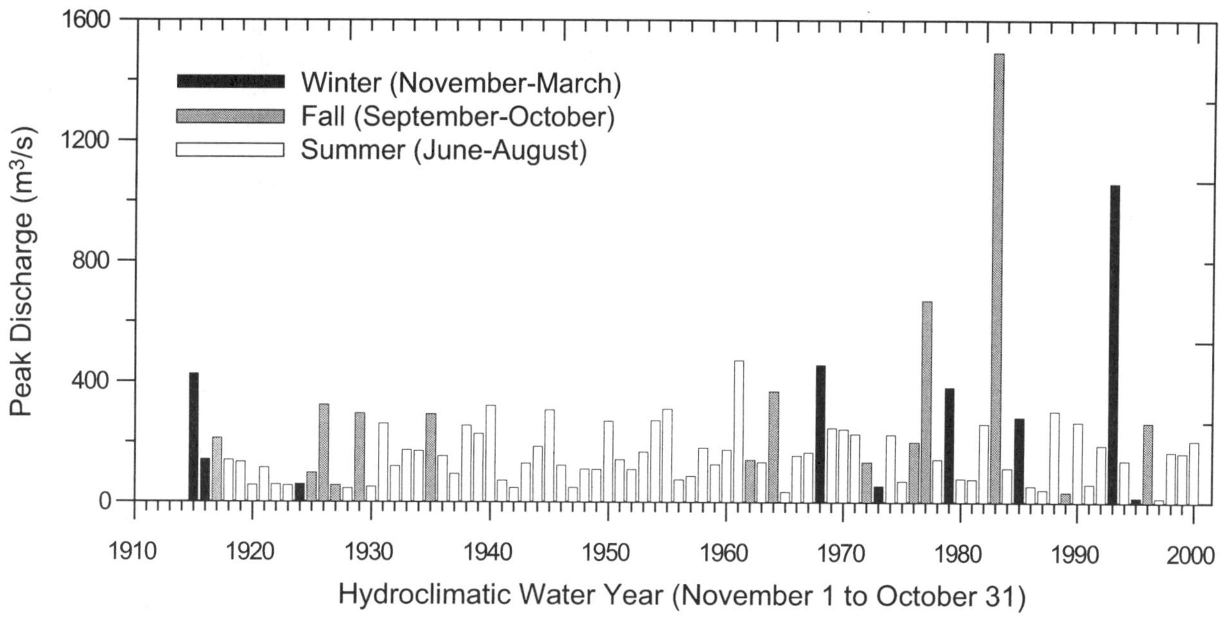

Figure 3. Annual peak instantaneous discharge of the Santa Cruz River at Tucson, Arizona, showing season of peak flow. Water years 1915-2000.

demonstrated a temporal association between specific late Holocene climatic episodes and changing frequencies of extreme storms and floods.

Building on established and new findings, we show first that 20th Century instrumented data strongly support the notion that large scale long-term climate exerts strong influence ("control") over the frequency of large floods in Arizona and southern California. We next discuss results from paleoclimate-paleoflood studies that demonstrate that climate can and does control flood regimes in these regions on longer time scales. After pointing to other flood studies that have shown similar patterns, we combine all of them to show that climatic variations have exerted similar controlling influence on decadal to millennial scales in the same regions.

Detection of Climate Effects on Flood Frequency Regimes

The ability to detect control by climate of large-flood frequency would allow us to understand better the advantages and limitations of using information on the largest modern, historical, and paleofloods for flood hazard mitigation and water resources management. We emphasize that the ability of climate to exert control over flood distributions will be strongly a function of location and season, and will therefore show geographic, hydroclimatic, and sometimes topographic preferences, factors which in turn obviously have repercussions for the ability to detect climate linkages. Detection requires both a sufficient record and a signal to detect. Some areas will have the requisite record but no signal to detect, others have the strong climate signal but no record to work with, and some have neither.

In some cases (e.g., southwest United States) the presence of a mixed population of flood-causing atmospheric conditions (winter cyclones, summer monsoon, autumn tropical storms) may complicate the detection of variability in floods through time in a specific region [Hirschboeck, 1985, 1987a; Webb and Betancourt, 1992] (Figure 3, also on CD). Analysis of just the extreme flood subset in basins that are demonstrably sensitive to particular flood-producing mechanisms or sets of mechanisms produces better insights into relations between climate and extreme floods [e.g., Hirschboeck, 1985; Ely et al., 1993]. Two sources of variability may mask each other by affecting different seasons, or by alternating between sources in a single season.

Because reliable long flood records are very difficult to find, we should also seek relationships between records of quantities relevant to floods (e.g., flood records themselves, heavy precipitation records, etc.) and records of patterns of regional and large scale climate, whose temporal variations are known further back in time. This often includes the need to establish correspondence between flood occurrence and extreme precipitation events during their coincident record. The latter may have longer records that can then be analyzed for modern "proxy floods." Stationarity of the flood-climate relationship will be explicitly assumed over the modern/paleoscale periods. That is, in a given region, the same climate state produces the same flood characteristics, perhaps not always a correct assumption, but one whose validity is even harder to establish or refute.

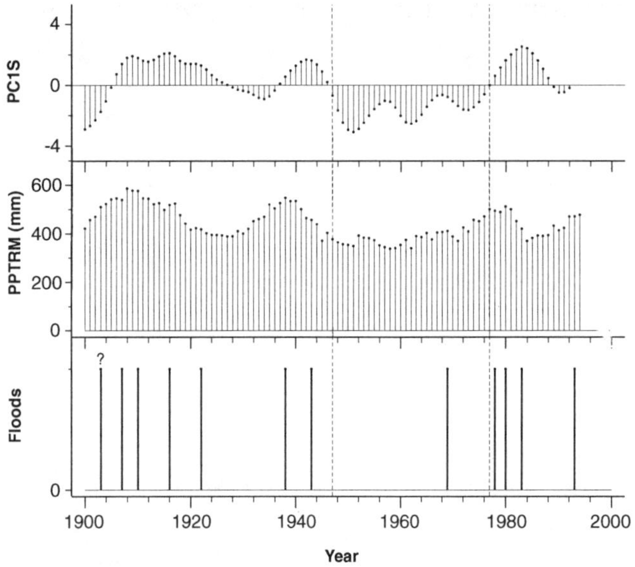

Figure 4. Relation between Mojave River floods, Southern California precipitation, and tree ring records during the 20th century. Years 1947 and 1977 are shown by dashed vertical lines. (top) Smoothed first principal component (PC1S) of annual tree-ring chronologies from Southern California [*Biondi et al.,* 2001]. A 10-year cubic smoothing spline [*Cook and Peters*, 1981] has been applied to the annual scores. (middle) Ten-year running mean of October-March precipitation, south coast climate division in California (PPTRM), 1895-2000, plotted in middle year. (bottom) Water years (ending year) of largest historical and gaged Mojave River floods [from *Enzel*, 1990] during the 20th century (vertical bars). The 1903 flood ("?") may be off by a year; the 1907 date is not quite certain.

Among the questions we wish to explore are: a) Should we expect any relationship at all between flood and climate? b) Is such detection possible (i.e., are there any examples of such detection)? c) At what temporal scales is it possible to detect such relationships? d) Do available chronologies of paleofloods reveal specific episodes that indicate climate change, and thus implied paleoflood change?

MODERN / LATE HOLOCENE FLOODS AND CLIMATE IN THE SOUTHWEST UNITED STATES

To help illustrate inference of changes in flood frequency via evidence from paleo studies, we have chosen two basins in the southwestern United States. In one of these basins dendrochronological data are emphasized, and in the other fluvial and lacustrine data are highlighted.

Mojave River, Southern California

The Mojave River in southern California provides an excellent example of such a linkage between specific basin floods and large scale climate. The Mojave River drains the San Bernardino Mountains northeast of Los Angeles and flows north and then east through the arid lowlands of the Mojave Desert to a terminal playa. Annual (July through June) precipitation in the headwaters can exceed 2000 mm during the wettest winter seasons.

A large fraction of the seasonal total rainfall is produced by a few extreme individual winter storms that can yield up to 1000 mm per storm [Enzel, 1990]. Such storms produce the largest floods in the continuous historical (since 1894) and gaged record [Stamos et al., 2001]. To reach the terminal playa in the desert, Mojave River floods must overcome large transmission losses along the 200 km long arid sand and gravel alluvial river bed [Enzel et al., 1989; Enzel, 1992; Enzel and Wells, 1997; Stamos et al., 2001; see Goodrich, 2000 for example from Arizona]. Therefore, only floods with very large peak discharges from the headwaters region reach the playa [Enzel, 1992; Lines, 1996; Stamos et al., 2001] (Figure 4, also on CD). These same floods also have the largest total flow volumes at upstream gages. Smaller floods merely infiltrate into the alluvium as in many other arid ephemeral streams [e.g., Greenbaum et al., this volume]. Figure 5 (also on CD) shows the relationship between annual peak instantaneous flow and annual total discharge. The terminal lakes formed as the result of the largest floods typically evaporate within 3-18 months, depending on the volume of floodwater that reached the playa, and to a lesser extent, on subsequent temperature and precipitation. In this manner the lacustrine stratigraphic re-

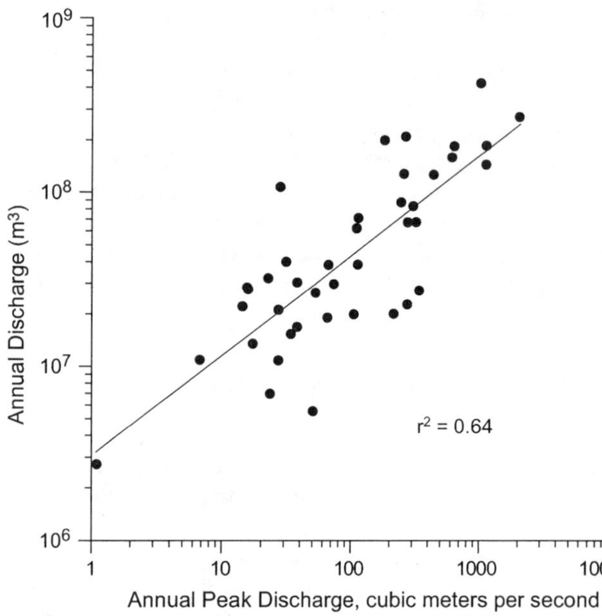

Figure 5. Relation between annual peak discharge and annual total discharge, Mojave River at The Forks gaging station, upstream from Hesperia California, and below confluence of West Fork and Deep Creek. Based on 41 unregulated years from within the interval 1906-1971. Data from *Corps of Engineers* [1986]. Correlation r = + 0.80.

cord of the playas indicates occurrences of past large floods.

Mojave River floods that reach the playa were selected from the historical/gage period of the past century. Their occurrence is compared in Figure 4, with the 10-year running mean of Southern California coastal precipitation during the October-March rainy season.

Very large floods (approximately 10-year return or more) are seen to cluster in decades that are wet. Monte Carlo tests show that the likelihood of such clustering by chance in wet 10-year intervals is less than 4 percent if the two indefinite ("?") dates are excluded, and less than 1 percent if they are included. The same conclusion holds for 5-year intervals, and for individual years, the respective probabilities are less than 0.05 and 0.4 percent respectively. An example of an exception is the individual winter of 1909-1910, with a major New Year's Day flood coming after a wet December, but during a dry winter overall.

On longer time scales during the middle and late Holocene, Enzel [1992], Enzel et al. [1992], and Enzel and Wells [1997] observed episodes where shallow perennial lakes (<8 m deep) formed in the terminal playas and persisted for up to ~100 years. Their appearance during the past few thousand years indicates frequent large floods in the Mojave River. Enzel [1992] placed both upper and lower constraints on the size of these paleofloods, concluding that they could not have been much larger than modern extreme floods (since the maximum elevation of the shorelines of these lakes is well established) nor much smaller, either. Evaporation differences are greatly insufficient to account for this persistence.

Lake persistence thus constitutes evidence of frequent "refreshment" by large floods that overcame infiltration loss into the sandy stream channel, reached the terminal playas, and prevented desiccation of the temporary lake [Enzel, 1992]. The most recent instance when this happened occurred ~400 years ago, and probably coincided with an extended period of cooler-than-normal summer temperatures [Biondi et al., 1999]. During such multidecade to century-long episodes, floods of the magnitude of the largest modern floods must have clustered over years and decades to form the observed perennial lakes. During extended episodes, the recurrence time for floods of this magnitude was a year or less [Enzel, 1992], compared with modern recurrence times of 15-30 years (depending on the specific gaging station analyzed).

Such physical settings as fluvial-lake systems, where the lakes are either formed or significantly affected by floods, provide persuasive evidence for changes in flood frequencies during the Holocene. The advantage of using joint fluvial and lake information instead of paleoflood chronologies is that even if one does not fully trust the exact ages assigned to these flooding episodes, the fact that episodes with high frequency of large floods did actually occur is not in question. Exclusively fluvial paleoflood chronologies, by contrast, depend on the quality and resolution of flood stratigraphy and age dating to show clustering [see discussions in House et al., this volume; and in Webb et al., this volume].

The inference drawn from these considerations is that an assumption of constant climate for this area is incorrect at decadal to centennial time scales. The observational flood record of the Mojave River is undoubtedly heavily influenced by climate variability.

We have searched for corroborative evidence from dendrochronological records in the headwaters of the Mojave. This evidence is suggestive but not quite persuasive. Because of the proximity to the coast, tree growth in years without major floods can occur because of the occurrence of at least several modest storms, or rainy episodes of longer duration and lower intensity, or snowfall which is more effectively absorbed into the soil. Conversely, major floods have occurred in winters that ended up with smaller growth rings, such as 1937-38. Annual precipitation here is not as strongly affected by single standout events as in the more arid region farther from the ocean. However, nearby tree ring records provide very useful information about climate variability on annual to decadal scales, which as shown earlier also relates to the likelihood of Mojave River large floods. Biondi et al. [2001] showed that the annual water year precipitation record for this area can be extended to the 1600s using tree-ring chronologies. Furthermore, there is regional structure to the correlation pattern (Figure 6, also on CD), extending maximally along a northeast axis toward Utah and also toward Arizona to the east (discussed next).

Verde River, Central Arizona

A different situation exists on the Verde River that allows judicious usage of dendrochronological records. Relationships between tree-rings, precipitation, and extreme floods can be expected in semiarid regions, where (a) tree growth is limited by effective precipitation and (b) very large fractions of seasonal precipitation are accumulated during a few relatively short (1 to 3 days) and intense large-scale storms. These heavy storms affect soil moisture, tree growth, and runoff, thereby providing the basis for the successful reconstruction of hydrological parameters from tree rings [e.g., Meko et al., 2001].

Historically, large floods in Verde River basin in central Arizona are associated with intense large scale winter storms (Figure 7, also on CD) [Hirschboeck, 1985; Ely et al., 1993; House and Hirschboeck, 1997; Cayan et al., 1999]. The Verde River demonstrates a reasonably high correlation between annual flow and peak discharge (Figure 8, also on CD). Applying various thresholds for annual flow, we note from Figure 8 (also on CD) that the largest annual total discharges are typically associated with a very

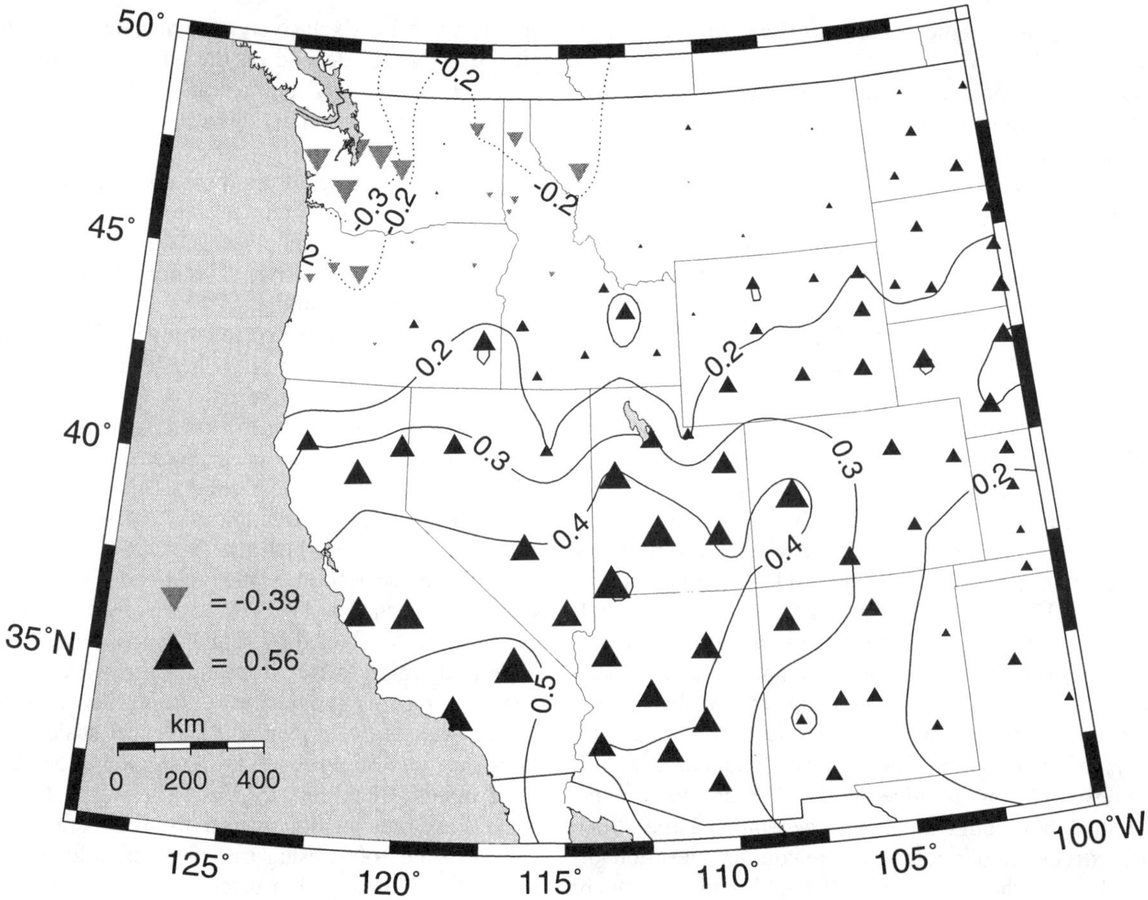

Figure 6. Correlation map between the spatial mean of a set of annual tree-ring chronologies from southern California / northern Mexico [*Biondi et al.*, 2001] and total annual precipitation over climate divisions of the western United States during the 20th century. Point correlations are represented by triangles (upward for positive values, downward for negative values; extremes are shown for scale), and interpolated values are represented by contour lines.

large annual peak discharge (i.e. with at least one very large flood in the winter rainy season). For example, with annual discharge thresholds in the range of $8 \cdot 10^8$ to $10 \cdot 10^8$ m^3, four out of the five years exceeding the threshold have annual flood peaks greater than 1500 m^3/s. For three cases exceeding a higher threshold ($12 \cdot 10^8$ m^3), all three have peak discharges larger than 1700 m^3/s. The floods in which we are interested are exclusively winter phenomena.

A recent dendrohydrological study [reported by Van West and Altshul, 1997, building on the work of Graybill, 1989] reconstructed Verde River annual discharge for the period AD 572 to 1985 (Figure 9, also on CD). House et al. [this volume] report a correspondingly long fluvial paleoflood record from the lower Verde River that provides an important point for comparison with the reconstructed flow time series. It is important to remember that for this exercise we are not interested in the high frequency portion of Figure 9, such as the annual peak (approximately 2-2.5 year return periods) or 5-year return floods, but rather in the top half dozen or so floods per century: the very largest events. The available paleo evidence becomes weaker and less conclusive the smaller and more frequent the flood considered. For annual discharges less than the very highest values, the possibility that peak discharge did not occur in a very large flood slowly begins to rise, which would compromise the validity of the underlying premise. As long as we are careful to restrict attention to this upper subset of the records, the correspondence between largest floods and highest annual discharge can be utilized to infer flood information from reconstructed annual discharge based on the tree-ring chronology shown in Figure 9. The reconstructed chronology indicates changes in the frequency of extreme annual discharge over periods spanning several decades to centuries. We see concomitant dramatic changes in the mean and variance of the annual flow at these temporal scales.

Guided by information in Figures 8 and 9, we can count the number of ring-index values that exceed different

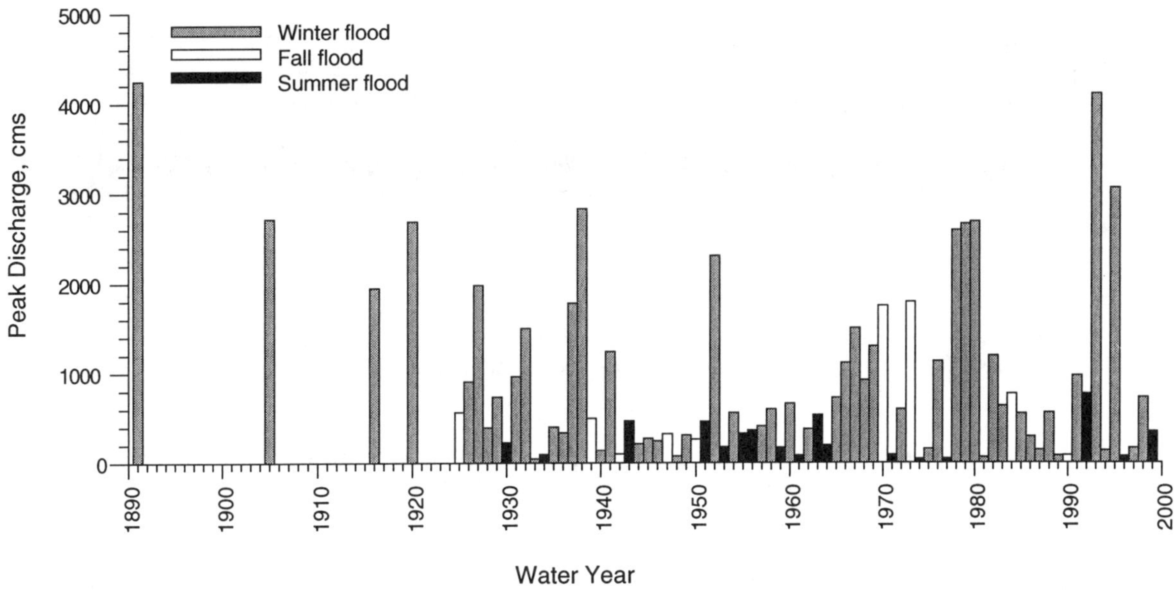

Figure 7. Annual peak instantaneous discharge of the Verde River below Tangle Creek, Arizona, showing season of flood peaks.

thresholds. These thresholds are selected roughly according to their average frequency of occurrence over the entire fourteen-century record. Then, a running count is obtained of the number of instances per 100 years that a particular threshold of interest is exceeded, for all possible 100-year-long periods in the record. We choose 100 years because that is the typical length of a long gaged record of streamflow, a period usually assumed to be quite adequate. Figure 10 (also on CD) shows running totals of the number of years per hundred years (for all possible 100-year intervals) with counts exceeding different thresholds (1.0, 1.5, 2.0 normalized units) from AD 572 through 1982. Hypothetically, a society could have formed and begun its century-long measuring program during any one of these years.

From AD 570 until about AD 1000 we see an extended period with numerous higher annual flows—and thus implied large floods—per 100 years (bottom trace on Figure 10). From about AD 1000 until the early AD 1100s an extended period of greatly reduced numbers of high flows, and (Figure 10) likely fewer floods also, is noted. Moderately high flows increase from AD 1200-1400, but very high flows again become relatively rare, indicating that at least winter floods (which are the largest) should have declined. Moderate annual flow years can still contain smaller tropical storm or monsoon thunderstorm floods. Hundred year periods from the mid-1400s to mid-1500s show a relative increase.

The highest 100-year frequency on the Verde River occurs during the period centered on the late 1800s. Concentrating especially on the least common and thus very large events, the last two centuries show the most events in the entire 1400-year record. Taken in its entirety, this analysis shows dramatic variations in the frequency of implied large floods. For this 1400-year-long period, reconstructed annual flows with an overall approximate 13-year return (7-8 per century, Z-scores of +1.5) fluctuate from 1-2 events per hundred years (50-100 year return interval) to as much as

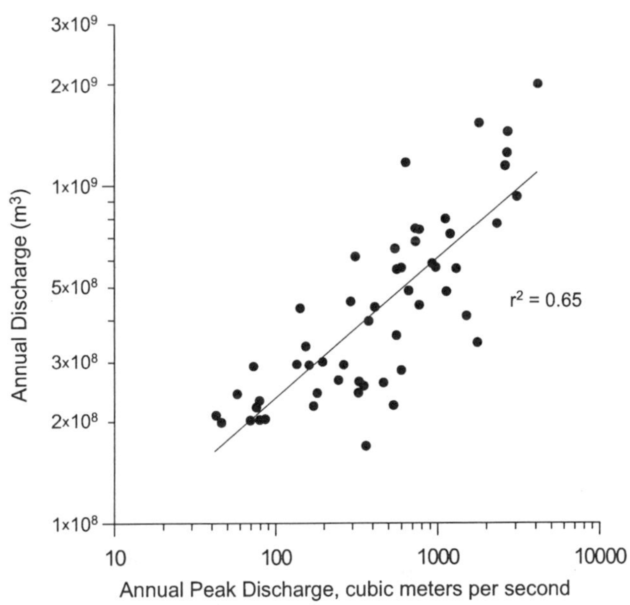

Figure 8. Relation between annual maximum peak discharge and annual total discharge, Verde River at Tangle Creek, Arizona (lower Verde River). Based on 54 years of data (1946-1999). Correlation r = + 0.81.

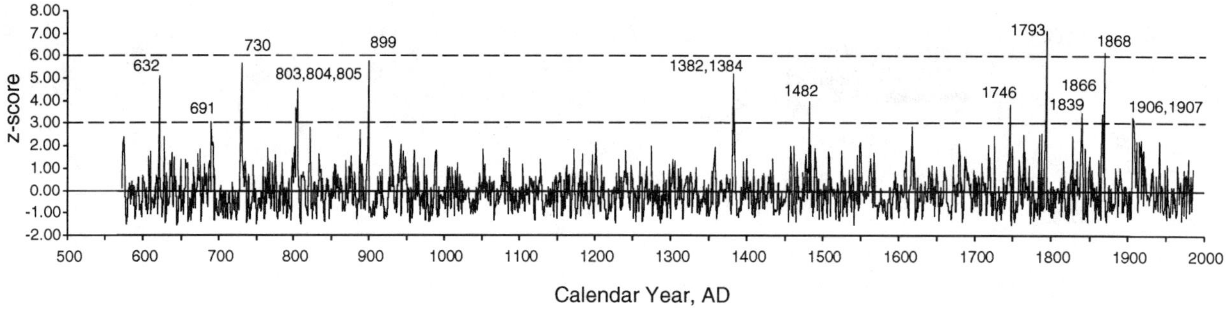

Figure 9. Tree-ring-based reconstruction of the total annual discharge volume on the Verde River [*Van West and Altschul*, 1997] (A.D. 572-1985) in Z-scores (average reconstructed discharge: $432 \cdot 10^6$ m^3, standard deviation: $228 \cdot 10^6$ m^3.

17 events per hundred years (6-year return interval). More rare flows with an overall 25-year return period (4 per century, Z-scores of +2.0) fluctuate from none per 200 years to eleven per 100-years (9 year return interval).

Although not central to our main point of the simple existence of such large fluctuations, it is nonetheless notable that the highest frequency of the rare flows is found in the past 150 years. Assuming a stable peak-to-total ratio, these past centennial scale episodes were also characterized by more frequent extreme floods. Independent evidence and studies show this same increase has also been observed in the paleoflood chronology for Arizona and southern Utah [Ely et al., 1993; Ely, 1997]. Moreover, House et al. [this volume] report that six of the 12 largest floods recorded in a 1600-year stratigraphic paleoflood record on the lower Verde River occurred within at most the last 150 years.

CLIMATE AND FLOOD VARIABILITY IN THE SOUTHWESTERN U.S.

The Climate System as a Flood Generator

The complete earth-atmosphere-ocean system, encompassing processes on all time and space scales, combines to produce a single weather or climate event at any given location. Thus, for this discussion, the following rather parochial approach can be adopted. In this constrained view, the

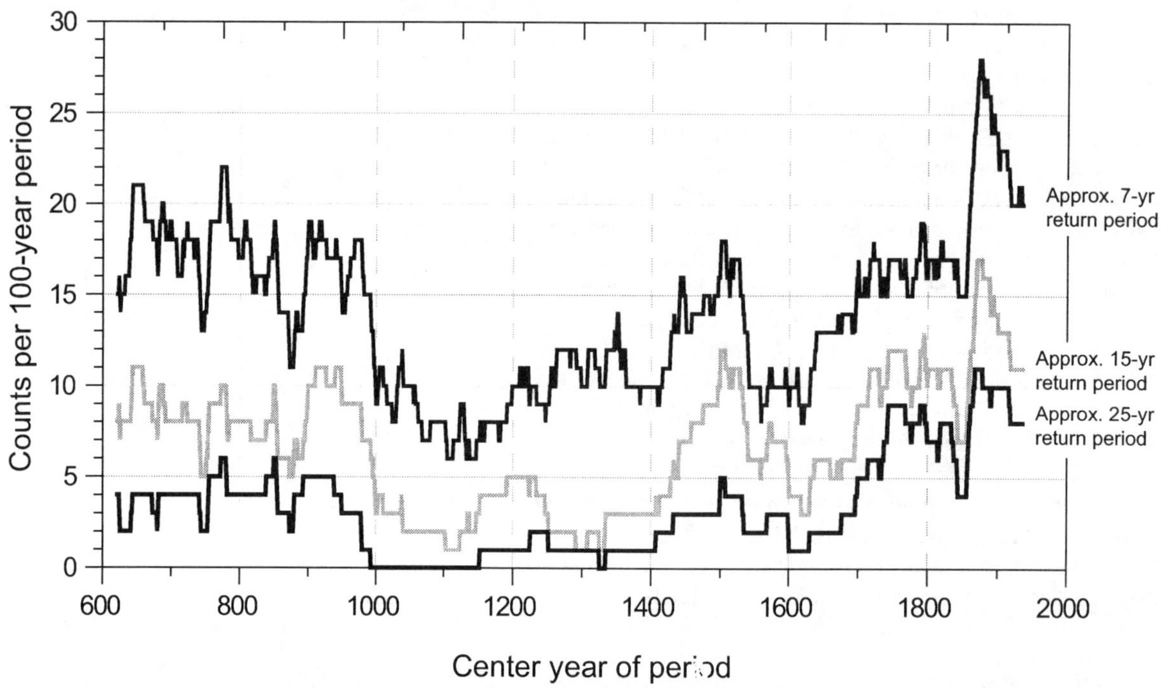

Figure 10. Number of years per running hundred years that Z-Score of reconstructed annual Verde River flow at Tangle Creek exceeded thresholds of +1.0, +1.5, and +2.0, A.D. 572-1985. Thresholds represent moderately rare flows.

entire climate system of the earth is conceived of as a device designed solely to act as a flood generator for a basin of interest. If the climate system were to remain in the same mode continuously, it would randomly produce floods obtained by drawing from an unvarying probability distribution.

Climate varies considerably at the monthly and seasonal scale, as monthly time series of any climatic parameter will show. In the general case, we have limited knowledge about what is responsible for such variations, particularly beyond the proximal causes. However, we do understand a few specific sources of variability at these time scales, and will devote attention primarily to one mechanism which is better understood, namely ENSO (El Niño/Southern Oscillation).

Interannual Time Scales: Climate and ENSO

At the outset, it should be noted that by focusing on ENSO, we do not wish to imply that this is the only, or even the most important, source of variability for this region; it is simply the most well-studied. For the points we wish to make, it is sufficient that it is at least a significant influence on Southwest climate, and on aspects of its flood regime. At a maximum, El Niño and La Niña account for only about one-third of the winter precipitation variance in the Southwest, and essentially none of the summer variance. Therefore other mechanisms are surely at work and ENSO alone cannot explain all of flood variability. The three main flood producers in the Southwest United States (winter cyclones, summer monsoon, autumn tropical storms) occur in different seasons with very different atmospheric and oceanic flow configurations. The combination of these produces the observed flood record. There is no particular reason to expect the year-to-year, decade-to-decade, or century-to-century variations of these different seasons to resemble one another. Hirschboeck [1985] and Webb and Betancourt [1992] have shown that flood frequencies in Arizona, when stratified by season (i.e., by causes), have all varied on multi-decadal time scales during the 20th century, but not in the same manner.

It has become quite well established that ENSO has major effects on climate throughout the world [e.g., Ropelewski and Halpert, 1986, 1987, 1989; Kiladis and Diaz, 1989; Dai and Wigley, 2000]. In fact, after the seasons, ENSO is the largest source of variability in the earth's climate system [e.g., Dai et al., 1997]. Studies of ENSO relationships have evolved toward regional scale behavior, and have increasingly emphasized predictability and practical effects, and elements other than temperature and precipitation.

One key area with significant associations has been the western half of the United States and North America. Variations in temperature, precipitation and snowfall, primarily in winter, are related to ENSO variations [Yarnal and Diaz, 1986; Andrade and Sellers, 1988; Kiladis and Diaz, 1989; Cayan and Peterson, 1989; Schonher and Nicholson, 1989; Redmond and Koch, 1991; Cayan and Webb, 1992; Cayan, 1996; Livezey et al., 1997; Kunkel and Angel, 1999; Wang and Fu, 2000]. Measures of the state of ENSO have included both atmospheric (Southern Oscillation Index, SOI), oceanic (Sea Surface Temperature, SST), or combined (e.g., Multivariate ENSO Index, MEI) [Wolter and Timlin, 1998] indices. In general the SOI relations to western U.S. climate are slightly stronger than those with SST and offer better predictability [e.g., Redmond and Koch, 1991] — why is not clear.

The warm phase of ENSO ("El Niño") typically brings wet winters to the southwestern U.S. and drier than usual winters to the Pacific Northwest. The cool phase typically brings wetter winters to the Pacific Northwest, and has so far been very consistent in bringing dry winters to the Southwest, probably the most reliable predictive climate relationship known in North America. Figures 11 and 12 (also on CD) illustrate this very well.

Winter precipitation and snowmelt is the primary source of annual streamflow in this region, so ENSO phase also affects the seasonal runoff. Studies on ENSO relations to snow pack and streamflow have focused primarily on the western United States [Redmond and Koch, 1991; Cayan and Webb, 1992; Kahya and Dracup, 1993; 1994; Dracup and Kahya, 1994; Redmond and Cayan, 1994; Cayan, 1996; Piechota et al., 1997; Pulwarty and Redmond, 1997;

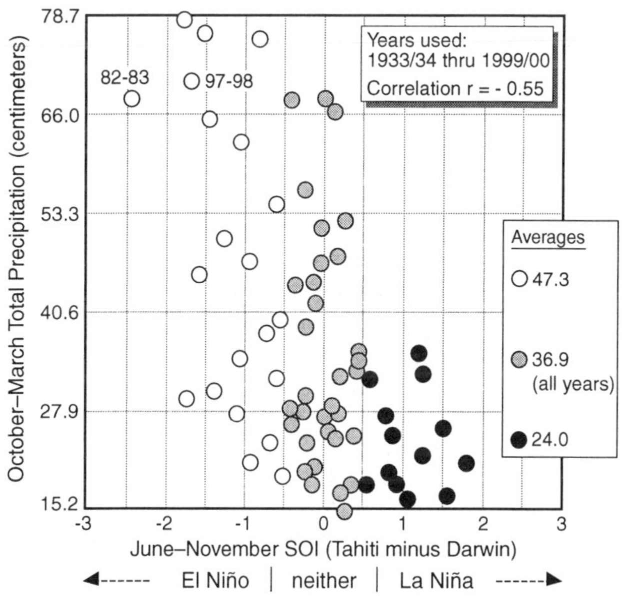

Figure 11. Scatterplot of June-November Southern Oscillation Index versus subsequent October-March average precipitation for South Coast climate division in California (Los Angeles to San Diego), for 1933/34 through 1999-00. Position of average symbols on Y-axis is proportional to value. Open: El Niño, Gray: All Years, Solid: La Niña, threshold used: +/- 0.50. Correlation is r = -0.55.

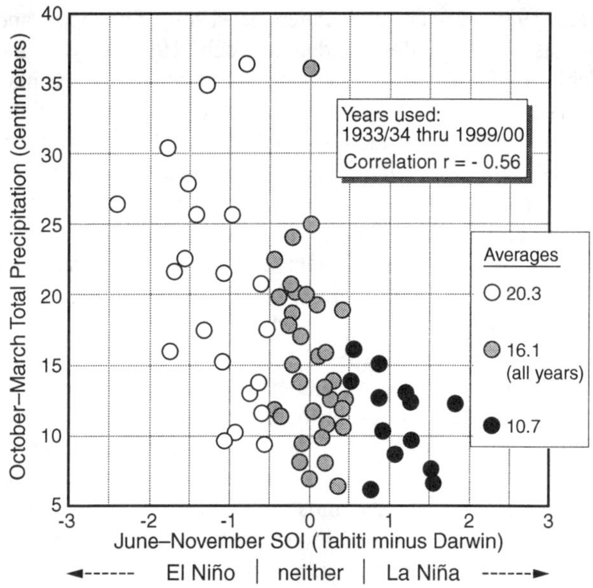

Figure 12. Scatterplot of June-November Southern Oscillation Index versus subsequent October-March statewide average Arizona precipitation, for 1933/34 through 1999-00. Position of average symbols on Y-axis is proportional to value. Open: El Niño, Gray: All Years, Solid: La Niña, threshold: +/- 0.50. Correlation is r = -0.56.

Kunkel and Angel, 1999; Dettinger et al., 1998, 1999, 2000a]. In the southwestern U.S., the streamflow relations to SOI look very much like the graphs in Figures 11 and 12. The very fact that the winter season precipitation distributions are so different from each other for the two ENSO phases means that the "i.i.d." assumption cannot be valid for climate on interannual time scales.

Woolhiser et al. [1993] perturbed a stochastic "weather generator" with SOI and found that in the Southwest in winter, ENSO phase affects both of the main precipitation processes: its occurrence and its amount (conditional on occurrence). During El Niño for example, a day with precipitation becomes more likely, and a wet day with El Niño will bring more precipitation than a wet day with La Niña. More recently, Cayan et al. [1999] showed that ENSO enlarges the upper tail of the daily precipitation distribution. Persistence is also enhanced. In central Arizona, there are many more occurrences of three consecutive wet days during El Niño than non-El Niño winters. Any factor which increases storm duration, frequency or magnitude, or causes lower temperatures, increases the efficacy of storms in wetting soil and producing immediate or delayed runoff. A number of other studies have examined the relation between atmospheric circulation patterns and winter flood hydroclimatology in the lower Colorado River drainage basin and southern California [Hirschboeck, 1985; Enzel et al., 1989; Cayan and Webb, 1992; Webb and Betancourt, 1992; Ely et al., 1993, 1994; House and Hirschboeck, 1997; Hirschboeck et al., 2000]. As Figures 11 and 12 make clear, El Niño winters are not consistently wet in the Southwest United States: many are average, and some are quite dry. However, nearly all of the wettest winters occur during El Niño conditions (7 of the 8 wettest Oct-Mar in Arizona from 1933-34 through 1999-2000). The converse is not true: for La Niña, there have been no wet winters, and thus floods are much less likely. We can conclude that ENSO phase has a major influence on the potential to produce large floods in this region. What is very clear is that the probability distribution is shifted, greatly, according to ENSO phase.

Causes of Southern California and Central Arizona Floods

Before proceeding on to longer time scales, we first briefly summarize factors associated with floods in the Southwest United States, and then consider associations between floods and ENSO.

The causes of major floods at various spatial scales in the southwest United States have been previously discussed by Maddox et al. [1980]; Hirschboeck [1985, 1987a, 1988, 1991], Enzel et al. [1989], Cayan and Webb [1992], Webb and Betancourt [1992], Ely et al. [1994], House and Hirschboeck [1997], Hirschboeck et al. [2000], and House and Baker [2001]. Explanations of the causative factors and associated physical phenomena for some of the very largest floods requires consideration of analogously large spatial and sometimes temporal scales of climate interactions [e.g., Cayan and Peterson, 1989].

The causes of large floods on the Mojave and Verde Rivers are generally similar, although the Verde has a slightly more diverse flood regime. The factors which result in large floods have been rather thoroughly described by Hirschboeck [1987a, 1988]. Over the past century, large floods in the Mojave River occur only during the winter and are associated with specific oceanic and atmospheric circulation patterns over the equatorial and north Pacific Ocean related to ENSO and other phenomena [e.g., Enzel et al., 1989; Ely et al., 1994; Enzel and Wells, 1997]. Similar, and occasionally the same, patterns are typical of floods on the Verde River [House and Hirschboeck, 1997; Ely et al., 1994]. One important distinction between the two rivers is that dissipating tropical storms have been responsible for moderate magnitude Verde River floods in the 20th century.

Sea-level pressure patterns associated with large Mojave River floods show a diminished eastern North Pacific high, along with a southward and eastward shift of the Aleutian low resulting in anomalous low pressure off the coast of central California and Oregon [Enzel et al., 1989]. Large floods on the Verde River are generally associated with a similar configuration, but the relative location of anomalous low pressure off the coast of California is less restricted latitudinally as is the case for the Mojave [Ely et al., 1994]. The Polar Jet during these floods is at a southerly position over southwestern United States with episodes of sustained and intensive moisture transport from the tropical central North Pacific Ocean (colloquially often referred to as the "Pineapple Express"). In some instances, merging of the Polar Jet with the more southerly Subtropical Jet can result in optimum conditions for large floods by directing alternatively cold and warm storms over the region [Enzel, 1990; Enzel et al., 1989]. The winter flooding in southern California and Central Arizona during 1993 was an excellent example of this phenomenon [House and Hirschboeck, 1997]. These atmospheric patterns are accompanied by warmer than usual sea surface temperatures offshore California and result in greatly increased rainfall in southern California and other parts of the southwestern United States. If they recur frequently, the seasonal storm track will be south of its regular position. If the sea-surface temperature anomalies persist and reemerge in the following winter season [e.g., Namias et al., 1988; Alexander et al., 1999] a setup for potential flood "clustering" over consecutive years is in place [Enzel, 1990; Enzel and Wells, 1997]. Therefore, other proxy-records of long-term, warmer water in the eastern North Pacific offshore from California can be indicative of similar oceanic and atmospheric patterns and storm tracks that increase the potential of large floods in the southwestern United States in general. The terminology about causes can be problematic, as it can encompass a range from proximal to "ultimate" causes (if such can be said to exist in a turbulent system which never comes to rest). The most satisfactory use of the word "cause" in the present context involves a chain of events in which the characteristics of probability distributions for specific locations and times can be traced back to large scale circulation features on the scale of oceans and continents, in the most favorable cases with lead times that allow a degree of predictability. When current conditions can be traced back arbitrarily far to prior initial conditions and boundary values, there is ample room for interpretation as to what constitutes a "cause." As a concrete example, the moist subtropical plume from the southwest (Pineapple Express) is often invoked as a "cause" of California floods, but what larger circumstances cause the probability of this configuration to increase?

Interannual Time Scales: ENSO and floods

As related above, ENSO has a major effect on the wintertime probability distribution of precipitation. To date there has not been a comprehensive examination of the relation between ENSO and floods. In Arizona, the flood regime is a complex function of elevation, basin orientation, drainage area, season, and location within the state. The strongest relationships to El Niño are found for those basins whose greatest flows occur in winter, thus the largest basins. The Mogollon Rim has a relatively large winter maximum in precipitation, and the basins with the best relations drain the south side.

To establish whether peak discharge is preferentially associated with one phase of ENSO, annual (water year) time series of this statistic were obtained from the U.S. Geological Survey for a number of larger streams in Arizona having minimal upstream storage. Each annual value was classified as monsoon, autumn (possible tropical storm, September-October), or winter. All winter peak discharges were then further classified as weak or strong La Niña or El Niño, or neutral, according the criteria of Redmond and Koch [1991]. Thus, ENSO effects were only allowed for November-May. Aggregate ranks were computed for El Niño, neutral, and La Niña conditions for winter peaks, and compared with ranks from Monte Carlo simulations where no relation is present (10,000 cases). All years were included in rankings regardless of season of peak discharge. For each case (El Niño La Niña) the probability of a chance relationship was determined ("p"). River gages with strong peak daily flow relationships to ENSO (particularly El Niño) are listed in Table 1. It should be noted that for La Niña, dry conditions generally prevail, and the largest flood can easily occur outside of winter, so that low-peak La Niña relationships can be readily covered up by other seasons whereas high-peak El Niño relationships cannot.

Many smaller basins do not show any particular preference toward positive or negative ENSO phase with respect to peak discharge. However, as the basin characteristics increasingly resemble those noted just above, El Niño becomes increasingly associated with the largest flows. That is, many basins in Arizona have nonexistent to weak to modest relationships with ENSO, but the very largest floods have strong relationships, in particular to El Niño.

Together, the variations in climatic patterns, in ENSO and in flood seasonality observed during the 20th Century provide clear evidence that our climate system flood generator has not been drawing from the same identical distribution during this time, and thus that the i.i.d. assumption is not correct for this region over the relatively short historical record.

The Past Century

What if El Niño or La Niña themselves become more or less frequent? Since the "1976 shift" [Ebbesmeyer et al., 1991], the number of "El Niño" winters (using the SOI criteria of Redmond and Koch, 1991) has exceeded the number of "La Niña" winters by 9 to 2 (through 1998). For the prior period from 1947-1976, there were 7 El Niños and 10 La Niñas, by the same criterion. In the Southwest, the post-1976 period was the wettest in the past century. A process which modulates the number of ENSO events (warm or cold), over years to decades at a time, will thus alter the likelihood that the climate system will generate a winter flood of a given magnitude at a given place. One mechanism is thus afforded for our flood generator (the global climate system) to be able to draw from different subpopulations for extended periods. Thus we have two

Table 1. Arizona basins with flood-ENSO relations. Significance level based on Monte Carlo tests (high values for El Niño, low values for La Niña) for peak instantaneous flow. Only winter floods are assigned an ENSO status, but all seasons are considered. For a special case on the Verde, all flows were winter peaks only, to avoid summer/autumn masking of La Niña low flows. Record lengths 47-79 years.

Gaging Station	Area (km^2)	El Niño	La Niña
Black River at Fort Apache, Arizona USGS no.09490500	3190	$p < 0.03$	$p < 0.015$
Salt River near Chrysolite, Arizona USGS no. 09497500	7376	$p < 0.013$	$p < 0.133$
Tonto Creek above Gun Creek, Arizona USGS no. 09499000	1747	$p < 0.038$	$p < 0.47$
Verde River below Tangle Creek, Arizona USGS no. 09508500	15,168	$p < 0.0019$	$p < 0.14$
Verde River below Tangle Creek, Arizona (winter only)	15,168	$p < 0.042$	$p < 0.039$

examples, at different time scales but within the period of systematic record, where a physical rationale exists for violation of the traditional "i.i.d." assumption.

McCabe and Dettinger [1999] have repeated the analysis of Redmond and Koch [1991], but using sliding 30-year segments during the 20th century. Their intriguing paper shows that some other (unknown) factor appears to modulate the strength of the relation between ENSO and western United States winter precipitation over multi-decade time spans. The Southwest relationship was always present, but the region of strongest correlation has slowly migrated from centering on New Mexico in the early 1900s toward centering on Arizona/Southern California in the late 1900s. In the Pacific Northwest, the relationship fades greatly near mid-century, then returns strongly toward its end. McCabe and Dettinger [1999] also note strong changes in the correlation between the SOI and SST in the equatorial Pacific, from moderate to weak to strong as the century progresses. No explanation has appeared in the literature. For this region, this behavior is another potential source of non-stationary behavior. The variations they describe do not correspond with those of the Pacific Decadal Oscillation (described below), especially its prominent change points. Dettinger et al. [2000b] also show slow changes in relationships elsewhere in the Americas.

The American River, discussed earlier, lies nearly on the nodal line of the western ENSO dipole between opposing effects in the Southwest and Northwest United States [e.g., Redmond and Koch, 1991]. Flood likelihood on this river shows greatly different probability distributions for the negative (El Niño), neutral, and positive (La Niña) values of the SOI (standardized Tahiti minus Darwin pressure), and signs of mixed distributions for neutral and positive SOI (Figure 13, also on CD). By contrast, no major Sierra Nevada floods have occurred with El Niño. For this discussion, we are less interested in the reasons for such differences in distributions [which appear to relate to the Intra-Seasonal Oscillation (ISO) near Indonesia] than we are in the simple existence of different probability distributions. Figure 13 is quite interesting, in that no overall correlation exists, and yet strong hints of significant structure are nonetheless present.

Evidence is accumulating that climate "regimes" are present with typical time scales of 20-30 years. The Pacific Decadal Oscillation (PDO) [Mantua et al., 1997], and the North Atlantic Oscillation – and associated Arctic Oscillation [Thompson and Wallace, 1998] – have thus far been identified at these time scales. Others may yet emerge. Efforts to forecast spring/summer runoff have uncovered statistically significant differences in the streamflow and flood probability distributions for the two phases of the PDO [Hamlet and Lettenmeier, 1999; Dettinger et al., 2000c].

These large-scale patterns shift the probability distributions of climate elements in different ways, again depending on location and time of year. They may have direct effects of their own, as well as interactions which modulate the effects of shorter-time-scale climate phenomena (such as the ENSO 2-9 year time scale). In cascading fashion, these in turn can modulate the likelihood of individual hydroclimatic events, once again with effects depending on season and on geographic location with respect to climatic patterns.

The Past Millennium

The time scales just discussed range from years to approximately half-centuries. The latter time scales are increasingly likely to be sampled by paleoclimatic proxy data, and furthermore are into the typical realm of most flood frequency analysis (50-200 years). An important question is whether the inclination toward "regime-like" behavior itself may fluctuate on longer, century-scale lengths of time. Biondi et al. [2001] developed a tree-ring proxy record of PDO, which shows significant spectral power at bidecadal time scales throughout the past four centuries. In the 1900s, however, much lower frequencies have emerged, and at the same time the number of large swings in a combined PDO-SOI time series increased. Especially striking in their time series (Figure 14, also on CD) are the longer regimes (two decades or more) separated by large and abrupt swings, a feature more prominent in the 20th century than during the previous three centuries. Other natural sources of variability can affect the operations of the global climate system in its function as our flood generator, such as solar output, volcanic dust, and internal dynamics of oceans and ice caps. New sources of climate variability, some long-lasting, have emerged: soot and other aerosols, land cover and vegetation changes, and radiatively active trace gases. The presence of these other causes of variability does not qualitatively change the conclusions thus far derived from primarily ENSO considerations; they simply add more possibilities.

Having established that factors in the Pacific Basin alone are sufficient to explain at least part of the inter-annual to century scale variability in climate characteristics in the southwestern United States, we proceed to ask whether analogous variations exist on the longer time scales.

Moderate and stronger El Niño events have been documented using historical information since the early 16th century [Quinn et al., 1987; Quinn and Neal, 1992; Ortleib, 2000]. Historical records from (primarily) the Americas do not conclusively establish the presence of a particular ENSO phase for individual years, but can be nonetheless quite useful in this regard [Deser and Wallace, 1987; Enfield, 1992]. Coral records from the world's oceans can also be employed [Cole et al., 1993; Dunbar et al., 1994; Urban et al., 2000, Cole et al., 2000]. Other extensions of the historical ENSO record, to the year 622 A.D., rely on a strong correlation with Nile River flood data [Quinn, 1992]. Anderson [1992] used the relation between Nile floods and ENSO effects on monsoon rains in East African highlands to show that the centennial variations in the frequency of ENSO can be dramatic, even recognizing the imperfect correspondence (e.g., the 1997-98 ENSO).

In recent years efforts have begun to systematically organize paleoclimatic records from around the earth into annual or subannual time series [e.g., Mann et al., 1995, 1998, 2000a], and place added emphasis on spatial patterns as well. These have so far focused more on temperature than on precipitation, but do show that not all regions behave similarly, consistent with what has been seen with precipitation during the last century in just a portion of North America. In addition, these approaches are trying to incorporate other sources of variability (solar and volcanic variations) extrinsic to the earth's climate system [Mann et al., 2000b and references therein]. Climate system behavior is so intimately linked across such a wide range of space scales that we must ultimately have access to such records to fully understand and explain flood history behavior at a particular point.

REGIONAL PALEOFLOOD CLIMATE RECORDS

There are only a few areas in the world where there are regional paleoflood records and independent paleoclimatic records at similar resolution and with adequate radiocarbon age-control to draw conclusions on the impact of climate on floods over centennial to millennial scales during the late Holocene. The most obvious place is coastal Peru where floods occur only during El Niño years [Wells, 1990; Wells and Noller, 1997; Fontugne et al., 1999]. This area represents the simplest case, with a single oceanic-atmosphere phenomenon that clearly controls occurrence of floods. Consequently it is the variations in the frequency of El Niño over the last several centuries [e.g., Quinn et al., 1987] that have controlled the frequency of floods there [e.g., Wells, 1987]. Another region where flood occurrence is strongly affected by ENSO behavior is northeastern Australia. For that area, Isdale et al. [1998] developed a 337-year chronology of high flows on coastal rivers determined from the presence of terrestrial organic compounds

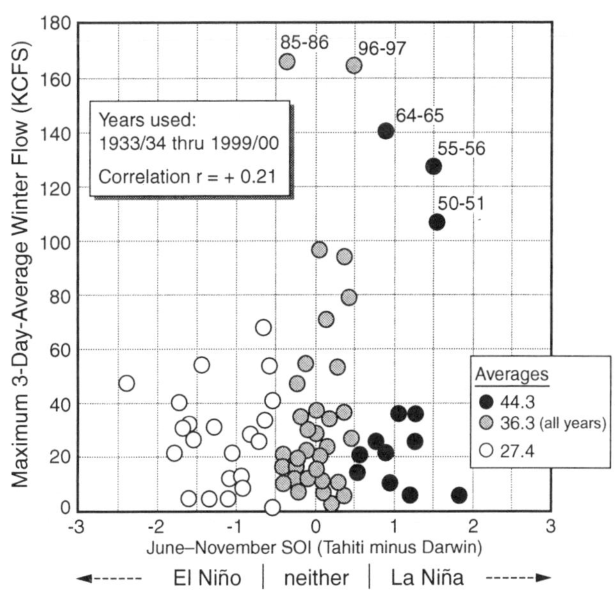

Figure 13. Scatterplot of June-November Southern Oscillation Index versus maximum 3-day average (reconstructed natural) flow during the following winter, for the American River at Fair Oaks gaging station, in Sacramento CA. 1933/34 through 1999-00. Position of average symbols on Y-axis is proportional to value. Open: El Niño, Gray: All years, Solid: La Niña, threshold used is +/- 0.50. Correlation r = + 0.21.

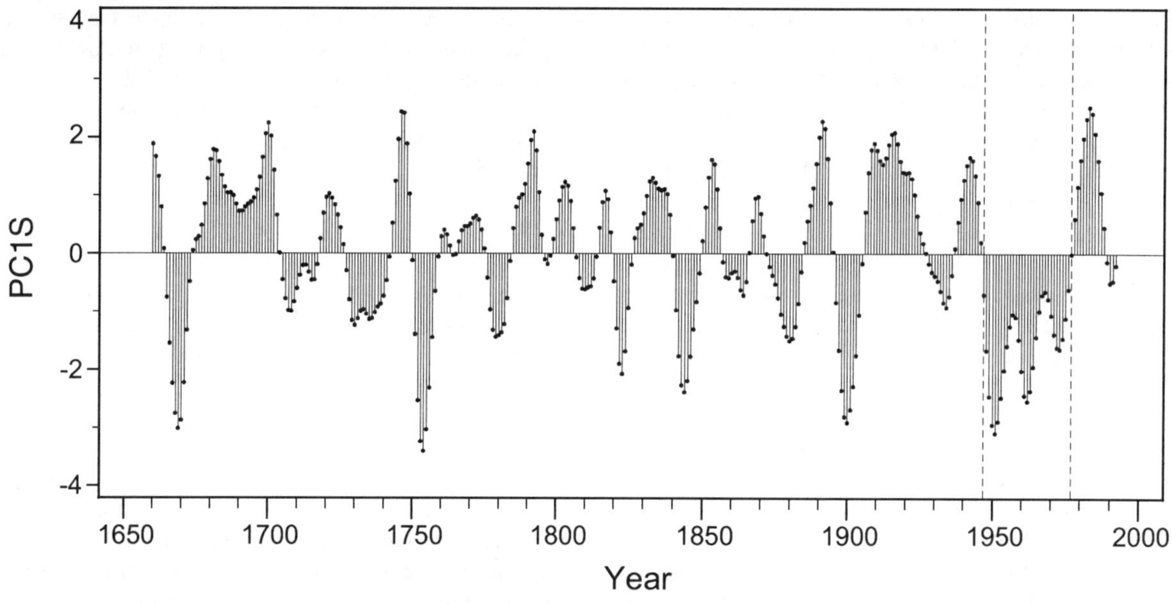

Figure 14. Decadally smoothed first principal component (PC1S) of the *Biondi et al.* [2000] annual tree-ring chronologies from Southern California spanning 1660-1992. The chronology in Figure 4 is the 20th century section of this record.

within the growth layers of coral reefs. These organic materials were derived from the Burdekin River drainage basin and reached the reefs during large floods. Their chronology shows that the mean and variance of the discharge of this tropical river varied dramatically at multidecadal to centennial scales.

There are two examples that involve the identification of individual large floods from specific regions in the United States. The regional paleoflood chronologies from the lower Colorado River drainage basin [Ely et al., 1993; Ely, 1997] and the Upper Mississippi Valley [Knox, 1993, 2000; Knox and Kundzewicz, 1997] are good examples of centennial-scale flood data, both of which strongly suggest non-stationarity in the occurrence of large floods over time. This is not a particularly surprising result in light of variability evident in modern and historical data as reviewed and discussed earlier. Aside from problems associated with age and event resolution of paleoflood stratigraphy [e.g., House et al. and Webb et al., this volume] these two paleoflood chronologies agree well with independent regional paleoclimate data [Knox, 2000; Hirschboeck et al., 2000]. This agreement is encouraging, since it provides independent confirmation of the interpretation of fluvial stratigraphy, whether the latter is related to large or moderate floods. For example, within the limitations of radiocarbon dating of the late Holocene, recent analysis of the paleohydrologic implications of cumulative lacustrine deposits in Lake Canyon, southeast Utah [Pederson, 2000] agrees reasonably well with the paleoflood chronology presented for the nearby region by Ely [1997]. In this setting, Pederson [2000] identified a) a 1000-900 year-old episode of increased flooding [c.f. Ely, 1997], which is also associated with one of the wettest Holocene episodes in the western United States [e.g., Leavitt, 1994; Petersen, 1994], and b) an episode immediately following of drier conditions, and a drop in flood frequency. This is in accordance with the interpretation of the regional fluvial paleoflood record [Ely, 1997]. The episode of frequent paleofloods centered around 3500-4300 years B.P. reported by Ely [1997] also appears in many records in the southwest United States and specifically in flood-lacustrine studies [Enzel and Wells, 1997; Pederson, 2000].

Additional evidence of extended episodes of diminished flood frequency is furnished by the presence of buried soils within flood deposit sequences. The formation of soils in alluvial deposits requires relatively prolonged episodes of subaerial exposure, or a period of minimal deposition of flood sediment. Deposition of thick flood deposits terminates soil-development at the pre-depositional level. Erosion by floods also precludes soil preservation or leave diagnostic stratigraphic evidence. Knox [2000] described buried soils in alluvial stratigraphy in the upper Mississippi Valley. At least one soil is well enough developed to indicate the absence of floods above a specific threshold for relatively long time. In central India, Ely et al. [1996] identified a sequence of flood deposits with a soil forming at their top, suggesting a break in the occurrence of large floods for a few hundreds of years. This soil was overtopped by 1-2 floods only during the 20th century [Ely et al., 1996]. Greenbaum et al. [2000] identified buried, incipient soils in a flood deposit sequence in hyper arid southern Israel. Again, whether or not the sequence is perfectly dated, the presence of buried soils indicates a significant time interval without any flood stage significantly high enough to interrupt soil development. The use of soils as indicators of episodes with relatively lower flood magnitudes and lacking any large flood is actually one of the earlier geological contributions to paleohydrology [e.g. Jahns, 1947], and is presently in a rapidly evolving stage of practical application for establishing constraints on flood-magnitude over time [e.g., Levish, this volume and references therein; Jarrett and Tomlinson, 2000; Klawon et al., 2000].

Another example of a regional paleoflood record has recently been reported from the hyper-arid Negev Desert, Israel. It too suggests episodes of increased flood frequency in the late Holocene [Greenbaum et al., 2000]. Such episodes are also synchronous with high stands of the Dead Sea, which are largely responses to wetter hydrologic regimes in headwater regions in northern Israel, Syria and Jordan. The periods of wetter conditions in the eastern Mediterranean during these high stands are also concurrent with periods of increased flood frequency in the adjacent northern Negev Desert [Frumkin et al., 1998]. A recent study [Kahana, 1999] demonstrates that in the modern record, the large floods in the northern Negev are caused by two types of synoptic-scale atmospheric patterns (an "anomalous" Active Red Sea Trough, and a Syrian Low). On other occasions such patterns produce short periods of heavy rain and snowfall in central and northern Israel, producing no flood but effectively contributing to large seasonal inflow into the Dead Sea. Therefore, the episodes of more frequent large paleofloods and Dead Sea rises in the late Holocene probably represent responses to variability in similar synoptic-scale patterns.

Cayan et al. [1989] describe the hydrologic causal chain in a given basin from precipitation to stream flow to lacustrine conditions and show that the century-long lake level record of Great Salt Lake, Utah, closely reflects the time series of precipitation and stream flow in its watershed. The lake level record can thus be used to infer decadal changes in precipitation and incoming streamflow. High-resolution (in this case up to multi-decadal) geological records would allow the extension of the lacustrine record and therefore streamflow into the late Holocene, indicating that these systems hold considerable promise for developing high-resolution paleoflood-paleoclimate records.

SUMMARY

A number of lines of evidence have steadily been converging on the conclusion that significant climatic variability occurs on time scales of decades to centuries over areas the size of typical river basins. A widely held view of climate as relatively static is giving way to a more realistic acceptance of climate as a dynamic entity exhibiting constant fluctuation.

The likelihood of extreme hydrologic events is strongly linked to climate. Fluctuations in climate on different time scales lead to changes in extreme event likelihood on those same scales. Variations in both mean conditions and in the degree of variability are of interest to a wide range of water specialists. Prior studies and additional observations have been employed here to emphasize that significant variability occurs on time scales important to flood-plain management, community planning, water-supply infrastructure, and engineering and design.

The short records of systematic data constrain our ability to obtain a longer-term perspective for analysis of risk. A particular facet of this analysis is a need to understand whether risk varies with time, and if so by how much and in what manner (i.e., the rate of change). In a kind of logical trap, the predominant focus on the short (typically sub-century) systematic records precludes the ability to learn whether this information is representative of longer periods. This makes the value of paleohydrologic data obvious for improving our understanding of climate variability over longer time frames. Systematic compilation of paleoflood and paleohydrologic data on a global scale would contribute immensely to our understanding of long-term climate variability.

In the practical decision framework the most important question is what characteristics the future is likely to have over some time scale of interest. It is useful and constructive to look upon flood-frequency analysis as, fundamentally, an exercise in prediction. Our research efforts to understand climatic and hydrologic variability, insofar as we wish them to improve the practice of flood-frequency analysis, will be more successful if we cultivate habitual awareness of the predictive nature of the problem and adopt a more flexible approach in general. This is particularly true in instances when basic assumptions are demonstrably false, or marginally appropriate at best.

Paleoflood and other paleohydrologic evidence, judiciously applied, can furnish unique and useful direct and indirect information about flood frequency and its temporal variability. It can also be shown that even in absence of direct age control on the occurrence of individual floods, certain types of geologic information attest to the occurrence of intervals with notably higher flood frequency than other periods of time. Also, it is possible to utilize paleoflood information to estimate maximum limiting values of flood discharge over discrete periods of time, or paleohydrologic bounds [Enzel et al., 1993; Levish, this volume]. This type of information does not speak directly to climate variability, but it does indicate limits on the magnitude of hydrologic events culminating from extreme climatologic and meteorologic conditions over discrete and typically long periods of time. In general, such information can help reduce uncertainty in flood estimation. It may be that there are cases where newer estimates bring more accurate mid-range values, but also larger uncertainty bounds if false confidence in former values is revealed. In all cases, paleohydrologic data provide the only link to hydroclimatologic phenomena that have actually occurred and, on this basis alone, should be given due consideration in any effort that seeks to better understand flood-frequency.

We caution that the use of paleohydrologic information must be carried out in full recognition of its limitations. Among these are its often-qualitative or imprecise nature, its inability in some cases to isolate specific events, occasionally poor temporal resolution, and difficulty in separating specific floods or flood episodes from generally moist climatic conditions. Some limitations are site or investigation specific, but in all cases it is important that the limitations are known. It is possible to develop quantitative methods for incorporating paleohydrologic data into statistical analysis while explicitly accounting for its potential uncertainty [e.g., Levish, this volume]. These and similar developments should help to reduce resistance to incorporating paleo data into conventional hydrologic practice.

Flood-frequency analysis is a consensus activity. Experience has shown that groups typically overestimate the confidence they place in their own forecasts [Nicholls, 1999]. The outcome of flood-frequency analysis is ultimately a decision or set of decisions. As pointed out by Sarewitz and Pielke [2000, p. 2-3], "...decision-making is a forward-looking process...[emphasis as in the original]" and therefore is essentially an act of prediction, a theme that we have strongly echoed, albeit in different form. They go on to say that "...the idea of a prediction as a disembodied number modified by an uncertainty is entirely too abstract to have any meaning in the real world." An elaboration of this sentiment is provided by [Baker, 1994], who argues that physical evidence of floods has considerably more influence on social perceptions of relative risk than do mathematical conceptions of risk derived from idealized, theoretical models. Further elaboration of this point is beyond the scope of this work, but the social, political, and scientific environments are all intimate parts of the context in which flood-frequency analysis takes place. The incorporation of paleoflood and paleohydrologic information into the process will be most effective when the intersection of these elements becomes more clearly understood.

It has been stressed that arid climates offer especially promising circumstances for the use of paleo data in comparison with other regions, because of clarity of evidence, better preservation of evidence, and growing human occupation of such lands. We should be taking advantage of these considerations to reconstruct past conditions in as many streams and washes in arid settings as possible. Knox [1993, 1999] has demonstrated that flood-climate linkages are apparent in humid regions and that there are data types appropriate to extending flood records in time. Other more recent studies in humid and subhumid environments have demonstrated that paleoflood records can be compiled in diverse settings [Jarrett and Tomlinson, 2000; Kite et al., this volume; Springer, this volume], but linkages with climate variability remain to be explored in detail.

Regardless of the region of study, there is a critical need to develop better methods of dating paleoflood events in both absolute and relative terms. This capability is greatly desired to improve the inferential power of paleoflood data when combined with other types of paleo evidence and interpreted with the assistance of physically-based climate models. The latter capability is expanding rapidly. Complementary advances in improving the age-resolution of paleoflood chronologies are less rapid, but more rigorous methods of sampling and extracting organic materials from flood deposits [e.g., Pohl, 1995; Levish et al., this volume; House et al., this volume] and improvements in the application of luminescence dating methods to fluvial sediments [e.g., Duller, 1996; Stokes, 1999] offer considerable promise.

The argument for multi-scaled linkages between climate and floods is not based on speculation. Rather, it is firmly rooted in the rich and mutually supportive body of observations of a highly interconnected, complex system involving the atmospheric, oceanic, terrestrial and hydrologic regimes. Consequently, it seems inescapable that variability in climate should be accompanied by variability in the spatial and temporal distribution of extreme hydrologic events. Abundant evidence exists that such variability occurs over a wide range of time scales. Our ability to discern and quantify the details grows more difficult with increasing time scale. This should serve to motivate integration of concepts originating in allied fields such as geology, paleobiology, and climatology. Their incorporation in the realm of practical hydrology and water resource management would require an expanded view of the pervasive presence of fluctuation and change. Such reorientation further entails recognition of the limitations of over-reliance on outmoded or untenable assumptions about natural systems whose complex behavior is not yet fully appreciated. This path depends on judicious selection and interpretation of evidence that is fully cognizant of both its inherent value and its limitations. A common goal should be the development of a better framework for flood-frequency analysis, water resource management, and related policy-making that is firmly grounded in science and is flexible enough to change in response to scientific discovery and technological advances.

Acknowledgements. K. Redmond would like to acknowledge support from NSF Grant EAR9996036 and from the NOAA Western Regional Climate Center. Y. Enzel received support from the Desert Research Institute and the University of Nevada, Reno. K House was also supported by NSF Grant EAR9996036, and by the Nevada Bureau of Mines and Geology (NBMG). F. Biondi participated with support of grants from the Paleoclimate Program of the National Oceanic and Atmospheric Administration and the Paleoclimate Program of the National Science Foundation. Jack Hursh (NBMG) prepared several of the figures. The authors greatly appreciate the comments of Katherine Hirshboeck and Henry Diaz.

REFERENCES

Alexander, M., A.C. Deser, and M.S. Timlin, The reemergence of SST anomalies in the North Pacific Ocean, *J. Clim.*, 12, 2419-2433, 1999.

Anderson, R.Y., Long-term changes in the frequency of occurrence of El Niño events, in *El Niño: Historical and paleoclimatic aspects of the Southern Oscillation*, ed. H.F. Diaz and V. Markgraf, pp. 193-200, Cambridge University Press, 1992.

Andrade, E., and W. Sellers, El Niño and its effect on precipitation in Arizona, *J. Climatol.*, 8, 403-410, 1988.

Baker, V.R., Paleoflood hydrology and extraordinary flood events, *J. Hydrol.*, 96, 79-99, 1987.

Baker, V.R., Magnitude and frequency of palaeofloods, in *Floods: Hydrological, Sedimentological, and Geomorphological Implications*, ed. K. Beven, and P. Carling, pp. 171-183, John Wiley, New York, 1989.

Baker, V.R., J.M. Bowler, Y. Enzel, and N. Lancaster, Late Quaternary palaeohydrology of arid and semi-arid regions, in *Global Continental Paleohydrology*, ed. K.J. Gregory, L. Starkel, and V.R. Baker, pp. 203-231, John Wiley, New York, 1995.

Benito, G., M.J. Machado, and A. Perez-Gonzalez, Climate change and flood sensitivity in Spain, in *Global Continental Changes: the Context of Paleohydrology*, ed. J. Branson, A. G. Brown, and K. J. Gregory, Geol. Soc. Spec. Pub., 115, London, 1996.

Benito, G., M.J. Machado, A. Perez-Gonzalez, and A. Sopefla, Paleoflood hydrology of the Tagus River, central Spain, in *Paleohydrology and Environmental Change*, edited by G. Benito, V. R., Baker, and K. J., Gregory, pp. 317-332, John Wiley, New York, 1998.

Biondi, F., D.L. Perkins, D.R. Cayan, and M.K. Hughes, July temperature during the second millennium reconstructed from Idaho tree rings, *Geophys. Res. Letters*, 26, 1445-1448, 1999.

Biondi, F., A. Gershunov, and D. R. Cayan, North Pacific decadal climate variability since AD 1661, *J. Clim. (Letters)*, 14, 5-10, 2001.

Brown, S. L., P. L. Bierman, A. Lini, and J. Southon, 10,000 yr record of extreme hydrological events, *Geology*, 28, 335-338, 2000.

Bryson, R.A., The paradigm of climatology: An essay, *Bull. Amer. Meteor. Soc.*, 78, 449-455, 1997.

Cayan, D.R., J.V. Gardner, J.M., Landwehr, J. Namias, and D.H. Peterson, Introduction, *in Aspects of Climate Variability in the Pacific and the Western Americas*, ed. D.H. Peterson, AGU Geophys. Monog. 55, p. xiii-xvi, 1989.

Cayan, D.R., and D.H. Peterson, The influence of North Pacific atmospheric circulation on streamflow in the West, *in Aspects of Climate Variability in the Pacific and the Western Americas*, edited by D.H. Peterson, AGU Geophys. Monog. 55, 375-397, 1989.

Cayan, D.R., and R.H. Webb, El Niño/Southern Oscillation and streamflow in the western United States, *in El Niño: Historical and paleoclimatic aspects of the Southern Oscillation*, Ed. H.F. Diaz and V. Markgraf, Cambridge University Press, 29-68, 1992.

Cayan, D.R., Interannual climate variability and snowpack in the western United States, *J. Clim.*, 9, 928-948, 1996.

Cayan, D.R., K.T. Redmond, and L.G. Riddle, ENSO and hydrologic extremes in the western United States, *J. Clim.*, 12, 2881-2893, 1999.

Chatters, J.C., and K.A. Hoover, Changing late Holocene flooding frequencies on the Columbia River, Washington, *Quat. Res.* 26, 309-320. 1986.

Cole, J.E., R.B. Dunbar, T.R. McClanahan, and N. Muthiga, Tropical Pacific forcing of decadal variability in the western Indian Ocean over the past two centuries, *Science*, 287, 617-619, 2000.

Cole, J.E., R.G. Fairbanks, and G.T. Shen, The spectrum of recent variability in the Southern Oscillation: results from a Tarawa Atoll coral, *Science*, 262, 1790-1793, 1993.

Cook, E R., and K. Peters, The smoothing spline: A new approach to standardizing forest interior tree-ring width series for dendroclimatic studies, *Tree-Ring Bulletin*, 41, 45-53, 1981.

Dai, A., I.Y. Fung, and A.D. Del Genio, Surface observed global land precipitation variations during 1900-1988, *J. Clim.*, 10, 2943-2962, 1997.

Dai, A. and T.M.L. Wigley, Global patterns of ENSO-induced precipitation, *Geophys. Res. Lett.*, 27, 1283-1286, 2000.

Deser, C., and J.M. Wallace, El Niño events and their relation to the Southern Oscillation, *J. Geophys. Res.*, 92, 14,189-14, 196, 1987.

Dettinger, M.D., D.R. Cayan, H.F. Diaz, and D.M. Meko, North-south precipitation patterns in western North America on interannual-to-decadal timescales, *J. Clim.*, 11, 3095-3111, 1998.

Dettinger, M.D., D.R. Cayan, and K.T. Redmond, United States streamflow probabilities based on forecasted La Niña, Winter-Spring 2000, *Experimental Long-Lead Forecast Bulletin*, 8(4), December 1999, 57-61, 1999.

Dettinger, M.D., D.R. Cayan, G.J. McCabe, and J.A. Marengo, Multiscale streamflow variability associated with El Niño/Southern Oscillation, in *El Niño and the Southern Oscillation, Multiscale Variability and Global and Impacts*, Ed. H. F. Diaz and V. Markgraf, Cambridge University Press, 113-146, 2000a.

Dettinger, M.D., D.S. Battisti, R.D. Garreaud, G.J. McCabe, and C.M. Bitz, Interhemispheric effects of interannual and decadal ENSO-like climate variations on the Americas, *in Present and Past Interhemispheric Climate Linkages in the Americas and Their Societal Effects*, Ed. Markgraf, V., Academic Press, 1-16, 2000b.

Dettinger, M.D., D.R. Cayan, G.J. McCabe, and K.T. Redmond, Winter-Spring 2001 United States streamflow probabilities based on anticipated neutral ENSO conditions and recent NPO status, *Experimental Long-Lead Forecast Bulletin*, 9(3), 55-60, 2000c.

Dracup, J.A., and E. Kahya, 1994, The relationships between U.S.

streamflow and La Niña events, *Water Resources Res.*, 30, 2133-2141, 1994.

Duller, G.A.T., Recent developments in luminescence dating of fluvial sediments, *Prog. Phys. Geog., 20*, 127-145, 1996.

Dunbar, R.B, G.M. Wellington, M. Colgan, and P.W. Glynn, Eastern Pacific sea surface temperature since 1600 A.D.: the Delta^{18}O record of climate variability in Galapagos corals, *Paleoceanography*, 9, 291-316, 1994.

Ebbesmeyer, C.C., D.R. Cayan, D. McLain, F. Nichols, D. Peterson, and K.T. Redmond, 1976 Step in the Pacific Climate: Forty environmental changes between 1968-1975 and 1977-1984, Proceedings, *Seventh Annual Pacific Climate (PACLIM) Workshop*, April 10-13, 1990, Asilomar Conference Center, Pacific Grove, California, pp. 115-126, 1991.

Eltahir, E. A. B., El Niño and the natural variabilty in the flows of the Nile River, *Water Resour. Res.*, 32, 131-137, 1996.

Ely, L.L., Y. Enzel, V.R. Baker, and D.R. Cayan, A 5000-year record of extreme floods and climate change in the southwestern United States, *Science*, 262, 410-412, 1993.

Ely, L.L., Y. Enzel, and D.R. Cayan, Anomalous North Pacific atmospheric circulation and large winter floods in the southwestern United States, *J. Clim.*, 7, 977-987, 1994.

Ely, L.L., Y. Enzel, V.R. Baker, V.S. Kale, and S. Mishra, Changes in the magnitude and frequency of late Holocene monsoon floods on the Narmada River, central India, *Geol. Soc. Am. Bull.*, 108, 1134-1148, 1996.

Ely, L.L., Response of extreme floods in the southwestern United States to climate variations in the late Holocene, *Geomorphology*, 19, 175-201, 1997.

Enfield, D.B., and S.L. Cid, Low-frequency changes in El Niño-Southern Oscillation, *J. Clim.*, 4, 1137-1146, 1991.

Enfield, D.B., Historical and prehistorical overview of El Niño/Southern Oscillation, *in El Niño: Historical and Paleoclimatic Aspects of the Southern Oscillation*, Ed. H.F. Diaz and V. Markgraf, Cambridge University Press, 95-117, 1992.

Enfield, D.B., Is El Niño becoming more common? *Oceanography Magazine*, 1, 23-27, 1988.

Enzel, Y., D.R. Cayan, R.Y. Anderson, S.G. Wells, Atmospheric circulation during Holocene lake stands in the Mojave Desert: Evidence of regional climate change, 1989, *Nature*, 341, 44-47, 1989.

Enzel, Y., *Hydrology of a large, closed arid watershed as a basis for paleohydrolgical and paleoclimatological studies in the Mojave River drainage system, southern California*. Ph.D. Dissertation, University of New Mexico, 360 p., 1990.

Enzel, Y., Flood frequency of the Mojave River and the formation of late Holocene playa lakes, southern California, *The Holocene*, 2, 11-18, 1992.

Enzel, Y., W.J. Brown, R.Y. Anderson, L.D. McFadden, S.G. Wells, Short-duration Holocene lakes in the Mojave River drainage basin, southern California, *Quat. Res.*, 38, 60-73, 1992.

Enzel, Y., L.L. Ely, P.K. House, R.H. Webb, and V.R. Baker, Paleoflood evidence for a natural upper bound to flood magnitudes in the Colorado River Basin, *Water Resour. Res.*, 29, 2287-2297, 1993.

Enzel, Y., and S.G. Wells, Extracting Holocene paleohydrology and paleoclimatology from modern records of extreme events: An example from southern California, *Geomorphology*, 11, 203-226, 1997.

Fontugne, M., P. Usselmann, D. Lavalle, M. Julien, and C. Hatt, El Niño variability in the coastal desert of southern Peru during the mid-Holocene, *Quat. Res.*, 52, 171-179, 1999.

Frumkin, A., N. Greenbaum, and A.P. Schick, Paleohydrology of the Northern Negev: Comparative evaluation of two catchments, Ed. Issar, A.S., and Brown, N., *Water, Environment and Society in Times of Climatic Change*, Kluwer, The Netherlands, 97-111, 1998.

Fuller, I.C., M.G. Macklin, J. Lewin, D.G. Passmore, and A.G. Wintle, River response to high-frequency climate oscillations in southern Europe over the past 200 k.y., *Geology*, 26, 3, 275-278, 1998.

Goodrich, D.C., Measured rainfall and channel transmission losses at Walnut Gulch, Figure 2, pp 52-54, Web summary, Hydrologic processes modeling workshop, Tucson AZ, extracted 23 June 2001, http://water.usgs.gov/wicp/acwi/hydrology/mtsconfwkshops/pdf/hydrolMod_workshop00.pdf, 2000.

Gottesfeld, A.S., British Columbia flood scars: maximum flood-stage indicators, *Geomorphology, 14*, 319-325, 1996.

Graybill, D.A., The reconstruction of prehistoric Salt River streamflow, in *The 1982 Excavations at Las Colinas: Environment and Subsistence*, edited by D.A. Graybill, D.A. Gregory, F.L. Nials, S.K. Fish, R.E. Gasser, C.H. Miksicek, and C.R. Szuter, p. 25-38. Archaeological Series No. 162, v. 5, Arizona State Museum, University of Arizona, Tucson, 1989.

Greenbaum, N., Schick, A.P., and Baker, V.R., The paleoflood record of a hyperarid catchment, Nahal Zin, Negev, Desert, Israel, *Earth Surf. Proc. and Landf.*, 25:951-971, 2000.

Hamlet, A.F., and Lettenmaier, D.P., Columbia River Streamflow Forecasting Based on ENSO and PDO Climate Signals, J. *Water Res. Plan. and Mgmt.*, 125 (6), pp. 333-341, Nov./Dec., 1999.

Hirschboeck, K.K., *Hydroclimatology of flow events in the Gila River basin, central and southern Arizona*, Ph. D. dissertation, University of Arizona, Tucson, 325 pp, 1985.

Hirschboeck, K.K., Hydroclimatically-defined mixed distributions in partial duration flood series, in *Hydrologic Frequency Modeling*, Ed. V.P. Singh, D. Reidel Publishing Company, 199-212, 1987a.

Hirschboeck, K.K., Catastrophic flooding and atmospheric circulation anomalies, in *Catastrophic Flooding*, Ed. L. Mayer and D.B. Nash, Allen & Unwin, 23-56, 1987b.

Hirschboeck, K.K., Flood hydroclimatology, in *Flood Geomorphology*, Ed. V.R. Baker, R.C. Kochel, and P.C. Patton, New York, J. Wiley & Sons, 27-49, 1988.

Hirschboeck, K.K., Climate and floods, in *National Water Summary 1988-1989, Floods and Droughts: Hydrologic perspectives on water issues*, U.S. Geol. Surv. Water-Sup. Paper 2375, 67-88, 1991.

Hirschboeck, K.K., L.L. Ely, and R.A. Maddox, Hydroclimatology of meteorologic floods, in *Inland Flood Hazards*, edited by E.E. Wohl, pp. 39-72, Cambridge University Press, 2000.

House, P.K., and K.K. Hirschboeck, Hydroclimatological and paleohydrological context of extreme winter flooding in Arizona, 1993, in *Storm-Induced Geological Hazards: Case Histories from the 1992-1993 Winter Storm in Southern California and Arizona*, Ed. R.A. Larson and J.E. Slosson, Boulder Colorado, Geol. Soc. Amer. Reviews in Engineering Geology, 9, 1-24, 1997.

House, P.K., and V.R. Baker, Paleohydrology of flash floods in small desert watersheds in western Arizona, *Water Resour. Res.*, 37, 6 , 1825-1840, 2001.

Isdale, P.J., B.J. Stewart, K.S. Tickle, and J.M. Lough, Paleohydrological variation in a tropical river catchment: A reconstruction using fluorescent bands in corals of the Great Barrier Reef, *The Holocene*, 8, 1-8, 1998.

Jahns, R.H., Geologic features of the Connecticut Valley, Massachusetts, as related to recent floods. *U.S. Geol. Surv. Water Supply Pap.* 996, 158 pp., 1947.

Jarrett, D.R., and E.M. Tomlinson, Regional interdisciplinary paleoflood approach to assess extreme flood potential, *Water Resources Res.*, 36, 2957-2984, 2000.

Kahana, R., *Synoptic hydroclimatology of major floods in the Negev and Arava, southern Israel*, MS. Thesis, Hebrew University of Jerusalem, 93 p. (in Hebrew with English abstract), 1999.

Kahya, E., and J.A. Dracup, U.S. streamflow patterns in relation to the El Niño/Southern Oscillation, *Water Resources Res.*, 29, 2491-2503, 1993.

Kahya, E., and J.A. Dracup, The influence of Type 1 El Niño and La Niña events on streamflows in the Pacific Southwest of the United States, *J. Clim.*, 7, 965-976, 1994.

Kale, V.S., S. Mishra, and V.R. Baker, A 2000 year paleoflood record from Sakarghat on Narmada, central India, *J. Geol. Soc. India*, 50, 283-288, 1997.

Kiladis, G., and H.F. Diaz, Global climate anomalies associated with extremes of the Southern Oscillation, *J. Clim.*, 2, 1069-1090, 1989.

Klawon, J.E., D.R. Levish, D.R.H. O'Connell, Paleofloods and paleosols on the North Platte River: Stratigraphic records of exceedance and nonexceedance, *Geol. Soc. Amer. Abstracts with Prog., 32, 7,* A-509, 2000.

Knox, J.C., Responses of river systems to Holocene climates, in *Late Quaternary Environments of the United States*, v. 2, The Holocene, Ed. H.F. Wright, Jr., University of Minnesota Press, Minneapolis, 26-41, 1983.

Knox, J.C., Fluvial Response to small scale climatic changes, in *Developments and Applications of Geomorphology*, Ed. J.E. Costa and P.J. Fleisher, Springer, Berlin, 318-342, 1984.

Knox, J.C., Large increase in flood magnitude in response to modest changes in climate, *Nature*, 361, 430-432, 1993.

Knox, J.C., and Z.W. Kundzewicz, Extreme hydrological events, palaeo-information and climate change, *Hydrol. Sci. J.*, 42, 765-779, 1997.

Knox, J.C., Long-term episodic changes in magnitudes and frequencies of floods in the Upper Mississippi River Valley, in *Fluvial Processes and Environmental Change*, Brown, A.G., and Quine, T.A., eds., Wiley, Chichester, 255-282, 1999.

Knox, J.C., Sensitivity of Modern and Holocene floods to climate change, *Quat. Sci. Rev.*, 19, 439-457, 2000.

Kunkel, K.E., and J.R. Angel, Relationship of ENSO to snowfall and related cyclone activity in the contiguous United States, *J. Geophys. Res.*, 104, D16, 19,425-19,434, 1999.

Lamb, H.H., *Climate: Present, Past and Future*, v. 2, Climatic History and the Future, Methuen & Co. LTD, London, 835 pp., 1977.

Leavitt, S.W., Major wet interval in White Mountains Medieval Warm period evidenced in ^{13}C of Bristlecone Pine tree rings, *Clim. Change,* 26, 299-307, 1994.

Levish, D.R., D.A. Ostenaa, and D.R.H. O'Connell, Paleoflood hydrology and dam safety, *Waterpower '97, Proceedings of the International Conference on Hydropower*, August 5-8, 1997, Atlanta, Georgia, 2205-2214, 1997.

Lines, G.C., Ground-water and surface-water relations along the Mojave River, Southern California, *USGS Water Resources Investigations Report 95-4189*, 43 p, 1996.

Livezey, R.E., M. Masutani, A. Leetma, H. Rui, M. Ji, and A. Kumar, Teleconective response of the Pacific-North American region atmosphere to large central equatorial Pacific SST anomalies, *J. Clim.*, 10, 1787-1820, 1997.

Maddox, R.A., F. Canova, and L.R. Hoxit, Meteorological characteristics of flash flood events over the western United States, *Mon. Weath. Rev.*, 108, 1866-1877, 1980.

Mann, M.E., J. Park, and R.S. Bradley, Global interdecadal and century-scale climate oscillations during the past five centuries, *Nature*, 378, 266-270, 1995.

Mann, M.E., R.S. Bradley, and M.K. Hughes, Global-scale temperature patterns and climate forcing over the past six centuries, *Nature*, 392, 779-787, 1998.

Mann, M.E., R.S. Bradley, and H.K. Hughes, Long-term variability in the El Niño Southern Oscillation and associated teleconnections, in *El Niño and the Southern Oscillation, Multiscale variability and global and regional impacts*, Ed. H.F. Diaz, and V. Markgraf, Cambridge University Press, 357-412, 2000a.

Mann, M.E., E. Gille, R.S. Bradley, M.K. Hughes, J. Overpeck, F.T. Keimig, and W. Gross, Global temperature patterns in past centuries: An interactive presentation, *Earth Interactions*, V. 4, Paper 4, 29 pp., 2000b.

Mantua, N.J., S.T. Hare, Y. Zhang, J.M. Wallace, and R.C. Francis, A Pacific interdecadal climate oscillation with impacts on salmon production, *Bull. Amer. Meteor. Soc.*, 78, 1069-1079, 1997.

McCabe, G.J., and M.D. Dettinger, Decadal variations in the strength of ENSO teleconnections with precipitation in the western United States, *Int. J. Climatol.*, 19, 1399-1410, 1999.

McCord, V.A.S., Fluvial process dendrogeomorphology: Reconstruction of flood events from the Southwestern United States using flood-scarred trees. in *Tree Rings, Environment, and Humanity*, Ed. J.S. Dean, D.M. Meko, and T.W. Swetnam, Radiocarbon Special Volume, Department of Geosciences, The University of Arizona, Tucson, pp. 689-699, 1996.

Meko, D.M., and C.H. Baisan, Pilot study of latewood-width of conifers as an indicator of variability of summer rainfall in the North American Monsoon Region, *International Journal of Climatology*, 21, 697-708, 2001.

Meko, D.M., M.D. Therrell, C.H. Baisan, and M.K. Hughes, Sacramento River flow reconstructed to A.D. 869 from tree rings, *J. Amer. Water Resour. Assoc.*, accepted 2001.

Namias, J., X. Yuan, and D.R. Cayan, Persistence of North Pacific sea surface temperature and the atmospheric flow patterns, *J. Clim.*, 1, 682-703, 1988.

National Research Council, *Improving American River flood frequency analyses, National Academy Press*, 120 pp., 1999.

Nicholls, N., Cognitive illusions, heuristics, and climate prediction, *Bull. Amer. Meteor. Soc.*, 80, 1385-1397, 1999.

Ortleib, L, The documentary historical record of El Niño events in Peru: An update of the Quinn record (sixteenth through nineteenth centuries), in *El Niño and the Southern Oscillation,*

Multiscale Variability and Global and Regional Impacts, Ed. H. F. Diaz and V. Markgraf, Cambridge University Press, 357-412, 2000.

Pederson, J.L., Holocene paleolakes of Lake Canyon, Colorado Plateau: Paleoclimate and landscape response from sedimentology and allostratigraphy, *Geol. Soc. Am. Bull.*, 112, 147-158, 2000.

Petersen, K.L., A warm and wet little climatic optimum and cold and dry Little Ice Age in the southern Rocky Mountains, U.S.A., *Clim. Change*, 26, 243-269, 1994.

Piechota, T.C., J.A. Dracup, and R.G. Fovell, Western U.S. streamflow and atmospheric circulation patterns during El Niño-Southern Oscillation, *J. Hydrol.*, 201, 249-271, 1997.

Pohl, M., Radiocarbon ages on organics from piedmont alluvium, Ajo Mountains, Arizona, *Phys. Geog., 16*, 339-353, 1995.

Pulwarty, R.S., and K.T. Redmond, Climate and salmon restoration in the Columbia river basin: The role and usability of seasonal forecasts, *Bull. Amer. Meteor. Soc.*, 78, 381-397, 1997.

Quinn, W. H., V.T. Neal, and S.E. Antunez de Mayolo, Niño occurrences over the past four and a half centuries, *J. Geophys. Res.* 92, 14449-14461, 1987.

Quinn, W.H., A study of SO-related climatic activity for A.D. 622-1990 incorporating Nile River flood data, in *El Niño: Historical and paleoclimatic aspects of the Southern Oscillation*, Ed. H.F. Diaz and V. Markgraf, Cambridge University Press, 119-149, 1992.

Quinn, W.H., and V.T. Neal, The historical record of El Niño events, in *Climate Since A.D. 1500*, [eds: R.S. Bradley and P.D. Jones], Routledge, New York, 623-648, 1992.

Redmond, K.T., and D.R. Cayan, El Niño/Southern Oscillation and western climate variability, in *Proceedings, Sixth Conference on Climate variations*, Nashville, Tennessee, Amer. Meteor. Soc., 141-145, 1994.

Redmond, K.T., and R.W. Koch, Surface climate and streamflow variability in the western United States and their relationship to large-scale circulation indices, *Water Resources Res.*, 27, 2381-2399, 1991.

Ropelewski, C.F., and M.S. Halpert, North American precipitation and temperature patterns associated with the El Niño/Southern Oscillation (ENSO), *Mon. Weath. Rev.*, 114, 2352-2362, 1986.

Ropelewski, C.F., and M.S. Halpert, Global and regional scale precipitation patterns associated with the El Niño/Southern Oscillation (ENSO), *Mon. Weath. Rev.*, 115, 1606-1626, 1987.

Ropelewski, C.F., and M.S. Halpert, Precipitation patterns associated with the high-index of the Southern Oscillation, *J. Clim.*, 2, 268-284, 1989.

Said, R., *The River Nile: Geology, Hydrology and Utilization*, Pergamon Press, 332 pp., 1993.

Sarewitz, D., and R.A. Pielke, Jr., Introduction, pp. 1-7, in *Prediction: Science, Decision-Making, and the Future of Nature*, Ed. D. Sarewitz, R.A. Pielke, Jr., and R. Byerly, Jr., 405 pp., Island Press, 2000.

Schonher, T., and S. Nicholson, The relationship between California rainfall and ENSO events, *J. Clim.*, 2, 1258-1269, 1989.

Stamos, C.L., P. Martin, T. Nishikawa and B.F. Cox, Simulation of ground-water flow in the Mojave River Basin, California, U.S. Geol. Surv. Water Resour. Invest. Rept. 01-4002, Version 1.1, Sacramento CA, 129 pp., 2001.

Stedinger, J.R., and T.A. Cohn, Flood frequency analysis with historical and paleoflood information, *Water. Resources Res.*, 22, 785-793, 1986.

Stockton, C. W. and G. C. Jacoby, Jr., Long-term surface-water supply and streamflow trends in the upper Colorado River Basin based on tree-ring analyses, *Lake Powell Research Project Bulletin 18*, 1975.

Stockton, C. W., Long-term streamflow reconstruction in the upper Colorado River basin using tree-rings, *in Colorado River Basin Modeling Studies*, Proceedings of a Seminar held at Utah State University, July 16-18, 1975, Ed. C. G. Clyde, D. R. Falkenberg, and J. P. Riley, pp. 401-41, Utah Water Research Lab, College of Engineering, Utah State University, Logan, 1975.

Stokes, S., Luminescence dating applications in geomorphologic research, *Geomorphology*, 29, 153-171, 1999.

Taylor, K.C., Rapid climate change, *American Scientist*, 87(4), 320-327,1999.

Thompson, D.W.J., and J.M. Wallace, The Arctic Oscillation signature in wintertime geopotential height and temperature fields, *Geophys. Res. Let.*, 25 (9), 1297-1300, 1998.

Torrence, C., and P.J. Webster, Interdecadal changes in the ENSO-Monsoon system, *J. Clim.*, 12, 2679-2690, 1999.

Tourre, Y.M., Y. Kushnir, and W.B. White, Evolution of interdecadal variability in sea level pressure, sea surface temperature, and upper ocean temperature over the Pacific Ocean, *J. Phys. Oceanogr.*, 29, 1528-1541, 1999.

U.S. Army Corps of Engineers, *Mojave River Dam - San Bernardino County, California, Final Feasibility Report*, Appendix 4, Evaluations of proposed modifications - Mojave River Dam, California, by The Hydrologic Engineering Center, Special Project Memo No. 85-2, 102 p., 1986,

U.S. Water Resources Council, *Guidelines for determining flood flow frequency*, Bulletin 17B, Hydrology Committee, Washington, D.C., 1981.

Urban, F.E., J.E. Cole, and J.T. Overpeck, Influence of mean climate change on climate variability from a 155-year tropical Pacific coral record, *Nature*, 407, 989-993, 2000.

Van West, C.R., and J.H. Altschul, Environmental variability and agricultural economics along the Lower Verde River, A.D. 750-1450, in *Vanishing River, Landscapes and Lives of the Lower Verde Valley: The Lower Verde Archaeological Project, Overview, Synthesis, and Conclusions*, Ed. S.M. Whittlesey, R. Ciolek-Torrello, and J.H. Altschul, Chapter 9, pp. 337-392, SRI Press, Tucson, Arizona, 1997.

Wang, H., and R. Fu, Winter monthly mean atmospheric anomalies over the North Pacific and North America associated with El Niño SSTs, *J. Clim.*, 13(19), 3435-3447, 2000.

Webb, R.H., and J.L. Betancourt, Climatic variability and flood frequency of the Santa Cruz River, Pima County, Arizona, *U.S. Geol. Surv. Water-Supply Paper 2379*, 1992.

Wells, L.E., An alluvial record of El Niño events from northern coastal Peru, *J. Geophys. Res.*, 92, 14,463-14,470, 1987.

Wells, L.E., Holocene history of the El Niño phenomenon as recorded in flood sediments of northern coastal Peru, *Geology*, 18, 1134-1137, 1990.

Wells, L.E., and J. S. Noller, Determining the early history of El Niño, *Science*, 276, 966-997, 1997.

Wohl, E.E., and Y. Enzel, Data for paleohydrology, in *Global Continental Paleohydrology*, Ed. K.J. Gregory, L. Starkel, and V.R. Baker, p. 23-59, John Wiley, New York, 1995.

Wolter, K., and M.S. Timlin, Measuring the strength of ENSO-How does 1997/98 rank? *Weather*, 53, 315-324, 1998.

Woolhiser, D.A., T.O. Keefer, and K.T. Redmond, Southern Oscillation effects on daily precipitation in the southwestern United States, *Water Resour. Res.*, 29(4), 1287-1295, 1993.

Yarnal, B., and H.F. Diaz, Relationships between extremes of the Southern Oscillation and the winter climate of the Anglo-American Pacific coast, *J. Climatol.*, 6, 197-219, 1986.

Kelly T. Redmond, Western Regional Climate Center Desert Research Institute, 2215 Raggio Parkway, Reno, Nevada 89512-1095, email: krwrcc@dri.edu

Yehouda Enzel, Institute of Earth Sciences, Hebrew University of Jerusalem, Jerusalem 91904, Israel

P. Kyle House, Nevada Bureau of Mines and Geology, University of Nevada, Mail Stop 178 Reno, Nevada 89557-0088

Franco Biondi , Department of Geography, University of Nevada, Mail Stop 154, Reno, Nevada 89557-0048

Paleoflood Reconstruction on Floodplains Using Geophysical Survey Data And Hydraulic Modeling

Geoff Pickup and Alan Marks

CSIRO Land and Water, Canberra City, Australia.

Mary Bourke

School of Geography, University of Oxford, England.

Arid zone floodplains may contain large-scale landforms created by paleoflood deposition. However, reconstruction of event magnitude using hydraulic modeling techniques may be difficult because of flow complexity and limitations of topographic data. In this chapter, we show how low-altitude airborne geophysical surveys provide terrain elevation data with sufficient resolution to describe the main elements of floodplain topography. Information from existing topographic maps cannot do this. We also present automatic methods for drainage net extraction and identification of cross-sections and flow patterns for use in gradually varied flow calculations on floodplains. Airborne geophysical surveys provide information on radioelement concentrations in near-surface soil and rock from gamma ray emission measurements. These data are used to identify sediment sources and large-scale sedimentary environments on the Todd River floodplain in arid central Australia. Gradually varied flow calculations are used to identify the extent of 50-10,000 year floods for comparison with the gamma ray and Landsat Thematic Mapper data. Patterns in the distribution of radioelements correspond closely with modeled extreme flood activity. They also show that, while much of the floodplain is active, there are older and larger sedimentary features dating from previous times and larger flood events than those experienced in the last 1,000 years.

INTRODUCTION

Paleoflood reconstruction is normally done from slack-water deposits in bedrock gorges [*O'Connor and Webb*, 1987] although there are some notable exceptions [e.g. *Bretz*, 1925; *Baker and Nummedal*, 1978; *Hupp*, 1987]. The usual practice is to survey gorge cross-sections and to carry out a flow profile analysis assuming gradually varied flow conditions. The calculated water surface elevations may then be related to slackwater deposit elevations in sheltered locations such as tributary entrances, caves, paleomeanders or areas on the margin of rapid flow expansions or contractions [*Kochel and Baker*, 1987]. These areas often have large eddies or areas of backed-up water where sediment deposition occurs.

Ancient Floods, Modern Hazards: Principles and Applications of Paleoflood Hydrology
Water Science and Application Volume 5, pages 47-60

While flow profile calculations are relatively easy to conduct in gorges, sediment deposits frequently do not survive in these locations for long periods in arid Australia. This limits the usefulness of paleoflood reconstruction techniques for long-term flood-frequency analysis and for studies of extreme events in climate change. For example, [*Baker et al*, 1988] describe slackwater deposits dating back only about 900 years in the Finke gorge in central Australia, yet this area has a much longer paleoflood history [*Bourke*, 1999]. Heavily truncated sediment deposits do survive in some parts of the gorge but the waterway seems to have been swept clean at some stage [*Pickup*, 1989]. Similarly, *Baker and Pickup* [1985] were only able to extend the flood history of the Katherine River back over about 150 years. Longer time series have been reported from the tropical cyclone belt to the north [*Gillieson et al*, 1991; *Wohl et al*, 1994; *Nott et al*, 1996]. However, paleoflood analysis has found few practical applications in Australia so far, in spite of being recognized as a viable technique in Australia's flood estimation manual [*Pilgrim*, 1987].

While paleoflood traces are limited in gorges, a much longer-term record may survive in sediment sinks such as floodplains and desert 'floodouts' [*Wells and Dohrenwend*, 1985; *Patton et al*, 1993; *Bourke*, 1999]. However, the deposits left behind by catastrophic floods do not form simple superpositional sequences in confined locations. Instead, they survive as discontinuous smears, spread over large areas, partially buried by subsequent events, and heavily-modified by commercial cattle grazing since European settlement [*Pickup, 1991*; *Patton et al*, 1993]. Dealing with this large scale, complex "horizontal stratigraphy" often defies traditional ground-based sediment sampling techniques. Indeed, it is better dealt with using remote sensing techniques as long as soil properties are not masked by vegetation, or if the vegetation reflects the soil properties closely enough [e.g. *Bourke*, 1999].

In the first part of this chapter, we examine how remote sensing may be used to identify sedimentary environments in arid central Australia and to infer the extent of extreme floods on the floodplain of the Todd River. While some data from Landsat Thematic Mapper (TM) are used, we concentrate on gamma-ray emission measurements obtained by airborne geophysical surveys. These provide information on the potassium (K), thorium (Th) and uranium (U) content of near-surface soil and rock and are largely unaffected by vegetation cover. Spatial patterns in these radioelements in alluvial environments may show paths of sediment transport, areas of deposition and the effects of sediment sorting. They may also reflect the extent of weathering and relative age of sediments. We use these data, together with information from field surveys, to build a picture of how extreme floods have affected large-scale floodplain morphology.

Even where paleoflood deposits can be identified on floodplains, paleoflood reconstruction is difficult. Channel cross-section surveys over large areas may be expensive and logistically difficult. Flood deposits from extreme events may extend for tens to hundreds of km and areas of inundation may be ten km wide or more. Also, because the terrain is relatively flat, it may be better to reconstruct event magnitude from the lateral extent of flooding rather than estimated water surface elevation.

In the second part of this chapter, we attempt to reconstruct the areal extent of extreme floods on the Todd floodplain using a combination of remote sensing, hydraulic calculations that estimate depth and location of inundation, and regional flood-frequency data. Existing topographic map data are inadequate for gradually varied flow calculations given that the contour interval is 20 m and only a few spot heights are available. Other sources of information on floodplain topography therefore must be sought. We therefore begin by describing techniques for combining widely available contour data with radar altimetry from airborne geophysical surveys to provide a useable digital elevation model (DEM) for both steep terrain and floodplain areas with low relief. Terrain analysis techniques are applied to extract flow directions from the DEM and to identify flow cross-sections. We use these cross-sections in a regional-scale flow profile analysis and estimate the lateral extent of floodplain inundation during extreme flood events with return periods of up to 10,000 years. The resultant patterns are then compared with the gamma-ray survey data to check their credibility.

METHODS

Airborne Gamma-ray Remote Sensing

Over the last decade, airborne geophysical surveys have been flown over about one-third of Australia using helicopters or fixed wing aircraft and are continuing. Most surveys are carried out at low altitude, with a flight line spacing of 200 m or 400 m. The gamma-ray data used in this chapter were supplied by the Northern Territory Department of Mines and Energy. Surveys were flown at 400 m line spacing and a mean terrain clearance of 60 m in a north-south direction. Radiometric and elevation data were sampled at 70 m intervals along each line. This resolution is suitable for mapping of large-scale sedimentary features, making these surveys a valuable but under-used tool in geomorphology.

Gamma-ray surveys use instruments mounted in low-flying aircraft to detect the abundance of potassium (K), thorium (Th) and uranium (U) in soils and near-surface rocks. They therefore offer potential for mapping paleoflood deposits at coarse resolution but on a regional scale. Relative K abundance is measured directly using the decay of the unstable isotope, ^{40}K, at 1.46 MeV. Relative U and Th abundance are measured indirectly from the gamma-emitting daughter products, ^{214}Bi and ^{208}Tl respectively [*Grasty*, 1979]. Most of the radiation is derived from shallow depths with approximately 90% coming from the top 30-45 cm for dry material with a density of 1.5 grams/cm^3 [*Grasty*, 1975]. Attenuation by vegetation is minimal except where there are dense stands of forest [*Dickson and Scott*, 1992]. In these forested areas, reductions in gamma ray counts of 14-22% have been recorded [*Aspin and Bierworth*, 1997]. Thus, there are some advantages over remote sensing in the visible-near infrared-middle infrared parts of the spectrum when the vegetation canopy obscures the soil.

Gamma-ray emissions largely reflect the geochemistry of the parent rock type and, to a lesser degree, the extent of weathering [e.g. *Dickson and Scott*, 1992; *Wilford*, 1995; *Dickson and Scott*, 1997]. Weathering effects vary with lithology but K is geochemically mobile and soluble under most weathering conditions, resulting in either loss in solution or adsorption onto clay minerals. U is readily leached and can be released from soluble materials under oxidizing conditions and precipitates under reducing conditions.

Differential rates of weathering may lead to variations in the gamma-ray signatures of different particle size fractions in soils. For example, radioelement analysis of size fractions from soils collected over granitic rocks show that U and Th tend to be retained in clay-rich or iron oxide-rich fractions [e.g. *Bierworth*, 1996]. K concentrations are greater in coarser factions where feldspar or biotite occur [*Dickson and Scott*, 1992]. The sorting that occurs during sediment transport may also affect gamma-ray signatures. For example, both K and Th vary with distance down hillslopes and with the intensity of erosion and deposition in both granites and metasediments [*Pickup and Marks*, 2000].

Transported sediments often reflect the characteristics of their parent material making it possible to track the path of material across the landscape and to identify deposited material from different sources [*Dickson and Scott*, 1992]. It may also be possible to identify depositional environments and surfaces with different ages from changes to source material gamma-ray signatures due to sediment sorting or pedogenesis [*Wilford*, 1995]. This opens the possibility of reconstructing paleoflood extents on floodplains and, perhaps, identifying relative ages of different surfaces. Indeed, preliminary work suggests that the pattern of K-rich floodplain sediments closely matches the extent of extreme floods on the Todd River floodplain [*Pickup and Marks*, 2001].

The gamma-ray data were gathered using two 256 channel airborne systems: a Picodas PDAS-1000 system and an Exploranium GR820 system. Data were calibrated and corrected using the processing techniques described by Grasty and Minty [1995] and Minty [1997]. Line survey data were interpolated to a 100 m grid using the bicubic spline algorithm of Akima [1970]. This level of interpolation is feasible because the survey instrument integrates emissions over an area rather than measures them at a point.

DEM Construction

High-quality DEM data are not routinely available over most of Australia. Until recently, the only digital elevation data available in our study area came from 1: 100,000 draft topographic mapping with a limited number of spot heights and a contour interval of 20 m. Most surface-fitting routines will not produce a DEM from these data that reproduces observed drainage patterns when flowpaths are calculated in terrain analysis routines. If a DEM is constructed using *ANUDEM* software [*Hutchinson*, 1989], with topographic low points forced to occur along mapped stream lines, observed drainage will be reproduced, but floodplain topography may still be poorly represented. It may also be distorted by the fitting process. This is hardly surprising given that variations in floodplain elevation may be 5 m or less and the source data only capture features with a height variation of 20 m or more.

Besides collecting information on magnetic properties and gamma-ray emissions from the underlying terrain, many geophysical surveys also record aircraft elevation above ground from the aircraft's radar altimeter. In central Australia, these data were recorded at intervals of 70 m along flight lines and, when coupled with information from the aircraft mounted differential global positioning system, allow calculation of terrain elevation. Tests of the accuracy of these derived elevations show that most points fall within 3 m of surveyed elevations [*Horsfall*, 1997]. However, the 400 m line spacing used in the central Australian geophysical surveys means that some topographic features are not recorded in the more rugged parts of the landscape (Plate 1, color version on CD). Conversely, geophysical survey elevation data may provide much more detailed representation of topographic variation than the contour data in the flatter parts of the landscape where contours may be more than 5 km apart. It might therefore be useful to combine both data sets when constructing a DEM.

Attempts to fit a DEM to combined contour and geophysical survey data using *ANUDEM* were not successful. Errors in contour location distort small variations in topography on floodplains recorded by the radar altimeter. Vertical accuracy of contour maps is also limited, with 10% of mapped elevations having errors greater than 10 m for this

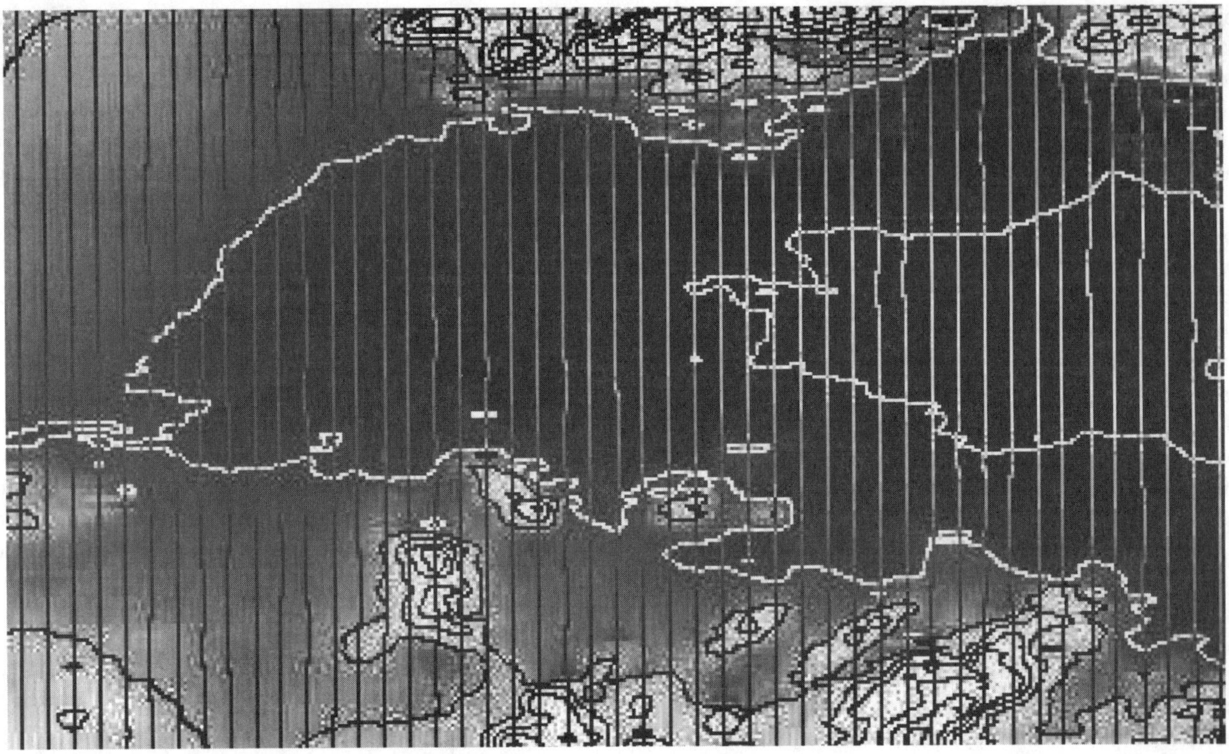

Plate 1. Geophysical survey flight lines overlaid on a DEM with 20-m contours marked. Note the wide contour spacing on the flatter areas. The survey lines are 400 m apart.

mapping series. Changes in aircraft attitude when crossing steep mountain ranges and, at the same time, attempting to maintain a constant altitude above the ground may mean that the altimeter does not always record distance to a point immediately below the aircraft. There may also be differences of up to 200 m in positional accuracy when combining GPS and topographic map data in central Australia though both reportedly use the same datum [*G. Priestley*, personal communication].

Our method for producing a composite DEM is to derive separate DEMs from the contour data and the geophysical survey data, and then to combine them. *ANUDEM* is used on the contour data but without drainage forcing since this distorts floodplain topography. Also, drainage line locations on topographic maps in central Australian are often incorrect or incomplete. The geophysical survey DEM is generated from flight line data using Akima's [1970] bicubic spline algorithm after first removing elevation offsets between individual surveys by microleveling [*Minty*, 1991]. The combination algorithm is based on the assumption that large differences between the DEMs occur because survey line spacing is too far apart to capture some elements of the topography. Where the difference is small, either DEM may be acceptable, but the geophysical survey data capture small topographic variations in more detail. We therefore weight the results towards the geophysical survey data where the difference between DEMs is small, and towards the contour data where the difference is greater. The combination procedure may be written as:

$$z = z_C \quad \text{for } zdiff \geq n \tag{1}$$

$$z = \sqrt{(zdiff / n)}\, z_C + \{1 - \sqrt{(zdiff / n)}\} z_R \quad \text{for } zdiff < n \tag{2}$$

where: z is the final elevation, $zdiff = |z_C - z_R|$, z_C is elevation from the contour-based DEM, z_R is elevation from the geophysical survey DEM, and n is the data range over which the correction is to be applied. In our case, n was set at half the difference between the 5th and 95th percentiles of the *zdiff* values.

While the composite DEM provides some improvement over individual DEMs, it is still less than ideal for paleoflood reconstruction. The central Australian airborne geophysical survey data were acquired with a line spacing of 400 m and there is little justification in generating a DEM with a grid cell size of less than 100 m. This is still large enough to miss many floodplain features. Equally, the reported accuracy of 3 m or less is enough to produce local errors in paleoflood reconstruction. In spite of these difficulties, we believe that it should be possible to reproduce the major features of many extreme floods and there is currently no other source of low-cost topographic data in the area of interest.

Flow Computations on Floodplains

Flow profile analysis in the bedrock gorges is usually carried out by surveying a series of channel cross-sections and carrying out gradually varied flow calculations using a computer package such as HEC-2 [*Hydrologic Engineering Center*, 1982]. Sub-critical flow is assumed and the calculations are carried out cross-section by cross-section in the upstream direction. Flow profile analysis on floodplains follows a similar procedure but if the source data come from a DEM, flow directions and channel locations must first be identified before extracting channel cross-section data. Cross-sections must also be sorted so the calculations proceed in the upstream direction in reaches where flow is sub-critical.

Several terrain analysis packages offer routines for calculation of slope and aspect, identification of flow directions, downstream accumulation of contributing areas and, if necessary, pit removal from DEM data. We have used the routines in *TNT MIPs* [*MicroImages*, 1999]. These routines produce a pitless DEM and route flow down slopes and across flats using an algorithm which passes the flow from a given grid cell into a single downstream neighbour based on the steepest direction of flow. This provides a data layer in which contributing drainage area increases progressively downstream and allows sorting of cross-sections into downstream or upstream sequences. Terrain analysis packages that use a dispersion approach and allow flow to be proportioned out among several downstream neighbors based on terrain curvature are unsuitable because downstream flow directions do not always result in progressively increasing drainage area.

The next step is to identify cross-section locations and to extract cross-section profiles. We first set a minimum drainage area that is used to define the stream network. Grid cells with a drainage area that exceeds this minimum are classed as channel centreline locations and sorted by decreasing drainage area. A second sorting pass orders channel centreline grid cells with a similar drainage area on the basis of increasing elevation. The procedures arrange channel centerline grid cells so that gradually varied flow calculations may be carried out in the upstream direction using the standard step method [*Chow*, 1959]. The calculations begin by assuming uniform flow in locations where the channel intersects the DEM boundary or where the channel terminates in a lake or floodout. Slopes for uniform flow calculations are taken from terrain analysis package results or, in the case of terminating drainage lines, assigned a starting value. Channel cross-sections are identified by calculating mean flow direction at the centerline pixel and constructing orthogonals on each side. Mean flow direction is estimated based on a 10-20 grid cell moving window that is passed along each stream channel link. The gradually varied flow calculations produce a water depth, a friction slope, and the location of the inundated area. Water depths in grid cells between cross-sections are calculated by interpolation.

The gradually varied flow algorithm, as described above, does not distinguish between active flow zones and areas of slackwater or backflow. These occur at expansions and contractions of the channel and of the flooded area. If the slackwater or backflow areas are included in the flow calculations, they will produce errors in cross-sectional area and mean velocity. We therefore conduct two passes through the calculations. The first pass makes no distinction between the active flow zone and other inundated areas but it does provide estimates of channel width. From this, we identify all constrictions along each channel. Active flow zones are then identified by restricting upstream and downstream flow expansions to a user-specified angle (usually 15-45 degrees) on each side of the channel (CD Plate 1). The gradually varied flow calculations are repeated with cross-sectional areas restricted to active flow zones. Inundated areas outside the active flow zone are also identified.

While this approach works reasonably well, there can be problems. No DEM is artifact-free, and in flat areas, small errors in flow direction can result in substantial misalignment of cross-sections. This is especially true with high-resolution DEMs, and it is sometimes better to re-sample to a lower spatial resolution before calculating channel aspect and setting the orthogonals that define each cross-section. A badly aligned cross-section may point upstream and downstream rather than across a valley. Where this happens, water depths continuously increase in the downstream arm of the cross-section. In some cases, they may even cross a watershed, resulting in impossible flow depths and preventing the gradually varied flow calculations from converging on a solution.

We use one or more of three options to restrict or remove this problem. Firstly, variations in flow direction may be smoothed to varying degrees by changing the size of the moving window. Secondly, a user-defined restriction may be placed on channel widths. This may involve limiting total width or setting a maximum expansion angle for the inundated area (usually 45-70 degrees, CD Plate 2). Thirdly, elevations in the cross-section may not fall below the channel centerline elevation minus a threshold value that is set by the user. The most successful of these approaches is to apply a maximum expansion angle coupled with flow direction smoothing.

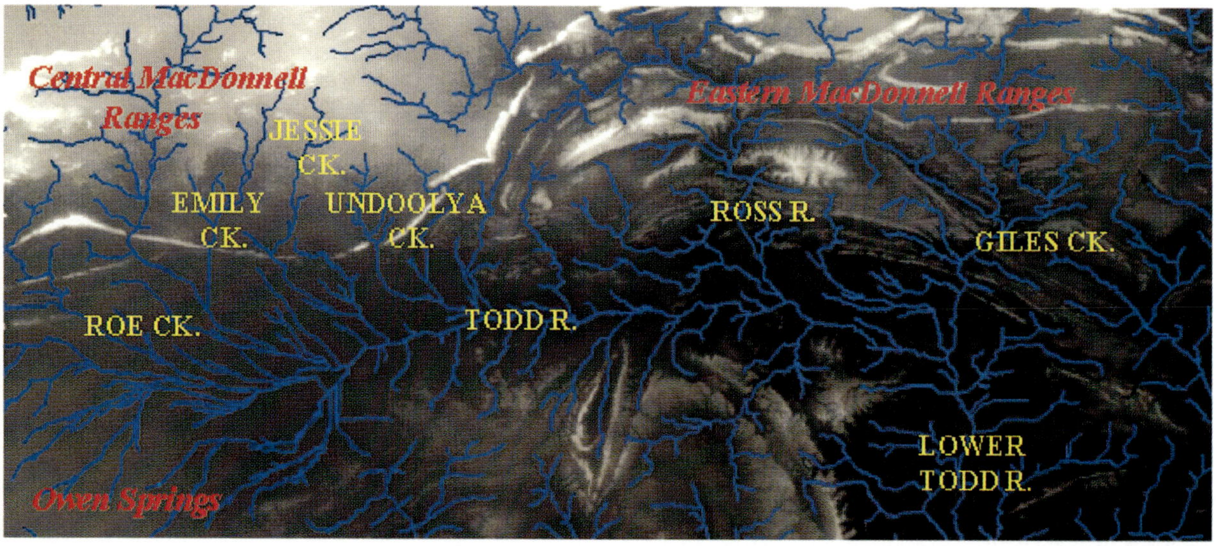

Plate 2. Digital elevation model and drainage systems of the study area. Area illustrated is 130 x 54 km.

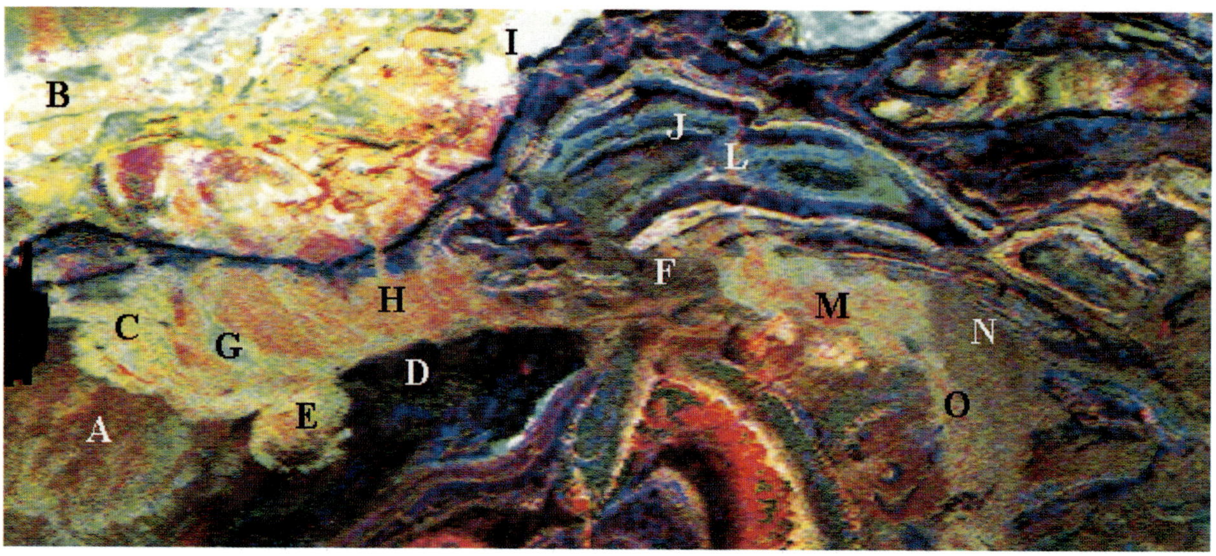

Plate 3. Radioelement concentrations in the MacDonnell Ranges and adjacent floodplains derived from an airborne gamma ray survey. Red is potassium, green is thorium and blue shows uranium. Area shown is 130 x 54 km. See text for an explanation of labels.

Plate 4. A Landsat TM image (copyright * Commonwealth of Australia, June 1988) overlaid with boundaries of the active flow area during a 50-year flood. Imagery is false color with TM band 2 = blue, TM band 3 = green and TM band 4 = red. Imagery provided by the Australian Centre for Remote Sensing, AUSLIG.

REMOTE SENSING OF FLOODPLAIN CHARACTERISTICS AND BEHAVIOR

The Todd River Drainage Basin

The Todd River is one the main ephemeral drainage systems of central Australia (Plate 2, also on CD) [*Bourke and Pickup*, 1999]. Its headwaters lie in the central and eastern MacDonnell Ranges where its main tributaries occupy narrow gorges (CD Plate 3). Beyond the ranges, the Todd occupies a broad alluvial plain (CD Plate 4), ending in a large floodout between the dunes of the Simpson Desert (CD Plate 5). The area has a mean annual rainfall of 274 mm but experiences heavy rainfalls when influxes of moist tropical air destabilise the predominantly high-pressure system of warm dry air. Indeed, there are examples of the mean annual rainfall being received in less than 24 hours.

Paleoflood records from bedrock gorges in the region extend back about 900 years [*Pickup et al*, 1988]. However, many large-scale fluvial features on the floodplains are older and result from much larger events [*Pickup*, 1991; *Patton et al* 1993; *Bourke* 1999]. The events in the 900-year record are also well below the envelope of maximum recorded discharges world wide for catchments of the same size [*Gillieson et al*, 1991]. Dating of flood deposits on the Todd floodplain as far back as 27,000 BP suggests that the largest events occurred in the late Pleistocene and early to mid-Holocene [*Bourke*, 1999]. Floods of the last 2,000 years are generally smaller in the Todd although the last century has been an extremely active period during which three of the four largest events in the 900-year record from the Finke have occurred.

Gamma-ray Data

Plate 3 (also on CD) shows radioelement concentrations in soils and near-surface rocks in the study area with the image enhanced to emphasize alluvial rather than bedrock areas. The main features are:

- The radioelement-rich gneiss and granite core of the MacDonnell Ranges;
- The folded metamorphic and sedimentary rocks on the southern and eastern fringe of this core; and
- The southern floodplain systems, including Roe Creek, the Todd and Ross Rivers, and Giles Creek.

Extensive folding is apparent in the sedimentary rocks that form a series of strike ridges developed on eroded anticlines and synclines. Transported sediment can be seen extending outwards from the mountain ranges on the southern floodplain systems. Indeed, some sediment can be tracked for long distances and gradual changes in their radioelement content observed.

The southern floodplain systems contain several major features. The Owen Springs area, which is bordered by the Hugh River in the west, is a relatively flat alluvial and limestone plain with extensive caliche deposits and a discontinuous dust mantle. The area has experienced substantial soil erosion since European settlement and much of the transported material has accumulated in a series of shallow washes and floodouts. This material is dark red-brown in colour on the image, indicating low radioelement concentrations but with K dominant. The exposed limestone, caliche and derived soils are green in colour, indicating relatively high concentrations of Th.

The shallow washes of Owen Springs (marked A on Plate 3) are bordered in the east by the radioelement-rich floodplain deposits of Roe Creek. This material is derived from the gneiss and granite core of the MacDonnell Ranges (B on Plate 3). The Roe Creek floodplain (C on Plate 3) sharply truncates the Owen Springs floodouts, suggesting that these drainages have been blocked off or backed up (CD Plate 6). This may have occurred because more sediment is coming down Roe Creek than from the Owen Springs area where the washes are choked by trees and shrubs. It is also possible that extreme floods are involved, with ponded water covering much of the Todd River floodplain. For example, the dark area on the southern edge of the Todd floodplain is a large dune field (D on Plate 3). A large embayment in this dune field at (E on Plate 3) contains an area of radioelement-rich sediment that appears to have come from Roe Creek and the Todd River. This suggests a large area of backflooding with ponded water and deposition of sediments. Sediments in this area are fine suggesting deposition from slow flowing or standing water. Floodplain surfaces closer to the active flow area of the Todd have large fields of low-amplitude gravel ripples.

The Todd floodplain extends eastwards to a narrow constriction created by the large dune field on its southern edge and several ridges on the north side that have blocked the passage of aeolian material, creating a discontinuous sand sheet. There is also a large floodout or alluvial fan (F on Plate 3) made up of sediments from the eastern MacDonnell Ranges that enter the floodplains immediately east of the gap. This may also partially block the gap and confine the Todd to a narrow reach on the southern side of the alluvial plain.

Upstream of the gap, the floodplain has two main elements: active floodplain reaches (e.g., G on Plate 3); and older, large-scale, paleoflood landforms (H on Plate 3). The active floodplain reaches are white, bright yellow or bright green in colour, indicating a high radioelement content and sediment sources in the gneiss and granite areas. The paleoflood landforms make up a series of slightly elevated surfaces, often covered by large-scale sedimentary features such as linear bars, diversion channels and fields of megaripples [*Pickup*, 1991] (CD Plates 7 and 8). These surfaces may have been continuous but are now breached by the radioelement-rich, active floodplains of a number of minor tributaries of the Todd River that drain the gneiss and granite area, notably Emily Creek, Jessie Creek and Undoolya Creek.

The paleoflood landforms become progressively redder to the east, suggesting that sorting has occurred leading to progressively higher K content. This may reflect differences in the lithology of gneiss and granite source areas drained by tributary creeks from the north. There also appears to be more gravel present than on paleoflood landforms further west, probably because sediment source areas are closer and there has been less downstream sorting.

Immediately west of the gap, the Ross River enters the Todd River floodplain (CD Plate 9). This system drains the eastern end of the gneiss and granite region of the MacDonnell Ranges (I on Plate 3) as well as a large area of folded sedimentary rocks (J on Plate 3). The K-rich (red) channel sediments can be seen right through the Ross River gorge as it cuts through the sedimentary rocks (L on Plate 3). They also spread across the Ross River floodplain (M on Plate 3). The rest of the Ross River floodplain consists of large paleochannel systems and megaripple fields (CD Plate 10) [*Patton et al*, 1993]. These are light green or yellow in colour in the gamma ray image indicating a high content of all radioelements. This suggests that much of the sediment is derived from the gneiss and granite region of the eastern MacDonnell Ranges. The Ross River drains this area as well as the limestones and folded sedimentary rocks to the immediate north of the floodplain. Much of this material is extremely fine, suggesting that ponded water occupied the area long enough to allow large-scale deposition of silts and clays.

Giles Creek marks the eastern end of the Ross River floodplain and brings in darker, radioelement-poor sediment derived from the sedimentary rocks (N on Plate 3). Reworked aeolian material also may be present since many of the valleys in this region contain extensive dunefields or sand sheets built from material blown in from the Simpson Desert immediately to the east. The Giles Creek sediments probably post-date and partly cover the paleochannel systems of the Ross River floodplain and extend further south to eventually flood out into the Simpson Desert [*Bourke*, 1999]. Even so, Ross River sediments can be seen on the left hand side of the flood plain (O on Plate 3) and extend right to the lower edge of the image. The changing colour of this material suggests mixing with Giles Creek sediments.

RECONSTRUCTION OF PALEOFLOOD EXTENTS ON THE TODD RIVER FLOODPLAIN

Extent of Flooding

We attempted to reconstruct the areal extent of flooding on the Todd River floodplain using the composite DEM, a regional flood frequency curve, and the flow computation techniques described above. The flood frequency curve was derived from gauging data in the Todd catchment [*MacQueen*, 1978], flood estimates for dam design [*Jackson and Paige*, 1979], and data from a hydraulic model and slackwater deposits from the Finke Gorge [*Pickup et al* 1988]. A Manning's n value of 0.125 [see *Chow*, 1959 for typical floodplain roughness coefficients] was used in the hydraulic calculations. Lateral expansion of the inundated area was limited to no more than 45 degrees from the main flow direction. Slackwater areas and backflow zones were designated by restricting expansion of the active flow zone to an angle of 15 degrees from the main flow direction both upstream and downstream of flow contractions. These angles were derived from observed changes in channel and floodplain width (e.g. CD Plates 1 and 2). Maps showing the water depth in inundated areas and/or active flow zones for flood return periods of 50 to 10,000 years are shown in Plates 4 and 5 (also on CD).

The 50-year flood reconstruction was used as a test of how well the DEM reproduced observed drainages and floodouts. Much of the drainage data on topographic maps in this area is of limited accuracy because well-defined channels often do not exist and the shallow washes that replace them are difficult to map. These features can be seen on Landsat TM imagery, particularly after heavy rains when vegetation response is greater along flow lines. We have therefore used Landsat TM imagery for June 1988 as "truth" and overlaid the boundaries of the active flow zone for the 50-year flood for comparative purposes (Plate 4). This image was acquired approximately two months after a major flood that inundated much of the Todd floodplain and other parts of central Australia. This flood had a return period of about 50 years on the Todd River upstream of Heavitree Gap. Its return period for downstream reaches was probably less than this as rainfall declined to the east.

The TM image is false colour with the red areas indicating vigorous plant growth, usually where inundation has added extra soil moisture. The dark red areas are floodplain plant communities developed on sandy deposits with river redgums (*Eucaplytus camaldulensis*), and an understorey of buffel grass (*Cenchrus ciliaris*) or couch grass (*Cynodon*

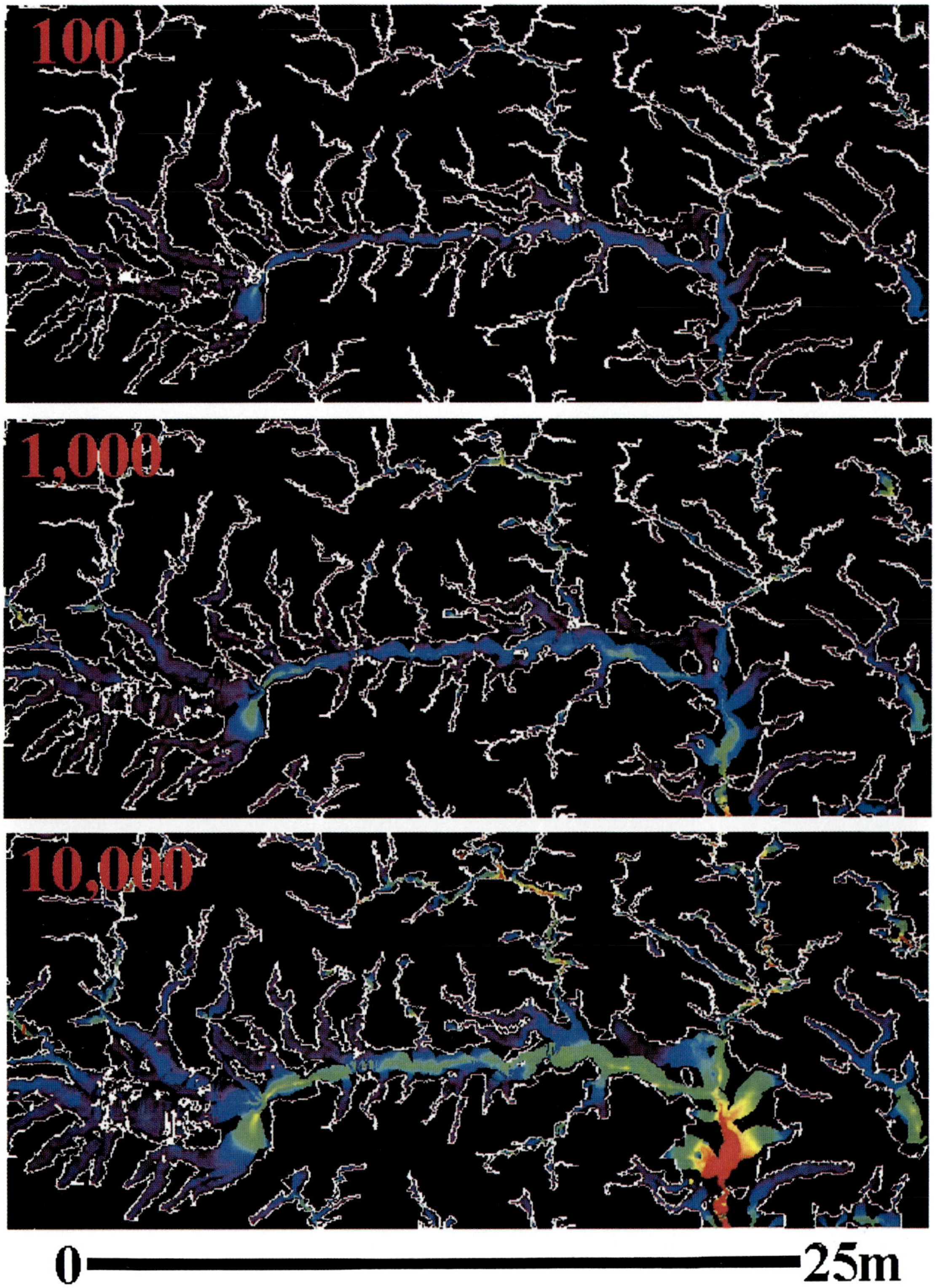

Plate 5. Depth of flooding in the study area during the 100-, 1000- and 10,000-year floods. White lines show boundary of inundated area. Colored area indicates water depths in m in active flow zone.

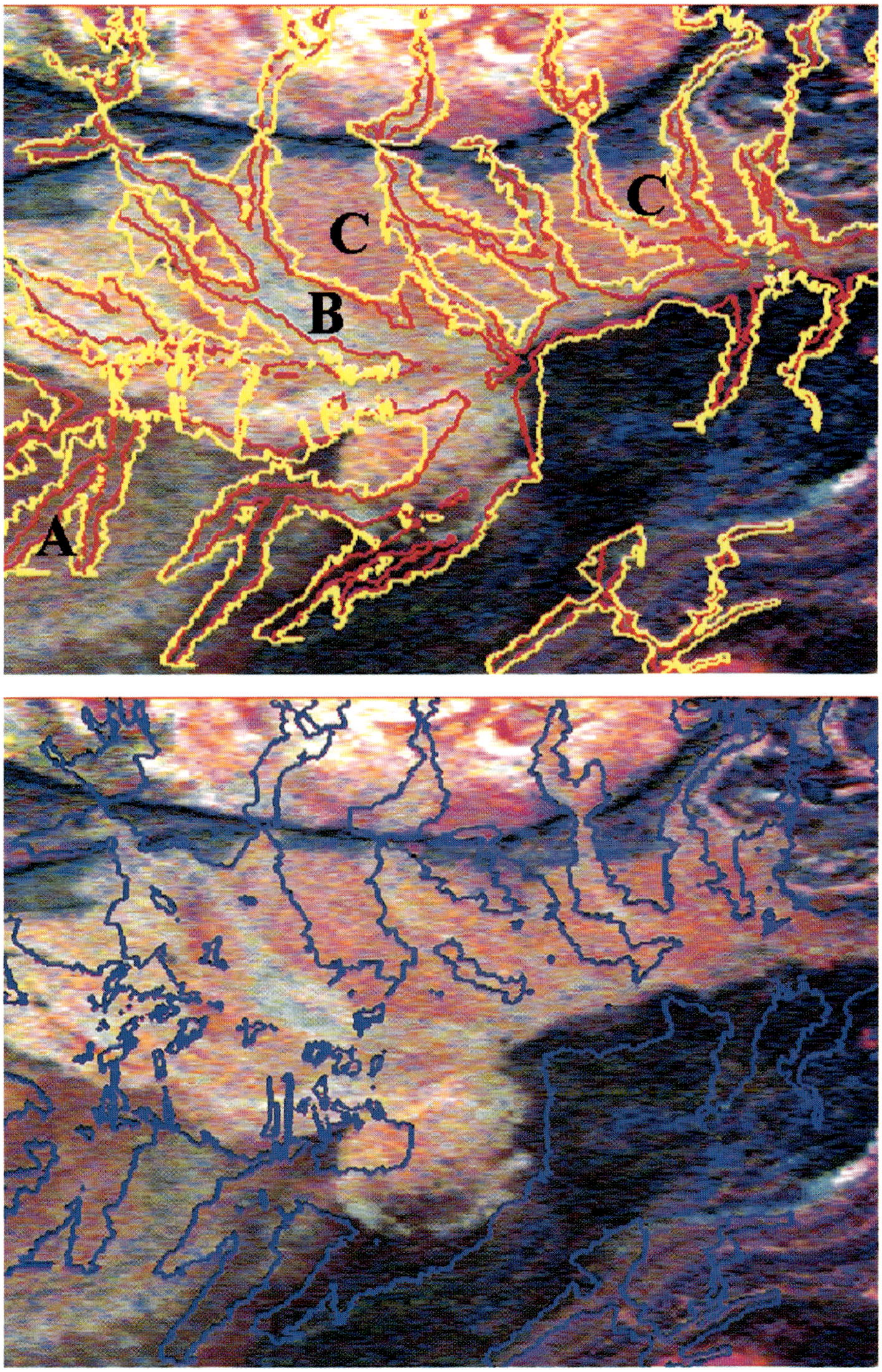

Plate 6. Radioelement concentrations and modeled floodwater extents during the 1,000-year (upper image) and 10,000-year (lower image) floods on the Todd River floodplain upstream of the Ross River junction. Red is potassium, green is thorium and blue is uranium on the gamma ray image. Red and yellow lines mark the extents of the active zone and total inundated area for the 1,000 year events respectively. Blue lines show the inundated area for the 10,000 year flood.

dactyledon). Both grasses are introduced species and are aggressive colonisers of floodplains and areas of sediment deposition. The pale red areas identify floodplains with either buffel grass or ephemeral native grasses and forbs, but fewer trees. These areas have clay or clay-loam soils in many cases, indicating deposition of finer material.

Comparison of the TM image and the active flow zones gives remarkably good results considering the quality of the elevation data and the fact that the DEM was fitted without forcing topographic lows on to mapped drainage lines. All channel systems occur within the flooded area and most areas of floodplain, as defined by the vegetation, are identified. Most of the dark red areas that lie outside the flooded area occur on alluvial fans at the base of the mountain ranges or cover floodplains of streams with drainage areas below cut-off. In the few cases where the TM data and the flooded area are mismatched, the elevation differences are 2m or less.

Given the level of success in reproducing the 50-year flood, we have calculated the extent and depth of inundation for the 100-, 1,000- and 10,000-year floods using discharges from the regional flood frequency curve (Plate 5). The curve includes data from a slackwater deposit record extending back about 900 years so the value for the 1,000-year flood is close to being within historical limits. The 10,000-year flood discharge was obtained by extrapolation and represents an extreme condition for purposes of comparison rather than a definitive value.

Many floodplain features are inundated by the 50-year event, suggesting that this floodplain is an active system. However, the small number of extreme flows in slackwater deposit record for this area suggests that change may be slow and that many features from earlier times should be preserved. Some areas remain above water during the flood. Many of these have been mapped as remnants of a Pleistocene floodplain surface [*Bourke*, 1999]. Other areas not inundated include alluvial fans, dunefields and aeolian deposits at the base of the mountain ranges. The fans would be active during most extreme events but have drainage basins too small to be included in the gradually varied flow calculations. The upstream sections of some of the largest paleoflood bars are not submerged, even by the 1000-year flood. However, their downstream sections are covered by 5-10 m of water during this event. Most of the paleoflood bars would be inundated by the 10,000-year event.

The water depths shown in Plate 5 were calculated assuming steady-state flow conditions. This means that all streams experience peak flow simultaneously and that peak flows are maintained long enough for the whole system to achieve steady state. This is unlikely given the size of the Todd River drainage basin. Some water depths therefore may be exaggerated, particularly where flow constrictions raise water surface elevation in upstream floodplain reaches. Water depths over the whole system may also be excessive in cases where localised storms generate extreme floods but infiltration into sandy riverbeds causes rapid depletion of flood discharges as flood waves progress downstream.

While many flood events in the slackwater deposit record seem confined to particular catchments, there is evidence for extreme rainfall events that occur across the region. There are common dates for events in the Finke Gorge, Dashwood and Derwent Creeks, the Todd River, and the Ross River [*Pickup et al*, 1988; *Patton et al* 1993; *Bourke*, 1999; *P. C. Patton*, personal communication]. The affected area has a 100 km radius so these rainfall events must have been very large. They may also have lasted long enough for all systems to experience large flows simultaneously although the occurrence of simultaneous flood peaks is unlikely. There is also anecdotal evidence for inundation of large areas in an early description of the site of Alice Springs in 1901:

"On the first survey flood marks were found which showed that the whole country to the north of the range must have been converted into a huge lake 15 feet deep and the blacks tell that before the white men came they had once to take refuge from the flood on the hill." [*White*, 1909, quoted by *Kimber*, 1996].

We conclude that the modeling results are reasonable for the regional scale events.

Comparison with Gamma-ray Imagery

The gamma-ray imagery provides information on floodplain sedimentary environments that may be compared with modeled flow conditions to see whether there is any similarity. Before making this comparison, we have divided the area inundated by the 1000-year flood into active flow zones and areas of backwater/slackwater since a different mix of particle sizes may be deposited in each. This process of sorting may affect radioelement content and give different gamma-ray signatures.

Plate 6 (also on CD) shows radioelement and flood extent data for the Todd floodplain upstream of the Ross River junction for the 1,000- and 10,000-year events. On the Owen Springs drainage systems, the inundated areas correspond reasonably well with the red-brown areas on the gamma-ray image (A on Plate 6) for both events. On the Todd River and Roe Creek floodplains, there is some similarity between active flow zones for the 1,000-year event and the green and yellow areas (B on Plate 6) but a better match was obtained with the 50-year flood (Plate 4). Most of these are mixed sand and gravel channel, avulsion and floodout deposits and are currently active to varying degrees.

The backwater/slackwater zones of the 1,000-year Todd floodplain correspond only partly with the red paleoflood features on the gamma-ray imagery (C on Plate 6). Many of these features lie outside the active flow zones and are only partly inundated by backwater/slackwater. We suggest that they do receive fine sediments deposited during extreme events, but are older than the more radioelement rich active flow areas. Also, the linear bars and megaripples on some of the red paleoflood surfaces (CD Plates 7 and 10) imply active flow transporting sediment in quantity rather than standing water. The inundated area for the 10,000-year event covers most, but not all of these features, so they may result from an even larger flood. Further dating and stratigraphic investigations are needed to produce a firm conclusion.

CONCLUSIONS

Terrain elevation data from airborne geophysical surveys can improve DEMs derived from topographic map information to the extent that they may be used for regional-scale modeling of extreme flood events using gradually varied flow algorithms. These surveys also provide data on radioelement content of surface and near-surface soil and rock that may be used to identify and map sedimentary environments. In central Australia, these techniques allow us to identify sediment sources and to discriminate between paleoflood deposits and sedimentary features associated with active floodplains.

Paleoflood modeling of the Todd River floodplain generates patterns of inundation similar to those identified on Landsat TM imagery. The results also generate patterns that match sedimentary environments in the gamma-ray imagery. This suggests that, while much of the floodplain has been active during the last 1,000 years, some floodplain features result from older and larger events. This conclusion matches the results of field studies that have identified floodplain landforms at several scales and concluded that the largest features are the product of episodic and catastrophic events.

Airborne geophysical survey data are becoming widely available and offer potential for regional scale investigation of both sediment transport and weathering patterns in a range of environments. They are deserving of much wider use by geomorphologists.

Acknowledgments. We are grateful to the Northern Territory Department of Mines and Energy for providing geophysical survey data.

REFERENCES

Akima, H., A new method of interpolation and smooth curve fitting based on local procedures, *J. Assoc. Computing Machinery, 17*, 589-602, 1970.

Aspin, S.J. and P.N. Bierworth, GIS analysis of the effects of forest biomass on gamma-radiometric images, Third National Forum on GIS in the Geosciences, *Australian Geol. Surv. Org. Rec.* 1997/36, 77-83, 1997.

Baker, V.R. and D. Nummedal, *The Channeled Scabland*, Planetary Geology Program, NASA, Washington, D.C., 186 pp, 1978.

Baker, V.R. and G. Pickup, Radiocarbon dating of flood events, Katherine Gorge, NT, Australia, *Geology* 13, 344-347, 1985.

Bierworth, P., Investigation of Airborne Gamma-Ray Images as a Rapid Mapping Tool for Soil and Land Degradation - Wagga Wagga, NSW. *Australian Geol. Surv. Org. Rec.* 1996/22, Australian Geological Survey Organisation, Canberra, 69pp, 1996.

Bourke, M.C.A., *Fluvial Geomorphology and Paleofloods in Arid Central Australia*, Ph.D. Thesis, Australian National University, Canberra, 208 pp, 1999.

Bourke, M. and G. Pickup, Variability of fluvial forms in the Todd River, Central Australia, in *Varieties of Fluvial Form* edited by A. Miller and A. Gupta, Wiley, Chichester, pp. 249-271, 1999.

Bretz, J.H., The Spokane flood beyond the Channeled Scabland, *J. Geol.*, 33, 97-115, 236-259, 1925.

Chow V.T. *Open-Channel Hydraulics*. McGraw-Hill Inc., New York, 680 pp, 1959.

Dickson, B.L. and K.M. Scott, Interpretation of Aerial Gamma-Ray Surveys, *CSIRO Div. Exploration Geoscience Restricted Report, 301R*, North Ryde, 144 pp, 1992.

Dickson, B.L. and K.M. Scott, Interpretation of aerial gamma-ray surveys – adding the geochemical factors, *Australian Geol. Surv. Org. J. Australian Geol. and Geophys.*, 17, 187-200, 1997.

Gillieson, D., D. Ingle Smith, and M. Ellaway, Flood history of the Limestone ranges in the Kimberley region, Western Australia, *Appl. Geog.* 11, 105-123, 1991.

Grasty, R.L., Atmospheric absorption of 2.62 MeV gamma-ray photons emitted from the ground, *Geophysics*, 40, 1058-1065, 1975.

Grasty, R.L., Gamma-ray spectrometric methods in uranium exploration - theory and operational procedures, *in Geophysics and Geochemistry in the Search for Metallic Ores, Geological Survey of Canada, Economic Geology Report* 31, edited by P. J. Hood, pp. 147-161, 1979.

Grasty, R.L. and B.R. Minty, A guide to the technical specifications for airborne gamma-ray surveys, *Australian Geol. Surv. Org. Record* 1995/60, 89 pp, 1995.

Horsfall, K.R., Airborne magnetic and gamma-ray data acquisition, *Australian Geological Survey Organisation Journal of Australian Geology and Geophysics* 17, 23-30, 1997.

Hupp, C. R., Plant ecological aspects of flood geomorphology and paleoflood history, in *Flood Geomorphology*, edited by V. R. Baker, K. C. Kochel and P. C. Patton, pp. 335-356, John Wiley, New York, 1987.

Hutchinson, M. F., A new method for gridding elevation and stream line data with automatic removal of pits, *Journal of Hydrology*, 106, 211-232, 1989.

Hydrologic Engineering Center, *HEC-2 Water Surface Profiles User Manual*, US Army Corps Of Engineers, Davis, California, 1982.

Jackson, D. and D. Paige, *Alice Springs Recreational Dam Hydrology Report*, Northern Territory of Australia Department of Transport and Works Hydrology Division, Darwin, 50pp, 1979.

Kimber, R.G., The dynamic century before the Horn Expedition: a speculative history, *in Exploring Central Australia: Society, the Environment and the Horn Expedition*, edited by S. R. Morton and D.J. Mulvaney, Beatty and Sons, Chipping Norton, Surrey, pp. 99-102, 1996.

Kochel, R.C. and V.R. Baker, Paleoflood analysis using slack-water deposits, in *Flood Geomorphology*, edited by V.R. Baker, R.C. Kochel and P.C. Patton, Wiley Interscience, New York, pp. 357-376, 1987.

MacQueen, A.D., *Flood Frequency in Central Australia, a Regional Model*, Northern Territory of Australia Dept. Transport and Works Hydrology Division, Darwin, 1978.

MicroImages Inc., *TNT Mips Reference Manual,* Lincoln, Nebraska, 1999.

Minty, B.R.S., Simple microlevelling for aeromagnetic data, *Explor. Geophys., 22*, 591-592, 1991.

Minty, B.R.S., Fundamentals of airborne gamma-ray spectrometry, *Australian Geol. Surv. Org. J. Australian Geol. and Geophys., 17*, 39-50, 1997.

Nott, J., D.M. Price, and E.A. Bryant, A 30,000-year record of extreme floods in tropical Australia from relict plunge-pool deposits. Implications for future climate change, *Geophys. Res. Let., 23*, 379-382, 1996.

O'Connor, J.E. and R.H. Webb, Hydraulic modeling for paleoflood analysis, in *Flood Geomorphology,* edited by V.R. Baker, R.C. Kochel and P.C. Patton, Wiley, New York, pp. 393-402, 1988.

Patton, P.C., G. Pickup, and D.M. Price, Holocene paleofloods of the Ross River, Central Australia, *Quat. Res., 40*, 201-212, 1993.

Pickup, G., Paleoflood hydrology and estimation of the magnitude, frequency and areal extent of extreme floods - an Australian perspective, *Civil Eng. Trans. Institution of Engineers, Australia*, 253-263, 1989.

Pickup, G., V.R. Baker, and G. Allan, History, Paleochannels and Paleofloods of the Finke River, central Australia, in *Australian Fluvial Geomorphology*, edited by R.F. Warner, Academic Press, Sydney, pp. 105-127, 1988.

Pickup, G., Event frequency and landscape stability on the floodplains of arid central Australia, *Quat. Sci. Rev., 10*, 463-473, 1991.

Pickup, G. and A. Marks, Identifying large-scale erosion and deposition processes from airborne gamma radiometrics and digital elevation models in a weathered landscape, *Earth Surf. Proc. Landf., 25*, 535-557, 2000.

Pickup, G. and A. Marks, Regional-scale sedimentation process models from airborne gamma-ray remote sensing and digital elevation data, *Earth Surf. Proc. Landf., 26*, 273-29, 2001.

Pilgrim, D. H., *Australian Rainfall and Runoff. A Guide to Flood Estimation*, Institution of Engineers, Australia, Barton, ACT, 374 pp, 1987.

Wells, S. G. and J.C. Dohrenwend, Relict sheet flood bed forms on late Quaternary alluvial fan surfaces in the southwestern United States, *Geology, 13*, 512-516, 1985.

Wilford, J., Airborne gamma-ray spectrometry as a tool for assessing relative landscape activity and weathering development of regolith, including soils, *Australian Geol. Surv. Org. Research Newsletter, 22*, 12-14, 1995.

Wohl, E.R., R.H. Webb, V.R. Baker, and G. Pickup, Sedimentary flood records in bedrock canyons in the monsoonal region of Australia, *Colorado State Univ. Wat. Resour. Pap.,* 107, 102 pp, 1994.

Geoff Pickup, CSIRO Land and Water, PO Box 1666, Canberra City, ACT 2601, Australia.

Alan Marks, CSIRO Land and Water, PO Box 1666, Canberra City, ACT 2601, Australia.

Mary Bourke, School of Geography and the Environment, University of Oxford, Mansfield Road, Oxford, OX1 3TB, England.

Mapping Flood Inundation in Southwestern Arizona using Landsat TM Data: A Method for Rapid Regional Flood Assessment Following Large Storms

Larry Mayer

Department of Geosciences, University of Arizona, Tucson, Arizona

Philip A. Pearthree

Arizona Geological Survey, Tucson, Arizona

Landsat data can be used to detect changes at the Earth's surface caused by flooding in the deserts of the southwestern United States. On September 25-26, 1997, Hurricane Nora made landfall across Baja California Norte and then dissipated across southwestern Arizona, where it caused widespread flooding. We introduce a method of orthogonalizing bands from Landsat data, obtained before and after the passage of Hurricane Nora, in order to perform radiometric normalization simultaneously with change detection. Using multi-spectral detection allowed us to separate two general kinds of change associated with floods: vegetation growth and sediment deposition. Landsat data from southwestern Arizona were analyzed using this method to detect the extent of the flooding caused by Hurricane Nora. These data reveal that detectable changes occurred along many washes in western Arizona as a result of the passage of this storm. Comparison of remote-sensing data with independent field mapping of flood inundation on the Tiger Wash alluvial fan in western Arizona allows us to assess the general usefulness of our approach. Tiger Wash experienced an extreme flood as a result of the passage of the remains of Hurricane Nora over the region. Field investigation indicates that change detection from space is a valid and effective approach to rapidly determining where flooding has occurred in southwestern Arizona, because there is a strong correspondence between detected changes and the extent of inundated areas on Tiger Wash.

INTRODUCTION

Hurricanes and other tropical storms represent one type of weather system that brings significant amounts of precipitation to the southwestern United States in the late summer and early autumn. Hurricanes that form in the eastern Pacific Basin can generate intense floods affecting southwestern Mexico, Baja California, and the southwestern United States. The eastern Pacific Basin is second only to the western Pacific in average annual production of hurricanes. Approximately 15 of the Earth's 80 tropical storms per year develop in the eastern Pacific; most of these storms track out into the Pacific Ocean toward Hawaii. Each year, typically, several of these tropical storms re-curve back towards Mexico or the southwestern United States, sometimes with devastating results [*Smith*, 1986].

Hurricanes have a long history in the region, dating from the time of Cortez and the Spanish conquest of Mex-

Ancient Floods, Modern Hazards: Principles
and Applications of Paleoflood Hydrology
Water Science and Application Volume 5, pages 61-75

ico. Cortez, whose assault on the Aztec stronghold of Tenochtitlan was slowed by flooding [*O'Hara*, 1995], lost one of his ships to a "terrible storm" in the Gulf of California. More recently, several hurricanes and dissipating tropical storms slammed into Los Angeles in 1939, killing 45 persons [*Gatewood, et al.*, 1946]. In 1976, Hurricane Liza devastated the city of La Paz, Baja California Sur, with strong winds, a 2 m storm surge, 160 mm of rain, and floods. As many as 1,000 people were killed [*Smith*, 1986]. Liza then dissipated over the southwestern United Sates generating flooding into the Death Valley region.

Many eastern Pacific hurricanes developed during the strong El Niño conditions that existed in the latter part of 1997, including Linda, Nora, and Pauline. Hurricane Linda was the strongest east-Pacific hurricane ever recorded. Flooding and landslides from Hurricane Pauline were responsible for at least 200 deaths in Mexico. Hurricane Nora (Plate 1; color version on CD) developed over the eastern Pacific and eventually tracked northward over Baja California, up the Gulf of California, and into Arizona, where it caused flooding across a relatively sparsely populated region.

This paper introduces a *new* method for using satellite remote sensing, specifically Landsat data, to assess the impact of hurricane-induced flooding by detecting the resultant landscape changes. Using the floods in south-western Arizona caused by dissipating Hurricane Nora as a case study, we introduce a rapid and relatively simple method for generating change images that detect the effects of flooding on the landscape. Field mapping over a small portion of the affected area provides a verification of the methodology. The method is well suited for use throughout the southwestern United States. With the introduction of Landsat 7, satellite data can be shared among government and educational agencies and relatively low costs permits their operational use.

HURRICANE NORA

Hurricane Nora formed on September 16, 1997, about 450 kilometers southwest of Acapulco, Mexico, over very warm water (29-30°C). On September 18 Nora had formed a distinct eye and had sustained wind speeds of over 250 km/hour that generated waves nearly 3 meters high when they hit the Mexican mainland. Nora stalled on the 20th of September and weakened quickly but then started to move again, finally merging with the track of Hurricane Linda. An omega-blocking pattern over the western United States then developed into a cutoff low during the last week of September, which steered Nora over warm waters where it regained strength just before making landfall on September 25. At Rancho Chapala, Baja California Norte, residents recalled the rapid filling of the playa at Laguna Chapala (Plate 1). According to eyewitness accounts, the lake was about "1 foot" deep. High water marks consisting of very clear vegetation debris mats about 15-20 cm wide formed a ring around the playa and allowed surveying of the lake level. Using a rough estimate of the lake area of 4 km^2, we estimate that about one million m^3 of runoff and direct precipitation accumulated in the lake as Nora passed on its way up to Yuma. Nexrad radar at Yuma documented an area of high rainfall, about 250 mm, near the northern end of the Gulf of California. Yuma reported about between 30 mm and 100 mm of rain, representing about 33% to 100% of its annual average rainfall. In the Harquahala Mountains of western Arizona, 300 mm of rain was recorded and over 100 mm fell on the adjacent Tiger Wash area in a 24-hour period (Flood Control District of Maricopa County, unpublished data). Although Nexrad data indicated locally high rainfall amounts in southwestern Arizona, finding where flooding actually occurred would have been a massive project using conventional field mapping. Satellite remote sensing, however, is ideally suited to this task.

SATELLITE-BASED FLOOD DETECTION

Satellite-based flood detection provides important information for government agencies monitoring disasters [*Rao et al.*, 1998] and supplies data to scientists seeking to understand the responses of surface processes to storm events at synoptic scales. The application of Landsat data to flood mapping is not new; rather it began with the first Landsat satellite, Landsat-1 [*Hallberg et al.*, 1973; *Deutsch and Ruggles*, 1974; *Rango and Salomonson*, 1974; *Morrison and White*, 1976]. Despite these early studies, the use of visible- and infrared-based remote sensing has not been extensively applied to flood mapping based on the published literature. One important reason for the lack of remote sensing flood applications over the visible-infrared bandwidth is that clouds obscure the view from space. Paradoxically, Landsat-based flood mapping requires that the view of the ground from space not be obscured by the same clouds that caused the flooding. The launching of synthetic aperture radar (SAR) satellites such ERS-1, ERS-2, JERS-1, and RADARSAT, has led to increased interest in flood mapping from space [cf. *Smith*, 1997]. SAR penetrates cloud cover and affords a view of the surface even during flooding, a requirement for direct observation of flood inundation. Both SAR and visible-infrared systems are useful, and in many ways complementary methods.

In many places in the world, large floods induce landscape changes that can be observed after clouds dissipate and before another event changes the landscape again. Unlike inundation mapping (common to SAR applications)

Plate 1. Shaded relief map of Mexico, and portions of Arizona and California, showing the general track of Hurricane Nora and selected geographic locations. Physiography is important because high elevation areas will generally correspond to areas of higher precipitation. The location of Tiger Wash in western Arizona is shown by TW.

which shows the direct observation of areas while still flooded, this paper describes a method to detect evidence of the effects of flooding after the floodwaters have receded.

Landsat Data and Change Detection

Our analysis is based on the comparison of two Landsat TM scenes for path-row 038-037. The first scene was recorded 15 July 1997 and represents the ground condition shortly before the passage of Hurricane Nora. The second scene was recorded 19 October 1997 and represents the ground condition after the passage of Hurricane Nora. These scenes were ordered from EOSAT through the EROS Data Center, were both projected in an Albers equal area projection and had terrain-level corrections applied. Image to image co-registration resulted in small registration errors on the order of one pixel or less. Due to the difference in the time of year that the scenes were acquired, there is an inherent difference in the solar illumination.

Landsat satellites acquire data from the visible, infrared, and thermal infrared bands, representing brightness across several electromagnetic wavelengths from descending orbits about 700 km above the Earth's surface. Landsat revisits the same area every 16 days, which defines the smallest period of time over which change can be documented. The information for each portion of the electromagnetic spectrum is collected in seven bands (Table 1). The spatial resolution of the visible and near infrared bands of Landsat satellites 4, 5, and 7 is about 30 meters.

Each band of Landsat is an eight-bit array of values referred to as Digital Numbers (DNs) or Brightness Values (BVs). The BVs vary from 0 to 255 and represent pixel brightness in a particular band. The visible color of an object on a satellite image is the result of its brightness in the three visible bands (3,2,1), which roughly represent red, green, and blue wavelengths. For example, if the BVs of the visible green band for a pixel are greater than zero, bright

Table 1. Spectral characteristics of the ETM+ sensor on the Landsat 7 satellite.

Band	Wavelength (micrometers)	Spatial Resolution (meters)
1 Visible	0.45 – 0.515	30
2 Visible	0.525 – 0.605	30
3 Visible	0.63 – 0.69	30
4 Infrared	0.75 – 0.90	30
5 Infrared	1.55 – 1.75	30
6 Thermal Infrared	10.40 – 12.50	60
7 Infrared	2.09 – 2.35	30
8 Panchromatic	0.52 – 0.90	15

objects have some green in them. If the BVs for this pixel in all the other visible bands are near zero, then the object would appear green in visible light. Using Landsat data we can also image the infrared response of an object (bands 7,5,4). The brightness of a pixel in any of these bands or wavelengths is a result of an object's brightness, or reflectance, in that wavelength. For example, leafy vegetation (Figure 1; also on CD) is very bright in the visible green wavelength (band 2), very dark in the red wavelength (band 3), and brightest in the near infrared wavelength (band 4).

Detecting the effects of flooding on the landscape from Landsat satellite data is based on a methodology referred to as *change detection*. The concept behind this methodology is simple. If a flood changes some aspect of the land surface, it should be detectable if the change (1) is large enough spatially, and (2) is reflected in a significant change in the brightness of a ground pixel in any of the sampled bands. Pixels may get either darker or brighter depending on the change, and a single pixel may get darker in one band and brighter in another. The application of change detection to a landscape is based on the careful comparison of precisely co-registered multi-temporal satellite images.

An inherent difficulty in change detection is that factors other than actual change of the land surface can also cause differences between multi-temporal images. For example, atmospheric conditions, haze, or the thickness of the atmosphere may be sufficiently different between the times the images were acquired that the brightness values are different between them. Studies of radar interferometry indicate that some atmospheric variation can be significant even within a single scene. Other sources of non-land surface change include differences in source illumination at different times of day or times of year, dust in the atmosphere, or sensor calibration. Often, even cursory examination of multi-temporal images indicates that they are not radiometrically equivalent. In other words, they appear different even though they are not. The correction of these differences in radiometric response is called radiometric normalization [*Schott et al.*, 1988; *Elvidge et al.*, 1995] or radiometric rectification [*Hall et al.*, 1991].

Many of the radiometric differences not related to land-surface changes could be corrected if targets with known spectral characteristics were deployed and used to adjust the responses. Lacking these installed targets, other, mostly empirical, methods are used to apply corrections to one or more of the multi-temporal images in order to make them radiometrically comparable to the others. In the case of two images, one image can be adjusted or normalized to the other in order to correct for non-land surface change differences between the two images. One can easily think of this correction as a linear transformation function:

$$Y = B_0 + B_1 X \tag{1}$$

where Y is the estimator of corrected or normalized image BVs based on a linear transformation of the original BVs in

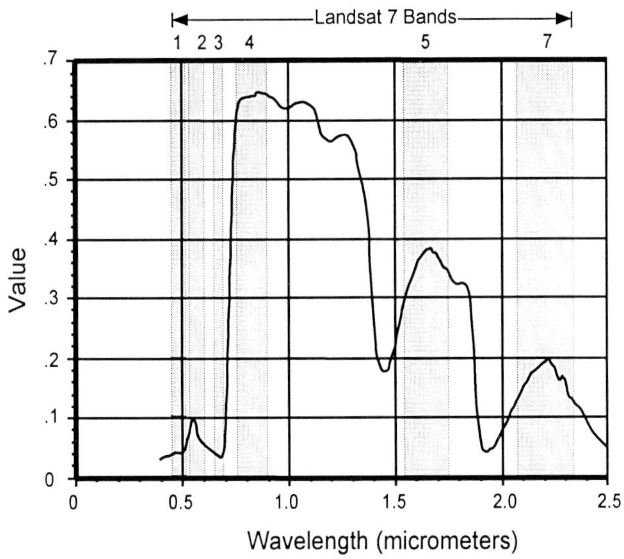

Figure 1. USGS laboratory spectral response of leafy vegetation (Maple leaves) over the wavelengths detected by Landsat satellites. The shaded areas correspond to the bandwidths measured by specific Landsat bands, indicated at the top of the figure. Notice that this response is similar to those of other brushy and grass vegetation types, with the brightest region being in band 4. Band 2 (visible green) is the brightest of all the visible bands. These response curves are used by remote sensing scientists to interpret the significance of the scene or analysis. For example, simultaneous brightness in bands 4 and band 3 for example, could not be the result of vegetation because vegetation is very dark in band 3.

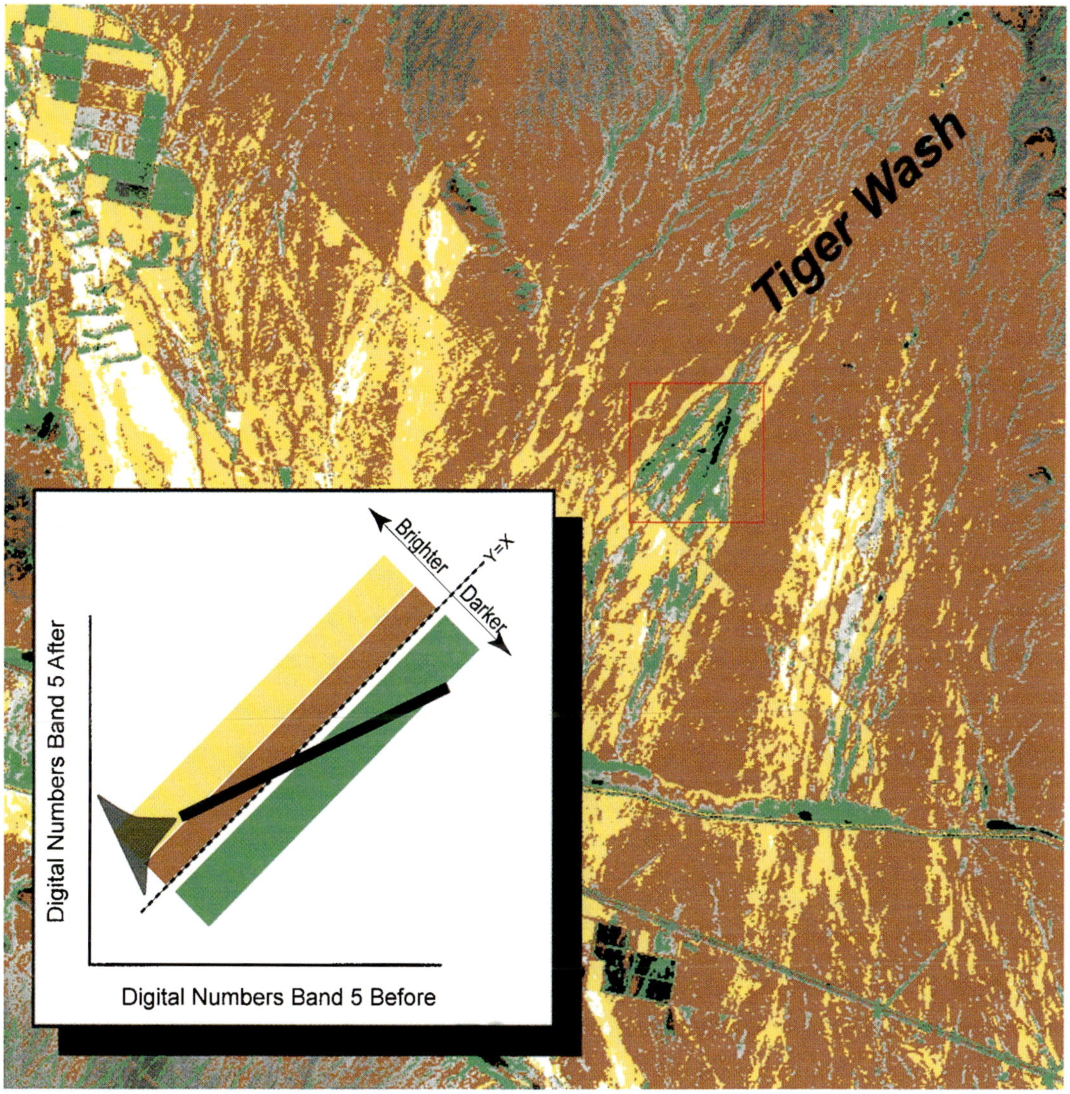

Plate 2. Changes in band 5 in the Tiger Wash area. Image processing programs allow you to see what areas are different in two scenes by comparing a single band on a scattergram. By selecting areas that are either below or above the line of equal response, one can highlight regions that became either darker or brighter. Note that the inset shows that without radiometric normalization, the differences would be skewed, as illustrated by the oblique cutting of the "actual radiometric response" by the ideal response.

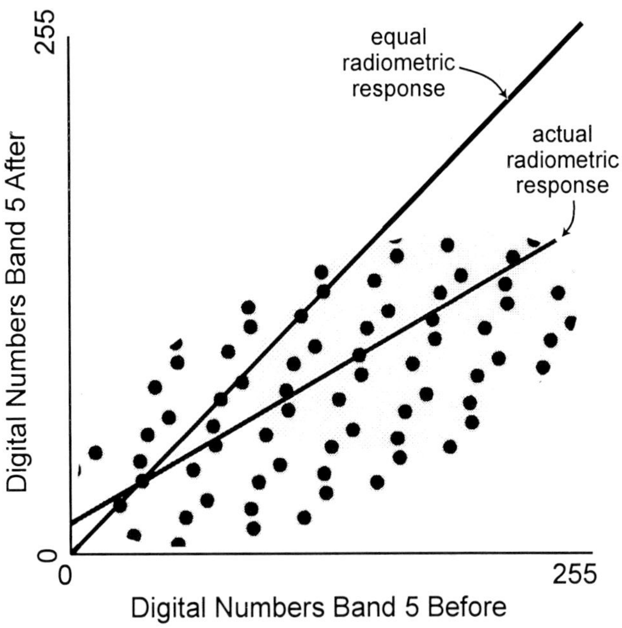

Figure 2. Bivariate scattergram for hypothetical data showing the importance of radiometric normalization in change studies using multi-temporal data. The scattergram plots band 5 [before] versus band 5 [after]. The line with slope = 1 represents the relationship where multi-temporal scenes are radiometrically equivalent. If they are not radiometrically equivalent, then the adjustments to one scene is made based on the regression line "actual radiometric response" (see text for discussion).

X. This operation assumes that there are clearly identifiable pixels that have not changed from one image to another. The regression is done for each band using a relatively small number of pixels. The regression coefficients are used for the coefficients of the radiometric transform operation. The pixels represent two kinds of objects in the images; the lightest and also the darkest objects. All other objects are linearly transformed to the radiometric response of the other image by means of the regression coefficients [*Hall et al.*, 1991]. Operationally, the linear transformation function (1) is estimated using linear regression. This method produces results comparable to histogram matching [*Schott et. al.*, 1988]. Other methods of radiometric normalization do not require the selection of only dark and bright objects, but assume that no change pixels can be identified throughout the brightness histograms of the bands [*Yuan and Elvidge*, 1996]. After the data are radiometrically normalized, image differencing, which subtracts the spectral response of one image from another, can identify significant changes between images.

To quickly review these relationships, consider the hypothetical Landsat data shown in Figure 2 (also on CD). Two images that are exactly the same should show a high correlation between them at all brightness levels. If there is a systematic radiometric difference between the two images, the slope of the regression line will not be one. Changes between the two images are represented by departures from the regression line. The larger the change between images, the larger the departure, either positive or negative. To identify where change has occurred, a difference image can be derived by subtracting one image from another or by highlighting areas of difference on the bivariate scattergram. For example, we can examine the bivariate scattergram for band 5 before and after the hurricane and select regions that depart from the line of equality (Plate 2; also on CD). One could directly highlight pixels that are very different, or only somewhat different, from very close to the line of equality. This difference image only shows changes in a single spectral band.

An alternative to direct and explicit radiometric normalization is orthogonalization of selected bivariate band pairs from before and after the flood (Figure 3; also on CD). This method relies on the correlation structure between spectral bands of the image. The spectral responses in Landsat TM bands are highly correlated to one another (Figure 4; also on CD). Principal components analysis can be used to determine the correlation structure and by use of the covariance matrix, calculate a new set of axes that minimize this correlation [*Gillespie*, 1980] and produce a new orthogonal reference frame. This characteristic of Landsat TM data led to the use of principal components analysis as a means of enhancing the color model display of Landsat images [*Williams*, 1983]. A related technique, also used to enhance the color model, is de-correlation stretching [*Rothery and Hunt*, 1990]. To perform a de-correlation stretch, one first calculates the principal components of the image, then stretches each of the bands in principal component space, and finally re-projects the data back into the original band coordinate system. The re-projection back to the original coordinate system allows the investigator to make interpretations from the spectral responses rather than from the more abstract principal components. Fortunately most image processing programs have a de-correlation stretch algorithm allowing the user to perform these calculations with a single command.

The goal of orthogonalization is to emphasize the changes in the two images by stretching the data in the orthogonalized reference frame and also to simultaneously accomplish radiometric normalization. The orthogonalization method is exactly equivalent to the de-correlation stretch method but with the specific input requirement that two bands are the same (from the "before" and the "after" images) in order to accomplish radiometric normalization. The advantages of this method over regression methods are that it uses more than a single band of information, is accomplished in a single automated step, and does radio-

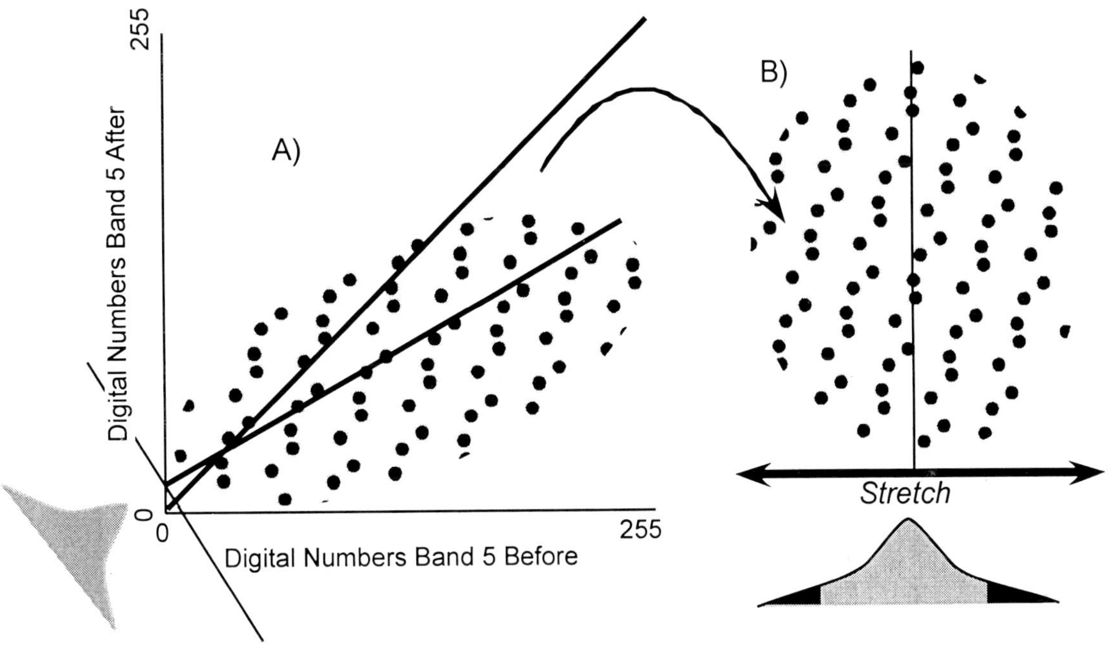

Figure 3. Sketch showing the goal of an orthogonal transform. Orthogonal transformations find a linear weighting of the data that in effect de-correlate them. A) The original data are correlated and also skewed because they are not radiometrically normalized. By finding the principle component, the axis around which the data are no long correlated, but represents the axis of major variation, we can determine the linear weighting required. B) Stretching the data following the linear transformation emphasizes the changes in the images.

metric normalization. The disadvantage is its unknown performance relative to regression techniques.

Specific Bands and Landscape Changes

The strategy for selecting the input bands is based on the kinds of changes expected from flooding. Band 4, which is excellent for detecting vegetation cover, is a good choice because is likely that greening or growth of vegetation following a flood will be evident on floodplains and along stream channels. Band 7 or band 5 can be used to look for non-vegetation changes, for example reworking of sediment in channels and other areas of relatively deep flow. We elected to use bands 5 and 4 as comparison bands, and we used band 5 for normalization. In this paper, we refer to band number and its timing relative to the change; band 4 [before] refers to band 4 on the image before the hurricane.

The RGB color model used to portray change between images is composed of de-correlation stretched bands:

Red channel – band 5 [after]
Green channel – band 4 [after]
Blue channel – band 5 [before]

If a region in the change image is any one of the primary channel colors, i.e., red, green, or blue, then the brightness is coming from the corresponding band. For example, if an area shows up as green on the change image, then the pixels are bright only in band 4 (the vegetation band) and therefore dark in the other channels. These areas likely had substantially more vegetation cover in the second satellite image. Regions that are simultaneously bright in the red and green channels (bands 5 [after] and 4 [after]) show up as yellow areas. These areas, typically located in or near stream channels, appear to represent the deposition of fresh sediment and some vegetation growth. Regions that are simultaneously bright in the green and blue channels (bands 4 [after] and 5 [before]) show up as cyan. Cyan areas in channels and floodplains appear to represent the greening of vegetation following the storm. Areas that appear blue on the change image are darker in band 5 after the flood. This may result from moist sediment or extensive dead vegetation (flotsam). Regions that are simultaneously bright in the red and blue channels (bands 5 [after] and 5 [before]) show up as purple or magenta. These colors represent no observable change.

SATELLITE-CHANGE IMAGERY FOR WEST-CENTRAL ARIZONA

Satellite change imagery was generated for Landsat 5 scenes bracketing the passage of dissipating Hurricane

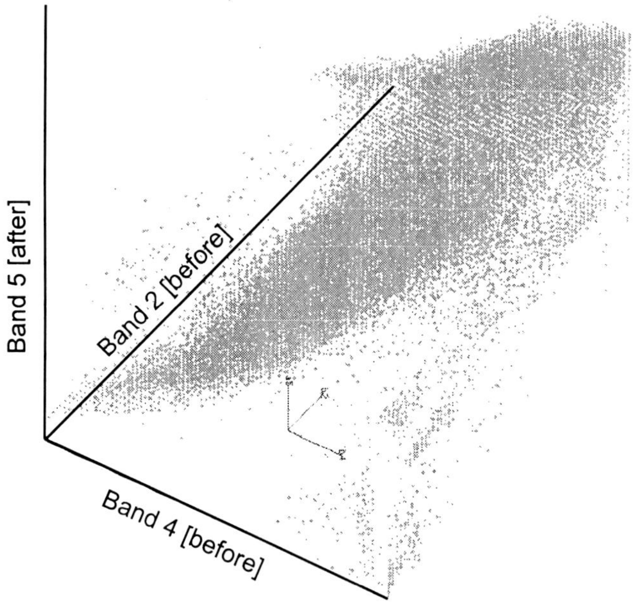

Figure 4. Three-dimensional scattergram showing the kind of correlation that exists within and between Landsat bands. These data represent a small subset of the Tiger Wash images. The points that lie high and low around the cloud, projected onto band 5, represent changes in the image.

Nora through western Arizona. The scenes date to July 1997 and October 1997, and they cover much of west-central Arizona (see Plate 3; also on CD). The images reveal widespread landscape change that we interpret as evidence for widespread flooding throughout the region. The change image shown in Plate 3 is a subset of one of the two larger portions of the Landsat scene we processed. As was discussed above, we believe that the yellow, cyan, and blue colors on the change image highlight areas where landscape change occurred between July and October 1997. Relict Pleistocene alluvial fans and other piedmont areas that were not seriously affected by flooding are represented by shades of magenta.

Colors representing landscape change are common on the piedmont areas in a northeast-trending belt through the study area (see Plate 3). This belt of substantial landscape change follows the approximate track of moisture associated with Hurricane Nora [*Flood Control District of Maricopa County,* unpublished data; *Klawon and Pearthree,* 2000]. Within this fairly broad belt, three general types of landscape change are represented. Relatively narrow bands of yellow and cyan within broader areas of magenta represent active channels and adjacent floodplains of limited extent that are incised into Pleistocene fans. Some of these washes may have experienced significant floods, but topographic confinement by higher, older fan deposits limited the lateral extent of inundation. More extensive yellow, cyan, and blue colors on piedmonts are associated with distributary drainage networks that experienced large floods. Finally, broad strips of cyan and yellow in the valleys are associated with larger axial drainages such as Bouse Wash and Centennial Wash. In this region, axial drainages typically have very broad, low relief floodplains and small, discontinuous channels. The satellite change data suggest that much of the axial floodplain areas of these larger drainages were inundated in September 1997.

The predominance of yellow and cyan hues along many smaller washes suggests that 1997 floods entailed mainly sediment reworking and deposition, and some vegetation growth. The post-flood satellite image was obtained less than three weeks after the event; the response of vegetation to flooding might have been more fully expressed in later scenes. Complex color patterns found in larger distributary drainage systems indicate that flood-induced changes in the landscape were more complicated along these systems. Changes probably included deposition of fresh channel sediment, silt and clay drapes, and organic litter, wet sediment and possibly standing water, and increased growth in vegetation after the flood. The complex spatial distribution of flood evidence on the change image illustrates how difficult it would be to know that a remote area was flooded without the remotely sensed data. The ability to determine geomorphic response at synoptic scales enables a new class of study, *synoptic geomorphology*, and its general importance is introduced in *Mayer* [2000].

FIELD VERIFICATION OF SATELLITE CHANGE DATA – TIGER WASH

Field mapping and analyses of the 1997 flood on Tiger Wash in western Arizona provide an opportunity to explore in detail the usefulness of satellite-based change analysis for flood detection. Inundation mapping of the 1997 flood was done primarily in the field in the spring of 1999 using relatively large-scale, post-flood aerial photos as a mapping base. Field mapping was later supplemented by interpretation of large-scale post-flood color aerial photographs [*Klawon and Pearthree*, 2000]. The inundation mapping was done completely independently of the satellite change analysis and *visa versa.* Thus, comparison of the field-based inundation map for the 1997 flood on Tiger Wash with the satellite change image provides an excellent test of the usefulness of flood detection mapping using satellite change methodology.

The 1997 Flood on Tiger Wash

Tiger Wash is one many drainages in the Gila River system in west-central Arizona that experienced floods as a result of the passage of the remnants of Hurricane Nora (Plate 3). The upper reaches of Tiger Wash are a ~300 km^2

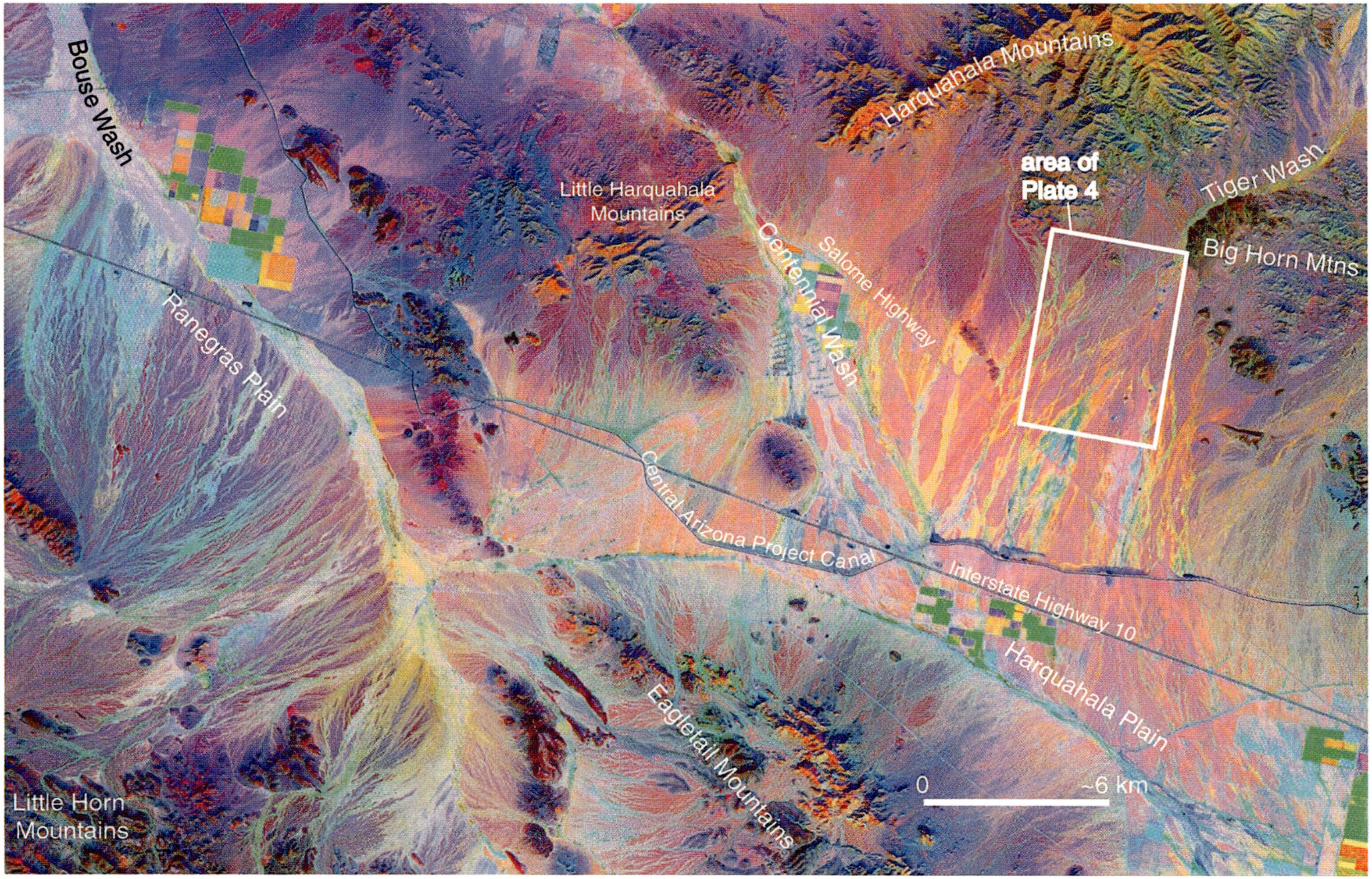

Plate 3. Results of image processing of the Landsat scene covering part of western Arizona, including the Tiger Wash area. The image shows change caused by the hurricane as different colors. The various colors result of different combinations or contributions from bands 5 and 4, before and after the storm. The yellow color highlights areas of the channels where flooding left fine grained sedimentary deposits. Green and cyan colors probably highlight areas of vegetation growth after the flood. See text for discussion.

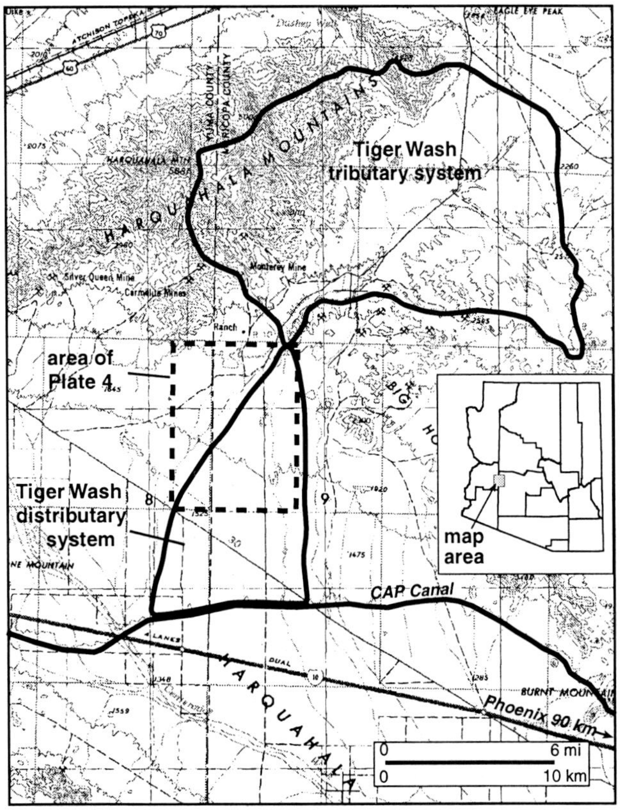

Figure 5. Location map for the Tiger Wash area. The tributary (upper) and distributary (lower) parts of the drainage basin are outlined. The approximate area covered by the satellite change image and the detailed inundation map in **Plate 4** is within the dashed line.

tributary system that drains parts of the Harquahala and Big Horn mountains and a moderately dissected basin between them (Figure 5; also on CD). As Tiger Wash enters the low-relief piedmont of the Harquahala Plain to the south, it changes from a tributary system to a large (~120 km^2) alluvial fan complex consisting of extensive Holocene deposits and a downstream-branching distributary drainage system.

The Tiger Wash drainage basin received upwards of 100 mm of precipitation on September 25 and 26, 1997, with as much as 300 mm falling at the top of the Harquahala Mountains [*Flood Control District of Maricopa County*, unpublished data]. A crest-stage stream gage on Tiger Wash operated by the U.S. Geological Survey was destroyed in the flood, but indirect discharge estimates made after the flood indicate peak discharges of about 230 cms at the gage site and about 250-300 cms at the head of the distributary portion of the system [*JE Fuller Hydrology / Geomorphology*, 2000]. The crest-stage gage on Tiger Wash was operated for most of the period from 1963 to 1997, and the 1997 peak discharge at the gage was almost twice as large as any previously recorded discharge [*Pope et al.*, 1998]. Based on statistical analysis done prior to the 1997 flood, the 100-year discharge estimate for the gage site was about 210 cms. Paleoflood analyses done at the stream gage site by P.K. House and J.E. Fuller in 1992 indicate that no floods larger than about 370 cms have occurred on Tiger Wash during the Holocene [*CH2MHILL*, 1992]. Clearly, the 1997 flood was a large event for this basin. In addition, peak flow in the branch of the distributary system that is west of Eagle Eye Road (Figure 5) was about 3 times larger than that of the eastern branch, so this was probably an extreme flood for the western part of the system [*Klawon and Pearthree*, 2000; *JE Fuller Hydrology / Geomorphology*, 2000].

Flood Inundation Mapping

Areas of inundation during the 1997 flood in the distributary portion of Tiger Wash were mapped in 1999 by a team from the Arizona Geological Survey as part of an effort to learn more about the character of alluvial-fan flooding in this region [*Klawon and Pearthree*, 2000; *JE Fuller Hydrology / Geomorphology*, 2000]. Based on the rain gage data available for the Tiger Wash drainage basin [*Flood Control District of Maricopa County*, unpublished data], there were probably no significant streamflow events on Tiger Wash between late September 1997 and January 1999, when we began our field mapping. Initial field mapping was done in early 1999 using a 1:15,000-scale black-and-white mosaic of aerial photographs taken in February 1998. Evidence of inundation is more distinctive on 1:9,600-scale color aerial photographs taken in late March 1999. These new photos were interpreted to revise and extend the inundation mapping. Inundation mapping was compiled on a new mosaic made from the 1:9,600-scale color photos. This detailed inundation map was then rectified and georeferenced so map patterns of inundation could be compared with the satellite change data. The satellite change imagery was not used in the inundation mapping, and the inundation map has not been revised based on the change image.

Several kinds of field evidence were used to map the extent of inundation. The most ubiquitous evidence of the flood is freshly deposited sediment. In areas of deeper inundation, fresh sediment typically is light-colored sand and gravel. This sediment partially buries vegetation of various sizes, and in many places sandbars were deposited on the lee sides of vegetation. In areas of shallow sheetflooding, fresh sediment typically is fine sand, silt and clay, with minor concentrations of coarser sand and fine gravel. The color of this fine-grained sediment is typically slightly redder than deposits that pre-date the flood. Although field mapping was done more than a year after the flood event, vegetation remnants observed on larger patches of fine sheetflood deposits suggest that plants, including grasses, grew in abundance after the flood. Flotsam (floated organic debris) is also a common indicator of flow. It is especially valuable around the margins of inundation because it floats and may mark the maximum extent of the water surface. Flotsam was also observed in trees and bushes in areas of deeper flow. Evidence of fresh scour and formation or modification of gullies of various sizes was common. In these areas, it was evident that the surface had recently been swept clean of loose sediment, and in many places small scour pits had developed. In areas of sheet flooding, scour commonly was focused between bushes or animal burrows.

Areas of the west branch of Tiger Wash that were inundated in the 1997 flood were subdivided into three categories based on character of flow. Flow in channels (typically > 0.3 m) was responsible for the transport of bedload sediment up to small boulders and erosion of channel beds and stream banks (Figure 6a; also on CD). Several large new channels formed during the flood, and numerous pre-existing channels were enlarged or extended. Relatively deep, unconfined flow occurred in broad overbank areas adjacent to main channels and downstream from them. This flow deposited as sediment as gravel and sand sheets and in streamlined pendant bars downstream of creosote bushes and other vegetation (Figure 6b; also on CD). Flow depths in these areas were quite variable, ranging from 1 meter or more in gullies to a few centimeters over pendant bars. Shallower sheetflooding (< 20 cm) occurred at the margins of flow, where water reached its maximum distance away from the main channels, and in broad sheets in the lower part of the distributary system. In areas of shallow sheetflooding, low sand deposits or "fingers" record flow and deposition around creosote bushes and areas between bushes typically are slightly to moderately scoured (Figure 6c; also on CD). At the very margins of flow, thin silt and clay layers typically cover the surface. Flow in the eastern branch of Tiger was not subdivided.

There are several uncertainties in the resulting flood inundation map (Plate 4; also on CD). There is uncertainty in

A)

B)

C)

Figure 6. Photographs of Tiger Wash: A) Ground photo illustrating a large channel that conveyed flow at least 1 m deep in the 1997 flood. The channel bed consists of sand and gravel, with little vegetation. Vegetation typically is quite dense (for this region) along the margins of sizable channels. B) Ground photo of an area that experienced deep unconfined flow in the 1997 flood. Water was at least 0.5 m deep in the swales (labeled "S"); flow was much shallower over sand bars (labeled "B"), which were deposited on the lee sides of vegetation. C) Ground photo of an area where the 1997 flood inundation was very shallow. The light surface color in the foreground reflects the thin veneer of fine sediment deposited by the flood.

the location of inundation limits where shallow flow covered young fine-grained deposits because of the minimal contrast in visible color. In addition, in featureless low relief areas, location on the aerial photo was less certain. There is complex distortion in the photos themselves, and our attempts to rectify and georeference the aerial photos and the inundation mapping have not removed all of this distortion.

The 1997 flood resulted in complex flow patterns and widespread inundation in the Tiger Wash distributary system (Plate 4). In the upper part of the distributary network, flow was mostly contained in channels with relatively minor areas of overbank flow. High-water marks are evident on channel banks and tributary channel mouths were backflooded. Flow on the eastern branch of Tiger Wash was complicated by the complete filling and spillover of several small gravel pits, and some flow from the eastern branch spilled back into the western branch (Plate 4). As flow became less confined by topography farther downslope, inundation was widespread and complex. Several sizable new channels developed during the flood. The largest of these is approximately 25 m wide, 1 m deep, and 1 km long

Comparison of Field-based Inundation Mapping with Satellite Change Imagery

Satellite-based flood detection analysis for the Tiger Wash area reveals a complex pattern of change that is strikingly similar to the field-based flood inundation mapping. The satellite change image highlights areas that were inundated in 1997 and provides some information about the dominant kinds of change found in the Tiger Wash area (Plate 4). As was discussed earlier, we believe that shades ranging from pink to magenta indicate areas where no change was detected. These colors cover much of the piedmont within and adjacent to the Tiger Wash distributary network, faithfully reflecting the fact that broad areas were not inundated in the 1997 flood. Within the distributary network of Tiger Wash, there are extensive relict Pleistocene alluvial fan surfaces as well as Holocene fan surfaces that are not now part of the active fluvial system and were not inundated in the 1997 flood [*Klawon and Pearthree*, 2000]. In addition, there are extensive dissected Pleistocene fans on the upper piedmont. In general, it appears that older alluvial surfaces are represented by a darker shade of magenta. This distinct color change from the low surfaces to the dissected ones may represent the difficulty in radiometrically normalizing areas that have shadowing or strong topographic texture.

Areas in the Tiger Wash distributary system that were inundated in the 1997 flood show up as yellow, green, cyan, or blue (Plate 4). The yellow color depicts a simultaneous brightening in bands 5 [after] and 4 [after], which we interpret to represent the deposition of fresh sediment and some increase in vegetation. Band 4 is sensitive to vegetation, but the yellow color cannot represent predominantly vegetation increase because while vegetation is bright in band 4, it is also dark in band 5 (see Figure 1). Thus to be simultaneously bright in both suggests a complex cause, which we attribute primarily to exposure of recently transported or uncovered sediment. The yellow color appears to represent areas that received extensive but shallow inundation in the 1997 flood, where fresh sediment is sand and finer. It is also possible that there was some ephemeral increase in vegetation in these areas that was not readily apparent in 1999. Green colors, which depict areas where increases in vegetation occurred, are found along some channels and in areas that experienced deep unconfined flow. The cyan color occurs in a few channels, but is most extensive in areas farther down in the distributary system where broad, deep and shallow unconfined flooding occurred. We suggest that this represents vegetation growth following infiltration of water from the flood. In the RGB color model, cyan is the mixture of green and blue, or band 4 [after] and band 5 [before]. If the pixel is bright in band 4 it could represent vegetation, but because the pixel was bright in band 5 [before], we know that there was little vegetation there before the flooding. Areas that appear blue in the satellite change image are extensive in the lower part of the distributary system where flow was quite shallow. The blue color indicates that band 5 was brighter in these areas before the flood. Field reconnaissance suggests that extensive mats of dead organic litter covered these areas after the flood, possibly decreasing the brightness of band 5. However, the large new channels that formed in the 1997 flood do not show very dramatically in the satellite change image (see Plate 4).

The combined areas of yellow, blue, green and cyan depicted on the satellite change image depict the extent of inundation in the Tiger Wash system extremely well. This is especially interesting given the rather large pixel size (~30 m^2) in the satellite image. In areas around the margins of the flood that experienced very shallow inundation, it is likely that the satellite change image represents the extent of inundation more completely than field mapping that was conducted more than a year after the flood. The fact that flood inundation is represented by a substantial range of colors indicates that a variety of landscape changes occurred as a result of the 1997 flood. It is clear that the colors shown on the satellite change image for Tiger Wash do not correspond in a simple manner with inundation depths and the character of flow in different parts of the system.

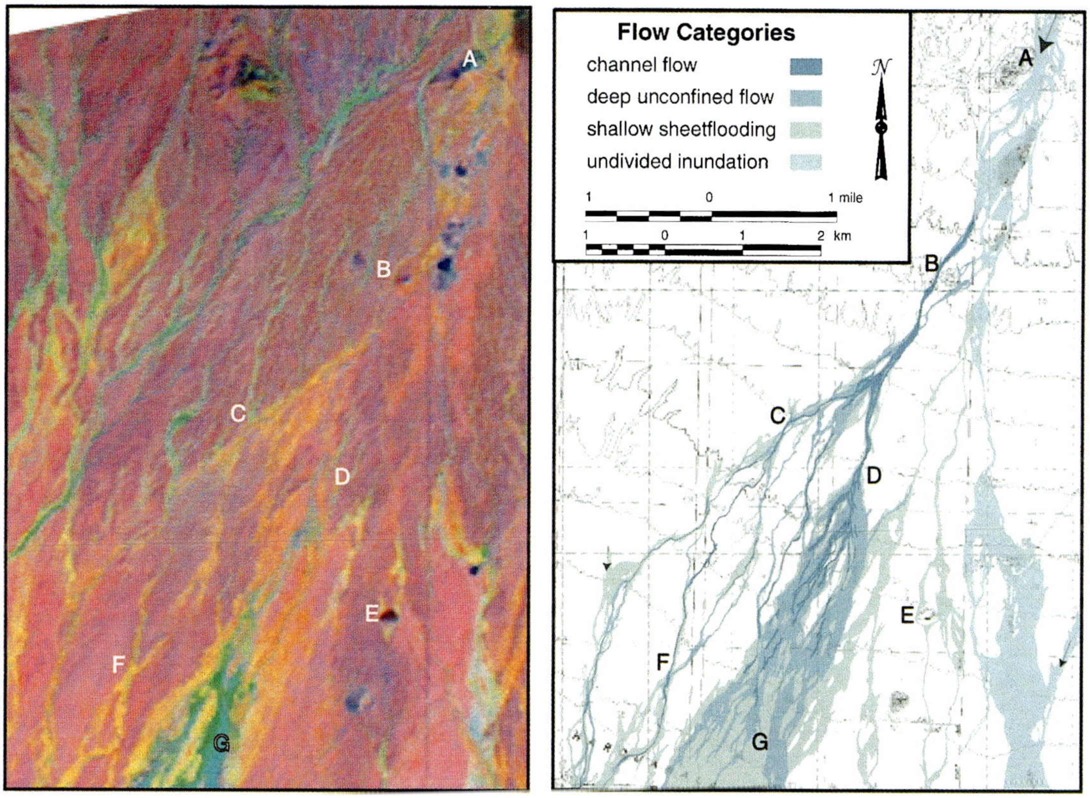

Plate 4. Map of inundation in the Tiger Wash distributary system in the 1997 flood based on detailed field mapping and the satellite change image for the same area. The inundation mapping was rectified and georeferenced as well as possible, but we were unable to completely remove the effects of distortion of the aerial photos.

CONCLUSIONS

We show that the analysis of medium resolution (30 m) satellite imagery to detect changes in the land surface is a viable means for detecting flooding and determining the extent of flood inundation after floodwaters have receded. We processed satellite data covering western Arizona for the period immediately before and after the passage of Hurricane Nora in late September 1997. Landsat data were analyzed to accentuate changes in infrared bands 4 and 5, which we believe reflect changes in sediment character and vegetation. The resulting satellite change image clearly highlights drainages that likely experienced floods as the hurricane dissipated over the region.

The extent of inundation resulting from the 1997 flood on Tiger Wash in west-central Arizona was mapped in the field completely independently of the satellite remote sensing data analysis. This field mapping reveals a complex pattern of inundation throughout the large distributary drainage system. Although the significance of each color on the satellite change image for Tiger Wash is not completely clear, the change image and the field-based inundation map are strikingly similar. This indicates that analysis of changes between before- and after-flood satellite images have great potential to locate those drainages that have experienced floods and eventually to help map the extent of flood inundation.

Acknowledgments. L. Mayer gratefully acknowledges the support of the U.S. Geological Survey, EROS Data Center for their support of this pilot project, which was initiated while he was a visiting scientist at EROS in 1998. This paper is an outcome of the Fundamentals through the Application of Space Technology (FaST) initiative that L. Mayer designed and launched. Research was supported by the U.S. Geological Survey, Agreement No. 99CRAG0009. This manuscript is submitted for publication with the understanding that the United States Government is authorized to reproduce and distribute reprints for governmental purposes. The views and conclusions contained in this document are those of the authors and should not be interpreted as necessarily representing the official policies, either express or implied of the US government. Funding for fieldwork in Baja during the summer of 1998 was provided through NASA's Mission to Planet Earth program in support of assessing the effects of Hurricane Nora. NASA's support was critical for documenting flooding in Baja California, along the track of the hurricane. Further fieldwork in Baja was supported through the National Science Foundation grant EAR0049053.

Many individuals contributed to the analysis of the 1997 flood on Tiger Wash. We are especially grateful for assistance from Jeanne Klawon, who contributed greatly to the Tiger Wash project and is responsible for much of the inundation mapping. Ted Lehman of JE Fuller provided the indirect discharge estimates reported here, and engaged in many discussions about this work. Tim Orr was responsible for map digitization and rectification and georeferencing of the inundation mapping. Funding for analysis of the 1997 flood was provided by the Flood Control District of Maricopa County and the Arizona Geological Survey. The work was conducted in cooperation with JE Fuller Hydrology & Geomorphology.

Thoughtful reviews by Nick Lancaster, Geoff Pickup and Kyle House substantially improved this manuscript. Ann Youberg proofed the manuscript and detected a few last imperfections.

REFERENCES

Blasco, F., M.F. Bellan, and M.U. Chaudhury, Estimating the extent of floods in Bangladesh using SPOT data, *Remote Sensing of Environment, 39*, 167-178, 1992.

CH2MHill, Alluvial fan data collection and monitoring study: A report to the Flood Control District of Maricopa County, 204 p., 4 plates, 1992.

Coppin, P.R., and M.E. Bauer, Digital change detection in forest ecosystems with remote sensing imagery, *Remote Sensing of Environment, 13*, 207-234, 1996.

Deutsch, M., and F. Ruggles, Optical data processing projected applications of the ERTS-1 imagery covering the 1973 Mississippi river valley floods, *Water Resources Research, 10*, 1023-1039, 1974.

Elvidge, C.D., D. Yuan, R.D. Weerackoon, and R.S. Lunetta, Relative radiometric normalization of Landsat multispectral scanner (MSS) data using an automatic scattergram controlled regression, *Photogrammetric Engineering & Remote Sensing, 61*, 1255-1260, 1995.

Gatewood, J.S., F.F. Schrader, and M.R. Stackpole, Notable local floods of 1939, Part I, Floods of September, 1939 in Colorado River Basin below Boulder Dam, *U.S. Geological Survey Water Supply Paper*, 967-A, 37 pp, 1946.

Gillespie, A.R., Digital techniques of image enhancement, in *Remote Sensing in Geology*, edited by B.R. Siegal and A.R. Gillespie, John Wiley, New York, 1980.

Hall, F.G., D.E. Strebel, J.E. Nickeson, and S.J. Goetz, Radiometric rectification: Toward a common radiometric response among multidate multisensor images, *Remote Sensing of Environment, 35*, 11-27, 1991.

Hallberg, G.R., B.E. Hoyer, and A. Rango, Application of ERTS-1 imagery to flood innundation mapping, *NASA Special Publication No. 327, Symposium on significant results obtained from Earth Resources Satellite 1, 1A*, 745-753, 1973.

JE Fuller Hydrology & Geomorphology, 2000, Inundation mapping of the September 1997 flood, surficial geologic mapping, and evaluation of the piedmont flood hazard assessment manual for portions of the Tiger Wash piedmont: Report to the Flood Control District of Maricopa County, Contract No. 98-48

Klawon, J.E., and P.A. Pearthree, 2000, Field guide to a dynamic distributary drainage system: Tiger Wash, western Arizona: Arizona Geological Survey Open-File Report 00-01, 34 p.

Mayer, L., 2000, Satellite remote sensing of flood related landscape change resulting from hurricanes in Baja California and the southwestern United States, in *The Hydrology-Geomorphology Interface:Rainfall,Floods, Sedimentation, Land Use*, edited by M. Hassan, O. Slaymaker, S. Berkowicz,

International Association of Hydrological Sciences Publication 261, pp. 201-214.

Morrison, R.B., and P.G. White, Monitoring flood innundation, in *ERTS-1 A New Window on our Planet*, pp. 196-208, U.S. Geological Survey Professional Paper 929, 1976.

Rango, A., and V.V. Salomonson, Regional flood mapping from space, *Water Resources Research, 10*, 473-484, 1974.

Rao, D.P., V. Bhanumurthy, G.S. Rao, and P. Manjusri, Remote sening and GIS in flood management in India, in *Flood Studies in India*, edited by V.S. Kale, pp. 195-218, Geological Society of India, Bangalore, 1998.

Rothery, D.A., and G.A. Hunt, A simple way to perform decorrelation stretching and related techniques on menu-driven image processing systems, *International Journal of Remote Sensing, 11*, 133-137, 1990.

Schott, J.R., C. Salvaggio, and W.J. Volchok, Radiometric scene normalization using pseudoinvariant features, *Remote Sensing of Environment, 26*, 1-16, 1988.

Smith, Walter, The effects of North Pacific tropical cyclones on the southwestern United States, *NOAA Technical Memorandum*, NWS-197, 229 pp., 1986.

Smith, L.C., Satellite remote sensing of river inundation area, stage, and discharge: A review, *Hydrological Processes, 11*, 1427-1439, 1997.

Williams, R.S., Geological applications, in *Manual of Remote Sensing*, edited by R.S. Williams, American Society of Photogrammetry, Falls Church, 1983.

Yuan, D., and C.D. Elvidge, Comparison of relative radiometric normalization techniques, *Journal of Photogrammetry & Remote Sensing, 51*, 117-126, 1996.

Larry Mayer, Department of Geosciences, University of Arizona, Tucson, Arizona, 85721, USA

Philip A. Pearthree, Arizona Geological Survey, 416 W. Congress St., Tucson, Arizona, 85701 USA.

Dendrochronologic Evidence for the Frequency and Magnitude of Paleofloods

Thomas M. Yanosky

U.S. Geological Survey, Reston, Virginia

Robert D. Jarrett

U.S. Geological Survey, Denver, Colorado

Tree-growth responses to flood damage can be used to document the frequency and magnitude of paleofloods, thereby extending the historical period of record and improving estimates of flood recurrence. Common responses include the formation of scars, sprouting from tilted stems, and eccentric ring growth; occasionally, ring-anatomical changes develop following damage to leaves and buds, or to prolonged root flooding that does not injure trees. The annual formation of rings permits flood dating to within a year, and sometimes to within several weeks in the case of anatomical responses. The average maximum height of scars provides a reliable indicator of paleoflood stages along both low- and high-gradient streams. Magnitude also can be estimated by recovering information of any kind from trees at successively higher flood-plain elevations. This paper discusses numerous paleoflood studies based on botanical evidence, and considers practical applications that include detection, sampling, and interpretation of botanical information.

INTRODUCTION

Background

In 1964, Robert S. Sigafoos, United States Geological Survey (USGS), published his classic treatise showing that riparian trees could be used to document the occurrence of past floods and depositional events (Figure 1). At the time, few gaged streams in the United States had continuous records longer than 30 years. Sigafoos argued that riparian trees sometimes live hundreds of years, form annual rings that can be dated precisely, and thus were an untapped resource for hydrologic inference. He studied tree-growth responses to flooding along reaches of the Potomac River near Washington, D.C., USA, verifying his findings against the systematic record of floods since 1930 (and since 1895 at Point of Rocks, Maryland, 67 km farther upstream). Sigafoos' primary intent was to develop a technique to improve estimates of flood-recurrence intervals, thereby more accurately defining flood-plain limits and facilitating the cost-appropriate construction or modification of bridges, dams, and levees.

Paleoflood hydrology originated in the United States during the early 19th century when geologists began to consider the role of floods in the evolution of landforms, and as a discipline began to expand exponentially after 1970 [*Costa*, 1987]. Studies are conducted to document flooding before the period of record, thereby extending flood histories, or to estimate the stages of known floods. In addition to providing information about floods, the

Ancient Floods, Modern Hazards: Principles
and Applications of Paleoflood Hydrology
Water Science and Application Volume 5, pages 77-89

Published in 2002 by the American Geophysical Union

broader field of paleohydrology also emphasizes hydrologic variability, most notably the frequency and magnitude of droughts, average monthly streamflow, and climate change. Paleohydrologic investigations often are interdisciplinary and may involve stratigraphy, chemical analysis of sediments, palynology, radiometric techniques, archeology, and dendrochronology [*Jarrett*, 1991; *Jarrett and Tomlinson*, 2000]. Of these, only dendrochronology provides proxy data at the resolution of individual years for reconstructing long-term hydroclimatic variables. However, ring-width chronologies used traditionally for reconstructions cannot be used to determine the occurrence or magnitude of individual paleofloods. Rather, the botanical basis for paleoflood inference results from predictable tree-growth responses to damage of the bark- and wood-forming tissues, buds, and leaves, and to radial growth following partial uprooting of the trunk. The remainder of this paper summarizes the kinds of botanical evidence most useful in paleofloods studies, with several illustrations of each and some considerations for practical applications.

Tree Growth and Tree Rings

Trees grow in height only from the division and elongation of cells at stem tips. Thus, a scar on a trunk or branch will remain at its original height as the tree grows taller. Growth in diameter results from the division of a thin layer of living cells (the cambium) just inside the bark, each year producing a new layer of wood, or ring. *Phipps* [1985] describes the three-dimensional nature of rings as concentric circles in transverse section and as parabolas when viewed longitudinally (Figure 2). The age of the tree can be approximated by counting the number of rings in the lower trunk, but a precise year can be assigned to each ring only by crossdating; that is, by matching ring-width patterns of a tree with those of nearby trees [*Fritts*, 1976]. Rings are composed of specialized tissues that conduct sap, provide support, and store organic materials. The relative size, amount, and arrangement of these components confer the typical appearance and physical properties of the various woods [see *Phipps*, 1985]. The structure of wood sometimes is modified by environmental factors and events that occur during the growth season, thus preserving anatomical evidence of past growing conditions.

The widths of tree rings are controlled in large part by environmental processes, most notably climate, and for this reason have been used extensively to reconstruct climatic variables where records are incomplete or lacking [*Fritts*, 1976]. Severe flood damage sometimes results in the production of a narrow ring, or even reduced radial growth for several consecutive years [*Yanosky*, 1982a]. Narrow rings most typically are associated with droughts, however, and thus have little value for paleoflood documentations. *Yanosky and Cleaveland* [1998] and *Yanosky* [1999] provide further discussions of the uses and limitations of tree rings in hydrology, whereas others (see below) consider the relations between vegetation and hydrology from a dynamic, process-oriented perspective.

Vegetation and Hydrogeomorphology

Fluvial ecosystems are four dimensional, consisting of longitudinal (upstream-downstream), lateral (floodplain-upland), vertical (hyporheic-phreatic), and temporal components [*Stanford and Ward*, 1992]. The establishment and growth of riparian vegetation is controlled in large part by hydrogeomorphic processes, human disturbance, land use, and climate [*Bedinger*, 1971; *Baker*, 1977, 1994; *Osterkamp and Hupp*, 1984; *Osterkamp and Costa*, 1987; *Hupp*, 1987, 1988; *Glavac et al.*, 1992; *Stanford and Ward*, 1992; *Shroder and Bishop*, 1995; *Hupp and Osterkamp*, 1996; *Scott et al.*, 1996; *Friedman et al.*, 1996a, 1996b]. Vegetation and hydrogeology are interrelated whether streams are ephemeral, intermittent, or perennial [*Wolman and Miller*, 1960; *Baker*, 1977, 1994; *Wolman and Gerson*, 1978; *Hedman and Oster-*

Figure 1. Robert Sigafoos (1920-1995), USGS research botanist, using an increment borer to sample a flood-damaged ash tree (*Fraxinus pennsylvanica* Marsh.) growing along the Potomac River. Photo by D. Usher.

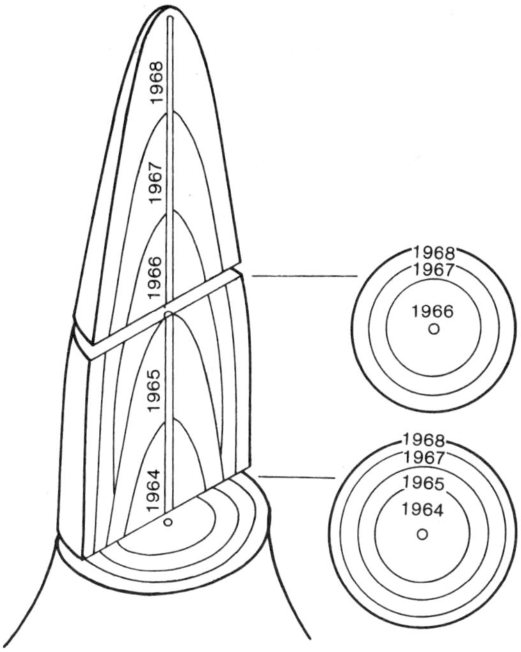

Figure 2. Stylized diagram of tree rings in transverse and longitudinal sections. From *Phipps* [1985].

kamp, 1982; Nilsson, 1987; *Everitt*, 1995]. The age, abundance, and species distributions of riparian vegetation are controlled in part by flow-duration regimes, flood frequency, magnitude, and velocity (intensity), shear stress, stream power, and sediment-transport dynamics. In turn, vegetation may alter flood flows by increasing roughness and promoting deposition. For example, *Osterkamp and Hupp* [1984] and *Hupp and Osterkamp* [1985] showed that riparian landforms support predictable assemblages of vegetation related to flow characteristics of three streams in the Virginia piedmont.

Most studies investigating the dynamic associations between vegetation and hydrogeomorphology are based on direct-gradient analysis [*Auble et al.*, 1994] and relate vegetation to specific ranges of hydraulic variables. These associations may permit predictions concerning changes in riparian ecosystems from land use, regulation, and climate [*Williams and Wolman*, 1984; *Richter and Richter*, 1990; *Stanford and Ward*, 1992; *Hupp*, 1992; *Johnson*, 1994; *Scott et al.*, 1996], and thus also are important to researchers attempting to recover evidence of paleofloods. *Shroder and Bishop* [1995] suggest that remote sensing provides a way to monitor subtle changes in riparian ecosystems, and noted that dendrochronological methods were particularly effective in comparative studies of ecosystems in the United States and Pakistan. Improvements in predictive models will result in the refinement of dendrochronologic methods and permit applications to a wider variety of hydrologic and botanical problems.

DENDROCHRONOLOGICAL TECHNIQUES

Tree Age and Growth Habit

Catastrophic flooding sometimes destroys all existing vegetation and creates a new surface for colonization by trees. If the flood occurred within the typical life span of colonizing trees, a forest of mixed age will develop in which the oldest trees are the same age. The lack of an age-synchronous cohort suggests that the flood likely occurred longer ago than the lifespan of trees, and thus ages of the oldest trees afford only a minimum estimate of the flood date [*Sigafoos*, 1964; *Everitt*, 1968, 1995; *Helley and LaMarche*, 1968; *Costa*, 1978; *Hupp*, 1988; *Gottesfeld and Gottesfeld*, 1990]. Generally, hydrologic inference from age-synchronous trees is most reliable along high-gradient streams prone to unusually catastrophic flooding. For example, *Hack and Goodlett* [1960] studied the aftermath of a cloudburst flood in June 1949 along Little River, Virginia, USA, and concluded from the age of flood-plain trees that floods of comparable magnitude had not occurred in at least 150 years. They also located numerous revegetated debris chutes, none of which supported trees established before 1949. *Yanosky* [1982b] found that most trees along inside-channel bends of the Potomac River survived the catastrophic flood of June 1972, as well as even larger floods in 1936 and 1942. However, most trees were destroyed along outside-channel bends exposed to the full force of floodwaters, and none of the few survivors had been growing prior to 1942. This study showed that the age and form of riparian forests was related to flood-flow velocities rather than to magnitude alone.

Trees with the greatest life spans, such as those in western North America, provide the longest potential record of paleofloods, particularly in remote areas where anthropomorphic disturbances are minimal. For example, *Helley and LaMarche* [1968] found that a large flood in northern California, USA, in 1964 eroded a stream terrace covering the stumps of redwoods (*Sequoia sempervirens* (D. Don) Endl.) killed by an earlier flood of approximately the same magnitude. Radiocarbon analysis of terrace deposits revealed that the original surface had been buried to a depth of ~6 m by a single flood that occurred ~A.D. 1560 $\pm$ 100 years. Thus, the investigators were able to assign a 400-year return period to the 1964 flood. Similarly, *Hupp* [1984] used the ages of trees on newly-created depositional landforms to date a series of debris flows along five creeks near Mt. Shasta, California. He was able to document nine debris flows not reported previously, including one that oc-

Figure 3. Growth habit ("Sigafoos tree") of a white oak (*Quercus alba* L.) on a flood-plain terrace along the Potomac River near Washington, D.C. USA. The vertical sprout at right started to grow from the tilted trunk in 1937. The flood of record occurred in March 1936, followed by the 4th highest flood of the century in April 1937. Photo by T.M. Yanosky.

curred in ~1670. This and similar studies require estimates of the time for vegetation to establish on newly-created surfaces

Flood-plain trees along high-gradient streams commonly are small and shrubby, particularly when flooded frequently. Partial uprooting results in the formation of vertical sprouts along a leaning trunk, imparting a characteristic growth habit sometimes referred to as a "Sigafoos tree" (Figure 3.). Trunk tilting or crown damage during dormancy results in the production of sprouts during the next growing season. Thus, the age of the sprout can be used to determine when the flood occurred [*Sigafoos,* 1964]. Some trees in high-damage zones bear sprouts that document numerous floods, the oldest typically forming along lowest parts of the trunk. For example, *Sigafoos* [1964] described one tree that formed sprouts documenting floods of the Potomac River that occurred in 1942, 1948, and 1955. If the crown and most of the bole are destroyed, two or more sprouts growing from the very base eventually impart a "split trunk" appearance similar to the multiple-stemmed trees growing from logged stumps.

Trunk tilting alters the widths of rings formed in subsequent years. Ring widths may be reasonably uniform along the circumference of vertical stems, whereas new growth becomes eccentric after a stem is tilted. That is, the width of a single ring may differ strikingly along the axis of lean. In hardwoods, rings are wider on the upper side than lower side of leaning trunks; rings in conifers are wider on the lower side. The number of eccentric rings can be used to determine how long the trunk has been tilted, and thus when the flood occurred.

Scars

Ice, logs, sediments, and other flood-borne debris sometimes abrade the bark and destroy part of the underlying cambium, resulting in the formation of scar tissue (callus) rather than new wood and bark. Callus is undifferentiated, often discolored tissue that forms a permanent record of localized cambial destruction. Eventually, new cambium grows over the scar and restores the production of wood and bark. Until this occurs, however, the scar is visible from the outside of the tree, sometimes for many years. Because trees grow in height only from shoot tips, the height of a scar above the low-water channel represents a minimum flood elevation (Figure 4), although the stage will be underestimated if the tree partly uproots after the scar forms. *Harrison and Reid* [1967] noted that the intensity of scarring does not necessarily reflect flood magnitude, but rather is a function of the amount of debris. Scarred trees most probably are found along stream reaches where bed gradients are high or flows are constricted, and are less frequent along streams with low gradients, wide flood plains, and small basin areas [*McCord*, 1996].

If abrasion destroys the bark and cambium but no underlying rings, an increment core [*Phipps*, 1985] can be taken directly into the scar face and through the entire trunk diameter. The flood year can be calculated by subtracting the number of rings along the scar-containing radius from that of the intact radius. However, the removal of a small wedge-shaped sample along the scar margin (Figure 4) is preferable when tissues beneath the cambium have been damaged or abraded away. Healed scars within the tree most likely are discovered when cross sections are examined (Figure 5), although this method is undesirable because it requires the sacrifice of the tree. Flood years are estimated by counting the number of rings between the healed scar and the outside ring. Subtle changes in bark morphology sometimes suggest the presence of a healed scar close beneath the surface, but deeper scars are detected only by hit-or-miss increment corings taken along the trunk. A series of corings at selected intervals (~30 cm) along the upstream- or channel-facing trunk increases the likelihood of detecting healed scars [*Waythomas and Jarrett*, 1994; *McCord*, 1996].

The most extensive use of scars to extend flood records has been that by *McCord* [1990; 1996]. He used increment corings and wedge cuts from coniferous trees to extend flood histories at eight sites in Arizona, New Mexico, and Utah, USA. The longest reconstruction was for Kanab Creek in southern Utah (Figure 6), where evidence was recovered for floods that occurred in 1471,

Figure 4. Impact scars on the trunk of a sycamore tree (*Platanus occidentalis* L.) growing along a stream in northwestern Virginia, USA. Scars document the occurrence of four floods and provide estimates of crest elevations. The uppermost scar is ~2.5 m above ground surface. The small wedge cut (arrow) along the margin of the bottom scar permits the counting of rings that formed since the cambium was damaged by flooding. Photo by C.R. Hupp.

1521, 1660, 1726, 1742, 1773, 1809, 1866, 1872, 1883, 1890, 1902, 1909, 1910, 1916, 1925, 1936, and 1949. Furthermore, McCord showed that floods were not random but rather clustered in the period 1866-1949. Climate change or land-use practices may account for the increased frequency of floods, which, in turn, probably initiated the formation of arroyos on Kanab Creek [*Webb et al.*, 1991].

Harrison and Reid [1967] compared a flood-frequency graph constructed from an analysis of ice-scarred trees to instrumented records for the Turtle River in eastern North Dakota, USA (Figure 7). Scars were recovered for 10 of the 12 floods during the 21-year period of record, despite that only three trees were studied. Evidence also was found for three floods that occurred before the period of record. More importantly, stage estimates from scar data generally were within 0.3-0.6 m of actual crest elevations recorded 32 km downstream. Thus, in addition to being the first to compare scar data to documented flows, *Harrison and Reid* [1967] showed that maximum scar heights corresponded to ice-laden flood crests along a low-gradient stream.

Scars on flood-plain trees were used by *Gottesfeld and Gottesfeld* [1990] to document 31 floods on the Morice River, British Columbia, Canada, during a 105-year interval. Log transport during peak stages was most likely to damage trees growing on stable banks, bedrock outcrops, and the upstream ends of bars. They also classified scars on the basis of their formation. "Impact" scars are small features caused by the collision of logs or

Figure 5. Impact scars healed by new growth. Three scars (arrows) formed along the upstream-facing radius. Life size. Photo by D. Usher.

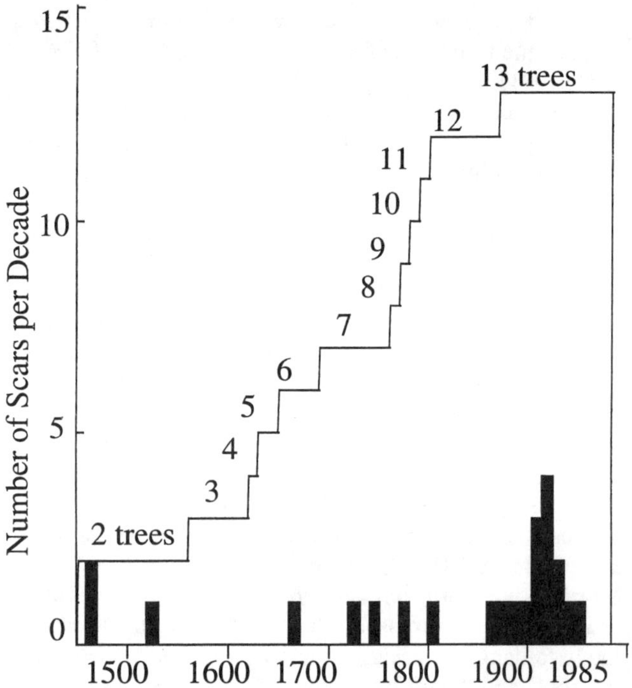

Figure 6. Flood-scar chronology by decade for Kanab Creek, Utah, USA. Note the clustering of floods during the 90-year interval beginning in the 1860s and the abrupt decrease in the number of study trees prior to the late 18th century. Modified from *McCord* [1996].

ice with upstream-facing trunks; "rubbing" scars occur when logs are lodged against the trunk and slowly abrade the bark and cambium, sometimes remaining in contact with the tree for more than one flood. Rubbing scars may be vertically extensive along upstream- and downstream-facing trunks. *Gottesfeld and Gottesfeld* [1990] concluded that their 105-year proxy flood record for the Morice River was "nearly complete."

Gottesfeld [1996] conducted a comprehensive study of scar heights following a nival (snowmelt) flood (recurrence interval ~9 years) in 1990 on the Skeena River, British Columbia. Flood damage typically results from log transport rather than from ice. Along a 170-m reach, he selected 48 scarred trees (nine species) likely to preserve a long-term record of the 1990 flood; that is, he excluded small, flexible trees, those growing on unstable surfaces, and those in which the outer bark was abraded but the cambium was not damaged. The height of scars (Figure 8) ranged from +9 cm to -80 cm relative to the flood crest (mean –20 cm). Most scars (88%) formed along a zone approximately four trees in width beginning at the channel edge. He concluded that impact scarring occurs instantaneously, whereas abrasion scars form during a short interval corresponding to the duration of maximum flood stage. Thus, as along a low-gradient stream susceptible to ice-laden flooding [*Harrison and Reid*, 1967], the heights of scars along the Skeena River were good indicators of peak flood-crest stages.

A NEW STUDY OF SCAR HEIGHTS AND FLOOD STAGE

Previous studies relating the heights of scars to peak flood stages were conducted at single locations along low-gradient streams and for relatively frequent floods [*Harrison and Reid*, 1967; *Gottesfeld*, 1996]. *Jarrett* [this paper] studied the heights of scars formed by an extraordinary flood in 1996 along Buffalo Creek, a high-gradient stream in the Colorado Rocky Mountains southwest of Denver, Colorado USA. The study was conducted along an 8-km reach (rather than at a single location) with bed gradients ranging from 0.01 to 0.04 m/m, thus extending the range of hydraulic conditions for documented tree-scar data.

Buffalo Creek is a perennial stream augmented by reservoir outflow (typically 0.3 to 2 m^3/s). The streambed contains pea gravel and boulders up to several meters in diameter, and average depth to bedrock is ~1 m; however, outcropping occurs locally along parts of the channel. Riparian trees include ponderosa pine (*Pinus ponderosa* Dougl. Ex Laws.), lodgepole pine (*Pinus contorta* Dougl. Ex Loud.), Douglas fir (*Pseudotsuga menziesii* (Mirb.) Franco), and cottonwood (*Populus fremontii* Wats.). Numerous ephemeral tributaries flowing into Buffalo Creek are fed primarily by melting snowpack and trans-basin flow diversions. Floods along Buffalo Creek and tributaries result from intense, localized thunderstorms, generalized rainstorms, and spring snowmelt [*Jarrett*, 1990]. Long-time residents reported no significant flooding in the study area during at least the previous 70 years.

On May 18, 1996, wildfire burned ~50 km^2 of forest within the Buffalo Creek watershed, destroying the organic litter of the shallow, moderately well-drained soils and increasing the susceptibility of the basin to flooding [*Colorado Water Conservation Board*, 1997]. A storm on July 12 generated 130 mm of rain within the burned area in slightly more than one hour [*Henz,* 1998]; according to *Miller et al.* [1973], the 100-year storm would be expected to produce ~70 mm of rain during a single hour. Peak discharges within parts of the burned basin were 20-40 times greater than those in nearby unburned areas. Some basins receiving maximum rainfall produced unit discharges (peak flow divided by drainage area) approximating 60 $m^3/s/km^2$, compared to maximum estimates of 40 $m^3/s/km^2$ for all known previous Colorado floods [*Jarrett*, 1990]. Along the reach of Buffalo Creek that tree scars were studied, the peak discharge ranged from ~85 m^3/s near the west (upstream) end of the burned area to ~500 m^3/s to the east (downstream), and in stage from ~1 to 5 m above bankfull.

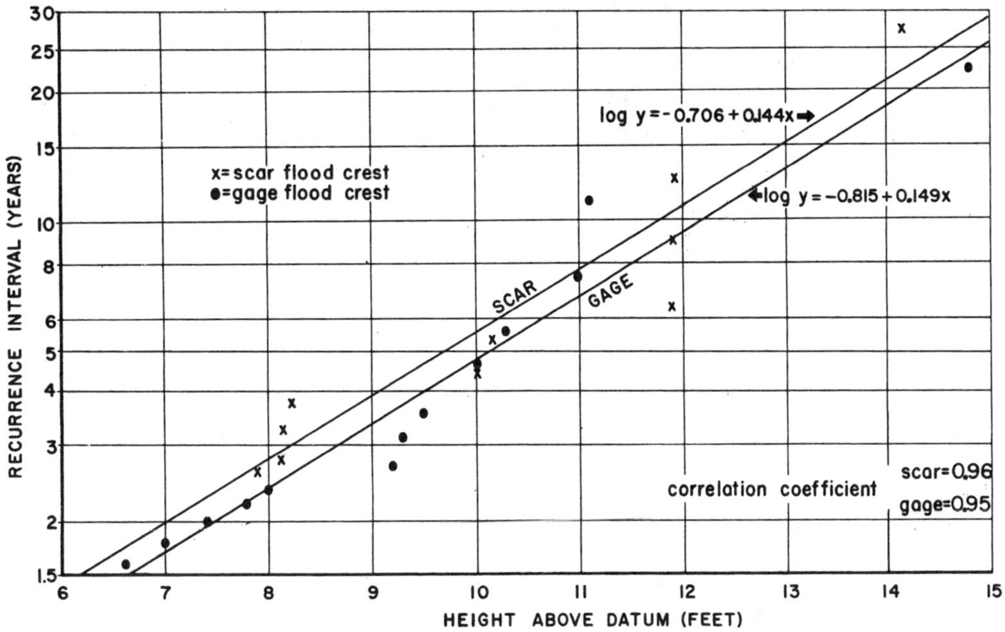

Figure 7. Flood-frequency graph for the Turtle River, North Dakota, constructed from instrumented records and from maximum scar heights. To convert stage heights to meters, divide by 3.281. From *Harrison and Reid* [1967].

Compared to the 100-year flood estimate of 20 m^3/s [*Federal Emergency Management Agency*, written communication, 1996], the flood was at least a 5,000-year event for pre-fire conditions [*Jarrett and Browning*, 1999]. Peak stages were reached within less than one hour and receded to within-bank flows one or two hours later. Well defined high-water marks on both streambanks consisted of deposition of black ash, charcoal, woody debris, and pines needles. Most trunk scarring is believed to have resulted from uprooted trees, bedload, and even by saltation of boulder-sized rocks.

Scar data were collected from 102 riparian trees along an 8-km reach of Buffalo Creek. The elevation of each high-water mark along the streambank was extended to the top of the highest scar on each tree with a string line and bubble level. No older flood scars were observed, a finding consistent with the observations of local residents. The heights of all scars ranged from –60 to +150 cm relative to high-water marks (mean +21 cm); 64 scars (63%) were within 20 cm (Figure 8). Scar heights relative to high-water marks were slightly greater along lower than upper parts of the study reach, possibly owing to increased flow depths, velocities, turbulence, and durations. Increased debris transport along lower reaches may increase the likelihood of scarring as large branches of uprooted trees protrude above the flood. A boulder ~75 cm in diameter was lodged 2 m above the streambank high-water mark in the branch forks of a cottonwood tree.

Scar data from the flood on Buffalo Creek also were analyzed according to the position of damaged trees within channel cross sections. Scar data were compared between streamside trees exposed to maximum flood-flow velocities ("main-channel" trees) and those more distant from the channel ("overbank" trees). The mean height of scars on 67 trees comprising the main-channel group was +23 cm relative to high-water marks, compared to +17 cm in 35 overbank trees. The lack of a significant difference between the two groups of scar heights probably reflects the extraordinary magnitude of the flood and the open-spaced nature of riparian forests in the semi-arid foothills of the Colorado Rocky Mountains. Additionally, some riparian areas along Buffalo Creek were burned severely, thus reducing even further the flow resistance typically provided by vegetation.

The mean height of scars relative to measured crest elevations along Buffalo Creek was ~40 cm greater than that determined along the lower-gradient Skeena River [*Gottesfeld*, 1996], and the range of scar heights also was greater (210 versus 90 cm). It is possible that extraordinary floods on low-gradient streams might possess sufficient energy to produce a tree-scar distribution similar to that of Buffalo Creek. Although additional tree-scar data are needed from a wider range of hydrologic and botanical conditions, the few studies conducted to date suggest that maximum scar heights provide acceptable estimates of paleoflood stages along low- and high-gradient streams, regardless of the kind of transported debris.

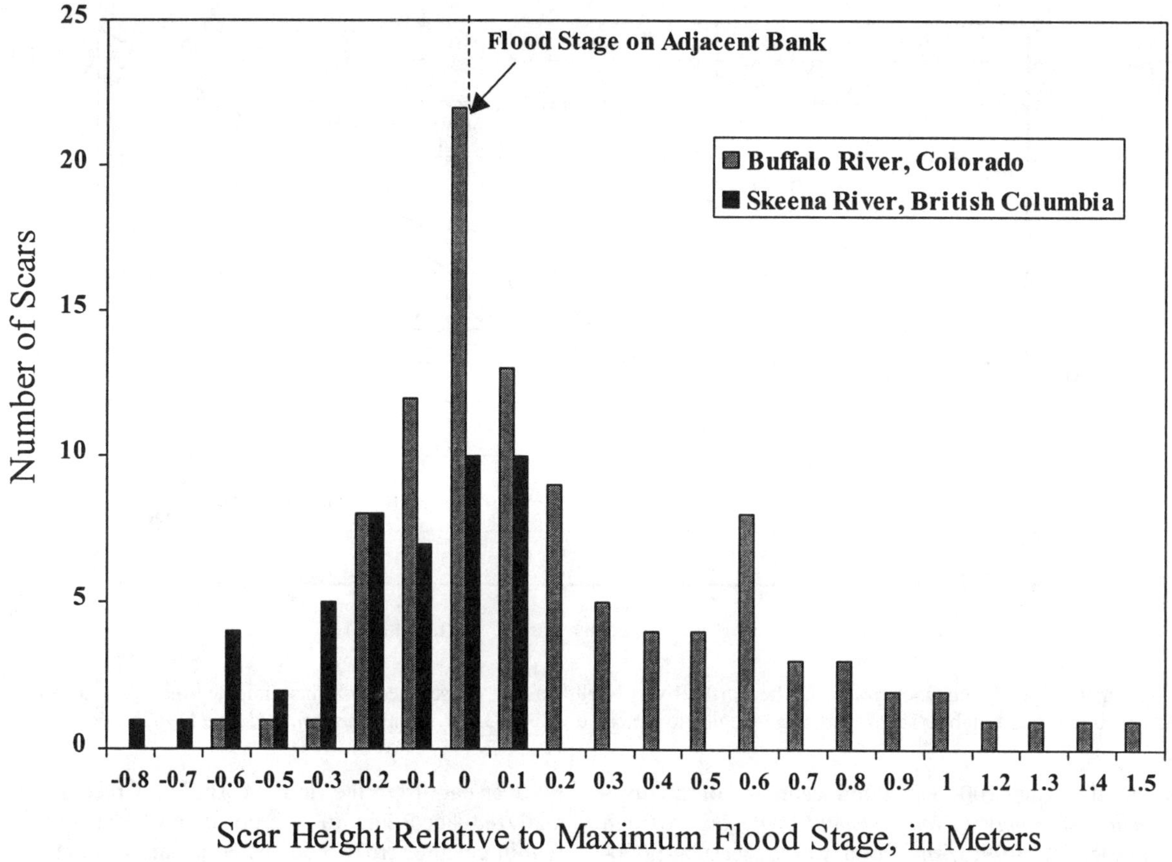

Figure 8. Maximum heights of scars on riparian trees relative to peak flood crests on the Skeena River, British Columbia (N=48 trees) [*Gottesfeld*, 1996] and on Buffalo Creek, Colorado (N=102 trees). Despite differences in gradient, flood magnitude, and the length of study reaches, the respective scar heights were reliable indicators of actual flood crests along both streams.

FLOOD-INDUCED ANATOMICAL RESPONSES

Although a scar technically is an anatomical response to flooding, callus growth is confined to small parts of trunks or branches. However, flooding sometimes alters the physiological processes that control radial growth, resulting in changes in the dimensions and proportions of tissues along much or the entire growth parabola of a ring; that is, along the circumference and height of the bole. Abnormal anatomical growth may be more difficult than a scar to interpret unambiguously, but nevertheless is easier to detect than small scars overgrown by new wood and bark.

Anomalous growth caused by flooding has been studied most intensely in ring-porous trees. Typically, ring-porous trees form one or several radial rows of large vessels (earlywood) during the expansion of leaves and elongation of twigs, and smaller vessels (latewood) thereafter for the remainder of the growth year. This striking difference in size between the two groups of vessels results in prominent ring boundaries. *Yanosky* [1983] found that ash trees (*Fraxinus pennsylvanica* Marsh.) growing along the Potomac River near Washington, D.C., sometimes formed rows of atypically large vessels ("flood rings") within the latewood (Figure 9a), and that anomalous vessels formed only during years when flooding occurred in late spring or summer. He observed that much or all of the leaf crop was stripped by flood currents, and that trees refoliated soon after flooding subsided. As a new crop of leaves developed, another set of spring-like vessels developed. The position of anomalous vessels relative to the total width of the ring could be used to estimate when the flood occurred, usually to within about two weeks. Floods that occurred during the interval of earlywood formation also resulted in anomalous growth, most commonly as jumbled or additional ranks of earlywood vessels. Yanosky (1983) found evidence for 12 floods that preceded the 1930 period of record, including three prior to 1895 (the inception of record keeping at the Point of Rocks gaging station).

Yanosky [1983] also noted that earlywood vessels were abnormally small and sparse in many trees subjected to catastrophic ice jamming in 1948 and 1968. Because vessel enlargement is controlled in part by hormones produced by expanding leaves, Yanosky concluded that ice damage to shoots and buds during dormancy resulted in a later deficiency of hormones during budbreak, and thus to a reduction in vessel diameters. Flood-ring vessels also are smaller than typical earlywood vessels, most likely because post-flooding refoliation also produces smaller amounts of growth hormones.

Growth responses involving the formation and position of atypical vessels depend on mechanical damage typically caused by flooding along high-gradient stream reaches. However, periodic flooding only of roots (that is, without inundation of branches or crowns) also may result in anomalous ring growth. For example, *Yanosky* [1984] studied the fibers (non-conducting support tissue) of ash trees subjected to episodic root flooding (1-3 days) along the Potomac River, and found that fibers often had thinner walls and larger lumens than those typically observed in latewood (Figure 9b). Rather, the dimensions of fibers were comparable to those produced during spring. Compared to typical latewood fibers, anomalous fibers appeared as light-colored bands ("white rings"), sometimes marking two or more separate episodes of root flooding within a single ring. *Yanosky* [1984] concluded that summer root flooding accelerated the rate of fiber production, resulting in the formation of spring-like fibers within the latewood. Similarly, *Yamamoto et al.* [1995] observed the formation of atypical latewood fibers following flooding of *Alnus japonica*.

More recently, *Astrade and Begin* [1997] studied changes in the vessel morphology of English oak (*Quercus robur* L.) related to flooding along the Saone River, France. They found that trees flooded during the early part of the 1983 growth year produced a typical first rank of vessels followed by an extensive zone of unusually small earlywood vessels. Apparently, the anomalous earlywood zones did not form in response to bud and leaf damage, but rather to physiological alterations from prolonged root flooding, and may be somewhat analogous to the "extended earlywood" variant of flood rings described by *Yanosky* [1983].

The effects of flooding upon ring growth of bur oak (*Quercus macrocarpa* (Michx.)) were studied by *St. George and Nielson* [2000] in the basin of the Red River, Manitoba, Canada. They constructed a 536-year oak chronology from living trees, buried logs, and architectural beams. Prolonged root inundation during May 1826 and 1852, the largest 19th century floods, resulted in the formation of rings with single ranks of diffuse, uniformly small earlywood vessels. Latewood vessels are scattered evenly throughout the remainder of rings, rather than in typical flame-like patterns, and the walls of fibers are atypically thin (Figure 10). As along the Saone River in France, it seems unlikely that trees in the Red River basin were damaged mechanically by floodwaters. St. George and Nielson also discussed the possibility of estimating the peak elevation of the 1826 flood, the century's largest, but most of their samples

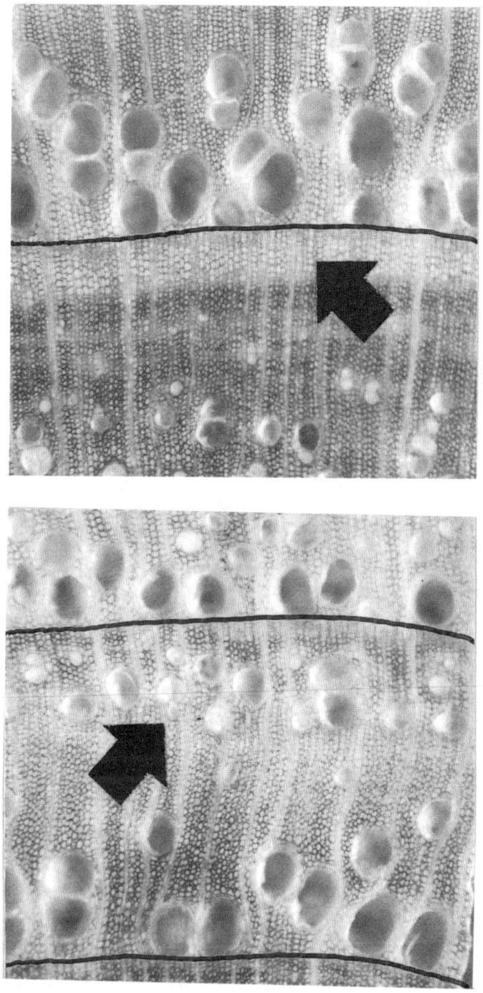

Figure 9. Ring anatomy of ash trees flooded during the growing season [*Yanosky*, 1983; 1984]. Direction of growth is from bottom to top, and ring boundaries are delineated by inked lines. Magnification X 60. (a) Large vessels ("flood ring") (arrow) within the outer part of this ring formed when the tree produced a new leaf crop following the stripping of leaves by a flood in late-May, 1942. Typically, large vessels form only within the first part (earlywood) of a ring and not within the remainder (latewood). (b) A zone of atypically large and thin-walled fibers ("white ring") (arrow) formed within the terminal latewood of this ring in response to late-summer root flooding (mid-August, 1942) that otherwise did not damage the leaf crop. Fibers produced before (below) the white ring are small and possess thick walls, thus imparting the dark hue typical of latewood. Photos by T.M. Yanosky.

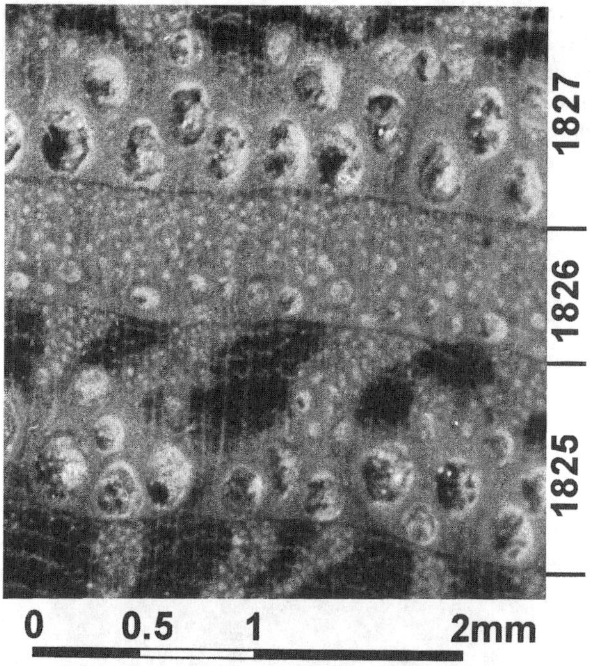

Figure 10. Ring anatomy of a bur oak (*Quercus macrocarpa* (Michx.)) inundated during the great flood of May 1826 flood on the Red River, Manitoba, Canada [*St. George and Nielson,* 2000]. Direction of growth is from bottom to top. Note the reduced width of the 1826 ring, the small, diffuse earlywood vessels, and the dearth of thick-walled fibers (the darkest tissues in adjoining rings). Photo by S. St. George.

were from architectural structures for which tree-growing locations could not be determined.

Estimates of maximum flood stage from ring-anatomical features generally are less precise than those from scars. A tree displaying anomalous growth related only to root flooding provides evidence that flows at least reached the base of the trunk; hence, a series of trees at successively greater elevations might be used to estimate the maximum stage. However, trees that produced anomalous vessels (flood rings) in response to flood damage must have been flooded to a depth sufficient to damage the canopy. Flood rings within the outer rings of tall trees, that is, in those unlikely to be bent over by flood flows, indeed suggest that anomalous growth developed in response to a large flood. *Yanosky* [1983] found that flood-ring vessels that formed after the great 1972 flood on the Potomac River had larger diameters than those that formed after smaller floods, presumably because the likelihood of crown damage is at least somewhat proportional to flood magnitude. If it is suspected that a flood temporarily tipped a small tree, thus resulting in an overestimation of stage, tree height in the year of the flood can be reconstructed by extracting cores at successive trunk heights. Even then, estimates of floodflow velocity near the tree also may be necessary to strengthen the contention that tipping did not occur.

ADDITIONAL CONSIDERATIONS AND CONCLUDING REMARKS

Botanical evidence used to recover hydrologic information is limited by the age and species of riparian trees. In the eastern United States, hardwoods (broad-leaved trees) commonly are plentiful along streams but do not attain the great ages typical of some western softwoods (coniferous trees). Unlike some softwoods, however, hardwoods commonly produce sprouts when tilted by floods, and ring-porous species such as the ashes and oaks sometimes produce anatomical responses to growing-season floods. Diffuse-porous hardwoods (the maples, gums, and willows, for example) typically form rings difficult to resolve, and have received little attention in paleofloods studies. Although scars and eccentric growth are the primary responses of conifers to flood damage, *Gottesfeld and Gottesfeld* [1990] noted the formation of traumatic resin canals in conifers flooded during late spring, but not in those flooded after the cessation of radial growth. Further studies are expected to demonstrate more fully the continuum of tree-growth responses to flooding, resulting in greater application and recovery of information.

The likelihood of locating botanical evidence of paleofloods increases if the researcher considers the seasonal distribution of floods and the probability of damage along various channel reaches. Floods that occur during predictable seasons are associated with predictable growth responses. For example, trunk scarring is predominant along streams that flood repeatedly during winter and early spring, whereas anatomical evidence commonly is preserved in trees likely flooded during the growing season. Nevertheless, scarring is easiest to detect and interpret unambiguously, and probably is the best evidence to seek initially. It is not uncommon to find individual trees scarred by numerous floods, sometimes manifesting other forms of evidence, as well, because some trees have an unusually high likelihood of sustaining damage on a frequent basis. For example, *Yanosky* [1983] found anatomical evidence of 18 floods in an ash tree growing in a frequently flooded, high-damage zone along the lower Potomac River.

Most paleoflood studies conducted to date analyzed from 20 to more than 200 trees, although the scar-based flood hydrograph by *Harrison and Reid* [1967] was constructed only from three trees, and *McCord*'s [1996] excellent flood extension of Kanab Creek, Utah, was based on a chronology of only 13 trees. Yet, regardless of the number of trees preserving evidence of paleofloods, the investigator invariably must decide the amount of evidence needed to document each flood with acceptable confidence. *Gottesfeld and Gottesfeld* [1990] considered evidence "significant" only when it is preserved in at least two different trees, although their reconstruction included evidence found only in single trees; *Hupp* [1987] based his reconstruction of Passage Creek, Virginia, USA, on three trees per flood. Several flood reconstructions (i.e. *Harrison and Reid*, 1967; *McCord*, 1990; 1996), however, routinely incorporated evidence only from single trees. The confidence of the researcher in such situations depends on the likelihood that the evidence indeed was caused by flooding. Flood scars typically differ from those caused by other agents [*McCord*, 1990; 1996], but trunks tilted by windthrow develop eccentric radial growth and vertical sprouts indistinguishable from flood-tilted trees, and defoliation by insects sometimes induces anatomical responses similar to flood rings. It seems reasonable that scarring along an upstream-facing trunk is reliable evidence of flood damage, particularly if additional scars or other flood-related features also are found in the same tree or in others nearby. That is, the prevalence of more recent evidence strengthens the contention that a single piece of older evidence also resulted from flooding. Similarly, anatomical abnormalities confined only to the rings of riparian trees likely resulted from defoliation by floods rather than by insects.

An additional consideration for the use of botanical evidence is that floods destroy evidence of earlier floods [*Sigafoos*, 1964], thus constantly reducing the number of trees still likely to preserve evidence of those floods. High-magnitude floods, however, also are more likely than small floods to damage large numbers of trees, increasing the likelihood that botanical evidence initially will be preserved [*Yanosky*, 1982b; *Hupp*, 1988]. A tree growing near the crest altitude of a great flood preserves evidence of the flood until obliterated by another great flood, or until the tree dies. Thus, chance mortality and the maximum-likely life span of trees probably control the persistence of evidence along terraces and the uppermost boundaries of flood plains. Evidence of large floods also may be preserved for many years along channel reaches where damage to trees is unlikely during small floods [*Yanosky*, 1982b]. Consequently, sheltered reaches sometimes support older trees, and thus evidence of older floods, than do reaches exposed to the full force of flood currents.

Little is known concerning the length of time that paleoflood evidence is likely to persist. *Sigafoos* [1964] found evidence of only two floods (1861 and 1889) on the Potomac River prior to the period of record at Point of Rocks; the earliest found by *Yanosky* [1983] was that for 1889 on the basis of flood-ring formation. The intent of both investigators, however, was to develop general methods to document paleofloods rather than to extend the flood history of the Potomac River. *Hupp* [1987] extended to 1720 the flood record of Passage Creek, but the earliest date derived from trunk scarring and/or sprout age was 1852. Thus, compared to the ~500-year scar record by *McCord* [1996] for a western stream, botanical evidence of paleofloods in the eastern United States is somewhat scanty, and the temporal limits of flood documentation await further study.

Geobotanical research during the previous three decades has built upon pioneering studies in the 1960s by *Sigafoos* [1964] and *Harrison and Reid* [1967]. Investigations have defined further the relation between scars and paleoflood crest stages, noted that flooding sometimes alters ring-anatomical structure, and applied digital imagery to the analysis and monitoring of hydrologic systems. Thus, dendrochronology may provide valuable proxy data where stream-gaging stations are widely scattered and flow records are short or unavailable. In addition to flood preparedness, these studies also may be useful to evaluate longer-term processes such as anthropogenic climate change and the effects of streamflow regulation upon riverine ecosystems. Model validations under a wider range of hydrologic and botanical conditions will continue to improve the acceptance of dendrochronologic studies within the water-resources community [*Jarrett and Tomlinson*, 2000].

Acknowledgments. The senior author thanks his mentors, the late Dr. Robert S. Sigafoos, and Dr. Richard L. Phipps, USGS. Dr. Phipps was the first to relate the formation of anomalous latewood vessels to flood damage, and graciously recommended further study to the senior author. Helpful review comments were provided by Drs. G. Richard Marzolf, Harvey E. Jobson, and Thomas F. Kraemer of the USGS; and by Drs. Franco Biondi, University of Nevada, and Connie Woodhouse, University of Colorado. Thanks are extended also to Kyle House, University of Nevada, senior editor of this Monograph.

REFERENCES

Astrade, L., and Y. Begin, Tree-ring response of Populus tremula L. and Quercus robur L. to recent spring floods of the Saone River, France, Ecoscience 4, 232-239, 1997.

Auble, G.T., J.M. Friedman, and M.L. Scott, Relating riparian vegetation to present and future streamflows, *Ecol. Applic., 4*, 544-554, 1994.

Baker, V.R., Stream-channel response to floods with examples from central Texas, *Geol. Soc. Amer. Bull., 88*, 1057-1071, 1977.

Baker, V.R., Geomorphological understanding of floods, *Geomorphology, 10*, 139-156, 1994.

Bedinger, M.S., Forest species as indicators of flooding in the lower White River valley, Arkansas, U.S. *Geol. Surv. Prof. Pap., 750-C*, C248-C253, 1971.

Colorado Water Conservation Board, Emergency response, flood hazard mitigation, and flood hazard awareness for residents of Buffalo Creek, Colorado, Dept. of Natural Resources, Denver, Colorado, 18 pp., 1997.

Costa, J.E., Colorado Big Thompson flood, geological evidence of a rare hydrologic event, *Geology, 6,* 617-620, 1978.

Costa, J.E., A history of paleofloods hydrology in the United States, 1800-1970, in *The History of Hydrology*, edited by E.R. Landa and S. Ince, pp. 49-53, American Geophysical Union, Washington, D.C., 1987.

Everitt, B.L., Use of the cottonwood in an investigation of the recent history of a flood plain, *Am. J. Sci., 266,* 417-439, 1968.

Everitt, B.L., Hydrologic factors in regeneration of Freemont cottonwood along the Fremont River, Utah, in *Natural and Anthropogenic Influences in Fluvial Geomorphology,* edited by J.E. Costa, A.J. Miller, K.W. Potter, and P.R. Wilcock, Am. Geophy. Union, Geophysical Monogr. 89, 197-208, 1995.

Friedman, J.M., W.R. Osterkamp, and W.M. Lewis, Jr., Channel narrowing and vegetation development following a Great Plains flood, *Ecology, 77,* 2167-2181, 1996a.

Friedman, J.M., W.R. Osterkamp, and W.M. Lewis, Jr., The role of vegetation and bed-level fluctuations in the process of channel narrowing, *Geomorphology, 14,* 341-351, 1996b.

Fritts, H.C., *Tree Rings and Climate*, 567 pp., Academic Press, London, 1976.

Glavac, V., C. Grillenberger, W. Hakes, and H. Ziezold, On the nature of vegetation boundaries, undisturbed flood plain forest communities as an example: a contribution to the continuum /discontinuum controversy, *Vegetation, 101*, 123-144, 1992.

Gottsfeld, A.S., British Columbia flood scars: maximum flood-stage indicators, *Geomorphology, 14,* 319-325, 1996.

Gottesfeld, A.S., and L.M.J. Gottesfeld, Flood-plain dynamics of a wandering river, dendrochronology of the Morice River, British Columbia, Canada, *Geomorphology, 3*, 159-179, 1990.

Hack, J.T., and J.C. Goodlett, Geomorphology and forest ecology of a mountain region in the central Appalachians, *U.S. Geol. Surv. Prof. Pap. 347*, 66 pp., 1960.

Harrison, S.S., and J.R. Reid, A flood-frequency graph based on tree-scar data, *Proc. N. D. Acad. Sci., 21,* 23-33, 1967.

Hedman, E.R., and W.R. Osterkamp, Streamflow characteristics related to channel geometry of streams in western United States, *U.S. Geol. Surv. Wat. Supply Pap., 2193,* 17 pp., 1982.

Helley, E.J., and V. C. LaMarche, December 1964, a 400-year flood in northern California, in *U.S. Geol. Surv. Prof. Pap., 600D*, D34-D37, 1968.

Henz, J.F., The Buffalo Creek flash flood of July 12, 1966, a reconstruction of rainfall and meteorology, Henz Meterological Services, Denver, Colorado, 28 pp., 1 appendix, 1998.

Hupp, C.R., Dendrogeomorphic evidence of debris flow frequency and magnitude at Mount Shasta, California, *Environ. Geol. Water Sci., 6,* 121-128, 1984.

Hupp, C.R., Botanical evidence of floods and paleoflood history, in *Regional Flood Frequency Analysis*, edited by V.P. Singh, pp. 355-369, D. Reidel, The Netherlands, 1987.

Hupp, C.R., Plant ecological aspects of flood geomorphology and paleofloods history, in *Flood Geomorphology*, edited by V.R. Baker, R.C. Kochel, and P.C. Patton, pp. 335-356, Wiley, New York, 1988.

Hupp, C.R., Riparian vegetation recovery patterns following stream channelization: a geomorphic perspective, *Ecology, 73*, 1209-1226, 1992.

Hupp, C.R., and W.R. Osterkamp, Bottomland vegetation distribution along Passage Creek, Virginia, in relation to fluvial landform, *Ecology, 66*, 670-681, 1985.

Hupp, C.R., and W.R. Osterkamp, Riparian vegetation and fluvial geomorphic processes, *Geomorphology, 14,* 277-295, 1996.

Jarrett, R.D., Paleohydrology used to define the spatial occurrence of floods, *Geomorphology, 3*, 81-95, 1990.

Jarrett, R.D., Paleohydrology and its value in analyzing floods and droughts, in *U.S. Geol. Surv. Water-Supply Pap., 2375*, 105-116, 1991.

Jarrett, J.D., and T.W. Browning, Geomorphic estimates of rainfall, floods, and sediment runoff: applied to the 1996 wildfire, Buffalo Creek, Colorado, *Geol. Soc. Am. Abstr. Programs, 31*, 313, 1999.

Jarrett, R.D., and E.M. Tomlinson, Regional interdisciplinary paleofloods approach to assess extreme flood potential, *Water Res. Rsch., 36*, 2957-2985, 2000.

Johnson, W.C., Woodland expansion in the Platte River, Nebraska: patterns and causes, *Ecological Monographs, 64*, 59-84, 1994.

McCord, V.A., Augmenting flood frequency estimates using flood-scarred trees, PhD thesis, U. of Arizona, Tucson, Arizona, 1990.

McCord, V.A., Fluvial process dendrogeomorphology: Reconstructions of flood events from the southwestern United States using flood-scarred trees, in *Tree Rings, Environment and Humanity*, edited by J.S. Dean, D.M. Meko, and T.W. Swetnam, pp. 689-699, *Radiocarbon*, U. of Arizona, Tucson, Arizona, 1996.

Miller, J.F., R.H. Frederick, and R.J. Tracy, Precipitation frequency atlas of the western United States, v. III—Colorado, National Oceanic and Atmospheric Administration, Silver Springs, Maryland, 67 pp., 1973.

Nilsson, C., Distribution of stream-edge vegetation along a gradient of current velocity, *J. of Ecology, 75*, 513-522, 1987.

Osterkamp, W.R., and J.E. Costa, Changes accompanying an extraordinary flood on a sandbed stream, in *Catastrophic Flooding*, edited by L. Mayer and D. Nash, Allen and Unwin, Boston, 201-224, 1987.

Osterkamp, W.R., and C.R. Hupp, Geomorphic and vegetative characteristics along three northern Virginia streams, *Geol. Soc. Amer. Bull., 95*, 1093-1011, 1984.

Phipps, R.L., Collecting, preparing, crossdating, and measuring tree increment cores, *U.S. Geol. Surv. Wat. Resour. Inv. Rpt., 85-4148*, 48 pp., 1985.

Scott, M.L., J.M. Friedman, and G.T. Auble, Fluvial

process and the establishment of bottomland trees, *Geomorphology, 14*, 327-339, 1996.

Shroder, J.F., Jr., and M.P. Bishop, Geobotanical assessment in the Great Plains, Rocky Mountains and Himalaya, *Geomorphology, 13*, 101-119, 1995.

Sigafoos, R.S., Botanical evidence of floods and flood-plain deposition, *U.S. Geol. Surv. Prof. Pap., 485A*, 35 pp., 1964.

Stanford, J.A., and K.V. Ward, Management of aquatic resources in large catchments: recognizing interactions between ecosystem connectivity and environmental disturbance, in *Watershed Management*, edited by R.J. Naiman, Springer Verlag, New York, 91-124, 1992.

St. George, S., and E. Nielsen, Signatures of high-magnitude 19th century floods in *Quercus macrocarpa* (Michx.) along the Red River, Manitoba, Canada, *Geology, 28*, 899-902, 2000.

Waythomas, C.F., and R.D. Jarrett, Flood geomorphology of Arthurs Rock Gulch, Colorado, paleoflood history, *Geomorphology, 11*, 15-40, 1994.

Webb, R.H., S.S. Smith, and V.A.S. McCord, Historic channel change of Kanab Creek, Southern Utah and Northern Arizona, Grand Canyon Natural History Association, *Monograph No. 9,* 91 pp., 1991.

Williams, G.P., and M.G. Wolman, Downstream effects of dams on alluvial rivers, *U.S. Geol. Surv. Prof. Pap., 1286*, 63 pp., 1984.

Wolman, M.G., and R. Gerson, Relative scales of time and effectiveness of climate in watershed geomorphology, *Earth Surf. Process.*, 3, 189-208, 1978.

Wolman, M.G., and J.P. Miller, Magnitude and frequency of forces in geomorphic processes, *J. of Geology, 68*, 54-74, 1960.

Yamamoto, F., T. Sakata, and K. Terazawa, Growth, morphology, stem anatomy, and ethylene production in flooded *Alnus japonica* seedlings, *Int. Assoc. Wood Anat. J., 16*, 47-59, 1995.

Yanosky, T.M., Hydrologic inferences from ring widths of flood-damaged trees, Potomac River, Maryland, *Envir. Geol., 4,* 43-52, 1982a.

Yanosky, T.M., Effects of flooding upon woody vegetation along parts of the Potomac River flood plain, *U.S. Geol. Surv. Prof. Pap., 1206*, 21 pp., 1982b.

Yanosky, T.M., Evidence of floods on the Potomac River from anatomical abnormalities in the wood of flood-plain trees, *U.S. Geol. Surv. Prof. Pap., 1296*, 42 pp., 1983.

Yanosky, T.M., Documentation of high summer flows on the Potomac River from the wood anatomy of ash trees, *Water Res. Bull., 20*, 241-250, 1984.

Yanosky, T.M., and M.K. Cleaveland, Tree rings in hydrological studies, in *Encyclopedia of Hydrology and Water Resources*, edited by R.W. Herschy and R.W. Fairbridge, pp. 650-655, Kluwer, 1998.

Yanosky, T.M., Tree ring analysis in hydrometry, in *Hydrometry: Principles and Practices,* 2d ed., edited by R.W. Herschy, pp. 265-289, Wiley, 1999.

Robert D. Jarrett, U.S.G.S, Box 25046, MS 412, Denver, Colorado, 80225 (rjarrett@usgs.gov)

Thomas M. Yanosky, U.S. Geological Survey, MS 430, Reston, VA 20192 USA (tyanosky@usgs.gov)

Reliability of Paleostage Indicators for Paleoflood Studies

Robert D. Jarrett

U.S. Geological Survey, Denver, Colorado

John F. England, Jr.

U.S. Bureau of Reclamation, Denver, Colorado

This paper documents the first systematic assessment for one aspect of paleoflood analysis: the relation between paleostage indicators (PSIs) and the peak stage of floods responsible for their emplacement. A main source of uncertainty in paleoflood reconstructions is maximum flood stage inferred from PSIs. Typically, the elevation of the top of the PSI is used as the minimum elevation of the flood that deposited the sediments, but little evidence supports this assumption. Data from recent large floods (median recurrence interval of 75 years), primarily in the western and west-central United States, were used for a comprehensive evaluation of the relation between flood-transported sediments (fresh PSIs) deposited as flood bars and/or slackwater deposits and flood high-water marks (HWMs). Surveys of flood-deposited sediments, HWMs, and channel geometry were made for 192 sites, primarily streams with gradients larger than 0.002 m/m and depths less than about 4.5 m. No statistically significant relation was identified with channel gradient, particle size, discharge, recurrence interval, expansion or contraction ratios, width, or depth variables. Analysis of the data indicates that the elevation at the top of the flood sediments (PSIs) is on average 15 mm higher than the HWMs; results slightly vary with type of deposit. The elevation at the top of flood-deposited sediments (new PSIs), preferably deposits nearest to channel margins, provides a reliable and accurate indication of the maximum height of the flood. Forty-six peak discharges that we estimated with the critical-depth and slope-conveyance methods also were compared with peak discharges independently made at streamflow-gaging stations, primarily determined using current-meter measurement; forty-two of the 46 estimates were within ±25 percent of the gage peak discharge measurements.

INTRODUCTION

"In the absence of sufficient rainfall data, it is fortunate that geology furnishes, in many instances, a clue to the probable heights of flood-waters. The clue referred to is afforded by flood-plains and terraces, which often give evidence of heights of water." *Fuller* [1917, p. 1270]

Paleoflood hydrology is the study of recent, past, or ancient floods [*Costa*, 1987; *Baker*, 1987; *Baker et al.*, 1988; *Jarrett*, 1991; *House et al.*, this volume]. Although most studies involve prehistoric floods, the methodology is applicable to historic or modern floods [*Baker*, 1987] pro-

Ancient Floods, Modern Hazards: Principles
and Applications of Paleoflood Hydrology
Water Science and Application Volume 5, pages 91-109

Published in 2002 by the American Geophysical Union

viding information about the number, magnitude, and age of floods not documented by standard hydrologic methods [*Jarrett*, 1991]. Paleoflood studies also provide important information that can be used for assessing the effects of climatic change on flooding and droughts [*Knox*, 1988, 2000; *Jarrett*, 1991; *Ely et al.*, 1993; *Ely*, 1997; *Jarrett and Tomlinson*, 2000]. Paleoflood data are particularly useful in providing evidence of the largest floods that have occurred in a river basin [*Enzel et al.*, 1993; *Parrett and Jarrett*, 2000; *Jarrett and Tomlinson*, 2000; *Jarrett*, 1990, 2000]. This paper documents an empirical relation for one aspect of paleoflood analysis: the relation between paleostage indicators (PSIs) and high-water marks (HWMs) of the peak stage (height) of floods.

Background

Accurate flood-frequency relations are necessary for floodplain management, hydrologic-risk assessments for high-hazard facilities such as dams, and related environmental investigations. In many river basins, streamflow-gaging station records generally do not contain a sufficient sample of infrequent floods to reliably estimate the frequency of extreme floods [*Costa*, 1978; *National Research Council (NRC)*, 1999]. Yet extreme or rare floods, with probabilities in the range of about 10^{-3} to 10^{-7} chance of occurrence per year, are of increasing interest to the hydrologic and engineering communities for purposes of planning, design, and analysis of structures such as dams [*NRC*, 1988].

Paleoflood investigations have been incorporated into a variety of water-resources investigations [*e.g.*, *Costa*, 1978; *Jarrett and Costa*, 1988; *Levish et al.*, 1994, 1997; *House et al.*, 1995; *Wohl*, 1995; *Pruess et al.*, 1998; *England, 1998*; *Jarrett and Tomlinson*, 2000; *Jarrett*, 2001], particularly in using paleoflood data for providing robust frequency estimates of extreme floods. Extensive research has been made during the past twenty years to incorporate paleoflood and historical data into flood-frequency analyses [*Stedinger and Cohn*, 1986; *Stedinger and Baker*, 1987; *Guo and Cunnane*, 1991; *Cohn et al.*, 1997; *O'Connell et al.*, 1998; *England*, 1998; *NRC*, 1999]. There also has been a trend to incorporate risk-based assessments for design-flood hydrology [*Levish et al.,* 1994, 1997; *NRC*, 1988, 1999; *O'Connell et al.*, 1998; *USBR*, 1999, *Levish*, this volume].

A primary objective of a flood study should be to generate as much information as practical about flooding at a site using multiple approaches and different sources of data [*NRC*, 1988; *Jarrett and Tomlinson*, 2000; *Parrett and Jarrett*, 2000]. The *NRC* [1988] emphasizes that accurately estimating the risk of extreme flooding is important. Risk-based hydrologic assessment of existing dams, which includes paleoflood data, is currently being developed in the dam-safety community, particularly within the Bureau of Reclamation [*Jarrett and Costa*, 1988; *Waythomas and Jarrett*, 1994; *Levish et al.*, 1994, 1997; *Levish and Ostenaa*, 1996; *Ostenaa et al.*, 1997; *O'Connell et al.*, 1998] and elsewhere [*Salmon et al.*, 1997; *Jarrett and Tomlinson*, 2000; *Jarrett*, 2000; *Parrett and Jarrett*, 2000]. In the mid-1990s, the Bureau of Reclamation began a program to use a risk-based assessment, which incorporates paleoflood investigations to provide estimates of the magnitude and frequency of extreme floods, to assist with dam safety decision-making [*Levish et al.*, 1994, 1997; *USBR*, 1999; *Levish*, this volume].

Parrett and Jarrett [2000] used a novel interdisciplinary approach to help assess dam safety for Tabor Dam near St. Ignatius, Montana. To help reduce uncertainty about plausible peak discharges and concerns regarding the ability of the Tabor Dam spillway to safely convey these discharges, the flood hydrology was evaluated on the basis of three hydrologic and geologic methods. These included: (1) regional paleoflood studies; (2) determining an envelope curve relating peak discharge and drainage area on the basis of regional flood and paleoflood data, and; (3) rainfall-runoff modeling in conjunction with regional precipitation information to determine plausible peak discharges for rainfall recurrence intervals up to 5,000 years. These three methods resulted in consistent plausible peak discharges that are substantially less than those predicted by the probable maximum flood technique. The U.S. Army Corps of Engineers also is implementing a risk assessment method to evaluate potential safety problems for its more than 550 dams to aid decision-makers prioritize investment decisions [*Foster*, 1999]. In 1999, the American Society of Civil Engineers began a task committee on paleoflood hydrology as it relates to dam safety and risk-based assessments.

There has been a misunderstanding that paleoflood techniques are only used for estimating very old and extreme floods. Paleoflood studies to obtain data for contemporary floods (recent to about 500 years) also are needed [*Hupp*, 1988; *Gottesfeld*, 1996; *NRC*, 1999; *Jarrett and Tomlinson*, 2000]. These studies help improve flood-frequency relations in the range of about 0.2 to about 0.002 chance of occurrence per year, assess the reliability of historic flood estimates [*Jarrett*, 1986, 1994; *House and Pearthree*, 1995; *Levish and Ostenaa*, 1996] and estimate convective rainfall amounts [*Jarrett*, 1990, 2001].

Jarrett and Tomlinson [2000], in one of many examples, demonstrates the value of incorporating paleoflood data in flood-frequency relations for the Yampa River at Steamboat Springs, Colorado, where data collection began in 1904. They compared the Federal Emergency Management Agency (FEMA) flood-frequency relation and gaged flood-

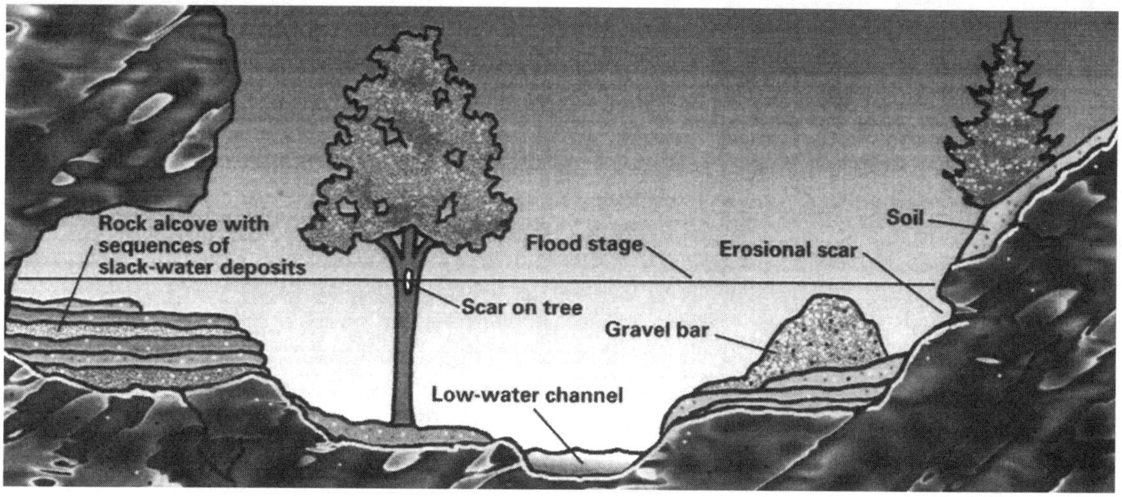

Plate 1. Diagrammatic section across a stream channel showing a peak flood stage and various paleoflood features used as paleostage indicators (PSIs) such as slackwater deposits (SWDs), cobble and gravel flood bars (FBs), tree scars, and erosional scars.

frequency; the two relations have good agreement to about the 50-year flood (CD Plate 1). For larger recurrence intervals, the FEMA relation increases sharply and does not fall within the 95-percent confidence limits for the flood-frequency relation based on analysis of annual peak-streamflow and paleoflood data using the Expected Moments Algorithm (EMA) [*Cohn et al.*, 1997]. In addition, the FEMA 500-year flood is almost double the largest paleoflood estimate of 311 m^3/s in about 9,000 years. FEMA is revising the flood hydrology for Steamboat Springs. In terms of extreme floods, the probable maximum flood for Stagecoach Reservoir located on the Yampa River upstream from Steamboat Springs [*Jarrett and Tomlinson*, 2000] also far exceeds a 10,000-year recurrence interval (CD Plate 1). Thus, efforts to identify if there may be uncertainty and bias in flood-frequency estimates are needed. Therefore, conducting research to improve paleoflood techniques and conducting studies to incorporate paleoflood information in water resources investigations are necessary and beneficial.

Paleostage Indicators

Floods leave distinctive sedimentary deposits [*Costa*, 1983; *Knox*, 1987, 1988; *Jarrett*, 1991; *Ely et al.*, 1993; *O'Connor et al.*, 1994; *England*, 1998; *Jarrett and Tomlinson*, 2000; *Webb et al.*, this volume; *House et al.*, this volume], botanical evidence [*Hupp*, 1988; *Gottesfeld*, 1996; *Yanosky and Jarrett*, this volume], erosional features on channel margins [*Baker*, 1987; *O'Connor*, 1993], and landforms [*Costa*, 1978; *Jarrett and Costa*, 1988; *Jarrett*, 1991, 2000; *Hupp*, 1988; *Levish et al.*, 1994, 1997; *Levish*, this volume] in streams and on floodplains. These features, termed paleostage indicators (PSIs), can be used to infer the maximum stage of past floods (Plate 1). Here, we document the appropriateness of the assumption that PSIs represent the maximum flood stage, by using recent flood HWMs against related stratigraphic and geomorphic evidence that has traditionally been used for estimating magnitudes of paleofloods.

In paleoflood studies, the most commonly used PSIs are slackwater deposits (SWDs) of silt and sand shown in Plate 2 [*Kochel and Baker*, 1988; *Baker*, 1987; *Webb et al.*, 1988; *Enzel et al.*, 1993; *Ely et al.*, 1993; *O'Connor et al.*, 1994; *Webb et al.*, this volume; *House et al.*, this volume]. SWDs are rapidly deposited from suspension in sediment-laden waters where flood flow velocities are minimal during the time that inundation occurs [*Kochel and Baker*, 1988]. Other types of PSIs used in paleoflood studies include bouldery flood bars (FBs) and coarse sandy deposits (also SWDs) shown in Plate 3 [*Jarrett*, 1990, 2000; *O'Connor*, 1993; *Waythomas and Jarrett*, 1994; *Wohl*, 1995; *Jarrett and Tomlinson*, 2000]. In addition, erosional features (Plate 1) such as truncated alluvial fans [*Costa*, 1978; *Jarrett and Costa*, 1988], eroded basalt benches [CD Plate 2], and modification of geomorphic surfaces such as terraces and floodplains [*O'Connor*, 1993; *Levish et al.*,

Plate 2. A good example channel for studying slackwater deposits (SWDs)—the Escalante River, Utah. The person to the left is standing on a typical sequence of SWDs that were deposited where the flow velocity decreased in the canyon of the Escalante River. (Source: Robert H. Webb, U.S. Geological Survey.) For color version see CD.

1994, 1997; *Jarrett*, 2000; *Levish*, this volume] also are used as PSIs.

Flood-deposited sediments composed of sand, gravel, and boulders can be preserved for many thousands of years [*Knox*, 1988, 1993, 2000; *Ely et al.*, 1993; *O'Connor*, 1993; *O'Connor et al.*, 1994; *Waythomas and Jarrett*, 1994; *Jarrett and Tomlinson*, 2000; *Parrett and Jarrett*, 2000; *Webb et al.*, this volume; *House et al.*, this volume]. Sites where flood deposits commonly are preserved include: (1) locations of rapid energy dissipation, where sediments would be deposited, such as tributary junctions, areas of decreased channel gradient, abrupt channel expansions, or areas of increased flow depth; (2) locations along the sides of valleys in wide, expanding areas where fine-grained sediments or slackwater deposits would likely be deposited; and (3) backwater areas upstream from channel-width constrictions. Optimal paleoflood study results are in bedrock channels that satisfy basic hydraulic criteria for discharge estimation [*Barnes and Davidian*, 1978; *Kochel and Baker*, 1988; *O'Connor*, 1993; *Waythomas and Jarrett*, 1994; *Webb and Jarrett*, this volume], and where alluvial channels can be shown to be stable for long time spans [*Jarrett*, 2000].

Research Objective

Although operational (applied) studies of PSIs for paleoflood reconstruction have been conducted for several decades, one critical issue for the reliability of data derived from paleoflood investigations has been representativeness of the height of PSIs as proxies for peak-flood stage. Generally, the elevation at the top of the flood-deposited sediments is assumed to represent a minimum estimate of the maximum paleoflood stage responsible for emplacing the sediments [*Baker*, 1987; *Kochel and Baker*, 1988; *Knox*, 1988; *O'Connor et al.*, 1994; *Webb and Jarrett*, this volume]. *Ely and Baker* [1985] documented that the heights of SWDs essentially were equal to HWMs for two floods in the Verde River, Arizona. Some investigators have calculated estimates of water above the top of flood-deposited sediment using flow-competence relations [*Costa*, 1983; *Knox*, 1987, 1988; *Levish et al.*, 1994; *Waythomas and Jarrett*, 1994], but these relations are not always accurate [*Costa*, 1983; *Komar*, 1987; *Wilcock*, 1992]. Documentation of fresh flood-deposited sediments on the Cheat River, West Virginia [*Springer and Kite*, 1997; *Kite et al.*, this volume], the Verde River, Arizona [*House et al.*, this volume], and Paria River, Arizona [*Webb et al.*, this volume] lend support to the assumption that SWDs are below HWMs. Based on the limited prior PSI-HWM documentation, the relation between the top of sediment deposits and peak stage is unclear.

In this paper, we conduct a comprehensive analysis for recent floods in streams located primarily in the western and west-central United States to provide quantitative information for the relation between fresh deposits of sediments (new PSIs) and flood height (HWMs). This can be viewed as a simple, empirical relation used to calibrate and verify river hydraulic models for extreme floods. *Oreskes et al.* [1994] suggest that, at present, verification, validation, and confirmation of numerical models of natural systems is

Plate 3. Downstream view toward inlet to Horsetooth Reservoir near Fort Collins, Colorado, of bouldery flood deposit, which resulted from a flood at least 5,000 years ago [Waythomas and Jarrett, 1994]. Backpack for scale is located near the upstream end of the deposit. For color version see CD.

impossible. Further, *Oreskes et al.* [1994] posit that models can be confirmed by the agreement between observation and prediction, but confirmation is partial and often can only be evaluated in relative terms, and their predictive value is always open to question. One approach to assess the validity of assumptions underlying paleoflood studies and models is to document and determine the relation between the height of PSIs and HWMs of recent large floods [*Ely and Baker*, 1985; *Kochel and Ritter*, 1987; *Kochel and Baker*, 1988; *House et al.*, 1995; *Springer and Kite*, 1997; *House et al.*, this volume; *Webb et al.*, this volume; *Kite et al.*, this volume; *Yanosky and Jarrett*, this volume]. Our approach is to collect and analyze data from recent floods and directly determine the relative effectiveness of the operational assumption. We used a "flood-chasing" approach to visit as many extreme flood sites as possible after a flood. Our hypothesis was that a relation of the difference between the height of PSI and HWM values would relate to hydraulic and sedimentary conditions in the channel reaches, particularly with respect to water-surface slope and particle size of the deposit. For example, one hypothesized relation was that as water slope increases, the PSI height minus HWM height (or PSI-HWM) value would decrease and approach a value of zero.

Plate 5. Downstream view of SWDs (type 1 deposit) from the July 1999 flood in Leopard Creek near Dallas Divide, Colorado. Survey rod is 2 m long. For color version see CD.

METHODS

The general strategy of a paleoflood investigation is to visit the most likely places where evidence of substantial flooding might be preserved. Our focus in this study is on flood depositional environments where a paleoflood reconstruction would normally be made. There are many factors that contribute to flood deposition including sediment sizes available for transport, hydraulic characteristic of the flow, and the geometry and roughness of local depositional environments [*Costa*, 1983; *Baker*, 1987; *Carling*, 1987; *O'Connor*, 1993]. We attempt to quantify as many of these characteristics as possible, but the most critical requirement was obtaining the actual peak flood HWMs at the location of associated flood-deposited sediments. HWMs are physical evidence such as bent grass, mud, silt, and seed lines, and flotsam, particularly fine organic debris (CD Plate 3), that documents the peak stage of a recent flood, but usually are not preserved for more than a few weeks to months in humid environments [*Williams and Costa*, 1988]. In semi-arid and arid climates, flood HWMs can be preserved for decades to, in some cases, over one hundred years [*Levish and Ostenaa*, 1996]. For example, HWMs of fine woody debris, pine needles, and bent grass for the 1964 flood along the Teton River near Farmington, Montana [*Boner and Stermitz*, 1967], the 1976 Big Thompson River flood, Colorado [*McCain et al.*, 1979], and for the 1982 Lawn Lake flood Colorado [*Jarrett and Costa*, 1986] still are clearly visible in numerous locations.

Plate 4. Upstream view of right overbank flood sediments deposited by the July 31, 1976, flash flood in the North Fork Big Thompson River near Drake, Colorado. For color version see CD.

After the flood, extensive post-flood surveys were made at a site. Usually, a site with fresh flood sediments and associated HWMs for a reach of river could be located in less than fifteen minutes. Our objective was to visually locate the highest deposits in a reach rather than all sediment deposits in a reach because these are the most relevant deposits for paleoflood reconstruction. At each site, we surveyed channel geometry for up to five cross sections, multiple HWMs to determine water-surface slope, deposit morphology, maximum bed-material size available for transport, and maximum particle size in the flood deposit. Standard post-flood surveying techniques and geomorphic measure-

Table 1. Field Measurements and Estimates Made at 192 Sites for Developing PSI-HWM Relations

Field Measurement or Estimate	Average	Minimum	Maximum
High-Water Mark (HWM) (m)	3.02	0.06	31.0
Paleostage Indicator (PSI) (m)	3.04	0.06	32.6
Maximum Flow Depth (m)	1.9	0.1	9.1
Water Surface Slope (m/m)	0.03	0.0001	0.35
Deposit Maximum Particle Size (mm)	250	0.003	4,570
Streambed Maximum Particle Size (mm)	1060	13	4,570
Channel Width (m)	56	1.2	792
Upstream or Downstream Channel Width (m)	42	0.6	792
Peak Discharge (m^3/s)	205	0.1	1930
Approximate Recurrence Interval (years)	75 (median)	2	10,000
Width-Depth Ratio (m/m)	27	3	235
Expansion-Contraction Ratio (m/m)	1.6	0.4	5.3

ments [e.g., *Jarrett*, 1985; *Williams and Costa*, 1988; *Webb and Jarrett*, this volume] were used.

We verified that observed flood-deposited sediments resulted from the recent flood. In most cases, there was little difficulty because fresh sediments were deposited over live brush or grass (Plates 4 and 5 and CD Plate 4). We restricted our analysis to flood-deposited sediments that were at least 100 mm thick and covered at least a 3 m^2 level surface because only accumulations of this magnitude are likely to be preserved in the stratigraphic record.

If the flood was not directly observed, deposit morphology and sedimentologic characteristics [*Williams and Costa*, 1988; *Waythomas and Jarrett*, 1994] were used to differentiate waterflow deposits from those that may have been emplaced by debris flows and hyperconcentrated flows, particularly in mountain basins. It is critical to differentiate between deposits formed by flooding and deposits formed by debris flows because paleoflood or flood reconstructions would be erroneous if made using deposits other than from water-dominated flows [*Williams and Costa*, 1988; *Waythomas and Jarrett*, 1994]. Generally, ranges of sediment concentration (by volume) are less than about 20 percent for water floods, about 20 to 50 percent for hyperconcentrated flows, and about 50 percent or larger for debris flows. Diagnostic post-event features of debris flows include: (1) poorly sorted, unstratified deposits of boulders, cobble, sand, and fines; (2) marginal levees of coarse clasts, the largest of which may be at the top of the levees; (3) terminal lobes of debris, which often may be partially reworked or removed by subsequent water-dominated flows; and (4) unusually large boulders, transported at the margins of the flow, which may have done little or no damage to vegetation or structures [*Williams and Costa,* 1988]. Flood deposits generally are well sorted and graded, although flash-flood deposits may have less sorting due to rapid changes in stage.

Onsite visits were made to recently flooded streams, primarily in the western and west-central United States during 1995-2000 to obtain the necessary data. Field measurements and estimates for the sites are summarized in Table 1. Notable recent floods that are part of our data set include the: June 1995 flood on the North Fork Flathead River near Columbia Falls, Montana that was larger than a 100-yr flood (C. Parrett, USGS, written commun., 1998); 1995 snowmelt runoff in Colorado, which produce up to 100-year floods in numerous rivers; July 1996 flash flood in Buffalo Creek basin, Colorado [*Jarrett*, 2001], ranging from about a 200- to 9,000-yr flood depending on location; January 1997 floods in east-central California and west-central Nevada, where several rivers had about 100-yr floods [*NRC*, 1999]; the July 1997 flood in Fort Collins, Colorado, at least 500-yr flood [*Grimm*, 1998]; the November 1995 flood in the Chelan River basin, Washington, was about a 125-yr flood, and; July 1999 floods in Saguache and Dallas Divide, Colorado, perhaps the largest during the Holocene for these two basins (unpublished data).

We expanded our analysis by looking at other recent extreme floods in semiarid and arid climates where HWM evidence can be preserved for many decades after a flood. These include the: June 1965 Plum Creek flood, Colorado [*Matthai*, 1969], about a 1,500-yr flood [*Jarrett and Costa*, 1988]; 1976 Big Thompson River flood (Plate 4) [*McCain et al.*, 1979], about a 10,000-yr flood [*Jarrett and Costa*, 1988]; 1982 Lawn Lake dam-failure flood (Plate 6) [*Jarrett*

Plate 6. Upstream view of July 15, 1982, Lawn Lake dam failure FB (type 3 deposit) in the Roaring River, Rocky Mountain National Park, Colorado, primarily 1 to 5 m diameter boulders (also see CD Plate 14). Logs within the deposit typically are about 300 mm in diameter and about 5 to 7 m long. For color version see CD.

and Costa, 1986] in Colorado, where the discharge exceeded the 500-yr flood by a factor of about 30; August 1990 flash flood near Opal, Wyoming (CD Plate 5), which is one of the largest known flash floods in Wyoming [*Corrigan and Vogel*, 1993]; and several streams in the Wasatch Mountains of Utah, which had discharges greater than 100-yr floods in 1983 [*Christensen et al.*, 1991].

We computed peak discharge using either the critical-depth or the slope-conveyance indirect methods [*Barnes and Davidian*, 1978; *Webb and Jarrett*, this volume]. For both methods, channels need to be as uniform (slope, depth, width, velocity) as possible. In many higher gradient streams, there may be few channels reaches that have the length necessary to satisfy requirements for hydraulic modeling (one- or two-dimensional; 1-D or 2-D). These include rapidly-varying and non-uniform flow, flow transitions from subcritical to supercritical flow, sharp bends, and boulder obstructions, which contribute to poor understanding of energy losses, hence, may produce large uncertainties when using 1-D and 2-D models.

For stream gradients of about 0.002 m/m or greater, flood discharge was estimated using the critical-depth method, which has been suggested for flood estimation, particularly for large floods [*Jarrett*, 1986; *Jarrett and Costa*, 1986; *Trieste and Jarrett*, 1987; *Grant*, 1997; *Jarrett and Tomlinson*, 2000]. In a comparison of the critical-depth and slope-conveyance methods with current meter derived peak discharges at eight streamflow-gaging stations in northwestern Colorado, *Jarrett and Tomlinson* [2000] noted an average of ±12 percent difference. Critical depth is assumed at a cross section for the respective flood height; peak discharge is the average of the critical-depth estimates for all cross sections in a reach. The slope-conveyance method was used to estimate flood discharge in lower-gradient rivers (less than about 0.002 m/m) where sub-critical flow generally prevails [e.g., *Jarrett*, 1985]. Estimates of flow-resistance coefficients for slope-conveyance method were made using criteria in *Jarrett* [1985]; *Webb and Jarrett* [this volume] provide additional references that provide techniques to help estimate flow resistance.

In some cases, PSI-HWM sites were located at or near streamflow-gaging stations operated by the U.S. Geological Survey or other water-resources agencies. For these sites, peak discharge using the critical-depth and slope-conveyance methods was compared with the peak discharge independently obtained for the gage. The gaged peak discharge was obtained by direct current-meter measurement at or near the peak and small rating curve extrapolation; in only a few cases were peak discharges independently estimated with the slope-area method [*Barnes and Davidian*, 1978; *Webb and Jarrett*, this volume]. Thus, our methodology provides additional information to: (1) validate the critical-depth and slope-conveyance methods, which we used for estimating peak discharge of floods; and 2) to assess the reliability when subsequently using these methods for paleoflood reconstructions for similar types of channel reaches.

Statistical analyses and modeling were used to investigate the relations between the difference of PSI height minus HWM height (PSI-HWM) and channel, flow, and geomorphic data (explanatory variables). The PSI-HWM data were analyzed using traditional statistical and graphical data analysis techniques [*Hirsch et al.*, 1993], including plotting, visualization, and multivariate, step-wise regression analysis. These techniques also were used to explore relations between peak discharge estimates at gaging stations and peak discharge estimated with the critical-depth and slope-conveyance methods as part of our study. Flood-frequency estimates were made using a variety of methods, including existing published values, regional flood relations available for the United States [*Jennings et al.*, 1994], and our own frequency analysis of annual peak-flow (gage) and paleoflood data using EMA [*Cohn et al.*, 1997].

RESULTS

Data documenting the relation between fresh, flood-deposited sediments (new PSIs) and HWMs responsible for emplacing the sediments were collected for 192 sites in thirteen states (number of sites); Alaska (4), Arizona (1), California (23), Colorado (126), Hawaii (1), Idaho (3),

Table 2. Deposit Types and PSI-HWM Results Summary for the 192 Sites

Type of Deposit	Type Number	Number of Observations	Average PSI-HWM (mm)	Minimum PSI-HWM (mm)	Maximum PSI-HWM (mm)
Overbank Slackwater Deposit	1	81	-32	-914	+305
(without Lawn Lake deposits)		(78)	(-33)	(-914)	(+305)
Main Channel Slackwater Deposit	2	6	0	0	0
(without Lawn Lake deposits)		(6)	(0)	(0)	(0)
Main Channel Flood Bar	3	70	+84	-183	+2,350
(without Lawn Lake deposits)		(67)	(+18)	(-183)	(+457)
Overbank Flood Bar	4	19	-14	-457	+183
(without Lawn Lake deposits)		(17)	(-26)	(-457)	(+30)
Delta	5	6	-26	-152	+122
(without Lawn Lake deposits)		(6)	(-26)	(-152)	(+122)
Fan apex	6	10	-3	-305	+152
(without Lawn Lake deposits)		(10)	(-3)	(-305)	(+152)
Total		192	+15	-914	+2,350
(without Lawn Lake deposits)		(184)	(-11)	(-914)	(+457)

Montana (2), Nebraska (1), New Mexico (6), Oregon (16), Utah (2), Wyoming (5), and Washington (4). The average and range for field measurements and estimates for the PSI-HWM sites are summarized in Table 1. Streams where PSI-HWM data were collected have drainage areas ranging from less than 0.1 km^2 to about 291,000 km^2. A summary for the 192 PSI-HWM observations and associated data are provided in the file "psi_hwm_table-cd.asc" on the CD. The file "psi_hwm_table-cd_descriptors.txt," which also is on the CD, is a text file describing the contents of the PSI-HWM data table. Data include the year of the flood, peak discharge estimated with the indirect methods, maximum flood depth, cross section width, upstream or downstream width, HWM and PSI elevation (arbitrary datum in most cases), water-surface slope, deposit form (new or enlarged), deposit type (Table 2), maximum particle size on the new flood deposit, and maximum particle size on the streambed or floodplain available for transport. The plot discharge "Q_{plot}" in the data table on the CD is our indirect discharge (Q_{site}), except where gage peak discharges (Q_{gage}) were available, which are more reliable. Reach type (expanding, contracting, or uniform) was estimated by comparing the PSI-HWM section width and the upstream or downstream channel width. The width to maximum depth ratio and expansion-contraction ratio of channel width were calculated from the field data.

For several large floods, primarily during the record snowmelt runoff in Colorado in 1995, maximum (or near maximum) flood stage was observed (*e.g.,* Plates 7-8 and CD Plate 6). Flood-deposited sediments were emplaced in a variety of locations in the channel and on the floodplain (Plates 5-10 and CD Plates 7-9). Numerous sites had multiple types and locations of sediment deposits (*e.g.*, Plate 7). After reviewing all PSI-HWM data, the data were stratified into six types based on the deposit location (Table 2). The classification is somewhat arbitrary in that all types of deposits result from a decrease of streamflow energy.

SWDs and FBs were differentiated based on the primary size of the deposit sediments and their location in the main channel or on the floodplain; these are types 1 to 4 (Table 2). Our broader definition of SWDs (silts and sands with isolated larger clasts) located on the floodplain were designated as a type 1 deposit, and if the deposits are located within the main channel were designated as type 2 deposits. Flood bars of cobble and boulders (FB) located within the main channel were designated as type 3 deposits and FBs emplaced on the floodplain were designated as type 4 deposits. In a few cases, a deposit was primarily sand and included a few isolated larger clasts, perhaps due to transitory extreme turbulence. For these cases, the deposit was classified as a SWD, but the maximum particle size is for the largest clast in the deposit (*e.g.*, Plate 5). Main channel

SWDs, in most cases, were sand deposits on mid-channel boulder bars that are many hundreds to thousands of years old, based on estimated ages using relative dating methods of *Jarrett and Tomlinson* [2000; Table 1]. The number of observations for each deposit type, the average and range (minimum and maximum) PSI-HWM values are provided in Table 2.

A Delta deposit (type 5), is a variation of main channel bouldery flood bars (type 3). These sediments were deposited where the water-surface slope dramatically decreased due to backwater effects (ponding) from a downstream constriction (*e.g.*, Plate 9). Thus, there is a dramatic decrease in particle size in a very short distance such as documented for the Roaring River flood produced by the 1982 Lawn Lake dam-failure flood [*Jarrett and Costa*, 1986; Figures 34 to 39]. In January 1997, flood flows from several high-gradient tributaries to the Merced River in Yosemite National Park, California, which was essentially ponded water due to width constrictions in Yosemite Valley, produced delta deposits corresponding with HWMs in the Merced River. Unfortunately, they were not recognized for their potential value and documented following the 1997 flood. These types of sediment deposits, usually cobble or boulder sized material, need to be documented in the future to assess their value.

Fan-apex deposits (type 6) are a variation of main channel bouldery flood bars and located where the channel width expands onto an alluvial fan and there usually are

Plate 7. Downstream view of FB (type 4 deposit) for Lefthand Creek near Boulder, Colorado (also see Plate 11), with 1995 flood waters about 200 mm below their peak stage. Note the sediments are deposited in the wake of 300 to 600 mm diameter trees. Peak discharge was 34 m3/s and the elevation of the top of flood bar (and a separate, type 1 SWDs of sand are visible in lower right center) equaled maximum flood elevation. Streamflow-gaging station is on right bank; the stage-discharge relation (rating curve) was defined by flow through a concrete flume. For color version see CD.

Plate 8. Upstream view of Avalanche Creek near Redstone, Colorado: a) with flood waters about 350 mm below their July 1995 peak stage (denoted by measuring tape); b) similar view of the new mid-channel FB (type 3 deposit) in November 1995. For color version see CD.

multiple channels. Like delta deposits, sediment size can dramatically decrease in short distances [*Jarrett and Costa*, 1986]. Cross sections need to be located at the fan apex where the largest clasts tend to be located and flow primarily is constrained to a single channel to obtain the most reliable peak discharge estimates.

Finally, flood frequency was estimated from a combination of as many as possible of the following methods: 1) published, if available; 2) EMA frequency analysis if the site is at or near a streamflow-gaging station; 3) regional flood-frequency relations developed by the U.S. Geological Survey for rivers in the United States [*Jennings et al.*, 1994]; 4) the relative height of the fresh flood deposits to maximum older PSIs at the site, and; 5) newspaper accounts of severity of flooding and associated precipitation.

Plate 9. Upstream view of bouldery, mid-channel delta (type 5 deposit) at Spring Creek near Fort Collins, Colorado, for the flood of July 28, 1997. Field assistant is holding survey rod at maximum flood height. For color version see CD.

Although estimates of flood frequencies are approximate, particular effort was made to estimate the frequency of extreme floods (>50-year recurrence interval). The Lawn Lake dam failure flood in the Roaring River [*Jarrett and Costa*, 1986] would exceed the magnitude of a meteorological-produced 10,000-yr flood. However, for ease of graphical viewing and statistical analyses, this flood was assigned a recurrence interval of 10,000 years.

Cross sections showing representative types of PSI data corresponding to the observed HWMs are provided in Plates 11-13 and CD Plates 10-12. The elevation at the top of the observed flood deposit (new PSI) minus the observed HWM provided the comparative difference (PSI-HWM) for the 192 observations (Tables 1 and 2; data on CD). Negative PSI-HWM values indicate the depth of the fresh sediments below the observed HWMs (*e.g.*, CD Plate 6). Positive PSI-HWM values (*e.g.*, Plate 9 and CD Plate 11) indicate the deposits protruded above the observed HWMs (see Discussion for explanation). The PSI-HWM differences, stratified by deposit type, were then plotted with available explanatory variables (Plates 14a-e and CD Plates 13a-b). Regression analyses were made between "PSI-HWM" and explanatory variables. The analyses were done for individual stratified deposit type (Table 2) and ungrouped data (all observations). There were no statistically significant relations (*e.g.*, slopes of regression lines were equal to zero), except for delta deposits. The regression for delta deposits only had five observations, and thus, was deemed insufficient for an acceptable statistical sample size.

During onsite visits, peak discharge was estimated using the flood HWMs and channel characteristics for 46 sites at or near streamflow-gaging stations. These estimates using the critical-depth and slope-conveyance methods were compared to the peak discharge obtained from gage records to help assess the accuracy of these two indirect discharge methods. The estimated peak discharge using the indirect methods ranged from -45 percent to +43 percent of gaged discharge (Plate 15), with an average of +1 percent for all sites. Additional confidence in paleoflood estimates is exhibited when multiple sites are used and results are similar (*e.g.*, North Fork South Platte River near Pine, Colorado).

DISCUSSION

This systematic documentation and analyses of recently emplaced flood-deposited sediments (new PSIs) provide important information for operational paleoflood discharge reconstructions. The study approach and results should be viewed as one possible framework for future research rather than an all-encompassing, conclusive study of the relation between PSIs and HWMs. The PSI-HWM data reflect a range of river hydraulics, sediments, and flood frequency that are representative of evidence typically preserved for extreme paleofloods. A lack of a statistical relation may reflect the approach we used of only documenting the highest fresh deposits in relation with the peak HWMs. For any recently flooded reach of channel, we quickly visualized and surveyed the highest flood-deposited sediments for PSI-HWM documentation. In most cases, fresh PSIs nearly equal to the height of the observed HWMs could be rapidly identified, even though many flood deposits sediment were well below HWMs. However, older PSIs covered with

Plate 10. Upstream view of Truckee River near Boca, California, following the January 1, 1997 flood. Extensive overbank sand splays (type 1 deposit) are located downstream from isolated trees and more commonly in moderate to thick brush. For color version see CD.

vegetation would take longer to identify. Our objective was to determine the relation between the top of the highest fresh PSI and the observed HWM for a reach of channel that meet hydraulic criteria [*Webb and Jarrett*, this volume] where for reliable paleoflood reconstructions could be made.

Perhaps, had we documented all fresh flood deposits in a study reach, analyses of such data may yield a statistical relation or predictive numerical model. One alternative approach is to randomly select river reaches at or near streamflow-gaging stations and collect pertinent pre-flood data, but would require a much greater investment of resources and longer time for onsite fieldwork. Then after a flood, an investigator could thoroughly document and map all sediment deposits in a reach, and obtain other pertinent information. This alternative approach certainly would provide a better understanding of the physics of large flood response and better predictive capabilities than our approach, but a pre-selected river could go many years without extreme flooding. However, in our "flood-chasing" approach, we were able to document the relation between HWMs and PSIs for many extreme floods in a relatively short period of time (1995-2000). Our research provides a large and comprehensive PSI-HWM database for our analyses and may stimulate discussion of research needs including optimizing a balance of "flood-chasing" and detailed approaches of PSI formation.

General Interpretations

The average value of PSI-HWM for the total 192 observations is +15 mm with a range of –914 mm to +2,350 mm (Table 2; data on CD). The graphical presentation of data (Plates 14a-e) helps visualize the ranges of hydraulic con-

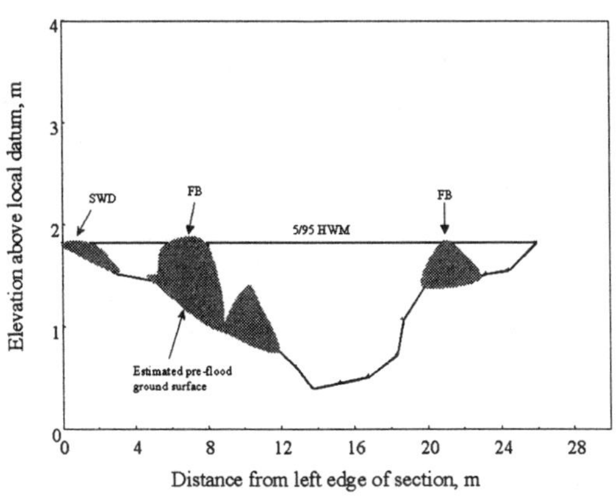

Plate 11. Cross section of Lefthand Creek near Boulder, Colorado (see Plate 7). The top of overbank FBs (type 4 deposit) and fine-grained SWDs (type 1 deposit) equals the 1995 HWMs. For color version see CD.

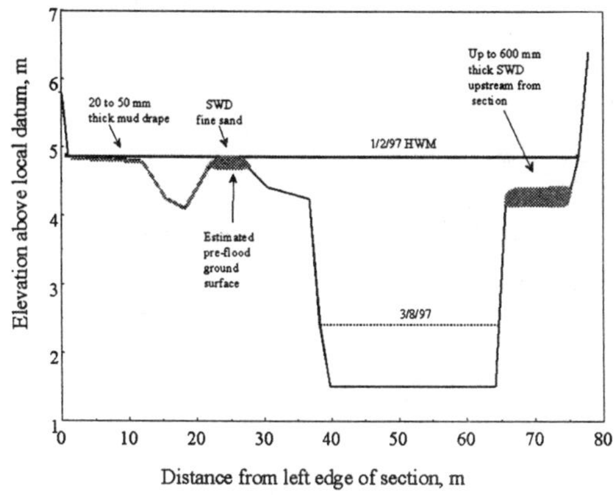

Plate 12. Cross section of Truckee River near Boca, California (see Plate 10) showing sand splays and mud drapes (type 1 deposit), which formed in backwater areas. For color version see CD.

ditions over which the observations were obtained and help identify where additional documentation is needed. Most of our data represent flows that were less than about 4 m depth (Plate 14a), and higher gradient (and energy) conditions (Plate 14b). Of particular importance is the number of observations for floods of 50 years or longer recurrence intervals (Plate 14e). For the wide range of values for each explanatory variable, results are relatively scale independent. For example, there is no definitive relation between PSI emplacement and expansion-contraction ratios in the reaches we documented (CD Plates 13a-b) as might be expected. As discussed in the next section, local-scale effects provide sufficient conditions for deposition of flood-transported sediments.

It is more meaningful to view the difference between PSIs and HWMs in terms of deposit type and location (Table 2). Deposits (silt to boulder size) located in the main channel (higher energy areas) tend to be emplaced slightly higher than those emplaced on floodplains. These deposits usually result from local-scale effects of velocity (conversion of kinetic energy to potential head) with the top of the deposit can approach the height of the energy-grade line (EGL). The EGL [*Jarrett and Webb*, this volume] equals the height of HWMs in quiescent water on the channel margins, which is the preferred location for obtaining HWMs (*e.g.*, Plates 7 and 10). The average difference of PSI-HWM for all observations within each deposit type is: -32 mm for overbank SWDs; 0 mm for main channel SWDs; +84 mm for main channel FBs; -14 mm for overbank FBs; -26 mm for delta deposits, and; -3 mm for fan apex deposits. Respective ranges for each deposit type are provided in Table 2.

Although no statistically significant relation for the PSI-HWM data was developed, as originally hypothesized, the

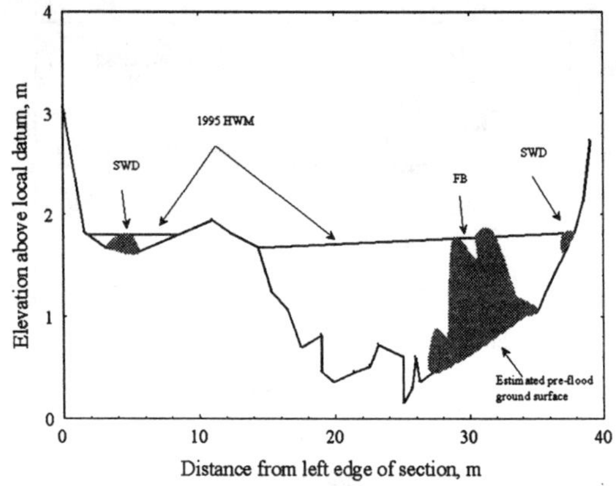

Plate 13. Cross section for Avalanche Creek near Redstone, Colorado (see Plate 8) showing FB (type 3 deposit). For color version see CD.

systematical documentation of fresh sediment deposits demonstrates that by documenting several PSIs in a channel reach, a reasonable estimate of maximum flood height associated with sediment (PSI) deposition can be made. Thus, for all deposit types we documented, the data indicate that to reconstruct the flood height for those respective deposit types in an operational paleoflood study, the HWM equals the top of the deposits plus the average value for PSI-HWM for the respective deposit type. For example, if using overbank FB deposits (type 4) to define the PSI, then the inferred paleostage would be 10 mm higher than the top of the highest FBs; for main channel FB deposits (type 3), the inferred paleostage would be 84 mm lower than the top of the highest FBs.

Local-Scale Effects

The 1982 Roaring River flood deposits were produced by the failure of Lawn Lake dam. Therefore, the summary PSI-HWM results were recomputed excluding these eight observations (Table 2); the average PSI-HWM changed from +15 mm to –11 mm and the range is –914 mm to +457 mm for the subset of 184 observations. In this light, the relation between PSIs and HWMs appear substantially improved and provide greater confidence in paleoflood discharge estimates. However, we believe the SWD and FB evidence for the Lawn Lake dam failure flood, which are preserved throughout most of the 10-km reach of the Roaring River [*Jarrett and Costa*, 1986], are an important set of PSI-HWM data.

The 1982 Lawn Lake flood produced the type of PSI evidence associated with the most extraordinary floods in mountain rivers increasingly being studied. The tops of boulder bar deposits (type 3) protrude well above HWMs (about 732 to 2,350 mm, Plate 14a) for the 1982 dam-failure flood in the Roaring River (*e.g.*, Plate 6). In reaches having channel slopes from about 0.07 to 0.35 m/m, clasts up to 8 m in diameter were mobilized and subsequently deposited in reaches where channel slope was less than about 0.07 m/m [*Jarrett and Costa*, 1986]. It is likely that such large boulders are mobilized by a combination of flow around the base of boulders, the force of flowing water on the boulders, and momentum transfer when struck by boulders already in transport. Boulder transport continues until flow depth decreases and sediments are deposited; for many deposits the height of the top of fresh bouldery deposits can be much higher than the EGL (CD Plate 14). Along the floodplain margins of the Roaring River, there are numerous isolated 2- to 3- m diameter boulders that continued rolling until the flow depth was less than 1 m (such isolated clasts were not used in our analysis). This suggests that using only isolated fluvial-transported clasts would be poor PSI evidence.

For less extraordinary floods (the subset of 184 observations), mid-channel and over-bank deposits (all but one site) correspond to boulder sized material transported in higher gradient streams and emplaced higher than HWMs. Our data suggest that when stream gradients exceeds about 0.04 m/m (Plate 14b), the tops of the sediments protrude up to 450 mm above the observed HWM elevations. These flood bars form either as new bars around an obstruction (boulders, trees, dense brush cluster) or when sediments ramp or roll up existing bars and cease motion when flow depth decreases. Record snowmelt flooding in 1995 from many streams in Colorado provided the opportunity to directly observe sediment transport and deposition at or very near peak-flow conditions (about 50- to 100-year floods). For example, 280-mm diameter boulders were observed being deposited at the peak flood stage, primarily on preexisting mid-channel bars, in Clear Creek upstream from Georgetown (data on CD). These clasts came to rest with their tops just protruding above the projected HWMs; nearby sand deposits in channel-margin areas equaled HWMs. Estimates of flood height from mid-channel PSIs downstream from reaches with large gradient (~ 0.04 m/m) and very turbulent, high-energy flow need to be lower than the tops of the boulders for discharge reconstruction.

Isolated obstructions in the flow such as trees (Plates 7, 9 and 10), brush (CD Plates 5 and 6), boulders (Plate 8), preexisting flood bars (CD Plate 10), or local-scale phenomena can have profound effects on the emplacement of flood sediments as reliable PSIs. For example, a July 1996 extraordinary flood in a small tributary to Sand Draw (0.08 km^2) near Buffalo Creek, Colorado, emplaced SWDs (5- to 20- mm diameter gravel) just downstream from a tree (CD Plate 15). This type of deposit was quite commonly ob-

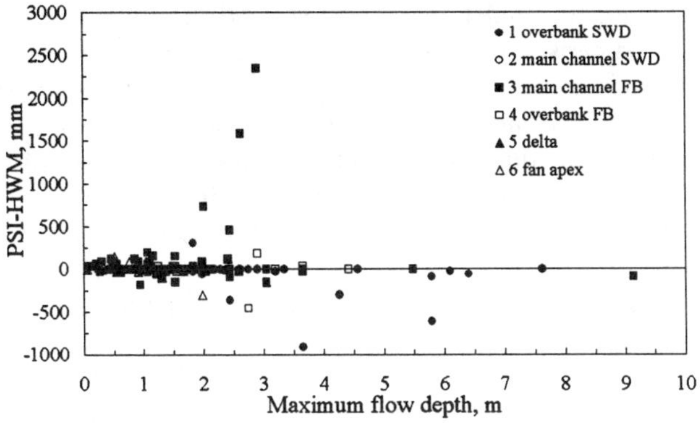

served. In addition, many similar, but much older SWDs are preserved in streams in the area and identified along many other streams we visited during our study.

Another commonly observed local-scale effect is cobble and boulder deposits associated with large woody debris located in channels or on floodplains (CD Plate 16). Although flow was partially ponded upstream from these trees, flood flows are forced under part of the trees producing locally accelerating flow, increasing flow competence. The transported cobble and boulders are immediately deposited as flow emerges on the downstream side of the tree; the top of the flood deposit (FB) usually was equal to adjacent HWMs.

In addition, when flows exceed the top of the vertical, main channel banks even by small flow depths (<50 to 100 mm) at the upstream point of overflow, bed material can be transported onto the floodplain and deposited (*e.g.*, CD Plate 4), particularly as shown in Plate 5. Extreme turbulence and upwelling of flow transported coarse streambed sediments over vertical banks 1.5-m high, yet flow depths were as shallow as 50 mm at the point where sediments were transported onto the floodplain (Plate 5 at survey rod). Extensive deposits of sand (type 1 deposit) including clasts up to 150 mm in diameter were emplaced on the floodplain at many locations along Leopard Creek. Thus, for higher gradient channels having coarse bed-material, a lack of coarse material on floodplain surfaces may indicate a lack of flood inundation.

For delta deposits (type 5) and fan-apex deposits (type 6), it is important to obtain the cross section, PSI-HWM and other data at the upstream end of the delta/fan, which usually coincided with the location of the maximum particle sizes for sites we documented. To assess the validity of using these deposits, peak discharge for delta deposit sites also was determined using the critical-depth method using tributary cross sections and HWMs from sites just upstream from the backwater affected reach. For the documented sites, the peak discharge computed with the critical-depth

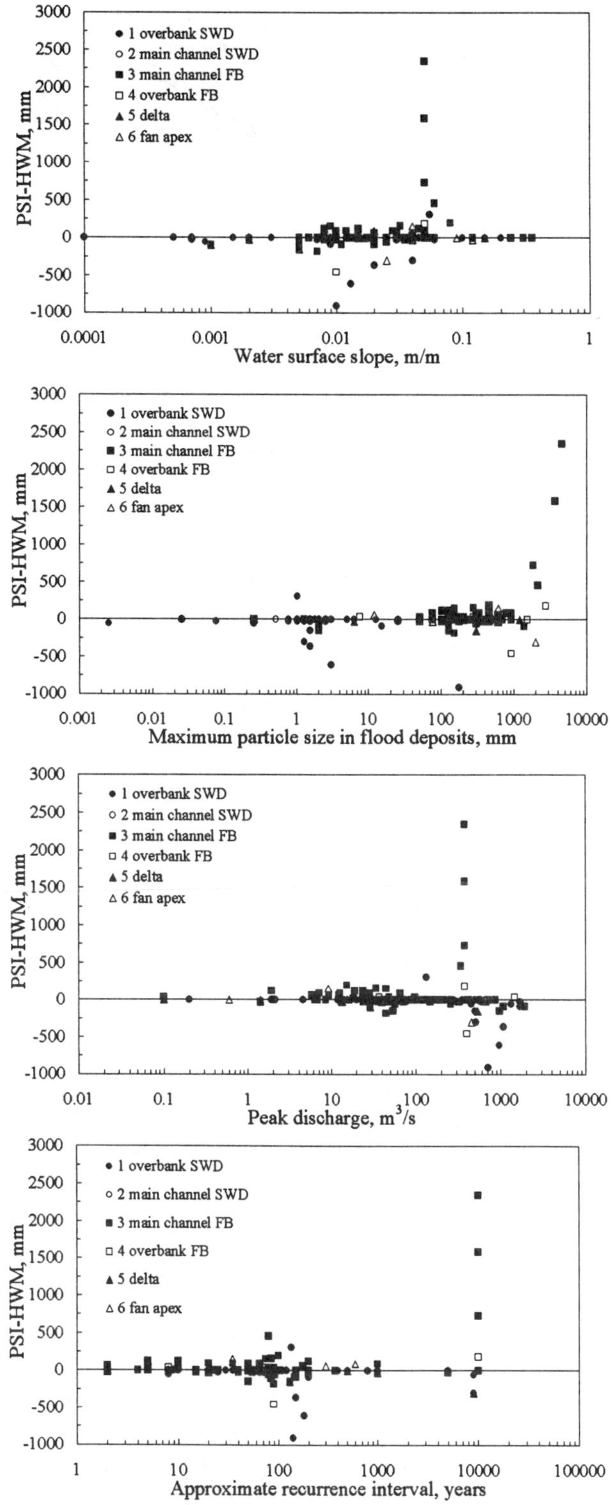

Plate 14. Graphs of difference between the elevation of new PSIs and HWMs stratified by deposit type with: a) maximum flow depth; b) water surface slope; c) maximum particle size in flood deposits; d) peak discharge; and e) approximate recurrence interval for recent floods. For color version see CD.

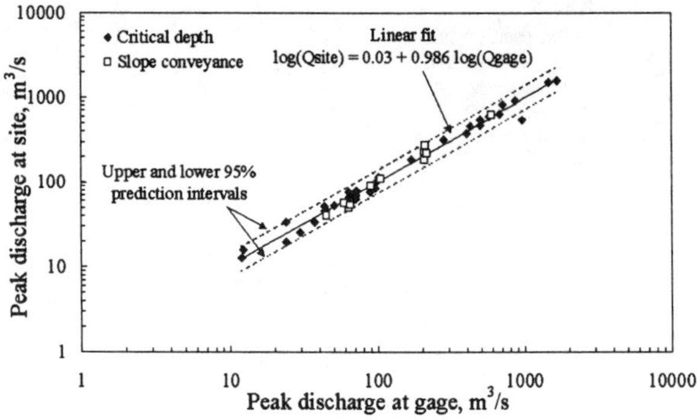

Plate 15. Relation of peak discharge between streamflow-gaging station values and peak discharge computed with the critical-depth and slope-conveyance methods. For color version see CD.

method at the deposit usually was within ~15% of the critical-depth discharge estimate at cross sections located in upstream, "channel control" reaches. This suggests that the largest clasts are rapidly deposited in near-critical flow, perhaps where rapid energy dissipation occurs. Such delta sites might prove fruitful for paleoflood and flood discharge reconstructions with additional study.

Evaluation of Indirect Discharge Methods

It is essential that validation of methods and results be used in an investigation where possible. In this study, we obtained 46 peak discharge estimates (Q_{site}) using the critical-depth (35 sites) and slope-conveyance (11 sites) methods for comparison with peak discharges obtained from gaged records (Q_{gage}) at or near our sites by others (Plate 15). Of these Q_{gage} floods, 22 peak discharges were determined using current-meter measurements, and 12 were estimated by using a modest rating-curve extension. Peak discharges were also derived using flow through a flume or over a diversion dam (6 sites), which we believe are more reliable than the indirect methods we used. In addition, Q_{gage} peak discharges were estimated using the slope-area method at six sites for comparison with our critical-depth discharge estimates. The range in the difference of the peak discharge using the critical-depth and slope-conveyance methods is –45 to +43 percent with an average difference of +1 percent (data on CD). Our results (Plate 15) have somewhat large uncertainty than noted by *Jarrett and Tomlinson* [2000, Table 2; ±12 percent average difference of Q_{gage}]. Their results have lower uncertainty primarily because low-frequency floods in northwestern Colorado result from relatively small magnitude, snowmelt peak flows.

These results for peak-discharge estimation are encouraging because of their ease of use when resources and time preclude use of 1-D (*e.g.*, step-backwater modeling) and/or 2-D hydraulic models used in paleoflood and flood studies. The comparison with gage data indicates the critical-depth and slope conveyance methods, when applied in hydraulically good channel reaches, provides reasonable flood estimates. In addition, the comparison indicates that the sites selected for documenting PSI-HWM relations are good hydraulic reaches for paleoflood discharge reconstruction provide credibility for use of the critical-depth and slope-conveyance methods for other streams with similar hydraulic characteristics. Certainly, if resources are available to use 1-D or 2-D hydraulic models, they could be expected to provide similarly reliable discharge estimates and additional flow characteristics depending on the model used. However, any method used to estimate discharge needs to be validated with flood data in the study reach or other study reaches with similar hydraulic and sedimentologic characteristics. It is important when using the critical-depth method to locate cross sections in stable channels with gradients preferably greater than 0.01 m/m and are relatively uniform (slope, width, depth), although the method may produce reasonable results for floods in somewhat lower gradients, perhaps as low as 0.002 m/m for sand-bed channels. *Jarrett and Tomlinson* [2000] noted that a relatively uniform reach length (one to two top-widths in length) usually can be identified within a longer, but primarily severely non-uniform reach. When using the slope-conveyance method, the main criterion is a reach is as uniform as possible. Both methods need to have relatively stable channel geometry. To reduce peak-discharge uncertainty for either method or a combination the methods, we recommend averaging results using three to six cross sections located in independent, uniform reaches.

Comparisons with Other PSI Documentation Studies

For eleven observations (five type 1, three type 3, and one each of deposit types 4, 5, and 6; data on CD), the tops of the highest flood deposits (new PSIs) were substantially lower (about 100 mm to 914 mm) than HWMs (Plate 14). These sites usually were located in relatively straight reaches confined by bedrock channels with very steep granite valley walls in streams, primarily in the Sierra Mountains of California. Sediment supply from decomposing granite generally is relatively small as compared to channels in softer rocks or alluvial channels. Perhaps the relatively small sediment supply combined with very large rainfall amounts and frequent flooding in these basins, which tend to flush most fine-grained sediment through these bedrock reaches, may explain the lower PSIs. We noted rainfall runoff also removed much of the thinner fine-grained deposits within a year after the January 1997 floods in California. These results are similar to the *Kite et al.* [this volume] PSI-HWM studies of the Cheat River in West Virginia, and emphasizes the need to better understand flood-sediment depositional processes. Additional PSI-HWM data are needed for these stream environments to develop predictive tools of the depth below HWMs for these SWDs with additional explanatory variables. Greater emphasis on using FB deposits may help estimate paleoflood depth in these canyon environments.

To date, we have not documented the commonly used massive SWDs of fine-grained sediments (*e.g.*, Plate 2) typical in rivers in the southwestern United States where most paleoflood studies have been done [*e.g.*, *Kochel and Baker*, 1988]. *Ely and Baker* [1985] documented SWDs equaled the height of HWMs for two recent floods for a site on the Verde River, Arizona. *Webb et al.* [this volume] provide data that SWDs ranged from 300 to 900 mm lower than HWMs for the Paria River, Arizona; *House et al.* [this volume] documented similar results. Our data and results appear to differ from other SWDs study results primarily because we document streams that generally have: (1) smaller basin size; (2) higher gradient (and higher energy); and (3) coarser sediments. Only 35 of our SWD sites had grain diameters less than about 1 mm (Plate 14c); typically, most other SWD comparisons with HWMs primarily have grain sizes less than 1 mm [e.g., *Webb et al.*, this volume]. For the subset of our data that have stream gradients less than about 0.002 m/m, which usually corresponds with finer grained deposits for larger flood depth and discharge, the top of SWDs is slightly (25 to 100 mm) lower than observed HWMs (Plate 14b). Our results are comparable with results of the other SWD studies (SWDs<HWMs), and also expand the range of environmental settings for PSI-HWM documentation. Additional study and documentation of PSI-HWM relations, particularly in lower gradient streams are needed.

With the increased emphasis on use of non-inundation surfaces to define bounds of maximum flooding, there is a need to determine relations between parameters such as shear stress and flood power with modification of geomorphic surfaces. We do not provide documentation between HWMs and non-inundation surfaces [*e.g.*, *Levish et al.*, 1994, 1997]. The objective of the non-inundation approach is to identify and assign ages to geomorphic surfaces adjacent to the river that serve as limits for the paleostage of large floods. In non-inundation studies, the maximum discharge that would not significantly inundate and modify a particular geomorphic surface is calculated and used as a bounding discharge. Further investigations to interpret long-term surface stability and PSI-HWM relations for these data are needed.

In some situations, non-inundation surfaces are difficult to interpret and flood preservation potential is poor. For example, on July 29, 1997, a thunderstorm produced about 400 mm of rain in four hours over a large area of Pawnee Creek basin located east of Fort Collins, Colorado. The rainstorm produced a peak discharge of about 2,125 m^3/s in Pawnee Creek years (Colorado Water Conservation Board, unpublished report, 2000), which had a flood recurrence interval in excess of 500 years. The Pawnee Creek channel has a gradient ranging from greater than 0.01 m/m in the uplands to about 0.005 m/m in downstream reaches. The channel and floodplain are covered with thick grass, which remained intact during the 1997 flood. Although there were ample thick (30 to 80 mm) HWMs of small organic material (grass, twigs, leaf litter) along channel margins, we were unable to locate flood-sediment deposits in the channel or on floodplain surfaces that might be preserved. We also did not see geomorphic expression of the flood (non-inundation surfaces). That such an extraordinary flood did not leave any recognizable evidence with long preservation potential is disconcerting to us. It is likely that there is some type of lasting evidence, perhaps mineralogical signatures of flooded compared with non-inundation surfaces. However, the evidence was not visibly apparent and may require very detailed analyses to interpret these flood signatures immediately after a flood, and particularly for studies of older paleofloods. It was beyond the objectives of our research to document these deposits in detail. However, *Costa and O'Connor* [1995] documented several extreme floods that were ineffective in causing significant (and lasting) landform changes in alluvial or bedrock channels. Research is needed to identify other types of PSIs, if they exist, and to better understand geomorphic effectiveness of floods. More detailed investigations, using field documentation, laboratory flume studies, and numerical modeling are needed, particularly for understanding local effects of deposition and flood conditions to support use of non-inundation surfaces.

Paleostage indicators (SWDs and FBs) and non-inundation surfaces have their respective advantages and disadvantages for operational paleoflood studies. For example, SWDs often contain many floods over long time periods (>5,000 years) and are particularly conducive to absolute and relative dating [*e.g.*, *Kochel and Baker*, 1988]. Because SWDs are usually fine-grained silt to sand-sized sediments, they may be eroded by lateral erosion, but SWDs emplaced by the largest flood is preferentially preserved for the longest time [*House et al.*, this volume]. Traditional SWDs (*e.g.*, Plate 2) may be limited to rivers where floods transport relatively high concentrations of fine-grained sediments.

FB deposits of cobble and boulder-sized material are less conducive to ^{14}C age dating, but ages can be determined using relative dating methods [*Waythomas and Jarrett*, 1994; *Grimm et al.*, 1995; *Pruess et al.*, 1998; *Jarrett and Tomlinson*, 2000]. Usually, there are fewer paleofloods preserved within FBs (a less complete flood record), but multiple floods within one deposit have been identified [*Waythomas and Jarrett*, 1994; *Grimm et al.*, 1995]. This is due in part to difficulty excavating deposits with cobble- and boulder-size material. Like SWDs, FBs from the largest flood are preferentially preserved, but are less erodible by subsequent floods than finer grained deposits. The geomorphic expression of bouldery FBs are relatively easily recognized and can be preserved for tens of thousands of years, particularly of the largest flood [*Jarrett and Tomlinson*, 2000]. The non-inundation approach has great potential for paleoflood reconstructions but it has had minimal validation with field studies or objective analysis with published flood studies. In an ideal river reach, maximum paleoflood discharge using non-inundation surfaces should be comparable with the estimate of PSIs (SWDs and FBs). Thus, we recommend using as many types and locations of PSIs (SWDs and FBs) as well as non-inundation surfaces to maximize advantages of each type of PSI and ensuring the most reliable operational paleoflood discharge estimates.

While our study provides important information about the relation of fresh sediment deposits and the maximum flood height, we are unaware of any studies documenting long-term preservation potential of PSIs. Can multiple extreme paleofloods separated by a short time (say within the reliability of about ±50 to 100 years using ^{14}C dating) in the early to mid-Holocene leave PSI deposits that can not be differentiated as distinct floods? What are the implications of incomplete flood record for flood-frequency analyses [*e.g.*, *NRC*, 1999; *Blainey et al.* this volume]? Also, are there post-depositional diagenesis and compaction effects, particularly of fine-grained sediment deposits, that affect PSI height and over what time scales might these effects be expected? *Kite et al.* [this volume] note that SWDs in moist environments (notably caves) rapidly alter deposit characteristics in short periods of time, including rapid soil profile development, and bioturbation. They suggest such modified flood deposits may not be distinguishable from colluvial material within a few hundred years.

SUMMARY AND CONCLUSIONS

In many past paleoflood studies, the elevation at the top of the flood-deposited sediments typically was assumed to represent a minimum estimate for the maximum paleoflood stage, or flow-competence equations were used to estimate the flow depth above the highest clast in the deposits. We evaluated this operational assumption by collecting and analyzing data from sites primarily in the western and west-central United States. Flood-transported sediments (fresh PSIs) deposited as flood bars (FBs) and/or slackwater deposits (SWDs) from recent large floods can be directly related to flood high-water marks (HWMs) at the same location. We used a "flood-chasing" approach to visit as many extreme flood sites as possible after a flood, and obtained PSI-HWM data from recent large floods at 192 sites. Our hypothesis was that a PSI and HWM height difference relation depended on hydraulic and sedimentary conditions in channels, especially water-surface slope and deposit particle size.

Analysis of the PSI-HWM data in this investigation indicates that the elevation at the top of some flood sediments (PSIs) in a short reach essentially equals the height of HWMs. The average PSI-HWM relation for all 192 observations is +15 mm with a range of –914 mm to +2,350 mm. Deposits were stratified into six deposit types: overbank slackwater deposits; main channel slackwater deposits; main channel flood bars; overbank flood bars; deltas; and fan apex deposits. Results slightly vary with type of deposit and flow energy in the depositional site. No statistically significant relations were found to relate PSIs and HWMs based on maximum flow depth, water-surface slope, maximum particle size, peak discharge, recurrence interval, width/depth ratio, or expansion/contraction ratio. Improved results were obtained by excluding eight out of 192 observations that were associated with the Lawn Lake Dam failure. By excluding just these eight values, the average PSI-HWM changed from +15 mm to –11 mm and the range was –914 mm to +457 mm for the 184 observations subset. The main result from this study is that use of the elevation at the top of flood-deposited sediments (new PSIs), preferably using deposits nearest to channel margins, provides a reliable and accurate indication of the maximum height of the flood. For the six deposit types we documented, the data indicate that to reconstruct flood height in an operational paleoflood study, the HWM equals the top of the deposits plus the average value for PSI-HWM for the respective deposit type.

This systematic documentation and analysis of recently emplaced flood-deposited sediments (new PSIs) provided a unique opportunity to study the physical evidence of floods in the United States, but it is not without limitations. Data were obtained from 13 states, including: Alaska (4), Arizona (1), California (23), Colorado (126), Hawaii (1), Idaho (3), Montana (2), Nebraska (1), New Mexico (6), Oregon (16), Utah (2), Wyoming (5), and Washington (4). Although most sites (65 percent) were located within Colorado, there were many extreme floods during the 1995-2000 period that were documented. Most of the sites were for streams with gradients larger than 0.002 m/m and for flow depths less than about 4.5 m. Data from sites with lower gradients and/or larger flow depths are needed; the commonly used massive SWDs of fine-grained sediments, observed in many areas of the southwestern United States, have not been documented in this study. Additional data collection and analyses are needed for these sites. The study approach and results should be viewed as one possible framework for future research rather than an all-encompassing, conclusive study of the relation between PSIs and HWMs. Perhaps if all fresh flood deposits were documented in a study reach, analyses of such data might have yielded a statistical relation or predictive model. Such a relation might be developed after additional data collection and research. Alternate data collections could be used to supplement observations reported in this paper. One alternative approach could be to randomly select river reaches at or near streamflow-gaging stations and collect pertinent pre-flood data; this would require a much greater investment of resources and longer time for onsite fieldwork.

Forty-six peak discharges that we estimated using the critical-depth and slope-conveyance methods were compared with peak discharges made for sites at or near streamflow-gaging stations. Peak discharges at these gages primarily were determined by current-meter measurement or rating-curve extension. Forty-two of the 46 estimates were within ±25 percent of the gage peak discharge measurements. These results provide important validation for the critical-depth and slope-conveyance methods for peak discharge estimation of large floods and paleofloods.

Our research provides a large and comprehensive PSI-HWM database for additional analyses. We hope to stimulate discussion of research needs, including optimizing a balance of "flood chasing" and detailed approaches of PSI formation. Based on the observations and analyses made as part of this study, we recommend investigations along several lines. Our results indicate further research is needed that better accounts for physical processes for sediment transport and deposition during flood conditions. There is much work to be done to improve our understanding of the distribution and location of sediment deposits and erosion thresholds of surfaces from extreme floods. This includes better understanding of the physics of large flood response and improved predictive capabilities. Research is needed to identify other types of PSIs, if they exist, and to better understand geomorphic effectiveness of floods. The importance of local-scale effects, especially in high-gradient reaches (exceeding 0.04 m/m) where large clasts can be mobilized, needs to be quantified.

Acknowledgments. We appreciate assistance with data collection from Tom Browning, Joe Capesius, Dan Cenderelli, Casey Clapsaddle, Jon Hortness, Ben Jarrett, Don Jensen, Doug Laiho, Chuck Parrett, Jeff Phillips, Jon Pruess, Ed Tomlinson, and particularly Dorothea Jarrett who helped on many weekends. We appreciate the thorough and helpful reviews provided by Vic Baker, Kyle House, Steve Kite, Jim Knox, Jim O'Connor, and Bob Webb.

REFERENCES

Baker, V.R., Paleoflood hydrology and extraordinary flood events, *J. Hydrol., 96*, 79-99, 1987.

Baker, V.R., R.C. Kochel, and P.C. Patton (Eds.), *Flood Geomorphology*, John Wiley, 503 pp., New York, 1988.

Barnes, H.H., Jr., and J. Davidian, Indirect methods, in *Hydrometry-Principles and Practices*, edited by R.W. Hershey, pp. 189-190, John Wiley, New York, 1978.

Boner, F.C., and F. Stermitz, Floods of June 1964 in northwestern Montana, *U.S. Geol. Surv. Water-Supply Pap., 1840-B*, 242 pp., 1967.

Carling, P.A., Hydrodynamic interpretation of a boulder berm and associated debris-torrent deposits, *Geomorphology, 1*, 53-67, 1987.

Christensen, R.C., D.D. Carlson, G.L. Ashcroft, and D.L. Anderson, Utah floods and droughts, in *National Water Summary 1988-89 – Hydrologic Events and Floods and Droughts, U.S. Geol. Surv. Water-Supply Pap., 2375*, 527-534, 1991.

Cohn, T.A., W.L. Lane, and W.G. Baier, An algorithm for computing moments-based flood quantile estimates when historical information is available, *Water Resour. Res., 33*, 2089-2096, 1997.

Corrigan, P., and J.L. Vogel, Meteorological analysis of a local flash flood near Opal, Wyoming 16 August, 1990, *13th Conference on Weather Analysis and Forecasting and Symposium on Flash Floods*, Vienna, Va., August 2-6, 1993, American Meteorological Society, 375-378, 1993.

Costa, J.E., Holocene stratigraphy in flood-frequency analysis, *Water Resour. Res., 14*, 626-632, 1978.

Costa, J.E., Paleohydraulic reconstruction of flash flood peaks from boulder deposits in the Colorado Front Range, *Geol. Soc. Am. Bull., 94*, 986-1004, 1983.

Costa, J.E., A history of paleoflood hydrology in the United States, 1800-1970, in *The History of Hydrology*, edited by E.R. Landa and S. Ince, pp. 49-53, AGU, Washington, D.C., 1987.

Costa, J.E., and J.E. O'Connor, Geomorphically effective floods, in *Natural and Anthropogenic Influences in Fluvial Geomorphology, The Wolman Volume, Geophysical Monograph 89*, edited by J.E. Costa, A.J. Miller, K.W. Potter, and P.R. Wilcock, pp. 45-56, AGU, Washington, D.C., 1995.

Ely, L.L., Response of extreme floods in the southwestern United States to climate variations in the late Holocene, *Geomorphology, 19*, 175-201, 1997.

Ely, L.L., and V.R. Baker, Reconstructing paleoflood hydrology with slackwater deposits: Verde River, Arizona, *Phys. Geog., 5*, 103-126, 1985.

Ely, L.L., Y. Enzel, V.R. Baker, and D.R. Cayan, A 5000-year record of extreme floods and climate change in the southwestern United States, *Science, 262*, 410-412, 1993.

England, J.F. Jr., Assessment of historical and paleohydrologic information in flood frequency analysis, M.S. thesis, 292 pp., Colo. State Univ., Fort Collins, 1998.

Enzel, Y., L.L. Ely, P.K. House, V.R. Baker, and R.H. Webb, Paleoflood evidence for a natural upper bound to flood magnitudes in the Colorado River Basin, *Water Resour. Res., 29*, 2287-2297, 1993.

Foster, J., Improving the state of the art in risk analysis, *Hydro Rev., 38*, 12-14, 1999.

Fuller, M.L., Discussion, Final Report of the Special Committee on Floods and Flood Prevention, *Trans. Am. Soc. Civ. Eng., 81*, Paper No. 1400, 1269-1278, 1917.

Gottesfeld, A.S., British Columbia flood scars: Maximum flood-indicators, *Geomorphology, 14*, 319-325, 1996.

Grant, G.E., Critical flow constrains flow hydraulics in mobile-bed streams: A new hypothesis, *Water Resour. Res., 33*, 349-358, 1997.

Grimm, M.M, E.E. Wohl, and R.D. Jarrett, Coarse-sediment distribution as evidence of an elevation limit for flash flooding, Bear Creek, Colorado, *Geomorphology, 14*, 199-210, 1995.

Grimm, M.M., Floodplain management, *Civil Engineering, ASCE, 68*(3), 62-64, 1998.

Guo, S.L. and C. Cunnane, Evaluation of the usefulness of historical and palaeological floods in quantile estimation, *J. Hydrol., 129*, 245-262, 1991.

Hirsch, R.M., D.R. Helsel, T.A. Cohn, and E.J. Gilroy, Statistical Analysis of Hydrologic Data. In *Handbook of Hydrology*, edited by D.R. Maidment, McGraw-Hill, New York, 1993.

House, P.K., and P.A. Pearthree, A geomorphologic and hydrologic evaluation of an extraordinary flood discharge estimate, *Water Resour. Res., 31*, 3059-3073, 1995.

House, P.K., P.A. Pearthree, and J.E. Fuller, Hydrologic and paleohydrologic assessment of the 1993 floods on the Verde River, Arizona, *Arizona Geol. Surv. Open-File Rept., 95-20*, 23 pp., 1995.

Hupp, C.R., Plant ecological aspects of flood geomorphology and paleoflood history, in *Flood Geomorphology*, edited by V.R. Baker, R.C. Kochel, and P.C. Patton, pp. 335-356, John Wiley, New York, 1988.

Jarrett, R.D., Determination of roughness coefficients for streams in Colorado, *U.S. Geol. Surv. Water Resour. Invest. Rep., 85-4004*, 54 pp., 1985.

Jarrett, R.D., Evaluation of the slope-area method of computing peak discharge, in *Selected Papers in the Hydrologic Sciences 1986*, edited by S. Subitzky, *U.S. Geol. Surv. Water-Supply Pap., 2310*, 13-26, 1986.

Jarrett, R.D., Paleohydrologic techniques used to define the spatial occurrence of floods, *Geomorphology, 3*, 181-195, 1990.

Jarrett, R.D., Paleohydrology and its value in analyzing floods and droughts, in *National Water Summary 1988-89 – Hydrologic Events and Floods and Droughts, U.S. Geol. Surv. Water-Supply Pap., 2375*, 105-116, 1991.

Jarrett, R.D., Historic-flood evaluation and research needs in mountainous areas, in *Hydraulic Engineering-Proceedings of the symposium sponsored by the American Society of Civil Engineers, Buffalo, New York, August 1-5, 1994*, edited by G.V. Cotroneo and R.R. Rumer, pp. 875-879, Am. Soc. of Civ. Eng., Reston, Va., 1994.

Jarrett, R.D., Paleoflood Investigations for Cherry Creek basin, Eastern Colorado, in *Proceedings, 2000 Joint Conference on Water Resources Engineering and Water Resources Planning and Management of the American Society of Civil Engineers, Minneapolis, Minn., July 30-Aug. 2*, edited by R.H. Hotchkiss and M. Glade, Am. Soc. of Civ. Eng., Reston, Va., 2000.

Jarrett, R.D., Paleohydrologic estimates of convective rainfall in the Rocky Mountains, in *Proceedings of the American Meteorological Society, Annual Meeting, Albuquerque, New Mexico, January 14-18, 2001*, p. J2.14-1-J2.14-4, 2001.

Jarrett, R.D., and J.E. Costa, Hydrology, geomorphology, and dam-break modeling of the Lawn Lake Dam and Cascade Lake Dam failures, Larimer County, Colorado, *U.S. Geol. Surv. Prof. Pap., 1369*, 78 pp., 1986.

Jarrett, R.D., and J.E. Costa, Evaluation of the flood hydrology in the Colorado Front Range using precipitation, streamflow, and paleoflood data, *U.S. Geol. Surv. Water Resour. Invest. Rep., 87-4117*, 37 pp., 1988.

Jarrett, R.D., and E.M. Tomlinson, Regional interdisciplinary paleoflood approach to assess extreme flood potential, *Water Resour. Res., 36*, 2957-2984, 2000.

Jennings, M.E., W.O. Thomas, and H.C. Riggs, (compilers) Nationwide summary of U.S. Geological Survey Regional Regression Equations for Estimating Magnitude and Frequency of Floods for Ungaged Sites, 1993, *U.S. Geol. Surv. Water Resour. Invest. Rep., 94-4002*, 196 pp., 1994.

Knox, J.C., Stratigraphic evidence of large floods in the upper Mississippi Valley, in *Catastrophic Flooding*, edited by L. Mayer and D. Nash, pp. 155-180, Allen and Unwin, Boston, 1987.

Knox, J.C., Climate influence on Upper Mississippi valley floods, in *Flood Geomorphology*, edited by V.R. Baker, R.C. Kochel, and P.C. Patton, pp. 279-300, John Wiley, New York, 1988.

Knox, J.C., Large increases in flood magnitude in response to modest changes in climate, *Nature, 361*, 430-432, 1993.

Knox, J.C., Sensitivity of modern and Holocene floods to climate change, *Quat. Sci. Rev., 19*, 439-457, 2000.

Kochel, R.C., and V.R. Baker, Paleoflood analysis using slackwater deposits, in *Flood Geomorphology*, edited by V.R. Baker, R.C. Kochel, and P.C. Patton, pp. 357-376, John Wiley, New York, 1988.

Kochel, R.C., and D.F. Ritter, Implications of flume experiments on the interpretation of slackwater paleoflood sediments, in *Regional Flood Frequency Analysis*, edited by V.P. Singh, pp. 365-384, D. Riedel, Boston, 1987.

Komar, P.D., Selective grain entrainment and the empirical evaluation of flow competence, *Sedimentology, 34*, 1165-1176, 1987.

Levish, D.R. and Ostenaa, D.A., Applied Paleoflood Hydrology in North-Central Oregon, *Guidebook for Field Trip 2, April 19-21, 1996, Cordilleran Section 92nd Annual Meeting, Geological Society of America, Portland, Oregon*, 182 pp., U.S. Bureau of Reclamation Seismotectonics Report 96-7, Denver, Co., 1996.

Levish, D.R., D.A. Ostenaa, and D.R.H. O'Connell, A non-inundation approach to paleoflood hydrology for the event-based assessment of extreme flood hazards, in *11th Annual Conference Proceedings*, pp. 69-82, Assoc. of State Dam Safety Off., Lexington, Ky., 1994.

Levish, D.R., D.A. Ostenaa, and D.R.H. O'Connell, Paleoflood

Hydrology and Dam Safety, in *WATERPOWER '97 Proceedings of the International Conference on Hydropower*, pp. 2205-2214, Am. Soc. of Civ. Eng., Reston, Va., 1997.

Matthai, H.F., Floods in June 1965 in the South Platte River Basin, *U.S., Geol. Surv. Water-Supply Pap., 1850-B*, 64 pp, 1969.

McCain, J.F., L.R. Hoxit, C.F. Chappell, and F. Caracena, Meteorology and hydrology in the Big Thompson River and Cache la Poudre River basins, Larimer and Welds Counties, Colorado, Part A of storm and flood of July 31-August 1, 1976, *U.S. Geol. Surv. Prof. Pap., 1115-A*, 1-85, 1979.

National Research Council (NRC), *Estimating Probabilities of Extreme Floods, Methods and Recommended Research*, 141 pp., National Academy Press, Washington, D.C., 1988.

National Research Council (NRC), *Improving American River Flood Frequency Analyses*, 120 pp., National Academy Press, Washington, D.C., 1999.

O'Connell, D.R.H., D.A. Ostenaa, , and D.R. Levish, 1998, Risk-based hydrology, basesian flood-frequency analysis using paleoflood data and data uncertainties, in *Proc. of the First Federal Interagency Hydrologic Modeling Conference, Las Vegas, NV*, by the Subcomm. on Hydrology of the Interag. Advis. Comm. on Water Data, 1998.

O'Connor, J.E., Hydrology, hydraulics, and geomorphology of the Bonneville Flood, *Geol. Soc. Am. Spec. Pap. 274*, 83 pp., 1993.

O'Connor, J.E., and R.H. Webb, Hydraulic modeling for paleoflood analysis, in *Flood Geomorphology*, edited by V.R. Baker, R.C. Kochel, and P.C. Patton, pp. 403-420, John Wiley, New York, 1988.

O'Connor, J.E., L.L. Ely, E.E. Wohl, L.E. Stevens, T.S. Melis, V.S. Kale, and V.R. Baker, A 4500-year record of large floods on the Colorado River in the Grand Canyon, Arizona, *J. Geol., 102*, 1-9, 1994.

Oreskes, N., K. Shrader-Frechette, and K. Belitz, Verification, validation, and confirmation of numerical models in earth sciences, *Science, 263*, 641-646, 1994.

Ostenaa, D.A., D.R. Levish, D.R.H. O'Connell, and E.A. Cohen, Paleoflood Study for Causey and Pineview Dams, Weber Basin and Ogden River Projects, Utah, 69 pp., U.S. Bureau of Reclamation Seismotectonic Report 96-6, Denver, Co., 1997.

Parrett, C. and R.D. Jarrett, Flood hydrology for Dry Creek, Lake County, Northwestern Montana, *U.S. Geol. Surv. Water Resour. Invest. Rep., 2000-4069*, 12 pp., 2000.

Pruess, J.W., E.E. Wohl, and R.D. Jarrett, Methodology and implications of maximum paleodischarge estimates for mountain channels, Upper Animas River basin, Colorado, U.S.A., *Arctic and Alp. Res., 30*, 40-50, 1998.

Salmon, G., D. Cattanach, and D. Hartford, Risk management at Wahleach Dam, *Civil Engineering, ASCE, 67*(2), 39-41, 1997.

Springer, G.S., and J.S. Kite, River-derived slackwater sediments in caves along Cheat River, West Virginia, *Geomorphology, 18*, 91-100, 1997.

Stedinger, J.R., and V.R. Baker, Surface water hydrology-Historical and paleoflood information, *Rev. Geophys., 25*, 119-124, 1987.

Stedinger, J.R., and T.A. Cohn, Flood frequency analysis with historical and paleoflood information, *Water Resour. Res., 22*, 785-793, 1986.

Trieste, D.J., and R.D. Jarrett, Roughness coefficients of large floods, in *Irrigation and Drainage Division Specialty Conference "Irrigation Systems for the 21st Century", Portland, Oregon, 1987*, edited by L.D. James and M.J. English, pp. 32-40, Am. Soc. of Civ. Eng., Reston, Va., 1987.

U.S. Bureau of Reclamation, A Framework for Characterizing Extreme Floods for Dam Safety Risk Assessment, Prepared by Utah State University and Bureau of Reclamation, 67 pp., 1999.

Waythomas, C.F., and R.D. Jarrett, Flood geomorphology of Arthurs Rock Gulch, Colorado, paleoflood history, *Geomorphology 11*, 15-40, 1994.

Webb, R.H., J.E. O'Connor, and V.R. Baker, Paleohydrologic reconstruction of flood frequency on the Escalante River, South-Central Utah, in *Flood Geomorphology*, edited by V.R. Baker, R.C. Kochel, and P.C. Patton, pp. 403-418, John Wiley, New York, 1988.

Wilcock, P.R., Flow competence, a criticism of a classic concept, *Earth Surf. Processes and Landforms, 17*, 289-298, 1992.

Williams, G.P., and J.E. Costa, Geomorphic measurements after a flood, in *Flood Geomorphology*, edited by V.R. Baker, R.C. Kochel, and P.C. Patton, pp. 65-77, John Wiley, New York, 1988.

Wohl, E.E., Estimating flood magnitude in ungauged mountain channels, Nepal, *Mtn. Res. and Devel., 15*, 69-76, 1995.

Robert D. Jarrett, U.S. Geological Survey, MS-412, Denver Federal Center, Denver, CO, 80225, rjarrett@usgs.gov; John F. England, Jr., U.S. Bureau of Reclamation, P.O. Box 25007, D-8530, Denver Federal Center, Denver, CO, 80225.

One-Dimensional Estimation Techniques For Discharges Of Paleofloods And Historical Floods

Robert H. Webb

U.S. Geological Survey, Tucson, Arizona

Robert D. Jarrett

U.S. Geological Survey, Denver, Colorado

Paleoflood data in the form of paleostage indicators (PSIs) are used as estimates for maximum flood stage. Discharges for paleofloods with PSI data can be estimated using one of the following one-dimensional flow techniques: (1) slope-conveyance method, (2) slope-area method, (3) step-backwater method, and (4) critical-depth method. The purpose of this paper is to review these methods, with an emphasis on their appropriateness in paleoflood studies. The step backwater and critical-depth methods are the most commonly used in paleoflood studies, although early paleoflood work relied on the slope-conveyance method. The underlying hydraulic theory behind these methods is similar, but each method uses different amounts of data and different solution schemes that affect their usefulness in paleoflood studies. Each method relies on the Manning's equation, and energy losses attributed to bed roughness are described with the Manning's n. Techniques for objectively determining Manning's n allow minimizing this source of error, although roughness changes at the time of the paleoflood cannot be objectively calculated but instead must be estimated. Discharge estimates in many previous paleoflood studies have been adequately documented, and such documentation should be made in future studies to aid evaluation of the accuracy of discharge estimates.

INTRODUCTION

Accurate measurement or estimation of flood discharges is an important goal for surface water hydrologists. In arid and semiarid regions, such as the southwestern United States, this goal may be difficult to achieve. Floods that occur on rivers with gaging stations may be measured directly with current meters, providing the most accurate type of discharge measurement, although this rarely occurs because of the vicarious nature of flash flooding in the region. More often, flood discharges are estimated from stage-discharge relations, which are anchored at high discharges by indirect discharge measurements. Discharges are estimated indirectly and post flood using several techniques described in this paper. Finally, for historical floods, or paleofloods that occurred before historical records were kept, the only way to estimate discharge is using indirect techniques.

Measurement or estimation of flow in natural channels—whether at a gaging station or using paleoflood evidence—is inherently uncertain. Even where discharge is directly measured with a current meter, uncertainties approaching 10% are common [*Pelletier*, 1988]. In the western United

Ancient Floods, Modern Hazards: Principles and Applications of Paleoflood Hydrology
Water Science and Application Volume 5, pages 111-125

Table 1. Types of paleostage indicators (PSIs) and their relevance to water-surface elevations of paleofloods.

Quality of data	Type of evidence	Problems with evidence
BEST	Organic drift	May not reflect main-channel flow (flow separation); preservation problems
	Silt lines	May not reflect main-channel flow (flow separation); preservation problems
	Flood deposits	Uncertain relation with water surface (water surface likely was higher than deposit
	Tree scars	Uncertain relation with water surface (water surface could be higher or lower than the scar)
	Terrace deposits	Uncertain depth of water over terrace
	Boulder deposits	Relation between deposits and water surface uncertain
WORST	Hillslope trimlines	Large uncertainty in water surface required to cut trimline in hillslope

States, where much of the research on discharge estimation for paleofloods has occurred, discharges for large floods typically are estimated using indirect methods similar to those used to estimate discharges for paleofloods and historical floods. *Glancy and Williams* [1994] discuss common problems with peak discharge estimation at gaging stations in arid and semiarid regions of the United States, including channel change, rapidly varying or nonuniform flow, high sediment transport rates, and supercritical flow.

The important difference between indirect discharge estimates at gaging stations and for paleofloods is in the amount of field evidence of the water-surface profile (WSP) and the types of reaches that can be used. Immediately after a flood, reaches with gradually varied flow can be chosen for discharge estimation because high-water marks (HWMs) are usually continuous. Evidence of paleofloods, however, is usually confined to specific reaches that may or may not have gradually varied flow, and evidence of WSP may be discontinuous or ambiguous. Information on the extent of scour or deposition during peak discharge is not readily reconstructed for either paleofloods or recent floods. Despite these uncertainties, estimates of paleoflood discharges provide constraints on the magnitude of floods in a river basin, which help minimize uncertainties in flood-frequency analysis [*Stedinger and Cohn*, 1986].

The purpose of this paper is to review the techniques for estimation of discharges for historical floods and paleofloods. In the late 1990s, hydrologists began using multidimensional flow models to estimate discharges for paleofloods [see *Levish*, this volume; *Denlinger et al.*, this volume]. Despite this recent advance in computational ability, we only consider one dimensional models because most studies completed before 1995 and most studies [e.g., *Kite et al.*, this volume; *Pickup et al.*, this volume; *Webb et al.*, this volume] use this type of analysis. Moreover, because of the relatively large expense of data collection for multidimensional models, one-dimensional models will likely continue to be the most commonly used techniques for estimating discharges for paleofloods. Readers are referred to the other papers in this volume as well as *Wohl et al.* [1999] for information on two-dimensional hydraulic analysis.

FIELD EVIDENCE FOR PALEOFLOOD DISCHARGE ESTIMATION

Paleostage Indicators (PSIs)

Discharges of paleofloods and historical floods are estimated using a combination of paleostage indicators (PSIs), modern channel geometry, and hydraulic models. PSIs are preserved along channel margins and have varying ages. PSIs range from flood deposits [termed "slackwater deposits"; *Kochel and Baker*, 1982], which typically are lower than the water-surface elevation; to silt lines, which are HWMs composed primarily of silt, clay, and organic matter plastered against bedrock walls [*Webb*, 1985]; to coarse, bouldery deposits [*Jarrett and Malde*, 1987; *Waythomas and Jarrett*, 1994; *Jarrett and Tomlinson*, 2000]. Other PSIs include scars on riparian trees, which may actually be higher than the water-surface elevation because of pileups; trimlines on hillslope soils, which are ambiguous HWMs and are difficult to date accurately; and terrace deposits, which represent minimum HWMs (Table 1).

The relation between PSIs and water-surface elevation has not been systematically studied until recently [*Jarrett and England*, this volume]. *Kochel* [1980] estimated that actual water depths were 7 to 10% greater than maximum water depths suggested by the tops of flood deposits long the Pecos River in Texas. *Webb* [1985] observed silt lines that graded into flood deposits along the Escalante River in Utah. Flood deposits are typically 0.5 to 0.9 m lower than associated silt lines along the Paria River in northern Arizona [*Webb et al.*, this volume]. Finally, *House et al.* [1996] found that peak discharge estimates based on flood deposits alone underestimated the actual peak discharge, estimated from driftlines, by 30%. These studies indicate that better documentation may be required to fully understand the relation between flood-deposit and water-surface elevations, and that driftlines or silt lines on bedrock walls more accurately reflect paleoflood stages, where this type of evidence is preserved.

Channel Geometry and Roughness

Paleoflood analysts typically use modern channel geometry in conjunction with PSIs to estimate discharges. Investigators generally select reaches that have bedrock or

Figure 1. San Juan River. A. (1921). In 1921, the U.S. Geological Survey explored the San Juan River in search of potential damsites. Eugene C. LaRue took this panoramic photograph about 3 km downstream from the entrance to the canyon to document the first potential damsite (shown drawn onto the photograph on the upstream side of the reach). Very little riparian vegetation is present in this reach. The lack of vegetation is similar to other photographs of this river taken between 1875 and 1945. B. (1998). Today, and partially because of the presence of dams upstream, the San Juan River is lined with riparian vegetation. Much of this vegetation is non-native tamarisk, which clearly was not present when paleofloods occurred and should not be included in roughness estimates. Note the persistence of rocks in the channel, which suggests long-term stability of channel roughness.

other types of stable control for paleoflood analysis. Alluvial channels in the southwestern United States are unstable and have fluctuated up to 30 m in base level over the last several thousand years [*Webb*, 1985; *Webb*, 1987; *Webb and Baker*, 1987; *Smith*, 1990; *Webb et al.*, 1991]. Even some reputedly "bedrock-controlled" channels are unstable; for example, scour depth of the Colorado River at Lee's Ferry, 4 km upstream from a paleoflood site [*O'Connor et al.*, 1994] has reached 10 m historically [*Burkham*, 1986]. Elsewhere in Grand Canyon, late Holocene bed fluctuations of as much as 30 m have been documented [*Webb et al.*, 1999]. Changes in alluvial channels during the period of paleohydrologic record have been estimated and used in discharge estimates for paleofloods [*Fuller*, 1987].

Channel stability can be evaluated using several techniques. Repeat photography has shown that in the absence of debris flows from side canyons, bedrock reaches of rivers on the Colorado Plateau are relatively stable with, in some cases, little discernable change in the channel bed over as long as 130 years [*Webb and Rathburn*, 1988; *Webb et al.*, 1991; *Webb*, 1996]. An example of repeat photography of a bedrock canyon used in paleoflood studies appears in Figure 1. Historical discharge measurements contain channel cross-section data, which are used to assess the stability of alluvial channels [*Topping*, 1997; *Allred and Schmidt*, 1999]. Few methods have been proposed to assess prehistorical channel changes, particularly in bedrock canyons, although *Enzel et al.* [1994] used aerial photography to reconstruct historical channel dimensions after construction effects and a large dam-break flood.

Little historic change has occurred in sediment storage along rivers with drainage areas greater than 10,000 km^2 in southern Utah [*Graf*, 1987], which suggests that historical aggradation and degradation may be minimal. *Webb et al.* [1988] used rudimentary photogrammetry to calculate the amount of channel change at the entrance of a bedrock channel apparent between photographs taken in 1872 and 1985. Other evidence of channel stability or instability can be obtained from the position and age of riparian trees [*Webb*, 1985], aboriginal structures, rock inscriptions, soil profiles, or river terraces. Most paleoflood sites in the southwestern United States have at least lateral bedrock control of channel width; many sites also have bedrock control of channel depth.

Assignment of roughness coefficients for paleoflood discharge estimates is complicated by many factors, including channel change. Riparian vegetation, one of the major sources for floodplain roughness [*Barnes*, 1967], has greatly increased along most rivers in the southwestern United States [*Turner and Karpiscak*, 1980; *Hereford*, 1993; see Figure 1]. Part of the reason for this increase is the invasion of non-native tamarisk along many rivers [*Webb*, 1996], but decreased historical flood frequency may also be important [*Webb et al.*, 1991]. For many studies in this region, and perhaps other regions where paleoflood studies

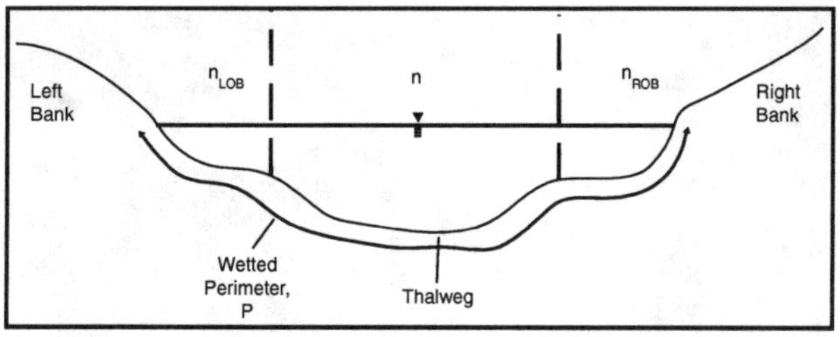

Figure 2. Schematic diagram showing definitions of terms in a channel cross section.

are conducted, riparian vegetation may be either greatly reduced or eliminated as a factor in channel roughness.

METHODS FOR ESTIMATING PALEOFLOOD DISCHARGES

Four methods have been used to estimate discharges from PSIs from paleofloods or HWMs from recent floods: (1) slope-conveyance, (2) slope-area, (3) step-backwater, and (4) critical-depth methods. All four of these methods use one-dimensional hydraulic theory as their basis. Of these methods, the slope-area method is primarily used for indirect-discharge estimates by the U.S. Geological Survey; only the step-backwater and critical-depth methods are widely used in paleoflood discharge reconstructions. Application of these methods is essentially the same for paleofloods and recent floods, except there may be a greater uncertainty in PSIs and channel geometry during paleofloods that needs to be considered.

Manning's Equation and Manning's n

Frictional losses caused by channel roughness are typically accounted for in one dimensional models using Manning's equation,

$$V = a/n \cdot R^{2/3} \cdot S_f^{1/2}, \qquad (1)$$

where V = velocity, a = a constant that is equal to 1 for metric units and 1.49 for English units, n = a roughness coefficient called Manning's n, and S_f = friction slope or energy gradient [*Chow*, 1959]. Manning's equation could be referred to as the Gauckler-Manning equation after its originators [*Williams*, 1970]. The hydraulic radius, R, is equal to A/P, where A = cross sectional area and P = wetted perimeter (see Figure 2). A, R, and P are usually surveyed, whereas n is either estimated or calculated from the modern channel cross section.

In most hydraulic computations, the accuracy of the computed discharge depends on the uncertainty in the selection of Manning's n for the reach in consideration. Manning's n is generally considered to reflect the effect of channel roughness on energy dissipation during flow [*Chow*, 1959]. In another sense, Manning's n is a coefficient that reflects the uncertainty in discharge related to the one-dimensional flow assumption and to factors not included in Manning's equation.

Several tools have been developed to assist with the selection of n values, which can vary longitudinally along the channel, with increasing stage, and seasonally where riparian vegetation is dense. *Chow* [1959] provides general criteria and photographs for selecting n values in channels. *Barnes* [1967] contains color photographs for a wide range of natural channels in the United States with n values computed from known water-surface elevations, channel geometry, and measured discharges; *Hicks and Mason* [1991] have a similar publication for New Zealand. *Limerinos* [1970] studied a range of natural channels in California, calculated n values for a range of flow depths, and developed an equation to estimate n values from bed-material size. Finally, *Brownlee* [1981] quantitatively linked n values to bedforms in a scour model, effectively providing a means for estimating hysteretic changes in channel roughness over the course of flood hydrographs.

Many states and some countries now have graphical publications to aid in assignment of Manning's n values. Using Arizona as an example, *Aldridge and Garrett* [1973] developed guidelines for selection of n values for ephemeral rivers in Arizona that are typical of sand-bed channels in arid regions. *Thomsen and Hjalmarson* [1991] updated the work for a single county (Maricopa), providing color photographs for reaches with verified Manning's n values. *Phillips and Ingersoll* [1998] extended work with verified Manning's n to 14 river and canal reaches. For gravel-bed rivers, they modified a relation from *Cowan* [1956]:

$$n = (n_b + n_1 + n_2 + n_3 + n_4) \cdot m, \qquad (2)$$

where n_b = a base value for a uniform channel, n_1 = correction for surface irregularities, n_2 = correction for variations in channel geometry, n_3 = correction for large roughness elements in the channel, n_4 = correction for riparian vegetation, and m = a correction factor for channel meandering. They provide values of n_b for selected reaches in Arizona, and n_1, n_2, and n_3 usually are 0 for reaches selected for indirect-discharge estimation. For riparian vegetation, *Phillips et al.* [1998] provide an adjustment equation:

$$n_4 = 0.0008 \cdot B - 0.0007, \qquad (3)$$

where B = the percentage of flow blocked by vegetation [also see *Gillen*, 1996, and *Coon*, 1998].

In addition to selection of Manning's n for the channel, overbank areas are considered separately (Fig. 2). The n values for the left and right overbank areas, termed n_{LOB} and n_{ROB}, respectively, are assigned separately from the main channel roughness. The areas of the left and right overbank areas are determined by trapezoidal or rectagonal polygons, as shown in Figure 2. As discussed in *Davidian* [1984, p. 21-22], improper subdivision of cross sections may result in overall roughness coefficients that differ substantially from the component n values.

Several attempts have been made to eliminate Manning's n from the Manning's equation because of the relations between n value, channel geometry, and slope. *Riggs* [1976] developed a simplified slope-area method that can be used to develop a formula for Manning's n in terms of area, hydraulic radius, and slope. *Arcement and Schneider* [1984] calculated n values and developed equations and a guide for selecting n values for densely vegetated flood plains in the southeastern United States. *Jarrett* [1985] studied a variety of channels in the mountains of Colorado, calculated n values for a range of measured flows, and developed an equation to estimate n values as a function of hydraulic radius and slope. *Jarrett* [1985] also summarized the current state-of-the-art methods for selecting n values in natural, urban, and agricultural channels and flood plains.

Slope-Conveyance Method

Manning's equation was developed for uniform flow, for which discharge, n, R, and S_f do not vary over a specified reach. Uniform flow is characterized by parallel elevations of the channel bed, water-surface profile, and energy-grade lines. S_f, the friction slope, is defined as the slope of the energy-grade line (Fig. 3); during uniform flow, S_f is also equal to the slope of the channel bed, S_0, and the water surface. If any of these variables change over a specified reach, the flow is nonuniform and additional relations are required to estimate discharge.

The first estimates of discharges for paleofloods in the southwestern United States [*Kochel*, 1980; *Kochel and Baker*, 1982] were obtained using the slope-conveyance method. This method requires a modified form of Manning's equation:

$$Q = \frac{C}{n} A R^{2/3} S^{1/2}, \qquad (4)$$

where S = a channel slope, which may be the bed slope (S_0) or a slope associated with flow energy (S_f), depending upon the application (see Fig. 3). To solve (4), one cross-sectional area is used with the top of the flood deposits commonly assumed to represent the water-surface elevation at peak discharge [*Kochel*, 1980]. For convenience, the channel conveyance, K, is defined as:

$$K = \frac{CAR^{2/3}}{n}. \qquad (5)$$

Use of only one cross section implies that the section used is representative of average channel geometry, which may not be the case. Because many of the flood deposits were in tributary mouths away from the main channel flow [*Kochel*, 1980], the heights of some of the flood deposits may have represented the water-surface elevation plus some fraction of the energy associated with flow velocity. S was first assumed to equal S_o (uniform flow), and then S was increased to 1.25 S_o after the initial discharge estimates were found to be much less than the discharges recorded for the same floods at gaging stations [*Kochel*, 1980]. Finally, a least-squares curve fitting revealed that discharges estimated from known water-surface elevations were 1.6 to 2.0 times larger than the discharges estimated using the height of flood deposits as the water surface elevation [*Kochel*, 1980]. The results of this study show that considerable uncertainty in discharge estimates can result from assuming uniform flow and use of the slope-conveyance method applied to only one PSI and one cross section. Better results may be achieved by averaging several cross sections before applying the slope-conveyance method (see discussion under Critical Depth Method below).

Flow in natural channels is usually nonuniform, hence large errors can be expected from the slope-conveyance method applied to PSI evidence. Methods of discharge estimation useful for nonuniform flow are derived from a combination of Manning's equation (1), conservation of mass, and conservation of energy. For a reach with no inflows or losses, Q is the same at the entrance and exit of the reach, which can be expressed as the equation of conservation of mass,

$$Q = A_1 V_2 = A_2 V_2, \qquad (6)$$

between cross sections 1 and 2.

Conservation of energy is calculated from the Bernoulli equation, which can be stated as

$$h_1 = h_2 + (h_L)_{1\text{-}2} + k(\Delta(h_v)_{1\text{-}2}), \qquad (7)$$

where h = the total flow head, h_L = friction head loss, and $\Delta(h_v)_{1\text{-}2}$ = the difference in velocity head between sections 1 and 2 (Fig. 3). The energy-loss coefficient, k, conventionally is 0.5 for expanding reaches and 0.0 for contracting reaches [*Davidian*, 1984]. An alternative view is given in *U.S. Army Corps of Engineers* [1998], which advocates 0.3 and 0.1, respectively, for k.

The total flow head is calculated from

$$h = z + \frac{\alpha V^2}{2g}, \qquad (8)$$

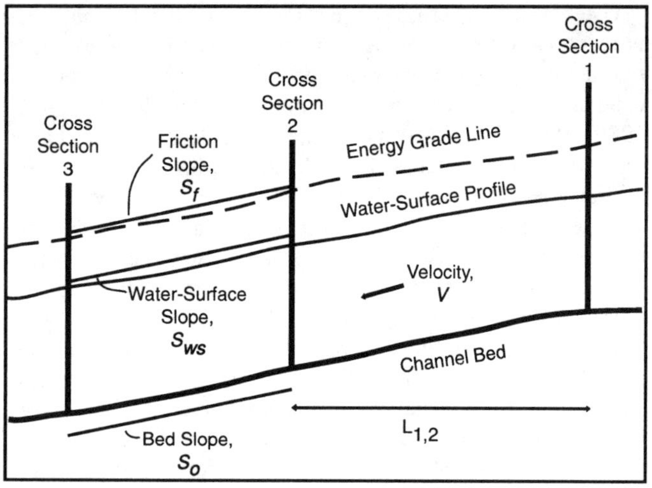

Figure 3. Schematic diagram of a longitudinal profile defining terms used in one-dimensional hydraulics.

where z = water-surface elevation; α = a kinetic-energy correction coefficient for variation in velocity across the channel (α is assumed equal to 1 if no subdivisions of cross sections are made [*Davidian*, 1984]); and g = gravitational acceleration (9.81 m/s^2). The head loss, $(h_L)_{1\text{-}2}$, is calculated from the Manning's equation using

$$(h_L)_{1-2} = L\frac{Q^2}{k_1 k_2}, \tag{9}$$

where L = the reach length between sections 1 and 2. The velocity head is calculated from

$$h_v = \frac{\alpha V^2}{2g}. \tag{10}$$

The energy state of flow is evaluated using the Froude number,

$$F = \frac{V}{(gD)^{0.5}}, \tag{11}$$

where D = mean flow depth, and D = A/T, where A = cross-sectional area and T = top width. Critical depth occurs where F = 1; for F > 1, flow is termed supercritical and for F < 1, flow is termed subcritical. WSPs may be grouped according to Froude number and changes in channel width or slope; the types of WSPs are given in *Henderson* [1966] and *Davidian* [1984]. The transition from supercritical to subcritical flow may cause a hydraulic jump, the size and location of which is a function of the relative flow energies [*Chow*, 1959; *Henderson*, 1966].

Supercritical flow is not sustainable for significant distances in most natural channels [*Jarrett*, 1984; *Trieste*, 1992], although it does occur in certain channels where drops are present [see discussion in *Martínez-Goytre et al.*, 1994]. The prediction of a hydraulic jump in one dimensional hydraulic models should prompt an investigation for the reasons and possible checking of hydraulic parameters. *Martínez-Goytre et al.* [1994] used a predicted n from the method of *Jarrett* [1984] to reduce estimated paleoflood discharges in 8 drainages from mostly supercritical to critical or subcritical. For possible ways of calculating hydraulic jumps in natural channels, the reader is directed to *Hager* [1992].

Slope-Area Method

The slope-area method is commonly used to calculate discharges indirectly after floods [*Dalrymple and Benson*, 1968]. The slope-area method uses the independent data of flow area, as constrained by surveyed HWMs, to estimate discharge. By use of PSIs or HWMs, water-surface elevations, and hence the areas and hydraulic radii, are determined. Using estimates of n, conveyances are calculated. Then, by substituting equations (8), (9), (10), and (6) into the Bernoulli equation (9), discharge is obtained by

$$Q = \sqrt{\frac{(z_1 - z_2)}{(L/K_1K_2) - (1-k)(\alpha_1/A_1^2 - \alpha_2/A_2^2)/2g}}. \tag{12}$$

This can be expressed as $Q = (K_1 K_2 S_f)^{0.5}$, where

$$S_f = \frac{(z_1 - z_2)}{1-(1-k)(K_1K_2/L)(\alpha_1/A_1^2 - \alpha_2/A_2^2)/2g}, \tag{13}$$

or

$$S_f = [\,(z_1 - z\,) + (1 - k)\,(h_{v1} - h_{v2})\,]\,/\,L\,. \tag{14}$$

Dalrymple and Benson [1968] also give expansions of (12) for three or more channel cross sections.

The slope-area method is rarely used in paleoflood discharge estimation because the relation between water-surface elevation and flood deposits is not exactly known. Flood deposits or other PSIs may not be adequate to define the water-surface elevation at all cross sections needed to evaluate nonuniform flow [see discussion in *House and Pearthree*, 1995]. Available PSIs may not be sufficiently numerous or of good enough quality to define a continuous water-surface profile through a reach; in this case, the slope-area method cannot be used. If sufficient PSIs are present in a reach but are not present in a specified cross section, the water-surface elevation for that cross section could be estimated by interpolation.

Quick [1991] provided evidence for high positive bias in slope-area estimates owing to four types of error in post-flood measurements: Manning's n value estimation, channel geometry, estimation of water-surface profiles, and the effects of erosion and sediment transport on hydraulic properties. *Jarrett* [1986, 1987, 1994] discusses the potential for errors in applying the slope-area method for computing peak discharge from HWMs of recent floods in mountain rivers. He found that discharges estimated using the slope-area method could be substantially overestimated because of measurement error of +60% and in some cases up to or greater than +100%. *O'Connor and Webb* [1988] found that use of a two-section slope-area method resulted in overestimation of discharges for paleofloods compared with results from a multi-section step-backwater analysis.

The slope-area method may be most appropriately used at sites with silt lines or other PSIs that are preserved in relatively long and straight reaches and closely represent the water surface elevation of the paleoflood [*Webb et al.*, this volume]. Another important use of the slope-area method is to obtain preliminary discharge estimates to be used in other hydraulic analyses, such as step-backwater analysis or the critical-depth method. By using the slope-area method, which is relatively easy in terms of computational requirements, one could constrain some variables, such as Manning's n, in the more intensive methods.

Step-Backwater Method

The step-backwater method [*Chow*, 1959; *Davidian*, 1984] is the most commonly used method for estimating discharges of paleofloods from PSIs. Numerous studies have relied on this method in the southwestern United States [*Webb*, 1985; *Ely and Baker*, 1985; *O'Connor et al.*, 1986a, 1986b; *Partridge and Baker*, 1987; *Fuller*, 1987; *Roberts*, 1987; *Webb et al.*, 1988; *Smith*, 1990; *O'Connor*, 1993; *Enzel et al.*, 1994] and Australia [*Baker et al.*, 1987; *Wohl et al.*, 1994]. In the step-backwater method, channel geometry and hydraulic properties are determined from field data and water-surface elevations are computed at specific cross sections for various given discharges.

The step-backwater method uses the same hydraulic equations as the slope-area method: the conservation of mass (6), energy (7), and Manning equations (1). Both methods use the same coefficients. However, the slope-area method uses known water-surface elevations to compute the discharge, whereas the step-backwater method uses discharge to compute stage. WSPs for various assumed discharges are compared by trial and error with the heights of PSIs until the profile that best approximates the PSIs is found. The corresponding discharge is adopted as the discharge of the paleoflood that emplaced the PSIs. *O'Connor and Webb* [1988] show applications of the step-backwater method to analyses of paleoflood discharges.

This difference in approach between the slope-area and step-backwater methods is essential to determining the most appropriate method for estimating discharges for paleofloods. To use the slope-area method, the relation between PSIs and water-surface elevation must be known precisely, and there must be a sufficient number of PSIs to define a WSP [*House and Pearthree*, 1995]. With the step backwater method, the WSPs associated with various discharges are computed independently from PSIs. Repeating this computation for a number of specified discharges defines a stage discharge relation at each cross section. This is a standard method of defining stage-discharge relations when direct measurements are not available [*Davidian*, 1984, p. 43].

The step-backwater method can be used even if the HWMs are too few or too poorly defined to determine the WSP required for the slope-area method. Flood deposits represent lower-limit estimates of water-surface elevations [*Kochel and Baker*, 1982], and use of flood deposits as HWMs results in minimal estimates of discharge. Use of silt lines or other PSIs that more closely approximate water-surface elevations yields more accurate estimates of discharge. However, by using the step-backwater method, additional cross sections can be used to account for nonuniform-flow conditions even though water-surface elevations are not known or PSIs are not present [see *O'Connor and Webb*, 1988].

Of several methods used in step-backwater analysis [*Henderson*, 1966], the standard step method is most commonly used for estimation of water-surface profiles in natural channels. Details of the iterative procedure are given by *Davidian* [1984]. Channel geometry and roughness data are obtained for the reach and a value of Q is chosen. Calculation proceeds in an upstream direction for subcritical flow and in a downstream direction for supercritical flow for pragmatic reasons, but there is no theoretical limitation that determines the computational direction. For subcritical flow, a water-surface elevation is chosen for the cross section at the downstream end of the reach (cross section 1) by assuming either critical depth, uniform flow, or by using a known HWM.

The water-surface elevation is computed for the nearest cross section upstream (cross section 2) by trial and error; a trial elevation is assumed and terms from both sides of the Bernoulli equation (5) are calculated. If the difference between the two sides of (5) is greater than a predetermined tolerance limit, a new water-surface elevation is assumed for cross section 2 and the computation is repeated until the difference is less than the tolerance limit. When this is achieved, the calculated water-surface elevation for cross section 2 is accepted, cross section 2 becomes cross section 1, the next cross section upstream becomes cross section 2, and the iteration begins again. Step-backwater analysis is usually performed using a computer model, and commonly used computer programs for step-backwater analysis are WSPRO [*Shearman et al.*, 1986], HEC-2 [*Feldman*, 1981], and HEC-RAS [*U.S. Army Corps of Engineers*, 1998]. The differences among these programs are mainly in user interface and specific routines to handle flow around floodplain structures.

Care must be used when choosing the starting water-surface elevation for step-backwater analysis. Because the starting water-surface elevation in the step-backwater method generally is unknown, convergence of multiple starting water-surface elevations is required to establish normal depth for each selected discharge used to define the stage-discharge relation [*Davidian*, 1984]. For subcritical flow, the channel geometry should be defined downstream from the most downstream HWM or PSI for a distance such that the computed stage at the HWM or PSI is not sensitive to the choice of downstream starting elevation. The necessary distance decreases as the channel slope increases; an approximate formula for an M1 profile is given by *Bailey and Ray* [1966] as:

$$(L \cdot S_o)/\, y_o = 0.8598 - 0.6398 \cdot (S_o \cdot C^2)/g\,, \qquad (15)$$

where C = the Chezy roughness coefficient. *Bailey and Ray* [1966] provide curves to make this estimate.

For each discharge, starting elevations should be chosen high enough and low enough to demonstrate convergence of M1 and M2 profiles to normal depth. The resultant normal-depth profiles should be consistent with the pattern of the HWM and PSI; if they are not, that is an indication that channel geometry may have changed, or the channel was partially dammed during the floods, or a hydraulic control downstream has affected the chosen reach. The convergence criterion requires additional cross sections to be defined at the downstream end of a subcritical flow reach for proper analysis. Since these sections are used primarily to eliminate the effects of the starting elevation, they do not need to be defined as precisely as those within the study reach. These cross sections might perhaps be scaled from available maps, but a field survey might still be necessary to establish the downstream channel-bed elevations and channel slope.

Critical-Depth Method

A special case of the step-backwater method is the critical-depth method [*Chow*, 1959; *Barnes and Davidian*, 1978; *Jarrett*, 1986]. This method relies on the occurrence of critical flow at a constriction or drop-off in the river downstream from recent HWMs or PSIs and requires the assumption of subcritical flow upstream. The starting water-surface elevation is known or assumed to be equal to critical depth at the construction or drop-off. This method is particularly advantageous if PSIs are available in the critical flow reach, documenting that the drop-off is not drowned out by conditions downstream. A particular advantage to this method is its insensitivity to roughness and energy loss coefficients.

Like the step-backwater method, the WSP is computed for each discharge and compared with HWMs and PSIs. In contrast to the step-backwater method, in which insensitivity to downstream starting elevation must be demonstrated by multiple computations, critical-depth computations begin at a single specified cross section (the critical-flow section) and at a precise starting elevation (critical depth). The occurrence of critical depth must then be demonstrated with HWMs from recent floods, from PSIs from recent, historic, or paleofloods, or by subsequent hydraulic analysis.

In natural rivers, critical depth occurs most commonly at channel constrictions, where the bed slope increases, or in water falls or drop-offs. Width contractions, which may be either natural or human-made constrictions such as bridge embankments, generally are used more on lower-gradient rivers, whereas natural slope increases and drop-offs are more commonly used for higher-gradient rivers [*Jarrett*, 1986]. Drop-offs also can be human-made, such as at weirs. At constrictions or slope increases, particularly for floods, flow goes from subcritical to critical or supercritical, but supercritical flow generally occurs only for short distances in natural channels [*Jarrett*, 1984, 1986; *Trieste and Jarrett*, 1987].

Preferred natural critical-depth sites are drops controlled by bedrock outcrops. In other pool-drop rivers, debris bars consisting of agglomerations of cobbles and boulders creates drops that may possibly create critical-flow conditions, if the drop is of sufficient height and the debris bars can be shown to be stable. The channel bottom typically is a series of steps, and the spacing and height are controlled by the bed slope [*Judd and Peterson*, 1969]. As the slope increases, spacing decreases and height increases. These steps develop as deposits of boulders across the river and seem to be formed during extremely large floods [*O'Connor et al.*, 1986b]. The steps usually remain stable during smaller floods; hence, they also may be used as drop-off controls for critical-depth sites.

The critical-depth method was used fairly extensively in the first half of the 1900s, but it was replaced by the slope-area method for computing flood discharges for recent floods [*Jarrett*, 1986]. Recently, however, the critical-depth method is being used more widely. Use of the critical-depth method is advocated for high-energy channel conditions, particularly where pool drop morphology is prevalent, by *Dobbie and Wolf* [1953], *Randle and Humphrey* [1984], *Jarrett* [1984, 1986], and *Trieste and Jarrett* [1987]. Examples of the critical-depth method include *Randle and Humphrey* [1984], *Jarrett* [1986], *Jarrett and Costa* [1986], *Jarrett and Malde* [1987], and *Jarrett and Tomlinson* [2000]. *Jarrett* [1986] and *Trieste and Jarrett* [1987] discuss the hydraulic and hydrologic justification for use of the critical-depth method for computing peak discharge of recent floods.

A drop through critical depth may not be a necessary criterion for use of the critical-depth method in rivers with slopes greater than ~0.01 m/m. *Jarrett and Tomlinson* [2000] and *Jarrett and England* [this volume] found that relatively long reaches in rivers with high bed slopes can be at critical flow, and that averaging 3-6 critical-depth estimates in such a reach will obtain a realistic discharge estimates within ±15% of discharges measured with current meters.

SUGGESTED FIELD PROCEDURES

Field-survey requirements for the estimation of paleoflood discharges are similar for all methods of discharge estimation. For the flow computation to be credible, it must be possible to assess the uncertainties in the application of the hydraulic equations and in the values of the various parameters. Therefore, it is essential for the investigator, during both field and office work, to anticipate as many questions as possible about the data reliability, to investigate those questions as thoroughly as possible, and to document the results explicitly and in detail. Although it may not always be possible to reduce the uncertainties in paleoflood discharge estimations, it is essential to document them as explicitly as possible.

Use of PSIs in discharge estimation requires care because of their uncertain relation to water-surface elevation. Many different types of PSIs may be present in a reach used for paleoflood analysis. Consequently, the type, location, and quality of PSI evidence must be carefully described in the field. For example, flood deposits in tributary mouths may not be representative of the water-surface profile in the main channel. The relations between PSI elevations and main-channel water-surface elevations depend on the configuration of the flow streamlines near the PSIs, and this in turn depends on the topography of the stream channel and bank near the PSIs. Therefore the topographic setting of each PSI should be documented, preferably with a scaled plan view. If the channel bed and banks are straight and have no irregularities near the PSI, that fact should be documented. Photographs may aid in evaluation or reinterpretation of PSIs during analysis.

Depending on the accuracy requirements, channel cross sections and their location can be obtained by field surveys [*Jarrett and Costa*, 1986], topographic maps [*Jarrett and Malde*, 1987], photogrammetric techniques [*Denlinger et al.*, this volume], and (or) stereo aerial photographs with field-surveyed ground control [*Enzel et al.*, 1994]. *Benson and Dalrymple* [1967] and *Davidian* [1984] give general information on collection of field data on HWMs and selection of location for cross sections. If the amount of channel change can be quantified, this information should be used to evaluate the uncertainty of discharge estimates. Steps taken to determine channel stability, scour, or fill should be documented, along with any evidence obtained. A definite conclusion as to stability, instability, or existence of uncertainty should be stated.

Channel cross sections are located in the field at major constrictions, expansions, or changes in bed slope (for examples of placement of cross sections, see *Benson and Dalrymple*, 1967; *O'Connor and Webb*, 1988; *Davidian*, 1984]. HWMs, PSIs, and inferred water-surface elevations are plotted and cross sections are located at major changes in the slope of the water surface profile [*Benson and Dalrymple*, 1967]. A profile showing the HWMs and PSIs should be prepared even if there are not enough marks to define a continuous water-surface profile. To help define breaks in channel slope, a low-water profile, a profile of the top of bank of the active channel, and profiles of other relevant geomorphic features may be plotted. Criteria for subdivision of channel cross sections are given in *Davidian* [1984].

Manning n values are selected in the field for subreaches using methods described above. Channel features on which the n-values are based should be documented, including particle size of bed- and bank-materials, channel-boundary irregularities, vegetation density, and other contributors to flow resistance. Photographs showing cross sections from upstream and downstream directions are highly desirable for substantiation of n values.

For the critical-depth method, special attention must be given to the control section at which critical flow is assumed and must be ascertained in subsequent hydraulic analysis. A cross section is required at the control, which commonly is the narrowest width in the constriction or at the brink of the increase in channel slope or drop-off. Although criteria have not been determined for exactly where the control section should be located, sensitivity analysis has indicated that a certain range in position is acceptable as long as the cross sections where HWMs or PSIs are located are at least 2.5 times the critical depth upstream of the control section [*Barnes and Davidian*, 1978; *Jarrett and Malde*, 1987]. Because flow just upstream of the control section is highly nonuniform, two to three cross sections (often synthesized from the control section) and spaced very close to the control section help to calculate nonuniform flow in the control section.

A cross section should be located a short distance downstream from the critical-depth control to assess whether the assumption of critical depth is correct in the hydraulic analysis. Additional cross sections at approximately channel width spacing upstream of the control and at the locations of the PSIs and HWMs also are needed. Averaging of cross sections helps reduce the uncertainty in discharge estimates from both the slope-conveyance and critical-depth methods.

COMPUTER ANALYSIS

Numerous computer programs are available to perform slope-area and step-backwater analyses. Step-backwater computer programs are routinely used to compute water-surface profiles for a range of specified discharges that are anticipated to cover the range of elevations of PSIs or HWMs. From this analysis, a stage-discharge relation can be determined for each cross section where the PSIs or HWMs are located and the flood discharge is determined. The model output provides hydraulic information such as Froude numbers for the specified discharges at each cross section to assess whether critical flow occurred for the discharge being assessed.

An essential component of any flood-discharge estimate is a sensitivity analysis of parameters that affect the reliability of the discharge so that uncertainty can be quantified given the constraints of field data. This includes making many step-backwater analyses to compute flood discharge with ranges of n values, contraction and expansion coefficients, assessments of scour and deposition, uncertainty in amount of ineffective flow area, and the uncertainty of the elevation of the PSIs for the reach of channel being analyzed [*Jarrett and Malde*, 1987; *O'Connor and Webb*, 1988]. A range of discharges may then be related to the PSIs or HWMs, and the most reasonable discharge can be chosen from the hydraulic analyses.

Use of computer models of the step-backwater and critical-depth methods, aided by improved methods of selecting n values (see previous section), has increased the accuracy of discharge estimates for recent and historical floods and paleofloods. These methods are flexible and can be used over a wide range of channel conditions. "Engineering judgment" may be required to evaluate the uncertainties associated with discharge estimation for paleofloods. Additional research will continue to improve the estimation techniques.

ASSUMPTIONS, ERRORS, AND QUALITY CONTROL

Assumptions concerning the effective flow area, hydraulic interpretation of PSIs, and choice of energy-loss coefficients cause uncertainties in discharge estimates. Use of the step backwater method requires the assumption of one-dimensional flow in the channel reaches. Although standard methods suggest choice of reaches with minimal hydraulic complications [*Benson and Dalrymple*, 1967; *Davidian*, 1984], the hydraulics of sites conducive to deposition of flood deposits or preservation of PSIs are usually complicated by ineffective flow areas or eddies. In addition, evidence may be preserved in channel bends, which might require calculation of superelevation [*Apmann*, 1973] in the discharge-estimation procedure.

Uncertainties in the sizes of ineffective flow areas and energy-loss coefficients may cause errors in the discharge estimates because the effective cross-sectional area is not known precisely. Flood deposits are assumed to represent the WSP of the mean channel flow. Use of flood deposits in tributary canyons, which are removed from the flow in the main channel, will introduce errors in discharge because these deposits are indicative of both the water-surface elevation and some fraction of the main-channel velocity head. Choice of n and k values introduces additional uncertainty in discharge estimates, although errors associated with reasonable values of n and k are much less than the errors created by uncertainties in cross sectional areas [*Jarrett and Malde*, 1987; *O'Connor and Webb*, 1988].

Kirby [1987] examined the effect of uncertainty in P, A, n, and k on the accuracy of discharges calculated using a general form of (12). He found that errors in discharge become large as the velocity head, h_v, and (or) head losses, h_L, exceed the change in water-surface elevation over the reach. Also, errors increase rapidly when lateral variation in velocity is large. The accuracy of discharge estimates is most affected by uncertainties in scour and fill [*Kirby*, 1987]. In the case of paleofloods, uncertainty in PSIs relative to the actual water-surface elevation increases potential errors. Although uncertainty in n values induces the same order-of-magnitude error in discharge, errors in flow depth caused by errors in water-surface elevation, scour, or fill induce larger errors in discharge.

Discharges estimated for paleofloods have inherent uncertainty because of the potential for channel change between floods, and channel cross sections are measured decades to thousands of years after the flood has occurred. For this reason, bedrock-controlled channels are the preferred sites for paleoflood-discharge estimates because the channel generally is stable and cross sections change little with time, although in extremely large floods plucking of bedrock blocks may create slight changes in channel cross sections. Discharge data estimated from PSIs, and any extreme flood, should be carefully evaluated for uncertainties before the data are used in flood-frequency analysis.

Paleoflood data should be subject to quality-assurance provisions similar to those for regular flood-discharge estimates. Standards for indirect-discharge procedures are presented in the relevant U.S. Geological Survey publications cited in this chapter. The special circumstances of paleoflood hydrology mean that certain kinds of evidence and assumptions sometimes may have to be accepted, even though they might be deemed unacceptable for studies of contemporary floods. For example, it may be necessary to work with fewer HWMs or PSIs than might be desired; it might be necessary to accept PSIs that are not as representative of the main channel velocity head as might be desired; it may be necessary to work in channels that have an unknown amount of ineffective flow area, high Froude numbers, or other adverse hydraulic characteristics; or it may be necessary to make questionable assumptions about channel stability and roughness characteristics. Many of these same issues are problems with indirect-discharge estimates made shortly after a flood has occurred. Despite the potential for higher uncertainties in paleoflood discharge estimates, the value of paleoflood discharge estimates is extremely high for many hydrologic studies, particularly those involving estimation of large recurrence intervals (floods > 100 year recurrence interval). Uncertainty in paleoflood estimates, if it can be reasonably quantified, can be used in flood-frequency analysis [*Blainey et al.,* this volume].

The necessity to sometimes work with less-than-optimal data in paleoflood discharge estimation makes it all the more imperative to document completely and in detail all aspects of the data collection and analysis so that the reliability and accuracy of the data and results can be assessed and evaluated as objectively as possible. The documentation needs to be made not only in the field notes, but also in any reports, informal or formal, whose purpose is to present the measurement results.

Table 2 lists the typical information that should be collected for paleoflood discharge estimates. The number of PSIs and locations of PSIs relative to the main flow streamlines should be listed for each paleoflood that is analyzed or shown graphically. The type of PSIs and their relation to the main channel water surface should be included, preferably in map form. Elevations of the PSIs and corresponding calculated water-surface elevations or ranges should be provided as well as the total fall of channel and water surface, length of the reach, and number and location of cross sections. Any unusual hydraulic conditions that are present should be listed, especially constrictions. Channel cross-section configuration and dimensions, including area, hydraulic radius, top width, and presence of ineffective flow areas or flood plains, should also be included if space allows. Characteristics used to assign n values, including particle size of bed and bank material and character of vegetation, should be discussed explicitly in reports. In addition, information about the computational procedures should be given, including identification of procedure (and computer program, if applicable), means used to establish downstream boundary conditions, and means used to ensure that the computations were representative of the hydraulic situation in the field. All of these matters are legitimate concerns of potential users of paleoflood data.

COMPARISON OF DISCHARGE ESTIMATES FOR LARGE FLOODS AT GAGING STATIONS AND PALEOFLOOD SITES

Few large floods in gaging records were directly measured with current meters. Instead, the discharges of these floods were estimated using the techniques described in this paper. If the methods described here are used correctly, this allows direct comparison of paleoflood discharges with historic floods. In past studies, discharges estimated for historic floods on the basis of PSIs generally compare favorably with discharges recorded at gaging stations or discharges estimated immediately after the flood [*Webb and Rathburn*, 1989; Table 3]. Similarly, *Jarrett and Tomlinson* [2000] and *Jarrett and England* [this volume] found that the critical-depth discharge estimates were ±15% of discharges estimated using current meters. Some notable discrepancies have been found between discharges estimated using paleoflood techniques and estimated at gaging stations for the same flood [*Partridge and Baker*, 1987; *O'Connor et al.*, 1986a; *Roberts*, 1987]. Discharges estimated using paleoflood techniques typically are lower than discharges estimated at gaging stations using indirect techniques. The discrepancies have been explained either by differences in discharge-estimation technique [slope-area versus step-backwater technique; *O'Connor et al.*, 1986a] or by flow from tributaries between the flood-deposit site and the gaging station [*Partridge and Baker*, 1987].

Several other methods can be used to assess the accuracy of paleoflood discharges. On many rivers in the southwestern United States, the largest flood in the gaging record may be two or more times larger than the second largest flood [*Webb et al.*, this volume]. Such discrepancies are the reasons for outlier tests in standardized flood-frequency analyses [*U.S. Water Resources Council,* 1982],

Table 2. Information needed to support discharge estimates for paleofloods or historical floods. These data should be appropriately archived for future use by other interested hydrologists.

BASIC DATA
Map of the entire reach with cross sections labeled
Plan map of local channel geometry at flood deposits or other PSI
Paleostage indicators (PSIs) or high-water marks (HWMs)
Number of PSIs or HWMs
Type of PSI or HWM
Relation between PSI or HWM and flow
Elevations of PSI or HWM
Plots of channel cross sections
Number of cross sections
Distance between cross sections (L)
How cross sections were located
Area and hydraulic radius of cross sections (A and R)
Amount of ineffective flow area and method of determination
Subdivision of cross sections
Manning-n values and method for determination
Expansion and contraction coefficients (k)
COMPUTATIONAL RESULTS
Type of hydraulic model
Comparison of computed and known water-surface profiles
Convergence of different profiles obtained in convergence tests (step-backwater method)
Froude number and velocity for each cross section
Method for determining and amount of scour or fill

Table 3. Summary of paleoflood-discharge estimates for rivers or creeks in the southwestern United States [modified from *Webb and Rathburn*, 1989].

Drainage name	Drainage area (km^2)	Number of paleofloods	Largest recorded paleoflood (m^3/s)	Date of [a] largest paleoflood (AD)	Length of paleoflood record (ka)	Method used [b] for estimating discharge	Number of cross sections	Reference
TEXAS								
Pecos River	9,500	30	27,400	1954	9.5	SC	1	*Kochel*, 1980
Devils River	7,100	0	16,700	1954	8.7	SC	1	*Kochel*, 1980
UTAH								
San Juan River	60,000	10	4,300	1911	1.0	SA, SBW	35	*Orchard*, 2001
Paria River	3,670	5	2,400	prehistoric	4.2	SBW	32	*Webb et al.*, this volume
Escalante River	810	18	720	950	2.1	SBW	10	*Webb et al.*, 1988
Boulder Creek	450	3	400	1650-1950	1.0-1.1	SBW	27	*O'Connor et al.*, 1986b
ARIZONA								
Colorado River	173,000	15	14,000	350-750	4.5	SBW	15	*O'Connor et al.*, 1994
Salt River	21,000	27	11,900	850	1.1	SBW	12	*Fuller*, 1987
Verde River	15,000	12	4300	<1200	>1.4	O[d]	n/a[d]	*House et al.*, this volume
Salt River	11,000	14	4,600	<950	>0.6	SBW	24	*Partridge and Baker*, 1987
Kanab Creek	5,370	10 - 12	600	1550	0.5	SBW	11	*Smith*, 1990
Tonto Creek	1,630	45	1,000	1980	0.28	SBW	13	*O'Connor et al.*, 1986a
Tonto Creek	1,630	46	1,500	1993	0.28	SBW	13	*Fuller et al.*, 1996
Aravaipa	1,340	6	970	unknown	1.1	SBW	28	*Roberts*, 1987
Tanque Verde	110	1	220	historic	10-100	SBW	6	*Martínez-Goytre et al.*, 1994
Buehman	103	1	240	historic	10-100	SBW	17	*Martínez-Goytre et al.*, 1994
Canada del Oro	80	1	105	historic	10-100	SBW	8	*Martínez-Goytre et al.*, 1994
Edgar	74	1	100	historic	10-100	SBW	8	*Martínez-Goytre et al.*, 1994
Sutherland	55	1	95	historic	10-100	SBW	6	*Martínez-Goytre et al.*, 1994
Alder	46	1	45	historic	10-100	SBW	8	*Martínez-Goytre et al.*, 1994
Youtcy	24	1	40	historic	10-100	SBW	10	*Martínez-Goytre et al.*, 1994
Pima	10	1	50	historic	10-100	SBW	7	*Martínez-Goytre et al.*, 1994

[a] Dates of floods are given in calendar years A.D. that were obtained in some cases from conversion of radiocarbon dates.
[b] SC—slope conveyance, SA—slope-area method, SBW—step-backwater method, CD—critical-depth method, O—other.
[c] Discharges are lower than maximum discharge reported by U.S. Geological Survey.
[d] see *House et. al*, [this volume] and *Denlinger et al.*, [this volume] for discussion

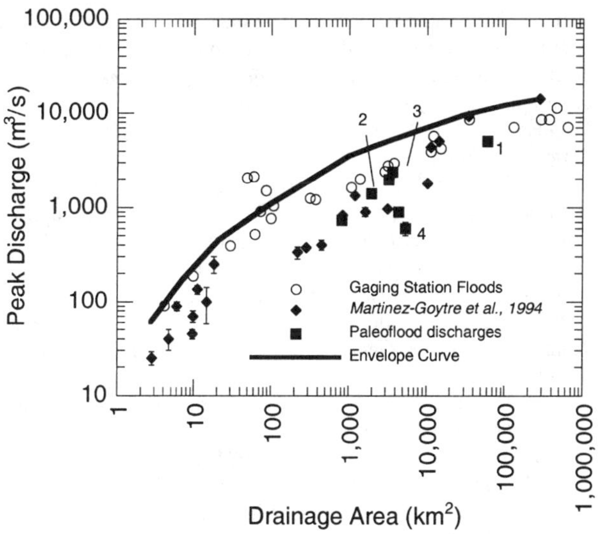

Figure 4. Enveloping curves for maximum floods, recorded at gaging stations and in paleoflood records, for the Colorado Plateau [modified from *Enzel et al.*, 1993]. (1) San Juan River at Bluff, Utah; paleoflood discharge from 1911 flood (Lynn Orchard, unpublished data, 2000); (2) Escalante River near Escalante, Utah; paleoflood discharge from a flood about 1.0 ka [*Webb et al.*, 1988]; (3) Paria River at Lee's Ferry; paleoflood discharge from large prehistoric flood [*Webb et al.*, this volume]; and (4) Kanab Creek downstream of Kanab, Utah; paleoflood discharge from a large prehistoric flood [*Smith*, 1990].

but for some rivers, the large floods may result from either unusual hydroclimatic conditions or from several populations of floods (*e.g.*, snowmelt versus rainfall floods). It is not unusual in paleoflood hydrology to obtain paleoflood discharges that are significantly higher than floods recorded in the gaging record, and probably for the same hydroclimatic reasons.

Enveloping curves are another method of checking whether paleoflood discharges are reasonable, particularly when they are significantly larger than floods recorded at gaging stations. *Enzel et al.* [1993] compared the discharges for the largest historical floods at gaging stations and paleofloods on the Colorado Plateau using an enveloping curve (Fig. 4). The largest drawback to this approach is that flood discharges are discriminated using potential contributing drainage area as the only independent variable, and other factors, including proximity to storm-producing moisture sources and drainage orientation, may influence peak discharges regionally. Despite these caveats, the technique is useful for assessing the reasonableness of paleoflood discharge estimates. For the Colorado River basin, paleoflood discharges are only slightly less than the enveloping curve of the largest floods in the region plotted by drainage area [*Webb and Baker*, 1987; *Enzel et al.*, 1993].

Acknowledgments. The authors thank John Costa, Jim O'Connor, and Ellen Wohl for critically reviewing this manuscript.

REFERENCES

Aldridge, B.N., and J.M. Garrett, Roughness coefficients for streams in Arizona, *U.S. Geological Survey Open-File Report*, 87 pp., 1973.

Allred, T.M., and J.C. Schmidt, Channel narrowing by vertical accretion along the Green River near Green River, Utah, *Geological Society of America Bulletin*, 111, 1757-1772, 1999.

Apmann, R.P., Estimating discharge from superelevation in bends, *Journal of the Hydraulics Division, Proceedings of the American Society of Civil Engineers*, HY1, 65-79, 1973.

Arcement, G.J., Jr., and V.R. Schneider, *Guide for selecting Manning's roughness coefficients for natural channels and flood plains*, Federal Highway Administration, Report No. FHSA TS-84-204, 62 p., U.S. Department of Transportation, Washington, D.C., 1984.

Bailey, J.F., and H.A. Ray, Definition of stage-discharge relation in natural channels by step backwater analysis, *U.S. Geological Survey Water-Supply Paper 1869-A*, 24 pp., 1966.

Barnes, H.H., Jr., Roughness characteristics of natural channels, *U.S. Geological Survey Water Supply Paper 1849*, 213 p., 1967.

Barnes, H.H., Jr., and Jacob Davidian, Indirect methods, pp. 149-204, in Hershey, R.W. (ed.), *Hydrometry -- Principles and practices*, John Wiley, New York, 1978.

Baker, V.R., G. Pickup, and R.H. Webb, Paleoflood hydrologic analysis at ungaged sites, central and northern Australia, pp. 321-334, in Singh, V. P. (ed.), *Regional flood frequency*, D. Reidel Publishing Company, Boston, 1987.

Benson, M.A., and T. Dalrymple, General field and office procedures for indirect discharge measurements, *U.S. Geological Survey Techniques in Water Resources Investigations*, Book 3, Chapter A1, 30 pp., 1967.

Brownlee, W.R., *Prediction of flood depth and sediment discharge in open channels*, Report No. KH-R-43A, W.M. Keck Laboratory of Hydraulics and Water Resources, Division of Engineering and Applied Science, California Institute of Technology, Pasadena, 232 pp., 1981.

Burkham, D.E., *Trends in selected hydraulic variables for the Colorado River at Lees Ferry and near Grand Canyon for the period 1922-1984*, U.S. Department of Commerce, National Technical Information Service, Report PB88-216098, GCES/07/87, Springfield, Virginia, 1986.

Chow, V.T., *Open channel hydraulics*, McGraw Hill, New York, 680 pp., 1959.

Coon, W.F., Estimation of roughness coefficients for natural stream channels with vegetated banks, *U.S. Geological Survey Water-Supply Paper* 2441, 133 pp., 1998.

Cowan, W.L., Estimating hydraulic roughness coefficients, *Agricultural Engineering*, 37(7), 473 475, 1956.

Dalrymple, T., and M.A. Benson, Measurement of peak discharge by the slope-area method, *U.S. Geological Survey Techniques in Water Resources Investigations*, Book 3, Chapter A2, 12 pp., 1968.

Davidian, J., Computation of water-surface profiles in open channels, *U.S. Geological Survey Techniques in Water Resources Investigations*, Book 3, Chapter A14, 48 pp., 1984.

Dobbie, C.H., and P.O. Wolf, The Lynmouth flood of August

1952, London, *Proceedings of the Institute of Civil Engineers*, 2, 522-588, 1953.

Ely, L.L., and V.R. Baker, Reconstructing paleoflood hydrology with slackwater deposits: Verde River, Arizona, *Physical Geography*, 6(2), 103-126, 1985.

Enzel, Y., L.L. Ely, P.K. House, V.R. Baker, and R.H. Webb, Paleoflood evidence for a natural upper bound to flood magnitudes in the Colorado River basin, *Water Resources Research*, 29, 2287-2297, 1993

Enzel, Y., L.L. Ely, J. Martinez-Goyre, and R.G. Vivian, Paleofloods and a dam-failure flood on the Virgin River, Utah and Arizona, *Journal of Hydrology*, 153, 291-315, 1994.

Feldman, A.D., HEC models for water resources system simulation, theory and experience, *Advances in Hydroscience*, 12, 297-423, 1981.

Fuller, J.E., *Paleoflood hydrology of the alluvial Salt River, Tempe, Arizona*, M.S. Thesis, University of Arizona, Tucson, 70 pp., 1987.

Gillen, D.F., Determination of roughness coefficients for streams in west-central Florida, *U.S. Geological Survey Open-file Report* 96-226, 39 pp., 1996.

Glancy, P.A., and R.P. Williams, Problems with indirect determinations of peak streamflows in steep, desert stream channels, *American Society of Civil Engineers Hydraulics Division 1994 National Conference on Hydraulic Engineering and Symposium on Fundamentals and Advancements in Hydraulic Measurements and Experimentation*, Buffalo, New York, 1994.

Graf, W.L., Late Holocene sediment storage in canyons of the Colorado Plateau, *Geological Society of America Bulletin*, 99, 261-271, 1987.

Hager, W.H., *Energy dissipators and hydraulic jump*, Kluwer Academic Publishers, Dordrecht, The Netherlands, 288 pp., 1992.

Henderson, F.M., *Open channel flow*, MacMillan, New York, 522 pp., 1966.

Hereford, R., Entrenchment and widening of the upper San Pedro River, Arizona, *Geological Society of America Special Paper 282*, 46 pp., 1993.

Hicks, D.M., and P.D. Mason, *Roughness characteristics of New Zealand rivers*, Water Resources Survey DSIR, Marine and Freshwater, Christchurch, New Zealand, 329 pp., 1991.

House, P.K., and P.A. Pearthree, A geomorphologic and hydrologic evaluation of an extraordinary flood discharge estimate: Bronco Creek, Arizona, *Water Resources Research*, 31, 3059-3073, 1995.

House, P.K., P.A. Pearthree, and J.E. Fuller, Hydrologic and paleohydrologic assessment of the 1993 floods on the Verde River, central Arizona, *Arizona Geological Survey Open-File Report* 95-20, Tucson, Arizona, 23 pp., 1995.

Jarrett, R.D., Hydraulics of high-gradient streams, *Journal of the Hydraulics Division, American Society of Civil Engineers*, 110(11), 1519-1539, 1984.

Jarrett, R.D., Determination of roughness coefficients for streams in Colorado, *U.S. Geological Survey Water-Resources Investigation Report* 85-4004, 54 pp., 1985.

Jarrett, R.D., Evaluation of the slope-area method for computing peak discharge, *U.S. Geological Survey Water-Supply Paper* 2310, 13-24, 1986.

Jarrett, R.D., Errors in slope-area computations of peak discharges in mountain streams, *Journal of Hydrology*, 96, 53-67, 1987.

Jarrett, R.D., Historic-flood evaluation and research needs in mountainous areas, pp. 875-879 in Cotroneo, G.V., and R.R. Rumer, R.R. (eds.), *Hydraulic Engineering, Proceedings of the American Society of Civil Engineers*, Buffalo, New York, August 1-5, 1994, New York, American Society of Civil Engineers, 1994.

Jarrett, R.D., and J.E. Costa, Hydrology, geomorphology, and dam-break modeling of the Lawn Lake Dam and Cascade Lake Dam failures, Larimer County, Colorado, *U.S. Geological Survey Professional Paper* 1369, 78 pp., 1986.

Jarrett, R.D., and H.E. Malde, Paleodischarge of the late Pleistocene Bonneville Flood, Snake River, Idaho, computed from new evidence, *Geological Society of America Bulletin*, 99, 127-134, 1987.

Jarrett, R.D., and E.T. Tomlinson, Regional interdisciplinary paleoflood approach to assess extreme flood potential, *Water Resources Research*, 36(10), 2957-2984, 2000.

Judd, H.E., and D.F. Petersen, *Hydraulics of large bed element channels*, Utah Water Research Laboratory, Report no. PRWG17-6, Utah State University, Logan, 115 pp., 1969.

Kirby, W.H., Linear error analysis of slope-area discharge determinations, *Journal of Hydrology*, 96, 125-138, 1987.

Kochel, R.C., *Interpretation of flood paleohydrology using slackwater deposits, lower Pecos and Devils Rivers, Southwestern Texas*, Ph.D. dissertation, University of Texas, Austin, 364 pp., 1980.

Kochel, R.C., and V.R. Baker, Paleoflood hydrology, *Science*, 215, 353-361, 1982.

Limerinos, J.T., Determination of the Manning coefficient from measured bed roughness in natural channels, *U.S. Geological Survey Water-Supply Paper* 1898-B, 47 pp., 1970.

Martínez-Goytre, J., P.K. House, and V.R. Baker, Spatial variability of small-basin paleoflood magnitudes for a southeastern Arizona mountain range, *Water Resources Research*, 30(5), 1491-1501, 1994.

O'Connor, J.E., Hydrology, hydraulics, and geomorphology of the Bonneville flood, *Geological Society of America Special Paper 274*, 83 pp., 1993.

O'Connor, J.E., J.E. Fuller, and V.R. Baker, *Late Holocene flooding within the Salt River basin, central Arizona*, Department of Geosciences paper, University of Arizona, Tucson, 84 pp., 1986a.

O'Connor, J.E., R.H. Webb, and V.R. Baker, Paleohydrology of pool-and-riffle pattern development, Boulder Creek, UT, *Geologic Society of America Bulletin*, 97, 410-420 1986b.

O'Connor, J.E., and R.H. Webb, Hydraulic modeling for paleoflood analysis, pp. 393-402 in Baker, V.R., R.C. Kochel, and P.C. Patton, (eds.), *Flood geomorphology*, John Wiley and Sons, New York, 1988.

O'Connor, J.E., L.L. Ely, E.E. Wohl, L.E. Stevens, T.S. Melis, V.S. Kale, and V.R. Baker, A 4500–year record of large floods on the Colorado River in the Grand Canyon, Arizona, *Journal of Geology*, *102*, 1-9, 1994.

Partridge, J. B., and V.R. Baker, Paleoflood hydrology of the Salt River, Arizona, *Earth Surface Processes and Landforms*, 12, 109-125, 1987.

Pelletier, P.M., Uncertainties in the single determination of river discharge: a literature review, *Canadian Journal of Civil Engineering*, 15, 834-850, 1988.

Phillips, J.V., and T.L. Ingersoll, Verification of roughness coefficients for selected natural and constructed stream channels in Arizona, *U.S. Geological Survey Professional Paper 1584*, 77 pp., 1998.

Phillips, J.V., D. McDoniel, J.P. Capesius, and W. Asquith, Method to estimate effects of flow induced vegetation changes on channel conveyances of streams in central Arizona, *U.S. Geological Survey Water-Resources Investigations Report* 98-4040, 43 pp., 1998.

Quick, M.C., Reliability of flood discharge estimates, *Canadian Journal of Civil Engineering*, 18, 624-630, 1991.

Randle, M.L., and J.H. Humphrey, Estimating peak flows in unstable channels using indirect methods, pp. 574-585, in Elliott, C.M. (ed.), *River meandering*, New Orleans, Louisiana, Conference Rivers '83 Proceedings, American Society of Civil Engineers, 1984.

Riggs, H.C., A simplified slope-area method for estimating flood discharges in natural channels, *Journal of Research of the U.S. Geological Survey*, 4, 285-291, 1976.

Roberts, L.K., *Paleohydrologic reconstruction, hydraulics, and frequency-magnitude relationships of large flood events along Aravaipa Creek, Arizona*, M.S. thesis, Department of Geosciences, University of Arizona, Tucson, 63 pp., 1987.

Shearman, J.O., W.H. Kirby, V.R. Schneider, and H.N. Flippo, *Bridge waterways analysis model*, Research Report, Federal Highway Administration, Report Number FHWA/RD-86/108, U.S. Department of Transportation, Washington, D.C., 112 pp., 1986.

Smith, S.S., *Large floods and rapid entrenchment of Kanab Creek, Southern Utah*, M.S. Thesis, Department of Geosciences, University of Arizona, Tucson, 82 pp., 1990.

Stedinger, J.R., and T.A. Cohn, Flood frequency analysis with historical and paleoflood information, *Water Resources Research*, 22(5), 273-286, 1986.

Thomsen, B.W., and H.W. Hjalmarson, *Estimated Manning's roughness coefficients for stream channels and flood plains in Maricopa County, Arizona*, Flood Control District of Maricopa County, U.S. Geological Survey report, Phoenix, Arizona, 126 pp., 1991.

Topping, D.J., *Physics of flow, sediment transport, hydraulic geometry, and channel geomorphic adjustment during flash floods in an ephemeral river, the Paria River, Utah and Arizona*, Ph.D Dissertation, University of Washington, Seattle, 404 pp., 1997.

Trieste, D.J., Evaluation of supercritical/subcritical flows in high-gradient channels, *Journal of Hydraulic Engineering*, 118(8), 1107-1118, 1992.

Trieste, D.J., and R.D. Jarrett, Roughness coefficients of large floods, pp. 32-40, in James, L.D., and M.J. English, (eds.), *Irrigation and Drainage Division Specialty Conference "Irrigation Systems for the 21st Century,"* Portland, Oregon, 1987, Proceedings, American Society of Civil Engineers, New York, 1987.

Turner, R.M., and M.M. Karpiscak, Recent vegetation changes along the Colorado River between Glen Canyon Dam and Lake Mead, Arizona, *U.S. Geological Survey Professional Paper 1132*, 125 pp., 1980.

U.S. Army Corps of Engineers, *HEC-RAS river analysis system, version 2.2*, Hydraulic Reference Manual CPD-69, U.S. Army Corps of Engineers, Hydrologic Engineering Center, Davis, California, variable pp., 1998.

U.S. Water Resources Council, *Guidelines for determining flood flow frequency*, U.S. Interagency Advisory Committee on Water Data, Hydrology Subcommittee, Bulletin 17B, Washington, D.C., 1982.

Waythomas, C.F., and R.D. Jarrett, Flood geomorphology of Arthurs Rock Gulch, Colorado, paleoflood history, *Geomorphology*, 11, 15-40, 1994.

Webb, R.H., *Late Holocene flooding on the Escalante, River, South-central Utah*, Ph.D dissertation, Department of Geosciences, University of Arizona, Tucson, 204 pp., 1985.

Webb, R.H., Occurrence and geomorphic effects of streamflow and debris flow floods in northern Arizona and Southern Utah, pp. 247-265, in Mayer, Larry, and D. Nash (eds.), *Catastrophic flooding*, Allen and Unwin, Boston, 1987.

Webb, R.H., and V.R. Baker, Changes in hydrologic conditions related to large floods on the Escalante River, South-central Utah, pp. 309-323, in Singh, V.P. (ed.), *Regional flood frequency*, D. Reidel, Boston, 1987.

Webb, R.H., and S.L. Rathburn, Paleoflood hydrologic research in the Southwestern United States, *Transportation Research Record*, 1201, 9-21, 1989.

Webb, R. H., J.E. O'Connor, and V.R. Baker, Paleohydrologic reconstruction of flood frequency on the Escalante River, South-central Utah, pp. 403-418, in Baker, V.R., R.C. Kochel, and P.C. Patton, (eds.), *Flood geomorphology*, John Wiley, New York, 1988.

Webb, R.H., S.S. Smith, and V.A.S. McCord, *Historic channel change of Kanab Creek, Southern Utah and northern Arizona*, Grand Canyon Natural History Association Monograph Number 9, 91 pp., 1991.

Webb, R.H., *Grand Canyon, a century of change*, University of Arizona Press, Tucson, 320 pp., 1996.

Webb, R.H., T.S. Melis, P.G. Griffiths, J.G. Elliott, T.E. Cerling, R.J. Poreda, T.W. Wise, and J.E. Pizzuto, Lava Falls Rapid in Grand Canyon, Effects of late Holocene debris flows on the Colorado River, *U.S. Geological Survey Professional Paper 1591*, 90 pp., 1999.

Wohl, E.E., R.H. Webb, V.R. Baker, and G. Pickup, Sedimentary flood records in the bedrock canyons of rivers in the monsoonal region of Australia, *Water Resources Papers*, 107, 102 pp., 1994.

Wohl, E.E., Uncertainty in flood estimates associated with roughness coefficients, *Journal of Hydraulic Engineering*, 124, 219-223, 1998.

Wohl, E.E., D.M. Thompson, and A.J. Miller, Canyons with undulating walls, *Geological Society of America Bulletin*, 111(7), 949-959, 1999.

Williams, G.P., Manning formula -- a misnomer? *Journal of the Hydraulics Division, American Society of Civil Engineers*, 96(HY1), 193-200, 1970.

Robert H. Webb, U.S. Geological Survey, 1675 W. Anklam Road, Tucson, AZ 85745, rhwebb@usgs.gov

Robert D. Jarrett, U.S.G.S, Box 25046, MS 412, Denver, Colorado, 80225.

Robust Determination of Stage and Discharge: An Example From an Extreme Flood on the Verde River, Arizona

Roger P. Denlinger

Cascades Volcano Observatory, Vancouver, Washington

D. R. H. O'Connell

U.S. Bureau of Reclamation, Lakewood, Colorado

P. Kyle House

Nevada Bureau of Mines and Geology, Reno, Nevada

Depth-averaged models can simulate nearly all aspects of three-dimensional flow in rivers, yet are more efficiently applied to large areas. The assumption that fluid pressure is hydrostatic is found to provide a good approximation for shallow water flow over complex three-dimensional terrain. Here we compare depth-averaged flow simulations with data from two large floods that occurred on the Verde River, Arizona, in January and February, 1993. Using a combination of photogrammetric and geodetic data, we construct a 6 km by 3 km grid of a reach of the river that contains a USGS stream gage and a narrow bedrock constriction. Elevation accuracy is 0.05 m and grid accuracy is 0.2 m. In the study area the channel is bedrock-controlled and forms a right-angle bend followed by a constriction at Sheep Bridge. Water dammed by this constriction during the 1993 floods created a 3 m drop in water-surface elevation over a distance of 50 m. High velocities generated by this drop scoured the channel and built a large bar of boulders and cobbles that stretched across the channel 150 m downstream. Indicators of minimum water surface elevation, including mudlines, debris stranded on hillsides, and grass wrapped around trees were mapped at 45 sites along the channel. Discharge was monitored through a spillway at Horseshoe Dam, 15 km downstream of the bridge, during the larger January flood, and stage and velocity were measured at the USGS Tangle Creek gage during the smaller February flood. High-water indicators near the gage are 0.07 m lower than the peak stage measured at the gage in January 1993. Peak stage and peak velocities exceeding 6 m/s lasted 4

Ancient Floods, Modern Hazards: Principles
and Applications of Paleoflood Hydrology
Water Science and Application Volume 5, pages 127-146

Published in 2002 by the American Geophysical Union

hours during the January flood, justifying the use of a steady-flow model for the 6 km long reach. The flow simulations for a discharge of 3398 m^3/s (120,000 ft^3/s) and a Manning's n value of 0.038 agree closely with direct velocity and stage measurements at the gage, a direct measurement of discharge through the spillway, and with the distribution of high-water marks along the channel. Other combinations of discharge and Manning's n result in larger residual errors in model fits to high-water indicators, and cannot fit both the high water marks and the observed combination of stage and peak velocity at the Tangle Creek gage for the 1993 floods. Thus discharge and bed roughness are uniquely constrained using profiles of maximum inundation along the channel margins extending upstream and downstream from the constriction. The inclusion of surface velocity measurements is not required, but improves resolution and statistical confidence in peak discharge estimates.

INTRODUCTION

Large scale simulations of river flow, for river channels tens of kilometers long or more, have traditionally used 1D models such as FLDWAV [*Jin and Fread*, 1997] and HEC-2 [*Hydrologic Engineering Center*, 1997], since multi-dimensional models are assumed to be too complex and computationally intensive to be practical. Yet the apparent simplicity of these 1D models is the result of ignoring momentum transport back and forth across the river and arbitrary and labor-intensive positioning of discharge-dependent channel thalweg positions in constructing the 1D grid. For example, islands occur in rivers at all scales (typically where the channel widens), yet are impossible to treat with 1D formulations in any rigorous way. Lacking cross-stream momentum transport, 1D models cannot fit observed variations in stage across sinuous constricted natural channels that give rise to differences in patterns of inundation commonly observed on right and left banks.

Multi-dimensional (2D and 3D) models are simpler to implement since they operate on a grid that is a three dimensional representation of actual channel topography. Variations in flow around islands and through constrictions, as well as stage variations across the channel, are obtained unambiguously by solving for flow on a grid. This is done using either the shallow-water flow equations in 2D, or using Reynolds-averaged Navier Stokes equations in 3D [*Miller and Cluer*, 1998]. Of these two choices, 3D formulations are more expensive to compute as they require an order of magnitude more degrees of freedom than 2D models, making them less practical for large-scale (i.e. million-cell) flow simulations. In this paper, we show that a depth-averaged 2D flow model can reproduce 3D variations in stage associated with large peak discharges on a sinuous bedrock-confined channel.

Depth-averaged, 2D solutions to the shallow water equations accurately represent the depth-averaged components of full 3D flow field as long as the flow is approximately hydrostatic [as defined in *Vreugdenhil*, 1997]. Flume studies [*Nelson et al.*, 1994] and field experiments [*Miller and Cluer*, 1998] show that gradients in depth-integrated averages of horizontal velocities may be used to estimate the full 3D flow field. For instance, *Wiele et al.*, [1996] calculated a 3D flow field from a depth- averaged model and used it to build a sediment transport model for a flood on the Little Colorado River, Arizona. What has not been reported is that a depth-averaged 2D flow model may be used to unambiguously constrain discharge and bed roughness and is particularly useful for constraining discharges of large floods. This more sophisticated approach to flow modeling requires more data than conventional modeling used in typical paleoflood studies [e.g., *O'Connor and Webb*, 1988; *Webb and Jarrett*, this volume], but it offers considerably more detailed information about hydraulic conditions that has applicability to both conventional flood studies and non-inundation studies.

Flow along any natural 3D river channel is disrupted by diversion of flow around obstructions or dammed behind constrictions [*Denlinger and Walters*, 1998], often resulting in eddies or switching of the current back and forth across the channel. These diversions or obstructions can create large variations in stage along and across the river channel that change markedly with discharge, providing a means to constrain both discharge and resistance to flow from bed roughness (friction). If stage or velocity is measured above and below a constriction, for example, then the solution for discharge and bed roughness that fit those data may be unique.

In this paper we test the ability of a 2D depth-averaged code (TRIMR2D) to reproduce stages and velocities associated with the two largest floods in the systematic record of the Verde River, Arizona [see discussion in *House et al.*, this volume]. The floods occurred six weeks apart, in January and February 1993, with the January flood leaving

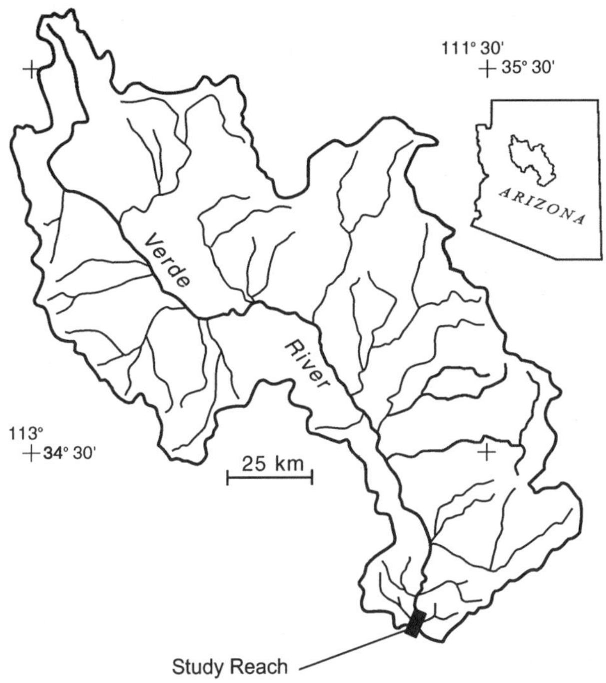

Figure 1. Map of the Verde River Basin, central Arizona. Approximate location of the study reach is indicated (see also *House et al.*, this volume, Figure 2).

high-water indicators approximately 0.2-0.6 m above those of the February flood. A detailed study of these floods [*House et al.*, 1995; *House et al.*, this volume], USGS gage records at selected sites along the river, a spillway to constrain discharge at the lower end of the drainage area, a dry climate that preserves flood deposits and debris, and a channel that is nearly dry in August make the Verde River an ideal candidate to study large floods. Here we use detailed measurements of the bedrock channel morphology along a 6 km long reach of the lower Verde River (Figures 1 and 2, also on CD), and present the position of high-water indicators along the reach, a direct measurement of stage, velocities, and discharge at a USGS gage, and locations of deposition and scour that occurred during the floods.

The Study Reach

In May, 2000, forty five unambiguous high-water indicators were measured along 4 km of the channel. These were combined with high-water data collected in the constricted portion of the reach between April, 1993 and April, 1998. Simulated stages along both sides of the channel are compared with the observed stages at the same locations to determine the discharge/bed roughness combination with the least residual error. Using the distribution of high-water indicators above and below the channel constriction at Sheep Bridge allows us to make a unique determination of bed friction and discharge. Once these flow parameters are constrained, flood power is derived from the best-fitting discharge model. Comparison of simulated power with observations of scour and deposition demonstrates the utility, stability, and precision of 2D, depth-averaged models for interpreting paleohydrologic bounds on inundation [*Levish et al.*, 1994] and on slackwater deposits.

Paleohydrologic observations by *House et al.* [1995] and *House et al.* [this volume] are concentrated in a tributary mouth immediately upstream from the constriction at Sheep Bridge. This site is near the downstream end of the unregulated drainage area, approximately 0.25 km upstream from the USGS Tangle Creek stream gage (U.S. Geological Survey gage no. 09508500), and 15 kilometers upstream from Horseshoe Dam.

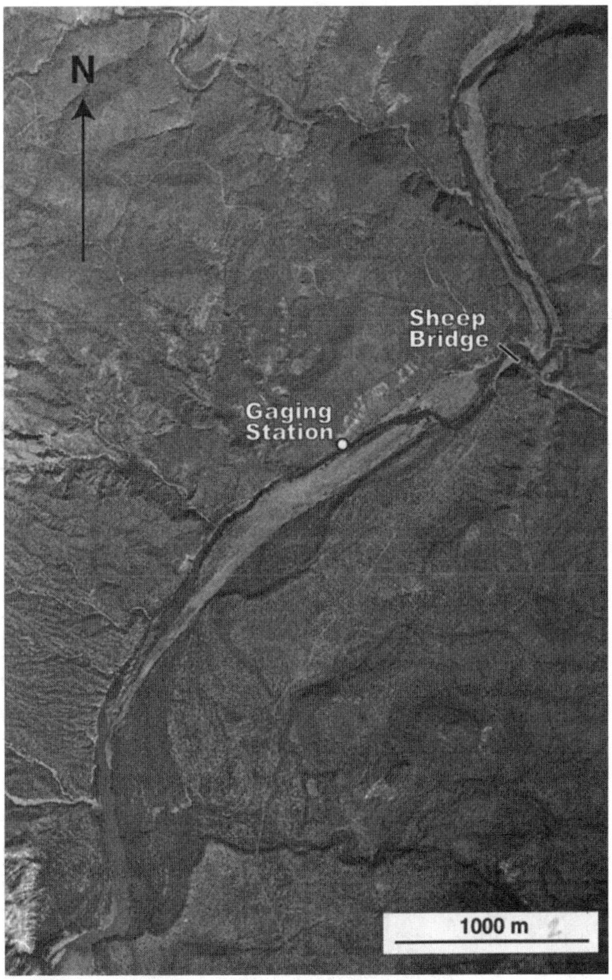

Figure 2. Aerial photo of the study reach, showing the location of Horse Creek, Sheep Bridge, and the Tangle Creek gage.

OBSERVATIONS OF THE 1993 FLOODS ON THE LOWER VERDE RIVER

Previous Studies and Reports

Two large floods occurred on the Verde River in January and February 1993. Comparing direct measurements of stage with paleoflood stages, *House and Hirschboeck* [1997], *House et al.* [1995], and *House et al.* [this volume] found these events to be comparable to the largest floods on the Verde River during the past millennium. Meteorological and hydrologic descriptions of the 1993 flood conditions are presented by *House and Hirschboeck* [1997]. Data from hydrologic measurements and observations of the floods are reported in *Smith et al.*, [1994].

Flood Conditions in the Study Reach

The bedrock channel of the Verde River makes a nearly right-angle bend at the confluence with Horse Creek at Sheep Bridge (Figure 2). Then the channel forms a narrow constriction at the bridge. Flood waters were dammed behind this constriction (Figure 3a, also on CD) during both floods in 1993, and a huge debris pile (Figure 3b, also on CD) accumulated in a back eddy that formed at the confluence with Horse Creek. The hydraulically-dammed water surface dropped 3 m through the constriction, generating high velocities, scouring the channel immediately downstream, and pushing the bed material into a large, natural bar across the channel. This bar, and the visible parts of the channel bed upstream and downstream are composed primarily of cobbles (0.1m) with large boulders (1 to 2 m) near the center of the channel.

Flow Data

There are two independent, direct assessments of flow conditions in the study reach during the 1993 floods. Flow over the dam spillway provided a direct measurement of peak discharge during the larger, January flood of 3105 m^3/s (109,657 ft^3/s) [Salt River Project, Bureau of Recla-

Figure 3. Photographic images of the study reach and effects of the 1993 floods. (a). Photo looking upstream towards Sheep Bridge during the flood. The bridge crosses the constriction near the upper end of the river in the photo (provided by USGS); (b). Photo of debris deposited in eddy at mouth of Horse Creek, with geologist for scale; (c). Photo of Tangle Creek gage during the January 1993 flood (provided by USGS).

Table 1. Direct measurements during the 1993 floods

Date	Location	Type	Value	Reference
Jan. 8, 1993	Horseshoe Dam spillway	discharge	3119 cms	Salt River Proj., unpub. data, 1993
Jan. 8, 1993	Tangle Creek gage, USGS station 09508500	water surface elevation	625.57 m	*Smith et al.*, 1994
Feb. 20, 1993	Tangle Creek gage, USGS station 09508500	water surface elevation	625.30 m	*Smith et al.*, 1994
Feb. 20, 1993	Tangle Creek gage, USGS station 09508500	current meter velocity	6.2 m/sec	USGS, unpub. data, 1993

mation, unpublished data]. A direct, near-surface measurement of flow velocity of 6.2 m/s [USGS, unpublished data] was obtained during the February flood at the USGS Tangle Creek gaging station located approximately 0.25 km downstream from the constriction (Figure 3c, also on CD; USGS gage no. 09508500). Discharge estimated from current and depth measurements of the February flood at the Tangle Creek gage was 3511 m^3/s [*Smith et al.*, 1994], which is 6 m^3/s larger than the Horseshoe Dam spillway discharge for the January flood. Yet the maximum stage adjacent to the gage in February was 0.26 m lower than the stage at the gage during the January flood. The duration of the peak discharge at the Tangle Creek gage for the January flood was 4 hours and sustained high flow was recorded for more than 10 hours [*Smith et al.*, 1994; *House and Hirschboeck*, 1997], whereas the sustained high flow was less than 2 hours for the February flood. No peak spillway discharge is available for the February flood. Table 1 contains direct observations made during the 1993 floods.

High-water Marks and Channel Topography

High-water indicators of the 1993 floods and post-flood channel topography were mapped using a combination of surveying methods (described below), and are shown in Figure 4 (also on CD) and listed in Table 2. Between 1993 and 1998, *House et al.*, [unpublished data] used a laser ranging electronic distance meter and theodolite to survey the elevation of high-water indicators from the 1993 floods (flotsam, mud, and debris). They also obtained the post-flood channel bathymetry in the vicinity of Sheep Bridge.

In other areas of the channel, large bedforms that were exposed at a low discharge were surveyed. In 1999,a kinematic GPS survey mapped 17,432 control points along a 2 km reach upstream and downstream of Sheep Bridge (Figure 4). To complete the data set, we subsequently surveyed high-water marks several kilometers upstream and downstream of Sheep Bridge (Table 2 and Plate 1, also on CD) and located sites which showed evidence of high values of flood power, such as scoured surfaces or large trees uprooted or sheared in half.

The accuracy and bias (deviation from zero mean discrepancy) of channel topography that was obtained with photogrammetry alone was improved using the GPS and total station data. The initial base topography of a 6 km long channel spanning Sheep Bridge was constructed by photogrammetric means using 1:6000 scale aerial photos, with enhanced elevation control determined along specific ('break') lines and at specific targets scattered through the

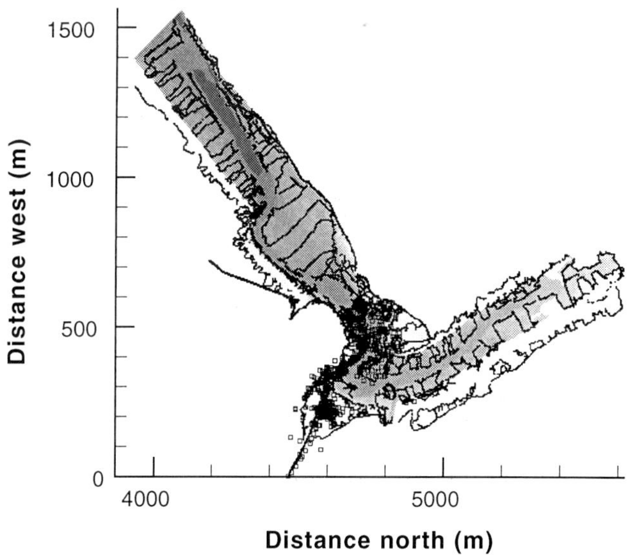

Figure 4. Distribution of geodetic control points used to refine the photogrammetric map along the study reach. The extent of this figure includes only the portion of the reach that was surveyed manually. The remaining extent of the study reach topography was constructed with photogrammetric methods.

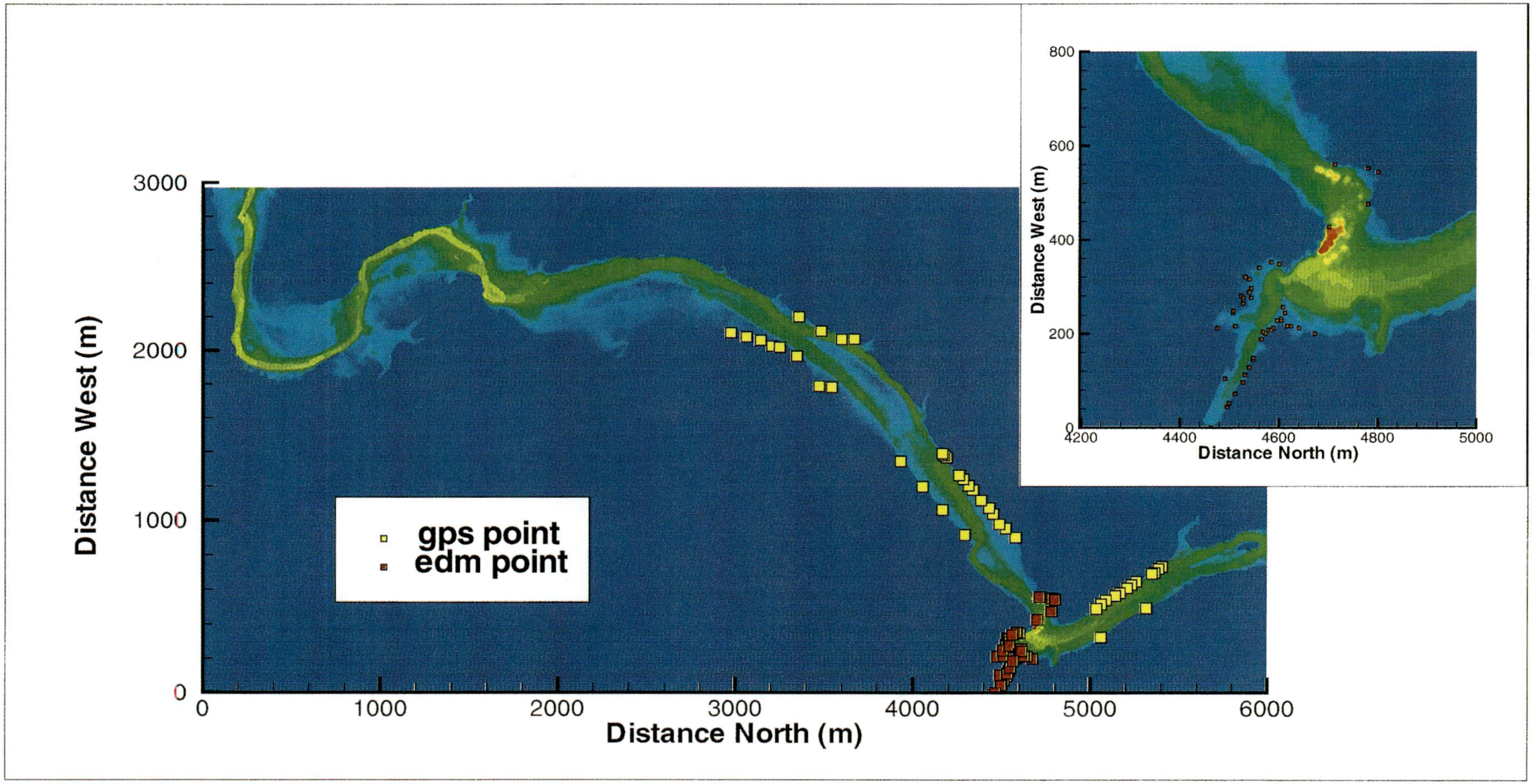

Plate 1. Distribution of geodetic control points at high water marks (Table 2)

Table 2. Description of high water marks measured with GPS

Station	North latitude	West longitude	Elevation (m)	station	Description
V110	34° 04' 29.85819"	111° 42' 49.05818"	626.323	V110	Flotsam on colluvium
V22	34° 04' 58.67978"	111° 42' 26.47713"	626.173	V22	Flotsam in tree
V33	34° 05' 01.52839"	111° 42' 35.83722"	631.764	V33	Debris pile in granite boulders
V34	34° 05' 00.93833"	111° 42' 35.36528"	631.197	V34	Debris pile in granite boulders
V35	34° 05' 00.19076"	111° 42' 34.45993"	631.332	V35	Flotsam on rock
V36	34° 04' 59.71429"	111° 42' 34.23410"	631.045	V36	Flotsam on boulder
V38	34° 04' 56.85148"	111° 42' 32.49025"	631.433	V38	Flotsam mat
V40	34° 04' 55.98146"	111° 42' 31.70314"	631.053	V40	Flotsam mat
V41	34° 04' 55.20735"	111° 42' 30.96031"	631.553	V41	Flotsam mat
V42	34° 04' 53.87807"	111° 42' 29.89284"	630.33	V42	Flotsam mat
V43	34° 04' 53.03917"	111° 42' 29.26335"	630.532	V43	Flotsam mat
V45	34° 04' 51.28029"	111° 42' 28.26351"	629.772	V45	Flotsam and sediment
V46	34° 04' 51.29265"	111° 42' 28.30580"	630.403	V46	Flotsam
V47	34° 04' 50.40157"	111° 42' 27.35732"	630.598	V47	Flotsam
V48	34° 04' 49.55002"	111° 42' 26.35492"	629.982	V48	Flotsam
V102	34° 04' 34.70350"	111° 42' 42.37064"	626.515	V102	Flotsam mat on flood sediment
V103	34° 04' 32.87210"	111° 42' 44.45169"	627.127	V103	Flotsam mat on colluvium
V107	34° 04' 31.73694"	111° 42' 45.43493"	625.899	V107	Flotsam near rock outcrop
V109	34° 04' 30.57647"	111° 42' 47.69127"	626.041	V109	Flotsam on colluvium
V110	34° 04' 29.85819"	111° 42' 49.05818"	626.323	V110	Flotsam on colluvium
V112	34° 04' 28.44889"	111° 42' 50.83211"	626.002	V112	Flotsam
V114	34° 04' 26.91282"	111° 42' 53.21978"	626.118	V114	High flotsam above debris pile
V116	34° 04' 26.13557"	111° 42' 54.17870"	625.398	V116	Flotsam mat on rock outcrop
V117	34° 04' 25.16996"	111° 42' 55.64528"	625.608	V117	Flotsam

Table 2. Continued.

Station	North latitude	West longitude	Elevation (m)	station	Description
V118	34° 04' 24.37178"	111° 42' 56.54740"	625.39	V118	Flotsam near gage
V122	34° 04' 22.06514"	111° 43' 00.50478"	624.811	V122	Flotsam below gage
V123	34° 04' 21.63931"	111° 43' 00.96363"	624.436	V123	Flotsam in colluvium
V124	34° 04' 21.30667"	111° 43' 01.40892"	624.55	V124	High flotsam on rock outcrop
V128	34° 04' 18.20983"	111° 43' 09.77159"	624.735	V128	Flotsam
V133	34° 04' 05.24915"	111° 43' 27.84433"	622.725	V133	Trim line
V134	34° 04' 02.90758"	111° 43' 27.77222"	622.943	V134	Flotsam on cliff
V137	34° 03' 59.12992"	111° 43' 29.70203"	622.212	V137	Flotsam on colluvium
V139	34° 03' 55.05771"	111° 43' 33.01143"	622.005	V139	Flotsam and flood sediment
V149	34° 03' 42.61794"	111° 43' 29.26961"	621.276	V149	Flotsam above scour pits
V150	34° 03' 45.45696"	111° 43' 28.35347"	621.354	V150	High flotsam
V151	34° 03' 48.02940"	111° 43' 27.55694"	621.523	V151	Flotsam
V152	34° 03' 50.14759"	111° 43' 26.17784"	621.237	V152	High flotsam
V153	34° 03' 51.55873"	111° 43' 25.93855"	621.915	V153	High flotsam
V154	34° 03' 53.58254"	111° 43' 24.89680"	622.71	V154	High flotsam (older than 93?)
V157	34° 03' 58.76805"	111° 43' 16.91757"	622.429	V157	High flotsam
V158	34° 04' 01.10702"	111° 43' 16.67645"	622.727	V158	Debris pile
V159	34° 04' 13.72767"	111° 42' 59.61729"	624.507	V159	Flotsam on terrace
V160	34° 04' 17.72883"	111° 42' 53.93238"	624.639	V160	Flotsam on bedrock
V161	34° 04' 21.34669"	111° 42' 48.67157"	625.379	V161	Flotsam on terrace riser
V163	34° 04' 25.44955"	111° 42' 42.96917"	624.565	V163	Flotsam on terrace riser

grid. The root mean square difference between all GPS elevations and the map elevations gave a standard deviation in accuracy of 0.74 m and a bias of 0.1 m prior to correction with the GPS data. The photogrammetric surface was corrected by using an inverse-square weighting applied to the distance between a surveyed point elevation and a photogrammetric grid point elevation. After correcting the photogrammetric elevations with the GPS elevations, the interpolated map elevation at a geodetic position was close to the measured elevation. A histogram of the final residual differences for the GPS data are shown in Figure 5 (also on CD). A standard error analysis of these residuals produces a bias of 0.05 m and a standard deviation in accuracy of 0.2 m; the residuals are normally distributed. The corrected surface was used as the topographic base for a 6 km by 3 km model grid of 4 m cells to simulate flooding of the channel in 1993.

MODEL FOR FLOODING

The TRIMR-2D flow model [*Casulli*, 1990] is based on the depth-averaged, shallow water equations, the derivation of which is described in *Vreugendhil*, [1994]. The calculation of the flow field consists of the numerical solution of these equations, which describe transport of mass and momentum that is averaged through the depth of the flow and conserved on a horizontal plane. These equations are written below for reference and as a basis for our subsequent discussion of the relative influence of inertia as well as for the calculation of the power dissipated by the flow on the bed. The equations have the same cartesian coordinate system used for the grid, with x and y defining the coordinate directions in the horizontal. Continuity is given by

$$\frac{\partial h}{\partial t} + \frac{\partial hu}{\partial x} + \frac{\partial hv}{\partial y} = 0 \qquad (1)$$

Momentum in the x direction is

$$\frac{\partial u}{\partial t} + u\frac{\partial u}{\partial x} + v\frac{\partial u}{\partial y} - \left(\frac{\partial}{\partial x}\varepsilon\frac{\partial u}{\partial x} + \frac{\partial}{\partial y}\varepsilon\frac{\partial u}{\partial y}\right) + g\frac{\partial \eta}{\partial x} + \frac{\tau_x}{\rho h} = \qquad (2)$$

and momentum in the y direction is

$$\frac{\partial u}{\partial t} + u\frac{\partial v}{\partial x} + v\frac{\partial v}{\partial y} - \left(\frac{\partial}{\partial x}\varepsilon\frac{\partial v}{\partial x} + \frac{\partial}{\partial y}\varepsilon\frac{\partial v}{\partial y}\right) + g\frac{\partial \eta}{\partial y} + \frac{\tau_y}{\rho h} = 0 \qquad (3)$$

where
x is east
y is north
z is bed elevation
η is water surface elevation $= z + h$
g is gravity
h is flow depth
u is the vertically average velocity in the x direction
v is the vertically average velocity in the y direction
ρ is the density of water
τ_x is the bed shear stress in the x direction
τ_y is the bed shear stress in the y direction
ε is the eddy viscosity

The numerical solution of these equations in TRIMR-2D is described in *Casulli* [1990] for shallow water flow. At each time step, the momentum equation is substituted into the continuity equation to produce an equation that can be implicitly solved to estimate water surface elevation. Then the water surface elevation is used to explicitly update flow velocities through the momentum equations. Discharge is conserved through areas of transcritical flow, though momentum is not. Steady state conditions were obtained after several hours of flow.

The eddy viscosity terms are included for completeness and occur in the model algorithm, but an eddy viscosity of zero is used here. Eddy viscosity large enough to affect momentum diffuses velocity gradients in the flow [*Miller and Cluer*, 1998], inhibiting eddy formation and size as well as sharp contrasts between separate currents in the river (current seams). Small to negligible values of eddy viscosity are required to match observed patterns of flow in the Colorado River for a discharge of 3500 m^3/s [*Miller and Cluer*, 1998]. In addition, eddy viscosity plays no direct role in stream power [*Vreugdenhil*, 1994]. Since the goal here is to model stage and power for a given discharge, eddy viscosity is omitted from the simulations.

Modeled discharges use a depth-averaged velocity, with the actual variation in velocity with depth represented by the source term for bed shear stress. The bed shear stresses assume logarithmic velocity profiles with depth found for fully turbulent flow over rough surfaces [*Keulegan*, 1938; *Schlicting*,1984]. The x and y components are

$$\tau_x = \rho c_f u V \qquad (4)$$

and

$$\tau_y = \rho c_f v V \qquad (5)$$

respectively, with $V = \sqrt{u^2 + v^2}$ and c_f representing the friction coefficient. This coefficient is defined as

$$c_f = \left[\frac{\kappa}{\ln\frac{h}{z_0} - 1}\right]^2 \left[\frac{h/z_0}{h/z_0 - 1}\right]^2 \qquad (6)$$

and may be derived assuming a logarithmic velocity profile in depth z given by

$$u(z) = \frac{\sqrt{\tau/\rho}\left(\ln\frac{z}{z_0}\right)}{\kappa} \qquad (7)$$

bulent, approximately hydrostatic flow, the flow is dominated by bed topography and the power dissipated on the bed depends upon the cube of the mean flow velocity.

FIELD VERIFICATION OF SIMULATED RIVER FLOWS

Observations of the hydraulic and geomorphic effects of the 1993 floods on the Verde River allow testing of 2D flow models by comparing observed and computed stage along the study reach and comparing computed and observed velocity for that stage at the Tangle Creek gage. The flow parameters with the lowest residual differences are used to calculate the distribution of flood power and compare it with the locations of scour and deposition during the flood. Interpretation of the high-water residuals is limited by the precision of the grid (Figure 5). Vertical resolution of stage exceeds the horizontal power resolution, which is based on 4 m wide cell averages.

The flow simulations assumed a Manning equation for bed friction with n values ranging from 0.030 to 0.038. The n value of 0.038 corresponds to drag on the river bed produced by obstacles with an average height ranging from 10% of the flow depth at shallow depths (~ 1 meter) to 7.5% of the flow depth for depths in the ten meter range (eqn. 9). These estimates are consistent with the large boulders (~1 m) that litter the nearly-dry, main channel of the Verde river in August, which in turn are flanked by cobbles. A Manning's n of 0.030 corresponds to a lower average form drag, corresponding to bed roughness values 2/3 of those above. Measurements of velocity profiles show that the addition of grass or weeds greatly increases the form drag and therefore the roughness [*Henderson*, 1966, p. 99, Table 4-2). However, as the 1993 floods on the Verde River removed soil and grass wherever flood power was high, we did not consider the additional drag of vegetation in our modeling.

Comparisons of Maximum Stage and Velocity

The positions of high-water marks, described in Table 2 and shown in Figure 4, provided the primary constraints for discharge estimation. The difference in meters between the measured elevation of the high-water indicator (Table 2) at each site and the calculated stage at that site are shown as histograms, one for each modeled peak discharge (Figure 6, also on CD). The uncertainty at each of the observed stage indicators is 0.05 m, whereas the elevation uncertainty in the grid is 0.2 m. Stage uncertainty is confirmed by direct observation at Tangle Creek gage, where gage measurements of high water during the January 1993 flood were 0.07 m higher than the maximum elevation of high water indicators we observed at the gage. It is possible that each high-water indicator, such as flotsam or driftwood, may not have been stranded at the peak or shifted downslope, and so is a lower bound. Collectively, the high-water indicators become increasingly inundated as simulated discharge increases. All the observed high-water indicators are submerged between a discharge of 3115 m^3/s (110,000 ft^3/s) and 3398 m^3/s (120,000 ft^3/s) for a Manning's n of 0.038 (Figure 6 and Plate 2a, also on CD). This compares favorably with the peak discharge of 3105 m^3/s (109,657 ft^3/s) that was measured passing through the spillway at Horseshoe dam during the January 1993 flood. A discharge of 4248 m^3/s and n = 0.030 also fits the stage at the gage, but not upstream at the bridge, and overall gives poorer residuals that reflect a flow that does not fit stage measurements (Plate 2b, also on CD). Removal of a gravel bar below the bridge only affected residuals in the immediate vicinity of the bridge. The best-fit result for this special bed configuration gave a slight increase in discharge, to 3450 m^3/s (125,000 ft^3/s), and the same distribution of stage residuals above and below the bridge. Thus the discharge that minimizes the high water residuals is robust: it is within a few percent of 3400 m^3/s, regardless of whether or not the gravel bar near Sheep Bridge was present during the 1993 floods.

Surface velocity can also be used to help constrain flood discharge when combined with stage measurements. The maximum near-surface velocity measured during the February flood was made at the Tangle Creek gage near the middle of the channel. The observed value of 6.2 m/s (22 ft/s) [USGS, unpublished data] is shown in Figure 7 (also on CD), along with the maximum velocities obtained with all the modeled peak discharges using a value of Manning's n of 0.038. A velocity profile for a Manning's n of 0.030, for which the discharge had to be increased to match the stage at the gage, is also shown and gave anomalously high velocities. Within the error of the measurements, the observed peak velocity is only achieved at a discharge/bed roughness combination of ~3400 m^3/s and n=0.038. This discharge is comparable to the direct estimate of 3200 m^3/s estimated from gage data taken on February 20, 1993. For lower values of Manning's n, it is possible to fit the stage at Tangle Creek by using a higher discharge or fit the velocity by using a lower discharge. However it is not possible to fit both the stage and velocity together (Plate 2b and Figure 7) for a Manning's n of 0.030, or any other value of n other than 0.038.

A similar constraint on both discharge and bed roughness can be obtained from the distribution of high-water indicators alone if the high-water indicators span a constriction. The position of high-water marks through the constriction at Sheep Bridge and along the straight reach

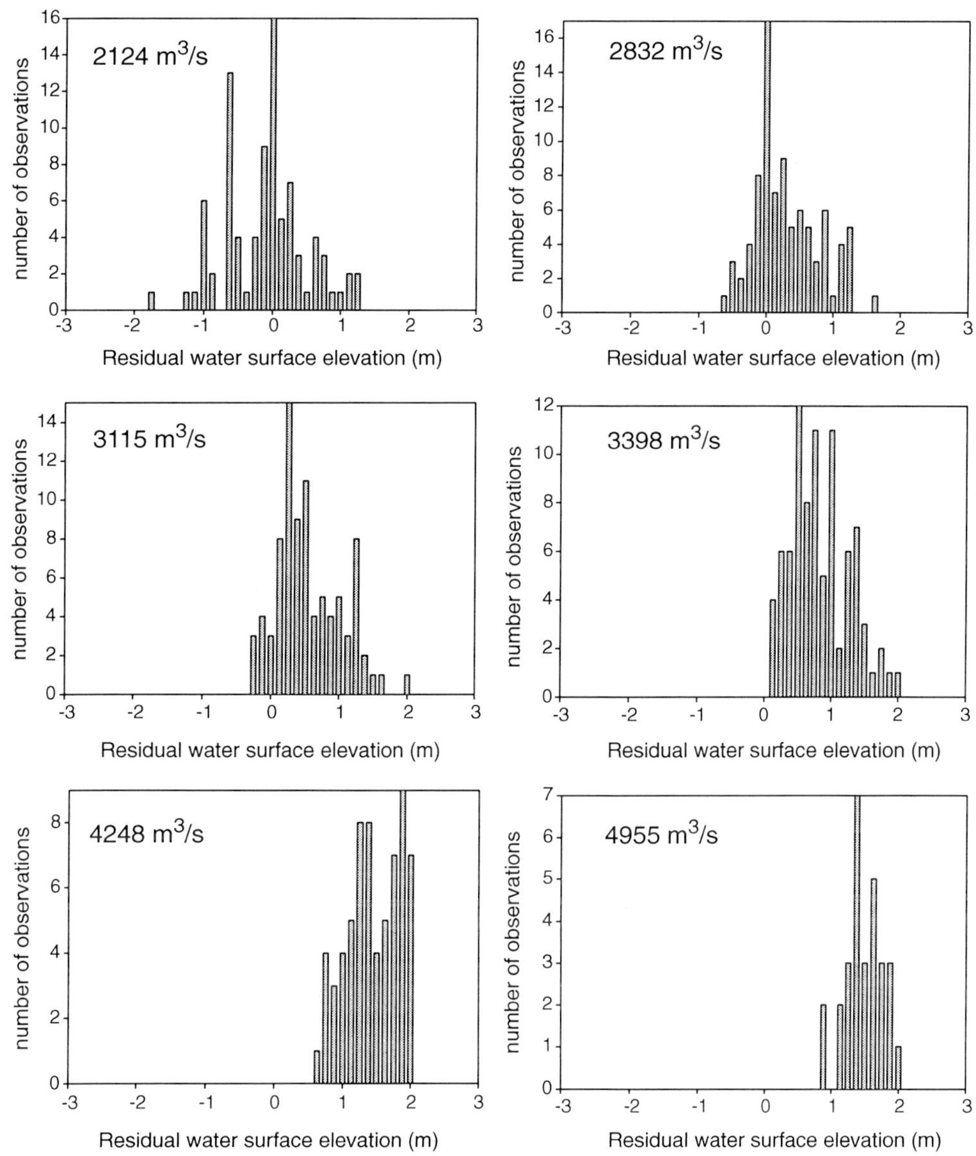

Figure 6. Histograms of the high water residuals for the suite of model discharges as labeled. Residuals are simulated stage minus observed stage indicator. The indicators are distributed over 0.2 m to 1. m of elevation at each site and are assumed to represent minimum high water values.

near Tangle Creek allows us to uniquely determine both discharge and bed roughness. The drop in water surface elevation through the constriction is sensitive to discharge, and constrains the discharge within narrow bounds. Values of bed friction corresponding to n less than 0.038 increase the modeled discharge to match the stage at the Tangle Creek gage, but in doing so increase the misfit of high water indicators above and below the constriction at Sheep Bridge. The same unique combination of discharge and roughness that minimizes high-water residuals (Plate 2a) is also the only one that provides a good fit for the combined stage/velocity data at Tangle Creek gage (Figure 7).

Flood Power Comparisons

Flow velocity and stage are constrained using field observations, and simulated stages and velocities are used to obtain the distribution of flood power. We selected and surveyed eight locations along the study reach with flood damage that included significant scour of the bedrock and

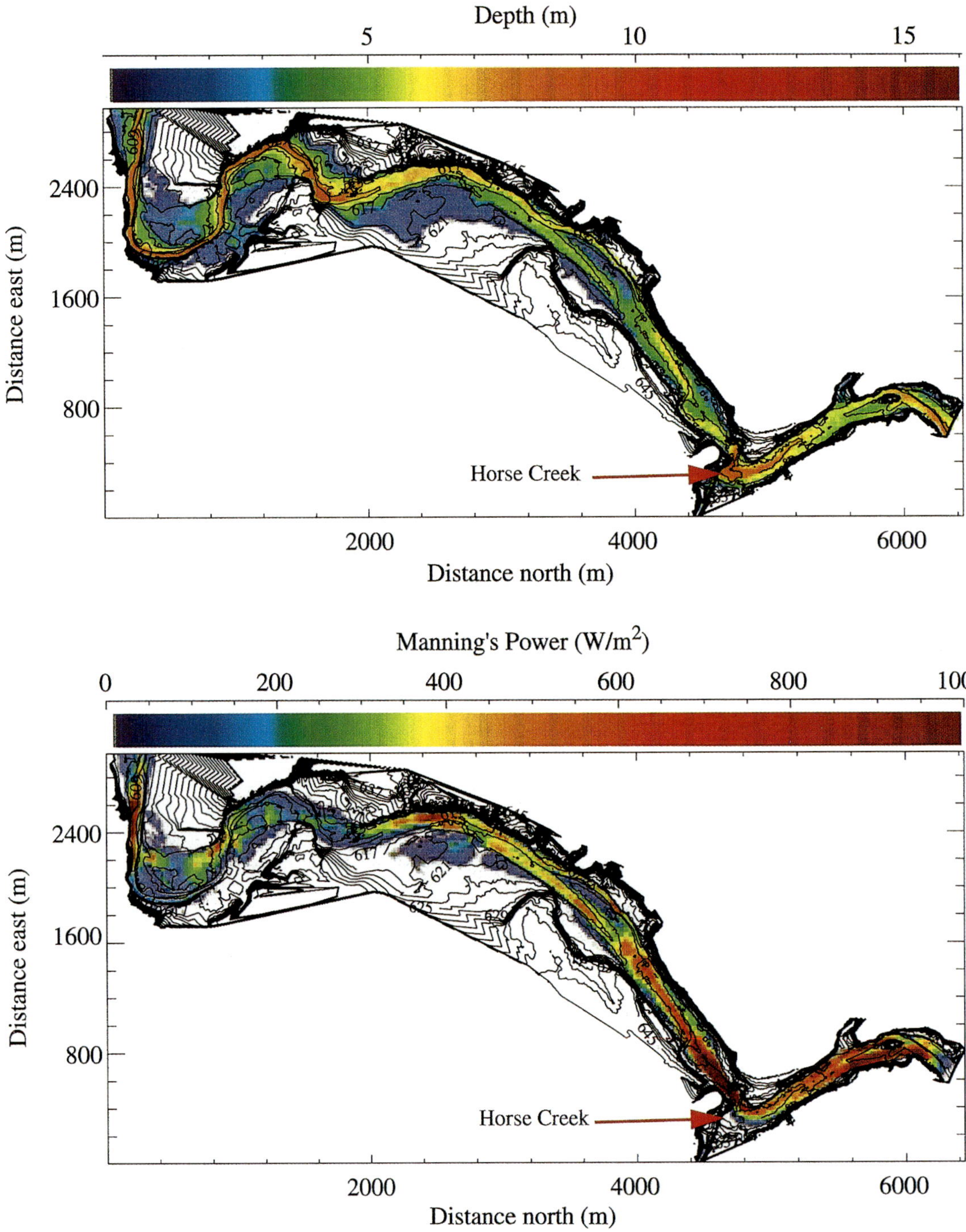

Plate 3. (a) Inundation and water depth along the channel at a discharge of 3398 m^3/s with 2m topographic contour interval and white velocity vectors. (b) Unit power distribution (W/m^2) for the same configuration as (a). Deposition of logs and debris occurred in all inundated areas shown in (a) that have low (<30) values of unit power, with a huge debris pile (Figure 3c) accumulating in the eddy at the mouth of Horse Creek at the location of the red arrow. These model output data may be found in the file verde_river_120k_n038.dat on the enclosed CD.

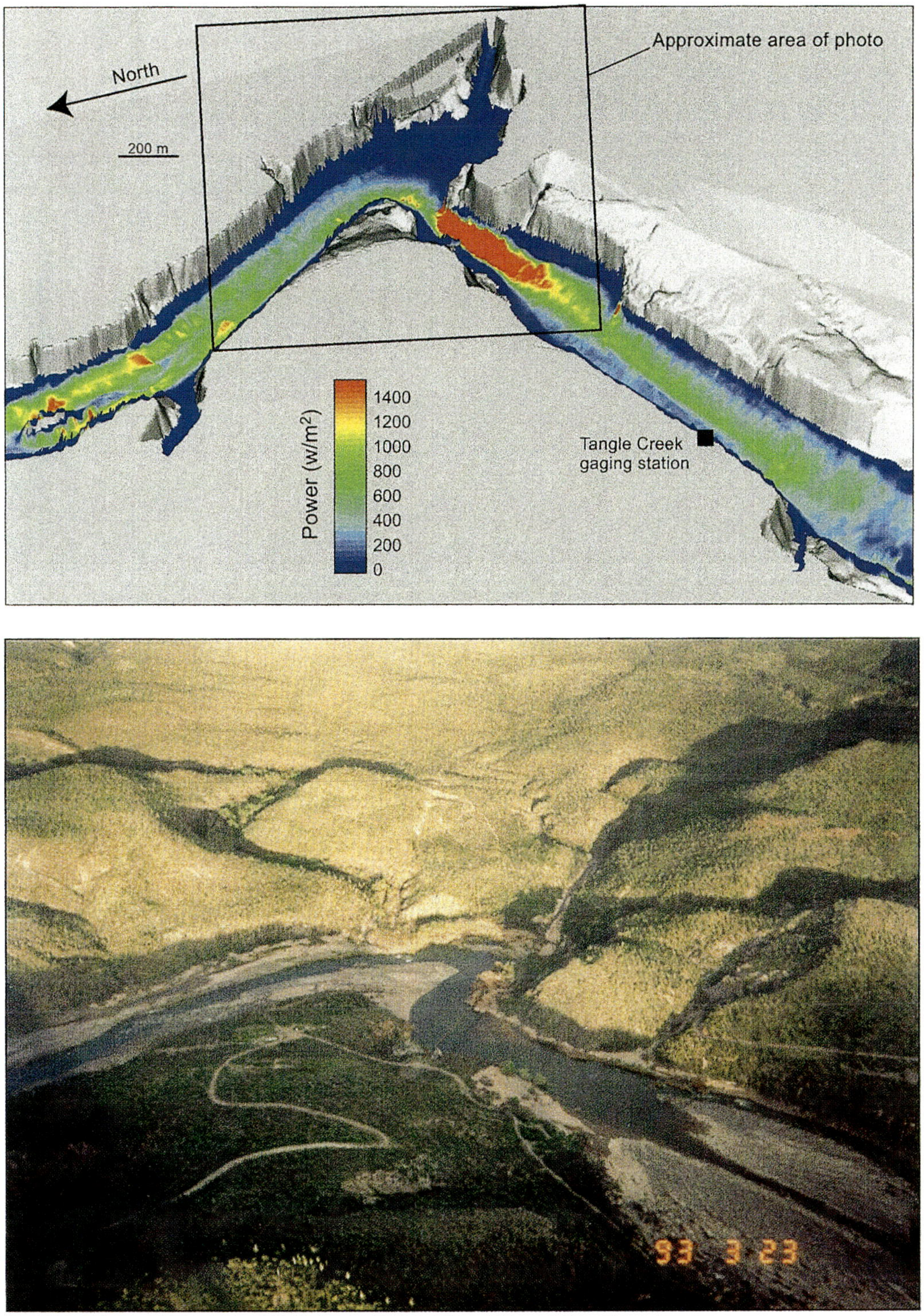

Plate 4. Comparison of the calculated power dissipated on the bed with the scour and deposition of the channel bed near Sheep Bridge, March 23, 1993 (photo provided by USGS).

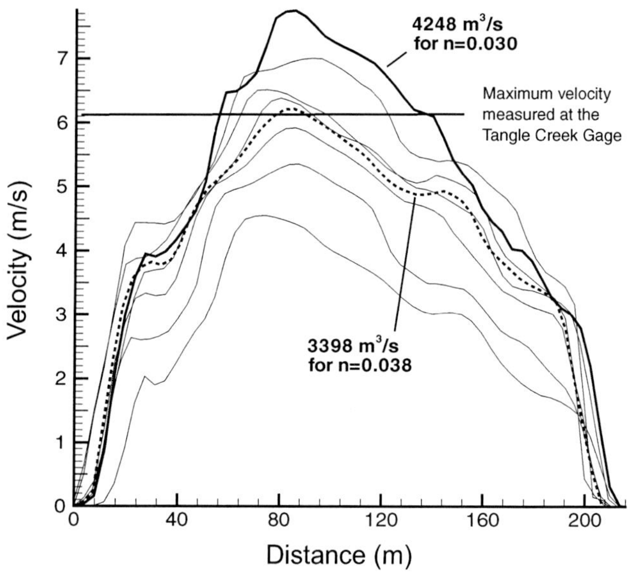

Figure 7. Computed variation of surface velocity across the channel at Tangle Creek, compared to the maximum surface velocity measured along that cross section, for various discharges and a Manning's 'n' of 0.038. The observed velocity of 6.2 m/s compares closely with the velocity of 6.22 m/s obtained with the model. Also shown in blue is the velocity obtained when the stage at the gage is matched with $n = 0.030$.

uprooted or sheared cottonwood trees (Table 3). Nearby high-water marks were also surveyed. Details of the power distribution around Sheep Bridge, as well as at one site upstream and downstream of the bridge are given in Plates 3 through 5 (also on CD). A data set of model output (verde_river_120k_n038.dat) is provided on the CD and gives details of the flow depths and velocities at a discharge of 3398 m3/sec (120,000 cfs) for a Mannings 'n' of 0.038. Power in these figures and in the data is normalized by cell area and is given as watts/m^2.

DISCUSSION

The transport of flow momentum back and forth across the channel and the disruption of flow by channel constrictions produces large variations in stage-discharge relations along the model reach. The stage at Tangle Creek gage, downstream of the Sheep Bridge constriction and along a straight reach of channel is 2.5 times less sensitive to discharge than at points just upstream of Sheep Bridge because of the hydraulic-damming of flood waters behind the constriction, as is illustrated by stage-discharge relationships (Figure 8, also on CD). Elevation of the water surface behind the constriction for the preferred discharge of 3398 m^3/s is illustrated in Plate 6 (also on CD), and the variation

Table 3. Description of selected high power sites

Station	North latitude	West longitude	Elevation (m)	Description	Power (watts/m^2)
V28	34° 04' 8.21600"	111° 42' 33.5733"	629.7	Destroyed tree	123
V44	34° 04' 52.55535"	111° 42' 28.5338"	627.4	Swept tree with new growth	75
V104	34° 04' 31.7382"	111° 42' 44.6698"	621.5	Scour hole adjacent to outcrop	280
V105	34° 04' 31.9961"	111° 42' 44.5989"	621.2	Deepest part of scour hole	280
V140	34° 04' 53.5982"	111° 42' 33.5816"	619.7	Scour and furrows in slack water deposits	236
V141	34° 04' 46.1887"	111° 42' 38.2514"	615.9	Scour hole near riverbank	355
V142	34° 04' 45.5841"	111° 42' 39.1425"	618.7	Scour line near river bank	231
V144	34° 04' 35.6908"	111° 42' 40.8369"	616.7	Huge overturned cottonwood	375

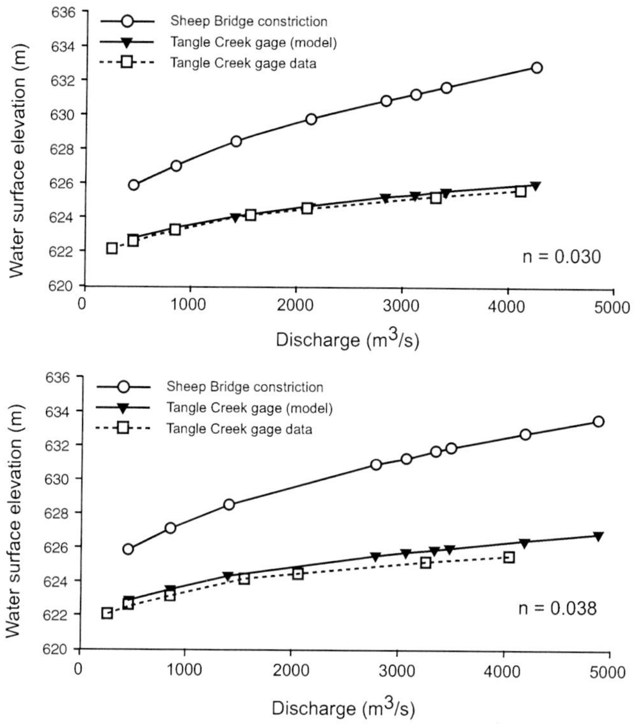

Figure 8. Variation of water surface elevation with discharge at two locations along the study reach—the Tangle Creek gage and the Sheep Bridge constriction—for two different values of bed friction ('n'). For low values of discharge and 'n' = 0.030, we find an excellent correspondence between direct measurements at the gage and model simulations of steady flow. Higher discharges require higher values of bed friction to match the observed variation of high water marks along the entire reach.

of stage with discharge is shown for two locations along the reach in Figure 8 for two different values of bed friction and discharge. Where ponding of flood waters occurs behind constrictions, it may be possible to exploit the increased sensitivity of stage to discharge to constrain both bed roughness and discharge for a particular flood. The comparisons in Plate 2 and Figure 8 show that two combinations of discharge and bed roughness produce the same stage at Tangle Creek gage. However those two combinations differ in stage by about a meter behind the Sheep Bridge constriction. Alternatively, this same optimum discharge and bed friction combination is obtained by matching both velocity and stage at the gage.

Flow modeling of the 1993 floods on the Verde River illustrates how a multi-dimensional pattern of high-water indicators can be used to uniquely constrain values of bed friction and discharge in a complex channel for a given flood. Thus, our analysis suggests that a robust method to determine peak discharge in the absence of direct velocity measurements is to measure stage variations above and below bedrock constrictions on both banks of a channel. The variation in water surface elevation along and across the channel may be well replicated by a depth-averaged, 2D simulation, providing the constraints needed to estimate flow discharge.

It is usually difficult or impossible (i.e. dangerous) to safely make direct velocity measurements during large floods. These results show that three-dimensional stage measurements upstream and downstream of constrictions, combined with depth-averaged 2D flow modeling, may uniquely constrain discharge in the absence of both direct velocity and roughness estimates. Slackwater sites behind significant constrictions may have discharge resolutions that exceed (by a factor of 2 or more) discharge resolution capabilities of typical gage sites along straight sections of channels for comparable stage measurement uncertainties. Consequently, peak discharge uncertainties for paleostage indicators may often be lower at sites behind significant constrictions than along straight sections of rivers.

Power computed using the discharge of 3398 m^3/s (Plate 3) is consistent with field observations. Inundated areas with low power (<30 watts/m^2) accumulated sediment and debris, including a huge debris pile in a back eddy behind the constriction at Sheep Bridge (Figure 3c). Some locations with high power, described in Table 3 and shown in Plates 4 and 5, are qualitatively consistent with the power distributions obtained at a discharge of 3398 m^3/s but not at discharges 500 m^3/s higher or lower than this value. Higher discharges scour the prominent bar formed downstream of Sheep Bridge and deposit it farther downstream. Lower discharges do not inundate areas where significant deposition occurred. Conversely, model simulations indicate that locations such as V28 (Table 2) where trees were severed are locations where current seams remain stationary for discharges greater than 2000 m^3/s.

Our simulations of stage, velocity, and power along the Verde River demonstrates that the solution of 2D, depth-averaged, shallow water equations can provide an excellent estimate of flow over complex topographic surfaces. Though the solution method is not strictly capable of balancing momentum in transcritical flow, it does conserve discharge, and accurately reproduces field measurements. There is excellent agreement between the model simulations and the observed high-water marks on the Verde River despite meters of water surface variation through a constriction at Sheep Bridge, as well as farther downstream in low-lying areas with complicated patterns of flooding. There is also a close correspondence between actual measurements of stage and discharge at Tangle Creek gage at low discharges, and model predictions of stage and discharge (Figure 8). Accurate simulation of these multi-dimensional aspects of flow allow us to make a robust and

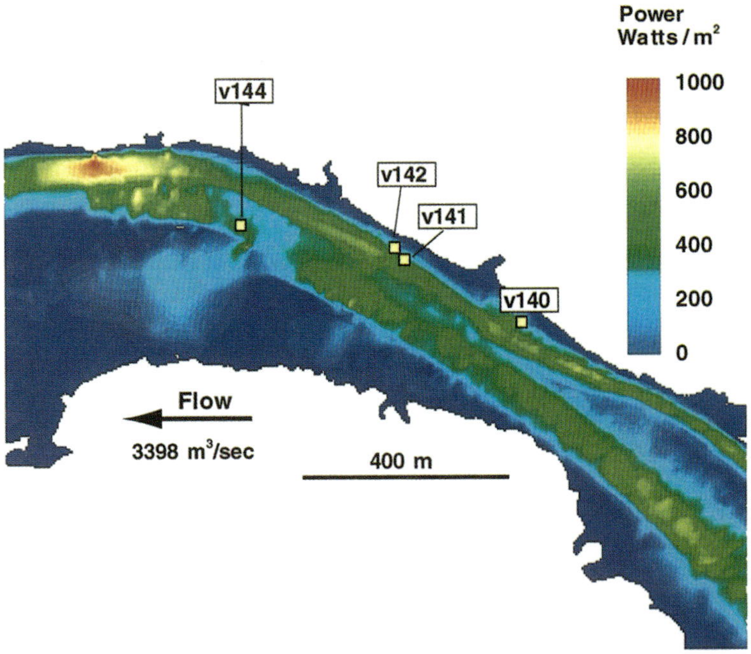

Plate 5. Calculated power distribution downstream of Tangle Creek gage, showing sites listed in Table 3. The colored area is inundated area, so deposits accumulated in much of the blue areas during the 1993 floods. Flow direction is south.

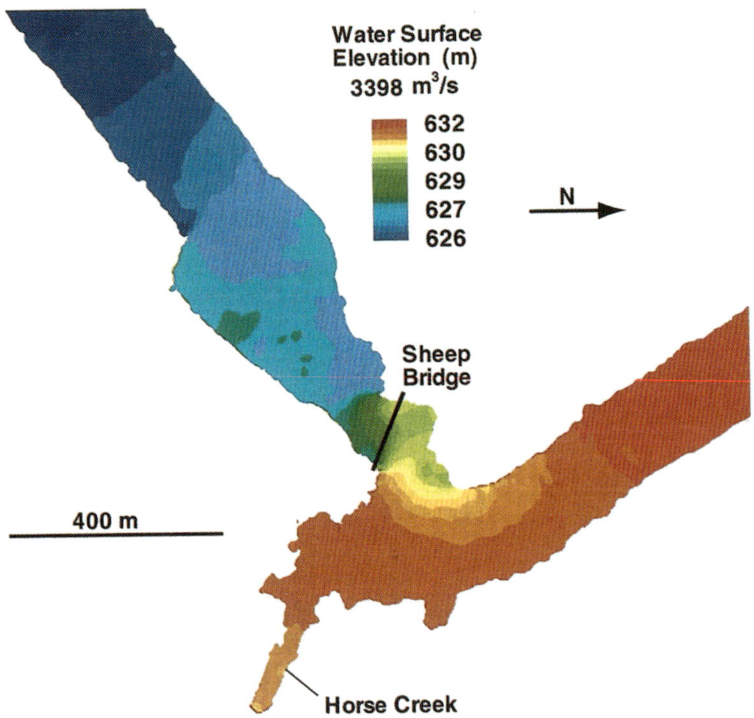

Plate 6. Variation of maximum stage for peak discharge of 3398 m^3/s near Sheep Bridge. The water surface falls nearly 3 meters over a distance of 100 meters.

unique interpretation of the peak discharges associated with the 1993 floods.

CONCLUSIONS

A 2D depth-averaged model for shallow water flow over three-dimensional topography reproduces observations of stage, discharge, and velocity from two large floods in the bedrock-confined channel of the Verde River. Erosional and depositional areas are accurately modeled for a given channel shape despite large concentrations of transported sediment; although flood power depends upon the cube of velocity, it varies linearly with bed friction, For the bedrock channels of the Verde River, with relatively high slopes and high Froude number flows, the primary factor affecting the distribution of stage, velocity, and flood power is channel topography. Bed friction can significantly impede velocities on the margins of the channel, and scales the magnitude of the power dissipated on the stream bed for a given flow.

For the January 8, 1993 flood on the Verde River, discharge and bed-friction are uniquely constrained with field data and flow modeling, either by combining stage and velocity measurements or by fitting the 3D distribution of high-water indicators upstream and downstream of a significant channel constriction. Field data for the 1993 flood, combined with flow modeling, constrain the discharge to be about 3398 m^3/s and the bed friction to be given by a Manning's n of 0.038. This is the single combination of discharge and bed friction in the range of 2000-5000 m^3/s and n = 0.03-0.04 that simultaneously matches the distribution of observed high water marks along the river as well as the velocity measured at the Tangle Creek gage.

Depth-averaged 2D flow models are useful for assessing flood hazards and for interpreting paleostage data. Our results may be used to establish probabilistic estimates of flood hazards using non-exceedance bounds for inundation and power [*Levish et al.*, 1994; 1997, *O'Connell et al.*, 1997, *Ostenaa et al.*, in press]. Estimates of both discharge and bed roughness (friction) are more robust for a single flood if a 3D distribution of high-water indicators along the channel are used rather than a few isolated measurements at gaging stations. Use of multidimensional flow models to interpret 3D patterns of inundation facilitates accurate interpretation of the damage and depositional record of past floods. Field observations of inundation and flow simulations combined with paleoflood constraints on maximum inundation limits through time provides a strong physical basis to quantify flood hazards.

Acknowledgments. This work was made possible by financial assistance from the Dam Safety Research Program (PALED and PHYLM) of the US Bureau of Reclamation, and field assistance from Dan Levish and Matt Jones. We also would like to thank Matt Narocki for his knowledgeable guidance and untiring assistance performing the GPS work that determined the channel topography. We would also like to thank Mike Charliebois and Dallas Reigle (SRP), and Jeff Phillips and Owen Baynham (USGS). Support for House's participation was provided by the National Science Foundation, Hydrologic Sciences Program Grant EAR-9996036.

REFERENCES

Bagnold, R.A., An approach to sediment transport problem from general physics, *U.S. Geol. Surv. Prof. Pap. 422-I*, 37 pp., 1966.

Casulli, V., Semi-implicit finite difference methods for the two-dimensional shallow water equation, *J. Comp. Phys.*, 86, 56-74, 1990.

Denlinger, R.P., R.A. Walters, D. Levish, and D. Ostenaa, Robust methods for flood routing over highly irregular terrain, *Proc. First Fed. Interagency Hydrologic Modeling Conf., April 19-23*, Las Vegas, Nevada, 2, 6.89-6.96, 1998.

Henderson, F.M., *Open Channel Flow*, 522 pp., MacMillan Publishing Co., New York, New York, 1966.

House, P.K., and K.K. Hirschboeck, Hydroclimatological and paleohydrological context of extreme winter flooding in Arizona, 1993, *Geol. Soc. Amer. Rev. Eng. Geol.*, *11*, 1-24, 1997.

House, P.K., P.A. Pearthree, and J.E. Fuller, Hydrologic and paleohydrologic assessment of the 1993 floods in the Verde River, central Arizona, *Ariz. Geol. Surv. Open- File Rep. 95-20*, 23 pp., 1995.

Hydrologic Engineering Center, HEC-RAS River Analysis System version 2.0, *Hydraulic Reference Manual*, U.S. Army Corps Eng., Davis, California, 1997.

Jin, M. and D.L. Fread, Dynamic flood routing with explicit and implicit numerical solution schemes, *J. Hydraul. Eng.*, 123, 166-173, 1997.

Keulegan, G.H., Laws of turbulent flow in open channels, *NBS Research Paper RP1151*, 21, 707-741, 1938.

Levish, D. R., D. A. Ostenaa, and D. R. H. O'Connell, A non-inundation approach to paleoflood hydrology for the event-based assessment of extreme flood hazards, in *Association of State Dam Safety Officials 1994 National Meeting Proceedings,* 69-82, 1994.

Levish, D.R., D.A. Ostenaa, and D.R.H. O'Connell, Paleoflood hydrology and dam safety, in *Waterpower 97, Proceedings of the International Conference on Hydropower*, edited by D. J. Mahoney, 2205-2214, Amer. Soc. Civ. Eng., New York, 1997.

Miller, A.J. and B.L. Cluer, Modeling considerations for simulation of flow in bedrock channels, in *Rivers over Rock: Fluvial Processes in Bedrock Channels*, *Amer. Geophys. Union Mono. 107*, edited by Ellen Wohl, 61-104, 1998.

Nelson, J.M., R.R. McDonald, and D.M. Rubin, Computational prediction of flow and sediment transport patterns in lateral separation eddies, *EOS Trans Amer. Geophys. Union*, Fall Meeting Supplement, 268, 1994.

O'Connell, D.R.H., D.R. Levish, and D.A. Ostenaa, Bayesian flood frequency analysis with paleohydrologic bounds for Late Holocene paleofloods, Santa Ynez River, California, in *Twenty years later: What we have learned since the Big Thompson flood, Special Publication No. 33*, edited by E. Gruntfest, 183-196, Natural Hazards Research and Applications Information Center, Univ. Of Colorado, Boulder, 1997.

Ostenaa, D.A., D.R.H. O'Connell, R.A. Walters, and R.J. Creed, Holocene paleoflood hydrology of the Big Lost River, western INEEL, Idaho, in *Geology, Hydrogeology and Environmental Remediation, Idaho, GSA Special Paper 354* edited by Link, P.K. and L.L.Mink, National Engineering and Environmental Laboratory, Eastern Snake River Plain, Idaho, in press.

Schlicting, Hermann, Boundary Layer Theory, 814 p., McGraw-Hill Publishing Co., New York, New York, 1979.

Sinha, S.K., Sotiropoulos and A.J. Odgaard, Three dimensional numerical model for flow through natural rivers, *J. Hydraul. Eng.*, 124, 13-24, 1998.

Smith, C.F., P.D. Rigas, L.K. Ham, N.R. Duet, and D.W. Anning, Water Resources Data, Arizona, Water year 1993, *U.S. Geological Survey Water-Data Report AZ-93-1*, 360 pp., 1994.

Steffler, P.M., *CDG2D Depth Averaged Hydrodynamic Model Reference Model*, University of Alberta, Edmonton, 1997.

Vanoni, V.A., Sedimentation Engineering, ASCE Manuals and Reports on Engineering Practice- No. 54, 745 pp., 1977.

Vreugdenhil, C.B., *Numerical methods for shallow water flow*, Kluwer Academic Publishers, 259 pp., 1994.

Wiele, S.M., J.B. Graf, and J.D. Smith, Sand deposition in the Colorado River in the Grand Canyon from flooding of the Little Colorado river, *Water Resour. Res.*, 32, 3579-3596, 1966.

Roger P. Denlinger, Cascades Volcano Observatory 5400 MacArthur Boulevard Vancouver, WA 98661, roger@usgs.gov

Daniel R.H. O'Connell, Geophysics, Paleohydrology, and Seismotectonics Group US Bureau of Reclamation P.O. Box 25007 D-8330, *Lakewood, CO 80225* geomagic@seismo.usbr.gov

P. Kyle House, Nevada Bureau of Mines and Geology University of Nevada, Mail Stop 178, Reno, NV, khouse@unr.edu

Initial Motion of Boulders in Bedrock Channels

Paul A. Carling

Department of Geography, University of Southampton, Highfield, Southampton, SO17 1BJ, UK.

Mark Hoffmann and Andrea S. Blatter

Institut für Wasserwirtschaft und Kulturtechnik, Universität Fridericiana zu Karlsruhe, Germany

Natural boulders situated on flat or rugose bedrock channel surfaces may be isolated or they may occur in simple or complex groups. Grouped boulders may be imbricated with flow-normal spans (B) comparable with the channel width (W). However, before the hydraulic stability of complex boulder bedforms can be studied, an evaluation of entrainment conditions for isolated boulders and simple groupings of boulders is required. The problem reduces to a consideration of (i) isolated boulders; (ii) stacks of a few boulders of limited span ($B<W$) and, (iii) stacks of many spanwise boulders ($B \cong W$). In this paper, preliminary results pertaining to the entrainment of isolated cubes on bedrock are reported. Latterly, data from two published studies of simple grouped blocks are re-evaluated and finally, examples of relatively simple natural boulder bedforms on bedrock are presented. Flume experiments quantify the force balance equation for initial motion of isolated smooth-faced cubes and oblongs. Flume experiments consisted of two parts: (1) Initial motion on a smooth concrete surface; (2) Initial motion on an artificially-roughened concrete surface simulating bedrock positive and negative irregular steps up to 2mm in height transverse to the flow. Three different cases of initial motion were recorded: (1) the block is emergent and pivots out of place; (2) the block is emergent and slides out of place; (3) the block is submerged and slides out of place. Thus initial motion by pivoting is induced by the flow characteristics associated with emergence. Due to the turning moment, the friction coefficient (μ_f) for the smooth bed was not constant, but on the simulated bedrock surface the friction coefficient was approximately constant. For the smooth concrete bed, with maximum contact between block and bed, μ_f was 0.80 for sandstone and 0.60 for slate. For the simulated bedrock the number of points of contact were reduced such that μ_f equaled 0.50 and 0.59 for the two rock types, respectively. In a natural smooth-bedded limestone

Ancient Floods, Modern Hazards: Principles
and Applications of Paleoflood Hydrology
Water Science and Application Volume 5, pages 147-160

channel μ_f was 0.81 for sandstone blocks, 0.73 for slate and 0.70 for natural limestone blocks derived from the channel bed. The variation in the drag coefficient (C_d) for conditions of initial motion was related to particle orientation (L/B) and relative water depth (h/d). For the fully submerged case $C_d = 0.95$, but for emergent conditions C_d increased rapidly as relative depth was reduced. For palaeohydraulic estimates of flow parameters it may be concluded that the friction coefficient is constant. Setting C_d equal to 0.95 is appropriate where blocks were deeply submerged ($(h/d)(L/B) > 2$) and entrained by sliding. However, where blocks were in shallow water or emergent ($(h/d)\ (L/B) < 2$), increasingly large values of C_d pertain as h/d decreases.

INTRODUCTION

Isolated or grouped large boulders are commonly observed in bedrock channels. Some of these boulders may be residual 'lag' deposits which have fallen from the side walls of the channels and are too large to have been transported by floods. Other boulders may have been transported and deposited by floods. Thus, in principle, knowledge of the forces required to move these latter boulders should allow estimation of the palaeodischarge, whilst the lag boulders should provide an upper limit to estimations of palaeoflow competence.

The question arises as to why competence studies particular to bedrock channels are required, when the entrainment of well-rounded cobbles from gravel beds consisting of mixed grain-size has been well-studied. The latter condition usually entails initial motion by a pivoting mechanism in relatively deep sub-critical flow [*Komar*, 1996] because gravel-bedded rivers are free to adjust their boundaries such that negative feedback ensures super-critical and critical flows occur only locally [*Grant*, 1997]. In contrast within bedrock channels the bed slope and geometry of many channel reaches may be adapted to maximize the unit discharge through a minimization of the mean specific energy [*Grant*, 1997; *Tinkler*, 1997; *Chanson*, 1999]. This condition occurs when the flow is critical (Froude number ~ 1.0). However, the boundaries of many bedrock channels adjust slowly, if at all, to the flow regime and so, unlike gravel-bedded streams, the largest residual boulders in a bedrock channel may be subject to sub-critical, critical or super-critical flows. Consequently the largest water-transported boulders may have been moved by near-critical or super-critical flows. Finally, the bed of bedrock channels may be physically-smooth and boulders may be tabular rather than rounded. Thus a consideration of the mechanics of entrainment by sliding (as well as pivoting) in critical and super-critical flow is important in bedrock systems.

PERSPECTIVE

Boulders often lie on extensive smooth or rugose bedrock surfaces. Boulders may be either well-rounded and entrained by pivoting or they are blocky with beveled edges. In the latter case, initial motion may be by pivoting or by sliding. Boulders immediately upstream of a positive bedrock step, which is higher than the height of the boulder, often can neither slide nor pivot from place. In this case initial motion only occurs if the lift force exceeds the gravitation force and the boulder is lifted into the flow. *Wende* [1999] termed this 'hydraulic plucking' and calculated the associated critical velocities. Boulders may be completely submerged or boulders may have dimensions comparable with, or greater than, the water depth. Thus the depth of flow above the obstacle may be small. The flow may exhibit free-surface undulations which generate quasi-periodic temporal variation in the bed shear stress [*Montes and Chanson*, 1998; *Chanson*, 1999], or the boulder may be emergent such that large standing waves occur around and above the obstruction (Figures 1 and 2). For the emergent case, air is entrained both on the upstream side, along the lateral margins and on the downstream side of the obstacle. In the extreme, on the downstream side, the rockbed surface is exposed within a horseshoe-shaped standing wave (Figure 3). Thus in near-emergent or emergent cases, the effects of sharp-edges of boulders, stagnation zones, wake, air entrainment and wave drag all may influence the over-all drag on the obstacle.

Boulders may be isolated or they may occur in groups. Often the grouped boulders are imbricated, especially if the source lithology favors the production of tabular blocks. Grouping may occur owing to the stacking of clasts against a step in the bedrock surface or against a stalled obstacle clast (Figures 4 and 5). In either case a small cluster may result, the total width of which is much less than the channel width (i.e. $B \ll W$) (Figure

5), or the stacking mechanism can generate stacked boulder units which together have flow-normal spans (B) comparable or equal to the channel width (W) (Figure 6). A wide range of boulder configurations may thus develop. For the purposes of assessing the mechanics of entrainment, the problem can be reduced to a consideration of (i) isolated boulders; (ii) stacks of a few boulders of limited span ($B<<W$) and, (iii) stacks of many spanwise boulders ($B \cong W$). However, before the stability of complex boulder bedforms can be studied, an evaluation of entrainment conditions for isolated boulders and simple groupings of boulders is required. In this paper preliminary results pertaining to the entrainment of isolated cubes and oblongs on bedrock are reported. Latterly, data from two published studies of simple groups of blocks are re-evaluated. Finally examples of relatively simple natural boulder bedforms on bedrock are presented.

THE FORCE BALANCE FOR ISOLATED BLOCKS

This paper primarily considers isolated boulders of cuboid form which are termed 'blocks'. The critical velocity (U_c) for entrainment of boulders by sliding may be expressed [*Carling and Grodek*, 1994];

$$U_c^{\,2} = \frac{2(\rho_s - \rho)gD_b}{\rho}\,\frac{\mu_f}{C_d + \left[C_l \mu_f (D_b / D_c)\right]} \quad (1)$$

where $(\rho_s - \rho)g$ is the immersed mass of a boulder with medium and short axes, D_b and D_c. The literature describing the parameters within the force balance equation for initial motion shows that the drag coefficient (C_d) and lift coefficient (C_l) of obstacles, and the friction coefficient (μ_f) between blocks and bedrock are poorly quantified for immersed or emergent boulders.

Dimensional analysis of the variables affecting the drag and lift over a single obstacle in a non-uniform flow, with free water surface effects, reduces variables to a consideration of the dimensionless parameters:

$$C_d = \phi\,(R_e, Fr, h/d, h/L, W/B, P) \quad (2)$$

where $R_e = Uh/\nu$ is the Reynolds number and $Fr = U/\sqrt{(gh)}$ is the Froude number. The water depth is h, d is the block height, g is the acceleration due to gravity, ν is the kinematic viscosity, here B is the width of the single block transverse to flow and L is the length of the block in the downstream direction. The velocity profile parameter P accounts for non-logarithmic profile effects. Other than selecting an appropriate reference level for U in the flow depth, the effect of P usually can be neglected [*Yen*, 1965; *Flammer et al.*, 1970]. The width of the unimpeded flow between channel walls and the obstacle is relevant. Generally, for negligible wake interference in experimental sub-critical flow, W/B should exceed 20 [*Flammer et al.*, 1970], which approximates in natural flows to a few percent of the bed area being occupied by obstacles [*Abrahams and Parsons*, 1994]. In supercritical flow W/B should exceed 2.0 [*Flammer et al*, 1970]. Reynolds number effects are slight for transitional flow conditions [*Flammer et al.*, 1970].

The Drag Coefficient

Reported values of the drag coefficient are dependent on the reference level selected for the value of U used to derive C_d [*Flammer et al.*, 1970] or, values equal to C_d/μ_f are reported as the putative drag coefficient [*e.g.*

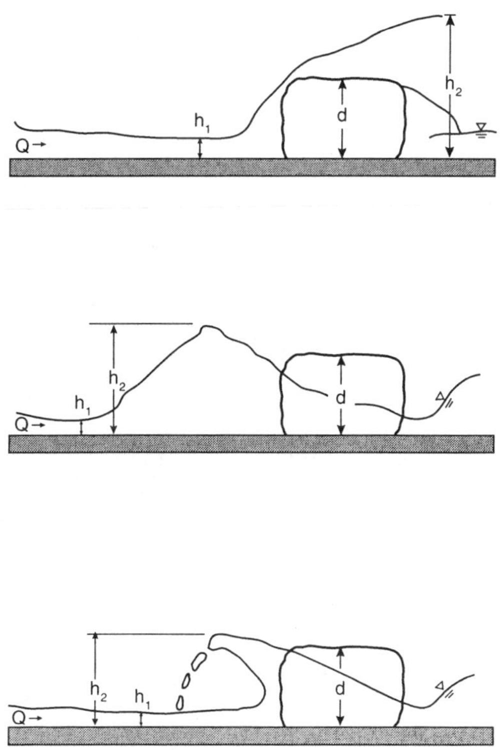

Figure 1. Examples of the style of waveforms associated with shallow flow past obstacle clasts on a plane non-mobile bed [after *Zgheib and Urroz-Aguirre*, 1990].

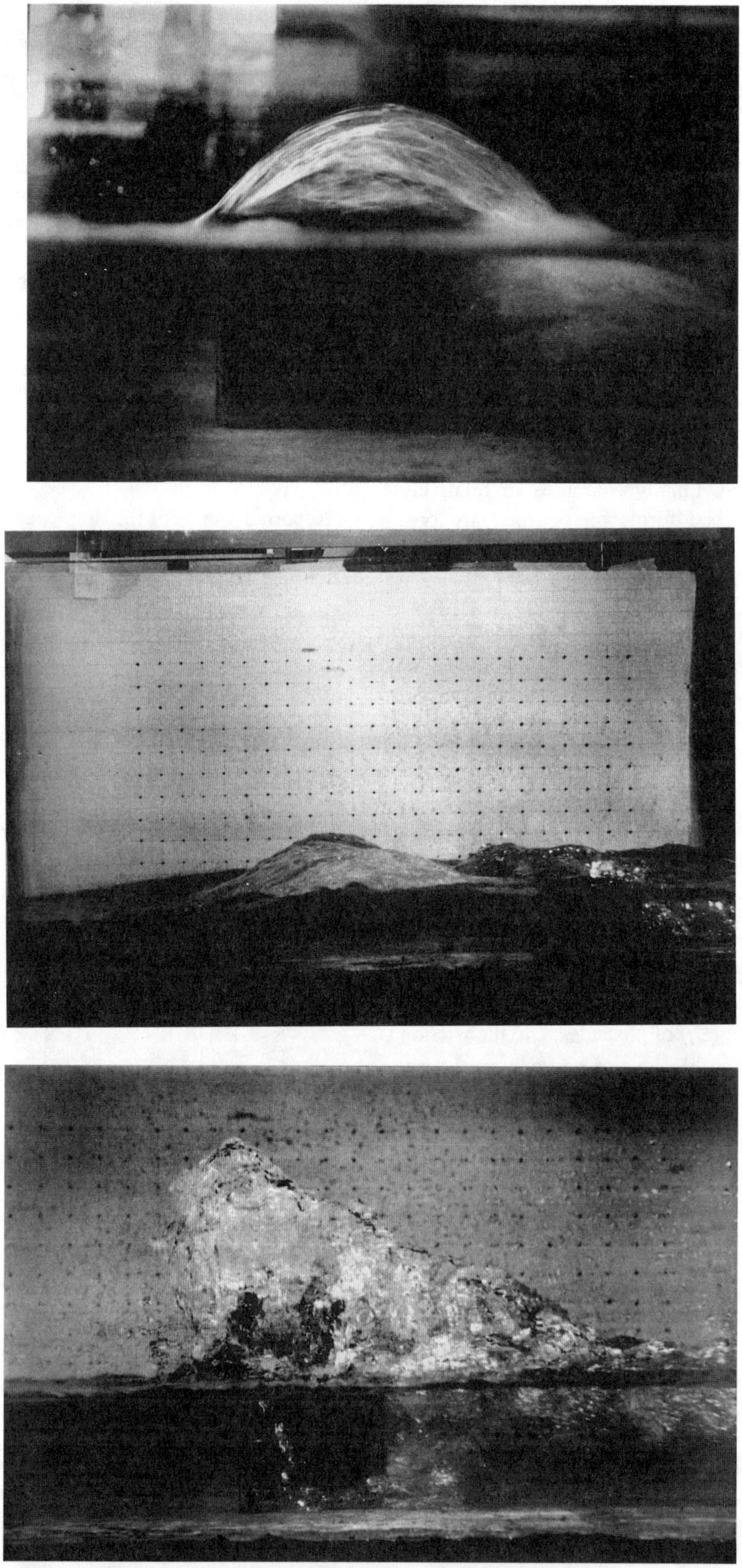

Figure 3. View obliquely upstream of block described in Fig. 2c.

Allen, 1942; 1947]. The drag coefficient C_d for submerged and emergent hemispheres has been evaluated in several studies [*e.g.* Flammer *et al.*, 1970] but less-streamlined shapes provide greater drag. *Pizzuto et al.* [1999] suggested a value of the drag coefficient of 1.36 ± 0.68 for angular boulders. For regular rectangular blocks and natural irregular-shaped rocks, values in the range 0.71 to 1.2 have been reported [*Hoerner*, 1965; *Engineering Sciences Data*, 1976; *Denny*, 1988; *Abrahams and Parsons*, 1994]. *Rajaratnam* [1967] argued that C_d varied along a downstream transect beneath an hydraulic jump, such as that which develops above near-emergent boulders. Thus C_d does not assume a constant value, but is flow dependent [*Herbich and Shulits*, 1964; *Yen*, 1965]. For example, initial motion data for plaster cubes on a smooth steel surface [*Ikeda*, 1988] showed that C_d/μ_f increased as relative depth decreased for emergent blocks in supercritical flows.

The Lift Coefficient

The lift coefficient C_l is poorly defined and existing data resulting from various studies are not consistent [*James*, 1990]. James reviewed existing data for spheroids perched on a bed of similar particles and concluded that, despite conflicting results, the lift coefficient could be regarded as equaling $0.5C_d$ for fully turbulent flow in the Reynolds number range ~ 10^3 to 10^4. For higher Reynolds numbers values for C_l in excess of $2.0C_d$ might apply [*James*, 1990]. *Renius* [1986] reported variation in the pressure coefficient above imbricated regular blocks (see Figures 8 and 9). Lift coefficients, as determined from Renius's data, range widely from positive values for supercritical flow above normally imbricated blocks to negative values for normal imbrication beneath hydraulic jumps. Negative values also pertained for supercritical flow above arrays of blocks tilted upstream. Positive values of C_l varied widely from 0.09 to 0.46 for Froude numbers typically ranging between 1 and 3.

The Friction Coefficient

The coefficient of friction (μ_f) is given by:

$$\mu_f = \frac{\tau_o}{\sigma} \qquad (3)$$

Figure 2. Styles of water surface instabilities for emergent block (150mm x 99mm x 50mm) on simulated bedrock. (a) Long axis parallel with flow: $h/d = 0.67$; $h/L = 1.0$; $W/B = 18$. The block causes minimal obstruction to the bulk flow, such that the regional water surface is little perturbed. The block is slightly emergent such that the leading edge causes a pronounced localized standing wave to break above the top of the block. Air is entrained along the full width of the upstream face and escapes along the full height of the lateral margins. Consequently an aerated space 5mm to 10mm in width causes the face in view to be 'dry' down to the local water surface which recovers to approximately half the block height at the downstream face. (b) Long axis transverse to flow: $h/d = 1.0$; $h/L = 0.5$; $W/B = 6$. The block causes substantial flow blocking such that the regional water surface is perturbed with a standing wave across the full flume width upstream of the block. (c) Long axis in the vertical, short axis transverse (see also Fig. 3): $h/d = 0.61$; $h/L = 0.4$; $W/B = 18$. The block causes minimal obstruction to the bulk flow, but being 50% higher than the water depth, reflects the standing wave on the upstream face. The water surface locally is drawn-down along the full length of the face in view.

Figure 4. Example of isolated 2m-long block (foreground) and a stack of a few boulders of limited span (*B*<*W*) imbricated against a bedrock step. Flow right to left. River Dee, Dentdale, Yorkshire, England.

Figure 5. Example of isolated 2m-long block (background) and a stack of limited span developed against a small bedrock step where one large block has formed a focus for cluster development. Note vertical imbrication of the blocks on the upstream side which is characteristic of such systems and the small shadow deposit of cobbles and pebbles in the leeside. Birk Beck, tributary to the River Lune, Cumbria, England.

Figure 6. Stacks of many spanwise boulders against a closely-spaced series of flow transverse bedrock steps constitute bedforms of span comparable to the channel width ($B \cong W$). Durack River, Kimberley, Australia. Flow bottom left to top right. Figure center-left provides scale.

where τ_o is the critical shear stress and σ is the effective normal stress resisting initial motion and for immersed blocks, $\sigma = (\rho_s - \rho)gD_aD_bD_c$. Thus for regular cuboids σ is readily determined and τ_o can be determined directly by pulling blocks using a force gauge. Despite the simplicity of equation 3 there are few determinations of the friction coefficient for different rock types and rock surface-roughness textures. *Carling and Grodek* [1994], *Wende* [1999] and *Carling and Tinkler* [1999] reviewed earlier studies and concluded that $\mu_f = 0.82$ to 0.89 applied to submerged tabular blocks on bedrock.

A full consideration of the parameters in Equation 2 will be published in due course. It is sometimes argued [*e.g. Yen*, 1965] that the drag coefficient is significantly correlated with the Froude number and to a lesser degree with the Reynolds number. However, in palaeoflow assessments these values are not known. Our detailed data (which are not reported here) show that apparent Froude and Reynolds number effects can be represented as relative depth (h/d) effects [see also *Flammer et al.*, 1970]. Here brief results on the effect of relative depth on the drag coefficient are reported, as within bedrock channels boulder size is measurable whilst depth estimates are possible using palaeostage indicators. Flume and river experiments quantify the force balance equation for initial motion of isolated smooth-faced cubes and rectangular blocks (Table 1). Semi-empirical equations were derived to demonstrate the variation of C_d as blocks, of varying dimensions (D_a, D_b, D_c) and orientations to the flow (L, B and d), were entrained as the relative depth of submergence (h/d) was varied. The study consisted of three parts: Initial motion on a smooth concrete surface; Initial motion on an artificially-roughened concrete surface simulating bedrock positive and negative irregular steps up to 2 mm in height roughly transverse to the flow; Initial motion on natural irregular bedrock surfaces in the field.

EXPERIMENTAL METHOD

Two bed conditions were investigated within a tiltable laboratory hydraulic flume. The flume has a working section of 10.56m and a width of 0.9m. Bed condition (i) consisted of lining the complete length of the flume with smooth-faced concrete paving stones 0.89m wide by 0.5m long. The paving stones were carefully grouted in place to provide a 'smooth' bed surface. Bed condition (ii) used simulated natural-stone paving stones with irregular surfaces that included 1mm to 2mm high flow-transverse rugosities and ridges.

Blocks of varying dimensions and weights were machine-sawn from sandstone or slate and placed in exactly the same place on the test bed for replicate test runs, although the block orientation was varied. By adjusting flume slope and tailgate height, flow depths were less than, equal to, or greater than the block height. Discharge was increased slowly until the blocks moved slightly or were swept away. A qualitative description of

Table 1: Summary of experimental materials. Machine-sawn slightly porous blocks 1 through 12 were weighed air-dried and after immersion in water for several days. The natural block was used to define the friction coefficient in a limestone bedrock channel and was not used in flume studies.

Block Number	Dry Weight M (g)	Wet Weight M (g)	Long Axis D_a (mm)	Int. Axis D_b (mm)	Short Axis D_c (mm)	Rock Type
1	288	295	50	50	50	Sandstone
2	583	596	100	50	50	Sandstone
3	874	894	150	50	50	Sandstone
4	1169	1196	199	50	50	Sandstone
5	1745	1786	150	99	50	Sandstone
6	2637	2696	150	150	50	Sandstone
7	2668	2689	230	99	50	Sandstone
8	4380	4420	123	123	123	Sandstone
9	19660	19720	266	177	177	Sandstone
10	397	397	51	51	51	Slate
11	5570	5570	124	124	124	Slate
12	16820	16820	178	178	178	Slate
Natural Block	39266	-	655	210	125	Limestone

the style of initial motion was recorded and then the block was removed from the flume. For each test, but with the block removed, the water surface slope (I_s), bedslope (I_b) and depth (h) were recorded. Vertical log-normal velocity profiles, measured using a miniature flow impellor or acoustic doppler velocimeter, were used to calculate the depth-average velocity ($\overline{U}$) and roughness length (z_0) of the flume bed. The critical velocity ($\overline{U_c}$) for initial motion impinging on the upstream side of the block was measured at a height above the bed equal to one-half the block height and, in addition, the depth-average velocity was calculated from logarithmic velocity profiles.

In all tests supercritical steady-flow pertained (Fr = 1.5 to 2.5) and, because of the physically-smooth beds, the flow approaching the obstacles was hydraulically transitional rough-turbulent. No measurable difference was recorded in the flow parameters comparing the smooth and roughened concrete surfaces. The roughness length z_0 for the simulated bedrock was of the order of 0.1mm.

The drag coefficient was obtained from the relationship:

$$C_d = \mu_f \frac{2g(K\gamma_s - A_b h\gamma)}{(\gamma A_d)U^2} \tag{4}$$

wherein γ_s and γ are the specific weight of the blocks and the water, respectively. K is the volume of the block, g is the acceleration due to gravity, and A_b and A_d are the basal and the flow normal submerged areas of the blocks, respectively. For the flume tests the friction coefficient μ_f was calculated for each rock type by first assuming C_d was known, using an engineering standard for regular blocks (settling through a water column) with varying combinations of the lengths: D_a,D_b,D_c [*Rosemeier*, 1976]. The average value of μ_f for each rock type was then used in Equation 4 to derive specific values of C_d for each test which included submerged and emergent conditions. The value of drag coefficient depends on the characteristic velocity used to calculate C_d [*Flammer et al.*, 1970]. The velocity ($\overline{U}$) integrated over the height of the obstacle will give a larger C_d value than one using $\overline{U}$ measured above, or integrated over, the height of the boundary layer. We used both approaches, but for our shallow flows there was no statistically significant difference in the results. Thus we report data pertaining to the depth-average velocity which is the parameter of interest for palaeodischarge estimates. The above procedure was justified in-as-much as the method obtained values of the drag coefficient for submerged blocks which are consistent with published literature on the value of the drag coefficient for regular blocks adjacent to a boundary (see Equation 5 below). Thus it can be reasonably assumed that the C_d values for emergent blocks are consistent and representative.

The friction coefficients reported below however were derived using Equation 3 and a strain gauge to pull blocks across the bed. These latter tests consisted of (i) sawn sandstone or slate blocks on the smooth and rough concrete in the flume; (ii) sawn sandstone and slate on natural limestone river bedrock; (iii) a natural limestone cubic block on natural limestone bedrock.

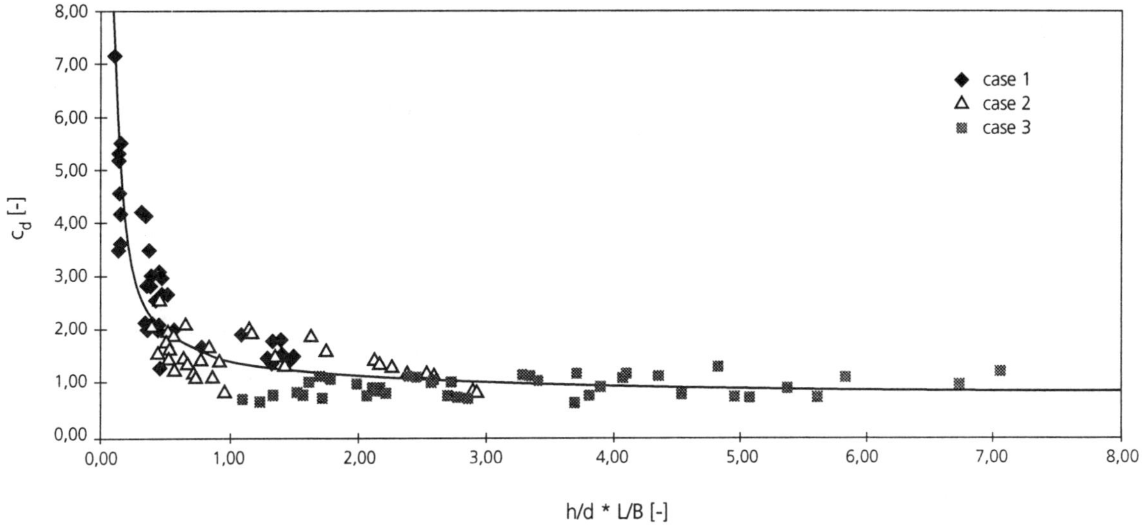

Figure 7. Variation of the drag coefficient as a function of the relative depth and block form-factor. Case 1: Block is emergent and pivoting. Case 2: Block is emergent and sliding. Case 3: Block is submerged and sliding.

RESULTS OF SINGLE BLOCK EXPERIMENTS

Three different cases of initial motion were recorded: (1) The block is emergent and pivots out of place; (2) The block is emergent and slides out of place; (3) The block is submerged and slides out of place. Initial motion by pivoting was always induced by flow characteristics associated with emergence.

Due to the influence of a turning moment, the friction coefficient for the smooth bed was not constant. In (1) and a strong turning moment the friction coefficient was higher than in (2). This effect did not occur on the rough simulated bedrock surface where the friction coefficient was approximately constant. Initial motion was not affected by μ_f if the block tilted as the turning moment is independent of μ_f. Thus μ_f had little influence on initial motion as, given the slightest irregularity in the surfaces, many blocks pivoted from position before sliding downstream. For the smooth concrete bed, with maximum contact between block and bed, μ_f was 0.80 for sandstone and 0.60 for slate. For the simulated bedrock surface the number of points of contact were reduced such that μ_f for blocks sitting proud on bedrock rugosities ranged between 0.59 and 0.50. In a natural 'smooth'-bedded limestone channel μ_f was 0.81 for sandstone blocks, 0.73 for slate and 0.70 for a natural limestone block derived from the limestone channel bed. Variability in these results, comparing smooth and rough surfaces, reflects both changes in contact area between bed and block, and the increase in critical shear stress required to push blocks up over bedrock rugosities. In the latter case the applied force must overcome friction and supply work to lift the blocks against the vertical force [*Atkinson and Bransby*, 1978].

For practical applications the drag coefficient for incipient motion was shown to be a function of relative submergence and block orientation; both factors being readily estimated from field data (Figure 7). A least squares empirical function applies equally well to both emergent and fully submerged blocks on the simulated bedrock surface:

$$C_d = \left(\frac{0.6}{(h/d)(L/B)} \right) + 0.9 \quad r^2 = 0.79 \tag{5}$$

where B is the block length transverse to the flow and L is the length parallel to the flow.

The uncertainty expressed by the data spread is described by limiting curves; $y = (0.8/x) + 1.3$, and $y = (0.4/x) + 0.6$. For relative depths of about 10, or greater, C_d assumes a constant value of about 0.95, which compares well with other recommended values [*Hoerner*, 1965; *Carling and Grodek*, 1994; *Wende*, 1999; *Carling and Tinkler*, 1998]. However, for shallow submergence the drag coefficient can be greater than 1 and increases very rapidly as water level falls below the block height. For the smooth test bed the values of the coefficient were generally about 30% greater than the simulated bedrock such that in deep flow ($h/d > 10$) C_d equalled 1.29.

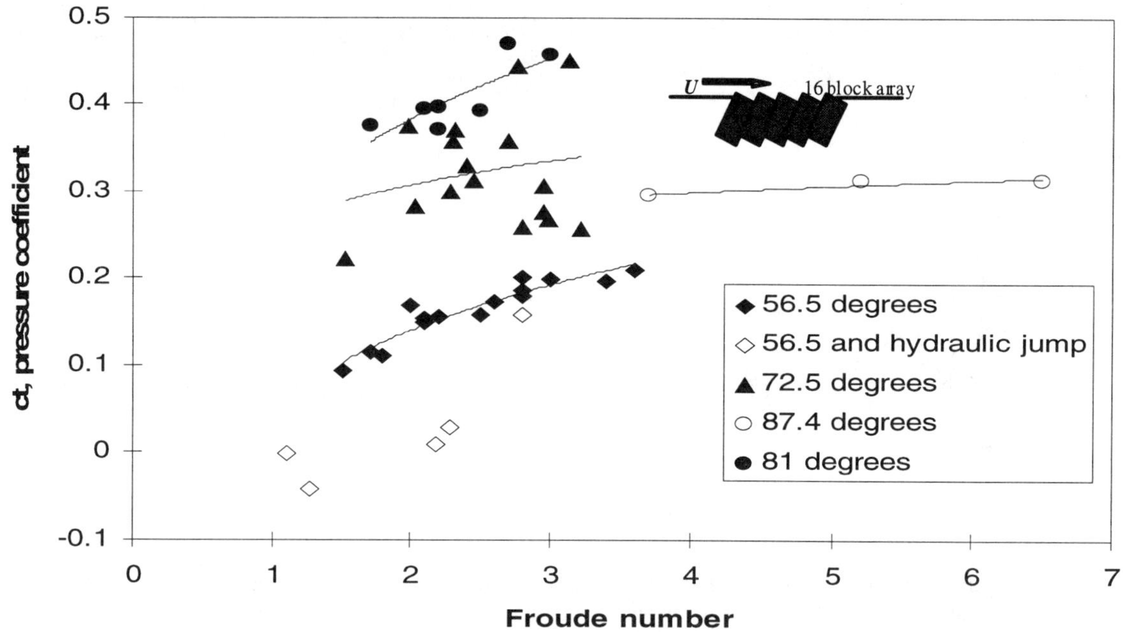

Figure 8. Variation of the pressure coefficient (C_t) above an array of 'normally' imbricated blocks as a function of the Froude number. The coefficient refers to the force acting normal to the top surface of rectangular blocks in fully turbulent flow parallel to the base of the flume. Data from Reinius (1986).

INITIAL MOTION OF MULTIPLE BLOCK BEDFORMS

This paper primarily has considered isolated blocks. However as further research will require a consideration of grouped blocks ($B<W$ and $B \cong W$), a brief consideration of these latter cases is warranted. There are few data in the literature which pertain to complex grouped blocks, but consideration of two studies demonstrates it is possible to define the nature of the parameter space. These studies provide some confidence that the range of entrainment conditions and the related discharge values can be determined for palaeoflow evaluation.

Renius [1986] provided a systematic study of the variation in the pressure coefficient (C_t) as a function of the Froude number for supercritical flow over a series of blocks. Sixteen identical blocks were imbricated such that the angle of imbrication for the group was varied. Imbrication angle (relative to the horizontal) was either 'normal', i.e. in the downstream direction, similar to the common shingling seen in platy river gravel, or 'inverse', *i.e.* in the upstream direction. In either case the data can be interpreted as representing the variation in the pressure above either fissured bedrock, or a simple stack of blocks ($B<W$). A number of main conclusions can be drawn. Firstly, the pressure coefficient for 'normal' imbrication is positive and increases both as the Froude number increases and as the angle of imbrication is increased (Figure 8). Secondly, for any given angle of imbrication (here 56.5°) the development of an hydraulic jump above the blocks will reduce the pressure coefficient, possibly to induce negative values of pressure, such that a significant lift force develops above the surface of the blocks. Inverse imbrication induces negative pressure, which is more pronounced as the degree of imbrication is reduced, but does not seem to be a function of the Froude number (Figure 9). Thus Renius's study demonstrates that a systematic study of hydrodynamic forces above grouped blocks is tractable.

Allen [1942 and 1947] investigated the initial motion of simple cuboid blocks. In particular, the initial motion of isolated concrete blocks on a cast-concrete bed in an hydraulic flume was investigated [see *Carling and Tinkler*, 1999]. Allen also considered the variation in initial motion criteria for nine different configurations of groups of identical blocks. At the simplest, blocks were placed side-by-side across the flume to form a simple flow-normal obstacle ($B = W$). Additional flow-normal rows of blocks were then added, as shown in Figure 10. Finally, blocks were stacked pell-mell to form a flow-transverse bedform and, for this condition, the flow conditions were recorded for both initial motion of one or two blocks and also for substantial break-up of the bedform. Thus Allen's artificial stacks of blocks may be

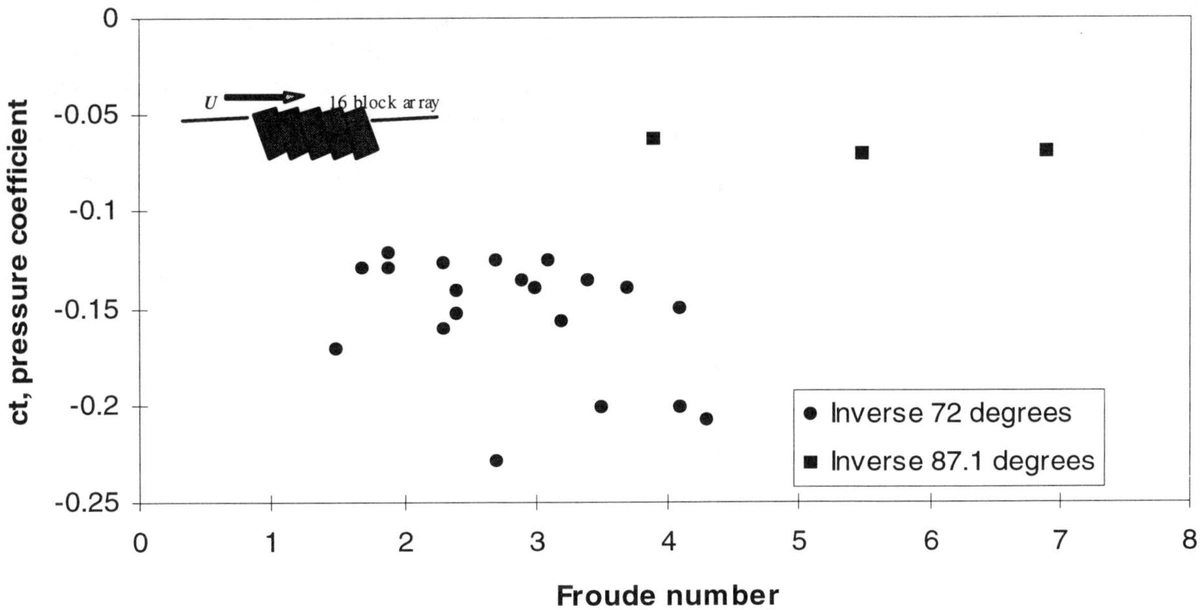

Figure 9. Variation of the pressure coefficient (C_t) above an array of 'inversely' imbricated blocks as a function of the Froude number. The coefficient refers to the force acting normal to the top surface of rectangular blocks in fully turbulent flow parallel to the base of the flume. Data from *Reinius* (1986).

used to evaluate the range of entrainment conditions associated with transverse boulder bedforms in bedrock channels. In Figure 11, Allen's data are plotted to show the variation in the non-dimensional critical entrainment velocity (U'/U_c) as a function of relative depth (h/d). Here velocity is the depth-mean value, U' is the critical entrainment velocity for the break-up of the grouped blocks, U_c is the critical entrainment velocity for an isolated block, h is the water depth and d is the height of the unit block.

A number of observations can be made with respect to Figure 11. Firstly, the range in values of the non-

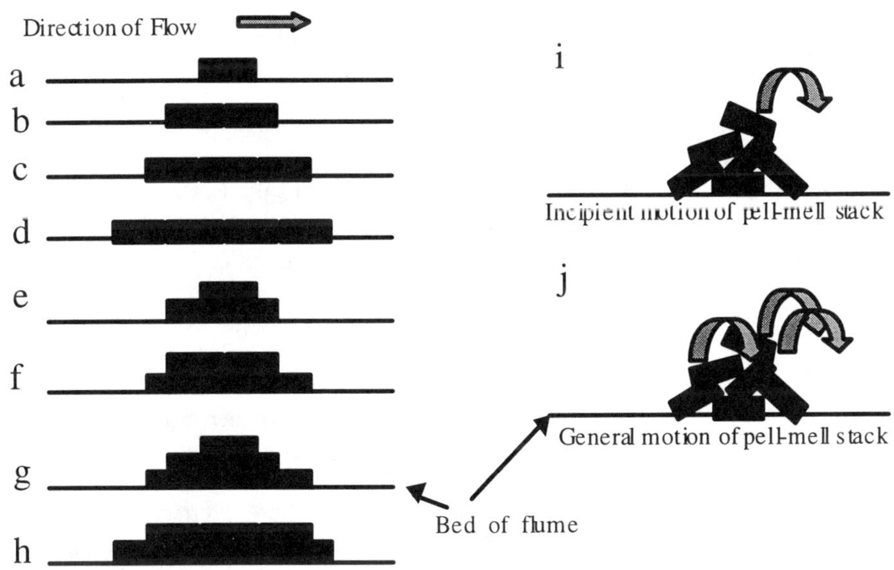

Figure 10. Definition diagram for the flume experiments of *Allen* (1942; 1947) depicting the geometry of the ten experimental configurations for which data is reproduced in Fig. 11. Each cartoon is a flow-parallel section through block configurations stacked across the full width of the channel.

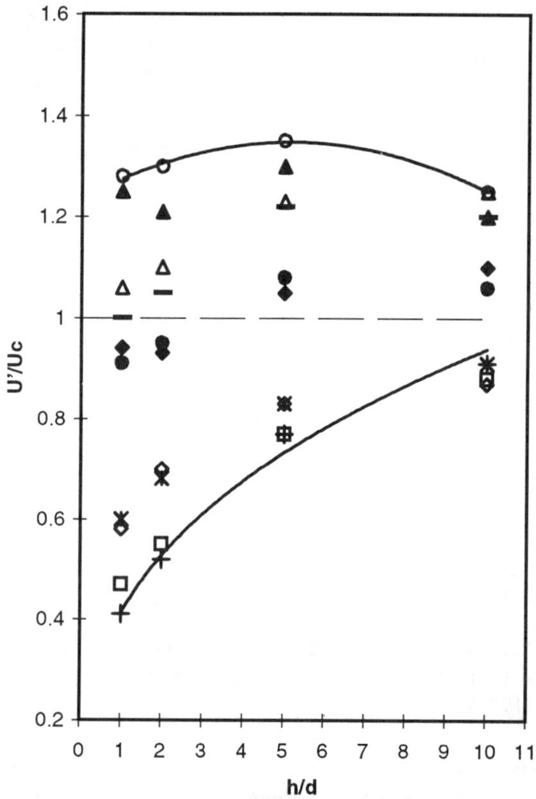

Figure 11. Data of *Allen* [1942; 1947] for transverse groupings of block configurations (see Figure 10 for definition of geometries). (a) Single row; (b) Two rows; (c) Three rows; (d) Four rows; (e) Three block stack; (f) Five block stack; (g) Six block stack; (h) Seven block stack; (i) Pell-mell initial motion; (j) Pell-mell general motion.

dimensional critical velocity is wide, especially in shallow flows, but this changes systematically as stack geometry is varied. Secondly, for relative flow depths greater than 10, the variation in U'/U_C is greatly reduced, converging within the range 0.9 to 1.2. Introducing a single row of blocks transverse to the flow (Series a) greatly reduces the critical velocity of the flow required to induce 'break-up', owing to the increase in flow-blocking when contrasted with an isolated single block. The simple addition of a second, third or fourth row (Series b, c, d) increases the group resistance to entrainment and reduces the dependence on relative depth. Further addition of additional tiers of blocks (Series e, f, g) reduces the resistance to entrainment, probably as a joint function of the relative ease of entraining isolated perched blocks and the increased flow blocking. However, low aspect stacks of several blocks (Series h) once again increase resistance to entrainment. For the case of more natural 'pell-mell' stacks, two pertinent observations may be made. Firstly, for the initial motion of only one or two blocks (Series i), the function is not dissimilar to that recorded for simple geometric stacks of a few blocks (Series a, e, g). For the condition of bedform break-up (Series j) the non-dimensional velocity is increased, but is once again similar to simple stack geometries. For a relative depth of *circa* 10 the critical velocity for both initial motion and general break-up of the 'pell-mell' bedform describes a limited range; 0.8 to 1.25 U'/U_C. Consequently, for $h/d > 10$ Allen's bedforms were subject to a flow with minor free surface effects. Thus in studies of more natural configurations of bedforms developed from blocks, it should be possible to relate the variation in flow parameters to reference values for isolated single or systematically-grouped blocks.

EXAMPLES OF MULTIPLE BLOCK BEDFORMS IN RIVERS

Examples of type-iii boulder bedforms have been described by *Pflüger and Seilacher* [1991] and by *Wende* [1999]. Using aerial photography, *Pflüger and Seilacher* [1991] identified coarse gravel- or boulder-bedforms in wadis near Rutba, Iraq (Figure 12). In the main channels these bedforms are isolated one from another by intervening regions of exposed bedrock. No site visit has been made to verify the following interpretation but it is hypothesised that flash floods mobilized, and then concentrated, a limited coarse sediment-fill in the wadis leaving the bedrock bare in between the bedforms. The rhomboid bedforms often display unusually sharp apices pointing upstream which are interpreted as indicative of super-critical flow. The distinctive seleniform thus may indicate a dominant hydraulic control on the planform of the deposits rather than any control exerted by joint patterns in the bedrock. In a similar context, *Ikeda* [1983] reported rhomboid gravel bedforms developed in transitional flow conditions, and *Allen* [1984, p398-405] reported allied rhomboid bed features within supercritical flows where oblique hydraulic jumps developed in steep shallow flows. *Wende* [1999] also reported rhomboid boulder bedforms, but additionally identified reticulate networks of boulder bedforms from flash-flood dominated flow regimes in the Durack River, N.W. Australia. Fieldwork confirmed that the distinct reticulate patterns reflect the control exerted by a pattern of bedrock joints on the plan view of flow-normal imbricated stacks of large boulders, whilst the rhomboid bedforms were ascribed to the spacing of standing waves developed during critical flow.

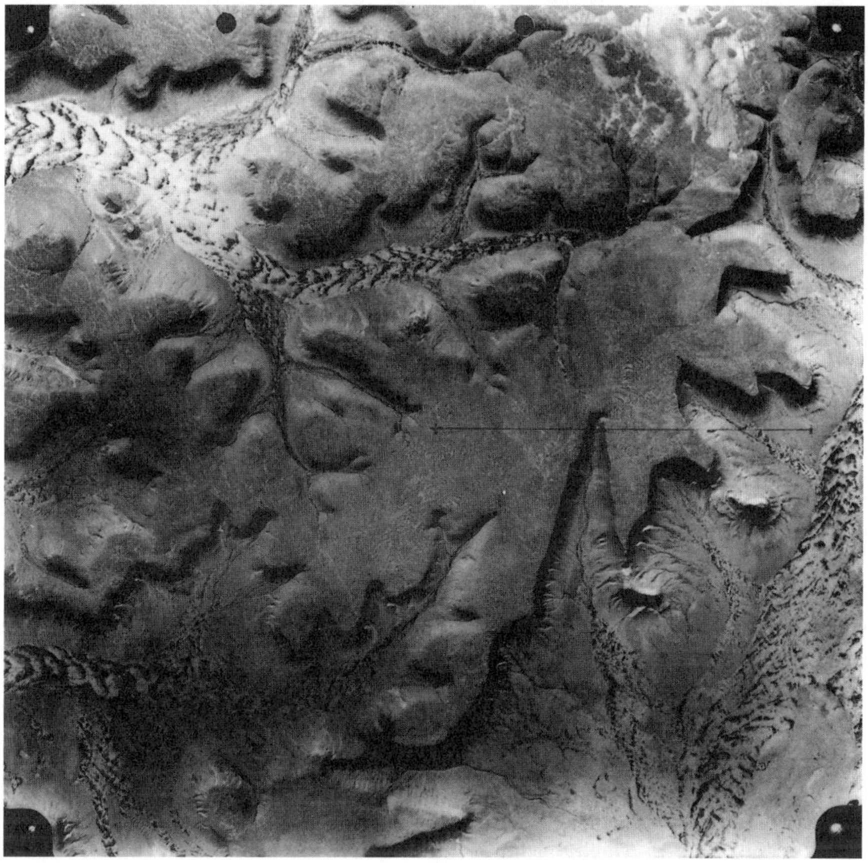

Figure 12. Seleniform boulder or large-cobble bedforms developed in the bed of wadis near Rutba, Iraq. Note that the bedforms are developed in parallel and not en echelon. The bedforms in the main channel truncate those emanating from the tributaries, indicating either prolonged run-off in the main valley, or separate flood events generated from within small sub-catchments. Field of view approx. 3km in width.

CONCLUSIONS

The particular flow conditions associated with isolated, or grouped boulders, in bedrock channels necessitate a special consideration of initial motion criteria. In particular, the degree of relative submergence of boulders in the flow influences the drag associated with flow around the obstacles. For assessing the mechanics of entrainment of seemingly complex groupings of boulders, the problem initially can be reduced to a consideration of (i) isolated boulders; (ii) stacks of a few boulders of limited span and, (iii) stacks of many spanwise boulders. For palaeohydraulic estimates of flow parameters it may be concluded that the friction coefficient (μ_f) between boulder and bedrock is constant as flow varies. However, an appropriate value needs to be selected for given bedrock and boulder roughness. Smooth bedrock against smooth boulder surfaces induces larger friction coefficients than rough bedrock adjacent to smooth boulders.

Careful consideration of the value for the drag coefficient is required when relative submergence is such that the water depth is less than about twice the block height. The variation in the drag coefficient for conditions of initial motion can be described by empirical equations relating relative water depth and particle orientation. For the fully submerged case C_d equals 0.95 for simulated bedrock surfaces, but for emergent conditions C_d increased rapidly as relative depth is reduced. Setting C_d equal to 0.95 is appropriate where blocks were probably deeply submerged ($(h/d)(L/B) > 2$) and entrained by sliding. However, where blocks were in shallow water or emergent ($(h/d)\ (L/B) < 2$), for both pivoting and sliding increasingly large values of C_d pertain as h/d decreases.

Acknowledgements. Rainer Wende and Friedrich Pflüger kindly provided Figure 6 and Figure 12 respectively, whilst Greg Springer thoughtfully directed us to pertinent literature. Andy Quin and Paul Williams prepared the flume bed and Charles Blakeley provided technical support. Paul Komar and Ellen Wohl are thanked for providing careful reviews of the manuscript.

REFERENCES

Abrahams, A.D. and Parsons, A.J., Hydraulics of interrill overland flow on stone-covered desert surfaces. *Catena*, 23, 111-140., 1994.

Allen, J., An investigation of the stability of bed materials in a stream of water. *J. Institution of Civil Engineers*, 5, 1-34. 1942.

Allen, J., *Scale Models in Hydraulic Engineering*, Longmans, London, 407pp., 1947.

Allen, J.R.L., *Sedimentary Structures: their Character and Physical Basis.* Elsevier, Amsterdam, 1, 593pp., 1984.

Atkinson, J.H. and Bransby, P.L., *The Mechanics of Soils.* McGraw-Hill, Maidenhead. 375pp., 1978.

Carling, P.A. and Grodek, T., Indirect estimation of unguaged peak discharges in a bedrock channel with reference to design discharge selection. *Hydrological Processes*, 8, 497-511.

Carling, P.A. and Tinkler, K., Conditions for the entrainment of cuboid boulders in bedrock streams: an historical review of literature with respect to recent investigations, pp 19-34 In: *Rivers Over Rock,* Edited by K.J. Tinkler and E. E. Wohl, Geophysical Monograph 107, American Geophysical Union, Washington, USA. , 1999.

Chanson, H., Comment on "Critical flow constrains flow hydraulics in mobile-bed streams: A new hypothesis" by G. E. Grant. *Water Resources Research*, 35, 903-905., 1999.

Grant, G.E., Critical flow constrains flow hydraulics in mobile-bed streams. *Water Resources Research*, 33, 349-358., 1997.

Denny, M.W., *Biology and the Mechanisms of the Wave-swept Environment.* Princetown University Press, Princetown, N.J., 1988.

Engineering Sciences Data, *Fluid Mechanics, External Flow*, 2, Engineering Sciences Data Unit, London., 1976.

Flammer, G.H., Tullis, J.P. and Mason, E.S., Free surface, velocity gradient flow past hemispheres. *J. Hydraulics Division, ASCE,* 96, 1485-1502., 1970.

Herbich, J.B. and Shulits, S., Large-scale roughness in open channel flow. *J. Hydraulics Division, ASCE*, 90, 203-230., 1964.

Hoerner, S.F., *Fluid Dynamic Drag*. Hoerner Fluid Dynamics, Bricktown, NJ, pp. 3-17. , 1965.

Ikeda, H., Experiments on bedload transport, bed forms, and sedimentary structures using fine gravel in the 4-metre-wide flume. *Environmental Research Center Papers* No. 2, Environmental Research Center, The University of Tsukuba, Ibaraki, Japan, 78pp., 1983.

Ikeda, H., Experimental study of heterogeneous sediment transport. *Environmental Research Center Papers* No. 12, Environmental Research Center, The University of Tsukuba, Ibaraki, Japan, 50pp., 1988.

James, C.S., Prediction of entrainment conditions for nonuniform, noncohesive sediments. *J. Hydraulic Research,* 28, 25-41., 1990.

Komar, P.D., Entrainment of sediments from deposits of mixed grain sizes and densities, in *Advances in Fluvial Dynamics and Stratigraphy,* edited by P.A. Carling and M.R. Dawson, pp 127-181, Wiley, Chichester. , 1996.

Montes, J.S. and Chanson, H., Characteristics of undular hydraulic jumps: experiments and analysis. *J. of Hydraulic Engineering*, 124, 192-204., 1998.

Pflüger, F. and Seilacher, A., Flash flood conglomerates, in *Cycles and Events in Stratigraphy* edited by G. Einsele, W. Ricken and A. Seilacher, pp 383-391, Springer-Verlag, Heidelberg. 955pp., 1991.

Pizzuto, J.E., Webb, R.H., Griffiths, P.G. Elliot, J.G. and Melis, T.S. , Entrainment and transport of cobbles and boulders from debris fans, in *The Controlled Flood in Grand Canyon, Geophysical Monograph* 110, pp 53-70, American Geophysical Union., 1999.

Rajaratnam, N., Hydraulic jumps, in *Advances in Hydroscience,* edited by V.T. Chow, 4, Academic Press, New York, pp. 197-280., 1967.

Reinius, E., Rock erosion. *Water Power and Dam Construction*, 38, 43-48., 1986.

Rosemeier, G., *Winddruckprobleme bei bauwerken*, Springer-Verlag, Heidelberg., 1976.

Tinkler, K., Critical flow in rockbed streams with estimated values for Manning's *n. Geomorphology*, 20, 147-164., 1997.

Wende, R., Boulder bedforms in jointed-bedrock channels, in *Varieties of Fluvial Form*, edited by A.J. Miller and A. Gupta, pp. 189-216, John Wiley & Sons Ltd, Chichester., 1999.

Yen, B-C., Discussion of: Large-scale roughness in open channel flow. *J. Hydraulics Division, ASCE*, 91, 257-262., 1965.

Zgheib, P. W. and Urroz-Aguirre, G.E., Flow transition around a single large bed element, in *Hydraulic Engineering* edited by H.H. Chang and J.C. Hill, 2, pp 1036-1041, Proceeding of the 1990 National Conference of the ASCE., 1990.

Paul A. Carling, Department of Geography, University of Southampton, Highfield, Southampton, SO17 1BJ, UK. (e-mail:p.carling@soton.ac.uk).

Mark Hoffmann, Institut für Wasserwirtschaft und Kulturtechnik, Universität Fridericiana zu Karlsruhe, Karlsruhe, Germany.

Andrea Silke Blatter, Institut für Wasserwirtschaft und Kulturtechnik, Universität Fridericiana zu Karlsruhe, Karlsruhe, Germany.

Bias and Information Content of Paleoflood Data in Flood-Frequency Analysis

J.B. Blainey[1], R.H. Webb[2], M.E. Moss[3], and V.R. Baker[4]

Paleoflood hydrology is commonly used to assess the magnitude and frequency of floods on rivers in bedrock-controlled reaches. Previous studies suggest that paleoflood data may either degrade or enhance the information content of gaging data in flood-frequency analysis, depending on analysis assumptions. In this study, we calculate the effect of combining synthetic paleoflood and gaging data on estimates of low-frequency flood quantiles. Using Monte Carlo simulation, we generate 5,000 realizations of a 4,000-year annual-flood series from a log-Pearson type III distribution. Using type I and type II censored models of paleoflood deposition and the last n-years (20-100 years) of the series as the gaging record, we fit the combined paleoflood and gaging data with the expected moments algorithm (EMA) and maximum-likelihood analysis (MAX). Our results are relevant to paleoflood records greater than 1,500 years in length and recurrence intervals ranging from 10 to 5,000 years. In general, the combination of paleoflood and gaging data improves the accuracy of flood-quantile estimates compared to gaging data alone when the three-parameter log-Pearson type III distribution is assumed to be the underlying parent distribution. Use of a two-threshold scenario, comprised of the largest flood as a single exceedance and the second largest flood as a period-of-record threshold, resulted in the best retrodiction of the parent population. Incomplete dating of paleoflood deposits, coupled with the assumption of a multiple-threshold exceedance scenario, may degrade the information content of paleoflood data in flood-frequency analysis compared to gaging data alone. Quantile estimates for gaging records with skew coefficients less than zero may not be improved with paleoflood data if the bias between paleoflood stage and maximum flood stage is not accounted for. Uncertainty of our results increases with increasing skew coefficient, suggesting limits to regionalization of our results.

[1] California State University Foundation, U. S. Geological Survey, Sacramento, CA
[2] U. S. Geological Survey, Tucson, AZ
[3] Retired, U. S. Geological Survey, Tucson, AZ
[4] Department of Hydrology and Water Resources, University of Arizona, Tucson, AZ

Ancient Floods, Modern Hazards: Principles and Applications of Paleoflood Hydrology
Water Science and Application Volume 5, pages 161-174

INTRODUCTION

Flood-frequency analysis is a critical element of many hydrological designs. Estimating the t-year flood is fundamental to risk analysis for design and retrofit of floodplain and flood-control structures [*Stedinger and Cohn*, 1986]. In addition to engineering applications, information about recurrence intervals of large floods is crucial to the study of channel morphology and development [*Kochel and Baker*, 1982]. The short length of most gaging records limits the accuracy of flood-quantile estimates [*National Research Council*, 1988].

Conventional approaches for estimating exceedance probabilities of large floods become less accurate as recurrence intervals exceed the length of gaging records, which typically are less than 100 years.

Paleoflood hydrology has often been used for incorporating geological and botanical information into flood-frequency analysis [*Kochel and Baker*, 1982; *Baker et al.*, 1983; *Baker*, 1987; *Kochel and Baker*, 1988]. It is emphasized that paleoflood hydrology merely extends the time period of observation of naturally occurring floods. Paleoflood hydrology provides new observational constraints on flood-discharge estimates required for use in procedures such as frequency analysis, rainfall-runoff modeling, regionalization, and probable-maximum-flood estimation. All these procedures benefit from the introduction of new data that can extend the period of observation in time and that increase spatial sampling. This new data provides constraints for model testing and provides a basis to reach engineering conclusions.

Some hydrologists have embraced the use of paleoflood data, developing new statistical techniques for efficiently using paleoflood information in flood-frequency analysis [*Stedinger and Cohn*, 1986; *Cohn and Stedinger*, 1987; *Cohn et al.*, 1997]. Other hydrologists have questioned the value of paleoflood data, suggesting that in some cases paleoflood data may degrade flood-frequency analysis in comparison to use of gaging data alone [*Hosking and Wallis*, 1986a, 1986b]. It is commonly asserted that paleoflood hydrological data (or "methods of flood estimation") contain more uncertainty than other data (or "other methods"). This uncertainty has been attributed to large errors presumed to characterize individual flood-magnitude estimates and to errors in supposed "proxy" relations between data and floods [*Yevjevich and Harmancioglu*, 1987]. Such statements are made as generalities, without attention to the great variety of paleoflood data sources. Discussion of these issues is beyond the scope of this study but is the subject of an extensive literature. The term "proxy" is inappropriate for the types of paleoflood data that will be considered in this paper [*Baker*, 1998], which are characterized by very high accuracy in comparison to other methods [*Baker*, 1987, 2000].

The purpose of this paper is to evaluate the information content of paleoflood data in flood-frequency analysis with variable length gaging records and with a variety of assumptions concerning the characteristics of the paleoflood data by using a series of Monte Carlo simulations.

STATISTICAL FRAMEWORK FOR FLOOD-FREQUENCY ANALYSIS

For regulatory purposes within the USA, the three-parameter log-Pearson type III distribution is mandated for all flood-frequency analyses in Bulletin 17B of *U.S. Water Resources Council* [1982]. Bulletin 17B is a regulatory document that also specifies a widely accepted method for including historical data in flood-frequency analysis by using a moments-based procedure [*U.S. Water Resources Council*, 1982]. This procedure has been criticized as an inefficient use of historic data [*Stedinger and Cohn*, 1986].

The use of historical data to improve the accuracy of flood-quantile estimates is not new. *Benson* [1950] used plotting positions to incorporate measurements of all flood flows exceeding a particular stage prior to installation of the gaging station. In addition to graphical methods, a statistical-weighting scheme incorporates historical data into flood-frequency analysis [*U.S. Water Resources Council*, 1982]. The Bulletin 17B historical procedure specifies a moments-based procedure that is most appropriate when the historical period can be described with a single threshold [*England*, 1998]. The historical-weighting adjustment was developed to incorporate pre-gaging information for two general cases: (1) a large flood in a short gaging record with community knowledge that the flood was the largest in a known time interval, and (2) knowledge of a large flood prior to establishment of the gaging record [*Cohn et al.*, 1997]. The critical assumption in this method is that each annual peak above the perception threshold is recorded either as a historic or gaging record peak. This assumption is inappropriate for long historical or paleoflood periods relative to the gaging record length [*Lane*, 1987].

The primary method of incorporating historical and paleoflood data into flood-frequency analysis is to assume that the data are censored. To preserve a record of a paleoflood, the stage of the resulting flood sedimentation unit must exceed the pre-existing stage of pre-existing flood deposits or other historical evidence. In other words, to be preserved, the stage of a flood must exceed the stages of all previous floods; consequently, many large floods may not be preserved in a historical or paleoflood record (censored data). In using paleoflood data, a monotonically non-decreasing trace of the discharge threshold, Q_0, can be defined over a time period during which sedimentation units of all floods greater than Q_0 are preserved. Although the preservation threshold can be altered by subsequent fluvial or geomorphic events, the number and time-duration of threshold exceedances are useful flood-frequency data.

Paleoflood data generally are a type I censored sample [*Cohn and Stedinger*, 1987] for which the points of truncation, or terminals, are known or can be determined, and the number of censored observations and the number of uncensored observations are random variables [*Cohen*, 1991]. The partial-duration series, recorded at most gaging stations, is a type I censored sample because all floods above a fixed threshold, known as the base discharge, are recorded. In type II censored samples, the total number of observations, the number of censored observations, and the number of non-censored observations are fixed values and the terminals are random [*Cohen*, 1991]. A well-known Type II censored sample is the annual-flood series; only the single largest flood in each year is known.

Previous Studies

In recent decades, new statistical procedures have been developed to incorporate historical and paleoflood data into flood-frequency analysis. *Stedinger and Cohn* [1986] fit two-parameter lognormal distributions, using maximum-likelihood analysis, to data generated from lognormal and log-Pearson type III distributions. Historical data consisted of all observations exceeding a known discharge threshold (type I censoring) over a period of 20 to 2,000 years. *Stedinger and Cohn* [1986] concluded "50 years of historical observations can provide as much information as 10 to 30 years of additional systematic flood record." *Cohn et al.* [1997] developed a method-of-moments algorithm to incorporate historical and paleoflood data in flood-frequency analysis more efficiently than the methods given in Bulletin 17B.

Hosking and Wallis [1986a, 1986b] studied the effect of unbiased measurement errors on the largest historical event (type II censoring) in 1,000 to 10,000 years using maximum-likelihood estimators for the Gumbel and Generalized Extreme Value (GEV) distributions and concluded the value of paleohydrologic information is much greater for three parameter distributions. The shape of the extreme tail of a three-parameter distribution is difficult to determine accurately from a short gaging record, and the paleohydrological maximum event, even if subject to errors on the order of ±50 %, can improve quantile accuracy. However, for a two-parameter distribution, "if the paleohydrological flood magnitude estimate is subject to errors of ±25 % or more, then its value to flood-frequency analysis is negligible."

Guo and Cunnane [1991] investigated the results of *Hosking and Wallis* [1986a, 1986b] and *Stedinger and Cohn* [1986] by comparing information gains with the addition of paleoflood data for type I and type II censoring. Using a Gumbel distribution in a maximum-likelihood framework, *Guo and Cunnane* [1991] concluded *Hosking and Wallis's* results "may be due to their use of type II censoring assumption." Their Monte Carlo experiments showed that quantiles estimated by type I censored data have smaller biases and root mean squared errors than those estimated by assuming type II censoring.

Pilon and Adamowski [1993] used a Monte Carlo analysis of a log-Pearson type III distribution to assess the accuracy of flood-quantile estimates for recurrence intervals ≤500 years. They employed a 100-year record with 49 years of gaging data and 51 years of historical record, and analyzed the data using maximum-likelihood estimators in both type I and type II censoring frameworks. In a type II censored framework, historical information increased the accuracy of flood-quantile estimates, even with large measurement errors.

Frances et al. [1994] assessed the value of historical/paleoflood information for type I and type II censoring with the two-parameter GEV distribution by analytically determining asymptotic variances in a maximum-likelihood framework. They concluded that statistical gains of type I censoring depend primarily on the quantile recurrence interval, the threshold recurrence interval, and on the ratio between the length of the historical/paleoflood record and the length of the gaging record. The value of historical/paleoflood information varies, depending on whether type I or type II censoring is assumed and on the relative values of the above-mentioned four factors.

One cannot assume type I or type II censoring *a priori*. The temporal order of deposition and stratigraphy of a particular paleoflood site controls whether type I or type II censoring exists at a given point in time. A major key to success in the use of paleoflood data is the determination of which type of censoring is appropriate. We attempt to simulate this by basing our modeling scenarios on commonly observed combinations of paleoflood data and censoring thresholds.

The second portion of the paper begins with a review of techniques used in paleoflood hydrology followed by a brief discussion of the uncertainties in the analysis of paleoflood data as they pertain to subsequent simulations. Commonly held assumptions regarding paleoflood methodology are discussed. In the next section, the stochastic model employed in the simulations is described along with additional assumptions about the nature of paleoflood data incorporated into the Monte Carlo simulations. The remaining two sections present the results, and the discussion and conclusions.

TECHNIQUES USED IN PALEOFLOOD HYDROLOGY

Paleoflood hydrology involves quantifying the dates and discharges of past floods by using geomorphic or botanical evidence such as sedimentary deposits, landforms, and riparian vegetation [*Costa*, 1986; *Baker*, 1987]. Evidence of paleofloods includes scars preserved in floodplain vegetation, flood bars and coarse-grained overbank deposits, landforms truncated by flood flows, and stratigraphy of flood deposits [*Kochel and Baker*, 1982; *Kochel and Baker*, 1988; *Ely et al.*, 1993]. For this study, we only consider paleoflood data in the form of slackwater deposit-paleostage indicator (SWD-PSI) evidence [*Baker*, 1987]. SWD-PSI evidence is dated by using radiocarbon dating of organic material entrained in paleoflood deposits and between flood units. Surveys of slackwater deposits and channel geometry, combined with one-dimensional hydraulic models, yield estimates of paleoflood stage and discharge. Paleoflood data are combined with gaging data to estimate recurrence intervals for large floods [*Kochel and Baker*, 1982; *O'Connor and Webb*, 1988].

The type of SWD-PSI evidence that we chose to investigate is flood deposits, comprised typically of coarse silts and fine sands, that accumulate in low-energy environments. A new sedimentary unit may be deposited for each subsequent flood with a stage exceeding the pre-existing elevation of sediment accumulation [*Kochel and Baker*, 1988]. If sediments accrue in sites protected from erosion by subsequent floods and runoff, these sites will continue to accumulate flood sediments from successive floods of greater magnitude [*Baker et al.*, 1983]. If one assumes that floods are highly sediment laden, controls on flood-deposit thickness include peak stage and duration of the flood, channel cross-section, channel gradient, and base elevation of the slackwater site above the channel [*Kochel and Baker*, 1988].

Sources of Error in the Analysis of Paleoflood Data

Errors in the analysis of paleoflood data may occur in both discharge reconstruction and event dating. Errors in discharge reconstruction include uncertainties in indirect-discharge estimates caused by channel change and/or mis-specification of roughness values [*O'Connor and Webb*, 1988], and differences between the actual maximum water-surface elevation and that inferred from flood deposit heights. The first type of error also occurs to some extent in gaging records when flood flows are not measured but are instead estimated by indirect-discharge techniques, a common practice in the southwestern United States.

Dating errors can be segregated into three types: (1) analytical, or laboratory counting errors; (2) incomplete dating of flood deposits; and (3) errors in associating the age of the carbon with the age of the flood. Radiocarbon dates of any type have an analytical error that, at best, typically is ±40 years; counting errors on small samples may be ±100 years or greater [*Webb and Rathburn*, 1989; *Ely et al.*, 1992]. Not all flood deposits may contain datable carbon, and therefore dates on some fraction of floods must be interpolated from dated deposits. Finally, radiocarbon ages of organic material in a slackwater deposit may differ from the actual age of the deposit depending on the residence time of the organic material in the environment prior to deposition in the slackwater deposit [*Baker et al.*, 1983; *Ely et al.*, 1992]. Association errors are difficult to synthesize appropriately in a stochastic model and are not considered here.

Paleoflood Methodological Assumptions

Common assumptions made in using slackwater deposits to calculate paleoflood discharges have been explicitly defined: (1) stable bedrock walls minimize lateral channel changes during floods, (2) the height of the flood deposit represents the minimum water-surface elevation of the flood, (3) regional aggradation and degradation of channels are insignificant, (4) scour and fill during floods are minimal, and (5) the paleofloods occurred in the modern hydrological regime [*Kochel et al.*, 1982]. During a large flood, stable channel walls in narrow bedrock canyons cause a rapid rise in stage. An obstruction or other feature along the channel, such as an alcove, cave, or tributary mouth, allows accumulation and preservation of flood deposits. Slackwater sediments are deposited relatively high above the channel and are protected, by a bedrock obstruction or other channel feature, from erosion and reworking by subsequent smaller magnitude floods.

For the purposes of our stochastic modeling, we assumed that no large climate shifts occurred in the late Holocene, or the last 4,000 years. Our assumption of climatic stationarity leads to the assumption that the modern hydrological regime has been approximately the same since the beginning of the late Holocene [*Ely*, 1985]. During the late Holocene, it is inferred that major atmospheric and oceanic circulation features of the present global atmospheric circulation were in place. *Ely et al.* [1993] claimed that large floods in this period are grouped, indicating climatic change. We interpret this grouping as regional response to large storm events, not necessarily climatic change.

STOCHASTIC MODELING OF PALEOFLOOD INFORMATION

A Monte Carlo analysis can provide insights into a variety of problems with both probabilistic and non-probabilistic content by performing sampling experiments on a computer. Monte Carlo simulations are used to assess the errors inherent in complex systems and are of particular use when a simple algebraic solution or single error analysis is impossible. Numerical methods that are known as Monte Carlo methods are statistical simulation methods that utilize sequences of random numbers to perform simulations. Statistical simulation methods may be contrasted to conventional numerical discretization methods, which employ equations that describe a physical system. In many applications of the Monte Carlo method, the physical process is simulated directly, by randomly sampling from the frequency distribution. Many simulations, or realizations, are then performed and the desired result is taken as an average over the number of observations.

Model Overview

Uncertainties inherent in flood-quantile estimates are unknown in most real-world flood-frequency analyses involving paleoflood data. To assess this uncertainty, we devised stochastic simulations of paleoflood deposition with the goal of retrodicting flood-frequency characteristics of a time-invariant parent population (Figure 1). In our model, we adopted a parent population equivalent to the station moments of the Salt

River near Chrysotile, Arizona (station number 09497500), a bedrock-constrained river in central Arizona with a 75-year gaging record [*Pope et al.*, 1998]. Paleoflood deposits occur downstream of this gaging station [*Partridge and Baker*, 1987], and paleoflood data from the Salt River have been used in flood-frequency analysis [*Stedinger et al.*, 1988]. No significant regulation occurs upstream from this gaging station, and we assumed a late Holocene stationarity in the mean, variance, and skew for the parent population.

In our Monte Carlo simulations, we generated an annual-flood series from a log-Pearson type III distribution to create 4,000 years of synthetic record [*Webb et al.*, 1994; *Blainey*, 2000]. We generated synthetic paleoflood data with several assumptions about depositional evidence and assembled a censored data set. We extracted the last n-years of record as a gaging record to evaluate the impact of gaging record length on flood-frequency analysis; we combined the paleoflood and gaging data in a log-Pearson type III flood-frequency analysis and calculated the 10, 50, 100, 500, 2,500, 4,000, and 5,000 year recurrence interval floods. Our analyses used two different algorithms that both incorporate censored data: (1) maximum-likelihood analysis (MAX) [*Stedinger et al.*, 1988] and (2) the expected moments algorithm (EMA) [*Cohn et al.*, 1997; *England*, 1999].

Stochastic Paleoflood Model Assumptions

In modeling paleoflood deposits, a few additional assumptions were made beyond the common ones made in paleoflood hydrology as previously described in the section titled paleoflood methodological assumptions. A 4,000-year paleoflood modeling record was selected as a conservative representative length of the best paleoflood records in the southwestern United States. For example, one of the longest records of large floods in the Southwest is the Colorado River near Lees Ferry, Arizona, with a 4,500-year stratigraphic record [*O'Connor et al.*, 1994]. The record for the Paria River in northern Arizona is 4,200 years long [*Webb et al.*, this volume]. In the stochastic model, paleoflood deposition is a function of the local stage-discharge relation. We compared stage-discharge relations for several gaging stations on bedrock-controlled rivers in Arizona and Utah and found the Salt River at Chrysotile, Arizona was representative of the relation between stage and discharge in this region.

In traditional flood-frequency analysis, data are from the annual-flood series. Two or more floods that could potentially deposit SWD/PSI information may occur in the same year on rivers in the Southwest (*e.g.*, the Gila River in Arizona had three large floods in January and February, 1993), but only the largest flood in each year is considered in flood-frequency analysis based on the annual-flood series. Consequently, annual floods are described with a single log-Pearson type III distribution with the traditional assumptions of stationarity, homogeneity, and interannual independence. Only SWD/PSI data of single, vertically accreting deposits [*Baker et al.*, 1988] were considered; inset deposits were not modeled, as they are difficult to synthesize in a stochastic model. In each paleoflood-deposit simulation, the initial preservation threshold was set at the 100-year flood stage.

Description of the Monte Carlo Simulations

Monte Carlo simulation begins by drawing a random number from a uniform probability distribution (Figure 1). This random number is transformed to a log-Pearson type III distribution by the moments of the gaging data from the Salt River at Chrysotile from 1924 to 1998. A sequence of 4,000 random numbers represents a 4,000-year synthetic annual flood series, censored to represent paleoflood data, with the last n-years representing an annual-flood series for a synthetic gaging record. After paleoflood censoring, the hybrid paleoflood data and annual-flood series were input to either MAX [*Stedinger et al.*, 1988] or EMA [*Cohn et al.*, 1997; *England*, 1999].

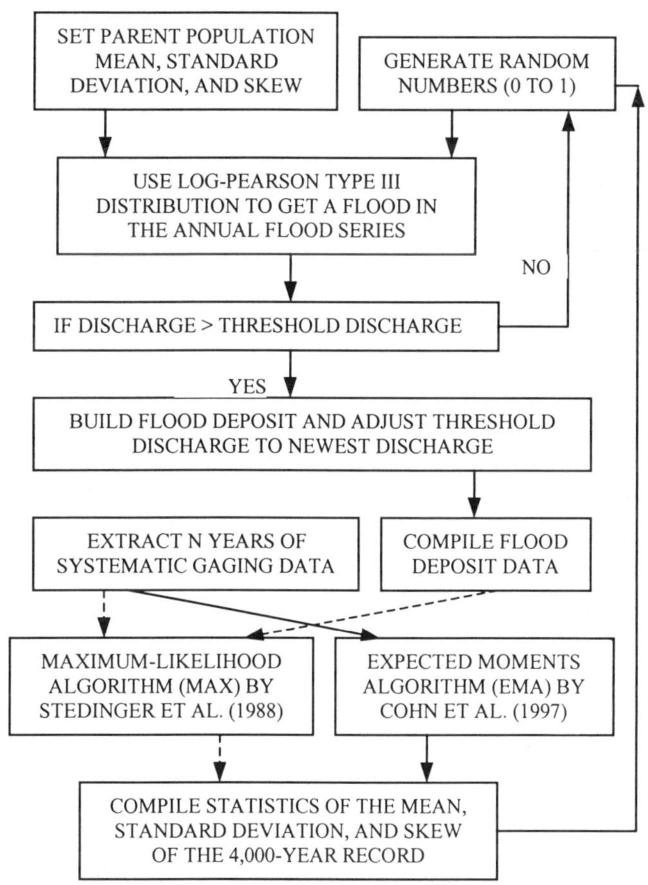

Figure 1. Schematic diagram showing the computational steps required for the rising-threshold scenario.

Using MAX, we only evaluated flood recurrence intervals equal to or less than 500 years; for EMA, we evaluated recurrence intervals up to and including 5,000 years. We evaluated 1,000 realizations of MAX fits and 5,000 realizations of EMA fits [*Blainey*, 2000]. Only 1,000 fits were used for MAX because of constraints on computer resources, whereas the EMA code was significantly faster, allowing more realizations. For both techniques, the sample averages of the three moments of the estimated distributions stabilized at the Chrysotile sample moments at about 1,000 realizations, but estimates for low-frequency quantiles made the larger number of realizations desirable for EMA.

For most of our analyses, we used a synthetic gaging record of length *n*, where $20 \leq n \leq 100$ years. Gaging records less than 20 years in length are generally unsuitable for flood-frequency analysis when calculating return periods greater than 100 years. Most gaging records in the southwestern United States that are not subject to the effects of flow regulation are less than 80 years in length [*e.g., Pope et al.*, 1998]. To allow direct comparisons among modeling scenarios, we used a synthetic gaging record length of 70 years.

Paleoflood Depositional Scenarios

For each synthetic data set, we simulated the following six scenarios: (1) a synthetic gaging record with no paleoflood data, which represents the least amount of flood data; (2) a largest flood scenario, an example of Type II censoring that results in the inclusion of the minimal amount of paleoflood data; (3) a one-threshold scenario that incorporates additional data from the second largest flood into the second scenario; (4) a rising-threshold scenario that incorporates all censored paleoflood data above a monotonically non-decreasing threshold, a scenario of Type I censoring; (5) a stage-bias scenario based on the data sets generated in scenario 4 that accounts for the potential differences between deposit heights and maximum flood stages; and (6) a scenario in which all paleofloods exceeding the 100-year recurrence interval are preserved, a best-case censored data scenario for Type I censored data.

Performance Measures

Recurrence intervals from 50 through 5,000 years were selected for performance evaluation. Smaller recurrence intervals (*i.e.*, 50 to 100 years) are commonly used for floodplain design, whereas the larger recurrence intervals (*i.e.*, 5,000 years) are used in dam-safety evaluation [*Levish et al.*, 1997]. Bias and relative root mean square error are the performance measures. For a single realization, the error, ΔQ, of a quantile estimate is the difference between the population value, Q_{pop}, and the average value determined from the realization, Q_{est}:

$$\Delta Q = Q_{est} - Q_{pop} \quad (1)$$

Bias is the average of the errors over a number of realizations in which the quantile being estimated and the conditions of the scenario are held constant. Root mean-square error, RMSE, is:

$$RMSE = [\Sigma \Delta Q^2/n\ ]^{0.5} \quad (2)$$

where n is the number of realizations (1,000 with MAX or 5,000 with EMA). For comparison among our scenarios, RMSE is normalized by dividing by the population value of the pertinent quantile to create the relative root mean-square error, RRMSE, using:

$$RRMSE = RMSE/Q_{pop} \quad (3)$$

RRMSE decreases to zero as the accuracy increases; therefore, a different measure is required to discriminate between relatively accurate scenarios. One such measure is relative information content (RIC), in which information is expressed according to the definition of *Fisher* [1960]:

$$RIC = (RRMSE)^{-2} \quad (4)$$

However, RIC does not discriminate well between imprecise models, and thus both RRMSE and RIC are used to evaluate the model performance of the full suite of scenarios.

RESULTS

Effect of Paleoflood Record Length (EMA)

A 4,000-year long record was used for most of our simulations because we assumed climate has been relatively invariant during the last 4,000 years. But what is the effect of shorter-length paleoflood records? We evaluated the effects of paleoflood record lengths that ranged from 250 to 4,000 years, in the largest flood, two-threshold, and all 100-year exceedances scenarios (Figure 2). Across this range, RRMSE generally decreases with increasing record length for each scenario and seems to approach asymptotes for each that are inversely related to the amount of paleoflood data included therein. In other words, the amount of information contained in each of the scenarios appears to increase as more data are included, both in terms of additional record length for a given scenario and across scenarios for a given record length.

Comparison of MAX and EMA

We compared the computer programs, EMA and MAX, for results at recurrence intervals less than 500 years to observe the effects of the different fitting algorithms on flood-frequency analysis. The largest differences were observed in the fit of gaging data of variable record length. The adjusted method-of-moments procedure in EMA resulted in positive

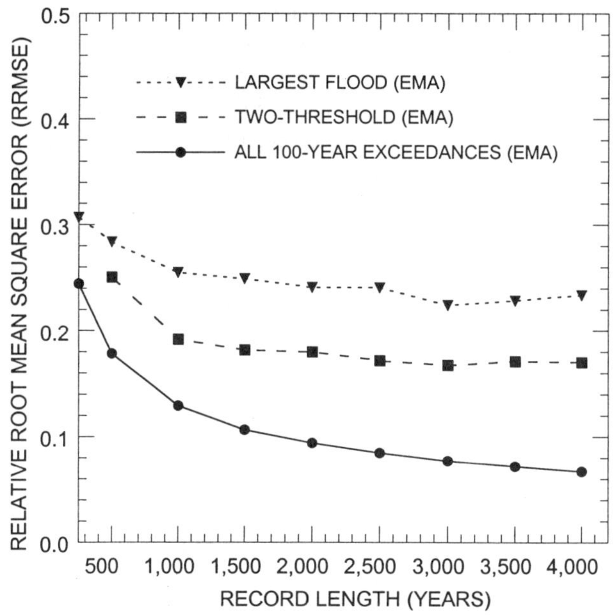

Figure 2. The effect of paleoflood record length on RRMSE of the 100-year quantile with a gaging record length of 70 years. The values were calculated using EMA.

biases, which would result in hydrologic overdesign, with absolute values that were significantly higher than those of the negative biases of the maximum-likelihood method, MAX, at recurrence intervals larger than 50 years (Figure 3). For the rising-threshold scenario, the biases introduced by the fitting algorithms are approximately the same for the two procedures (Figure 3).

In general, for large recurrence intervals, MAX produces less-biased quantile estimates than EMA (Figure 3). For example, bias obtained by using EMA is approximately 100 m^3/s larger (44% greater) than that from MAX at the 500-year recurrence interval for the rising-threshold scenario; differences in bias between the two fitting algorithms increased for the larger recurrence intervals of 100 and 500 years. Interpretation of the results based on RRMSE is essentially the same. Note the essentially overlying curves for the individual paleoflood scenarios in Figure 4, which shows the results for the 100-year recurrence interval. Comparison of Figures 3 and 4 indicate that differences in flood quantile retrodiction results between MAX and EMA are evident in the bias only.

Results from both MAX and EMA indicate that paleoflood data significantly improves flood-quantile estimates compared to gaging data alone. For the 100-year recurrence interval with a 70-year gaging record, RRMSE of paleoflood data is less than half that of the gaging record alone (Figure 4a). The marginal improvement of RRMSE is minimal with increased amounts of gaged record given a 4,000-year paleoflood record.

Results for Paleoflood Data with no Uncertainty (EMA)

With a constant gaging record length of 70 years, we observed large differences in bias between the scenarios that incorporate paleoflood data and that using the gaging record alone for recurrence intervals of 500 years or more (Figure 5a). Inclusion of paleoflood data significantly lowers the biases inherent in the use of that of the gaging record alone; at the 5,000-year recurrence interval, the bias of the gaging record alone is more than three times that of the rising-threshold scenario (Figure 5a). The biases for the paleoflood-deposit and the all 100-year-exceedances scenarios are similar for all recurrence intervals. Among the paleoflood scenarios, the two-threshold scenario performs best with small, negative biases at all recurrence intervals investigated (Figure 5a). Although the bias of the largest flood scenario is almost 2,000 m^3/s for the 2,500-year recurrence interval, this difference is only 15% of its comparable parent population quantile.

RRMSE revealed even larger differences between the results including paleoflood data and those with gaging data alone (Figure 5b). Differences in RRMSE between some paleoflood scenarios are significant; for example, RRMSE of the two-threshold scenario is approximately two-thirds that of the largest flood scenario for the 2,500-year recurrence interval (Figure 5b). For each recurrence interval greater than 2,500-years, RRMSE of the two-threshold scenario model is closer to the best-case value of the all 100-year exceedances scenario than the other two paleoflood scenarios.

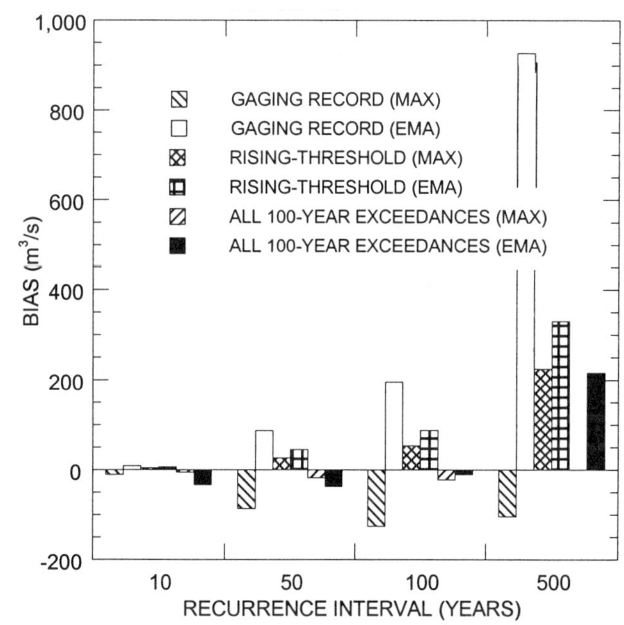

Figure 3. Comparison of the biases for flood-quantile ≤500 yrs with EMA and MAX for the gaging record, rising-threshold, and all 100-year exceedances models. The gaging record length is 70 years.

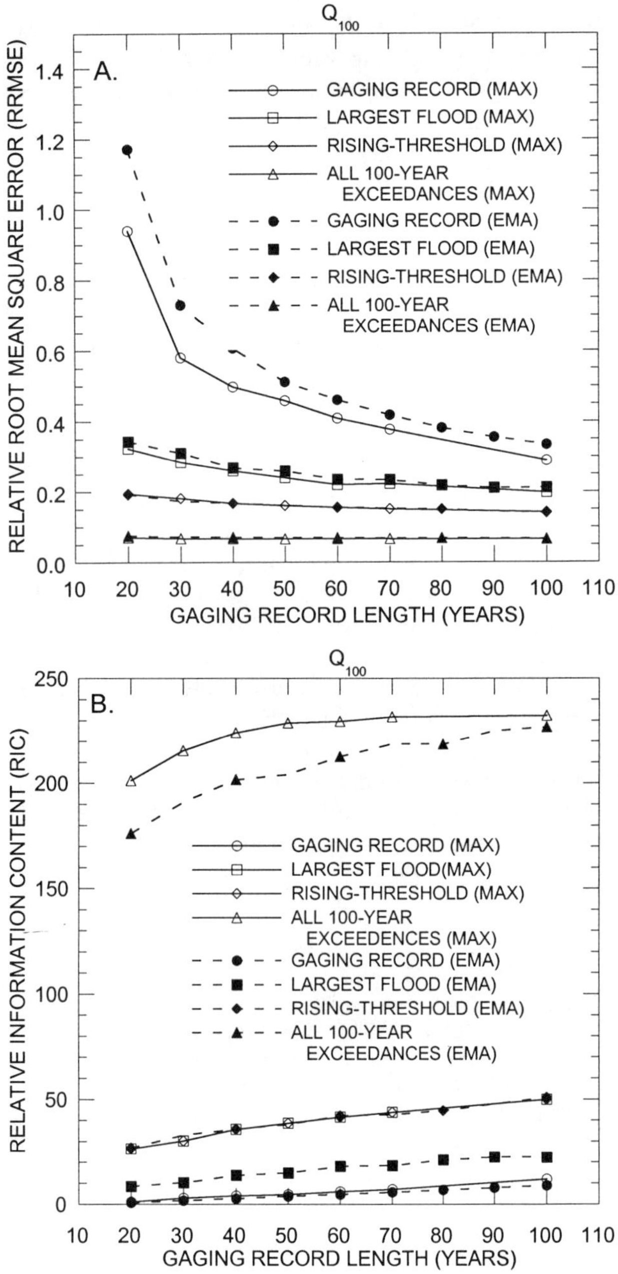

Figure 4. Comparison of results for the gaging record, largest-flood, rising-threshold, and all 100-year exceedances models for gaging record lengths of 20-100 years. A. Using RRMSE. B. Using RIC. The flood quantile is 100 years, and both EMA and MAX were used.

Effects of Errors in Paleoflood Data (MAX)

Using MAX, *Webb et al.* [unpublished data, 1994] evaluated the impact of various types of errors inherent in both paleoflood and gaged data on flood-frequency analysis. They considered discharge measurement errors, dating errors, errors associated with partial dating of floods, and the combination of these three errors (Figure 6). Discharge measurement errors were assumed to be unbiased; at gaging stations, measurement error is modeled as log-normally distributed with $2\sigma = \pm 20\%$, and at paleoflood sites, measurement error is modeled as log-normally distributed with $2\sigma = \pm 30\%$. Similarly, analytical radiocarbon dates have normally-distributed uncertainty that typically is ±40 years. A more significant error is the inability to determine a date for all flood units in a paleoflood deposit, which occurs because sources of stable carbon are somewhat rare in arid environments. Finally, *Webb et al.* [unpublished data, 1994] combined all errors in their Monte Carlo simulations.

These common uncertainties in the analysis of paleoflood data have insignificant effects on flood frequency compared to differences between the paleoflood data and the gaging data alone (Figure 6). RRMSE curves for models with and without measurement error essentially overlie each other with the exception of the gaging record model for gaging records 20 and 30 years in length. With a dating error of ±40 years, the average increase in RRMSE for the 100-year flood quantile in the rising-threshold scenario is 1.7%. Increasing the dating uncertainty to ±100 years also increased RRMSE only minimally [*Webb et al.*, unpublished data, 1994]. Combining uncertainty in measurement error and analytical radiocarbon dates minimally impacted RRMSE.

In addition to uncertainty in known data, an additional source of error is an incomplete paleoflood chronology. If only one-third of the deposits in a stratigraphic sequence can be dated, but statistical analyses are completed by bracketing undated layers by dated ones, the RRMSE curve intersects that of the gaging record alone curve at a gaging record length of 100 to 110 years (Figure 6). In the case of incomplete dating in the rising-threshold scenario, more information is gained by using the largest-flood scenario instead of the rising-threshold scenario. The largest-flood and one-threshold scenarios have a smaller RRMSE than the rising-threshold scenario for all gaging record lengths where only one-third of the deposits can be dated.

Stage-Bias Scenario (EMA)

The stage-bias scenario evaluates the impact of assuming flood deposits build to the maximum stage of the flood. We evaluated the effect of this assumption for the Salt River at Chrysotile by modeling a random differential elevation of 30 to 90 cm, uniformly distributed, between the height of the deposit and the maximum flood stage based on data from the Paria River [*Webb et al.*, this volume]. Bias caused by disregarding this differential stage ranged from -6% for the 100-year recurrence interval to -5% for the 5,000-year recurrence interval (Table 1). For the 5,000-year recurrence interval, the absolute bias of the paleoflood stage-bias scenario is one-half

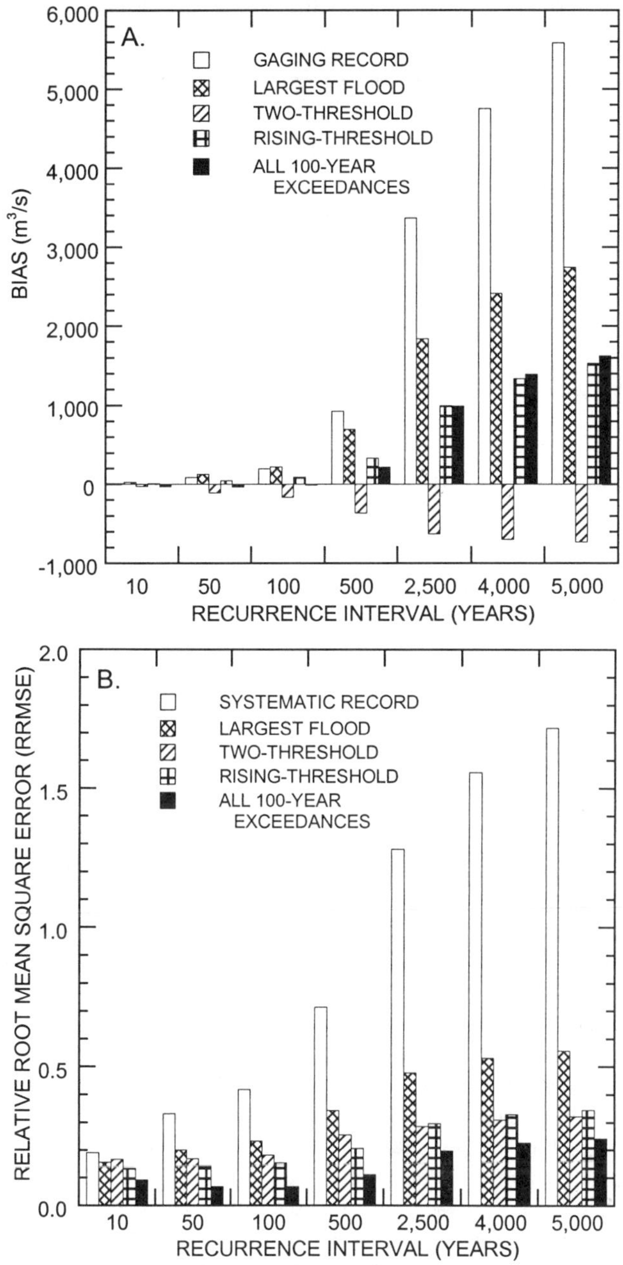

Figure 5. A. Biases for a range in flood quantiles for the gaging record, largest flood, two-threshold, rising-threshold, and all 100-year exceedances models with a gaging record of 70 years. The values were calculated using EMA. B. RRMSE of a range in flood quantiles for the gaging record, largest flood, two-threshold, rising-threshold, and all 100-year exceedances models with a gaging record of 70 years. The values were calculated using EMA.

that of the rising-threshold scenario for the Salt River near Chrysotile. Negative bias introduced in simulation of the differential elevation between the height of the deposit and the maximum flood stage, improves the overall model performance in the rising-threshold scenario by partially offsetting for the positive biases introduced by the moments-fitting algorithm in EMA (Figure 5a).

Clearly, offsetting biases might not occur for every river. The factors that might affect this outcome are the slope of the stage-discharge relation and the variance and especially the skew coefficient of the parent population of annual floods. Stage-discharge relations with high slopes (*e.g.*, in confined bedrock canyons) produce small changes in discharge with increasing stage, and the stage-bias inherent in paleoflood deposits is not as important. Conversely, the stage-bias problem is greatest for wide rivers with stage-discharge relations of low slope. Rivers with flood-frequency characteristics of high variance and especially high skew creates a higher probability for large floods. The interaction of flood-frequency characteristics and stage-discharge relation cannot be predicted using only one parent population and stage-discharge relation, so we tried to regionalize our results using data from other gaging stations in the region [*Blainey*, 2000].

To better extrapolate our results regionally, we plotted the standard deviations and skew coefficients for 8 gaging stations on bedrock-controlled rivers in Arizona and Utah (Figure 7). Four gaging stations represent the range in standard deviations and skew coefficients in the region: the Paria River near Lee's Ferry, the Little Colorado River near Cameron, the Virgin River near Springdale, and the Verde River below Tangle Creek (Table 2). Using normalized stage-discharge relations for these gaging stations and the Salt River near

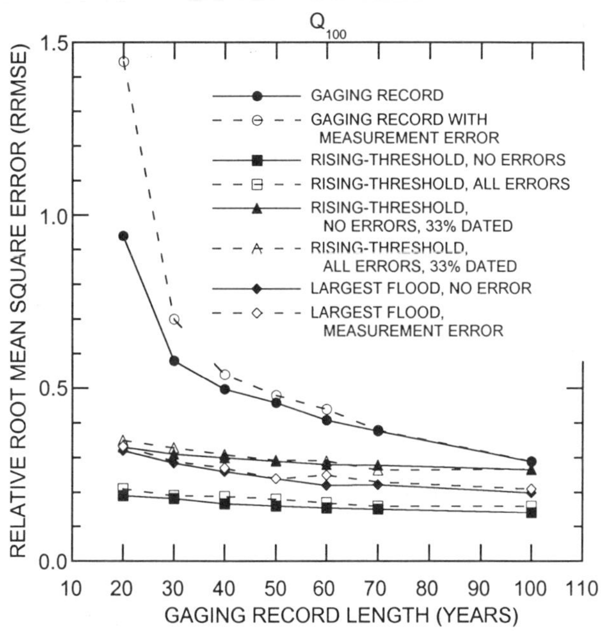

Figure 6. The effect of errors in the gaging record, rising-threshold, and largest-flood models on RRMSE using MAX for gaging record lengths of 20-100 years. The flood quantile is 100 years.

TABLE 1. Bias in paleoflood stage-bias model for a synthetic gaging record length of 70 years. The bias is normalized to the parent popoulation value (% deviation)

	BIAS (%)				
Recurrence Interval (years)	100	500	2,500	4,000	5,000
Paria River[a]	-15	-19	-22	-23	-23
Little Colorado River[b]	-4	-2	3	4	5
Virgin River[c]	-10	-12	-14	-15	-15
Salt River[d]	-6	-6	-5	-5	-5
Verde River[e]	-15	-18	-19	-20	-20

[a]09382000 is the Paria River near Lee's Ferry, Arizona (n=75 years).
[b]09402000 is the Little Colorado River at Cameron, Arizona (n=53 years).
[c]09405500 is the Virgin River near Springdale, Utah (n=73 years).
[d]09497500 is the Salt River near Chrysotile, Arizona (n=73 years).
[e]09508500 is the Verde River near Tangle Creek, Arizona (n=76 years).

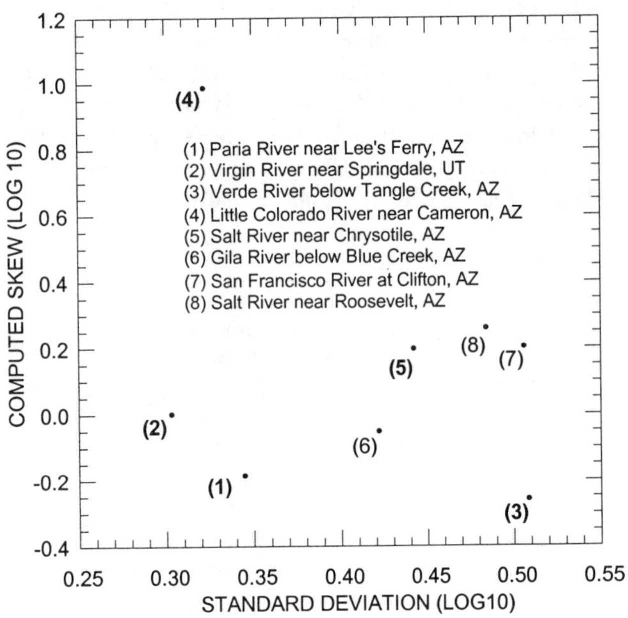

Figure 7. Regional variability of station standard deviation and skew coefficients for eight bedrock controlled rivers in the southwestern United States.

Chrysotile, we calculated the bias introduced by the stage differential.

The bias of both the rising-threshold scenario and the stage-bias scenario is strongly dependent on the skew coefficient of the log-Pearson type III distribution, permitting limited regional extrapolation of our results. The bias increases with the skew coefficient (Figure 8a). As the recurrence intervals increase, the impact of the skew coefficient becomes more significant. The highest correlation between recurrence interval and skew coefficient occurs for large recurrence intervals (*e.g.*, at the 5,000-year recurrence interval, $R^2 = 0.83$). For the 5,000-year recurrence interval, the bias resulting from the stage differential ranged from -2% for the Little Colorado River to -19% for the Paria River (Table 1). Most of the biases are negative (Table 1), unlike the mostly positive biases associated with EMA (Figure 8a). In comparison to the rising-threshold scenario, at the 500-year quantile the introduction of stage bias actually decreased the overall bias for the Little Colorado River because the negative bias offsets the positive bias (Figure 9). For the other four gaging stations, absolute bias increased with consideration of stage bias, indicating that stage bias detrimentally affects flood-frequency estimates using paleoflood data (Figure 8a).

Considerable variability arises in bias, as a percent, for the 500-year recurrence interval with a 70-year gaging record across all the models. Over all the gaging stations, plotted in order of increasing skew coefficient, the two-threshold, rising-threshold, and all 100-year exceedances scenarios have a small range in bias, which was approximately 5% (Figure 9).

RRMSE for the stage-bias scenario varied less among the five gaging stations than the bias (Figure 8b). RRMSE of the stage-bias scenario was lower than in the rising-threshold scenario for the Salt River and the Little Colorado River, the two stations with the largest positive skew coefficients [*Blainey*, 2000]. The difference in RRMSE between the rising-threshold scenario and the stage-bias scenario is minor (on average 0.05) for all benchmark recurrence intervals [*Blainey*, 2000].

TABLE 2. Station statistics (log10) for five gaging stations in Arizona and Utah (see Table 1 footnotes for full station names)

Gaging station	Mean	Standard deviation	Skew coefficient	Slope of stage-discharge relation
Paria River	1.9799	0.3446	-0.1841	0.26652
Little Colorado River	2.3475	0.3221	0.9883	0.44350
Virgin River	1.6930	0.3029	0.0022	0.56430
Salt River	2.4694	0.4420	0.1973	0.43124
Verde River	2.7104	0.5083	-0.2582	0.24799

Our results show that RRMSE for the stage-bias scenario is lower than for the gaging data alone, indicating that the stage-bias differential does not negatively impact flood-quantile estimates.

DISCUSSION AND CONCLUSIONS

Five analysis scenarios of paleoflood deposition and one of gaging data alone were developed to evaluate the effect of paleoflood data on flood-frequency analysis. Using Monte Carlo simulation, we have shown that paleoflood data generally improves the bias, RRMSE, and RIC of flood-quantile estimates compared with gaging data alone for gaging record lengths less than 100 years. In only one case — where only one-third of paleoflood deposits were dated — did use of paleoflood data potentially degrade information contained in the gaging data. Extrapolating from our results, we anticipate that paleoflood data would degrade flood-frequency analysis only if gaging records exceed 100 years in length, which currently does not occur in the southwestern United States. Bias and RRMSE of flood quantiles decrease with increasing positive skew coefficients, and paleoflood data may degrade flood-frequency analysis for rivers with large, positive skew coefficients in log space. Finally, our results appear to apply to paleoflood records with lengths greater than 1,500 years.

Use of paleoflood data in flood-frequency analysis significantly improves estimates of low-frequency flood quantiles, in particular for recurrence intervals greater than 100 years. For the 100-year recurrence interval, the bias and RRMSE of gaging data alone is more than twice that of the rising-threshold scenario. This difference increases as recurrence intervals become larger. At the 5,000-year recurrence interval, the bias of the gaging-record scenario is 3.6 times that of the rising-threshold scenario and RRMSE is five times larger. The bias of the rising-threshold scenario is not dependent on the length of the gaging record. Flood-quantile estimates based on gaging data alone, with record lengths of 20 to 40 years, have a RRMSE greater than three times that of the rising-threshold scenario for a recurrence interval of 100 years; flood frequency for gaging records of these record lengths, in particular, can be considerably improved with the addition of paleoflood data.

Even use of an incomplete paleoflood chronology improves flood-quantile estimates. If only the largest flood in 4,000 years is known, paleoflood data significantly improves flood-frequency analysis over use of gaging data alone. For example, for the 500-year recurrence interval, the largest flood scenario has a RRMSE half that of the gaging record alone. For recurrence intervals greater than 2,500 years, differences in RRMSE between the scenarios with paleoflood information greatly increase. The bias of the two-threshold scenario is one-third that of the rising-threshold scenario and the bias of the largest flood model is almost twice that the rising-threshold scenario for the Salt River at Chrysotile.

Our preliminary work showed that consideration of errors in paleoflood data does not significantly degrade flood-frequency analysis in comparison to use of gaging data alone [*Webb et al.*, unpublished data, 1994]. Measurement and radiocarbon-dating errors, considered separately or combined, do not significantly degrade flood-quantile estimates. However, if only one-third of the deposits in a paleoflood sequence can be dated, the RRMSE of the rising-threshold scenario intersects that of gaging data alone at a point between 100 and 110 years of gaging data. Even in this case, use of incompletely dated paleoflood data improves flood-frequency estimates for short (<50 yr) gaging records.

Unless corrected by additional stage data, such as silt lines [*Webb et al.*, this volume], paleoflood discharges are minimum estimates because the height of a flood deposit is assumed to be equal to the maximum flood stage. In some cases, deposit heights are lower than maximum flood stage, leading to a negative bias in discharge estimates. Despite this negative bias, use of paleoflood data improves the overall estimation of flood-quantiles for quantiles ≥500-year recurrence interval for the Salt River at Chrysotile. For four additional bedrock-controlled gaging stations in the region, the bias ranged from

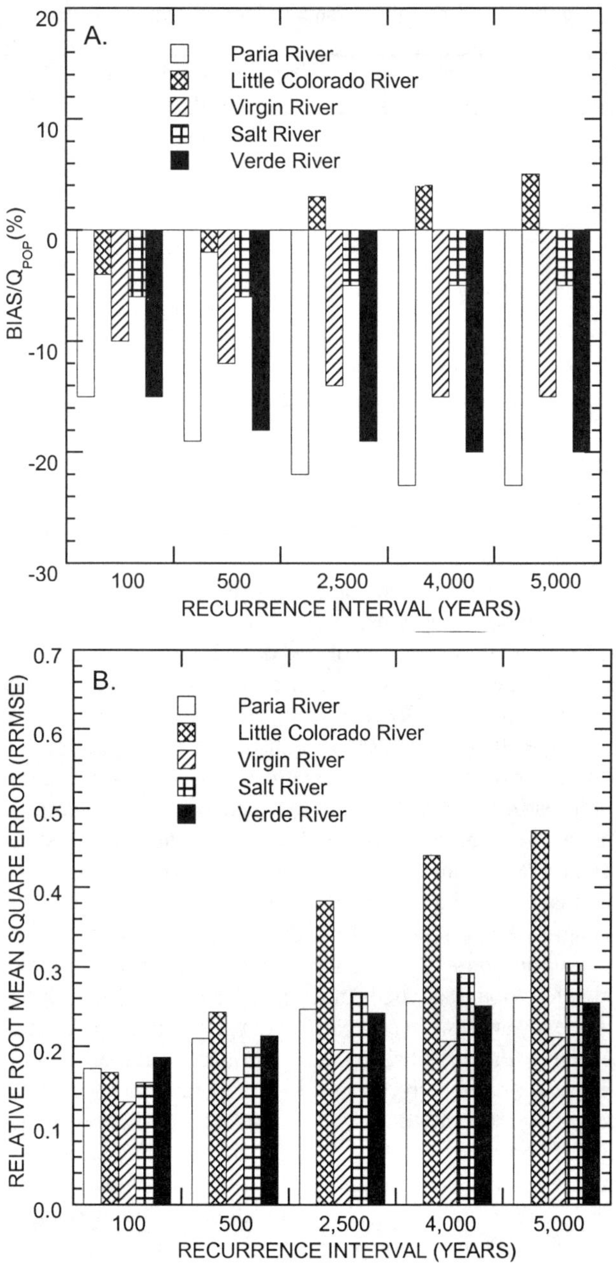

Figure 8. A. Biases (% deviation) of the stage-bias scenario of flood quantiles ranging from 100 to 5,000 years for 5 gaging stations (Table 2). The gaging record length is 70 years. The values were calculated using EMA. B. RRMSE of the stage-bias scenario of flood quantiles ranging from 100 to5,000 years for 5 gaging stations. The gaging record length is 70 years. The values were calculated using EMA.

2% to 9% for the 500-year recurrence interval. In addition, the paleoflood stage-bias scenario produced a smaller RRMSE than the rising-threshold scenario for the Salt River. The positive bias produced by fitting flood quantiles with the moments-based procedure in EMA is roughly balanced by the negative bias introduced by the underestimation of peak discharge. A small skew coefficient can degrade the value of the paleoflood stage-bias scenario compared to gaging data alone.

Because of its uncertainties, real or perceived, some hydrologists believe that paleoflood data should degrade gaging-station data in flood-frequency analysis. These uncertainties include discharge-estimation errors for paleofloods, uncertainty associated with radiocarbon dates, the number of datable deposits, and a systematic bias introduced by differences between water-surface elevation and associated deposit height. In one of the previous Monte Carlo analyses, paleoflood data were mis-specified in flood-frequency analysis as type II censored data [*Hosking and Wallis*, 1986a, 1986b]. Our results show that the correct specification of type I censored data, plus realistic uncertainty in discharge and flood dating, results in a favorable conclusion regarding the value of paleoflood data in flood-frequency analysis, even for rivers with long (*e.g.*, 70 year) gaging records.

Results obtained with the largest-flood scenario and consideration of measurement error conform with the three-parameter results of *Hosking and Wallis* [1986a, 1986b], who used a measurement error of up to ±100% in the GEV distribution. *Hosking and Wallis* concluded that even when the estimate of the largest flood is subject to an error of ±50%, a "worthwhile improvement in the accuracy of extreme flood estimates is still obtained." It is difficult to compare our re-

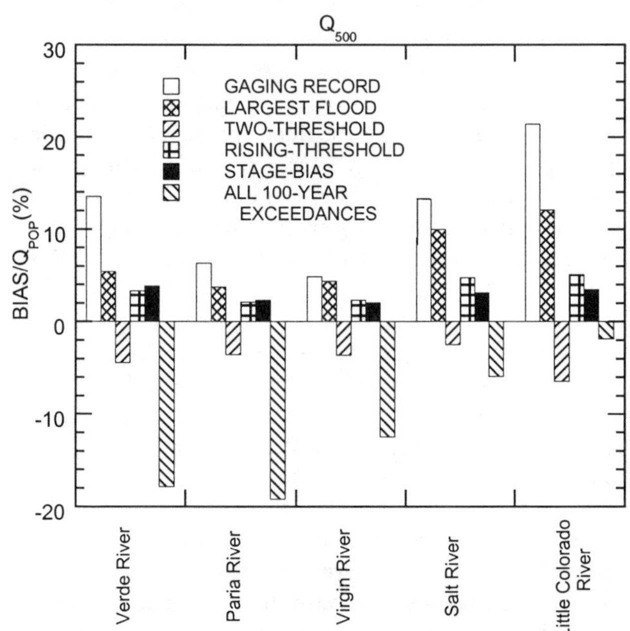

Figure 9. Biases (% deviation) of the 500-year flood quantile for all six models applied to five gaging records from the southwestern United States (Table 2). The gaging record length is 70 years. The values were calculated using EMA.

sults to other studies, which typically use two-parameter distributions. Paleoflood data are much more effective when using a three-parameter distribution [*Hosking and Wallis*, 1986a, 1986b].

Generally, our results show the factors affecting the value of paleoflood data in flood-frequency analysis agree with factors previously discussed in other studies [*Stedinger and Cohn*, 1987; *Guo and Cunnane*, 1991; *Pilon and Adamowski*, 1993; *Frances et al.*, 1994]. These factors include: (1) the flood quantile of interest, (2) the threshold-recurrence interval, (3) the length of available gaging data, (4) the length of the historical/paleoflood period, (5) moments of the parent population, and (6) consideration of uncertainty in paleoflood data.

Acknowledgments. Partial funding for this research was obtained through the U.S. Bureau of Reclamation Dam Safety Office Research Program. The authors thank Daniel Levish for his support of Blainey's research. David Pratt of the U.S. Geological Survey performed MAX simulations in 1993-1994 under the direction of authors Webb and Moss. The authors thank Jery Stedinger and Eduardo Martins for their critical reviews of the manuscript in addition to peer reviewers Robert Jarrett, Daniel O'Connell, and Upmanu Lall.

REFERENCES

Baker, V.R., R.C. Kochel, P.C., Patton, and G. Pickup, Paleohydrologic analysis of Holocene flood slackwater sediments, *Special Publication of the International Association of Sedimentologists*, 6, 229-239, 1983.

Baker, V.R., Paleoflood hydrology and extraordinary flood events, *J. Hydrol.*, 96, 79-99, 1987.

Baker, V.R., Paleohydrology and the hydrogeological sciences, in *Palaeohydrology and Environmental Change*, edited by G. Benito, V.R. Baker, and K.J. Gregory, pp.1-10, John Wiley and Sons, New York, 1998.

Baker, V.R., Paleoflood hydrology and the estimation of extreme floods, in *Inland Flood Hazards: Human, Riparian, and Aquatic Communities*, edited by E.E. Wohl, pp. 359-377, John Wiley and Sons, New York, 2000.

Benson, M.A., Use of historical data in flood frequency analysis, *EOS, Trans. Am. Geophys. Union*, 31(3), 419-424, 1950.

Blainey, J.B., Monte Carlo simulation of paleofloods: Information content of paleoflood data in flood- frequency analysis, 90 pp., M.S. Thesis, University of Arizona, Tucson, 2000.

Cohen, A.C., *Truncated and Censored Samples: Theory and Applications*, 312 pp., Marcel Dekker, Inc., New York, 1991.

Cohn, T.A., and J.R. Stedinger, Use of historical information in a maximum-likelihood framework, *J. Hydrol.*, 96, 215-223, 1987.

Cohn, T.A., W.L. Lane, and W.G. Baier, An algorithm for computing moments-based flood quantile estimates when historical information is available, *Water Resour. Res.*, 33(9), 2089-2096, 1997.

Costa, J.E., A history of paleoflood hydrology in the United States, 1800-1970, *EOS, Trans. Am. Geophys. Union,* 67(17), 428-430, 1986.

Ely, L.L., *Reconstructing paleoflood hydrology with slackwater deposits: Verde River, Arizona*, 49 pp., M.S. Thesis, University of Arizona, Tucson, 1985.

Ely, L.L, R.H. Webb, and Y. Enzel, Accuracy of post-bomb ^{137}Cs and ^{14}C in dating fluvial deposits, *Quat. Res.*, 38, 196-204, 1992.

Ely, L.L., Y. Enzel, V.R., Baker, and D.R., Cayan, A 5000-year record of extreme floods and climate change in the southwestern United States, *Science*, 262, 410-412, 1993.

England, J.F., *Assessment of historical and paleohydrologic information*, 292 pp., M.S, Thesis, Colorado State University, Fort Collins, Colorado, 1998.

England, J.F., *Draft users manual for EMA: at-site flood frequency estimation with historical/paleohydrologic data*, U.S. Bureau of Reclamation, Denver, Colorado, 1999.

Fisher, R.A., The generation of null hypotheses, in *The Design of Experiments*, Hafner Publishing Co., New York, 1960.

Francis, F., J.D. Salas, and D.C. Boes, Flood frequency analysis with systematic and historical or paleoflood data based on the two-parameter general extreme value models, *Water Resour. Res.*, 30(6), 1653-1664, 1994.

Guo, S.L. and C. Cunnane, Evaluation of the usefulness of historical and palaeological floods in quantile estimation, *J. Hydrol.*, 129, 245-262, 1991.

Hosking, J.R., and J.R. Wallis, The value of historical data in flood frequency analysis, *Water Resour. Res.*, 22(11), 1606-1612, 1986a.

Hosking, J.R., and J.R. Wallis, Paleoflood hydrology and flood frequency analysis, *Water Resour. Res.*, 22(4), 543-550, 1986b.

Jin, J., and J.R. Stedinger, Flood frequency analysis with regional and historical information, *Water Resour. Res.*, 25(5), 925-936, 1989.

Kochel, R.C., and V.R. Baker, Paleoflood hydrology, *Science*, 215, 353-361, 1982.

Kochel, R.C., V.R. Baker, and P.C. Patton, Paleohydrology of southwestern Texas, *Water Resour. Res.*, 18, 1165-1183, 1982.

Kochel, R.C., and V.R. Baker, Paleoflood analysis using slackwater deposits, pp. 357-376, in *Flood Geomorphology*, edited by V.R. Baker, R.C. Kochel, and P.C. Patton, John Wiley and Sons, New York, 1988.

Lane, W.L., Paleohydrologic data and flood frequency estimation, pp. 287-298, in *Regional Flood Frequency Analysis*, edited by V.P. Singh and D. Reidel, 1987.

Levish, D.R, D.A. Ostenaa, and D.R. O'Connell, A non-inundation approach to paleoflood hydrology for the event-based assessment of extreme flood hazards, in *Proceedings of the Association of State Dam Safety Officials 1994 Annual Conference*, pp. 69-82, Boston, Massachusetts, , 1994.

Levish, D.R., D.A. Ostenaa, and D.R. O'Connell, Paleoflood hydrology and dam safety, *Proceedings of the International Conference on Hydropower, Atlanta, GA, August 5-8, 1997*, pp. 2205-2214, 1997.

National Research Council, *Estimating probabilities of extreme floods: methods and recommended research*, 141 pp., National Academy Press, Washington D.C., 1988.

O'Connor, J.E., and R.H. Webb, Hydraulic modeling for paleoflood analysis, pp. 393-402 in *Flood Geomorphology*, edited by V.R. Baker, K.C. Kochel, and P.C. Patton, John Wiley and Sons, New York, 1988.

O'Connor, J.E., L.L. Ely, E.E. Wohl, L.E. Stevens, T.S. Melis, V.S.

Kale, and V.R. Baker, A 4500-year record of large floods on the Colorado River in the Grand Canyon, Arizona, *J. Geol.*, 102, 1-9, 1994.

Partridge, J., and V.R. Baker, Paleoflood hydrology of the Salt River, Arizona, *Earth Surf. Proc. Landf.*, 12, 109-125, 1987.

Pilon, P.J., and K. Adamowski, Asymptotic variance of flood quantile in log Pearson type III distribution with historical information, *J. Hydrol.*, 143, 481-503, 1993.

Pope, G.L., P.D. Rigas, and C.F. Smith, Statistical summaries of streamflow data and characteristics of drainage basins for selected streamflow-gaging stations in Arizona through water year 1996, 907 pp., *U.S. Geol. Surv. Water Resour.Invest. Rep. 98-4225*, 1998.

Stedinger, J.R., and T.A. Cohn, Flood frequency analysis with historical and paleoflood information, *Water Resour. Res.*, 22(5), 273-286, 1986.

Stedinger, J.R., and V.R. Baker, 1987, Surface water hydrology: historical and paleoflood information, *Rev. Geophys.*, 25(2), 119-124, 1987.

Stedinger, J.R., R. Surani, and R., Therivel, *Max users guide, a program for flood frequency analysis using systematic-record, historical, botanical, physical paleohydrologic and regional hydrologic information using maximum-likelihood techniques*, 51 pp., Cornell University, Department of Environmental Engineering, Ithaca, NY, 1988.

U.S. Water Resources Council, Guidelines for determining flood flow frequency, U.S. Interagency Advisory Committee on Water Data, Hydrology Subcommittee, Bulletin 17B, Washington, D.C., 1982.

Webb, R.H. and S.L. Rathburn, Paleoflood hydrologic research in the Southwestern United States, *Transp. Res. Rec.*, 1201, 9-21, 1989.

Yevjevich, V., and N.B. Harmancioglu, Research needs on flood characteristics, pp. 1-21, in *Applications of Frequency and Risk in Water Resources,* edited by V.P. Singh and D. Reidel, Boston, 1987.

Joan B. Blainey, U.S. Geological Survey, 6000 J Street, Placer Hall, Sacramento, CA 95819; email jblainey@usgs.gov.

Robert H. Webb, U.S. Geological Survey 1675 W. Anklam Road, Tucson, AZ 85745.

Marshall E. Moss, Retired, U.S. Geological Survey, 520 N. Park Avenue, Suite 221, Tucson, AZ 85719.

Victor R. Baker, University of Arizona, Department of Hydrology and Water Resources, P.O. Box 210011, Tucson, AZ 85721.

Paleohydrologic Bounds: Non-Exceedance Information for Flood Hazard Assessment

Daniel R. Levish

U.S. Bureau of Reclamation, Denver, CO

The need for widely applicable techniques to acquire paleoflood information for the Bureau of Reclamation in the Western United States prompted the development of the concept of paleohydrologic bounds. A paleohydrologic bound is defined as a time interval during which a particular discharge has not been exceeded. Paleohydrologic bounds are limits on paleostage provided by stable geomorphic surfaces adjacent to a stream. The age of the surface establishes the duration of the bound. The discharge associated with the bound is estimated from hydraulic model predictions of the discharge required to erode or significantly modify the geomorphic surface. The implementation of paleohydrologic bounds in paleoflood studies spurred the refinement of tools used in previous paleoflood studies as well as the development and application of new tools and concepts. These innovations include improvements in the use of soil stratigraphy and radiocarbon analysis, topographic mapping, two-dimensional hydraulic modeling, and statistical analysis. At present, it appears that the best approach is to combine information about the largest paleofloods and paleohydrologic bounds to constrain potential flood frequency behavior at low probabilities. As the use of risk analysis grows for flood hazard assessment, the need for paleoflood information will increase. In the future, large information gains will result from combining paleoflood information with more conventional hydrologic methods.

INTRODUCTION

At the Bureau of Reclamation (Reclamation) in the early 1990s, there was a pressing need to understand the magnitude of low probability floods at many sites in the Western United States. This resulted from the fact that numerous Reclamation dams cannot accommodate the Probable Maximum Flood (PMF). The PMF is a hypothetical maximum flood derived from the Probable Maximum Precipitation based on rainfall-runoff modeling [*Cudworth*, 1989]. Modifying a dam to the PMF standard can be very expensive, and in some cases the expense is difficult to justify [*Lave et al.*, 1990; *Graham*, 2000]. In part, this led Reclamation to employ quantitative risk analysis for the evaluation dam safety [*Reclamation*, 1997; 1999a]. Quantitative risk analysis for dam safety frequently requires information about floods with probabilities on the order of 1 in 10,000 [*Reclamation*, 1999b]. The paleoflood record documents the actual magnitude and history of low probability floods providing a basis for comparison with conventional hydrologic approaches [e.g., *Baker et al.*, 1987; *Baker* 1994; 2000; *Jarrett and Tomlinson*, 2000]. This spurred the development of a widely applicable strategy for collecting paleoflood information for application to dam safety [*Levish et al.*, 1994; 1997].

Most previous paleoflood studies have focused on reconstructions of individual floods [*Baker*, 2000]. This is generally accomplished through identification of paleostage

Ancient Floods, Modern Hazards: Principles and Applications of Paleoflood Hydrology
Water Science and Application Volume 5, pages 175-190

Published in 2002 by the American Geophysical Union

indicators and hydraulic reconstruction of the flood peak discharge [*Baker et al.*, this volume; *Denlinger et al.*, this volume; *Jarrett and England*, this volume; *Webb and Jarrett*, this volume]. These studies have advanced the science of paleoflood hydrology tremendously in the past 30 years [*Baker et al.*, this volume]. For flood hazard analysis the use of this type of flood history reconstruction can present some problems, particularly when trying to estimate the magnitude of very low probability floods (e.g., 1 in several 1,000 to 1 in 10,000). These problems arise from the preservation of flood records, the identification of individual floods, the uncertainty in reconstructing peak discharge, and the use of reconstructed flood magnitude in flood frequency analysis. Alluvial flood records can be difficult to interpret because of complex stratigraphy and variable preservation of flood records [*House et al.*, this volume]. Reconstructed individual floods can be difficult to use alone in flood frequency analysis because they may only provide minimum discharge estimates. Although this information is invaluable for understanding large floods, the uncertainty in this information may be magnified at low probabilities due to a lack of constraint on flood frequency extrapolation. Typically in flood hazard analysis, the key question is what discharge will not be exceeded at a given probability. For dam safety these probabilities can be very small [*Reclamation*, 1999b].

In general, it had been concluded that the use of slackwater records, fine-grained sand and silt deposited from suspension during floods, could be restricted to certain geologic settings [e.g., *Kochel and Baker*, 1988]. It was also clear that the engineering community had doubts about the use of this type of information [e.g., *Hosking and Wallis*, 1986; *Lane*, 1987], although some welcomed the ideas [e.g., *Dawdy and Lettenmaier*, 1987]. The initial concerns have since been addressed [*Baker et al.*, this volume]. A new, widely applicable strategy was required to apply geomorphic and stratigraphic information to flood hazard assessment. These techniques would have to apply to most rivers in the Western United States, and produce limits on maximum flood magnitude over long periods of time.

The development of this strategy began as a comparison of mapped patterns of inundation from large hypothetical floods (e.g., PMF) to geomorphic surface age. The strategy rapidly evolved to one of limiting the magnitude of paleofloods with geomorphic information. The initial approach was to estimate the age of terrace surfaces that would be significantly inundated by large floods of interest to dam safety. During the course of the first major study, this progressed rapidly to a non-exceedance approach. The goal of the non-exceedance approach is to limit flood paleostage through time with geomorphic and stratigraphic evidence [*Levish et al.*, 1994]. The term paleohydrologic bound refers to these limits on flood paleostage and therefore magnitude over time [*Levish et al.*, 1997]. The formalization of this new concept required the refinement of tools used in previous paleoflood studies, and the development and application of new tools and techniques. This includes refinement of soil stratigraphy and radiocarbon sampling, construction of high-resolution topographic maps of long reaches of river, refinement of two-dimensional hydraulic models particularly to display two-dimensional stream power distribution, and Bayesian statistical analysis with paleoflood and non-exceedance data.

Non-exceedance information can be effectively used in flood hazard studies for critical structures and can be rapidly collected for most streams in the Western United States. For the past eight years, Reclamation has used paleohydrologic bounds as input to flood-frequency analysis to estimate the magnitude of low probability floods (e.g., < 1 in 100) for dams in the Western United States. Paleohydrologic bounds allow paleoflood information to be developed in many settings where it was not previously thought possible. The collection of non-exceedance information can be tailored to the intended use. It can be collected quickly as a preliminary indicator of the flood potential of a basin or a region, or more comprehensive information can be developed for use in a detailed flood frequency analysis for hydrologic risk analysis of critical structures.

RELATION TO CONVENTIONAL PALEOFLOOD HYDROLOGY

Paleoflood hydrology includes the study of the geomorphic and stratigraphic record of past floods [e.g., *Baker*, 1989; *Jarrett*, 1991; *Baker et al.*, this volume]. There is a long history of paleoflood hydrology in a wide variety of settings throughout the world [e.g., *Costa*, 1987; *Patton*, 1987; *Baker et al.*, 1988]. The inclusion of paleoflood information can effectively lengthen a short data record and, therefore, reduce the uncertainty in hydrologic hazard analysis [*Jarrett*, 1991]. Paleoflood records are a direct, long-term measure of the ability of a stream or a region to produce large floods, and these records are often 10 to 100 times longer than the conventional record of annual peak discharge estimates. Obviously, the addition of paleoflood information allows for a higher degree of assurance for hydrologic hazard assessment of critical structures where floods with low probabilities or long return periods are the focus. In addition, paleoflood information can offer a long-term perspective on exceptional peak discharge estimates from stream gages, ungaged streams, and historical records [*Costa*, 1978a; *Levish et al.*, 1994].

The majority of paleoflood information comes from studies of slackwater deposits, where the stratigraphy of fine-grained sediment that accumulates in backwater areas is used to reconstruct the history of past floods [e.g., *Patton et al.*, 1979; *Baker et al.*, 1983; *Kochel and Baker*, 1988]. In this type of paleoflood study the goal is to reconstruct a complete record of all floods larger than a particular magnitude. Primarily this type of analysis formed the basis of the science of paleoflood hydrology and has added a wealth to the science of understanding floods [*Baker et al.*, this volume]. However, these reconstructions can be complex. This complexity arises from problems with preservation, lateral accretion, recognition of individual floods, discerning the precise relation between paleostage and flood deposits, and the progressive self-censoring of flood stratigraphy in vertically accreting deposits [*House et al.*, this volume; *Webb et al.*, this volume]. This type of information may also be limited to specific types of streams in certain environments [*Kochel and Baker*, 1988]. For the assessment of low probability flood hazards, it can be difficult to employ the reconstructed flood history from flood deposits. In some cases, paleostage indicators may accurately record the peak flood stage and therefore accurately document flood magnitude [*Jarrett and England*, this volume]. Typically, flood deposits are only capable of providing exceedance information, that is, minimum estimates of paleostage and paleoflood magnitude.

When this type of record of past floods is solely used in frequency analysis it amplifies the need to have accurate information about the magnitude, number, and timing of past floods. However, this information on minimum flood magnitude is a powerful complement to non-exceedance information, effectively bracketing the range of possible flood behavior at low probabilities. Combining exceedance and non-exceedance information reduces the necessity of having to reconstruct the record of past floods in great detail [*O'Connell et al.*, in press]. Each strategy has different strengths and weaknesses, but in general the information derived from non-exceedance and exceedance techniques is complementary.

Some paleoflood studies have noted non-exceedance information although they have not explicitly employed this information in analysis of flood frequency. However, it had been recognized that this type of paleoflood information could be useful in flood frequency analysis [*Stedinger et al.*, 1988]. Trim lines in soil or colluvium have been used as an indicator of maximum flood stage [e.g., *Baker*, 1987; *Baker et al.*, 1987; *Partridge and Baker*, 1987]. *Jarrett* [1990] demonstrated that the lack of evidence of large floods, as shown by landscape stability, is as important as the geomorphic record of floods in defining the spatial distribution of flooding. Recently some non-exceedance techniques, coupled with evidence of paleofloods, have also been applied to the evaluation of flood hazards for dam safety [*Jarrett and Tomlinson*, 2000; *Levish et al.*, 2000; *Jarrett and England*, this volume].

PALEOHYDROLOGIC BOUNDS

Background

In order to provide information on low probability flood magnitude in a wide variety of environments in the Western United States, a new strategy for collecting paleoflood data was required. The original approach for developing information for Reclamation dams was to compare existing downstream inundation maps to geomorphic surfaces of various ages. For many Reclamation dams inundation maps are available to show the predicted extent of flooding from hypothetical dam failures, flood discharges from maximized runoff such as the PMF, and maximum outlet-works capacity. These maps are used for emergency action planning and to evaluate potential economic damages from the failure of a structure or from a large spillway release. Initially, the goal was to simply look for substantial depth of inundation on identifiable geomorphic surfaces of a particular age. This depth of inundation would be anticipated to result in obvious modification of the geomorphic surface, such as the geomorphic impacts resulting from other large historical floods [e.g., *Shroba et al.*, 1979; *Baker and Costa*, 1987; *Brooks and Lawrence*, 1999].

The approach is similar to studies by *Mansfield* [1938] on the Ohio River and *Jahns* [1947] on the Connecticut River. They recognized that historical floods had overtopped sites not previously inundated in hundreds or thousands of years. The basic premise is, that lacking evidence of recent flood modification, erosion or deposition, the age of a stable geomorphic surface is an estimator of the minimum return period of a flood that could significantly alter that surface [*Costa*, 1978a; 1978b *O'Connell et al.*, 1997]. This comparison of predicted inundation and potential resulting surface modification, and geomorphic surface age could be conducted for flows equivalent to some key capacity of the dam or for some hypothetical flood like the PMF. The result would be frequency information about a flood magnitude of interest to dam safety or dam operation.

During the course of the initial study it became clear that, in some cases, these sorts of measures of hydrologic dam safety were well beyond any historical floods or paleofloods [*Levish et al.*, 1994; 1997]. It also became clear that this type of non-exceedance strategy would work with the geomorphic surfaces present along streams if discharges could be modeled that would produce predicted geomorphic modification. Additionally, it was evident that this non-exceedance information is of significant value in flood frequency analysis for defining the upper end of the frequency curve [*Levish et al.*, 1994; *O'Connell et al.*, 1997; *O'Connell et al.,* in press]. Trials based on preliminary field data and hypothetical data with MAX

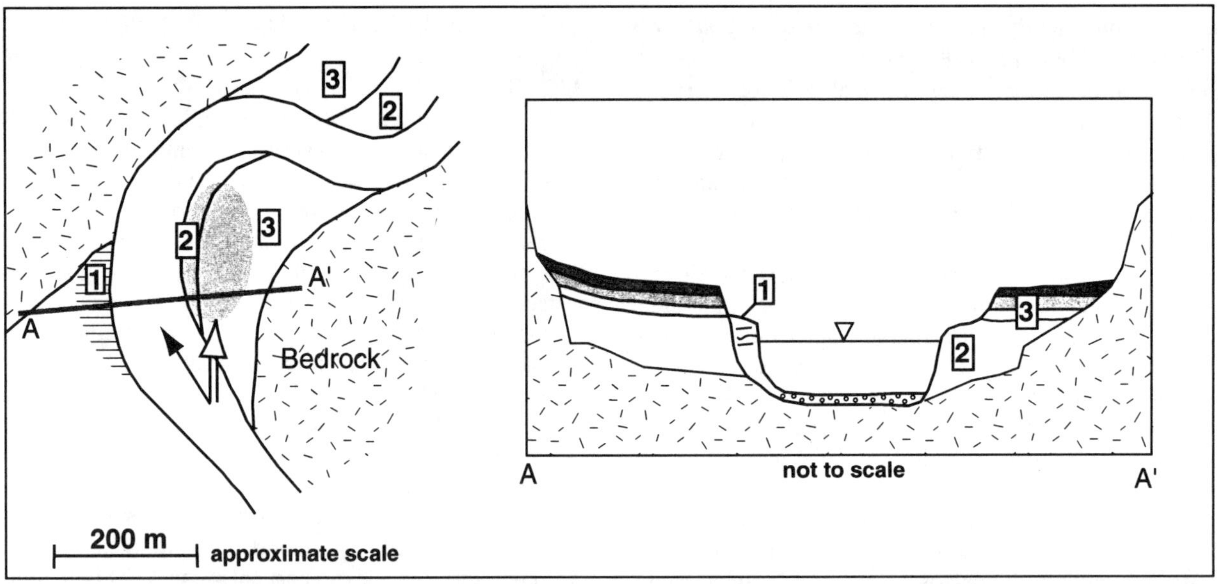

Figure 1. Schematic map and cross section of a bend in the North Platte River near Douglas, Wyoming. Unit 1 consists of vertically accreted flood deposits shown as GG4 in Figure 6. Unit 2 represents an eroded section of the stable geomorphic surface shown as Unit 3. At this site Unit 3 has been stable for about 2400 to 2800 years. The solid arrow represents the flow direction at low flows. The hollow arrow shows the hydraulic model predicted flow direction during large floods. The gray shaded area shows the region of high stream power predicted by the two-dimensional model. In this case, flows of this magnitude would be predicted to modify a surface that is more than 2400 years old. In the cross section note the close correspondence between the vertically accreted deposits on the outside of the bend and erosional scar on the inside.

[*Stedinger et al.*, 1988], showed that the best way to proceed with this type of study was to try to narrow the range of non-exceedance information. At the time, MAX was the only available method to effectively incorporate paleoflood data into flood frequency analysis. From the results of these initial trials, it appeared to be more important to attempt to bound the maximum size of floods rather than to reconstruct a detailed record of actual floods when estimating the magnitude of floods with probabilities less than 1 in several 100 for the purpose of risk analysis [*Ostenaa et al.*, 1996].

Definition

This iteration between field data collection and flood frequency calculation resulted in the formulation of the concept of a paleohydrologic bound. A paleohydrologic bound is a time interval during which a given discharge has not been exceeded. The term is not 'paleoflood bound' because it is not a bound on a particular past flood. In this way, the concept is somewhat different as the focus of most previous studies had been on the stratigraphic evidence of floods. The development of paleohydrologic bounds focuses on the positive evidence of landscape stability. Landscape stability is demonstrated by the presence of time-dependent stratigraphy such as soils developed on fluvial geomorphic surfaces adjacent to a stream. Such a record of stability provides positive evidence of no floods capable of modifying this record. This concept shares some similarities with paleoseismic studies where one of the important components is mapping geomorphic surfaces that have not been faulted [e.g., *McCalpin*, 1996]. The parallel in paleoflood studies is looking for geomorphic surfaces that have not been flood-modified. The next step is to predict what paleostage or discharge it would take to create a discernible and lasting record of geomorphic modification. Envisioning this approach can be difficult since one essential component is to model hypothetical large floods that by definition have not happened or at least have not happened in a very long time. However, it rapidly becomes apparent with hydraulic modeling of excessively large, hypothetical floods that they are not consistent with the geomorphology (Figure 1).

Geomorphic surfaces adjacent to streams are reliable indicators of maximum flood stage through time because fine-grained soils and stream-transported floodplain sediments that can be easily modified by floods typically underlie these surfaces. Preserved, non-modified surfaces of known age form paleostage limits that can be input into a hydraulic model to calculate the maximum discharge that would significantly modify the surface (Figure 1). This maximum discharge, together with the age of the surface, forms a bound on flood discharge through time that furnishes input data for flood frequency analysis. Paleohydrologic bounds are not actual floods, but rather are

Figure 2. Upstream view of the Medicine Bow River near Hanna, Wyoming. The stable terrace surface to the right in the foreground is about 7,000 years old based on six radiocarbon ages. A large magnitude flood, with a high enough stage, would flow over and eventually modify this surface. The river flows on bedrock and there is bedrock exposed along the left side of the channel.

limits on flood magnitude over a measured time interval. In this way, paleohydrologic bounds represent stages and discharges that have not been exceeded during the period of time since the geomorphic surface stabilized—the time since the surface ceased to be subject to active fluvial processes. The geomorphology, stratigraphy, and soils present on a terrace surface are positive evidence for stability of the surface and the lack of floods larger than a specific magnitude.

Identification

The objective of flood hazard studies based on paleohydrologic bounds is to identify and assign ages to geomorphic surfaces adjacent to a stream that serve as limits for the potential paleostage of large floods (Figures 1 and 2). The field expression of a paleohydrologic bound, a stable geomorphic surface, is a floodplain that has been abandoned. Once abandoned, the characteristics of a geomorphic surface change with time. Two of the most easily recognized changes involve the modification of surface morphology and the development of soil. Through time, slope processes and weathering mute the expression of surface irregularities related to fluvial erosion and deposition. Once a surface has stabilized, that is, it is no longer subjected to episodic erosion and deposition, soils form in a predictable sequence [*Birkeland*, 1999].

Disruptions in soil profiles and geomorphic features, such as eroded channels, that result from significant inundation by large floods, are generally easily recognized (Figures 3 and 4) [e.g., *Shroba et al.*, 1979; *Brooks and Lawrence*, 1999]. This is why these former floodplain surfaces have good potential to be reliable indicators of flood stage through time. Geomorphic surfaces adjacent to a stream define a maximum possible channel width, over the time period represented by the surfaces, through which a maximum discharge can be modeled. The ages associated with the geomorphic surfaces that form bounds for flood magnitude are almost always minimum ages because of the problems related to determining the precise time when a particular surface was abandoned. The result is an estimate of the maximum discharge during the minimum time interval since stabilization. These estimates are often larger

Figure 3. Inside of a bend on the North Platte River near Saratoga, Wyoming. The river flows from left to right. As stage increases flow will exit the channel farther upstream. Once the water leaves the main channel it flows across the surface to return to the channel downstream. The highest channels cut across the surface record floods with about a return period of about 50 to 100 years. The highest parts of the surface inside the bend have been stable for about 2,000 to 4,000 years.

than actual past floods due to vertical and lateral erosion, subsequent to the time of surface stabilization, resulting in apparently larger cross sections and discharges. Therefore, it is important to locate sites within stable reaches with the minimum channel capacity adjacent to geomorphic surfaces that place limits on paleostage (Figure 2). Paleohydrologic bounds provide the most utility when defined, based on physical evidence, as close as allowable to the magnitude of the largest paleofloods. It is possible to formulate paleohydrologic bounds where discharge is not tightly constrained. This may result in data that is not useful in flood frequency analysis.

Formation

Paleohydrologic bounds may form in several different ways. In stable areas with sufficient suspended sediment supply, flood sediment may vertically accrete until flood discharges no longer go out of bank [*Hereford*, 1984]. A subsequent larger flood could overtop and erode these vertically accreted deposits [*Nanson*, 1986]. Following a large flood, vertical accretion can take place relatively rapidly [*Schumm and Lichty*, 1963; *Moody et al.*, 1999]. In areas that are tectonically active, paleohydrologic bounds can form from down cutting in response to uplift making the floodplain accessible only to larger and larger magnitude floods. In areas of active subsidence paleohydrologic bounds could be difficult to establish because the lowering of the floodplain surfaces and aggradation of the channel could make these surfaces more accessible to floods. Other geomorphic features with time-dependent stratigraphy can also be used as paleohydrologic bounds, such as tributary fans that would be eroded by large floods [*Costa*, 1978b; *Levish and Ostenaa*, 1996].

Values Associated with Paleohydrologic Bounds

Both properties of the paleohydrologic bound—time and discharge—are independently determined from objective criteria in the field. Time, represented by the stability and age of a surface, can be determined based on soil stratigraphy and radiocarbon analysis. The depth of significant inundation can be calibrated based on the physical evidence in a particular reach and comparison with the

Figure 4. Soils and flood deposits on the North Platte River near Rawlins, Wyoming. Soils and organic material interbedded with fine sand and silt flood deposits create the striped appearance. From radiocarbon chronology at other sites it would appear that the floods and soils represent the last 500 years.

geomorphic impact of extreme floods [e.g., *Shroba et al.*, 1979; *Baker and Costa*, 1987; *Lapointe et al.*, 1998; *Brooks and Lawrence*, 1999].

Shear stress or stream power can be calculated and compared with information from other large floods to determine the values required to erode a deposit of known age or modify a geomorphic surface [e.g., *Baker and Costa*, 1987; *Magilligan*, 1992; *Lapointe et al.*, 1998; *Brooks and Lawrence*, 1999]. Two-dimensional models can be used to portray the spatial distribution of stream power to delineate areas of probable erosion and employ more complicated geometry in the analysis (Figure 1) [*Miller*, 1995; *Denlinger et al.*, this volume]. The exact value of stream power required to modify a surface may vary with the geomorphic effectiveness of a flood [*Costa and O'Connor*, 1995; *Jarrett and England*, this volume]. For this reason, the potential flood-producing mechanism and associated range of hydrographs should be kept in mind when assessing possible surface modification. Every site along a stream does not necessarily have equal potential to capture the record of a particular flood. Therefore, it is also important to have a good understanding of the fluvial geomorphology and geomorphic history of a stream in order to understand the likely geomorphic effects of large floods.

The level of effort expended to determine paleohydrologic bounds can be tailored to the economics of the project or the consequences of failure of a critical structure (Figure 5). For dam safety risk analysis, in situations of high consequences, the goal is to develop consistent paleohydrologic bounds, and age and magnitude estimates for the largest paleofloods, in multiple reaches. This is generally accomplished through selecting detailed study reaches with geometries and stratigraphy appropriate to develop bounds. In general, reaches are selected upstream and downstream of a dam and sometimes on major reservoir tributaries. The goal is to attempt to minimize the influence of other fluvial and geologic factors such as local aggradation, local bedrock controls on channel slope, or active tectonics. For large rivers this can often be accomplished by selecting reaches in different geologic settings.

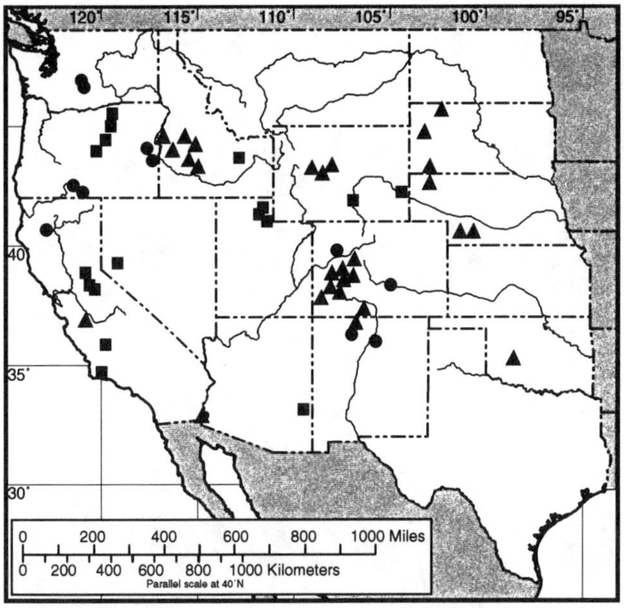

Figure 5. Map of the Western United States showing locations where Reclamation has collected paleoflood information. The different symbols represent different levels of effort. The squares show sites of the highest level of effort generally consisting of detailed mapping, hydraulic modeling, and radiocarbon chronology. The circles show sites of intermediate levels of effort generally consisting of less detailed modeling and less chronologic control. The triangles show sites of the lowest level of effort generally consisting of simple measurements of channel geometry, simplified discharge estimates, and chronology based on soil development and regional correlation.

In general, the reaches selected for study should be long enough to include several bends. Depending on the consequences, the scale of the river, the geometry of adjacent geomorphic surfaces, and the geomorphic and stratigraphic information available, a reach could extend from less than a kilometer to tens of kilometers. Within a reach multiple stratigraphic sites are described. Results for bounds are derived independently for each site based on the aggregate of evidence and then compared between reaches. The objective is to develop consistency between sites to ensure the accuracy of the results.

INNOVATIONS

The strategy of using paleohydrologic bounds to place limits on the magnitude of past floods requires the refinement of tools employed in previous paleoflood studies and the development and application of new tools and techniques. The new focus on quantifying landscape stability requires detailed understanding of fluvial erosional and deposition processes, the genesis of Holocene soils and radiocarbon methodology. Trying to predict the patterns of erosion that would result from large floods may require detailed topographic mapping and hydraulic modeling of long river reaches, generally including one or more bends. Bends usually afford good opportunities to limit discharge because stable surfaces present in these bends are commonly overtopped at high flows producing localized high stream power and lasting flood evidence [e.g., Shroba et al., 1979; Lapointe et al., 1998; Brooks and Lawrence, 1999] (Figures 1, 2 and 3). Bend geometry can be difficult to model with one-dimensional hydraulic models [e.g., *Denlinger et al.*, this volume]. Two-dimensional hydraulic models that can display the spatial distribution of stream power can be advantageous for more precisely estimating paleohydrologic bound discharges. Finally, in order to have utility in hydrologic risk analysis, paleohydrologic information needs to be accurately portrayed in a flood frequency analysis [*Stedinger and Baker*, 1987]. A faithful way to incorporate this paleoflood information in flood frequency analysis is through a Bayesian analysis that includes the uncertainty in the geologic and hydraulic information [*O'Connell et al.*, 1997; *O'Connell et al.*, 1998].

Soil Stratigraphy and Radiocarbon

One of the two components in the assignment of bounds on flood paleostage is to delineate and estimate numerical ages for stable geomorphic surfaces. Fine-grained sediments commonly deposited on floodplains, and the soils that subsequently form on them, are reliable indicators of surface stability. This is because fine-grained flood deposits in the Western United States are generally noncohesive and highly erodible. After an alluvial surface stabilizes, the soil profile will develop through time-dependent processes in a predictable sequence [e.g., *Birkeland*, 1999]. Radiocarbon analysis of organic material provides the most reliable method for assigning numerical ages to Holocene surfaces.

The age control for most slackwater studies comes from organic material recovered from flood deposits [*Baker*, 1987]. The primary concern is that radiocarbon datable material may possess some inherited age [e.g., *House et al.*, this volume; *Webb et al.*, this volume]. Dating stable alluvial surfaces is a little more difficult. A stable surface may not only be composed of a flood deposit, but also the soil that records stability since its deposition. Radiocarbon analyses of organic material in soils can yield a wide variation in ages that are considered minimum ages because young carbon is continuously added to the soil profile [e.g., *Gilet-Blein et al.*, 1980; *Hsieh* 1992; 1993; *Brown*, 1997; *Birkeland*, 1999]. The strategy developed for deriving ages from this type of deposit involves several key steps. Of fundamental importance is not to obtain radiocarbon ages from bulk organic material. This material

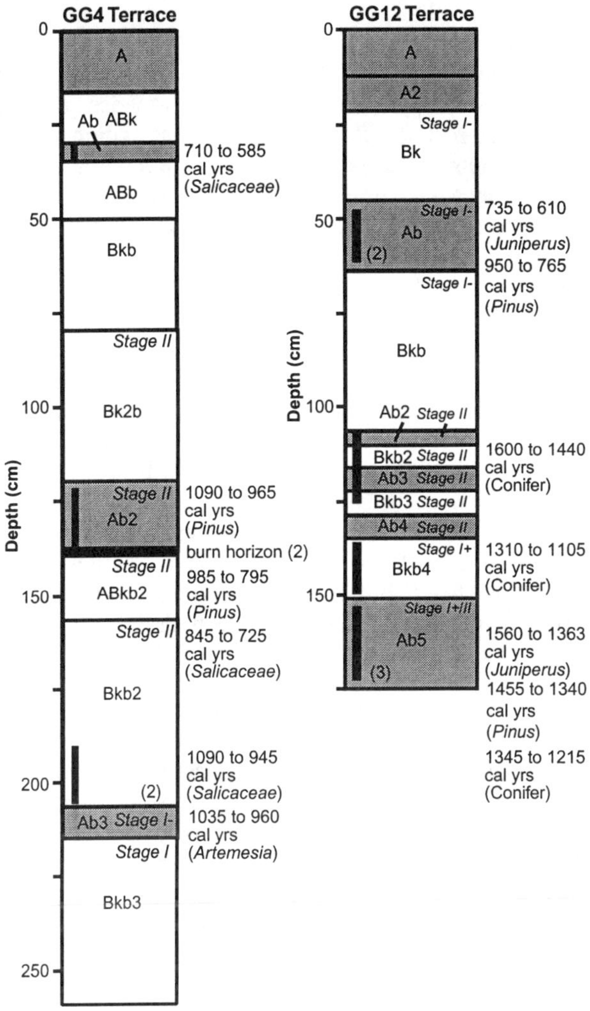

Figure 6. Vertically accreted flood deposits from sites about 10 km apart on the North Platte River near Douglas, Wyoming. The soils are described in accordance with *Birkeland* [1999]. The Ab horizons were A horizons once at the surface buried by later vertical accretion. The age from the buried soil represents a maximum age for the overlying flood deposit. The vertical black rectangles show zones where bulk samples were taken. The stage notation denotes the field morphology of pedogenic calcium carbonate accumulation. On the right of each column are calibrated radiocarbon ages in years before the year 2000. Shown in parentheses is the type of charcoal used in radiocarbon analysis. The ages of the last flood recorded correspond well.

may contain a wide range of material of very different ages [e.g., *Gilet-Blein et al.*, 1980; *McWeeney*, 1989]. It has proven successful to collect bulk material and have it separated by flotation. The floated macrofloral material can then be identified, usually to the genus level [e.g., *Puseman et al.*, 1999]. Analysis of bulk samples from some recent studies revealed more than a dozen different types of charcoal as well as roots, burned roots, small mammal bones, various types of mollusk shells, fruit, seeds, leaves, grass, monocot stems, insect parts and pupa, and rodent feces [*Levish and Ostenaa*, 1996; *Levish et al.*, 2000]. The material in these bulk samples is obviously composed of material washed in with flood sediment and material added later. Small pieces of charcoal also migrate downward in soil profiles and may yield anomalously young ages [e.g., *Stafford and Tyson*, 1989; *Stafford*, 1998] The most reliable ages appear to come from discrete fragments of charcoal, particularly from short-lived riparian species or from burned horizons [*Levish and Ostenaa*, 1996]. The best results are gained from multiple ages in stratigraphic order. In vertically accreting floodplains, buried soils and burn horizons can provide accurate ages (Figure 6). Material from the top of buried soils commonly provides reliable maximum ages for overlying deposits.

Many Holocene fluvial deposits contain an assemblage of terrestrial and aquatic snails [e.g., *Levish and Ostenaa*, 1996]. These snails can often be identified to the genus or species level [E. Evanoff, writt. commun., 1996]. Once identified the snails environmental preference can be inferred [E. Evanoff, writt. commun., 1996]. Depending on the goal, aquatic or terrestrial snails can be selected for radiocarbon analysis. Aquatic snails provide accurate ages of flood deposits and terrestrial snails can provide minimum ages for buried soils. Ages from snails should be corrected for radiocarbon reservoir effects [e.g., *Trumbore*, 2000].

In order to be of use in flood frequency analysis radiocarbon ages must be calibrated to calendar years. This is accomplished through the use of existing calibration curves [e.g., *Stuvier and Reimer*, 1993]. Calibrated radiocarbon ages are presented in years before 1950 by convention [e.g., *Bowman*, 1990]. Therefore, in order to accurately portray the period of time encompassed by a paleohydrologic bound in flood frequency analysis, the time spanned between 1950 and the year of the analysis must be added to the calibrated age.

The stratigraphy of vertically accreting floodplain deposits from the North Platte River near Douglas, Wyoming demonstrates the strategy followed for radiocarbon sampling and analysis (Figure 6). Both columns show typical floodplain stratigraphy found along many rivers in the Western United States with flood deposits separated by buried soils. The buried soils represent periods of surface stability between floods. In some situations, charcoal from short-lived riparian plants may not be available. The ages from the same horizon on shorter-lived plants are younger. The ages from site GG12 show the uncertainty introduced by ages from charcoal types that likely possess an inherited age. The ages from the uppermost buried soil also demonstrate the range of results introduced by radiocarbon ages from different types of charcoal.

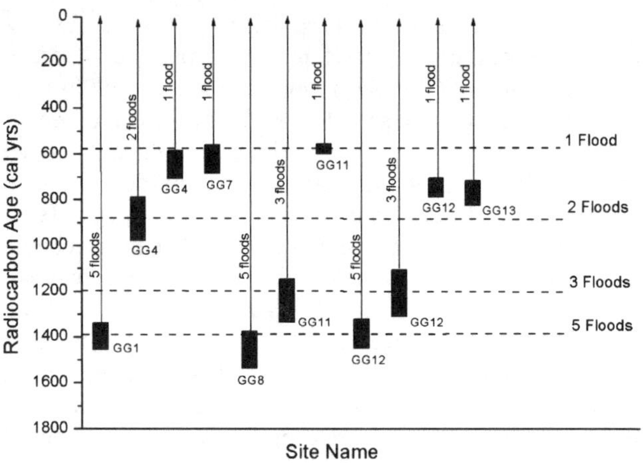

Figure 7. Composite diagram showing the flood history for seven sites along a 10 km reach of the North Platte River near Douglas, Wyoming. The ages are based on 29 radiocarbon ages from buried soils overlain by the number of flood deposits shown. In each case the floods are separated by distinct soils as shown in Figure 6.

Figure 7 shows a compilation of the vertically accreted flood records of seven sites and 29 radiocarbon ages described in an approximately 10 km reach of the North Platte River [*Levish et al.*, 2000]. The floods are plotted as younger than the buried soil they overlie. Although the floods fall into broad groups, numerous radiocarbon ages from multiple sites show that the best that can be done is to bracket floods to periods of several hundred years. However, when input into a frequency analysis that captures this range of uncertainty, this is valuable information on the timing of paleofloods. This information on paleofloods was collected as a complement to paleohydrologic bound data.

Topographic Mapping

The most consistent way to establish estimates of discharge for paleohydrologic bounds is by hydraulic modeling over a long reach, generally from one to tens of kilometers. The length of the reach depends on the scale of the stream, but generally should include several bends. Many of the best sites for establishing paleohydrologic bounds are located in these bends, and thus require flow modeling through these bends (Figures 1, 2, and 3). Two-dimensional hydraulic models are better equipped to model flow over the inside of bends, but require a topographic grid [*Denlinger et al.*, this volume]. To some extent, the accuracy of hydraulic modeling depends on the accuracy of the topography. In order to make this type of study economical, various methods can be used that are capable of producing accurate topographic maps depending on the size of rivers and floodplains. These include making very large-scale maps of short reaches using an electronic theodolite/rangefinder (total station), survey-grade global positioning satellite (GPS) equipment, and traditional photogrammetry.

For longer study reaches, computerized photogrammetry, or softcopy photogrammetry, provides a cost-effective alternative. Recent advances in computerized photogrammetry and in the use of survey-grade GPS measurements for ground control have substantially reduced the cost of producing high resolution, less than 0.5 m contour interval, topographic maps. Computerized photogrammetry also has the benefit of producing orthophotographs and three-dimensional terrain models that can in turn be used to produce geomorphic maps and other digital graphics. The ground control can be rapidly set with Real Time Kinematic GPS that provides sub-centimeter scale accuracy in a very short time. The use of this type of GPS allows the maps to tied to a local coordinate system or into a standard coordinate system. With this system it is possible to produce maps for many kilometers of river with an accuracy of 0.25 m or more. The resolution of the mapping is fixed by economics and the amount of information that can be handled by the hydraulic model.

Hydraulic Modeling

The best sites for the establishment of bounds on paleostage are sites that will be modified during large floods. Frequently sites where large floods generate high power at the lowest discharges are on the inside of bends overtopped by large floods (Figures 1, 2, and 3) [e.g., *Shroba et al.*, 1979; *Brooks and Lawrence*, 1999]. The calculation of the spatial distribution of stream power in these types of geometries requires the use of a two-dimensional model [*Miller* 1995; *Denlinger et al.*, this volume]. The model used and refined for these studies is discussed in detail in *Denlinger et al.* [this volume]. *Denlinger et al.* [this volume] also present some example model output.

If stable reaches with applicable stratigraphy and geometry for setting paleohydrologic bounds can be identified during reconnaissance field studies, topographic mapping and hydraulic modeling can be conducted prior to detailed fieldwork for stratigraphy and radiocarbon sampling. This will allow the most economical use of field time by focusing fieldwork on reaches or bends that will provide the most useful paleohydrologic bound information. In a reach that contains more than one good site for developing paleohydrologic bounds, the stratigraphic and chronologic information can be collected and compared for internal consistency between sites.

Statistical Methods

Currently for paleoflood information to be applied to engineering problems, it is necessary to put the paleoflood information in a frequency context. Although there are numerous contradictions between physical processes of floods and their portrayal in flood frequency [e.g., *Baker*, 1994; *Baker et al.*, this volume], at the present hydrologic

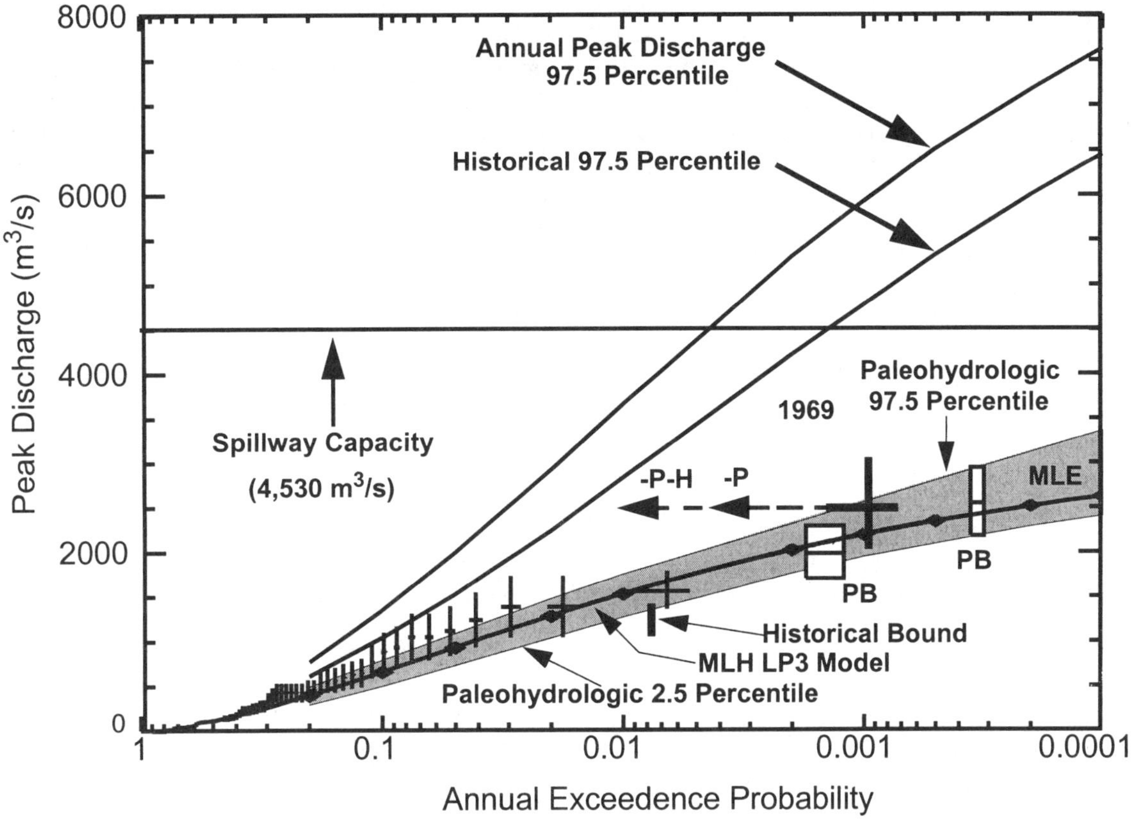

Figure 8. Flood frequency for the Santa Ynez River at Bradbury Dam [*O'Connell et al.*, in press]. Annual peak discharges are shown as horizontal lines for plotting position ranges and vertical lines for 2σ peak discharge estimate uncertainties. Paleohydrologic bounds (PB) are boxes denoting geologic age and discharge uncertainties. The 0.95 flood-frequency probability region using all data is shown in gray. The 97.5 percentile based only on the record of annual peak discharge estimates is also shown. The historical 97.5 percentile is the result of the calculation made using only historical information and the record of annual peak discharge estimates. Paleohydrologic 97.5 percentile that includes paleohydrologic bound information is much lower than the other two. The 1969 peak discharge is shown using thick lines and the dashed line arrows show the change in its plotting position when paleohydrologic (-P) and historical information (-P-H) are ignored.

risk analysis still requires the use of flood frequency information [e.g., *Reclamation*, 1999a]. This is true because safety of hydrologic structures is often based on a specific probability of failure, and quantitative information about hydrologic hazards is necessary input for basic risk analysis calculations. One of the most important effects of the use of paleoflood information in statistical analysis is that it can narrow the range of potential flood behavior at long return periods (Figure 8) [*O'Connell et al.*, 1997].

Incorporation of long-term paleohydrologic information in flood-frequency studies does not depend on being able to reconstruct the complete record of all past floods. Statistical techniques that can incorporate paleohydrologic bounds are a useful way to take advantage of paleohydrologic information [*Stedinger et al.*, 1988]. In this way, it is not important if floods of a specified recurrence are not recorded or included in the frequency analysis, although information on paleofloods helps constrain the location of possible flood frequency curves. What is important is that limits on flood magnitudes over time intervals can be identified. Sensitivity analyses show that the addition of only one or two paleohydrologic bounds that span a range of hundreds to thousands of years have a significant impact on the shape of the flood-frequency curve (Figure 8) [*Ostenaa et al.*, 1996; *O'Connell et al.*, 1997].

The statistical method developed and used in flood hazard studies at Reclamation is discussed in detail in *O'Connell et al.* [1997; 1998; and in press]. It employs a Bayesian approach and modified maximum likelihood methods of *Stedinger and Cohn* [1986] and *Stedinger et al.* [1988]. For flood frequency calculations, the information from the paleohydrologic bounds is combined with historical information, and annual peak discharge estimates. This approach allows for the accurate portrayal of the uncertainty associated with the establishment of values for discharge and duration of paleohydrologic bounds. Therefore the calculated statistical confidence reflects the confidence in field stratigraphic, numerical age, and hydraulic modeling information. At present, the best approach appears to be to obtain information on the magnitude and age of the largest paleofloods [e.g., *Blainey et al.*, this volume], and to develop bounds that closely bracket the limit on flood magnitude over a specific time interval [e.g., *O'Connell et al.*, in press]. This combination places strong constraints on the possible flood frequency functions at low probabilities.

LESSONS LEARNED

Just as important as the technique of developing paleohydrologic bounds is the experience gained from trying to apply this concept to numerous rivers of widely varying scales, from less than 10 km^2 to more than 40,000 km^2, in a variety of geographic settings in the Western United States (Figure 5). When trying to develop paleohydrologic bounds, geomorphic and stratigraphic limits on flood magnitude over time, the strict definition of what a paleohydrologic bound represents must be kept in mind. This may seem obvious and an oversimplification, but it is easy to go astray when trying to limit flood magnitude with stable surfaces. The development of a paleohydrologic bound requires the ability to demonstrate landscape stability over a time interval. This means there must be a time-dependent physical stratigraphic record that spans the time interval. Soil stratigraphy is probably the most powerful tool to establish stability. A collection of archeological artifacts or a collection of radiocarbon ages does not make a paleohydrologic bound. There must be stratigraphy indicative of stability that spans the interval of time. Keep in mind the basic concept of a paleohydrologic bound is to find a discharge that would erode or significantly modify an otherwise stable site that has time dependent soil stratigraphy.

As an outgrowth of these basic factors, paleohydrologic bounds are most easily developed on relatively high gradient streams in arid to semiarid regions. High-gradient streams work better because relatively shallow inundation of a terrace surface results in relatively higher stream power. This raises confidence in the evidence of landscape stability. Arid to semiarid regions are good locations for use of paleohydrologic bounds because the rates of pedogenesis are comparatively slower than more humid regions [*Birkeland*, 1999]. Where rates of pedogenesis are comparatively slow, the ability to recognize truncations in the soil profile is better (Figure 4). This is why the technique is highly effective for the majority of the Western United States.

In some situations, it is tempting to try to turn the concept of a paleohydrologic bound into one of a lack of deposition. There are a series of complexities introduced when the absence of deposits is used as a surrogate for a paleohydrologic bound. Basing paleohydrologic bounds on the potential thickness of nonexistent deposits is a difficult proposition at best. In order to develop a bound in this manner, the depositional site must be a nearly perfect stratigraphic recorder and the flood stratigraphy must be incontrovertible. The paleohydrologic bound concept is based on the predicted erosion or modification of a discharge-limiting site with time-dependent soil stratigraphy.

It is important to collect the right information for the conditions at the site and the intended use. The most important information for developing paleohydrologic bounds is the basic geomorphic and stratigraphic information. Before any other extensive analysis takes place, it must be clear how the geomorphic and stratigraphic information limits the magnitude of floods or documents the evidence of past floods. If the basic information is missing, the results of the analysis will always be open to question. In this way, it is important to be sure that there is geomorphic information to limit the magnitude of floods at a site before constructing detailed topographic maps and investing additional effort in two-dimensional hydraulic modeling and frequency analysis.

The primary benefit of two-dimensional modeling is to develop the spatial distribution of stream power and document predicted erosion of a stable discharge-limiting site with time dependent soil stratigraphy. For sites where erosional limits are not the goal, it is probably not cost-effective to do extensive topographic mapping, grid creation and manipulation, and two-dimensional hydraulic modeling. One-dimensional models work well for matching stage in geometrically simple reaches particularly if calibrated to large historical floods. One-dimensional modeling, critical depth calculations, or rating curve extrapolation may be all that is required at depositional sites with simple channel geometry [*Webb and Jarrett*, this volume].

USE IN ENGINEERING ANALYSIS

The driving force at Reclamation is to provide accurate information on the magnitude of low probability floods for use in assessing the safety of dams with respect to large floods. However this strategy is widely applicable. The paleohydrologic bound concept makes it portable to most streams in the Western United States. This type of study can be accomplished at different levels of effort depending on the goal (Figure 5). Simplified regional techniques may be useful indicators of regional flood potential [e.g., *Jarrett and Costa*, 1988; *Jarrett and Tomlinson*, 2000]. Most of the more complex tools have been developed for situations where millions of dollars are at stake. However, simplified analysis can be accomplished quickly. For example, reconnaissance paleohydrologic bound information can be used with a graphical extension of flood frequency. It is important to keep in mind the pitfalls and complications that can lead to spurious conclusions from a single site or an over-simplified analysis.

Paleoflood analysis only provides information on peak discharge. This may be all that is required for hydrologic structures that need to be designed or assessed for a specific discharge. However, in many engineering analyses particularly where there is some component of water storage, hydrographs are needed. The information derived from paleohydrologic bounds can be used to scale hydrographs, or it can be used to calibrate rainfall–runoff models.

SCIENTIFIC VALUE

Paleohydrologic bounds furnish direct insight into the time and magnitude of the largest floods on a stream, in a drainage basin, or in a region. When this information is combined with paleoflood records that provide ages and minimum discharge information, the two types of data bracket a river's possible flood history. This minimizes the influence of potential problems with flood record completeness, and alleviates many of the questions regarding the interpretation of individual floods. This maximizes the information content of the paleoflood record for understanding flood magnitude and frequency. Information on maximum flood magnitude provides insight into how rivers work, and how river geomorphology relates to flood magnitude. Paleohydrologic bounds also furnish valuable information on past flood magnitude where it was previously thought difficult to derive.

This knowledge of maximum flood magnitude may be of utility in climate change studies and analysis of time-dependent variations in flood frequency [e.g., *Baker et al.*, this volume; *Redmond et al.*, this volume]. Paleohydrologic bounds may span many cycles or episodes of climate change. Therefore, they offer an indication of potential maximum basin response even under different climatic conditions. This is a key question for studies of climate variability and flood magnitude.

The information paleohydrologic bounds provide should be directly applicable to the continued development of envelope curves of basin area versus discharge [*Baker et al.*, this volume] such as that by *Enzel et al.* [1993]. The information from paleohydrologic bounds can push envelope curves down from the top to help better position the curve. Instead of simply enveloping the data, the curve can be drawn between exceedance and non-exceedance information. Paleohydrologic bounds can also be developed rapidly over a large area, similar to *Jarrett* [1990]. This could facilitate the comparison of maximum flood magnitude between regions and help define the response of these regions to changes in climate.

FUTURE DEVELOPMENT

The formalization of the paleohydrologic bound technique is relatively new. The chief application has been for analysis of hydrologic dam safety. Often modification of a dam for larger design floods can be quite expensive. This expense justifies the cost of detailed paleoflood studies and other hydrologic studies [e.g., *Levish*, 1994; *Reclamation*, 1999b]. As awareness grows of this technique, it will gain wider acceptance in the engineering community. In addition, with advances in technology, the cost of this type of study will go down. As computerized photogrammetry and other methods for rapidly and inexpensively producing accurate topographic maps become more widely used they will also become less expensive and more readily available. As two-dimensional hydraulic models gain acceptance they will become more user-friendly, and creating and manipulating topographic grids will become easier. Additional studies in a wide variety of environments will help better define the criteria for setting bounds.

The results of studies based on paleohydrologic bounds should be directly compared with studies based on slackwater deposits to determine the relation between the results of different techniques. It may be most profitable to combine all types of data to bracket the Holocene flood record. In this way, paleohydrologic bounds can compensate for common difficulties associated with slackwater records by placing a cap on flood magnitude.

More studies on a regional level similar to *Jarrett and Tomlinson* [2000] and *Jarrett and Costa* [1988] will help with the regionalization of paleoflood information [e.g., *Stedinger and Baker*, 1987]. It is also important to understand the relative accuracy and precision of results based on different levels of effort. That is, do the results from reconnaissance-level paleoflood and paleohydrologic bound studies, where discharges are based on one or several cross sections and ages are based on regional soil correlations, differ substantially from the results of more detailed studies? In more detailed studies discharges are based on topographic mapping and two-dimensional modeling and the ages are based on radiocarbon analysis of stratigraphic sequences.

For engineering applications the most important issues for the near future are to find ways to effectively combine the results of paleoflood and paleohydrologic bound studies with more conventional hydrologic methods. This may include devising methods to scale paleoflood peak discharge information as input to hydrograph construction. For analysis of hydrologic risk, it is important to reconcile paleoflood frequency information with model constructed flood frequency such as the output from stochastic rainfall-runoff models. The consistency of results or the ability to combine results will yield a higher degree of assurance about the input to, and therefore the output of, risk analysis.

CONCLUSIONS

Dam safety risk analysis requires accurate information on the probability and magnitude of extreme floods. The most reliable way to obtain probability estimates of extreme floods is to study their geomorphic and stratigraphic record. Paleoflood hydrology effectively extends the length of the flood record in order to make realistic estimates of the magnitude of low probability floods. An expeditious and widely applicable technique to gain paleoflood information is through the use of paleohydrologic bounds. A paleohydrologic bound is a time interval during which a particular discharge has not been exceeded. The information derived from paleohydrologic bounds can be input directly into flood frequency analysis. This approach is appropriate for hazard assessment because paleohydrologic bounds provide a description of the limits of magnitude and timing of the largest floods. This description improves estimates of the magnitude of low probability floods. At present, it appears that the best approach is to combine information about the largest paleofloods and paleohydrologic bounds to constrain potential flood frequency behavior at low probabilities. As the engineering community continues to accept paleoflood techniques and information, these types of studies will become more widespread and less expensive.

Acknowledgments. The development of these concepts has benefited greatly from numerous discussions with Vic Baker, Kyle House, and Bob Jarrett over the past 8 years. Thought provoking discussions with Lisa Ely and Jim O'Connor helped refine the demonstration of landscape stability. Matt Jones made the innovations in topographic mapping possible. Jeanne Klawon prepared many of the figures. Ralph Klinger has provided many vital insights into fluvial geomorphology and Holocene geology. Dan O'Connell developed the Bayesian analysis, and refined the two-dimensional hydraulic model. I thank John England, Kyle House, Bob Jarrett, Jeanne Klawon, Ralph Klinger, and Jim O'Connor for their valuable review comments. The Dam Safety Office of the Bureau of Reclamation generously funded this research and development of these tools and techniques.

REFERENCES

Baker, V.R., Paleoflood hydrology and extraordinary flood events, *J. Hydrol., 96*, 79-99, 1987.

Baker, V.R., Magnitude and frequency of palaeofloods, in *Floods, Their Hydrological, Sedimentological and Geomorphological Implications* edited by K. Beven and P. Carling, pp. 171-183, John Wiley, Chichester, U.K., 1989.

Baker, V.R., Geomorphological understanding of floods, *Geomorphology, 10*, 139-156, 1994.

Baker, V.R., Paleoflood hydrology and the estimation of extreme floods, in *Inland Flood Hazards*, edited by E.E. Wohl, pp. 359-377, Cambridge University Press, Cambridge U.K., 2000.

Baker, V.R., and J.E. Costa, Flood power, in *Catastrophic Flooding*, edited by L. Mayer and D. Nash, pp.1-21, Allen and Unwin, Boston, 1987.

Baker, V.R., L.L. Ely, J.E. O'Connor, and J.B. Partridge, Paleoflood hydrology and design applications, in *Regional Flood Frequency Analysis*, edited by V.P. Singh, pp. 339-353, D. Reidel, Boston, 1987.

Baker, V.R., R.C. Kochel, and P.C. Patton, editors, *Flood Geomorphology*, 503 pp., John Wiley and Sons, New York, 1988.

Baker, V.R., R.C. Kochel, P.C. Patton, and G. Pickup, Paleohydrologic analysis of Holocene flood slack-water sediments, in *Modern and Ancient Fluvial Systems*, pp. 229-239, *Int. Assoc. Sed. Spec. Publ. 6*, 1983.

Birkeland, P.W., *Soils and Geomorphology*, 430 pp., Oxford University Press, New York, 1999.

Bowman, S., *Radiocarbon Dating*, 64 pp., University of California Press, Berkeley, 1990.

Brooks, G.R., and D.E. Lawrence, The drainage of the Lake Ha! Ha! Reservoir and downstream geomorphic impacts along Ha! Ha! River, Saguenay area, Quebec, Canada, *Geomorphology, 28*, 141 – 168, 1999.

Brown, A.G., *Alluvial Geoarchaeology: Floodplain Archaeology and Environmental Change*, 377 pp., Cambridge University Press, Cambridge, 1997.

Costa, J.E., Holocene stratigraphy in flood frequency analysis. *Water Resour. Res., 14*, 626-632, 1978a.

Costa, J.E., Colorado Big Thompson flood: geologic evidence of a rare hydrologic event, *Geology, 6*, 617-620, 1978b.

Costa, J.E., A history of paleoflood hydrology in the United States, 1800-1970, in *History of Hydrology,* edited by E.R. Landa and S. Ince, pp. 49-53, Amer. Geophys. Union, History of Geophysics, Vol. 3,Washington, D.C., 1987.

Costa J.E. and J.E. O'Connor, Geomorphically effective floods, in *Natural and Anthropogenic Influences in Fluvial Geomorphology*, edited by J.E. Costa, A.J. Miller, K.W. Potter, and P.R. Wilcock, pp. 45-56, Amer. Geophys. Union, Washington D.C., 1995.

Cudworth, A.G., *Flood Hydrology Manual*, 243 pp., U.S. Government Printing Office, Denver, 1989.

Dawdy, D.R., and D.P. Lettenmaier, Initiative for risk-based flood design, *J. Hydraul. Eng*., 113, 1041-1051, 1987.

Enzel, Y., L.L. Ely, P.K. House, V.R. Baker, and R.H. Webb, Paleoflood evidence for a natural upper bound to flood magnitudes in the Colorado River basin, *Water Resour. Res., 29*, 2287-2297, 1993.

Gilet-Blein, N., G. Marien, J. Evin, Unreliability of 14C dates from organic matter of soils, *Radiocarbon, 22*, 919-929, 1980.

Graham, W.J., Should dams be modified for the probable maximum flood?, *J. Amer. Water Resour. Assoc., 36*, 953-963, 2000.

Hereford, R., Climate and ephemeral-stream processes: twentieth-century geomorphology and alluvial stratigraphy of the Little Colorado River, Arizona, *Geol. Soc. Amer. Bull., 95*, 654-668, 1984.

Hosking, J.R., and J.R. Wallis, Paleoflood hydrology and flood frequency analysis, *Water Resour. Res.*, *22*, 543-550, 1986.

Hsieh, Y.P., Pool size and mean age of stable soils organic carbon in cropland, *Soil Sci. Soc. Amer. J.*, *56*, 460-464, 1992.

Hsieh, Y.P., Radiocarbon signatures of turnover rates in active soil carbon pools, *Soil Sci. Soc. of Amer. J.*, *57*, 1020-1022, 1993.

Jahns, R.H., Geologic features of the Connecticut Valley, Massachusetts, as related to recent floods, *U.S. Geol. Surv. Water Supply Pap. 996*, 158 pp., 1947.

Jarrett, R.D., Paleohydrologic techniques used to define the spatial occurrence of floods, *Geomorphology, 3*, 181-195, 1990.

Jarrett, R.D., Paleohydrology and its value in analyzing floods and droughts, *U.S. Geol. Surv. Water Supply Pap. 2375*, 105-116, 1991.

Jarrett, R.D., and J.E. Costa, Evaluation of the flood hydrology in the Colorado Front Range using precipitation, streamflow, and paleoflood data, *U.S. Geol. Surv. Water Resour. Invest. Rep. 87-4117*, 37 pp., 1988.

Jarrett, R.D., and E.M. Tomlinson, Regional interdisciplinary paleoflood approach to assess extreme flood potential, *Water Resour. Res.*, *10*, 2987 – 2984, 2000.

Kochel, R.C., and V.R. Baker, Paleoflood analysis using slack-water deposits, in *Flood Geomorphology,* edited by V.R. Baker, R.C. Kochel, and P.C. Patton, pp. 357-376, John Wiley and Sons, New York, 1988.

Lane, W.L., Paleohydrologic data and flood frequency estimation, in *Regional Flood Frequency Analysis*, edited by V.P. Singh, pp. 287-298, D. Reidel, Boston, 1987.

Lapointe, M.F., Y. Secretan, S.N. Driscoll, N. Bergeron, and M. Leclerc, Response of the Ha! Ha! River to the flood of July 1996 in the Saguenay Region of Quebec: large-scale avulsion in a glaciated valley, *Water Resour. Res., 34*, 2383 – 2392, 1998.

Lave, L.B., D. Resendiz-Carrillo, and F.C. McMichael, Safety goals for high-hazard dams: Are dams too safe?, *Water Resour. Res.*, *26*, 7, 1383-1391, 1990.

Levish, D.R, J.E. England Jr., J.E. Klawon, D.R.H. O'Connell, *Flood Hazard Analysis for Glendo and Seminoe Dams, Kendrick Project and Pick Sloan Missouri Basin Program, Wyoming,* Bureau of Reclamation Draft Report, Denver, 2000.

Levish, D.R., and D.A. Ostenaa, *Applied Paleoflood Hydrology in North-Central Oregon*, 182 pp., Guidebook for Fieldtrip 2, Geol. Soc. Amer. Cordill. Sect., Portland, OR, 1996.

Levish, D., D. Ostenaa, D. O'Connell, A non-inundation approach to paleoflood hydrology for the event-based assessment of extreme flood hazards, *Amer. Soc. Dam Safety Offic. Ann. Conf. Proc., 11*, 69-82, 1994.

Levish, D.R., D.A. Ostenaa, and D.R.H. O'Connell, Paleoflood hydrology and dam safety, in *Waterpower '97*, pp. 2205-2214, Proc. Intl. Conf. Hydropower, Aug. 5-8, 1997, Atlanta, GA, 1997.

Magilligan, F.J., Thresholds and the spatial variability of flood power during extreme floods, *Geomorphology, 5*, 373-390, 1992.

Mansfield, G.R., Flood deposits of the Ohio River, January-February 1937, a study of sedimentation, pp. 693-736, *in* Grover, N.C., *U.S. Geol. Surv. Water Supply Pap. 838,* 1938.

McCalpin, J.P, editor, *Paleoseismology,* 583 pp., Academic Press, San Diego, CA, 1996.

McWeeney, L., What lies lurking below the soil: beyond the archaeobotanical view of floatation samples. *North Amer. Arch., 10*, 227-230, 1989.

Miller A.J., Valley morphology and boundary conditions influencing spatial patterns of flood flow, in *Natural and Anthropogenic Influences in Fluvial Geomorphology*, edited by J.E. Costa, A.J. Miller, K.W. Potter, and P.R. Wilcock, pp. 57-81, Amer. Geophys. Union, Washington D.C., 1995.

Moody, J.A., J.E. Pizzuto, and R.H. Meade, Ontogeny of a flood plain, *Geol. Soc. Amer. Bull., 111*, 291-303, 1999.

Nanson, G.C., Episodes of vertical accretion and catastrophic stripping: a model of disequilibrium flood-plain development, *Geol. Soc. Amer. Bull., 97*, 1467-1475.

O'Connell, D.R.H., D.R. Levish, and D.A. Ostenaa, Bayesian flood frequency analysis with paleohydrologic bounds for late Holocene paleofloods, Santa Ynez River, California, in *Twenty Years Later: What Have We Learned Since the Big Thompson Flood*, edited by E. Gruntfest, pp. 183-196, Natural Haz. Res. Appl. Info. Center Spec. Pub., 33, University of Colorado, Boulder, CO, 1997.

O'Connell, D.R.H., D.R. Levish, D.A. Ostenaa, Risk-based hydrology: Bayesian flood-frequency analyses using paleo-flood information and data uncertainties, in *Proc. First Fed. Interagency Model. Conf.*, pp. 4-101 – 4-108, Subcomm. on Hydrol., 1998.

O'Connell, D.R.H., D.A. Ostenaa, D.R. Levish, and R.E. Klinger, Bayesian flood frequency analysis with paleohydrologic bound data, *Water Resour. Res.*, in press.

Ostenaa, D.A., D.R. Levish, and D.R.H. O'Connell, *Paleoflood Study for Bradbury Dam, Cachuma Project, California*, 86 pp., Bureau of Reclamation Seismotectonic Report 96-3, Denver, CO,1996.

Partridge, J., and V.R. Baker, V.R., Paleoflood hydrology of the Salt River, Arizona, *Earth Surf. Proc. and Landf., 12*, 109-125. 1987.

Patton, P.C., Measuring the rivers of the past: A history of fluvial paleohydrology, in *History of Hydrology,* edited by E.R. Landa and S. Ince, pp. 55-67, American Geophysical Union, History of Geophysics, No. 3, Washington, DC, 1987.

Patton, P.C., V.R. Baker, and R.C. Kochel, Slackwater deposits: A geomorphic technique for the interpretation of fluvial paleohydrology, in *Adjustments of the Fluvial System*, edited by D.D. Rhodes and G.P. Williams, pp. 225-252, Kendall-Hunt, Dubuque, IO, 1979.

Puseman, K., L. Ruggerio, and C.N. Thornberry, Examination of bulk sediment and detrital charcoal samples from study sites in the North Platte River project, Wyoming, 62 pp. *Paleo Res. Labs Tech. Rep. 99-92*, 1999.

Reclamation, *Guidelines for Achieving Public Protection in Dam Safety Decision Making,* 19 pp., Unpub. report dated April 4, 1997, Bureau of Reclamation, Denver, 1997.

Reclamation, *Dam safety risk analysis methodology*, 52 pp., Unpub. report dated September, 1999, Bureau of Reclamation, Denver, 1999a.

Reclamation, *A framework for characterizing extreme floods for dam safety risk assessment,* 67 pp. Unpub. report dated November, 1999, Bureau of Reclamation, Denver, 1999b.

Stafford, T.W. Jr., Radiocarbon chronostratigraphy, in *Wilson-Leonard: An 11,000-year Archeological Record of Hunter-Gatherers in Central Texas*, edited by M.B. Collins, pp. 1039 – 1066, Texas Archeological Research Laboratory Studies in

Archeology 31, University of Texas, Austin, TX, 1998.

Stafford, T.W. Jr., Accelerator radiocarbon dates on charcoal, shell, and human bone from the Del Mar Site, California, *Amer. Antiq., 54*, 389-395.

Schumm, S.A, and R.W. Lichty, Channel widening and flood-plain construction along the Cimarron river in southwestern Kansas, *U.S. Geol. Surv. Prof. Pap. 352-D*, 88 pp., 1963.

Shroba, R.R., P.W. Schmidt, E.J. Crosby, and W.R. Hansen, Geologic and geomorphic effects in the Big Thompson Canyon area, Larimer County, *U.S. Geol. Surv. Prof. Pap. 1115*, pp. 87-152, 1979.

Stedinger, J.R., and V.R. Baker, Surface water hydrology: historical and paleoflood information, *Rev. Geophys., 25*, 119-124, 1987.

Stedinger, J.R., and T.A. Cohn, Flood frequency analysis with historical and paleoflood information, *Water Resour. Res., 22*, 273-286, 1986.

Stedinger, J.R., R Surani, R. Therivel, *Max user's guide: a program for flood frequency analysis using systematic-record, historical, botanical, physical paleohydrologic and regional hydrologic information using maximum-likelihood techniques*, 51 pp., Dept. Environ. Engin., Cornell Univ., Ithaca, NY, 1988.

Stuvier, M., and P.J. Reimer, Extended C-14 data base and revised calib 3.0 C-14 age calibration program, *Radiocarbon, 35*, 215-230, 1993.

Trumbore, S.E., Radiocarbon geochronology, in *Quaternary Geochronology: Methods and Applications*, edited by J.S. Noller, J.M. Sowers, and W.R. Lettis, pp. 41-60, American Geophysical Union, Washington D.C., 2000.

D.R. Levish, Bureau of Reclamation, Flood Hydrology Group, P.O. Box 25007, D-8530, Denver, CO 80225.

Cosmogenic ^{3}He Ages and Geochemical Discrimination of Lava–Dam Outburst-Flood Deposits in Western Grand Canyon, Arizona

Cassandra R. Fenton, Thure E. Cerling, and Barbara P. Nash

University of Utah, Salt Lake City, UT

Robert H. Webb

U. S. Geological Survey, Tucson, AZ

Robert J. Poreda

University of Rochester, Rochester, NY

Quaternary terrace gravels almost entirely composed of basalt clasts are preserved on basalt ledges in western Grand Canyon between river miles 185 and 222. Sedimentological evidence preserved in these gravels suggests that they were emplaced by the catastrophic failure of lava dams, leading to deposition at elevations that decreased abruptly with increasing distance from potential damsites near the Uinkaret volcanic field (between river miles 179 and 188). At least two lava dams on the Colorado River may have failed in the last 160 ka, and deposits that correlate on the basis of exposure age and chemistry are preserved up to 26 river miles (43 km) downstream from the damsites in western Grand Canyon. These possible outburst–flood deposits range in height from 53 – 200 m above current river level. The deposits generally consist of a poorly–sorted basal hyaloclastite ash overlain by poorly–sorted, sub–angular to well-rounded basalt boulders, cobbles, and pebbles. Clast size and deposit thickness decrease with increasing distance from the damsites. Cosmogenic ^{3}He exposure ages and chemistry of whole rock basalt, basalt glass, and olivine phenocrysts in the deposits document at least two distinct flood deposits. The best preserved deposit has an average surface exposure age of 159±2 ka at river miles 189.5 and 202; the second flood deposit, at river mile 204.4, has a problematic exposure age of 66±6 ka and is likely older than 160 ka. A terrace cut into a pre–existing gravel deposit at river mile 204.6 has an exposure age of 100±6 ka. In the Uinkaret volcanic field, the

Ancient Floods, Modern Hazards: Principles
and Applications of Paleoflood Hydrology
Water Science and Application Volume 5, pages 191-215

Whitmore Cascade (177±9 ka), Younger Cascades (108±11), and Bar Ten (88±6 ka) lava flows were investigated as potential lava dams correlative with the outburst–flood deposits. Channel morphology and clast size can be used for future analysis of the flood wave that propagated downstream after the dam failures.

INTRODUCTION

Basaltic lava flows have dammed several rivers in the western United States. Intracanyon lava dams have been studied in Idaho and Oregon [*Malde*, 1982; *Howard et al.*, 1982; *Stearns*, 1931; *Peterson and Groh*, 1970], in Washington [*Smith*, 1903] and in western Grand Canyon in Arizona [*McKee and Schenk*, 1942; *Hamblin*, 1994]. Although several studies mention lava–dam failure in the western Grand Canyon [*Dutton*, 1882; *McKee and Schenk*, 1942; *Hamblin*, 1994; *Lucchitta et al.*, 2000], none of them report outburst–flood deposits resulting from catastrophic failure.

The Uinkaret volcanic field covers a large area mainly on the north rim of western Grand Canyon and was the source of basalt flows that cascaded into the Colorado River (Figure 1). *Alibert et al.* [1986], *Best and Brimhall* [1974], *Fitton* [1989], *Leeman*, [1974], and *Wenrich et al.* [1995] provide petrologic studies of basalts that broadly range a continuum of compositions in western Grand Canyon. *Wenrich et al.* [1997] mapped the distribution of basalt in this field but did not differentiate flow events; however, it has been shown that basalt flows become progressively younger towards the northeast side of this field [*Wenrich et al.*, 1995; *Fenton et al.*, in review].

Basalt flows and associated lava dams have been previously dated using K-Ar and Ar-Ar techniques [*McKee et al.*, 1968; *Hamblin*, 1994; *Wenrich et al.*, 1995; *Dalrymple and Hamblin*, 1998; *Lucchitta et al.*, 2000]. Few lava flows younger than 200 ka have been identified and dated using K-Ar or $^{39}Ar/^{40}Ar$ methods in the Uinkaret volcanic field. Young basalts may have anomalously old K-Ar or $^{39}Ar/^{40}Ar$ ages because of excess ^{40}Ar associated with magmatic fluid inclusions, glassy groundmass and/or low potassium concentrations [*Damon et al.*, 1967; *Dalrymple and Hamblin*, 1998]. Anomalously old ages for young basalts in other volcanic fields have been attributed to excess argon and low potassium concentrations [*Dalrymple and Hirooka*, 1965; *Laughlin et al.*, 1994]. The youngest, previously dated lava cascade is the Esplanade Dam with a K-Ar age of 110±53 ka, which *Dalrymple and Hamblin* [1998] do not consider a reliable date. Most K–Ar dates of western Grand Canyon lava dams range from 400 to 800 ka [*Dalrymple and Hamblin*, 1998]. Cosmogenic ^{3}He dating provides an alternative to K-Ar and $^{39}Ar/^{40}Ar$ dating in western Grand Canyon, particularly for relatively young (<200 ka) basalt flows [*Fenton et al.*, in review].

The most recent volcanism in the Uinkaret volcanic field created the lava dams in Grand Canyon. According to *Hamblin* [1994], lava flows cascaded into the inner gorge of western Grand Canyon and dammed the Colorado River at least 13 times in the last 1.2 Ma. Most of these dams formed between Toroweap Valley (river mile 179) and Whitmore Canyon (river mile 188), which are the major fault–controlled drainages leading from the Uinkaret volcanic field to the Colorado River (Figure 1). Lava filled the drainages to the level of the Esplanade sandstone; the lavas also cascaded over the rim into the inner gorge between these two drainages. According to *Hamblin* [1994], these dams ranged from 60 to 600 m in height, formed reservoirs of substantial water volume, and based on rates of waterfall migration at Niagara Falls, were stable for up to 40 ka. For destruction of the dams, *Hamblin* [1994] proposed a conceptual model that relied on waterfall-induced headward erosion through the dam.

Basalt–rich Quaternary terrace gravels in western Grand Canyon [*Huntoon et al.*, 1981] differ significantly from typical Pleistocene Colorado River gravels upstream of the Uinkaret volcanic field. We present sedimentological and chemical evidence that suggests these western Grand Canyon terrace gravels may be outburst–flood deposits resulting from the catastrophic failures of the most recent lava dams in western Grand Canyon. We hypothesize that basalt poured down drainages and/or over the canyon rim onto talus slopes and poorly sorted alluvium. This should have formed inherently unstable dams. The possibly unstable foundation and abutments, combined with hydrothermally induced fracturing of the basalt upon contact with the river, may have created optimal conditions for catastrophic dam failure. In this paper, we review the sedimentological evidence for catastrophic floods caused by failures of lava dams in western Grand Canyon and use geochemical correlations and cosmogenic-dating techniques to distinguish two outburst floods that occurred in the last 160 ka.

LAVA–DAM OUTBURST–FLOOD DEPOSITS

We investigated several deposits mapped by *Huntoon et al.*, [1981] as Quaternary terrace gravels between river miles 188 and 222, west and downstream of Lava Falls Rapid (river mile 179) (Figures 1 and 2). At river miles 189.5, 202, 204.4 and 204.6, we identified well–preserved deposits that are almost entirely composed of basalt boul-

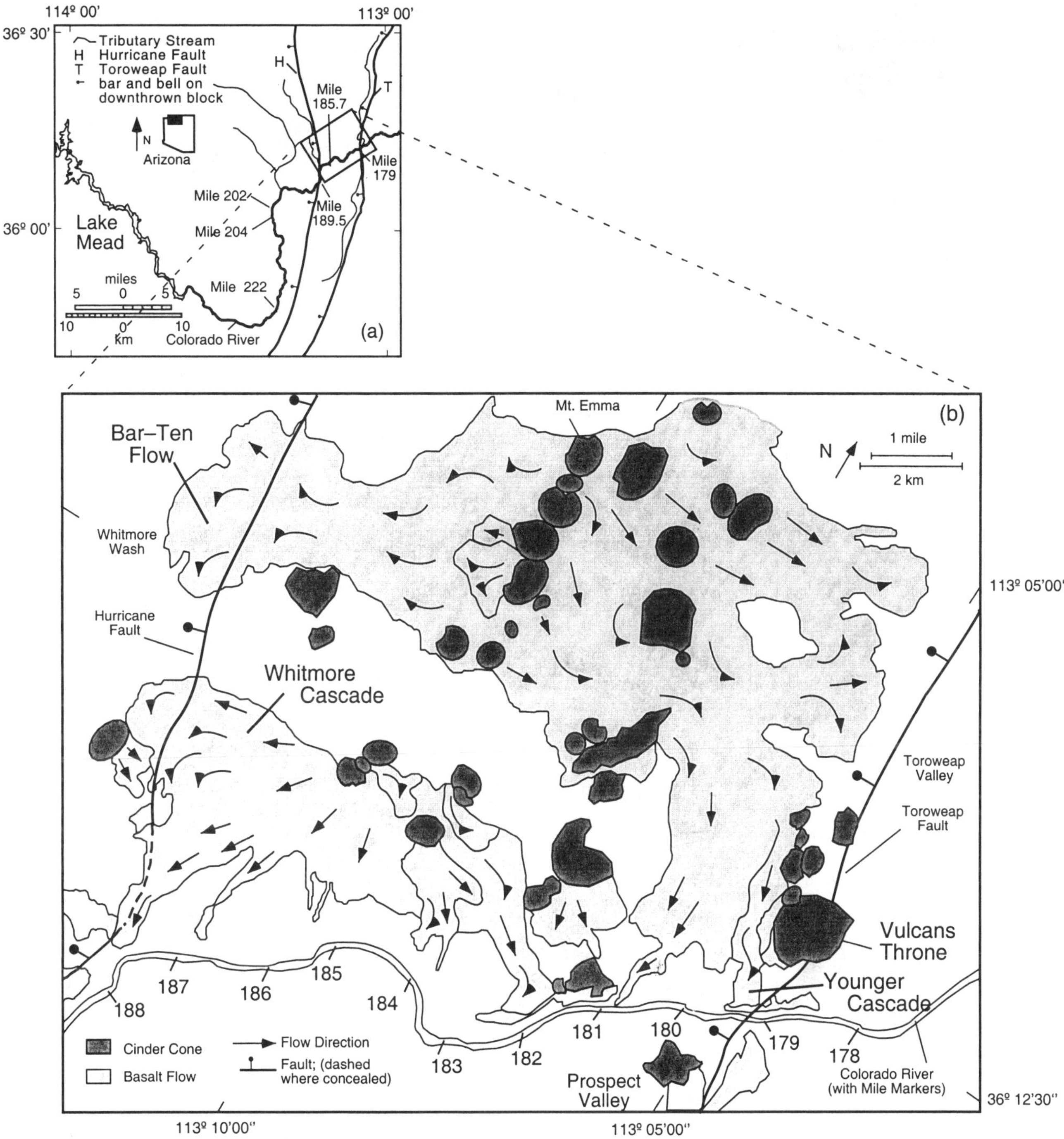

Figure 1. (a) Outburst–flood deposits are found along the Colorado River in northwestern Arizona between river miles 185 and 222. Lava–dam sites are mostly in the vicinity of the Uinkaret volcanic field. (b) Regional map of basalt flows produced in the southernmost portion of the Uinkaret volcanic field between Toroweap Valley (river mile 179) and Whitmore Wash (river mile 188) [adapted from *Hamblin*, 1994]. Arrows indicate the flow direction preserved in basalt flows; the prevalent flow direction is toward the Inner Gorge and the Colorado River.

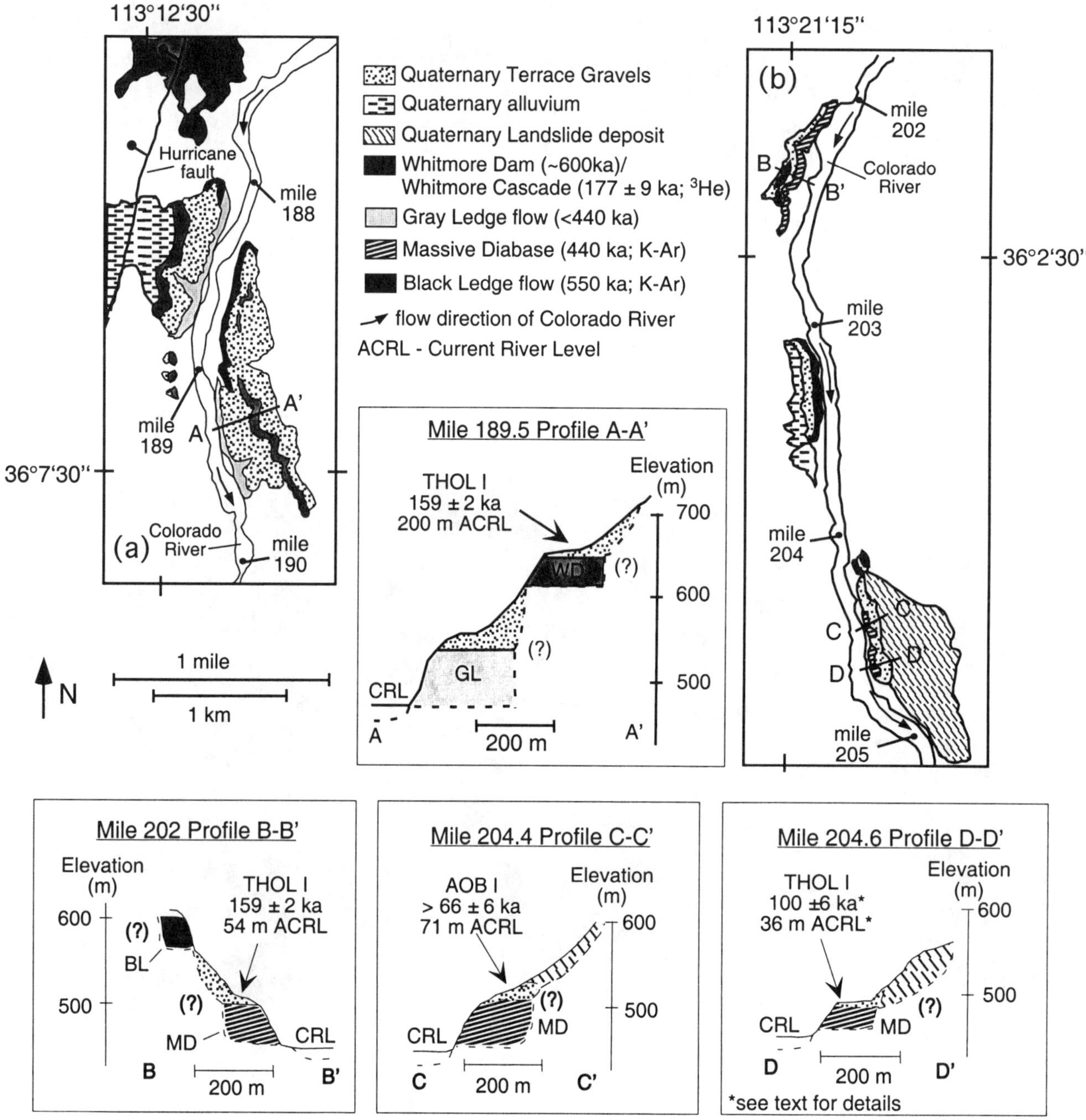

Figure 2. Maps of lava flows and Quaternary terrace gravels [adapted from *Huntoon et al.*, 1981 and *Hamblin*, 1994] (a) between Whitmore Wash (river mile 188) and river mile 190 and (b) between river miles 202 and 205. The scale and north arrow apply to both maps. Outburst–flood deposits were identified at river miles 189.5, 202, 204.4 and 204.6. Topographic profiles of flood deposits are given for transects A–A', B–B', C–C' and D–D'. In the profiles, the dashed lines and question marks represent the unknown lateral extent of the basalt flows and outburst–flood deposits. Due to the nature of the deposits and flows, it is difficult to determine contacts. The average ^{3}He exposure ages, chemistry of basalt glasses found in the deposits, and the maximum height above current level (ACRL) are listed beside each profile. AOB I=Alkaline Olivine Basalt I; THOL I = Tholeiitic Basalt I; see Figure 8 for details. The K–Ar ages and names of lava dams are those taken from *Dalrymple and Hamblin* [1998].

ders and cobbles and thick deposits of hyaloclastite tuffs. Geochemical and cosmogenic analyses are reported for these four deposits. The extant stratigraphy is preserved in point-bar or eddy deposits that overlie flat surfaces formed by the older Whitmore Dam, Black Ledge, Gray Ledge, and Massive Diabase lava flows [flows named by *Hamblin*, 1994] (Figure 2; Plate 1, for color version see CD).

Point counts were conducted on the surfaces and natural vertical exposures, where possible, of the basalt–rich gravels; the basalt content in these deposits decreases downstream from 98% near the damsites at river mile 189 to 82% in a deposit 18 river miles downstream (Figure 3). With increasing distance from the lava dams near the Uinkaret volcanic field, the basalt-rich deposits contain an increasing percentage of local Paleozoic clasts that were likely eroded from the canyon walls and the riverbed. Typical Colorado River gravels (e.g. extra–local porphyry and other igneous lithologies, quartzites, and gravels and clasts from local Paleozoic section) are found in western Grand Canyon [*Lucchitta et al.*, 2000]; however, these lithologies are excluded from the monolithological deposits except for small percentages of Muav limestone and Esplanade sandstone, which dominate the Paleozoic section of western Grand Canyon. Modern Colorado River gravels have a maximum of 60% basalt in gravel bars proximal to the Uinkaret volcanic field and basalt content decreases exponentially with increasing distance from the volcanic field (Figure 3).

Elevations of the basalt–rich deposits in western Grand Canyon were either measured using a total station or estimated using 1:24,000 topographic maps (Tables 1 and 2; Figure 4) and are reported as heights above current river level (ACRL). The maximum heights of the deposits decrease abruptly near potential damsites and continue to decrease with increasing distance downstream (Figure 4). The maximum heights range from 200 to 53 m ACRL, between river miles 189.5 and 205, respectively. These elevations are signigicantly different than Pleistocene river gravels found upstream. Typical Pleistocene main–stem Colorado River gravels comprise terraces preserved between river miles 66 and 72. The gravels are found at 36, 50, and <60 m ACRL, and have uranium–trend dates of 75±15, 150±30, and 210±25 ka, respectively [*Machette and Rosholt*, 1989].

Clast size of basalt–rich deposits decreases with increasing distance from the sites of dam failure (Figure 5). A deposit at river mile 188.5 has scattered basalt boulders up to 35 m in b-axis diameter. At river mile 204.6, the largest basalt boulder found in a monolithological deposit has a 4 m b–axis diameter. In deposits proximal to the Uinkaret volcanic field Large subangular to subrounded basalt boulders and angular limestone blocks up to 8 m in b–axis diameter are imbricated in a downstream direction. This indicates emplacement by a large–scale flood. In contrast, the largest boulders found in modern Colorado River gravels in western Grand Canyon were between 3-5 m in b–axis diameter at river miles 188 and 205.6, near Whitmore and 205–Mile Canyons, respectively. It is likely these large boulders were reworked into modern main–stem gravel bars after debris–flow deposition from the side canyons [*Webb et al.*, 1989]. The boulder size of modern Colorado River gravels does not decrease with increasing distance downstream (Figure 5), due to contributions of coarse sediment from side canyons along the length of the Colorado River [*Webb et al.*, 1989].

Basalt–rich gravel deposits closest to the Uinkaret volcanic field are at least 40 to 50 m thick, dominantly consisting of layers of clast–supported coarse angular limestone and subrounded basalt boulders and cobbles. The base of a deposit at river mile 188.5 has an angular matrix–rich basalt pebble and basalt glass lapilli breccia. The matrix contains basalt sand and basalt glass ash. *Hamblin* [1994; Figure 64D] describes large–scale foresets in one of these coarse gravel deposits at river mile 187.5 and states that this is the most convincing evidence for catastrophic flooding in the canyon during this period of time.

In distal deposits, there is commonly several meters of poorly–sorted hyaloclastite ash and lapilli with scattered pillow–basalt fragments, preserved as reworked hyaloclastite tuffs with large–scale foresets (Plate 2, for color version see CD). Well–rounded sand–size quartz grains are present in the tuffs but rare. Hyaloclastite (fragmented basalt glass) forms when basaltic lava is quenched in cooler water [*Naydu*, 1964]. Ideal conditions for formation of hyaloclastite were present when basalt flows encountered the Colorado River. *Hamblin* [1994] observed hyaloclastite deposits but referred to them as laminated tephra.

The hyaloclastite tuff is overlain by clast–supported and intermittently matrix–supported rounded to well–rounded basalt boulders, cobbles, and pebbles (Plates 1 and 2). The surfaces of distal deposits consist of clast-supported, rounded to well–rounded well–varnished basalt boulders and cobbles with interstitial hyaloclastite in a well–developed desert pavement (Plate 1).

The thickness of each basalt–rich deposit decreases with increasing distance from the Uinkaret volcanic field. The deposits range from 50–22 m in *minimum* thickness between river miles 190 and 204.6, respectively (Figure 4). A paleo–Colorado River channel at river mile 194 is filled with hyaloclastite ash overlain by basalt boulders and cobbles for a total thickness of 113 m. The lowest exposure of the deposit is 17 m above current river level (Figure 4), indicating the maximum elevation of the riverbed at the time of deposition. This is the thickest continuous sequence of hyaloclastite tuff and basalt boulders and cobbles identified in western Grand Canyon to date.

Plate 1. Downstream view of the Colorado River and the surface of an outburst–flood deposit preserved at river mile 202. The deposit has an exposure age of approximately 160 ka. The photographer is standing on the surface of the deposit, approximately 54 m above the river. The boulder in the foreground is approximately 1 m long. The prominent basalt flow in the distance is the Black Ledge flow of *Hamblin* [1994] near river mile 203. The top of the flow is roughly 45 m above the river. For color version see CD. (Photograph by Dominic Oldershaw.)

Stratigraphy of the basalt–rich deposits alone offers no clues for distinguishing different deposits of different origin and age. There is no correlation between a particular underlying lava flow and the flood deposit that caps it; for example, one outburst–flood deposit overlies both *Hamblin's* [1994] Whitmore Dam and Massive Diabase basalt flows at river miles 189.5 and 202, respectively.

In summary, basalt–rich flood deposits found between river miles 185 and 205 have characteristics atypical of normal Colorado River gravels. They (1) exclude extra–local lithologies; (2) are dominantly composed of basalt boulders, cobbles, pebbles and glass (hyaloclastite tuffs); (3) have scattered, imbricated large angular limestone blocks and large subrounded basalt boulders up to 8 and 35 m in b–axis diameter, respectively, in proximal deposits; (4) have distal deposits that contain poorly–sorted reworked hyaloclastite tuff with large–scale, randomly–oriented foresets overlain by a unit composed of either well–rounded basalt boulders or cobbles. We hypothesize that these types of deposits, local to western Grand Canyon, represent deposition during large–scale flood events produced by the failures of lava dams near the Uinkaret volcanic field. Hereafter, we refer to these deposits as lava–dam outburst–flood deposits.

CONCEPTUAL MODEL OF LAVA DAM FAILURE

Unlike *Hamblin's* [1994] model of lava dam construction, we propose that some lava dams were unstable because the bases and abutments rested on loose talus and debris-flow deposits, probably similar to those present today at Lava Falls Rapid [*Webb et al.*, 1999]. In addition, each structure likely had numerous macro- and microfissures that resulted from hydrothermal fracturing of basalt that cooled rapidly as it interacted with rising reservoir levels. We propose the following scenario to explain the sequence of deposition of the outburst-flood deposits discussed in this paper.

Basaltic lava cascaded into Grand Canyon, interacting with Colorado River water. Initially, most of the basalt was rapidly quenched by the river, forming hyaloclastite and pillow basalts that were formed and deposited in the dam's forebay. Presumably, significant amounts of hyaloclastite would not form downstream of the dam because the river's flow would be impounded behind the rising dam. As the volume of basalt increased, its discharge rate into the canyon must have been high enough to create an impoundment that had an asymmetric longitudinal profile, as originally proposed by *Hamblin* [1994]; the downstream end of the lava dam would have been thinner than the upstream end. The heights of the impoundments are unknown.

Failure began either by undermining caused by high pore pressures and flow under the base or abutments of the dam or by piping through fracture planes in the basalt. The first evidence of failure is preserved in outburst–flood deposits as reworked hyaloclastite tuffs with large–scale cross–bedding. The cross–bedding indicates large flow rates capable of entraining high sediment volumes from the riverbed and/or reservoir forebay. Hyaloclastites may have been previously deposited in the river corridor and may have been resuspended and deposited during the onset of a lava–dam failure.

A large percentage of each lava dam had to fail catastrophically to produce coarse, poorly–sorted deposits, almost debris–flow in appearance in places, that consist almost entirely (98%) of basalt particles near the damsites. Overtopping of a dam, which was the proposed failure mechanism of *Hamblin* [1994], would not rapidly mix water and sediment sufficiently to create a deposit of this type, nor would lapilli or ash be present in the matrix. The presence of angular limestone boulders in proximal outburst–flood deposits suggest failure may have begun on the abutments. Dam failures generally carry heterogeneous sediment loads of dam material [*Singh*, 1996]; in the case of Grand Canyon lava dams, 82–98% of the particles in the outburst–flood deposits are basalt. The coarsest, most poorly–sorted deposits occur proximally to the Uinkaret

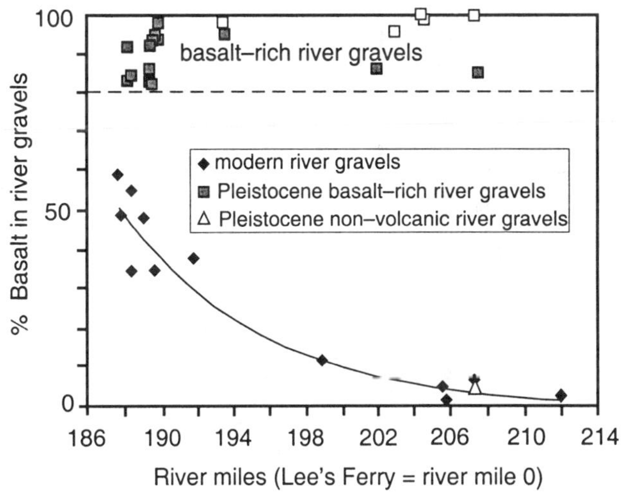

Figure 3. Basalt content of typical Colorado River gravels (both modern and Pleistocene) and basalt–rich river gravels in western Grand Canyon. The basalt–rich river gravels have greater than 80% basalt. Modern river gravels have a maximum of 60% basalt proximal to the Uinkaret volcanic field (river miles 179-188). Solid and hollow squares indicate point counts of vertical exposures and surfaces of basalt–rich gravel deposits, respectively. The Pleistocene non–volcanic river gravels, counted at river mile 207.3 are those described in *Lucchitta et al.* [2000] as Cycle III, with a Stage II carbonate development calibrated between 80 and 100 ka.

Table 1. ^{3}He exposure ages and locations, elevations and sample types of lava-dam outburst-flood deposits and lava flows in western Grand Canyon.

Event	Sample	Latitude (N)	Longitude (W)	Elevation/ Height ACRL[a] (m)	Phase[b]	Mass (g)	R/R_A[c] crush	R/R_A[c] fusion	^{4}He □ analytical error[d] (10^9 atoms/g)	$^{3}He_c$ □ analytical error[d,e] (10^6atoms/g)	%$^{3}He_c$	Production rate[f] (atoms/g/yr)	Cosmogenic $^{3}He_c$ age □ uncertainty[g] (ka)	Average $^{3}He_c$ age □1σ[h] (ka)
MILE 189.5 AND MILE 202 OUTBURST–FLOOD DEPOSIT: 157±2 ka														
Mile 202 outburst–flood deposit														
	96-AZ-238-GCRT[i]	36.05	113.35	511 / 54.2	ol	0.5161	5.04	79.31	151.81±7.59	16.68±0.83	95.7	145	115±8[j]	158±1[k]
	96-AZ-239-GCRT[i]	36.05	113.35	511 / 54.2	ol	0.4802	5.04	59.93	266.39±13.32	22.95±1.15	99.5	145	159±11	
	96-AZ-240-GCRT[i]	36.05	113.35	511 / 54.2	ol	0.3145	5.04	244.64	65.94±3.30	22.93±1.15	98.4	145	158±11	
	96-AZ-240-GCRT[i]	36.05	113.35	511 / 54.2	ol	0.4951	5.04	296.24	53.69±2.68	22.67±1.13	98.3	145	157±11	
	96-AZ-241-GCRT[i]	36.05	113.35	511 / 54.2	ol	0.5228	5.04	87.91	188.07±9.40	22.91±1.15	95.7	145	158±11	
Mile 189.5 upper outburst–flood deposit														
	99-AZ-623-GCRT[i]	36.13	113.28	685 / 200	ol	0.3206	5.04	275.75	75.96±3.80	29.87±1.49	98.8	185	161±11	161±11[l]
	99-AZ-624-GCRT[i]	36.13	113.28	685 / 200	ol	0.2913	5.04	213.99	85.03±4.25	25.78±1.29	98.1	185	139±10[j]	
	99-AZ-625-GCRT[i]	36.13	113.28	685 / 200	ol	0.3241	5.04	104.16	159.77±7.99	23.03±1.15	95.8	185	125±9[j]	
													AVERAGE	159±2[m]
MILE 204.6 OUTBURST–FLOOD DEPOSIT: 99±6 ka														
	93 LVA FLS 26[i]	36.01	113.35	490 / 36	ol	0.4495	5.79	193.47	85.78±4.29	13.51±0.68	97.9	142	95±7	100±6[k]
	93 LVA FLS 27[i]	36.01	113.35	490 / 36	ol	0.4954	5.79	126.01	88.81±4.44	14.32±0.72	96.2	142	101±7	
	93 LVA FLS 28[i]	36.01	113.35	490 / 36	ol	0.4655	5.79	69.96	71.71±3.59	12.97±0.65	92.5	142	91±6	
	93 LVA FLS 29[i]	36.01	113.35	490 / 36	ol	0.4030	5.79	77.58	29.69±1.48	10.46±0.52	93.8	142	74±5[j]	
	93 LVA FLS 29[i]	36.01	113.35	490 / 36	ol	0.4551	5.79	85.53	51.21±2.56	10.29±0.51	94.6	142	72±5[j]	
	93 LVA FLS 30[i]	36.01	113.35	490 / 36	ol	0.5377	5.79	263.00	33.03±1.65	15.04±0.75	97.9	142	106±7	
	93 LVA FLS 31[i]	36.01	113.35	490 / 36	ol	0.4754	5.79	48.41	90.37±4.52	14.80±0.74	88.9	142	104±7	
	93 LVA FLS 32[i]	36.01	113.35	490 / 36	ol	0.4323	5.79	158.05	94.19±4.71	15.18±0.76	97.2	142	107±7	
	93 LVA FLS 33[i]	36.01	113.35	490 / 36	ol	0.4329	5.79	139.41	29.28±1.46	15.27±0.76	96.6	142	107±7	
	93 LVA FLS 34[i]	36.01	113.35	490 / 36	ol	0.4683	5.79	39.84	181.25±9.06	13.08±0.72	86.8	142	92±7	
	93 LVA FLS 35[i]	36.01	113.35	490 / 36	ol	0.4669	5.79	124.07	116.17±5.81	14.47±0.65	95.8	142	102±7	
	93 LVA FLS 36[i]	36.01	113.35	490 / 36	ol	0.5190	5.79	37.61	52.29±2.61	13.42±0.67	85.7	142	94±7	
MILE 204.4 OUTBURST–FLOOD DEPOSIT														
	99-AZ-660A-GCRT[i]	36.02	113.35	526 / 71.4	ol	0.3533	6.94	18.02	488.09±24.40	8.09±0.40	63.7	138	59±4	66±6
	99-AZ-660C-GCRT[i]	36.02	113.35	526 / 71.4	ol	0.2323	6.94	44.63	165.48±8.27	9.36±0.47	87.8	138	68±5	

Table 1. Continued.

99-AZ-660E-GCRT[i]	36.02	113.35	526 / 71.4	ol	0.3762	6.94	7.91	862.70±43.13	9.75±0.49	99.5	138	71±5	
BASALTIC LAVA FLOWS													
Whitmore Cascade lava flow													
97-AZ-315A-WDA[n]	36.17	113.21	1012/NA	ol	0.4898	7.22	436.18	71.02±3.55	44.13±2.21	98.8	240	184±13	177±9[k]
97-AZ-316-WDA[o]	36.17	113.21	1012/NA	ol	0.4518	7.22	403.59	59.13±2.96	34.10±1.71	99.1	240	142±10[j]	
97-AZ-316A-WDA[o]	36.17	113.21	1012/NA	ol	0.4055	7.22	281.31	53.94±2.70	21.42±1.07	97.9	240	89±6[j]	
97-AZ-317-WDA[o]	36.17	113.21	1049/NA	ol	0.5182	7.22	228.35	101.09±5.05	32.09±1.60	96.8	247	130±9[j]	
97-AZ-317A-WDA[o]	36.17	113.21	1049/NA	ol	0.4831	7.22	247.27	100.84±5.04	35.07±1.75	97.3	247	142±10[j]	
97-AZ-320-WCB[o]	36.18	113.21	1122/NA	ol	0.4460	7.22	358.14	86.20±4.31	43.76±2.19	98.0	262	168±12	
97-AZ-323-WCD[o]	36.20	113.22	1122/NA	ol	0.4484	7.22	111.13	234.62±11.73	35.28±1.76	93.9	262	135±10[j]	
97-AZ-332-WCD[o]	36.20	113.21	1317/NA	ol	0.5462	7.22	197.90	200.14±10.01	54.88±2.74	96.5	303	181±13	
00-AZ-924-WC[n,p]	36.19	113.23	989/NA	ol	0.1306	7.22	128.96	15.97±0.80	30.98±1.55	96.1	247	125±9[j]	
00-AZ-925-WC[n,p]	36.19	113.23	989/NA	ol	0.1465	7.22	85.72	48.82±2.44	43.53±2.18	92.6	247	176±12	
00-AZ-926-WC[o,p]	36.19	113.23	989/NA	ol	0.1876	7.22	191.89	20.64±1.03	39.77±1.99	97.2	247	161±11	
00-AZ-935-WC[n,p]	36.15	113.21	747/NA	ol	0.2754	7.22	89.31	71.05±3.55	33.97±1.70	92.7	195	174±12	
00-AZ-936-WC[o,p]	36.15	113.21	747/NA	ol	0.2313	7.22	138.16	33.60±1.68	33.98±1.70	95.7	195	174±12	
00-AZ-937-WC[n,p]	36.15	113.21	747/NA	ol	0.2355	7.22	299.38	10.51±0.53	34.46±1.81	98.7	195	177±12	
00-AZ-938-WC[o,p]	36.15	113.21	747/NA	ol	0.2053	7.22	231.69	13.46±0.67	36.12±1.81	98.1	195	185±13	
00-AZ-940-WC[n,p]	36.15	113.21	759/NA	ol	0.1744	7.22	193.73	14.44±0.72	38.02±1.90	97.8	196	194±14	
Younger Cascade lava flow													
96-AZ-148-VT[n]	36.21	113.09	1258 /NA	ol	0.2321	6.63	75.61	278.17±13.91	27.76±1.39	91.3	290	96± 7	108±11[k]
96-AZ-148-VT[n]	36.21	113.09	1258 /NA	px	0.4843	6.63	72.47	296.23±14.81	29.34±1.47	92.0	290	101±7	
96-AZ-150-VT[o]	36.21	113.09	1122 /NA	ol	0.4651	6.63	38.43	650.80±32.54	35.12±1.76	97.2	293	120±8	
96-AZ-150-VT[o]	36.21	113.09	1122 /NA	ol	0.4269	6.63	103.63	248.74±12.44	35.08±1.75	94.1	293	120±8	
96-AZ-151-VT[n]	36.21	113.09	1122 /NA	ol	0.4994	6.63	134.77	175.25±8.76	33.03±1.65	96.6	293	113±8	
97-AZ-302-HYC[o]	36.20	113.08	732 / NA	ol	0.4849	6.63	21.12	756.81±37.84	15.81±0.79	68.1	183	86±6[j]	
97-AZ-303-HYC[o]	36.20	113.08	732 / NA	ol	0.3768	6.63	33.30	470.12±23.51	18.09±0.90	79.9	183	99±7	
Bar Ten lava flow													
97-AZ-326-YBT[o]	36.23	113.22	1268/NA	ol	0.4804	7.22	233.28	82.02±4.10	27.02±1.35	97.1	293	92±7	88±6
97-AZ-327-YBT[o]	36.23	113.22	1268/NA	ol	0.4904	7.22	244.96	76.27±3.81	26.43±1.32	97.3	293	90±6	
97-AZ-328-BT[o]	36.23	113.23	1256/NA	ol	0.4423	7.22	281.88	59.04±2.95	23.65±1.18	98.6	293	81±6	
97-AZ-330-BT[o]	36.23	113.22	1271/NA	ol	0.5286	7.22	266.84	64.84±3.24	24.40±1.22	97.5	295	83±6	

Table 1. Continued.

97-AZ-331-BT[o]	36.22	113.22	1280/NA	ol	0.5056	7.22	273.70	71.30±3.57	27.29±1.36	97.4	295	93±7

[a] ACRL=above current river level.
[b] ol = olivine, px = pyroxene.
[c] R_A=1.38 x 10^{-6}; a R/R_A crush value of 5.04 was determined for sample 99-AZ-627-GCRT and was used for the Mile 189.5 and Mile 202 outburst–flood deposits. A R/R_A crush value of 6.94 was determined for sample 99-AZ-660E-GCRT and was used for the Mile 204.4 outburst–flood deposit. R/R_A crush values of 5.18, 6.42, and 5.76 were determined for samples LVA FLS 28, 31, and 33, respectively, and an average value of 5.79 was used for the Mile 204.6 outburst–flood deposit. Sample 96-AZ-151-VT yielded an R/R_A crush value of 6.63, and this value was used in the calculation of exposure ages on the Younger Cascade. Sample 96-AZ-241-GCRT yielded an R/R_A crush value of 7.22 which was used for the calculation of exposure ages on the Whitmore Cascade and Bar Ten lava flows.
[d] Analytical error ≤ 5%; includes corrections for precision of mass spectrometer, blanks and standards.
[e] ^3He_c atom concentration corrected to the surface of each sample, accounting for self–shielding.
[f]Production rate is corrected for skyline shielding resulting from surround topography.
[g]Uncertainty includes error propagation of analytical error and production rate uncertainty (5%).
[h]Average and standard deviation of exposure ages on surface.
[i] Sample collected from the top of basalt boulders from surface of outburst–flood deposits.
[j]Younger ages are excluded from the average exposure age because they may represent surfaces that were either eroded, exhumed, and/or buried since deposition/emplacement.
[k] Ages excluded from calculation of average age and standard deviation of surface. See [j] for details.
[l]Oldest exposure age thought to be a better representation of ageof deposit. See [j] for details.
[e] Average of exposure ages of basalt boulders in outburst–flood deposits at river miles 189.5 and 202.
[m]Average exposure age of ouburst–flood deposits at river miles 189.5 and 202.
[n]Sample collected from a desert pavement.
[o]Sample collected from a lava flow top (i.e. aa, pahoehoe, tumuli and other primary flow surfaces).
[p]Samples analyzed on the University of Utah's noble gas mass spectrometer. All other samples analyzed at the University of Rochester.

Table 2. Olivine phenocryst and basalt glass samples and locations of their associated lava flows or outburst–flood deposits.

Location[a]	Sample	Latitude (N)	Longitude (W)	Maximum Elevation (m) ACRL[b]	Sample Type[c]	Olivines Number of data points[d]	Basalt glass size - Number of data points[d]
River Mile 185.7 - L	99-AZ-611-GCRT	36.16	113.17	174[g]	Tuff	--[j]	ash – 13
River Mile 189.5 - L	99-AZ-622-GCRT	36.13	113.20	200[g]	Tuff	--[j]	ash – 10
River Mile 195.2 - R	99-AZ-645BS2-GCRT	36.10	113.27	15[g,h]	Tuff	--[j]	ash – 9
River Mile 202 - R	99-AZ-645-GCRT	36.20	113.07	54.2[f]	Tuff	--[j]	ash – 10
						--[j]	1 Lapilli – 10
	99-AZ-653-GCRT	36.05	113.35	54.2[f]	Tuff	--[j]	1 lapilli – 10
	96-AZ-238-GCRT	36.05	113.35	54.2[f]	Boulder	10	--[j]
	96-AZ-239-GCRT	36.05	113.35	54.2[f]	Boulder	9	--[j]
	96-AZ-240-GCRT	36.05	113.35	54.2[f]	Boulder	10	--[j]
	96-AZ-241-GCRT	36.05	113.35	54.2[f]	Boulder	9	--[j]
River Mile 204.4 - L	99-AZ-681A-GCRT	36.01	113.34	71.4[f]	Tuff	--[j]	4 lapilli – 10 each
River Mile 204.6 - L	99-AZ-654-GCRT	36.01	113.35	36[f]	Tuff	--[j]	3 lapilli – 6, 7, and 9 each
	99-AZ-655-GCRT	36.01	113.35	36[f]	Tuff	--[j]	ash – 9
					Tuff		1 lapilli – 9
	LVA FLS 37	36.01	113.35	36[f]	Tuff	11	--[j]
	LVA FLS 26	36.01	113.35	36[f]	Boulder	10	--[j]
	LVA FLS 27	36.01	113.35	36[f]	Boulder	10	--[j]
	LVA FLS 34	36.01	113.35	36[f]	Boulder	10	--[j]
River Mile 222 - L	99-AZ-699A-GCRT	35.81	113.33	22[g]	Tuff	--[j]	3 lapilli – 8, 10, and 12 each
Bar Ten	96-AZ-325-YBT	36.23	113.22	NA[I]	flow	10	--[j]
basalt flow	96-AZ-326-YBT	36.23	113.22	NA[I]	flow	10	--[j]
	96-AZ-327-YBT	36.23	113.22	NA[I]	flow	10	--[j]
Younger Cascade basalt	97-AZ-302-HYC	36.20	113.08	NA[I]	flow	10	--[j]
flow	97-AZ-303-HYC	36.20	113.08	NA[I]	flow	9	--[j]
Whitmore Cascade	97-AZ-315A-WDA	36.17	113.21	NA[I]	flow	8	--[j]
basalt flow	97-AZ-316A-WDA	36.17	113.21	NA[I]	flow flow	10	--[j]
	97-AZ-317A-WDA	36.17	113.21	NA[I]	flow flow	8	--[j]
	97-AZ-320-WCB	36.18	113.21	NA[I]	flow	10	--[j]
	97-AZ-323-WCD	36.22	113.22	NA[I]		3	--[j]
	97-AZ-332-WCD	36.21	113.21			10	

[a]L=river left; R=river right for locations of outburst–flood deposits
[b] ACRL=above current river level.
[c] tuff = basalt glass ash or lapilli collected from a hyaloclastite tuff; boulder = basalt boulder collected from the surface of an outburst–flood deposit; flow = sample collected from the surface of a lava flow.
[d] number of probe analyses on same number of olivines from a given sample.
[e]number of probe analyses on either basalt glass ash or lapilli as indicated.
[f]Measured with a total station.
[g]Estimated from a 7.5 minute U.S.G.S. quadrangle.
[h]Minimum stage.
[I] not applicable.
[j] no data.

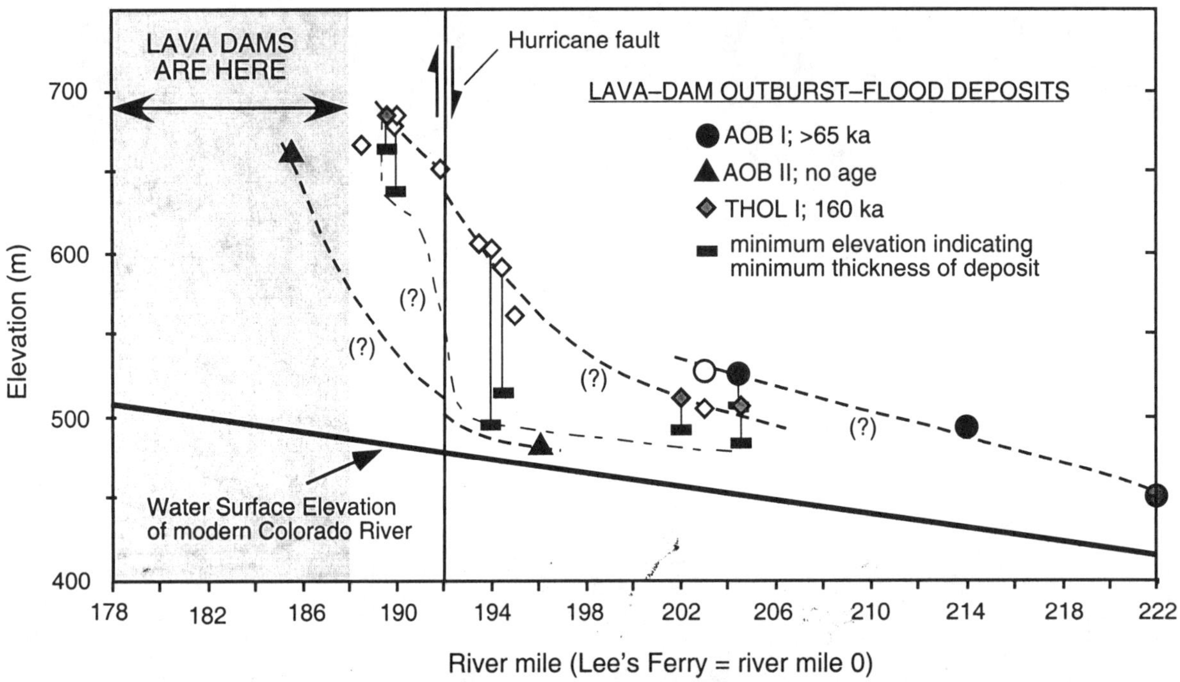

Figure 4. Elevations of two outburst–flood deposits and one uncorrelated hyaloclastite tuff between river miles 185 and 222 of the Colorado River. The water surface elevation of the modern Colorado River is depicted by the thick black line and is based on a best–line regression of elevations taken from *Stevens* [1983]. The filled in circles, triangles, and squares indicate deposits with basalt glass chemical signatures AOB I, AOB II, and THOL I, respectively (AOB = Alkali Olivine Basalt; THOL = Tholeiitic Basalt). The unfilled markers indicate tentative correlations between deposits (dashed lines) determined on the basis of elevation and mapping. AOB II (triangles) represents an uncorrelated deposit that consists of hyaloclastite ash; no basalt boulders or cobbles have yet been found associated with this deposit. Potential lava dam sites (shaded area with arrow) are found mainly between Lava Falls Rapid (river mile 179) and Whitmore Canyon (river mile 188) [*Hamblin*, 1994]. The lightly shaded area indicates the possible extent of the THOL I outburst–flood deposit based on measured minimum thickness. The question marks symbolize that the curves and slopes of dashed lines are approximate. The Hurricane fault crosses the Colorado River near river mile 191. This fault has an average vertical displacement rate of 79□7 m/Ma (meters per million years) [*Fenton*, in review] and is downthrown to the west. Based on this rate, 13 m of vertical displacement is expected between the 160 ka outburst–flood deposits at river miles 189.5 and 202. The AOB II tuff has an unknown age, thus the amount of displacement between the deposits at mile 185.7 and 195.2 is unknown.

volcanic field, suggesting probable downstream dilution into a hyperconcentrated flow [*Pierson and Scott*, 1985]. In general, dam failures create this type of deposition downstream [*Carling and Glaister*, 1987; *Cenderelli and Cluer*, 1998].

During typical dam failures, the bulk of large clasts is transported as bed load [*Singh*, 1996], and irregularly shaped particles, which were likely the product of dam failure, could become rounded within a short distance from the damsite. Clasts of local Paleozoic lithologies, previously river–reworked basalts and previously deposited hyaloclastites were likely entrained from the walls and bed in a manner similar to breaching of other natural dams [*Carling and Glaister*, 1987]. The high stages of peak flow probably resulted from complete rapid reservoir drainage following complete dam failure. The rapid decline in flood stage with distance from the damsites (Figure 4) suggests a highly unsteady flow rate, as would be expected from a rapidly draining reservoir [*Walder and O'Connor*, 1997]. No non–volcanic material was present in the outburst–flood deposits that could be attributed to deltaic or lacustrine deposits, which would have accumulated if the reservoirs had a long lifespan [*Hamblin*, 1994]. Because of the antiquity of the extant evidence, this scenario of lava-dam failure has many uncertain aspects. Estimation of discharges for paleo–outburst floods, such as occurred in the last 160 ka in western Grand Canyon, requires many assumptions concerning the nature of flood hydrographs and channel geometry.

The channel geometry in the modern canyon may be sufficiently similar to the channel geometry 75-150 ka to be used to estimate and route flow downstream from the dam failure. In locations where paleochannels are preserved, such as the previously mentioned channel at river mile 194, the paleo–geometry can be estimated by mapping the thickness and width of the fill sequence. In addition, many basalt flows in western Grand Canyon are near, at, or below the water level of today's Colorado River. The upper and lower Black Ledge flows yield $^{39}Ar/^{40}Ar$ ages of 603±8 and 524±7 ka, respectively, and the older of these flows caps river gravels that lie 30 m above the current river level near river mile 209 [*Lucchitta et al.*, 2000]. The base of the Massive Diabase unit [440 ka, K–Ar age; *Dalrymple and Hamblin*, 1998] at river mile 202 is below the water surface of the modern Colorado River. *Wenrich et al.* [1995] report K–Ar ages of 140 (no error) and 174±39 ka, respectively, for unspecified lava flows along the Colorado River at river miles 192 and 206, respectively. According to mapping of flows by *Hamblin* [1994] in these localities, the heights of the base of the lava flows are within 15 m of today's Colorado River level and some are below water level. *Lucchitta et al.* [2000] assert that the river has gone through at least four cycles of aggradation and degradation in the last 525 ka. We estimate that, within the accuracy of K–Ar dating of lava flows, the Colorado River in western Grand Canyon may have been somewhere near the same position for the past 600 ka.

METHODS

Individual outburst-flood deposits and lava flows near the Uinkaret volcanic field are identified on the basis of cosmogenic exposure ages and chemical compositions of basalt glass, whole rock basalts, and olivine phenocrysts. Although *Wenrich et al.* [1995] found that basalts in the Uinkaret volcanic field could not be distinguished chemically, our results indicate that individual basalt flows have distinct geochemical signatures that should allow tracing of clasts from lava flow to outburst-flood deposit.

Cosmogenic ^{3}He Dating Techniques

Cosmogenic ^{3}He is ideal for dating young basalts because it is a stable isotope with the highest production rate of any cosmogenic nuclide [*Kurz*, 1986; *Cerling*, 1990] and leakage of ^{3}He from olivines is minimal at surface temperatures [*Cerling*, 1990]. This method is particularly useful for dating basalt flows in western Grand Canyon because they contain abundant olivine phenocrysts [*Cerling et al.*, 1999]. Cosmogenic ^{3}He has been used extensively to determine exposure ages of Quaternary surfaces, basalt flows, and associated desert pavements [*Anthony and Poths,* 1992; *Cerling,* 1990; *Cerling and Craig,* 1994b; 1999; *Fenton et al.,* in review; *Laughlin et al.,* 1994; *Poreda and Cerling*, 1992; *Wells et al.*, 1995].

Theory. The amount of *in situ* cosmogenic ^{3}He is directly related to the length of time a rock has been exposed to cosmic rays [*Craig and Poreda*, 1986; *Kurz*, 1986; *Lal*, 1987]. The production rate of cosmogenic ^{3}He is controlled by elevation, latitude, and shielding of a sample. Shielding results from surrounding topography (i.e. hillslopes, cliff ledges), burial of a surface by sediment, or self-shielding. Production of cosmogenic ^{3}He decreases exponentially with depth, but is thought to be relatively constant in the top 4 cm of a surface [*Cerling and Craig*, 1994b]. As elevation and latitude increase and as shielding decreases, the production rate of *in situ* cosmogenic ^{3}He increases. Variations in the geomagnetic field also affect the production rate; the production rate is higher during periods of a weaker magnetic field [*Cerling and Craig*; 1994b. Four main processes contribute to the total helium mass in a rock: (1) air contamination, (2) radioactive decay and associated nuclear reactions, (3) mantle-gas contamination, and (4) cosmic-ray interaction [*Mamyrin and Tolstikhin*, 1984].

The cosmogenic component is calculated by subtracting the mantle, air, and radiogenic values from the total ^{3}He in a sample using:

$$^{3}He_{c}={}^{3}He_{tot}-{}^{3}He_{m}-{}^{3}He_{a}-{}^{3}He_{r}, \qquad (1)$$

where the subscripts *c*, *tot*, *m*, *a*, and *r* refer to the cosmogenic, total, mantle, air and radiogenic ^{3}He components, respectively. The contribution of radiogenic $^{3}He/^{4}He$ is significant in old rocks with young exposure ages (< 10–20

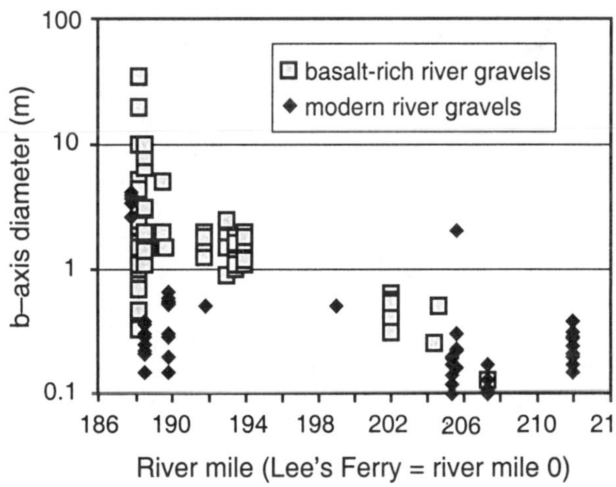

Figure 5. Size of largest clasts found in Pleistocene basalt–rich river gravels at various elevations and modern Colorado River gravels versus increasing distance from the Uinkaret volcanic field (found between river miles 179-188) in western Grand Canyon. As distance increases away from the lava dams, the size of boulders (>256 mm) in the basalt–rich gravels decreases, whereas the size of clasts in the modern gravels (generally cobbles and small boulders) does not appear to change dramatically over the same distance. There are large boulders found in modern river gravels at river miles 188 and 205.6. These gravel bars are near side canyons and appear to have local debris–flow contributions of these larger clasts. Clasts measured included basalts, limestones and sandstones.

Plate 2. Cross–sectional view of the Mile 204.6 outburst–flood deposit just below the 100 ka terrace. The section at this location is approximately 6 m thick and the top of the terrace is approximately 36 m above the river. Boulders related to this deposit are found as high as 53 meters above the river. The basal unit in this view is composed of basalt glass ash and lapilli preserved as a reworked hyaloclastite tuff with large–scale randomly oriented foresets. The tuff is overlain by well–rounded basalt boulders, cobbles and pebbles. This whole sequence overlies Hamblin's [1994] Massive Diabase flow. For color version see CD. (Photograph by Dominic Oldershaw.)

ka) but is not significant for young basalts from the Uinkaret volcanic field [*Cerling et al.*, 1999].

Sampling strategy. We sampled exposed surfaces of basalt boulders on the surfaces of outburst–flood deposits and primary flow structures from basalt flows. Primary flow structures include pahoehoe and aa flow morphologies preserved on a basalt flow surface as well as pressure ridges and tumuli. Rocks comprising desert pavements were also collected from both outburst–flood deposits and basalt flows for dating. Desert pavements in the southwest are typically inflationary surfaces that have been continuously exposed since the time of formation of the underlying landform and associated deposits [*Wells et al.*, 1995]. In this study, desert pavements and primary flow surfaces were collected for each basalt flow where both sample types were available.

Laboratory analysis. Mineral separates were prepared in accord with methods described in *Cerling et al.* [1999]. Olivine separates were analyzed for $^3He/^4He$ content on the VG 5400 or MAP–251 noble gases mass spectrometer at either the University of Rochester or the University of Utah, respectively (Table 1). Several mineral separates were crushed under high vacuum to release and measure mantle helium contained in inclusions. Remaining powders and uncrushed phenocrysts were melted at 1800°C in a double-walled modified Turner furnace. The released gas was purified using getters and cryogenic traps. Isotopic measurements were made on mass spectrometers each fitted with an electron multiplier and pulse counting electronics. $^3He/^4He$ ratios were standardized against the SIO-MM standard at 16.45 R_A (where R_A is the air ratio). All values were corrected for interference peaks, instrumental and extraction blanks [*Poreda and Cerling*, 1992].

The amount of cosmogenic ^3He in the sample, referred to hereafter as ^3He_c, was then calculated from:

$$^3He_c = {^4He}\left[\frac{^3He}{^4He_m} - \frac{^3He}{^4He_i}\right] \tag{2}$$

where the subscripts c, m, and i represent the cosmogenic, melted fraction, and inherited values, respectively [*Cerling et al*, 1994b]. Inherited values do not include radiogenic helium. *Cerling and Craig's* [1994a] absolute production rate of 115±4 atoms/g/yr for ^3He_c in olivine corrected to high latitude and sea level is used in this study. *Dunai and Wijbrans* [2000] verify this production rate back to 150 ka, indicating that the production rate is fairly constant on the 100 ka time scale and that the uncertainty in the ^3He production rate is <5% [*Ackert et al.*, 1998].

Electron Microprobe Analyses of Olivine and Basalt Glass

We performed electron-microprobe analyses of olivine phenocrysts and basalt glasses from basalt flows and outburst–flood deposits. All olivines were analyzed for Si, O, Ca, Mn, Ni, Fe, and Mg content. Basalt glasses from several different hyaloclastite tuffs were analyzed for Si, Ti, Al, Fe, Mn, Mg, Ca, K, Na, Cl, F, P, and O (Table 2). Basalt glass from hyaloclastite tuffs and interstitial basalt glass from the matrix–supported component were collected from various locations in each outburst–flood deposit. Table 2 lists the locations of samples from which the olivines or basalt glasses were separated.

Hyaloclastite tuffs were disaggregated, cleaned with dilute nitric acid, and ash-sized glass grains were mounted using an epoxy slurry. Basalt glass ash and lapilli were analyzed and each fragment was probed in different locations to account for chemical variation within a single grain. We used a Cameca SX-50 electron microprobe equipped with four wavelength-dispersive spectrometers. Analyses were made at 15 keV accelerating voltage, a beam current of 30 nA, and a beam diameter of 10 μm, with a mixture of natural minerals and synthetic compounds used as standards. Concentrations were calculated from relative peak intensities using the $\phi(\rho z)$ algorithm of *Pouchou and Pichoir* [1991].

ICP-MS Analyses Of Whole Rock Basalt And Basalt Glass Samples

Whole rock basalt samples were hand–crushed in a silica mortar and pestle. One hundred milligrams of each sample was microwave acid–digested using concentrated hydrofluoric and nitric acids. Rare earth and trace element concentrations of these solutions were determined by inductively coupled plasma–mass spectrometry (ICP–MS) at the University of Rochester. The ICP–MS data were checked against the BCR–2 (U.S. Geological Survey, Columbia River basalt) rock standard. Results for the BCR–2 standard were within 2% of the U.S. Geological Survey's reported rare earth element values for this standard.

RESULTS

Cosmogenic ^3He Exposure Ages of Outburst–Flood Deposits and Lava Flows

Table 1 lists the details of cosmogenic ^3He analyses and exposure ages of samples collected from the surfaces of outburst–flood deposits and lava flows.

Basalt Glass Chemistry

For the purposes of this study, we have classified basalt glasses collected between river miles 185.7 and 222 into three major groups—Alkali Olivine Basalt I and II (AOB I and AOB II) and Tholeiitic Basalt I (THOL I)—based on total alkali ($K_2O + Na_2O$) and silica (SiO_2) weight percents of the basalt glasses (Figure 6). Basalt glass ash and lapilli in hyaloclastite tuffs and in the matrix–supported

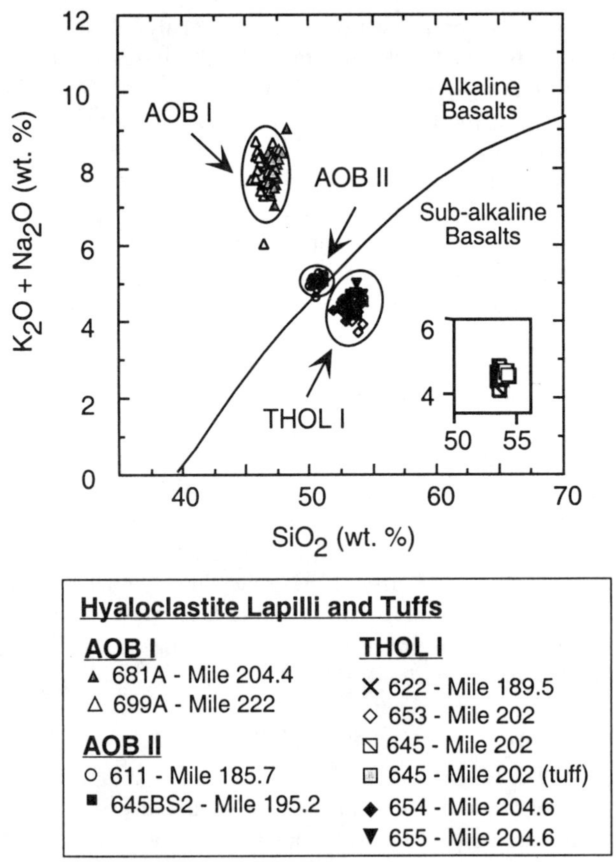

Figure 6. Alkali content of basalt glass lapilli and hyaloclastite ashes from lava–dam outburst–flood deposits. Basalt glasses are either alkaline or subalkaline basalts [after *Irvine and Baragar*, 1971], grouping into three distinct chemistries, AOB I, AOB II, and THOL I, where AOB and THOL indicate alkaline olivine and tholeiitic basalts, respectively. Basalt glass lapilli within desert pavements and associated hyaloclastite tuff have identical compositions (inset). Presumably, these basalt glasses were produced by the interaction of lava cascades from the Uinkaret volcanic field and the Colorado River.

basalt boulder components all exhibit the same chemical signature for a given outburst–flood deposit (Figure 6, inset). Thus, both basalt glass ash and lapilli from hyaloclastite tuffs are used for chemical characterization of the lava dam outburst–flood deposits.

Basalt glasses from the hyaloclastite tuff at river mile 204.4 and basalt glass lapilli overlying *Hamblin's* [1994] Black Ledge lava flow at mile 222 have the highest total alkali content and are grouped as AOB I (Figure 6). Alkalic basalt glasses found at river miles 185.7 and 195.2, overlying Black Ledge and Massive Diabase lava flows, respectively, have less total alkali and more total silica content than AOB I and are classified as AOB II. The third chemical group is referred to as THOL I which has the lowest total alkali and highest silica contents of all the basalt glasses. Samples from the outburst–flood deposits at river miles 189.5, 202, and 204.6 are grouped as THOL I.

Olivine Compositions

The forsterite component of olivines from lava flows, hyaloclastite tuffs, and outburst–flood basalt boulders ranges from Fo_{77} to Fo_{83}. Calcium and fayalite values are used to geochemically distinguish lava flows and outburst–flood deposits with surface exposure ages younger and older than 150 ka (Figure 9).

Rare Earth Element Patterns

Figure 8 illustrates the REE patterns, normalized to chondrite, of whole rock basalt samples from the Whitmore Cascade lava flow and flood deposits at miles 189..5, 202 and 204.4. Table 3 lists the La/Sm, Gd/Lu, and La/Lu ratios and the ΣREE concentrations of the samples. All REE concentrations are normalized to chondrite values reported in *Taylor and McClennan* [1985].

DISCRIMINATION OF OUTBURST–FLOOD DEPOSITS AND LAVA FLOWS

Outburst-Flood Deposits at river miles 189.5 and 202

Two outburst–flood deposits are preserved on ledges provided by *Hamblin's* [1994] Whitmore Dam and Massive Diabase lava dam remnants at river miles 189.5 and 202, respectively (Figure 2; Plate 1). These deposits are informally named the Mile 189.5 and Mile 202 outburst–flood deposits. They are found at maximum heights of 200 and 54 m above current river level, respectively (Table 2; Figure 4). A total of seven samples were collected from the surfaces of boulders greater than 50 cm in b–axis diameter on the deposits' surfaces. Three basalt boulders from the Mile 189.5 deposit yielded exposure ages of 125±9, 139±10, and 161±11 ka. The maximum exposure age of the flood deposit at mile 189.5 is 161±11 ka; the two younger ages may represent exhumation of the boulders. Slope wash may have washed away some of the fine–grained material on the surface of the flood deposit, resulting in slightly lower cosmogenic ^{3}He values for these two basalt boulders. Three of four boulders from the Mile 202 deposit range in age from 157±11 to 159±11 ka; one boulder yielded an anomalously young age of 115±8 ka. The average cosmogenic exposure age for the Mile 202 outburst-flood deposit is 158±1 ka, which is calculated by averaging the maximum exposure ages of the basalt boulders (Table 1). The anomalously young exposure age (115±8 ka) was discarded from the average on the basis that the boulder may have been rolled or exhumed, perhaps also by erosion of fine–grained sediments by slopewash.

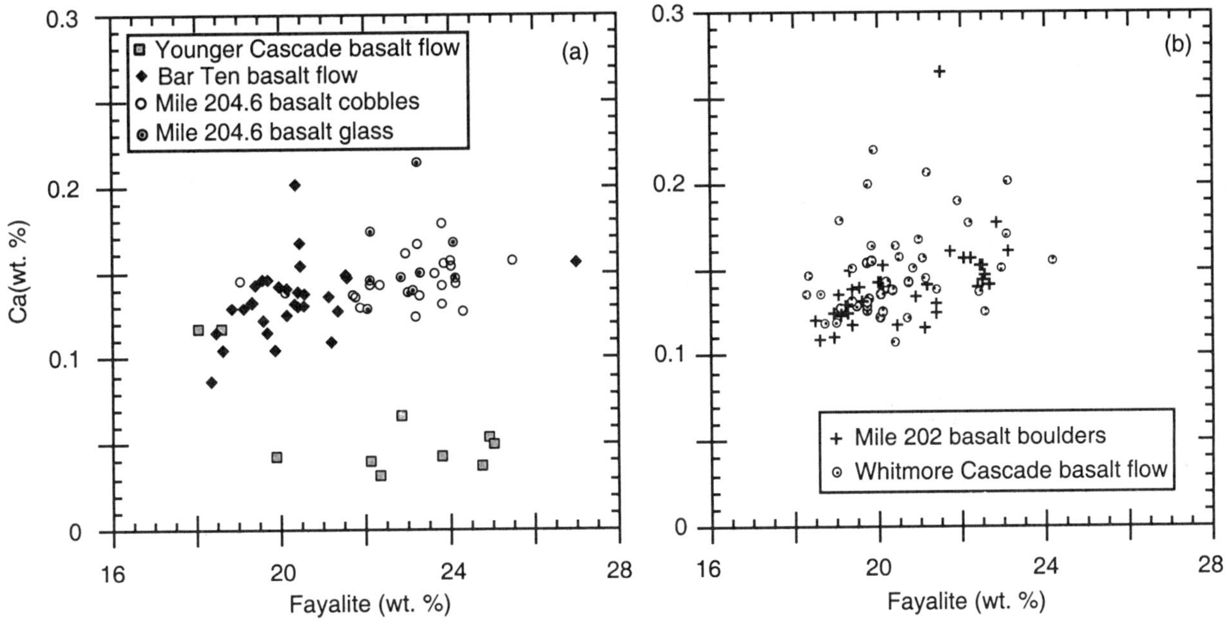

Figure 7. Calcium and fayalite contents in olivines phenocrysts (forsterite content 77-83%) from the Bar Ten, Younger Cascade and Whitmore Cascade lava flows and from outburst–flood deposits at river miles 202 and 204.6. (a) Although the ages of the Bar Ten (88±6 ka) and Younger Cascade (108±11 ka) lava flows and the Mile 204.6 outburst–flood deposit (100±6 ka) overlap within the error of the ^{3}He ages, the compositions of their olivines strongly suggest that they are unrelated. Olivines in the hyaloclastite and basalt cobbles of the Mile 204.6 outburst–flood deposit have very similar compositions, thus indicating that they may have been produced from the lava flow. (b) The Whitmore Cascade (177±9 ka) basalt flow and the Mile 202 outburst–flood deposit (159±2 ka) are unrelated, yet they have very similar olivine compositions.

Basalt glass ash and lapilli from both flood deposits were analyzed for major element concentrations. Sample 99-AZ-622-GCRT was collected from the Mile 189.5 deposit's basal hyaloclastite tuff. Two samples were collected from the Mile 202 hyaloclastite; sample 99-AZ-645-GCRT included ash-size and lapilli-size basalt glass collected from the base of the basal hyaloclastite tuff and sample 99-AZ-653-GCRT was a lapilli–sized glass fragment taken from the top 1 m of the matrix–supported basalt boulder component. The basalt glass analyses from these samples tightly cluster in group THOL I (tholeiitic basalt) in the subalkaline field of Figure 6. Rare earth element concentrations in basalt glass from the Mile 202 deposit also exhibit a tholeiitic chemical signature (Figure 8).

The whole rock rare earth element chemistry of a basalt boulder (sample 99-AZ-623-GCRT) from the Mile 189.5 deposit is identical to that of a basalt boulder (sample 96-AZ-240-GCRT) from the Mile 202 deposit. The exposure ages of these two boulders are essentially identical (Table 1). It is likely that the basalt boulders came from the same lava flow and were deposited during the same flood that occurred 159±2 ka. Both the Mile 189.5 and 202 outburst–flood deposits consist of reworked tholeiitic basalt glass ash and lapilli and alkali olivine basalt boulders. The major element chemical signatures of the basalt glasses are identical at both deposits; the chemical signatures of the interstitial basalt glasses are identical to the hyaloclastite tuffs.

There is a compositional difference between tholeiitic basalt glasses in the hyaloclastite tuffs and alkali olivine basalt boulders on the surfaces of deposits at river miles 189.5 and 202; however, we believe they may have been deposited 159±2 ka during the same outburst flood that followed the failure of an alkalic basalt lava dam. We hypothesize that the tholeittic basalt glass, created by previous interaction between the river and a tholeiitic basalt flow, was already in the river corridor. Later, an alkalic basalt dam was formed and at the onset of its failure, the rapidly draining reservoir resuspended the tholeiitic basalt glass and redeposited it as a reworked hyaloclastite tuff with large–scale foresets. Upon complete dam failure, basalt boulders and cobbles intermixed with the basalt glass and were deposited at the top of the outburst–flood sequence.

An alternative hypothesis is that the cross–bedded tholeiitic hyaloclastite tuff was deposited first and later covered with alkali olivine basalt boulders and cobbles with interstitial basalt glass resuspended from the older hyaloclastite tuff. The large–scale foresets, the stratigraphy,

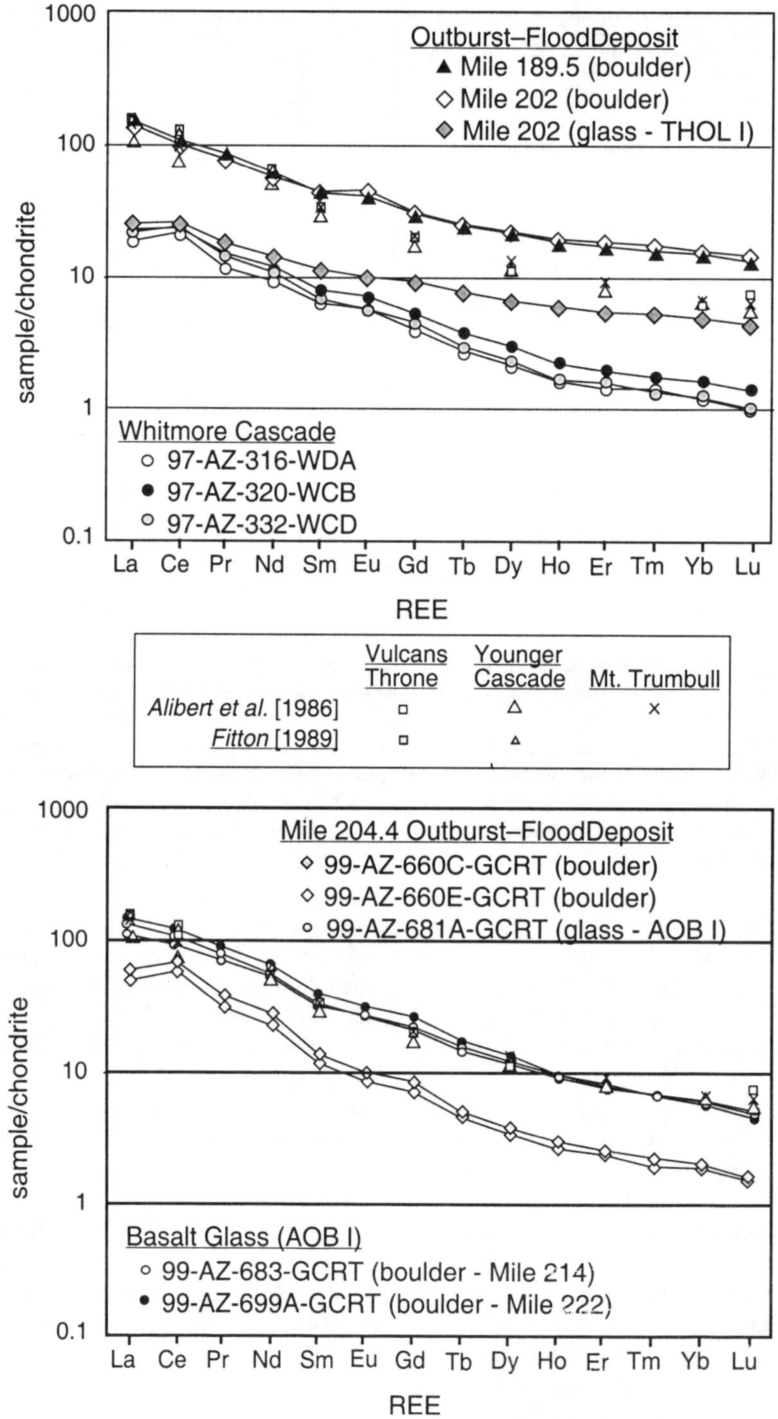

Figure 8. Chondrite–normalized rare earth element patterns of whole rock basalt samples from the Whitmore Cascade lava flow and from the surfaces of the Mile 189.5, 202 and 204.4 outburst–flood deposits. *Alibert et al* [1986] reports data for an alkali olivine basalt labeled as Tuweep and *Fitton* [1989] reports data for a basanite flow west of Vulcans Throne that contains abundant peridotite xenoliths. We believe these flows are the same and are equivalent to our informally named Younger Cascade. *Fitton* [1989] and *Alibert et al* [1986] also reports rare earth element data for Vulcan's Throne and Mt. Trumbull, a 3.5 Ma alkali olivine basaltic vent located approximately 5 miles north of Vulcan's Throne and the Colorado River. These data give typical rare earth element values for alkali olivine basalts in western Grand Canyon. All rare earth element concentrations in this diagram are normalized to chondrite values reported in *Taylor and McClennan*, [1985].

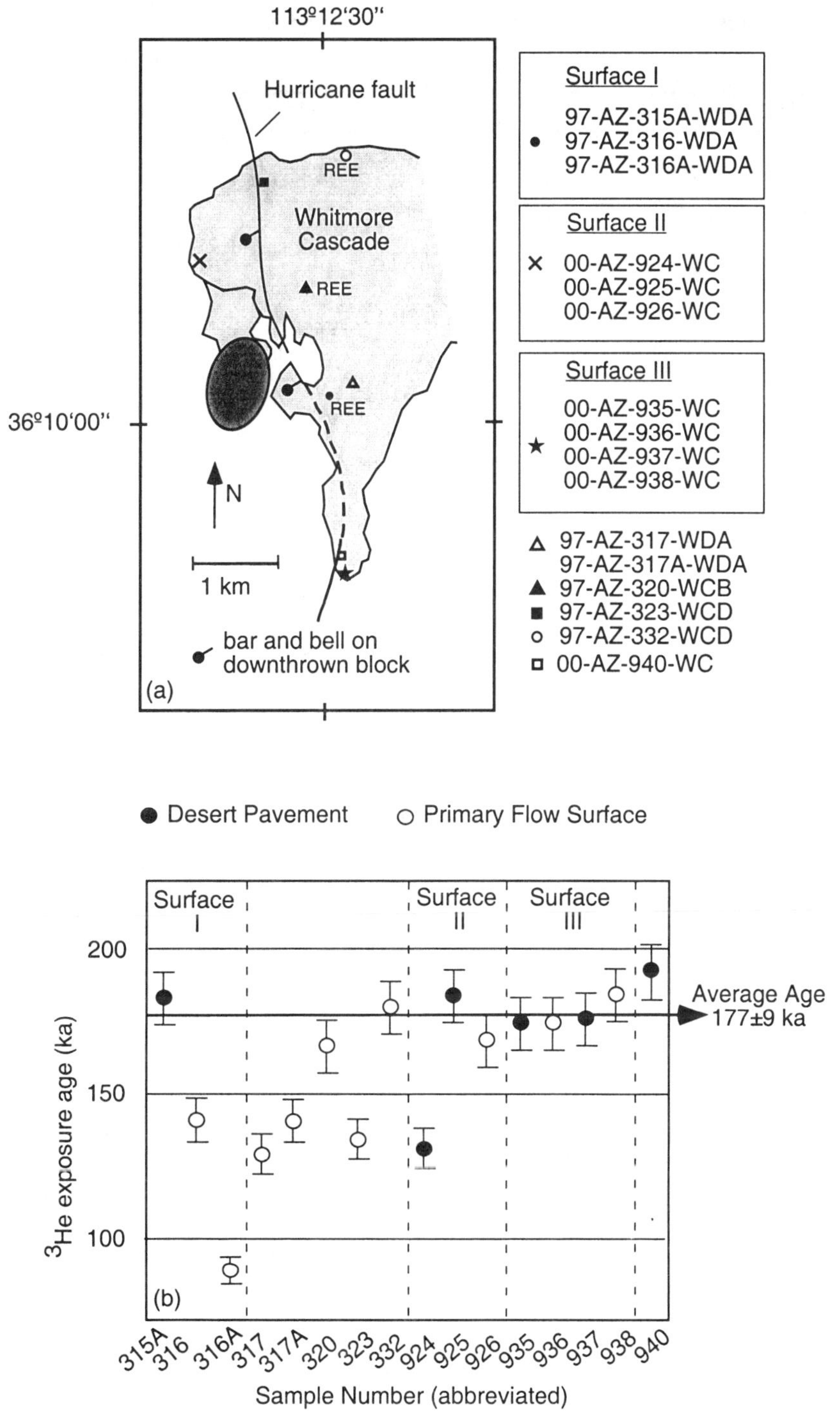

Figure 9. (a) Locations of cosmogenic and whole rock basalt samples taken from the surface of the Whitmore Cascade. Groups labeled as Surfaces I, II and III indicate those samples that were collected within a 10 m radius on one surface. REE shows where whole rock samples were collected for rare earth element analyses. (b) Exposure ages of desert pavements and primary flow surfaces collected at various locations on the Whitmore Cascade. Error bars (~7%) include 5% analytical error and 5% uncertainty in the production rate.

Table 3: Select chondrite–normalized ratios of the whole rock basalt samples from Whitmore Cascade and the 160 ka outburst–flood deposit.

Sample	ΣREE [a]	La/Sm [a]	Gd/Lu [a]	La/Lu [a]
Whitmore Cascade				
97-AZ-316-WDA [b]	90	4.71	32.26	19.05
97-AZ-320-WCB [b]	110	4.49	30.49	16.52
97-AZ-332-WCD [b]	101	5.05	35.87	21.09
160 ka Outburst–Flood Deposit				
99-AZ-623-GCRT [c] (river mile 189.5)	656	5.62	18.48	11.47
96-AZ-240-GCRT [c] (river mile 202)	636	5.12	16.82	9.58
99-AZ-653-GCRT [c] (river mile 202)	157	2.26	2.11	5.78
>65 ka Outburst–Flood deposit				
99-AZ-660C-GCRT [c] (river mile 204.4)	237	4.25	5.11	34.50
99-AZ-660E-GCRT [c] (river mile 204.4)	199	4.13	4.49	29.95
99-AZ-681A-GCRT [c] (river mile 204.4)	449	3.41	4.66	23.40
99-AZ-683-GCRT [c] (river mile 214)	494	3.90	4.10	24.79
99-AZ-699A-GCRT [c] (river mile 222)	581	3.63	5.08	27.67
Younger Cascade				
Tuweep [*Alibert et al.*, 1986]	NA [d]	3.73	3.02	18.83
Flow west of Vulcan's Throne [*Fitton*, 1989]	NA [d]	NA [d]	NA [d]	NA [d]

[a] chondrite–normalized
[b] whole rock samples from the surface of the basalt flow.
[c] Whole rock samples from boulders on the surface of the outburst–flood deposit.
[d] data not available.

the identical exposure ages, and identical basalt glass and whole rock chemistry strongly suggest that the hyaloclastite tuffs and basalt boulders were deposited at both river miles 189.5 and 202 during the same event.

Because of the strong agreement between the olivine compositions of the Whitmore Cascade and the boulders at the Mile 202 outburst–flood deposit (Figure 6b), the Whitmore Cascade was considered as a lava–dam candidate that may have failed to produce the outburst–flood deposits at miles 189.5 and 202.

Whitmore Cascade lava dam

Exposure ages of the Whitmore Cascade range from 89±6 to 194±14 ka (Table 1; Figure 9). Six samples from the Whitmore Cascade yielded exposure ages less than 150 ka; five of these samples were primary flow surfaces, one sample was a desert pavement (Figure 9b). These six ages were discarded from the average exposure age of the Whitmore Cascade on the basis that the ages represent degradation of primary flow surfaces or exhumation of clasts during the desert pavement formation process. The remaining ten surface samples yielded exposure ages greater than 150 ka; five of these samples were desert pavements, five were primary flow surfaces (Figure 9b). The average exposure age of the Whitmore Cascade is 177±9 ka, calculated by averaging these ten ages (Table 1). REE patterns of three whole rock basalt samples taken from the north end, middle, and south end of the flow clearly illustrate that the surface of the Whitmore Cascade is one tholeiitic basalt flow (Figures 8a and 9a). Variations in exposure ages do not result from sampling of younger, unidentified lava flows on the surface of the Whitmore Cascade.

Desert pavements yielded three of the oldest surface exposure ages on the Whitmore Cascade, whereas five of the youngest ages were yielded by primary flow surfaces. Each group of samples from Surfaces I, II, and II were collected within a 10 m radius on the Whitmore Cascade (Figure 9). Surfaces I and II illustrate that for a given surface, desert pavements often produce the oldest exposure age, most likely due to degradation of primary flow surfaces. Desert pavements can yield an anomalously young age (Surface II, Figure 9b), most likely due to exhumation of the clast during the desert pavement formation process. Surface III illustrates a very strong correlation between the exposure ages of desert pavements and primary flow surfaces on a given surface of a lava flow. Despite the ability of desert pavements to exhume underlying clasts and the degradation of primary flow surfaces, there is generally a 1:1 correlation (within error) between the ^{3}He exposure ages of the two surface types (Figure 9b). Therefore, desert pavements on late Pleistocene lava flows can give minimum ages of original eruptions, in accord with the findings of *Wells et al.* [1995], and on older lava flows, desert pavements may provide more reliable dates than intact flow surfaces.

Based on field mapping, we believe that our informally named Whitmore Cascade is the same as the Whitmore Dam basalt unit (993±97 ka; K–Ar age) mapped by *Hamblin* [1994]. The Whitmore Dam/Cascade contains abundant glass and thus may have a high atmospheric argon contribution that could result in an anomalously old K–Ar age [*Dalrymple and Hamblin*, 1998]. *Dalrymple and Hamblin* [1998] conclude that this date may not be reliable and they estimate that Whitmore Dam formed approximately 600 ka based on stratigraphic relationships with other K–Ar dated lava dams. We believe, however, that the ^{3}He exposure age of 177±9 ka more accurately reflects the age of the Whitmore Cascade [*Fenton et al.*, in review], and, thus, *Hamblin's* [1994] Whitmore Dam.

The REE patterns and exposure ages of Whitmore Cascade (177±9 ka) and basalt boulders from the outburst–flood deposits at river miles 189.5 and 202 (158±1 ka) do not agree. The Whitmore Cascade dammed the Colorado River approximately 180 ka (Figure 2), but it is not related to the these outburst–flood deposits.

Outburst-Flood Deposits at River Miles 204.4 and 204.6

Two deposits, informally named the Mile 204.4 and Mile 204.6 outburst–flood deposits, are preserved at river miles 204.4 and 204.6 on ledges formed by *Hamblin's* [1994] Massive Diabase lava dam, at maximum heights of 71 and 53 m ACRL, respectively (Figures 2 and 4; Plate 1). A lower erosional terrace is cut into the pre–existing Mile 204.6 deposit at a height of 36 m above current river level. There is a Quaternary landslide upslope of both the Mile 204.4 and Mile 204.6 deposits (Figure 2), but there is very little landslide material found in desert pavements on the surfaces of the outburst–flood deposits. This leads us to surmise that the flood deposits are inset against and younger than the landslide. Three basalt boulders were collected from the informally named Mile 204.4 deposit that yielded ages of 59±4, 68±5, and 71±5 ka. The average exposure age of 66±6 ka for this deposit is problematic because there is a large variation in only three exposure ages. The range in exposure ages of the Mile 204.4 deposit may result from burial by landslide material and then re–exhumation, possibly due to slopewash eroding away the fine–grained material. The lower terrace cut into the Mile 204.6 outburst–flood deposit yielded an average exposure age of 100±6 ka. Eleven basalt boulders were sampled from the terrace and ten of those samples ranged in age from 91±6 ka to 107±7 ka. Sample 93 LVA FLS 29 produced exposure ages of 72±5 ka and 74±5 ka, under duplicate analysis. This sample was excluded from the average exposure age because it likely represents exhumation or rolling of the boulder during the desert pavement formation process.

The composition of olivine phenocrysts suggest that olivines within the hyaloclastite tuff and basalt boulders in the Mile 204.6 outburst–flood deposit may have been produced from the same lava flow (Figure 7a). This conclusion is tentative because olivine composition can vary within a single phenocryst and it can vary among phenocrysts from the same basalt flow; however, variations in chemical composition of olivines from the hyaloclastite and from the boulders in the Mile 204.6 deposit cluster enough to indicate that they may be related.

Basalt glass in the Mile 204.6 deposit has a tholeiitic chemical signature, similar to the chemical signature of basalt glasses in the Mile 189.5 and Mile 202 outburst–flood deposits (159±2 ka). The composition of olivines from basalt boulders collected from the surfaces of both the Mile 202 and Mile 204.6 deposits also strongly agree; however, major element analysis of basalt glasses and olivines offer only crude distinctions of basalt and olivine type, chemistry alone is inconclusive. It is possible that either (1) two separate lava–dam failures of different ages emplaced outburst–flood deposits at miles 202 and 204.6, with chemically similar olivines and hyaloclastite tuffs, or (2) the outburst–flood deposits at miles 189.5, 202, and 204.6 were produced from the same lava–dam failure, and the surface at mile 204.6 (CD Plate 3) represents a planation surface that is 100 ka.

Major element chemistry indicates that basalt glasses from the Mile 204.4 deposit are classified as alkaline olivine basalt (AOB I; Figure 6). REE signatures of whole rock basalts and basalt glass from the surface of the Mile 204.4 deposit both exhibit REE patterns typical of alkali olivine basalts, but the glass and basalt boulders appear to be chemically unrelated to each other and to boulders at river miles 189.5 and 202. REE patterns of basalt glasses collected on top of basalt ledges at river miles 213 and 222 (Figure 4) are identical to the REE pattern of the glass collected at river mile 204.4 (Figure 8). The basalt glass may have formed from an older alkalic lava flow that cascaded into the Colorado River. The glass may have been resuspended and redeposited, forming large–scale cross-beds, during the later outburst–flood event that deposited the boulders at Mile 204.4. The flood discharge may have been large enough to distribute the basalt glass as far downstream as river mile 222. Alternatively, the hyaloclasite tuff may have been deposited in the river corridor during a different period, unrelated to the basalt boulders and cobbles that cap the tuff.

The REE pattern of basalt glass collected from the Mile 204.4 hyaloclastite is essentially identical to REE values reported by *Fitton* [1989] and *Alibert et al.* [1986] for the informally named Younger Cascade (Table 3; Figure 10b). The Younger Cascade (108±11 ka) may have produced the basalt glass preserved in the hyaloclastite tuff at river mile 204.4. Thus, the age of the Mile 204.4 deposit could range between 100±6 and 108±11 ka.

On the basis that the exposure ages and the chemical signatures of their basalt glasses differ significantly, we conclude that the Mile 204.4 and Mile 204.6 deposits (AOB I and THOL I, respectively) were deposited by two unrelated lava–dam outburst–floods. The Mile 204.4 deposit has a maximum exposure age of 71±4 ka, but is it is likely that the Mile 204.4 deposit was emplaced before the Mile 204.6 deposit (minimum age of 100 ka). If the Mile 204.6 deposit does in fact correlate with the outburst–flood deposits at river miles 189.5 and 202, the Mile 204.4 deposit could be older than 160 ka. The problematic younger age of the higher Mile 204.4 surface may result from burial and re–exhumation. If the Mile 204.4 deposit was emplaced after the 204.6 deposit, we would expect more scatter in the exposure ages of the lower deposit; however, ten of eleven basalt boulders on the Mile 204.6 surface yield exposure ages greater than 90 ka. The landslide predates both outburst–flood deposits and thus, is likely older than 160 ka.

The Younger Cascade and Bar Ten lava flows

Exposure ages for the informally–named Younger Cascade range from 86±6 to 120±8 ka, with an average exposure age of 108±11 ka (Table 3). Sample 96-AZ-150-VT was collected from the top of a tumulus approximately 7 m high; this sample yielded the oldest exposure age on the Younger Cascade. The other primary flow surfaces (97-AZ-302-HYC and 97-AZ-303-HYC) yielded younger exposure ages of 86±6 and 99±7 ka, most likely indicating degradation of the surfaces at those sample locations. The younger ages yielded by sample 96-AZ-148-VT may indicate that the clasts were exhumed during the desert pavement formation process. Five samples collected from primary flow surfaces on the informally–named Bar Ten basalt flow range in age from 83±6 to 92±7 ka, yielding an average exposure age of 88±6 ka (Table 3). All five samples were primary flow surfaces and the variation in ages may be due to differential degradation of the Bar Ten surface.

The Younger Cascade entered the Inner Gorge of the Colorado River and may have formed a dam. The Bar Ten lava flow never reached the river (Figure 1). The Younger Cascade has a source area near that of the Bar Ten; however, olivine compositions of the Younger Cascade and Bar Ten lava flows differ significantly (Figure 7a), indicating that the flows are not magmatically related. Although the exposure ages of the Younger Cascade (108±11 ka) and the Mile 204.6 outburst–flood deposit (99±6 ka) are similar, the compositions of their olivines indicate they are unrelated.

Uncorrelated Hyaloclastite Ash

Two locations have been identified that preserve a hyaloclastite tuff that is uncorrelated with any outburst–flood deposit. A uniquely yellow and greenish-yellow hyaloclastite tuff with a unique chemical signature is found at river miles 185.7 and 195.7, approximately 174 and 20 m above current river level, respectively. The colors result from chemical alteration of the basalt glass. This is common of basaltic glass, which is highly susceptible to palagonitization [*Nayudu*, 1964]. Palagonite has been described as a hydrated sideromelane, an alteration product, with a vitreous or resinous appearance varying in color from yellowish-brown, orange or reddish-brown to dark brown or light gray [*Hay and Iijima*, 1968]. Analyses of basalt glasses from river miles 185.7 and 195.7 tightly cluster as alkali olivine basalts in group AOB II (Figure 6). No basalt boulders or cobbles nor dateable surfaces have yet been found associated with this uncorrelated deposit. The lava that produced these deposits most likely entered the Inner Gorge upstream of river mile 185.7, probably near Toroweap Valley (river mile 179).

DISCUSSION AND CONCLUSIONS

We distinguished Pleistocence basalt–rich terrace gravels, identified as lava–dam outburst–flood deposits, in western Grand Canyon, that are significantly different than typical Pleistocene and modern Colorado River gravels in eastern and western Grand Canyon, respectively. The outburst–flood deposits are composed almost entirely of basalt boulders, up to 35 m in b–axis diameter, and cobbles, and hyaloclastite tuffs. Elevations of Pleistocene gravels in eastern Grand Canyon are lower than the highest elevations of outburst–flood deposits in western Grand Canyon. Despite the similarities in ages, the difference in lithologies and elevations indicate that the outburst–flood deposits are not related to the Pleistocene main–stem Colorado River gravels in eastern Grand Canyon. Basalt glasses, whole rock basalt, and olivine phenocrysts within the outburst–flood deposits exhibit distinct chemical signatures and help to distinguish outburst–flood deposits.

We believe that a lava dam near the Uinkaret volcanic field failed catastrophically and produced an outburst-flood deposit that extended from river mile 189.5 to at least mile 202; the dam has not yet been identified. Deposits recording this event are well–preserved on *Hamblin's* [1994] Whitmore Dam and Massive Diabase basalt flows at river miles 189.5 and 202, respectively, and have a combined average exposure age of 159±2 ka. The strong correlation among basalt boulder exposure age, basalt glass chemistry and whole REE rock patterns leads us to conclude that the outburst–flood deposits at river miles 189.5 and 202 were syndepositional. This outburst flood had the highest maximum height ACRL (200 m) of any flood for which we have found evidence. The Hurricane fault has a vertical displacement rate of 79±7 m/Ma [*Fenton et al.*, in review] and it crosses the river near river mile 191. Based on this displacement rate and the exposure age of the deposits, the outburst–flood deposit at river mile 202 has been downthrown approximately 13 m, relative to the Mile 189.5 deposit, since the time of deposition.

Olivines and basalt glass in the Mile 204.6 outburst flood deposit have chemical signatures similar to those of olivines and basalt glass in the Mile 189.5 and Mile 202 flood deposits. The maximum elevation of the Mile 204.6 deposit is 53 m ACRL. A terrace at 36 m ACRL was cut into the pre–existing flood deposit has an average exposure age of 100±6 ka. It is probable that this flood deposit was emplaced contemporaneously with the outburst–flood deposits at river miles 189.5 and 202, and then an erosional event created the lower terrace 100 ka. Olivines within the hyaloclastite tuff and basalt boulders in the Mile 204.6 outburst-flood deposit have indistinguishable characteristics and may have been derived from the same lava flow. The chemistry of olivines extracted from the Younger Cascade (108±11 ka) and Bar Ten (88±6 ka) lava flows indicate they are not magmatically related. The Younger Cascade may have formed a lava dam on the Colorado River, but it is not related to deposition of the Mile 204.6 outburst–flood deposit.

Hamblin's [1994] Whitmore Dam unit is the same basalt flow as the informally Whitmore Cascade; thus, we believe that the Whitmore Cascade dammed the Colorado River 177±9 ka. Whole rock REE patterns indicate that the outburst–flood deposits at miles 189.5 and 202 were not produced by the failure of the Whitmore Cascade.

Chemical signatures of whole rock basalt and basalt glass indicate that the Mile 204.4 deposit is not related to the outburst–flood deposits at river miles 189.5, 202, or 204.6. Basalt boulders and basalt glass in the Mile 204.4 deposit are chemically unrelated but both yield REE patterns typical of alkali olivine basalts. The Younger Cascade (108±11 ka) may have formed basalt glasses that are found at river miles 204.4, 213 and 222; REE values of the Younger Cascade do not correlate with those of whole rock basalt boulders in the Mile 204.4 deposit. The Mile 204.4 outburst–flood deposit has a problematic average exposure age of 66±6 ka. Because it contains basalt glass chemically similar to the Younger Cascade lava flow, the Mile 204.4 deposit may be younger than 110 ka, but it is more likely that the deposit is older than the Mile 204.6 deposit, which is 18 m lower in elevation.

Reconstruction of an event as catastrophic as a dam failure is difficult immediately after an recent event, much less for dam failures that occurred 160 ka. In western Grand Canyon, considerable depositional and chemical evidence supports the hypothesis that some lava dams failed catastrophically, each resulting in an unsteady flood wave that moved through western Grand Canyon depositing characteristic outburst–flood sediment packages.

Several unknown factors limit reconstruction of discharges in the flood wave. At present, the heights of the lava dams are unknown, and dam geometry is a necessary aspect of reconstruction of the outflow hydrograph [*Walder and O'Connor*, 1997]. Many models of dam failure, particularly failures in earthen dams [*Walder and O'Connor*, 1997], rely on allometric relations for rapid erosion of trapezoidal breaches through the dam's crest. In the case of lava dams, the available data suggests piping as the primary mechanism, followed by large-scale collapse of the structure. Dam geometry, channel geometry, and clast size can be used in the future to estimate the outflow hydrograph from the lava-dam failures.

Acknowledgments. The authors acknowledge the helpful assistance of Clair Roberts, National Park Service Ranger at Tuweep. Peter Griffiths, Ted Melis, Mimi Murov, Sue Phillips, Alan Rigby, Ben Passey, Jon Pelletier, Dave Marchetti, Elliot Lips and Lauren Hammack helped with the field work. Discussions with George Billingsley and Keith Howard, U.S. Geological Survey in Flagstaff and Menlo Park, respectively, helped tremendously. Dominic Oldershaw provided the photography in this paper. Andy Hunt and Carolyn Dowling provided valuable technical expertise at the University of Rochester. We thank the reviewers, Keith Knudsen, Joel Pederson and William Philllips for the critical reviews of early drafts of this manuscript. Funding for this project is from the Geological Society of America, Sigma Xi, National Science Foundation and the U.S. Geological Survey. We thank the Grand Canyon Monitoring and Research Center for logistical support.

REFERENCES

Ackert, R.P., B.S. Singer, H. Guillou, and M.D. Kurz, Cosmogenic ^{3}He production rates over the last 125,000 years:

calibration against $^{40}Ar/^{39}Ar$ and unspiked K–Ar ages of lava flows, Geol. Soc. Amer. Abst., 29, A–299, 1998.

Alibert, D., A. Michard, and F. Albarede, Isotope and trace element geochemistry of Colorado Plateau volcanics, *Geochim. Cosmochim. Acta, 50*, 2735–2750, 1986.

Anthony, E.Y., and J. Poths, ^{3}He surface exposure dating and its implications for magma evolution in the Potrillo volcanic field, Rio Grande Rift, New Mexico, USA, *Geochim. Cosmochim. Acta*, 56, 4105–4108, 1992.

Best, M.G., and W.H. Brimhall, Late Cenozoic alkalic basaltic magmas in the western Colorado Plateaus and the Basin and Range Transition Zone, U.S.A., and their bearing on mantle dynamics, *Geo. Soc. Amer. Bull., 85*, 1677-1690, 1974.

Carling, P.A., and M.S. Glaister, Reconstruction of a flood resulting from a moraine-dam failure using geomorphological evidence and dam-break modeling, in *Catastrophic flooding*, ed. L. Mayer and D. Nash, Allen & Unwin, Winchester, Massachusetts, pp. 182-200, 1987.

Cenderelli, D.A. and B.L. Cluer, Depositional processes and sediment supply in resistant-boundary channels: examples from two case studies, in *Rivers Over Rock: Fluvial Processes in Bedrock Channels*, ed. K.J. Tinkler and Wohl, E.E., pp 105-131, American Geophysical Union, Geophysical Monograph 107, Washington, D.C., 1998.

Cerling, T.E., Dating geomorphologic surfaces using cosmogenic ^{3}He, *Quat. Res.*, 33, 148–156, 1990.

Cerling, T.E., and H. Craig, Cosmogenic ^{3}He production rates from 39° N to 46° N latitude, western USA and France, *Geochim. Cosmochim. Acta*, 58, 249–255, 1994a.

Cerling, T.E., and H. Craig, Geomorphology and in-situ cosmogenic isotopes, *Ann. Rev. Earth Plan. Sci.* 22, 273–317, 1994.

Cerling, T.E., R.H. Webb, R.J. Poreda, A.D. Rigby, and T.S. Melis, Cosmogenic ^{3}He ages and frequency of late Holocene debris flows from Prospect Canyon, Grand Canyon, U.S.A., *Geomorphology*, 27, 93-111, 1999.

Craig, H. and R.J. Poreda, Cosmogenic ^{3}He in terrestrial rocks: the summit lavas of Maui, *Proc. Natl. Acad. Sci.*, 83, 1970-1974, 1986.

Dalrymple, G.B., and K. Hirooka, Variation of potassium, argon, and calculated age in a Late Cenozoic basalt, *J. Geophys. Res.*, 70 (20), 5291–5296, 1965.

Dalrymple, G.B., and W.K. Hamblin, K-Ar of Pleistocene lava dams in the Grand Canyon in Arizona, *Proc. Natl. Acad. Sci. USA*, 95, 9744–9749, 1998.

Damon, P.E., A.W. Laughlin, and J.K. Percious, The problem of excess argon-40 in volcanic rocks, 24 pp., in *Symposium on Radioactive Dating and Methods of Low-Level Counting*, 2-10 March 1967, Paper SM-87/45, International Atomic Energy Agency, Monaco, 1967.

Dunai, T.J. and J.R. Wijbrans, Long-term cosmogenic ^{3}He production rates (152 ka – 1.35 Ma) from $^{40}Ar/^{39}Ar$ dated basalt flows at 29° N latitude, *Earth Plan. Sci. Lett.*, 176, 147-156, 2000.

Dutton, C.E., Tertiary History of the Grand Canyon District, 264 pp., 1882.

Fenton, C.R. , R.H. Webb, P.A. Pearthree, T.E. Cerling, and R.J. Poreda, Displacement rates on the Toroweap and Hurricane faults: Implications for Quaternary downcutting in Grand Canyon, submitted to *Geology*, in review.

Fitton, J.G., Petrology and geochemistry of late Cenozoic basalt flows, western Grand Canyon, Arizona, in *Geology of Grand Canyon, Northern Arizona (with Colorado River Guides)*, edited by D.P. Elston, G.H. Billingsley, and R.A. Young, pp. 186–189, American Geophysical Union, Washington, D.C., 1989.

Hay, R.L., and A. Iijima, Nature and origin of palagonite tuffs of the Honolulu Group on Oahu, Hawaii, in *Studies in volcanology-A memoir in honor of Howel Williams*, pp. 331–376, Geol. Soc. Amer., 1968.

Hamblin, W.K., *Late Cenozoic lava dams in the western Grand Canyon*, 135 pp., Geol. Soc. Amer. Memoir 183, 139 pp., 1994.

Henderson, P., *Rare Earth Element Geochemistry*, 510 pp., Elsevier, Amsterdam, 1984.

Howard, K.A., J.W. Shervais, and E.H. McKee, Canyon-filling lavas and lava dams on the Boise River, Idaho, and their significance for evaluating downcutting during the last two million years, in *Cenozoic Geology of Idaho*, ed. B. Bonnichsen and R.M. Breckenridge, pp. 629–644, Idaho Bureau of Mines and Geology, Moscow, Idaho, 1982.

Huntoon, P.W., G.H. Billingsley, and M.D. Clark, *Geologic map of the Hurricane fault zone and vicinity, western Grand Canyon, Arizona*, Grand Canyon Natural History Association, Grand Canyon, Arizona, 1981.

Irvine, T.N. and W.R.A., A guide to the chemical classification of the common volcanic rocks, *Canad. J. Earth Sci.*, 8, 523–548, 1971.

Kurz, M.D., In-situ production of terrestrial cosmogenic helium and some applications to geochronology, *Geochim. Cosmochim. Acta, 50*, 2855–2862, 1986.

Lal, D., Production of ^{3}He in terrestrial rocks, *Chem. Geol. (Isotope geoscience section)*, 66, 89–98, 1987.

Laughlin, A.W., J. Poths, H.A. Healey, S. Reneau, and G. WoldeGabriel, Dating of Quaternary basalts using the cosmogenic ^{3}He and ^{14}C methods with implications for excess ^{40}Ar, *Geology*, 22, 135-138, 1994.

Leeman, W.P., Late Cenozoic alkali-rich basalt from the western Grand Canyon area, Utah and Arizona: isotopic composition of strontium, *Geol. Soc. Amer. Bull., 85*, 1691–1696, 1974.

Lucchitta, I., G.H. Curtis, M.E. Davis, S.W. Davis, and B. Turrin, Cyclic aggradation and downcutting, fluvial response to volcanic activity, and calibration of soil–carbonate stages in the western Grand Canyon, Arizona, *Quat. Res.*, 53, 23–33, 2000.

Machette, M.N., and J.N. Roshold, Quaternary terraces in Marble Canyon and eastern Grand Canyon, Arizona, in *Geology of Grand Canyon, Northern Arizona*, edited by E.P. Elston, et al., pp. 205-211, American Geophysical Union, Washington, D.C., 1989.

Malde, H.E., The Yahoo Clay, a lacustrine unit impounded by the McKinney Basalt in the Snake River canyon near Bliss, Idaho, in *Cenozoic Geology of Idaho*, edited by B. Bonnichsen, and R.M. Breckenridge, pp. 617–628, Idaho Bureau of Mines and Geology, 1982.

Mamyrin, B.A., and I.N. Tolstikhin, *Helium isotopes in nature*, 273 pp., Elsevier, Amsterdam, 1984.

McKee, E.D. and W.K. Hamblin, and P.E. Damon, K-Ar age of lava dams in Grand Canyon, *Geol. Soc. Amer. Bull.*, 79, 133–136, 1968.

McKee, E.D., and E.T. Schenk, The lower canyon lavas and related features at Toroweap in Grand Canyon, *J. Geomorph.*, 5 (4), 245–273, 1942.

Nayudu, Y.R., Palagonite tuffs (hyaloclastites) and the products of post-eruptive processes, *Bull. Volcan.*, 27, 391–410, 1964.

Peterson, N.V., and E.A. Groh, Geologic tour of Cove Palisades State Park near Madras, Oregon, *The Ore Bin*, 32 (8), 141–168, 1970.

Pierson, T.C., and K.M Scott, Downstream dilution of a lahar: Transition from debris flow to hyperconcentrated streamflow, *Water Res. Res.*, 21, 1511-1524, 1985.

Poreda, R.J., and T.E. Cerling, Cosmogenic neon in recent lavas from the western United States, *Geophys. Res. Let.*, 19 (18), 1863–1866, 1992.

Pouchou, J.L., and F. Pichoir, Quantitative analysis of homogenous or stratified microvolumes applying the model "PAP", in *Electron probe quantification*, ed. K.F.J. Heinrich and D.E. Newbury, pp. 31–75, Plenum Press, New York, 1991.

Singh, V.P., *Dam breach modeling technology*, 244 pp., Kluwer Academic Publishers, Boston, 1996.

Smith, G.O., Ellensburg folio, *U.S. Geol. Surv. Atlas, folio no. 86*, 1903.

Stearns, H.T., Geology and water supply resources of the middle Deschutes River basin, Oregon, *U.S. Geol. Surv. Water Supply Paper*, 637, 125–212, 1931.

Stevens, L., *The Colorado River in Grand Canyon*, 110 pp., Red Lake Books, Flagstaff, 1983.

Taylor, S.R., and S.M. McClennan, *The Continental Crust: its composition and evolution*, 312 pp., Blackwell Scientific Publications, London, 1985.

Walder, J.E., and J.E. O'Connor, Methods for predicting peak discharge of floods caused by failure of natural and constructed earthen dams, *Water Res. Res.,* 33, 10, 2337-2348, 1997.

Webb, R.H., T.S. Melis, P.G. Griffiths, J.G. Elliott, T.E. Cerling, R.J. Poreda, T.W. Wise, and J. Pizzuto, Lava Falls Rapid in Grand Canyon: Effects of Late Holocene debris flows on the Colorado River, *U. S. Geol. Surv. Prof. Paper 1591*, 90 pp., 1999.

Webb, R.H., P.T. Pringle, and G.R. Rink, Debris flows from tributaries of the Colorado River, Grand Canyon National Park, Arizona, *U. S. Geol. Surv. Prof. Paper 1492*, 1989.

Wells, S.G., L.D. McFadden, J. Poths, and C.T. Olinger, Cosmogenic ^{3}He surface-exposure dating of stone pavements: Implications for landscape evolution in deserts, *Geology*, 23 (7), 613–616, 1995.

Wenrich, K.J., G.H. Billingsley, and B.A. Blackerby, Spatial migration and compositional changes of Miocene-Quaternary magmatism in the western Grand Canyon, *J. Geophys. Res.*, 100 (B7), 10,417–10,440, 1995.

Wenrich, K.J., G.H. Billingsley, and P.W. Huntoon, P.W., Breccia-pipe and geologic map of the northeastern part of the Hualapai Indian reservation and vicinity, Arizona: U.S. Geol. Surv. Misc. Invest. Series Map I-2440, 2 sheets, p. 19 + map., 1997.

Wenrich, K.J., G.H. Billingsley, and P.W. Huntoon, P.W., Breccia-pipe and geologic map of the northeastern part of the Hualapai Indian reservation and vicinity, Arizona: U.S. Geol. Surv. Misc. Invest. Series Map I-2440, 2 sheets, p. 19 + map., 1997.

Cassandra R. Fenton, Thure E. Cerling, and Barbara P. Nash, 135 S 1460 E, Rm 719, Dept. of Geology and Geophysics, University of Utah, Salt Lake City, UT 84112, crfenton@mines.utah.edu;

Robert H. Webb, U.S. Geological Survey, 1675 W. Anklam Road, Tucson, AZ 85745.

Robert J. Poreda, Hutchison Hall 227, Dept. of Earth and Environmental Sciences, University of Rochester, Rochester, NY 14627.

Evaluation of Glacial Outburst Flood Hypothesis for the Big Lost River, Idaho

Keith L. Knudsen[1] and Janet M. Sowers

William Lettis & Associates, Inc., Walnut Creek, CA

Dean A. Ostenaa and Daniel R. Levish

U.S. Bureau of Reclamation, Denver Federal Center, Denver, CO

Geomorphic evidence along the Big Lost River in Idaho suggests that a catastrophic flood likely occurred earlier than previously thought. Previous researchers described flood-related features in the Box Canyon area and suggested that a large flood might have originated about 20 to 30 ka from Glacial Lake East Fork, about 100 km upstream of Box Canyon. Geomorphic mapping of glacial till and outwash terraces near the location of Pleistocene Glacial Lake East Fork reveals problems with this model and raises new questions about the viability of 20 to 30 ka glacial outburst flooding as the mechanism responsible for forming the features in Box Canyon. Smooth low hills of older till downstream from the former lake show no evidence of overtopping or scouring. Outwash terraces estimated to be 20 to 30 ka also show no evidence of having been overtopped by a large flood. The incised valley between these terraces has insufficient channel capacity to pass a large flood. Between the glaciated headwaters and Box Canyon, geomorphic features suggestive of outburst flooding are few and most are explicable by other processes. A 5-m granitic boulder on an outwash terrace about 15 km downstream of Glacial Lake East Fork may have been ice rafted in an earlier outburst flood. Limestone walls of narrow canyons, through which the Big Lost River flows, show no evidence of abrasion, and their abundant ledges and hollows contain no fluvial sediment or exotic clasts. Scarps cut into the toes of middle to late Pleistocene alluvial fans near Mackay could have been produced by a large flood or by a laterally migrating stream. Upstream of Box Canyon, near Arco, fluvial terraces estimated to be at least 50 ka appear not to have been overtopped since their formation, providing a minimum age for large floods on this reach of the Big Lost River. Downstream of Box Canyon, soil development on alluvium that should have been modified substantially or may have been deposited by an outburst flood is similar to or greater than soil development on late Pleistocene terrace surfaces near

Ancient Floods, Modern Hazards: Principles and Applications of Paleoflood Hydrology
Water Science and Application Volume 5, pages 217-235

[1] Present address: California Division of Mines and Geology, San Francisco, CA

Arco, suggesting a minimum age of 50 to 90 ka for any large floods. Thus, the features near Box Canyon must be significantly older than previously suggested.

INTRODUCTION

This evaluation presents geomorphic evidence that bears on the flood history of the Big Lost River. The hypothesis proposed by *Evenson et al.* [1982], *Rathburn* [1989, 1993] and *Cerling et al.* [1994] that a large flood affected the Big Lost River about 20 to 30 ka is reexamined in the light of geomorphic evidence from reaches upstream and downstream of Box Canyon. Previous workers have been puzzled by the lack of strong evidence for this flood outside of the Box Canyon area. This study is a systematic examination of geomorphic features that might be expected to provide some evidence, positive or negative, for the occurrence and timing of the flood. Three types of evidence are identified herein: (1) direct stratigraphic and geomorphic evidence of a flood; (2) the absence of evidence for an outburst flood at locations where such evidence should be preserved; and (3) direct stratigraphic and geomorphic evidence precluding the occurrence of an outburst flood or constraining the timing of a large flood. Reevaluation of the evidence of flood features and discharge estimates made by *Rathburn* [1989, 1993] in the Box Canyon area was not part of this investigation.

The Big Lost River is one of the principal drainages of the Pioneer Mountains and Lost River Range in eastern Idaho. It flows northeastward out of the Pioneer Mountains from four main tributary watersheds: the North and East Forks of the Big Lost River, Wildhorse Creek, and Summit Creek (Figure 1, also on CD). Multiple times during the late Quaternary each of these watersheds has contained large alpine glaciers. At their maximum extent, the glaciers reached no farther downstream than the confluence of the North Fork and East Fork. Below this confluence, the Big Lost River flows through and over alluvium that is primarily reworked glacial outwash and till. In the Thousand Springs area (Figure 1), the mainstem Big Lost River turns to the southeast and flows along the base of the bajada of the Lost River Range to Arco. From the North and East Fork confluence to about the location of Arco, the Big Lost River is an alluvial stream. Downstream of Arco, the river flows across and percolates through basalt bedrock of the Eastern Snake River Plain. In the Box Canyon area, the river has incised a narrow canyon into the basalt. Farther downstream the river emerges from the canyon and terminates in a closed alluvial basin.

The motivation for this study was concern on the part of the Idaho National Engineering and Environmental Laboratory (INEEL) about flood hazards along the Big Lost River. The investigation and results presented herein are part of a larger probabilistic flood hazard assessment for INEEL [*Ostenaa et al.*, 1999]. In this investigation, an array of geomorphic methods and reasoning were used to constrain the timing and magnitude of past large floods on the Big Lost River. This investigation might be considered a case study for the approach outlined by *Levish* [this volume].

The Outburst Flood Hypothesis

Rathburn [1989, 1991, 1993] conducted a geomorphic and hydrologic study in the Box Canyon area (Figure 1) and described features that she interpreted as evidence of catastrophic flooding. Such features include scabland topography, elongate hills parallel to the direction of flow, a loess scarp, and flood-transported boulders. Using step-backwater modeling, *Rathburn* [1989, 1993] estimated that a discharge of about 60,000 m^3/s (about 2.1 x10^6 cfs) was required to produce the flood features identified near Box Canyon.

Granitic boulders found in the Box Canyon area are of a lithology thought to occur only in one place in the Big Lost River watershed—the headwaters of Wildhorse Creek. Granitic boulders found elsewhere along the Big Lost River, including a boulder found near Swensen Butte (Plate 1, color version on CD), are of similar lithology. Ice-rafted boulders near the Wild Horse Creek/East Fork confluence are thought to be derived from the same parent rock [*Brugger et al.*, 1983]. *Evenson et al.* [1982] and *Rathburn* [1989] concluded that the most likely mechanism for transport of these boulders to downstream reaches is an outburst flood.

Age of the Proposed Outburst Flood

Flood features near Box Canyon have been interpreted to be cut into the Quaking Aspen Butte lava flow, dated at about 95 ka [*Kuntz et al.*, 1994; *Forman et al.*, 1994], placing a possible maximum age on the flood [*Rathburn*, 1991]. *Rathburn* [1989] noted that middle to late Pleistocene alluvial fans were truncated (perhaps by a large flood) near Mackay (Figure 1), whereas the latest Pleistocene fans in this area grade to the modern channel. Additionally, she described probable outburst-flood-related terraces as having "a moderately well developed surface soil", calcic horizons, and 1-mm-thick carbonate rinds on clasts. She also described well developed weathering rinds on flood-related basalt boulders, and identified erratic, presumably flood-transported boulders that are buried by loess.

Cerling et al. [1994] used cosmogenic ^{3}He and ^{21}Ne to estimate the age of flood-deposited basalt boulders and

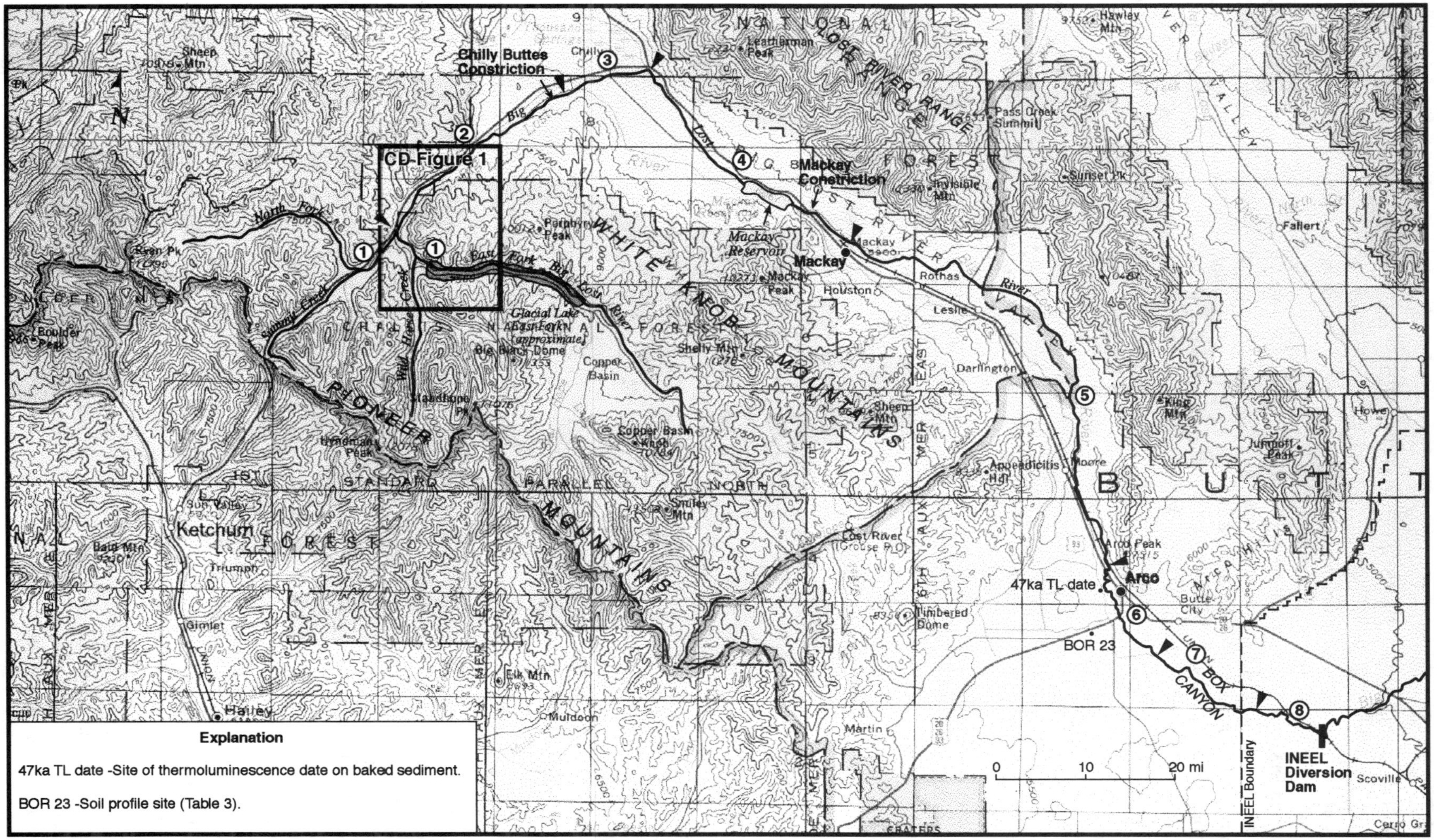

Figure 1. Location map, upper Big Lost River, and identification of reaches 1 through 8.

scoured-bedrock surfaces in the Box Canyon area. *Cerling et al.* [1994] estimated the age of five samples, three from boulders derived from less than 200 ka basalt flows [*Kuntz et al.*, 1994] within two separate boulder bars. A fourth sample was from a flood-transported boulder thought to be derived from an approximately 520 ka basalt flow [*Kuntz et al.*, 1994]. The fifth sample was collected from the bedrock lip of a feature interpreted to be a cataract formed during an outburst flood. Cosmogenic exposure ages for the boulders sampled downstream of the cataract ranged from 19,000 to 43,000 cal yr B.P., with the discrepancy in ages attributed to pre-flood exposure of some of the boulders to cosmogenic radiation. The sample from the bedrock lip of the cataract yielded an age of 92,800 cal yr B.P. The relatively older age for this sample was explained by erosion of the bedrock lip during an earlier flood event. *Cerling et al.* [1994] did not report an age for the fourth sample. Based on the two youngest age estimates, *Cerling et al.* [1994] concluded that the Big Lost River flood occurred at about 20,500 cal yr B.P.

The work of *Evenson et al.* [1982], *Brugger et al.* [1983], and *Brugger* [1996] provides an overview of the glacial history in the area of Pleistocene Glacial Lake East Fork. *Evenson et al.* [1982] were the first to document evidence for Glacial Lake East Fork, a possible source for the flood thought to have formed some of the features near Box Canyon. *Brugger et al.* [1983] described the provenance of glacial deposits, and reconstructed glacial flow paths for four glacial advances. *Brugger* [1996] used till provenance studies to describe flow paths and dynamics of the two principal tributary glaciers in Wildhorse Creek, the source of ice and debris responsible for damming the East Fork and forming Glacial Lake East Fork. These studies did not include numerical age estimates for the Wildhorse Creek glacial advances. Thus, additional studies conducted within the greater region of central and eastern Idaho also were used in this study as a basis for estimating ages of late Quaternary deposits and landforms (Table 1).

The ages of the Wildhorse I and II glaciations are not well constrained, but they are thought to correlate with the Pinedale glaciations of the Rocky Mountains (Table 1). *Brugger et al.* [1983] suggested that Wildhorse I correlates with the early Pinedale, the late Wisconsin glacial maximum, and Wildhorse II correlates with the late Pinedale. Nowhere in the literature on the upper Big Lost River is there a suggestion that any of these moraines date from the early Wisconsin, or about 60 to 80 ka. However, glacial deposits of this age (oxygen isotope stage 4) are documented elsewhere in the Rocky Mountains [e.g., *Forman et al.*, 1993], and *Gianniny* [1998] documents a high stand of Lake Terreton, located at the downstream end of the Big Lost River, at 88 ka.

Source of the Proposed Outburst Flood

A source for the flood responsible for forming the features in Box Canyon has not been identified conclusively. *Rathburn* [1989, 1993] pointed out that Glacial Lake East Fork [*Evenson et al.*, 1982] could have failed catastrophically resulting in an outburst flood, but that the discharge from this lake would not have been sufficient to produce the required discharge downstream. Based on a survey of ice-rafted erratics and planimetric calculations, *Rathburn* [1989, 1993] estimated the volume for Glacial Lake East Fork to be on the order of 6.3 km^3, with a mean depth greater than 130 m (Table 2). Using this calculated lake volume and published regression relationships relating lake volume to peak discharge for historic large outburst floods, she estimated a peak outburst discharge of 26,000 to 38,000 m^3/s. These values are significantly less than the 60,000 m^3/s estimated based on step-backwater modeling at Box Canyon. More recent work by *Brugger* (personal communication) also showed that catastrophic failure of Glacial Lake East Fork could not have produced flows of the magnitude thought to have occurred at Box Canyon (Table 2).

Rathburn [1989] also discussed the attenuation of a flood peak between the site of Glacial Lake East Fork and Box Canyon. She estimated that a discharge of 200,000 m^3/s would be necessary at Glacial Lake East Fork to produce the modeled discharge of 60,000 m^3/s downstream at Box Canyon. No other sites of glacial lakes or other sources for the proposed flood have been identified in the Big Lost River watershed.

METHODS

This study included aerial photograph interpretation, field and aerial reconnaissance, reconnaissance-level soil/geomorphic evaluations in the area of Glacial Lake East Fork, and more detailed soil/geomorphic investigations in downstream reaches near Box Canyon. Geomorphic maps of fluvial terraces and glacial deposits for key reaches of the Big Lost River were prepared from field reconnaissance and interpretation of aerial photography. Ages of geomorphic surfaces and late Quaternary geologic deposits were qualitatively assessed by: (1) relative geomorphic position; (2) reconnaissance-level evaluations of soil profile development; (3) degree of surface modification; and (4) thickness of carbonate coatings on clasts. Deposit /geomorphic surface age is important because a non-inundation approach [*Levish*, this volume] is used to constrain the timing and magnitude of past floods in some reaches. The relations between fluvial terraces and glacial deposits of different age, and between fluvial deposits, eolian deposits, and basalt flows were examined. The walls of bedrock constrictions were searched for flood deposits, er-

Table 1. Tentative correlation chart for deposits mentioned in this study. Age estimates (in ka) given when provided by previous workers. Arrows suggest possible alternate correlations.

Glacial Deposits									Loess Deposits		Lacustrine Deposits		
Big Lost River [this study]			Wildhorse Canyon [Evenson et al., 1982]		Big Lost River [Evenson et al., 1982]	North Yellowstone [Sturchio et al., 1994]	McCall Idaho [Colman and Pierce, 1986]	West Yellowstone [Colman and Pierce, 1981]	Eastern Snake River Plain [Forman et al., 1993]	Southeastern Idaho [Pierce et al., 1982]	Lake Terreton [Gianniny et al., 1999]	Marine Oxygen Isotope Stage	
Alluvial Fans	Outwash Terraces	Glacial Till											
									10	10		13	1
Qf4	Qt4		Wild-horse Canyon	IV			12.5	12.5					
Qf3	Qt3	Qgw II		III, II	Potholes	15.5 Deckard Flats 19.5	Pinedale	Deckard Flats					
Qf2	Qt2	Qgw I		I		22.5 Pinedale 30	Pinedale 22	22	Pinedale	Loess Unit A	~22 high stand		2
↓	↓	↓		↓		34 Early Pinedale 47		35 Pinedale 43	35 to 40			35	3
							58 Intermediate 70		~60 Early to Mid. Wisconsin ~80	70 Loess Unit B	~85-90 high stand	65	4
												79	5a, 5b, 5c, 5d
Qf1	Qt1	Qgd	Devils Bedstead	II, I	Copper Basin		130 Bull Lake 150 Outer Bull Lake ↓	130 Bull Lake 140		150	120 high stand 160	132	5e, 6
												198	7, 8, 9, 10, 11
												480	12, 13
		Qgd+	Pole Creek		Pioneer		Pre-Bull Lake						

ratics, evidence of abrasion, ice rafted rocks, or other evidence of past large floods.

ASSESSMENT AND IDENTIFICATION OF EVIDENCE RELATED TO THE GLACIAL OUTBURST FLOOD HYPOTHESIS

In the following sections, evidence observed in eight reaches of the Big Lost River is described. Several of the reaches provide little information relevant to this study because of the wide valleys and abundant Holocene and late Pleistocene valley-fill sediment.

Big Lost River Reach 1 - Glaciated Headwaters Upstream of the East Fork/North Fork Confluence

There is abundant evidence supporting the existence of Pleistocene Glacial Lake East Fork as proposed by *Evenson et al.* [1982]. The most conclusive evidence for this lake is a number of large granitic and migmatitic boulders perched on the canyon walls of the East Fork for about 10 km upstream of the terminus of the Wildhorse Glacier (CD Figure 1). There is no source for rocks of these lithologies in the watershed of the East Fork [*Brugger et al.*, 1983]; rather, they are thought to be derived from the Wildhorse Creek watershed. These boulders likely were carried up the East Fork to their present locations by icebergs that calved off the Wildhorse Creek glacier into the lake. Two of these boulders were found on the north wall of the East Fork canyon, located at approximately 2240 m (7350 feet) elevation, or 82 m (270 feet) above the present stream (elevation about 2158 m [7,080 feet]) (CD Figure 1). These granitic boulders have a few small patches of glacial polish preserved on their surfaces, as do some of the boulders observed on the surface of Wildhorse I till deposits.

The existence of Glacial Lake East Fork is further supported by the distribution of till from the Wildhorse I gla-

Table 2. Previous hydrologic estimates for Glacial Lake East Fork volumes and discharges.

Author	Maximum Water Surface Elevation (m and ft)	Volume (m^3)	Peak Discharge and Method (m^3/s)
Evenson et al. (1982)	"lake depths of up to 120 m" (a)	NA	NA
Rathburn (1989)	2,280 m - 7,480 ft	6.3×10^9 (mean depth in excess of 130 m)	26,000-38,000, based on relations between lake volume and maximum discharge
Brugger (personal communication.)	2,225 m - 7,300 ft	1.95×10^8 4.0×10^8 (maximum)	900-2,200, subglacial tunnel model
			5,800-6,000, subaerial breach model
	2,286 m -7,500 ft (about 60 m higher than paleoshorelines)	1.06×10^9	~30,000, subaerial breach model, maximum lake levels

(a) Brugger (personal communication) reports that the 120 m lake level is high; he believes a maximum depth of about 90 m is more appropriate.

cier, which can be used to evaluate the geometry of the ice dam. This till, of Wildhorse Creek provenance [*Brugger et al.*, 1983], is found on both the south and north sides of the East Fork near the confluence (QgwI on CD Figure 1). Thus, the Wildhorse I glacier must have crossed the East Fork, potentially damming it. The elevations of the highest Wildhorse I moraine crests on the south side of the confluence (7,400 to 7,600 feet) are consistent with the elevation of the ice rafted boulders (7,350 feet) on the wall of the canyon of the East Fork. Thus, the Wildhorse I glacier surface and ice dam was high enough to impound a lake to at least the elevation of the ice-rafted granitic boulders.

Other features support the past presence of Glacial Lake East Fork, including: (1) subtle strand lines or benches on alluvial slopes on the south side of the East Fork (CD Figure 1); and (2) the grading of these alluvial slopes to a base level above the present channel (CD Figure 1). The strand lines or benches are apparent on aerial photographs. In the field, the strand lines are most apparent when viewed from the north side of the East Fork, near the boulders shown on CD Figure 1.

As pointed out by *Evenson et al.* [1982], older glaciers of equal or greater size also may have dammed the East Fork and created glacial lakes in its valley. For example, based on the distribution of till, the Devil's Bedstead glaciation (Table 1) was more extensive than the Wildhorse I glaciation; thus, there also may have been a larger lake during Devil's Bedstead time. Based on the distribution of till (CD Figure 1), the Wildhorse I glacier is the youngest of the glaciers capable of damming the East Fork.

There is not conclusive evidence for a catastrophic outburst flood emanating from Glacial Lake East Fork during Wildhorse I time. Previous workers [*Evenson et al.*, 1982; *Brugger*, personal communication] identified two lines of evidence for outburst flooding from this lake. The first is the presence of large (2- to 5-m-diamter) granite and migmatite boulders on outwash terrace surfaces up to 15 km downstream in Reach 2 (see reach 2 discussion). The second line of evidence for catastrophic flooding is incision of the East Fork into bedrock directly downstream of the ice dam. Although this channel could have been incised during an outburst flood, such a channel can be cut in a normal fluvial or glaciofluvial system. This bedrock reach of the stream runs along the north margin of the roughly east-west trending valley (CD Figure 1) against the bedrock valley margin where bedrock is likely to be shallow. More importantly, the current channel is incised into outwash terrace Qt2 of Wildhorse I age, thus the incision postdates the Wildhorse I glaciation. As discussed below, any glacial outburst flood must predate the Qt2 outwash terraces because the cross-sectional area between remnants of Qt2 is insufficient to pass a large flood without overtopping and leaving a record on these smooth, unaltered surfaces. Therefore, incision of the East Fork into bedrock has taken place primarily since the early Pinedale, likely during the Wildhorse I recession, and is the result of the position of the stream at the north margin of the valley, where bedrock is shallow.

No other evidence of catastrophic failure of Glacial Lake East Fork was found in this reach during this study. If the lake failed catastrophically, the floodwaters would be expected to have affected the outside of bends in the river, such as the prominent bend to the northeast where the East Fork joins the North Fork (CD Figure 1). The hillslope on the northwest side of this bend was examined in a search for ice rafted boulders, scour marks, or fluvial sediment. No such features were observed on these slopes. However, old, probably middle Pleistocene till was found in patches on the top of one bedrock knoll (CD Figure 1); this till apparently was not removed or altered by a younger flood.

Reconnaissance-level field observations of boulder weathering and soil development were made in order to derive an independent estimate for the age of moraines and outwash deposits of Wildhorse I and II age in Reach 1. The surfaces of moraines of Wildhorse I age are littered with weathered granitic boulders. The boulders are rough, with single mineral grain relief, and weathering pedestals 5 to 12 cm high are present on a small percentage of the boulders. Glacial polish is rare, and typically occurs as a remnant patch on very quartz-rich veins or inclusions. The percentage of weathered boulders with single mineral grain relief on Wildhorse I age moraines was qualitatively compared with that documented by *Burke and Birkeland* [1983] for moraines in the Sierra Nevada of California. The percentage of weathered granitic boulders on Wildhorse I moraines (95%) is more similar to that on the Tahoe moraines of oxygen isotope stage 6 (Table 1) (70 to 80%) than the Tioga moraine of oxygen isotope stage 2 (10 to 20%). Although weathering rates may vary between the two sites due to differences in climate and granite composition, these observations suggest a pre-late Wisconsin age for the Wildhorse I moraines.

Multiple soil exposures in Reach 1 were examined at a reconnaissance level, including two roadcuts in terrace deposits inset into the Wildhorse II moraine, and one stream cut exposure in outwash of the Wildhorse I glaciation. Based on accumulation of carbonate, development of soil structure, degree of horizonation, and carbonate coatings on clasts, soil profile development of the soils examined was consistent with at least a 25 to 50 ka age for Wildhorse I and a younger age (about 15 to 25 ka?) for Wildhorse II. These age estimates are consistent with, or slightly older than, the ages inferred from the correlation of *Evenson et al.* [1982] to other better constrained glacial chronologies (Table 1).

Big Lost River Reach 2 - East Fork/North Fork Confluence to Chilly Buttes Constriction

In Reach 2, outwash terraces flank the river for about 20 km (CD Figure 1). Study of this reach included a search for evidence that a large flood had affected the surface morphology, composition, grain size, or structure of the terraces, or had left deposits or erosional features on adjacent bedrock valley walls. Four terraces (Qt1, Qt2, Qt3 and Qt4, in order of decreasing age) are mapped on CD Figure 1. These terraces grade to upstream deposits related to distinct glacial advances and are used to constrain the timing of past large floods.

Terrace Qt4 is found directly adjacent to the present stream and usually less than 5 meters above the stream water surface. In places, it shows evidence of having been overtopped by flooding in the past several years. Because of its geomorphic position, it is assumed that terrace Qt4 stabilized during the late Holocene. Terrace Qt3 is higher above the present stream than Qt4, and is therefore older. Terrace Qt3 is probably no younger than early Holocene to latest Pleistocene in age and, based on its geomorphic position, tentatively is correlated with the latest Pleistocene Wildhorse II glaciation.

Terrace Qt2 is the most extensive, laterally continuous terrace surface found downstream of the North Fork/East Fork confluence. The Qt2 surface is about 18 m above the present stream channel near the confluence, and about 7 m above the channel at Chilly Buttes, based on interpretation of 7.5-minute topographic maps. The surface of terrace Qt2 is smooth with no evidence of post-deposition surface modification. Surface soils formed on the Qt2 surface are characterized by continuous calcium carbonate coatings 1 to 3 mm thick on the bottoms of clasts in a Bk horizon. Because this terrace is graded to Wildhorse I moraines this terrace is correlated to the Wildhorse I glaciation.

Boulders were observed in road cuts and ditches within the Qt2 deposits, but relatively few boulders were found at the surface of the deposit. In several places, boulders and large cobbles were observed in the terrace riser separating terrace Qt2 from terrace Qt3. These boulders likely were left behind during the incision event that eroded Qt2 prior to deposition of the Qt3 terrace. The boulders and coarse cobbles indicate that the stream responsible for depositing Qt2 was a higher energy system than the present system. However, the smooth surface of Qt2, lack of large bedforms, and, where deposits are exposed, no sedimentologic evidence of large floods (e.g. large-scale cross bedding), argues against the terrace deposits resulting directly from an outburst flood.

Remnants of terrace Qt1 and alluvial fan Qf1 are preserved along the margins of the valley (CD Figure 1). The surface of these terrace and fan remnants is generally smooth, although there is greater post-stabilization dissection than on terrace Qt2. Terrace Qt1 and fan Qf1 are tentatively correlated with the Devil's Bedstead glacial advance (Table 1).

The outwash terraces Qt2 and Qt1 show no evidence of being overtopped by an extreme flood. No erosional scours, large boulder bars, cataracts, meander scars, mega ripples, or other evidence of anomalously high stream power was observed on the surface of the terraces. Additionally, exposures of terrace Qt2 and Qt1 have well developed soil profiles, indicating prolonged post-depositional surface stability and little erosion. Thus, neither of these terrace surfaces appear to have been overtopped by significant depths of high velocity water.

If an outburst flood passed through this reach, it must predate the Qt2 terrace. The channel cross-sectional area between remnants of Qt2 is insufficient to pass a large flood without overtopping the terrace surface (cross sections A-A' and B-B' on CD Figure 1). Assuming an average velocity of 5 m/s, and using cross-sectional areas measured from cross sections A-A' and B-B' (CD Figure 1), the maximum discharge that can flow through the valley without overtopping the Qt2 surface is about 6,400 m^3/s at cross-section A-A' and about 15,000 m^3/s at cross-section B-B'. Using higher velocities in these rough estimates would increase the discharges (e.g. assuming an average velocity of 10 m/s would double the discharge estimates). Discharges estimated in this fashion are much lower than the 60,000 m^3/s thought to have reached Box Canyon. The age of terrace Qt2 is constrained by its association with the Wildhorse I glacial moraines, and by reconnaissance-level observations of soil-profile development in the field. As noted previously, the Wildhorse I advance and terrace Qt2 are estimated to be no younger than about 25,000 years.

An outburst flood that post-dated stabilization of terrace Qt1 and pre-dated the Qt2 terrace may have occurred. Cross-sections A-A' and B-B' on CD Figure 1 show the estimated discharges that might have been responsible for forming the steep scarps separating Qt2 from Qt1, without significantly overtopping terrace Qt1. The discharge estimates assume that the channel was scoured to bedrock, which is thought to be near the base of the active channel on the basis of the presence of bedrock in the channel bed near the East Fork/North Fork confluence. The discharge estimates assume average velocities of 5 m/s and 10 m/s. Using 5 m/s, discharges of about 60,000 m^3/s and 83,000 m^3/s could be routed between the Qt1 and Qfl surfaces, and thus leave no record on these surfaces.

Significance of anomalously large boulders. Perhaps the strongest evidence to support past extreme flooding in the Big Lost River watershed outside of the Box Canyon area is the presence of anomalously large exotic boulders on terrace surfaces within Reach 2 [*Evenson et al.*, 1982; *Rathburn*, 1989]. In Reach 2, the largest such boulder is located 15 km from the glacial terminus, on outwash terrace Qt2 near Swenson Butte (Plate 1). This approximately 5x3x2 m granite migmatite boulder is thought to originate from the Wildhorse Canyon watershed. This boulder is subrounded, rough, and weathered over most of its surface except for a small patch of glacial polish preserved on the top of the boulder. A 7- to 9-cm-high erosional pedestal was found at the top of the boulder. Ice rafting during a glacial outburst flood has been suggested as a possible mechanism for emplacement of this boulder [*Evenson et al.*, 1982; *Rathburn*, 1989].

The ice rafting mechanism deserves further examination. In a discussion of the effective density of ice encased boulders, *Carling and Tinkler* [1998] state that a volume of ice

$$V_i \geq V_c\,[(\gamma_s-\gamma)/(\gamma_i-\gamma)]$$

is required to "float" an ice encased boulder, where V_i is the ice volume, V_c is the boulder volume, γ_i is the specific weight of ice, γ_s is the specific weight of the boulder, and γ is the specific weight of water. Assuming $\gamma_i = 0.9$ g/cm^3, $\gamma_s = 2.65$ g/cm^3, and $\gamma = 1$ g/cm^3, a volume of ice approximately 16.5 times the boulder volume is required to "float" a boulder. Using *Rathburn's* [1989] measurements for the Swensen Butte boulder (5.2x3.3x2.6 m), a boulder volume of about 30 m^3 is assumed. A volume of ice 16.5 times this size (approximately 495 m^3) would be necessary to "float" this boulder. This volume of ice would be equivalent to a rind of ice approximately 3 m thick. In order to suspend such an ice-encased boulder, the water would have to be about 10 m deep (combined thickness of boulder and ice). If this 2.5-km-wide, flat valley were filled with 10 m of water moving at a velocity of 5 m/s, the resulting discharge would be 125,000 m^3/s, significantly greater than the maximum discharge of about 30,000 m^3/s for failure of the ice dam at Glacial Lake East Fork (calculated by *Rathburn* [1989] and *Brugger* [personal communication]). Assuming the maximum discharge of 30,000 m^3/s and the above dimensions and velocities, the calculated water depth is 2.4 to 4 m, insufficient to float this boulder, but perhaps enough to slide or roll it short distances.

The hypothesized extreme discharges needed to ice raft these large boulders are inconsistent with the lack of geomorphic evidence of extreme discharges flowing over the surface of the terraces upon which the boulders sit. The terrace surfaces are smooth, planar, and gently sloping, not scoured or rippled. Mechanisms for transport of large boulders that do not require extreme discharges of a glacial outburst flood should be explored. First, the boulder may have been encased in an iceberg calving from the toe of the glacier, but was not floated downstream. Any thickness of ice frozen to a boulder will decrease the effective density of the boulder, making its transport as bedload by fast moving water more likely. If the boulder sat on an active outwash plain for hundreds of years, glacial melt flows may have been capable of transporting the boulder short distances. Second, the boulder may be much older than the terrace surface on which it sits. It may be a "lag" from an earlier deposit or flood that was reworked and remobilized to form terrace Qt2. Third, the boulder may have been moved by a debris flow, although its location many kilometers downstream of its source argues against this possibility.

Bedrock constrictions. Two bedrock constrictions, the Chilly Buttes constriction between Reaches 2 and 3 (Figure 1; Plates 1 and 2, color versions on CD) and the Mackay constriction in Reach 4 (Figures 1 and 2, also on CD) were identified as locations that might have preserved evidence of an outburst flood. In both cases, the river flows through a relatively narrow gap, 0.8 and 0.5 km wide, respectively, between steep slopes underlain by limestone bedrock. At each of these constrictions, the depth of water associated with a discharge of 60,000 m^3/s was estimated given the cross-sectional area of the constriction (measured from 7.5-minute topographic maps), and an assumed velocity of about 3 to 6 m/s. Lower velocities are used for these calculations than for calculations upstream because the slope in these reaches is lower. Stages of at least 20 m would be reached by 60,000 m^3/s floods at the two constrictions (Plate 2 and Figure 2). The bedrock slopes upstream and downstream of the constrictions were searched up to and above that height for evidence of past floods. Macroturbulence within the narrowest part of the constriction would likely prevent deposition; however, areas upstream and downstream might preserve evidence of a large flood. Such a flood likely would have left evidence on the bedrock walls in the form of clasts of exotic lithologies, fluvial deposits in the abundant ledges and hollows on the slope, or abraded rock surfaces. Loose colluvial deposits on the bedrock slopes likely would have been eroded during such a flood.

No evidence of catastrophic flooding is preserved on the bedrock slopes on either side of the Chilly Buttes constriction or upstream and downstream of the constriction. Also, no clasts of granitic, volcanic, or metamorphic origin were found, and no fluvial deposits were identified. No abraded bedrock surfaces were found, although such surfaces could have become pitted and roughened by chemical weathering of the soluble limestone since the latest Pleistocene. The exotic granitic and metamorphic clasts that could be expected to have been deposited would not have been similarly weathered. No expansion bars are present downstream of the constriction; expansion bars have been used elsewhere as indicators of past extreme floods [e.g., *O'Connor*, 1993]. In addition, direct evidence may preclude catastrophic flooding since the latest Pleistocene. Thick deposits of colluvium sit at the foot of the bedrock slope at Chilly Butte. The age of these colluvial deposits is not known, but their thickness (5 to 7 m), pedogenic reddening, and weathered clasts suggest they may be at least 20 to 30 ka. These colluvial deposits likely would have been eroded away during a 20 to 30 ka outburst flood.

Big Lost River Reach 3 - Chilly Buttes to Thousand Springs

Downstream of the Chilly Buttes constriction, the river flows across a large alluvial fan into the basin at the foot of the Lost River Range. The gradient of the alluvial-fan surfaces and the present stream is much less than that of the stream and terraces upstream of the Chilly Buttes constriction. The large alluvial fan likely correlates to terrace Qt3 on the basis of the similarity of surface morphology, geomorphic position, and soil profile development. The soil developed on the fan surface indicates a latest Pleistocene to early Holocene age for the alluvial fan. This fan surface is separated from a higher fan surface by a < 3-m-high scarp. The higher surface may be correlative with the Qt2 terrace mapped upstream. The alluvial fans extend across the valley, likely burying any evidence of 20 to 30 ka outburst flooding.

Big Lost River Reach 4 - Thousand Springs to Mackay

In Reach 4 (Figure 1), the Big Lost River flows southeast, parallel to the Lost River Range, and is confined between large Pleistocene alluvial fans [*Scott*, 1982] that coalesce to form continuous sloping bajadas on either side of the valley. In many places, these tributary fans are truncated abruptly at their toes by the active channel of the Big Lost River. Fluvial scarps resulting from this truncation typically are 'scalloped' in plan form (concave toward the channel) with each scallop resembling in size and curvature meander bends in the present river. These scalloped scarps clearly are the result of lateral migration of the Big Lost River during the Holocene.

Some of the older fans, however, have linear or broadly curving scarps (Figure 2). These linear scarps may have formed during a large flood in which the valley filled with water and the smaller-scale meander pattern was overwhelmed; alternatively, the scarps may have formed by migration of the active channel against the downslope end of the alluvial fans. Because these scarps have been eroded to gentler slopes than the more recent scarps and are partly buried by younger alluvium, the scarps likely are no younger than late Pleistocene in age.

Near Mackay there is a bedrock constriction that is similar to the Chilly Buttes constriction in reach 2. No evidence of catastrophic flooding is preserved on the bedrock slopes of the Mackay constriction (Figure 2). The depth of water that would be associated with a 60,000 m^3/s discharge passing through the 0.5-km-wide constriction is estimated to be 20 to 40 m based on the cross-sectional area of the constriction and an assumed velocity of about 3 to 6 m/s (Figure 2). The bedrock slopes up to and above that height were searched for evidence of an extreme flood, along with slopes upstream and downstream of the constriction. Hollows and ledges, plentiful in the limestone bedrock, are devoid of any overbank or slackwater deposits. No clasts of granitic, volcanic, or metamorphic origin were found, and no deposits with fluvial bedding were observed.

(a)

(b)

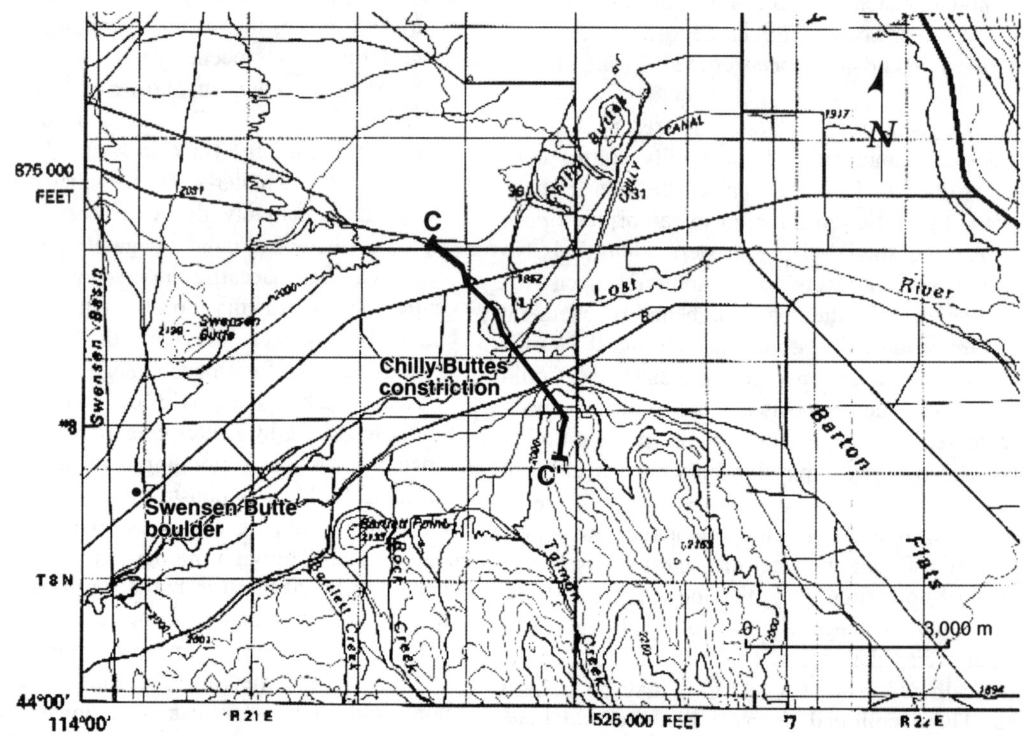

Plate 1. (a) "Swensen Butte boulder," view to southwest (upstream); granitic boulder is approximately 2x3x5 meters. **(b)** Location of "Swensen Butte boulder," and "Chilly Buttes constriction." Base map U.S. Geological Survey Borah Peak, Idaho, 1:100,000 quadrangle. Chilly Buttes cross section is shown on Plate 2.

(a)

(b)

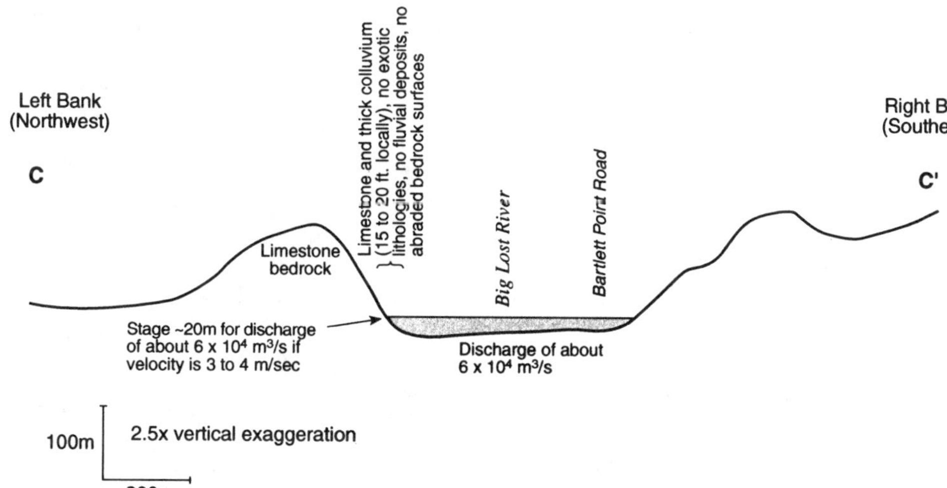

Plate 2. (a) Chilly Buttes constriction, aerial view to southeast across constriction; Big Lost River flows from right (southwest) to left (northeast). **(b)** Chilly Buttes constriction. Location of C-C' is shown on Plate 1b.

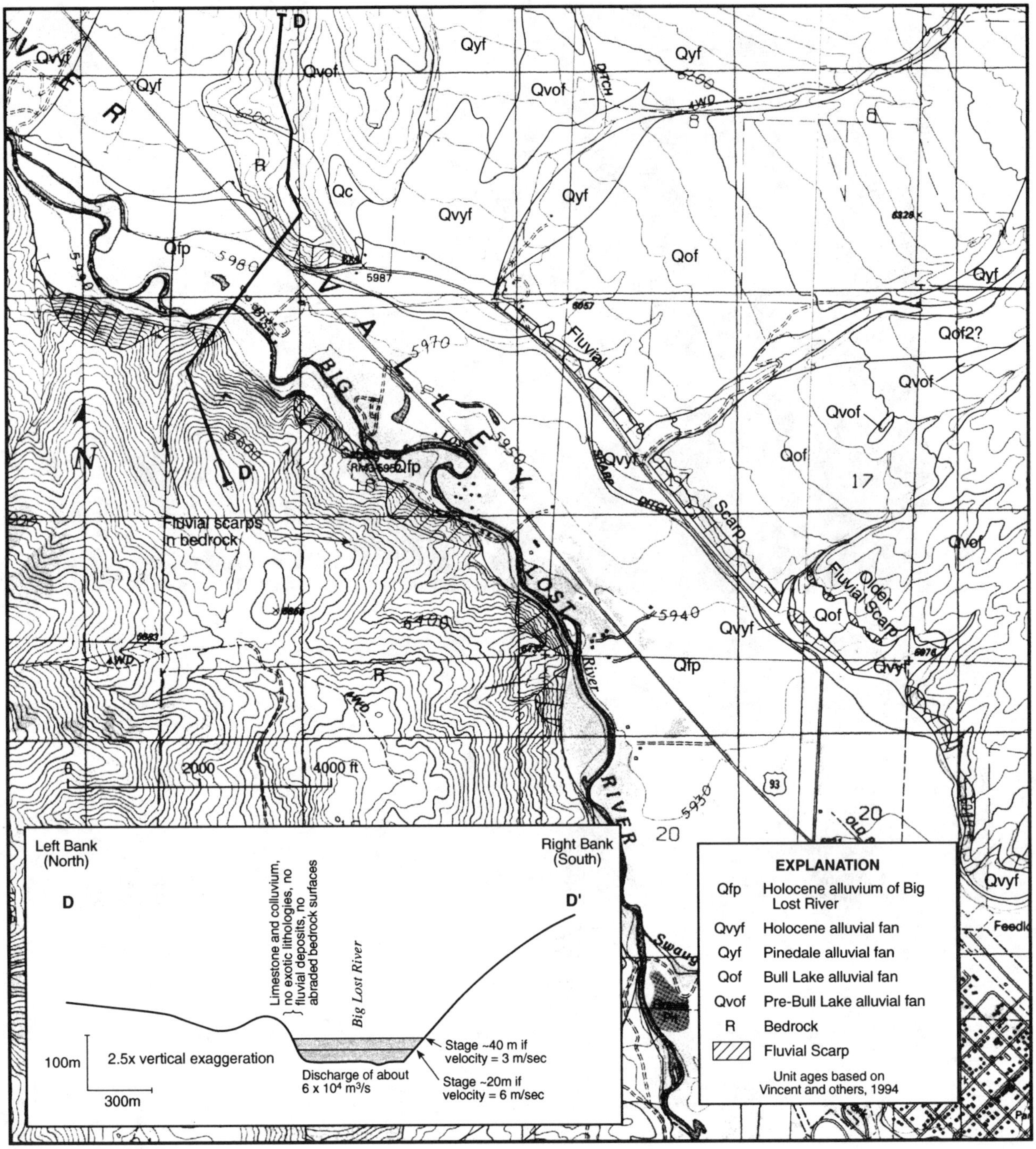

Figure 2. Quaternary geology and cross section, Mackay constriction. Base maps: U.S. Geological Survey Mackay and Mackay Reservoir 7.5-minute quadrangles.

Big Lost River Reach 5 - Mackay to 3 km North of Arco

In this reach of the Big Lost River, the late Holocene flood plain is 2 to 4 km wide and is confined between large Holocene and late Pleistocene alluvial fans [*Scott*, 1982]. Many of the tributary alluvial fans are graded to the present valley floor. However, there are reaches along the river where 2- to 3-m-high scarps are cut into the bordering alluvial fans. The scarps are arcuate and in many cases parallel to the modern channel. They are interpreted to have formed within the last several thousand years when the Big Lost River migrated across its flood plain and cut into the tributary alluvial fans. The Holocene and late Pleistocene valley fill sediment and tributary alluvial fan sediment likely blankets any evidence of late Pleistocene outburst flooding, if present.

Big Lost River Reach 6 - 3 km North of Arco to Arco Gaging Station

Near Arco the Big Lost River is flanked by a wide area of Holocene to late Pleistocene alluvium [*Kuntz et al.*, 1994] of low relief. Older Pleistocene alluvium underlies surfaces that are more than 5 to 10 m above the present stream level. A large complex of basalt flows overlies an older fluvial surface [Qmo of *Kuntz et al.*, 1994] west of Arco. *Kuntz et al.* assigned an age of 30 to 50 ka to the Qmo units. Exposures of well developed soils in gravel pits near the Arco Airport indicate that portions of the Qmo surface are significantly older than 30 to 50 ka (Table 3) [*Ostenaa et al.*, 1999]. Northwest of Arco, the edge of this basalt is exposed slightly below the general level of the main Qmo surface because at this location the basalt flowed on a slightly lower, younger surface than the Qmo surface. At this site, Forman [cited in *Olig et al.*, 1997] obtained a TL age of 47 ka on baked sediment beneath the basalt. Because this site is inset below the level of the main Qmo surface, this age provides a minimum age for stabilization of the Qmo surface. There is no evidence that floods have overtopped the basalt in this area. Correlative Qmo surfaces southeast of Arco near the Butte City scarps (Figure 3) were interpreted as older than 30 to 50 ka by *Olig et al.* [1997] and *Woodward-Clyde Federal Services* [1997]. Numerous soil pits excavated for these studies exposed no evidence of younger flood erosion on the alluvial surfaces in this area. Extensive areas of alluvium correlative to Qmo in this area have well developed soils [*NRCS* unpub. data on file in Arco] that are indicative of an age of at least 50 ka (Table 3, Stan variant site).

Big Lost River Reach 7 - Arco Gaging Station to Box Canyon

South of Arco, the Big Lost River flows onto basalt of the Snake River Plain and is confined to a relatively narrow and shallow canyon for several kilometers (Figure 4). Basalt flows in this area are interpreted by *Kuntz et al.* [1994] to be several hundred thousand years in age. The narrow canyon would have presented substantial hydraulic obstructions to a 20 to 30 ka outburst flood on the Big Lost River. The upstream portion of this reach (Figure 3) was evaluated in a related study of the Big Lost River [*Ostenaa et al.*, 1999]. Two-dimensional flow simulations of this reach indicate that the entrance to Box Canyon is a significant hydraulic constriction that will cause backwater upstream at larger flows [*Ostenaa et al.*, 1999]. The simulation results indicate that for discharges of 800 m^3/s and larger, there is significant backwater, > 7 m deep in the Big Lost River channel at the mouth of Box Canyon, that begins to extend onto the > 50 ka Qmo surfaces upstream. Any flood that overtops the divides on either side of Box Canyon would require flow depths of at least 13 m, and possibly > 20 m, near the mouth of Box Canyon. The highest Holocene flood deposits in this reach are less than 3 m above the river [*Ostenaa et al.*, 1999] and there is no accumulation of slackwater sediment farther upslope. While some thin beds of slackwater sediment conceivably may be masked within the loess that overlies the alluvium and basalt upstream of Box Canyon, if such deposits are present, they must be quite thin as the loess in this area is generally 1 to 1.5 m thick (Table 3). Soil development in loess overlying the basalt and alluvium suggests ages of >30 to 40 ka for the underlying fluvial deposits and basalts (Table 3, site BOR 21). The Box Canyon features are downstream of this detailed study reach and are located atop the basalt surfaces into which the canyon is cut. If an outburst flood produced the Box Canyon features, such a flood must have substantially inundated the sites upstream of the canyon.

Evidence for a large flood near Box Canyon. We did not specifically reexamine the evidence described by *Rathburn* [1993] in the Box Canyon area as part of this study. However, these features are distinctly different from evidence found at other sites along the Big Lost River. Among the evidence of a large flood are cataracts, scabland topography, a loess scarp, a streamlined loess hill, boulder bars up to 300 m in length and 15 m wide, and erratic clasts derived from the Pioneer Mountains, nearly 100 km upstream [*Rathburn*, 1989, 1993]. Box Canyon is narrow and sinuous with an average top width and depth of 40 m and 23 m, respectively [*Rathburn*, 1993]. The flood features near Box Canyon are distributed across a 3- to 4-km-wide basalt upland into which the canyon is incised and require complete inundation of Box Canyon with substantial flow on the adjacent upland. Hydraulic modeling by *Rathburn* [1989, 1993] indicates calculated flow depths of 9 to 31 m for her preferred discharge estimate. Estimates of unit stream power calculated from the modeling suggest that the flood had sufficient power to erode bedrock and transport boulders up to 4.5 m in diameter.

Big Lost River Reach 8 - Downstream of Box Canyon

Downstream of Box Canyon, there are extensive areas of Big Lost River alluvium that bury the 95 ka and older Quaking Aspen basalt flows [*Kuntz et al.*, 1993]. The degree of soil development indicates that large areas of this alluvium are no younger than 30 to 40 ka, and possibly as old as 95 ka (Table 3). The alluvial surfaces in these areas are relatively smooth, even adjacent to basalt outcrops that

Table 3. Soil profile characteristics and estimated ages.

Site	Location	Surface morphology and site characteristics [(a)]	Soil characteristics						Estimated age	References for soil and age data [(b)]
			Loess			B-Horizon	Gravel			
			Loess thickness (cm)	Carbonate stage in loess	Maximum Carbonate (weight %)		Gravel Carbonate stage	Grussified stones		
BOR21	Upstream of Box Canyon	up	125	II+	21	Bw, Bk	na	na	> 50 ka	1
BOR23	Arco Airport	a, ag	40 – 150	II	36	Bt, Btk, K	III	yes	>> 100 ka	1
BOR25	Downstream of INEEL diversion dam	a, pat, lm	65 – 140	II+ to III	29 – 34	Bw, Bk, K	I- to II-	few	>30-40 ka to <90 ka	1,4,5
WCC	Butte City	a, ag	5 – 96	I to II	not described	not described	I to III	not described	> 50 ka	1,2,3
Stan variant, typical pedon location	Near Arco gaging station	a, ag	68 – 107 (?)[(c)]	not described?	20	Bw, Bk	not described; pendants 1-2 mm thick on underside of gravel	not described	> 50 ka	1,6

Notes: na – not applicable
(a) surface morphology abbreviations: a – alluvium with loess; pat – mound pattern on aerial photographs; lm – large mounds 1-2 m high; up – upland basalt with loess; ag – agricultural, field or modified surface
(b) references: 1 – Ostenaa et al., 1999; 2 – Olig et al., 1997; 3 – Olig, 1997; 4 – Forman et al., 1994; 5 – Dechert et al., 1994; 6 – NRCS, Arco
(c) estimated loess thickness interpreted from NRCS descriptions

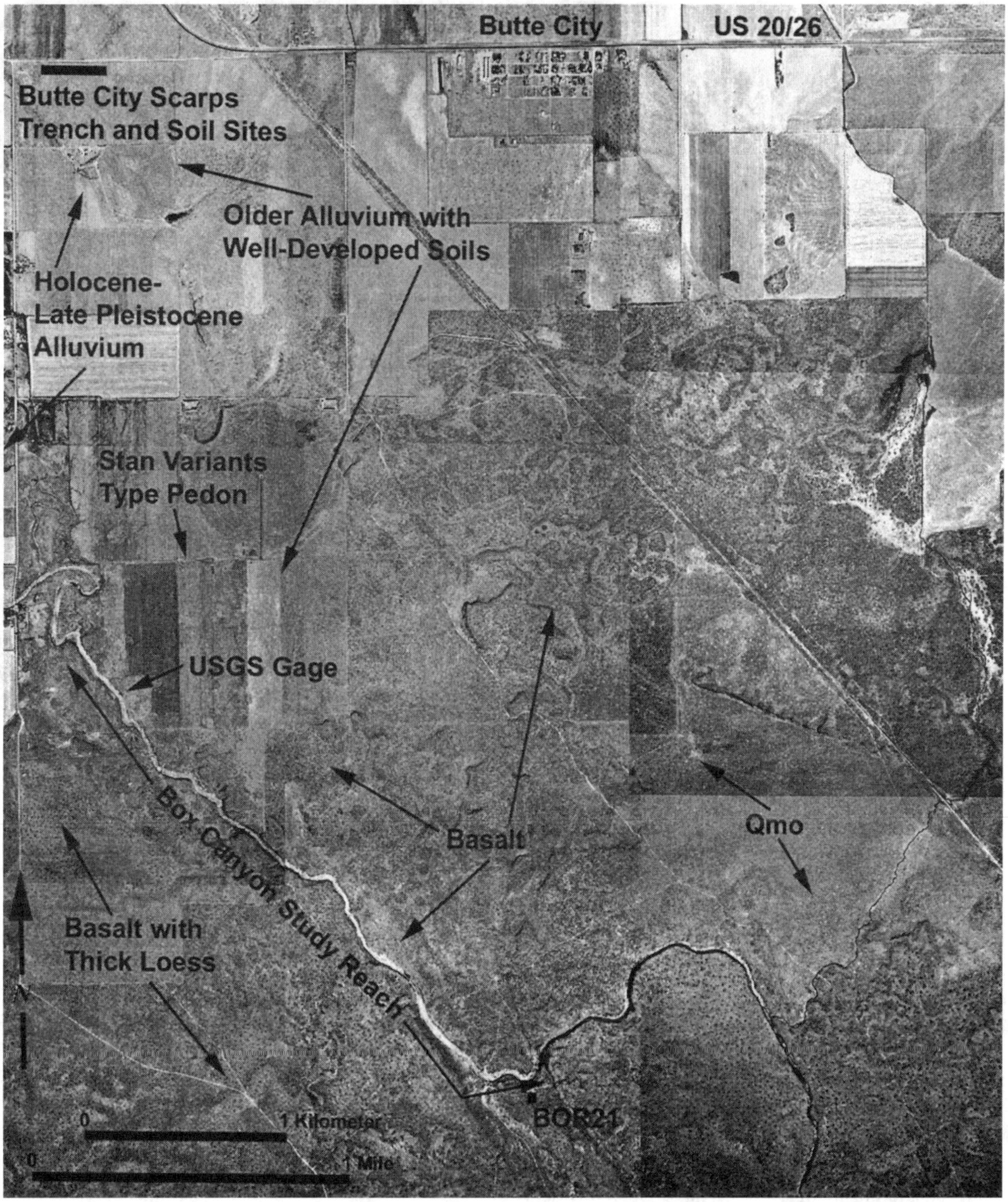

Figure 3. Photomosaic of the Butte City to upper Box Canyon area. Previous investigations of the Butte City scarps [*Olig et al.*, 1997; *Woodward-Clyde Federal Services*, 1997] concluded that these surfaces are at least 30 to 50 ka in age. A similar age is assigned to unit Qmo [*Kuntz et al.*, 1994]. The Stan Variants soil series has characteristics that suggest a minimum age of 30 to 40 ka as do soils at BOR21 (Table 3). Similar soil development appears to characterize the thick loess overlying basalt in the southwest corner of the photograph. Flow simulations for the Box Canyon reach of *Ostenaa et al.* [1999] predict that backwater occurs for discharges as low as 800 m^3/s on these surfaces due to hydraulic constriction at Box Canyon.

Figure 4. Oblique aerial view looking southeast along the Big Lost River at upstream end of Box Canyon modeling reach of *Ostenaa et al.* [1999]. Buildings in foreground sit on Holocene to late Pleistocene alluvium. River is confined by basalt beginning at sharp bends near the center of photograph. Field to left is alluvial surface; soil development on correlative surfaces suggests an age > 50 ka.

are buried by the alluvium and stand in relief above the surface. No large scours or bedforms indicative of high-energy flow are preserved on this surface. Surface morphology suggests a braided stream system that was deflected around basalt outcrops. Limited reconnaissance performed for this study and by *Dechert et al.* [1994] did not locate exotic clasts on the higher areas of the 95 ka basalt. Many shallow pits excavated in the alluvium in this area exposed Big Lost River gravel with maximum clast sizes that are less than 0.5 m, and usually less than 0.25 m. Conspicuously absent are large, angular basalt clasts that might have been eroded from the basalt in the Box Canyon area.

DISCUSSION

In studying the late Quaternary flood history of the Big Lost River, it is convenient to think in terms of three kinds of evidence: (1) direct stratigraphic and geomorphic evidence supporting the occurrence of a large late Pleistocene flood; (2) the absence of evidence for a catastrophic flood at locations where such evidence might be preserved; and (3) direct stratigraphic and geomorphic evidence precluding the occurrence of a large flood or constraining its age.

Possible Evidence for a Large late Pleistocene Flood

In addition to the features in the Box Canyon area, several features have been identified in this and earlier studies that suggest the possible occurrence of a large flood along the Big Lost River. These include: (1) boulders in Box Canyon and near Swensen Butte (reach 2) that are lithologically similar to bedrock in the Wildhorse Creek watershed; (2) truncated tributary alluvial fans near Mackay and elsewhere along the Big Lost River; (3) the presence of exposed bedrock in the eroded channel downstream of the confluence of Wildhorse Creek with the East Fork; (4) the presence of a glacially dammed lake in the headwaters of the Big Lost River (Glacial Lake East Fork), and (5) valley margin scarps in bedrock and separating terrace Qt1 from Qt2 in the reach downstream of the Wildhorse I glacial terminus. However, most of these features do not require an outburst flood for their formation. Rather, they may be explained by lower energy fluvial or glaciofluvial processes.

The most compelling evidence for a large flood outside of Box Canyon is the presence of large granitic boulders in reach 2.

Absence of Evidence for a Glacial Outburst Flood

No direct evidence was observed that a catastrophic flood occurred as a result of failure of the ice dam at Glacial Lake East Fork. Except for the features described near Box Canyon, geomorphic evidence observed at sites of other extreme floods (e.g. Missoula flood [*Baker*, 1973; *Waitt*, 1980], Bonneville flood [*Malde*, 1968; *O'Connor*, 1993] is not present downstream of the site of Glacial Lake East Fork, or elsewhere along the Big Lost River. No deposits or geomorphic features upstream or downstream of Box Canyon directly relate to, or are uniquely formed by, glacial outburst flooding. Two constrictions in the Big Lost River Valley, at Chilly Buttes and Mackay, have no evidence within, upstream or downstream of them that suggest a large flood flowed through these constrictions. The lack of obvious catastrophic flood features in the Big Lost River Valley between Box Canyon and the headwaters has been mentioned by previous workers [e.g. *Rathburn*, 1991].

Evidence of no Flood in the Latest Pleistocene

Direct stratigraphic and geomorphic evidence indicates that a catastrophic flood did not occur in the last 20 to 30 ka along the Big Lost River. Evidence includes four primary observations: (1) terrace surfaces adjacent to the Big Lost River directly downstream of the Wildhorse Creek glacier's terminus, which correlate with the Wildhorse I glacial advance, have not been overtopped by a large flood. Remnants of these at least 25 ka Qt2 terrace surfaces exist on both sides of the creek and the cross-sectional area between the remnants is not sufficient to pass a 60,000 m^3/s flood. Simple estimates indicate that no more than about 15,000 m^3/s would flow between these terrace surfaces. (2) Colluvial deposits are preserved within the Chilly Buttes constriction that would have been eroded in a 20 to 30 ka flood. (3) Near Arco, greater than 50 ka, low-lying fluvial terraces have not been overtopped since their stabilization. (4) Downstream of Box Canyon, alluvial deposits that would not have survived being overtopped by a large flood without showing evidence for that flood are no younger than 30 to 40 ka, and possibly as old as 95 ka.

Features in the Box Canyon Area - Alternative Hypotheses

The results of this study indicate that the minimum age for a large outburst flood along the lower Big Lost River is at least 50 ka, and possibly 95 ka. Thus, the features in Box Canyon must be significantly older than the previously suggested age of about 20 to 30 ka. If these features are older than 95 ka, then Box Canyon, which was at least in part formed by incision, may have been much smaller at the time of formation of the features. The features in Box Canyon may need to be evaluated in a framework of the long-term evolution of the Big Lost River drainage that considers the potential for episodes of volcanic activity, tectonic deformation, and erosion by floods. In this framework, possible outburst flood magnitudes could be consistent with the estimated volume(s) associated with catastrophic failure of Glacial Lake East Fork. Any assessment of flood magnitude in the Box Canyon area needs to consider the timing of these floods relative to the formation of Box Canyon as well as any subsequent volcanic or tectonic deformation. Because the channel geometry may have changed since formation of the features in Box Canyon, the 60,000 m^3/s estimate of *Rathburn* [1989, 1993] may have considerable uncertainty associated with it.

CONCLUSIONS

Direct evidence exists for a late Pleistocene glacial lake (Glacial Lake East Fork), as described by *Evenson et al.*, [1982] where the East Fork Big Lost River was dammed by ice flowing out of Wildhorse Creek. Ice-rafted granitic boulders from Wildhorse Canyon perched on the canyon wall of the East Fork upstream of the dam can only have been emplaced via ice rafting within the lake. The most recent damming of the East Fork occurred during the Wildhorse I glacial advance, which is probably no younger than 25 to 30 ka. However, there is no compelling evidence for catastrophic failure of this ice dam. In the reach of the Big Lost River directly downstream from the former ice dam, the hypothesized 60,000 m^3/s flood would have had great velocity and very high stream power. Yet no evidence was found of large scale erosion, high energy deposits, or large bedforms.

Outwash terrace "Qt2" surfaces, which grade to glacial deposits of the Wildhorse I glacial advance, show no evidence of having been overtopped by a catastrophic flood. The terrace surfaces are fairly continuous, smooth, planar landforms without scours, boulder bars, or other indications of anomalously high stream power. The incised valley between terrace Qt2 remnants has insufficient cross-sectional area to pass an outburst flood of the magnitude thought to have reached Box Canyon. Thus, if an extreme flood flowed down this reach of the upper Big Lost River, it must have occurred prior to stabilization of the Qt2 terrace, at least 25 to 30 ka. Large exotic boulders on these terraces may be inherited from earlier deposits.

At two locations, Chilly Buttes and Mackay, the river passes through constrictions between steep limestone hills. Rough calculations show that a 60,000 m^3/s flood would have risen 20 to 40 m up the bedrock canyon walls. At both constrictions, and upstream and downstream of the constrictions, there are (1) no fluvial deposits or clasts of exotic lithology on bedrock ledges, in crevices, hollows, or interbedded with colluvium, and (2) no abraded bedrock surfaces. Also, a weathered, approximately 5-m-thick colluvial mantle within the Chilly Buttes constriction is undisturbed and precludes a large flood since its emplacement.

Fluvial terrace surfaces adjacent to the Big Lost River near Arco also appear not to have been overtopped since their formation, yet should have been overtopped in a flood of 60,000 m^3/s. Soil development, geomorphic studies, and correlation to a thermoluminescence-dated basalt flow, suggest that some of these surfaces are at least 50 ka, providing a minimum age for extreme floods on this reach of the Big Lost River.

Downstream of Box Canyon, alluvium that overlies the 95 ka Quaking Aspen Butte basalt flow would be flooded by any catastrophic flood. However, both the surface morphology and the sedimentology of this alluvium suggest that it must post-date any catastrophic flood. Soil development on this alluvium and stratigraphic relationships to the Quaking Aspen Butte basalt flow indicate an age for the alluvium of 30 to 40 ka to 95 ka.

In conclusion, there is no direct or conclusive evidence along the Big Lost River for the occurrence of a 20 to 30 ka catastrophic flood. To the contrary, evidence precludes the occurrence of a catastrophic flood in the past 25 to 30 ka in upstream reaches of the Big Lost River, in the past approximately 50 ka near Arco, and the past 30 to 40 ka to possibly 95 ka in the area downstream of Box Canyon. The features in the Box Canyon area merit reexamination in the light of these results.

Acknowledgments. This study was undertaken at the request of R.J. Creed, DOE Idaho Operations Office, for the Idaho National Environmental and Engineering Laboratory as part of a broader flood hazard characterization. Messrs. Creed and R.P. Smith, LMTCO, Idaho Falls, ID provided extensive assistance in obtaining background information and in coordinating various logistical aspects of the study. We gratefully acknowledge the willingness of researchers familiar with the area to share their insights into the geomorphic history of the Big Lost River, including: Edward Evenson (Lehigh University), Glen Thackray (Idaho State University), Kirk Vincent (U.S. Geological Survey), and Keith Brugger (University of Minnesota at Duluth). This paper benefited from the reviews of Scott Lundstrom, Jim O'Connor and Robert Webb.

REFERENCES

Baker, V.R., Paleohydrology and sedimentology of Lake Missoula flooding in eastern Washington, *Geol. Soc. Amer. Spec. Pap.* 144, 1973.

Baker, V.R., Flood sedimentation in bedrock fluvial systems, in *Sedimentology of gravels and conglomerates*, edited by E. H. Koster, and R. J. Steel, Can. Soc. Pet. Geol., Memoir 10, 87-98, 1984.

Brugger, K.A., Implications of till-provenance studies for glaciological reconstructions of the paleoglaciers of Wildhorse Canyon, Idaho, U.S.A.: *Annals of Glaciology*, 22, 93-101, 1996.

Brugger, K.A., E.B., Evenson, and R.A. Stewart, Chronology and style of glaciation in the Wildhorse Canyon area, Idaho, in *Tills and Related Deposits, Genesis/Petrology/Application/ Stratigraphy, Proceedings of the INQUA Symposia on the Genesis and Lithology of Quaternary Deposits/ USA 1981/Argentina 1982,* edited by E.B. Evenson, C. Schuchter, and J. Rabassa, A.A. Balkema, Rotterdam, 151-175, 1983.

Burke, R.M., and P.W. Birkeland, Reevaluation of multiparameter relative dating techniques and their application to the glacial sequence along the eastern escarpment of the Sierra Nevada, California, *Quat. Res.*, 11, 21-51, 1983.

Carling, P., and K. Tinkler, Conditions for the entrainment of cuboid boulders in bedrock streams: An historical review of literature with respect to recent investigations, in *Fluvial Processes in Bedrock Channels*, edited by K. Tinkler, and E. E. Wohl, Amer. Geophys. Union, Geophysical Monograph 107, 19-34, 1998.

Cerling, T.E., R.J., Poreda, and S.L. Rathburn, Cosmogenic ^{3}He and ^{21}Ne age of the Big Lost River flood, Snake River Plain, Idaho: *Geology*, 22, 227-230, 1994.

Colman, S.M., and K.L Pierce, Weathering rinds on andesitic and basaltic stones as a Quaternary age indicator, Western United States: *U.S. Geol. Surv. Prof. Pap.* 1210, 1981.

Colman, S.M., and K.L. Pierce, Glacial sequence near McCall, Idaho: Weathering rinds, soil development, morphology, and other relative-age criteria, *Quat. Res.*, 25, 25-42, 1986.

Dechert, T.V., P.A. McDaniel, and A.L. Falen, Aggradational and erosional history of the Radioactive Waste Management Complex at the Idaho National Engineering Laboratory, EGG-WM-11049, Prepared for EG&G, Idaho, Inc., Subcontract No. C85-110544, Task Order 82, by Univ. of Idaho, College of Ag., Soil Science Div., Moscow, ID, 26 p, 1994.

Evenson, E.B., J.F.P. Cotter, and J.M. Clinch, Glaciation of the Pioneer Mountains: A proposed model for Idaho, in *Cenozoic Geology of Idaho,* edited by B. Bonnichsen and R. Breckenridge, Idaho Bur. Mines Geol. Bull. 26, 653-665, 1982.

Forman, S.L., J. Pierson, R.P. Smith, W.R. Hackett, and G. Valentine, Assessing the accuracy of thermoluminescence for dating baked sediments beneath late Quaternary lava flows, Snake River Plain, Idaho, *J. Geophys. Res.*, 99, 15,569-15,576, 1994.

Forman, S.L., R.P. Smith, W.R. Hackett, J.A. Tullis, and P.A. McDaniel, Timing of late Quaternary glaciations in the Western United States based on the age of loess on the Eastern Snake River Plain, Idaho, *Quat. Res.*, 40, 30-37, 1993.

Gianniny, G.D., Report on temporal recharge variability for the Eastern Snake River Plain aquifer, based upon osctracode species composition in cores from both sub-basins of Quaternary Lake Terreton with implications for planning and remediation at Teat Area North (TAN) and the Idaho National

Engineering and Environmental Laboratory (INEEL), unpublished report, 1 p. plus figures, tables, 1998.

Hacket, W.R., J.A. Tullis, R.P. Smith, S.J Miller, T.V. Dechert, P. A. McDaniel, and A.L. Falen, Geologic processes in the RWMC area, Idaho National Engineering Laboratory: Implications for long-term stability and soil erosion at the Radioactive Waste Management Complex, INEL-95/0519, Idaho Nat. Eng. Lab., Lockheed Martin Idaho Technologies, Idaho Falls, ID, 49 p, 1995.

Kuntz, M.A., B. Skipp, M.A. Lamphere, W.E. Scott, K.L. Pierce, G.B. Dalrymple, D.E. Champion, G.F. Embree, W.R. Page, L.A. Morgan, R. P. Smith, W. R. Hackett, and D. W. Rodgers, Geologic map of the Idaho National Engineering Laboratory and adjoining area, eastern Idaho, *U.S. Geol. Surv. Misc. Invest. Map*, I-2330, scale 1:100,000, 1994,

Malde, H.E., The catastrophic late Pleistocene Bonneville Flood in the Snake River Plain, Idaho, *U.S. Geol. Surv. Prof. Pap.* 596, 1968.

O'Connor, J.E., Hydrology, Hydraulics, and Geomorphology of the Bonneville Flood, *Geol. Soc. Amer. Spec. Pap.* 274, 1993.

Olig, S.S., A.E Gorton, R.P. Smith and S.L. Forman, Additional geologic investigations of the Southern Lost River fault and northern Arco rift zone, Idaho, unpub. consulting report submitted to Lockheed Martin Idaho Technologies, 15 p. plus figures, tables, appendices, 1997.

Ostenaa, D.A., D.R. Levish, R.E. Klinger and D.R.H. O'Connell, Phase 2 Paleohydrologic and Geomorphic Studies for the Assessment of Flood Risk for the Idaho National Engineering and Environmental Laboratory, Idaho, Report 99-7, U.S. Bur. Rec., Denver, CO, 112 p. plus appendices, 1999.

Pierce, K.L., Fosberg, M.A., Scott, W.E., Lewis, G.C., and S.M. Colman, Loess deposits of Southeastern Idaho: Age and correlation of the upper two loess units, in *Cenozoic Geology of Idaho*, edited by B. Bonnichsen and R. Breckenridge, Idaho Bur. Mines Geol. Bull. 26, 717-725, 1982.

Rathburn, S.L., Pleistocene glacial outburst flooding along the Big Lost River, East-Central Idaho, M.S. thesis, University of Arizona, Tucson, Arizona, 1989.

Rathburn, S.L., Quaternary channel changes and paleoflooding along the Big Lost River, Idaho National Engineering Laboratory: TriHydro Corp. report for EG&G Idaho, Inc., subcontract no. C90-132903, 33 p., 1991.

Rathburn, S.L., Pleistocene cataclysmic flooding along the Big Lost River, east central Idaho, *Geomorphology*, 8, 305-319, 1993.

Scott, W.E., Surficial geologic map of the eastern Snake River Plain and adjacent areas, 111° to 115° W., Idaho and Wyoming, *U.S. Geol. Surv. Misc. Invest. Map,* I-1372, scale 1:250,000, 1982.

Sturchio, N.C., K.L. Pierce, M.T. Murrell, and M.L. Sorey, Uranium-series ages of travertines and timing of the last glaciation in the northern Yellowstone area, Wyoming-Montana, *Quat. Res.*, 41, 265-277, 1994.

Vincent, K.R., W.B. Bull, and O.A. Chadwick, Construction of a soil chronosequence using the thickness of pedogenic carbonate coatings, *J. Geol. Ed.*, 42, 316-324, 1994.

Waitt, R.B., About forty last-glacial Lake Missoula Jokulhlaups from glacial Lake Missoula, *Geol. Soc. Amer. Bull.*, 95, 1271-1286, 1980.

Woodward-Clyde Federal Services, Trench investigation of the Butte City scarps, unpub. consulting letter report to Lockheed Martin Idaho Technologies, 4 p. plus figures, tables, 1997.

Keith L. Knudsen, William Lettis & Associates, Inc., 1777 Botelho Dr., Ste 262, Walnut Creek, CA, 94596, (present address: California Division of Mines and Geology, 185 Berry St., Ste 210, San Francisco, CA, 94107)

Janet M. Sowers, William Lettis & Associates, Inc., 1777 Botelho Dr., Ste 262, Walnut Creek, CA, 94596

Dean A. Ostenaa and Daniel R. Levish, U.S. Bureau of Reclamation, P.O. Box 25007, Building 67, Denver Federal Center, Denver, CO, 80225

Watershed Scale and the Stratigraphic Record of Large Floods

J. C. Knox and J. M. Daniels

Department of Geography, University of Wisconsin, Madison, Wisconsin

The sedimentological characteristics of overbank alluvium deposited by large floods vary systematically with size of the contributing upstream watershed. The effects of watershed scale result from the strong association between areal extent of flood-producing meteorological events and watershed size, and from the tendency for floods to experience downstream decreasing power to transport coarse sediment. Stratigraphic evidence of large overbank floods that have occurred during historical and Holocene time in the Upper Mississippi Valley includes: (1) large clasts from which former flood depths can be estimated and (2) fine-texture (sand, silt, and clay) slackwater deposits from which flood depths can sometimes be estimated but more often provide only flood frequency information. The use of clasts to compute depths of flow competent for their transportation is mainly restricted to watersheds smaller than a few hundred km^2 in the Upper Mississippi Valley, but use of slackwater sediment for flood frequency analyses extends across all sizes of watersheds including the main channel upper Mississippi River. A 1 - 2 cm vertical sampling resolution for grain size analyses is ordinarily sufficient for detecting sedimentological evidence of large floods, although higher resolution sampling is necessary when large floods have occurred in close temporal succession. The widespread distributions of the fine-texture slackwater sediments on floodplains and low terraces provide relatively long and nearly continuous records of large overbank floods that have occurred in the Upper Mississippi Valley. Recognition of a relatively complete record of large post-glacial floods is supported by radiocarbon ages that indicate the Holocene upper Mississippi River along western Wisconsin has been dominated by progressive aggradation. Heavy summer rainfalls are responsible for the largest floods on watersheds smaller than several hundred km^2 in the Upper Mississippi Valley, but on larger watersheds, snowmelt is a principal cause of the largest floods. The scale-related different origins for large overbank floods provide a framework from which long-term seasonal behavior of large floods can be evaluated. Results show that rainfall and snowmelt floods in the region have for certain extended periods of the late-Holocene behaved in similar ways while at other times may have differed significantly. These findings indicate that more

Ancient Floods, Modern Hazards: Principles
and Applications of Paleoflood Hydrology
Water Science and Application Volume 5, pages 237-255

detailed comparisons of paleoflood histories from wide-ranging sizes of watersheds in the Upper Mississippi Valley can provide relatively specific information concerning how climate changes influence large snowmelt and rainfall floods. In turn, the results provide a more sound basis to foresee how magnitudes and recurrence frequencies of large floods might respond to future climatic changes.

INTRODUCTION

Modern instrumental records are too short to adequately capture the possible range of flood magnitudes and the variability in flood recurrence probabilities that might occur with changing environmental conditions. Paleoflood hydrology offers assistance for this dilemma because stratigraphic and morphologic records of past floods can be correlated with other past environmental information to better anticipate flood responses that are possible with potential future environmental changes [*Kochel and Baker*, 1982; *Jarrett*, 1991; *Jarrett and Tomlinson*, 2000; *Knox*, 1985, 1993; *Ely et al.*, 1993].

Reconstructions of floods from stratigraphic records involve a variety of methods [*Kochel and Baker*, 1982; *Williams*, 1983; *Costa*, 1987; *Jarrett*, 1991; *Baker*, 1994]. Applications of these methods, however, are often limited by geological conditions that characterize specific localities [*Baker*, 1987]. Furthermore, the applicability of paleoflood methodologies may change with geographic location in drainage systems. Surprisingly, there has been relatively little attention given to understanding how watershed size influences the type of stratigraphic record left by large floods. The causal mechanisms for floods change in importance as environments shift from small headwater tributaries to moderate size large alluvial valleys to floodplains associated with extremely large watersheds of tens of thousands of square kilometers. The general objective of this paper is to examine how watershed scale influences morphologic and stratigraphic evidence of past floods in the humid-continental environment of the Upper Mississippi Valley (UMV) and to illustrate how the appropriate methodologies for paleoflood estimations change with watershed scale.

Paleoflood investigations normally involve evaluations of flood frequencies and magnitudes. We show that long and continuous stratigraphic records important for frequency analyses become more common in the downstream direction in many UMV watersheds, although frequency analyses are normally possible in watersheds of all sizes. Paleoflood magnitudes require either the preservation of high-water marks or preservation of coarse clasts that can be used to estimate former depths of flow necessary for downstream transportation of clasts. High-water marks or their proxies are easily destroyed in the humid UMV and the availability of coarse clasts suitable for estimating depths of past flows decreases rapidly in the downstream direction. Consequently, estimates of paleoflood magnitudes in UMV watersheds are most easily achieved in small and intermediate scale systems and they become difficult to achieve in large watersheds that exceed several hundred square kilometers. Our procedure here is to use examples from both recent historical and older Holocene alluvial stratigraphy in small and large watersheds. The historical record is a particularly useful analogue for interpreting older alluvial sequences because it presents a high-resolution image of alluvial activity where age control is usually quite good.

Paleoflood frequency information is more abundant in the alluvial stratigraphic record than is paleoflood magnitude information. This is because frequency information can be derived from alluvial sequences that span the full range of grain size diameters. Furthermore, frequency information can be extrapolated from alluvial deposits that occur in a wide range of topographic settings ranging from floodplains to terraces and other locations where flood deposition might occur. The principal information required for frequency analysis includes: 1) sedimentological evidence that a flood has occurred and 2) an estimate of its age. The method requires an ability to distinguish discrete floods within a stratigraphic sequence that may or may not contain obvious indications of a hiatus in flow. Quantification of relict flood magnitudes on the other hand requires estimation of former flood stages. Two principal approaches have been used to extrapolate paleostage information from the stratigraphic record. Use of relative heights of slackwater sediments above former channel bed positions is the most commonly practiced method, especially in arid and semiarid regions [*Kochel and Baker*, 1982, 1988; *Ely et al.*, 1993]. The other commonly used method to estimate former stages of relict floods involves computing the competent depth of flood waters necessary to transport the coarsest cobbles and boulders found in alluvial deposits [*Baker and Ritter*, 1975; *Costa*, 1983; *Williams*, 1983; *Komar*, 1987; *Knox*, 1987a, 1993]. Applications of the slackwater and competency methods involve several assumptions, including presumed stability of cross sections and availability of sediment.

WATERSHED SIZE AND GENERATION OF HIGH-ENERGY FLOODS

Watershed scale influences characteristics of the alluvial stratigraphic record of large floods in two principal ways. The first is related to the strong association between areal extent of flood-producing meteorological events and watershed size. The second is related to the association between flood power and drainage area. These two factors induce downstream systematic changes in the competency to erode, transport, and deposit sediments available to a river system.

Meteorological Influence on Downstream Distribution of Flood Energy

Large floods on small and moderate size watersheds normally require meteorological conditions that produce rapid rates of surface runoff because excess water usually is efficiently transported downstream in headwater locations. Consequently, snowmelt, even in association with light rain, is seldom responsible for the largest floods in small and moderate size watersheds of the UMV [*Knox*, 1988]. Figures 1, 2, and 3, for example, show stream flow responses to light rain on melting snow for three sites in the Platte River system of Grant County, Wisconsin. Note that the contributing watershed area increases approximately by an order of magnitude between each site. In the steep headwater tributary of Figure 1, the accumulated volume of meltwater from the 0.03 km^2 watershed is sufficiently small that the depth of flow does not submerge even the small boulders on the channel bed. Farther downstream, where drainage area has increased to about 3.21 km^2, accumulated runoff from many upstream small tributaries fills the channel to about one-third of its full capacity and all bed sediment is submerged (Figure 2). Still farther downstream where drainage area is increased to about 230 km^2, the accumulated runoff has filled the channel to nearly full capacity (Figure 3).

The annual recurrence of snowmelt runoff in this region is highly consistent. Therefore, snowmelt is responsible for many small floods of high recurrence frequency, but its capability to produce large and extreme floods is mainly restricted to watersheds of large size. This relationship is illustrated by Figure 4, which shows the monthly distributions of annual maximum floods for three UMV watersheds of widely ranging size. The basins are represented by a 4.4 km^2 headwater tributary, a 368 km^2 moderate size tributary, and a site on the main channel upper Mississippi River where the drainage area is 153,325 km^2. The annual maximum flood occurs in nearly every month of the year, but such maxima are rare during the winter months of December and January when precipitable water vapor content of regional air masses is low, and low temperatures minimize snowmelt. The two smaller watersheds have experienced their largest floods during the months of June, July, and August in association with heavy rainfalls. Although excessive rainfall can produce extreme floods on very large watersheds of the Mississippi River scale, as was illustrated by the great flood of 1993 on the upper Mississippi River [*Parrett et al.*, 1993; *Wahl et al.*, 1993; *Pitlick*, 1997], such occurrences are infrequent compared to the role of snowmelt runoff (Figure 4). Note that the four largest floods on the Mississippi River at

Figure 1. Snowmelt runoff seldom produces large floods on headwater tributaries as shown here for this small, steep gradient channel in southwestern Wisconsin's Little Platte River watershed. Average thalweg water depth is about 15 cm. (Photo: J. C. Knox).

Figure 2. The downstream progression to larger and larger drainage areas gradually produces deeper and higher energy flows as snowmelt runoff accumulates from many small tributaries of the type shown in Figure 1. Average thalweg water depth is about 50 cm. The above channel, located in southwestern Wisconsin's Little Platte River watershed, is about 0.25 km from the channel of Figure 1. (Photo: J. C. Knox).

Winona, Minnesota occurred in conjunction with April snowmelt runoff.

Excessive rainfall is the most common source of large floods on small and moderate size watersheds. *Hirschboeck* [1988] and *Hirschboeck et al.* [2000] note that isolated or small thunderstorms and squall line disturbances commonly produce the type of high intensity rainfalls that lead to large floods on small watersheds. The large floods shown for June, July, and August on Rocky Branch watershed are examples of this process (Figure 4). *Hirschboeck* [1988] and *Hirschboeck et al.* [2000] have shown that larger mesoscale systems involving severe thunderstorms and multiple squall lines are typically the cause of the large floods on moderate side watersheds. The large floods shown for June and July in the 368 km^2 Platte River watershed are examples (Figure 4). These extreme rainfalls are responsible for floods that may cover entire valley floors to depths of 2-3 m or greater as illustrated by the 1978 flood of about 100-year recurrence probability on the Kickapoo River in southwestern Wisconsin (Figure 5). These large rain-generated floods on headwater and moderate size watersheds have sufficient power to transport large clasts onto upper positions of point bars and onto floodplain surfaces. Former large floods are often easily recognized in vertical alluvial sequences by gravel units or by texture reversals in the normal fining-upward trend (Figure 6). Extreme floods may transport exceptionally large clasts, if available, to and along floodplain surfaces in small and moderate size watersheds. Figure 7 represents a site farther downstream from the Figure 6 site, where a July 16, 1950 flood deposited large imbricated boulders on the floodplain of the Platte River. Standard Log Pearson Type III [*U. S. Water Resources Council Hydrology Committee*, 1981] analysis indicates this July 1950 flood to have a 500-year recurrence probability. Eighteen of the largest boulders on the floodplain shown in Figure 7 were found to have an average intermediate-axis dimension of 309 mm and ranged from 175 mm to 440 mm. The boulders rest on a bed of sandy gravel that was deposited by the 1950 flood. The depth of water competent to transport the large boulders down-valley on the Platte River floodplain was estimated by application of equation (1) [*Knox*, 1987a]:

$$D = 0.0001A^{1.21}S^{-0.57} \tag{1}$$

where D is the minimum depth of water in meters competent to transport the largest clasts, A is the clast intermediate axis in mm, and S is an approximation of the energy slope in m m^{-1}. The equation assumes a constant of 9800 N m^{-3} for specific weight of the water and sediment load. The equation, which was calibrated for particle diameters ranging between 50-3300 mm, is statistically significant at the 0.01 probability level and the two variables explain 74% of the statistical variance in depth

Figure 3. The simultaneous inputs of snowmelt runoff from the numerous smaller tributaries have filled the main channel of southwestern Wisconsin's Little Platte River to nearly full capacity. The water depth in the thalweg is about 3 m deep. This site is 0.1 km distance from the channel shown in Figure 1. (Photo: J. C. Knox).

for the calibration data set. Use of equation (1) requires that: (a) the maximum particle size transported should be smaller than the maximum size of particles that were available for transportation, and (b) the particles were transported by normal fluvial processes rather than by other processes such as ice-rafting or mass wasting. Application of the competent depth method also requires that lithologies underlying the watershed are capable of producing a range of clast sizes that are suitable for testing the competent depth limits of large floods. Crystalline rocks such as quartzite and unweathered granite and sedimentary rocks such as dolomite and limestone commonly are sufficiently resistant to breakdown that they provide a wide-ranging array of clast sizes. Less resistant crystalline rocks such as many types of schist or sedimentary rocks such a shale and poorly cemented sandstone often readily breakdown into their constituent compositional grain textures and normally are unsuitable for application in competent depth analyses. For the paleoflood studies in the UMV discussed here, the mean intermediate axis dimensions of the coarsest 10 and the coarsest 5 carbonate boulders were used to calculate the depth of water competent to transport the boulders along the floodplain. Their application in equation (1) indicated respective water depths of 2.7 and 3.1 m on the floodplain. Depth estimates of these magnitudes were in close agreement with elevations of flood scars on valley floor trees and with maximum stage height recorded for the July 1950 flood at a U. S. Geological Survey gaging station located 1.4 km downstream.

As watershed size increases, larger meso- and macro-scale weather systems involving slow-moving or stalled major fronts are a common source of intense or long-duration rainfalls that produce large floods [*Hirschboeck*, 1988; *Hirschboeck et al.*, 2000]. Nevertheless, the probability that a given flood-producing weather system will cover the entire watershed decreases as watershed size increases. Consequently, basin-wide snow cover is more important than widespread heavy or long-duration rainfall as a cause of large and extreme floods in watersheds that are thousands of square kilometers in size or larger.

The systematic differences in causality of large floods with changing size of the contributing watershed can be expected to produce differing frequencies of overbank alluvial sedimentation in the downstream direction. Overbank sedimentation on floodplains of small headwater tributaries tends to occur more intermittently over time than overbank sedimentation on floodplains of large downstream river systems. The high relative variability in flood magnitudes of headwater and moderate size tributaries compared to the low relative variability in flood magnitudes on large downstream rivers implies that the depositional records of overbank floods probably differ significantly with regard to temporal-scale resolution. Large floods generally are well-represented in overbank alluvium of small and moderate size tributaries, but high-frequency small floods are less well-represented there because typical high-frequency

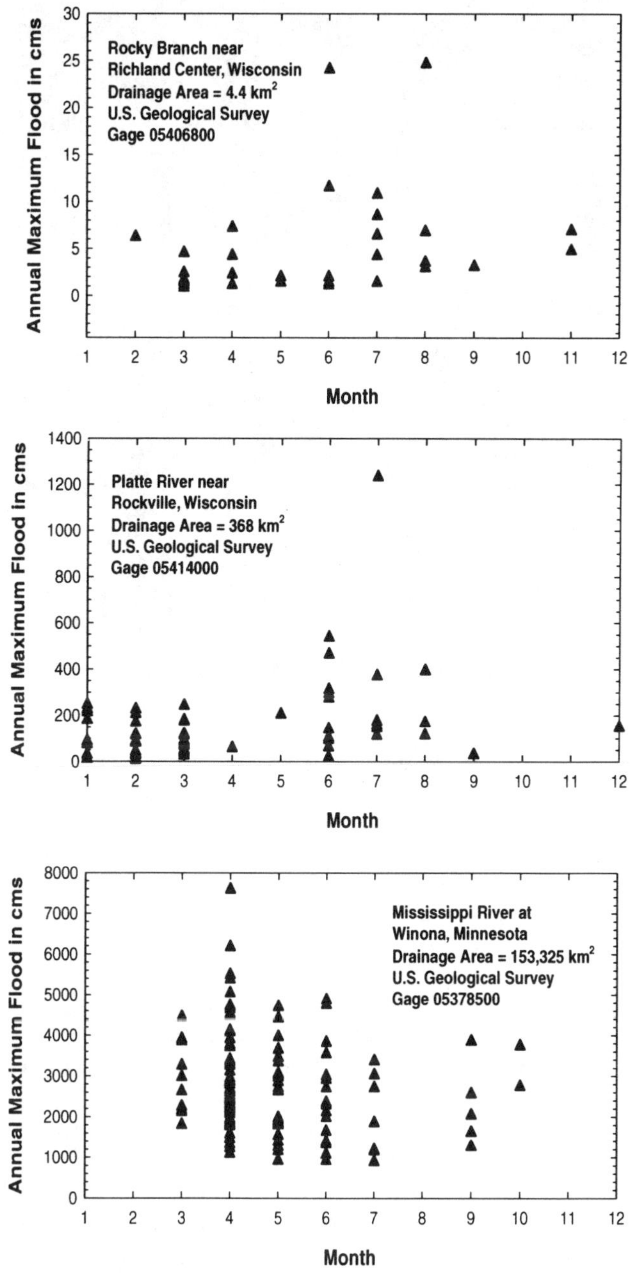

Figure 4. The largest floods on small and intermediate size watersheds in the Upper Mississippi Valley tend to be produced by heavy rains that occur during the months of June, July, and August. However, as watershed size increases beyond a few thousand km^2, as shown here for the upper Mississippi River, the largest floods tend to be dominated by snowmelt runoff. Note: January = 1; December = 12. (Data source: U. S. Geological Survey).

Figure 5. Heavy rainfall that included 24.1 cm falling in a 4 day period at a headwater monitoring station produced the July 1978 flood on southwestern Wisconsin's Kickapoo River [*Hughes et al.*, 1981]. A Log Pearson Type III analysis indicates this flood is slightly smaller than the expected 100 years recurrence probability flood [*U. S. Water Resources Council*, 1981]. The drainage area for the above site is approximately 1300 km^2. (Photos: J. C. Knox).

floods of moderate and small magnitude are often confined within either the channel or active meander belt. Furthermore, downstream spatial differences in stream power are likely to cause the total long-term Holocene flood record in the small and intermediate size high-energy tributaries to be relatively shorter than for downstream sites of relatively low-energy. *Mandel* [1995], for example, examined chronostratigraphy representing more than 65 sites in the Central Great Plains of the U. S. and found that long-term erosion had removed most early and middle Holocene alluvium from bottoms of small, high-energy valleys. On the other hand, he found that large downstream valleys contained most of the Holocene record either in alluvial fans or in deeply-buried valley fills. Consequently, there is a tendency for the continuity of overbank alluvial records of floods to increase in the downstream direction where low-energy floods of a wide range of recurrence frequencies overtop the valley bottom.

In light of the issues noted above, the possibility must be considered that an unrecognized erosional unconformity may exist in any alluvial deposit. Such unconformities are especially difficult to identify in fine-grained alluvium where bioturbation has occurred. Nevertheless, the possibility of erosional stripping of overbank alluvium by large floods is of extremely low probability for the vegetated floodplains of the UMV. No erosional hiatus was apparent when the detailed overbank alluvial stratigraphy representing the floodplain of the 700 km^2 Grant River watershed in southwestern Wisconsin was compared against a U. S. Geological Survey local 50-year daily stream flow record [*Paramore*, 1992]. Here, all moderate and large floods were represented in the alluvium and no evidence for an erosional hiatus was present [*Paramore*, 1992]. The present senior author has observed floodplains of UMV tributaries and main valleys for more than 50 years during which several floods have occurred with magnitudes that exceeded the 100-year recurrence probability (*e.g.* see Figure 7). Stripping of floodplain alluvium by floods has been a rare exception. In the few places where stripping occurred, it was associated with valley reaches where anomalous constrictions of flow occur as in narrow rock gorges. *Costa* [1974] observed local floodplain scouring occurred in similar reaches of valley constrictions during tropical storm Agnes flooding in the Appalachian Piedmont in 1972. *Magilligan* [1985] modeled environmental factors that controlled historical floodplain sedimentation on the Galena River system of the

Figure 6. Sandy gravel layers, marked by numbers, denote six reversals in the normal fining-upward sedimentology common to overbank alluvium. These gravels represent deposition during former large overbank floods that occurred on this 25 km^2 tributary watershed in the upper Platte River system of southwestern Wisconsin. (Photo: J. C. Knox).

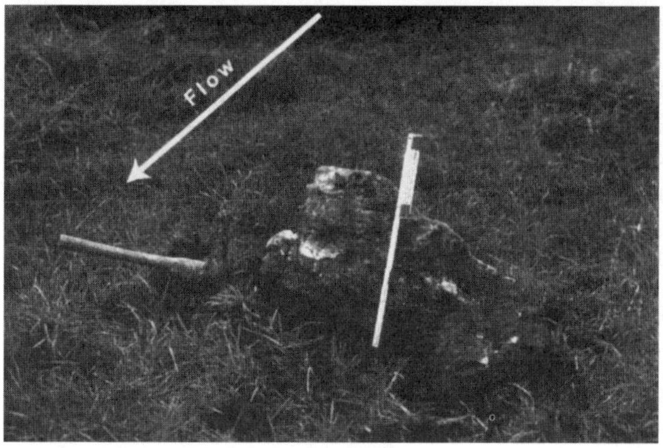

Figure 7. Large imbricated boulders were deposited on the floodplain of southwestern Wisconsin's Platte River during the July 16, 1950 flood. These boulders are gradually being buried by fine-texture overbank alluvium of smaller floods that have occurred since 1950. A log Pearson Type III analysis indicates the 1950 flood has an expected recurrence probability of about once per 500 years [*U. S. Water Resources Council*, 1981]. (Photo: J. C. Knox).

UMV and showed that floodplain width plays a dominant role in explaining downstream variations in depths of overbank historical alluviation. He attributed relatively shallow depths and small volumes of historical overbank floodplain deposition in reaches of anomalously narrow valley widths to occurrences of deep, high velocity flows in those reaches during large floods. *Miller and Parkinson* [1993] observed similar associations between constricted flows and floodplain stripping following the great flood of November 1985 in the South Branch Potomac River watershed of West Virginia. *Miller* [1995] modeled spatial patterns of flow energy associated with the 1985 flood and demonstrated that areas of erosional stripping were indeed associated with concentrated exceptionally high stream power. Typical impacts of large floods on unconstricted floodplains in tributary valleys of the UMR are bank erosion and overbank deposition as shown in Figures 8 and 9. Channel incision does not occur in most moderate and large size tributaries to the upper Mississippi River because Holocene channel movements have been mainly restricted to lateral migration on an armored coarse boulder gravel of late glacial age [*Mason and Knox*, 1997]. Even for the occurrences of accelerated floods during the period of historical agriculture, floodplain erosional stripping in agricultural fields is relatively rare and is typically isolated to unique locations where deep swift flows are associated with old paleochannels or other anomalous locations of constricted deep flows. These examples show there is a possibility for erosional hiatuses to occur in areas where landform morphology favors constricted deep flows. Extra caution is necessary when interpreting the stratigraphic record from these types of localities.

Physiographic Influence on Downstream Distribution of Flood Energy

The second principal way that watershed scale influences the type of stratigraphic record left by large floods is related to systematic downstream variability in the power of floods to erode, transport, and deposit sediment. Channel gradient and flow magnitude are the dominant factors controlling the power of flood waters [*Baker and Costa*, 1987]. Average diameters for sediment particles tend to decrease in the downstream direction. The downstream-fining pattern has been attributed to particle abrasion with distance transported and to downstream reduction in stream power. Stream power probably is the dominant factor because large reductions in the energy gradient occur in the downstream direction and this reduces the competency of the fluvial system to transport coarse particles. Total stream power per unit length of channel, Ω, represented in watts per meter ($W\ m^{-1}$) is defined by equation (2):

$$\Omega = \gamma Q s \qquad (2)$$

where γ is the specific weight of the water and sediment load (approximately 9800 $N\ m^{-3}$), Q is water discharge in

Figure 8. The bright surface on the floodplain represents 2 - 3 cm of sandy silt that was deposited by a June 29, 1990 flood in southwestern Wisconsin's Platte River system. The site is approximately at the same location of the bank exposure shown in Figure 6. Water depth on the floodplain was 25 -50 cm deep. The recurrence probability of the flood is unknown, but extrapolation from U. S. Geological Survey records for a downstream gaging station suggests it may be between about 2.0 and 2.5% on an annual basis. (Photo: J. C. Knox).

Figure 9. This illustration represents a close-up view of the June 29, 1990 sandy silt deposition that occurred on the floodplain shown in Figure 8. Later, these types of sandy silt units become buried by finer sediment related to overbank floods of smaller magnitude as shown in Figure 10. (Photo: J. C. Knox).

$m^3\ s^{-1}$, and *s* is the dimensionless energy gradient. Stream power per unit boundary area ω, expressed in watts per square meter ($W\ m^{-2}$), is defined by equation (3):

$$\omega = \gamma Q s / w \qquad (3)$$

where γ, Q, and *s* are defined as in (2) and w is defined as water-surface width. Several studies have confirmed that maximum stream power tends to occur in watersheds of moderate size rather than of large size [*Graf*, 1983; *Baker and Costa*, 1987; *Magilligan*, 1992; *Lecce*, 1997; *Knighton*, 1999]. The results of *Lecce* [1997] are particularly relevant to the present discussion because his study area is contained within and is quite representative of the UMV region of focus here. *Lecce* [1997] found that both cross-sectional and mean stream power increased to a maximum for drainage areas between about 10 and 100 km^2 and then decreased rapidly with distance downstream. He explained this relationship as principally determined by channel slope.

STRATIGRAPHIC RECORDS OF LARGE FLOODS

Small and Intermediate Scale Watersheds

Large floods in small and intermediate scale watersheds transport and deposit both coarse- and fine-texture sediment onto UMV floodplains. The gravel horizons shown in the eroding channel bank of Figure 6 and the large boulder on the floodplain of Figure 7 are illustrative of overbank coarse-sediment responses to high-energy floods that characterize small and intermediate scale watersheds in the region. More common expressions of large floods, however, are represented by the fine-grained overbank deposition illustrated by the 2 - 3 cm thick silty sand that was deposited during a June 1990 flood which over-topped the upper Platte River floodplain (Figures 8 and 9). The recurrence probability of the 1990 flood at this 25 km^2 upper Platte River site is unknown. Farther downstream, where the drainage area increases to 368 km^2 at a gaging site, application of Log Pearson Type III frequency analysis indicates the flood has a recurrence probability of about 2.5% in a given year [*U. S. Water Resources Council*, 1981]. High-water marks indicated the depth of flow responsible for the 2 - 3 cm deposition of silty sand was between 25 and 50 cm deep on the floodplain in the area of Figure 8. If channel alluvium of sandy texture is available in a fluvial system, then large overbank floods will normally produce floodplain sedimentation that is enriched in sand. Sandy horizons in overbank alluvial sequences therefore can be a proxy of former large floods if source conditions have not changed over time.

Overbank depositional units represented by fine sediment of sand, silt, and clay usually are inadequate for quantitative determinations of former flood magnitudes. On the other hand, their relative thickness and their fractions of sand, silt, and clay provide qualitative evidence of former flood magnitudes and semi-quantitative evidence for frequencies of past large floods [*Knox*, 1987a]. Figure 10 shows typical fine-texture historical overbank deposits in the upper half of a stream bank along southwestern Wisconsin's Grant River where the upstream drainage area is about 695 km^2. The vertical rod in the left center of the photo is 1.6 m length and denotes the magnitude of agriculturally-accelerated overbank flood sedimentation since July 1940. A nearby, continuously recording stream gage, in operation since October 1934, permitted correlation of individual flood depositional units with streamflow [*Paramore*, 1992]. The light-colored units in the exposure represent deposition of silt and sandy silt from singular large floods or from complexes of large floods that occurred close together within a single season. Depositional units associated with the large floods of 1940, 1950, 1954, and 1978 are benchmarks that facilitated easier age estimations for other flood units represented in the section. The wide range of fine sediment sizes within the sand through clay fractions at the Grant River Site of Figure 10 are representative of most UMV watersheds. For these grain sizes, overbank deposition from a large flood typically begins with a high silt fraction that shifts to a high sand fraction as a large flood reaches its peak magnitude.

Figure 10. The light colored units exposed at the Yager Site on southwestern Wisconsin's lower Grant River represent sandy silt and coarse silt that were deposited during historical large overbank floods. Vertical variability in fine-grained overbank flood depositional units of this type in the tributaries of the UMR typically coarsen upward toward the middle or upper part of the unit and then commonly show a fining upward tendency toward the top of the unit [*Knox*, 1987a]. These units represent the buried equivalent of sandy silt and coarse silt shown on the floodplain surface in Figures 8 and 9. The prominent deposits of 1940, 1950, 1954, and 1978 all were associated with extreme summer rainstorms in the upstream watershed of about 695 km^2.· These deposits serve as key beds that are traceable from site to site in the alluvium of the lower Grant River valley. A U. S. Geological Survey continuous recording stream gage has been in operation since the mid-1930s near the Yager Site. The gaging record allowed reasonably accurate age assignments for nearly all of the distinctive units shown in the top 1.5 m of the 3.4 m of overbank alluvium deposited since about 1835 when Euro-American agriculture became significant in the region. The length of the wooden (light-colored) section of the hoe handle is 1.4 m and the length of the mattock is 44 cm from its base to the end of the handle. Deposition associated with the largest floods, for example, 27 July 1940 and 16 July 1950 ranged between 15-17 cm, while other numerous smaller floods deposited 2-4 cm or less overbank sediment. (Photo: J. C. Knox).

Subsequently, the high sand zone is commonly capped with a thin unit of silt or clay as a large flood recedes [*Knox*, 1987a].

Age-controlled flood sedimentology at or near sites with long stream gaging records provides a basis for interpreting flooding history from longer stratigraphic records at those sites or other sites in the region. However, long-term interpretation of sedimentological changes in vertical sedimentary sequences must be evaluated in the context of possible changes in other environmental factors that might also influence sediment transport. The Stelpflug Site 2, with a drainage area of 422 km^2 on the lower Platte River of southwestern Wisconsin, illustrates the type of detailed chronology of large floods that can be extracted from fine-texture overbank alluvial sequences and extended to a longer pre-instrumented record (Figure 11). The Stelpflug Site is geographically adjacent to the Grant River watershed of Figure 10 and therefore displays a history of large floods similar to that shown in Figure 10. The section at Stelpflug Site 2 includes both a historical and late-Holocene stratigraphy. The bottom of the historical section corresponds with the beginning of intense Euro-American settlement about 1830. The Holocene section probably is not older than about 1000 - 1500 years

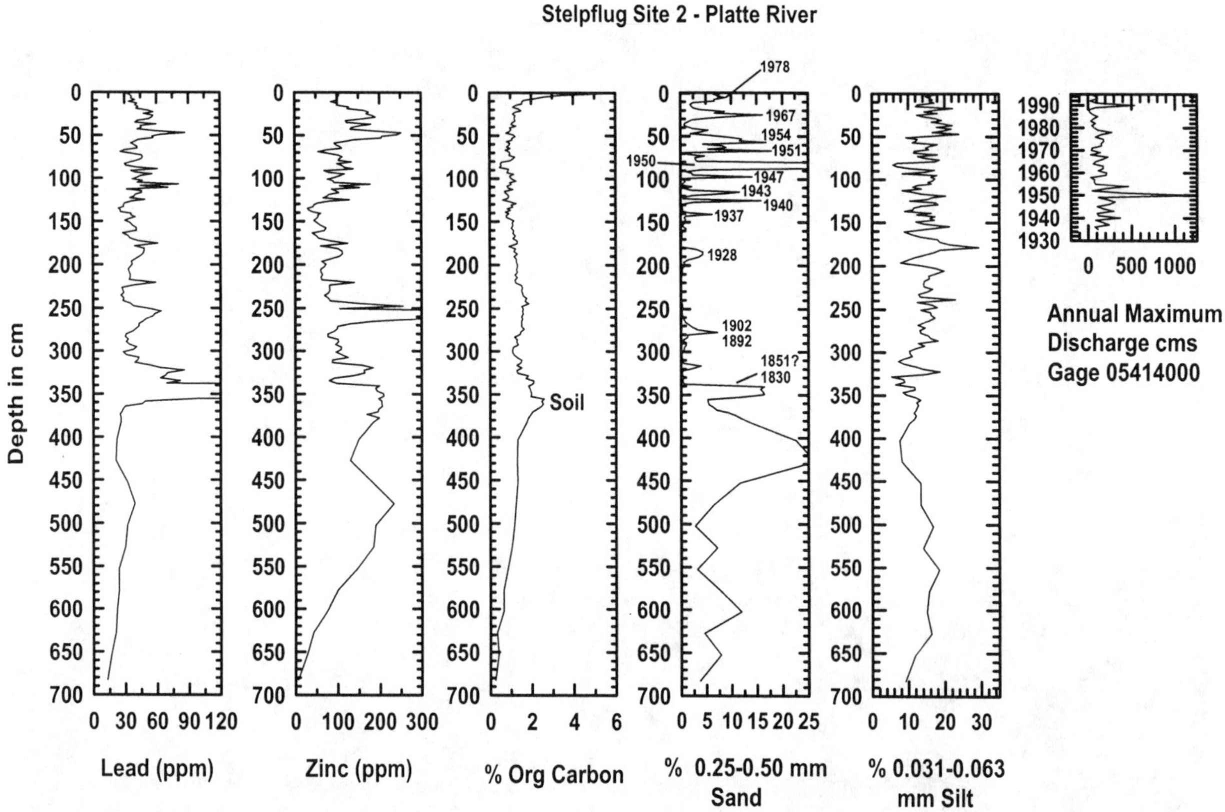

Figure 11. The stratigraphic record of historical large floods to A. D. 1989 at Stelpflug Site 2 on southwestern Wisconsin's lower Platte River is nearly identical to that shown for the Yager Site in Figure 10. The size of the upstream watershed is 422 km^2. Age assignments for units representing deposition associated with large overbank floods were based on correlation of the sedimentological record with a continuous record of stream flow since October 1934 at a gaging site 9 km upstream, and on correlation with trace metals generated by mining activity of known age. The high level of organic carbon at 360 cm depth denotes a buried Mollisol that was the floodplain surface prior to introduction of agriculture by Euro-Americans. The increased level of sand in the upper section of the pre-1830 horizon is interpreted as representing an episode of large floods that characterized the very late Holocene. (Discharge data: U. S. Geological Survey).

because it is stratigraphically and sedimentologically similar to an alluvial sequence with a basal age of A.D. 735 ± 50 years at the Hinderman Site in an adjacent tributary of the Platte River system [*Knox*, 1999].

Age control is relatively good for the period since the beginning of intense Euro-American settlement in the 1830s at Stelpflug Site 2. The early settlers were both miners and farmers and their disturbance of the upland land cover of prairie and forest led to a nearly immediate acceleration of upland soil erosion followed by accelerated downstream floodplain sedimentation [*Knox*, 1977; 1987b]. The accelerated sedimentation has resulted in very distinctive stratigraphic evidence of former large floods and other stratigraphic discontinuities (Figure 11). The pre-settlement floodplain horizon is easily recognized in the field as a dark Mollisol. An increased level of organic carbon at about 360 cm depth represents this buried soil on Figure 11. Further confirmation that the pre-settlement floodplain of about 1830 occurs at the 360 cm horizon is indicated by the abrupt increase in lead from a background concentration of less than about 30 ppm to maximum concentrations of 490 - 560 ppm between 340 and 355 cm depth. The original Federal land survey field notebooks and maps of 1832 and 1833 indicate active lead mines in small tributaries located about 5 km upstream and a later map by *Freeman* [1842] shows that a lead smelter was in operation there in 1842. These tributaries were associated with several episodes of well-documented mining from the early 19th century until about 1960 [*Heyl et al.*, 1959]. Between about 1915 and 1917 two mines alone produced more than 60,000 tonnes of zinc and lead ore and this production is represented by the elevated levels of heavy metals occurring between about 240 - 270 cm depth (Figure 11). A large flotation mill processed zinc and lead ores in the area from 1954 to about 1960 [*Heyl et al.*, 1959]. Downstream transport of sulfide metals from tailings associated with this mill is responsible for the sharp increase in metal concentrations occurring at about 60 cm depth at Stelpflug Site 2. In addition to the trace metal concentrations and buried soil as age marker horizons, the extreme flood of 1950 also produced a very prominent stratigraphic horizon. These age controls were combined with a continuous record of Platte River flow since October 1934 from a U. S. Geological Survey stream gage located 9 km upstream. The combined records were then used to assign ages to stratigraphic horizons indicative of other large floods in the Stelpflug alluvial sequence (Figure 11).

The coupling of age-assigned floods with the alluvial record at Stelpflug Site 2 shows that floods with a magnitude equal or greater than approximately the 10-year flood are associated with highly recognizable stratigraphic signatures. Medium sand (0.25-0.50 mm diameter) and coarse silt (0.031-0.063 mm) were selected to represent the effects of floods on transport and deposition of different grain sizes. Medium sand is selected over other sand fractions and over total sand because it is coarse enough to be a sensitive indicator of flood energy, but fine enough to be transported in reasonable abundance even in the small magnitude floods. Suspended transport of sand coarser than medium sand normally occurs in the lower portion of the water column, whereas transport of sand finer than 0.125 mm commonly occurs at all levels in the water column [*Colby*, 1963]. Coarse silt was selected because it is a sensitive indicator of upland erosion in the loess-covered watersheds of the Upper Mississippi Valley. Large rainstorms capable of producing sheet, rill, or gully erosion favor large inputs of silt to the river system because the loess-dominated parent material is typically represented by 60-70% silt composition. Coarse silt also provides a good proxy of variability in small floods when large floods are absent from the chronology. The percentage of coarse silt also normally varies inversely with the amount of fine silt and clay. Therefore, the anomalously low values of coarse silt occurring at about 160 and at about 195 cm depth at Stelpflug Site 2 correspond with anomalous high percentages of fine silt and clay. The high magnitudes of fine sediment here are possibly related to deep gully incision that penetrated through the loess cover and into underlying clayey residuum [*Knox*, 1987b]. Possible changes in sedimentology of overbank alluvium due to abrupt shifts in channel positions, as with avulsions, can be ruled out at Stelpflug Site 2. Comparisons of the modern channel locations on township section lines in the reach of the lower Platte River containing the Stelpflug Site with their surveyed location in 1832-33 shows lateral movement has generally not exceeded 5-8 m [*Knox*, 1977].

The use of sand, silt, and clay grain size variations for estimating qualitative variations in flood magnitudes and frequencies requires that other potential environmental changes also be considered. The relatively modest frequency of flood-deposited sand units before 1940 and their dramatic increase after about 1940 at Stelpflug Site 2 gives the impression that flood magnitudes and frequencies have increased sharply since about 1940. Although the record of magnitudes for annual maximum floods at the U. S. Geological Survey gaging site located 9 km upstream supports the idea that the stratigraphic record provides a good proxy of large floods since the mid-1930s, an instrumental record is not available for earlier historical stratigraphy. The relatively short period of instrumental record is unfortunate because it begins during a dry period of few floods and only about 3 years before the Corps of Engineers raised the normal low-water base level in 1937 at the mouth of the Platte River 9.5 km downstream. The rise in base level resulted in aggradation of the channel bed and it accelerated overbank flooding. The rising river bed made it easier to elevate sand onto the floodplain surface thereby magnifying the stratigraphic imprint of large floods. Natural aggradation or degradation of the channel bed also contributes to changes in overbank flood deposition, but this has not been a significant factor during the Holocene for the Upper Mississippi Valley sites that are the focus of this paper. Surveyed cross-valley drilling transects in tributary valleys of southwestern Wisconsin have shown that channel bed elevations have remained remarkably stable during the Holocene as these channels migrated laterally on an armored surface of late-Pleistocene cobble and boulder gravels [*Knox*, 1985; *Mason* and *Knox*, 1997]. The anomalous post-1937 aggradation of the channel bed at the Stelpflug Site resulted directly from downstream dam closure and related baselevel rise. Comparing elevations of the modern channel bed with the elevation of the former pre-dam and Holocene channel bed elevations preserved under Holocene alluvial deposits indicates the amount of post-dam channel bed aggradation ranges between 2 and 3 m. This magnitude agrees reasonably well with a nearby Little Platte River site located 13.5 km upstream of the Platte River mouth at the Mississippi River. Comparison of 1989 survey results with a 1939 survey by the Corps of Engineers indicates that the bed of the Little Platte River had aggraded about 2 m since 1939. The rapid aggradation since the late 1930s is anomalous for the entire historical period because radiocarbon age control of the Little Platte River alluvial stratigraphy showed that the channel bed had previously been vertically stable since at least about A.D. 735 [*Knox*, 1999].

During the post-1937 period of accelerated aggradation of the channel bed, floods as small as those of 10-year recurrence probability are represented by high sand fractions in the overbank alluvium at Stelpflug Site 2 (Figure 11). The notable decreased frequency of beds with high sand fractions for the pre-1937 historical component of overbank alluvium occurs mainly because larger floods were required to elevate sand to the floodplain level then. Comparisons of daily and monthly local rainfall records since 1890 with magnitudes of floods and associated rainfalls since 1937 indicate that the large floods responsible for the units with high sand fractions in the pre-1937 historical alluvium probably equaled or exceeded modern floods of 25-yr to 50-yr recurrence probabilities. The 1928 flood occurred when 16.3 cm of rain fell during a four-day interval with 10.7 cm of the rainfall occurring near the end of the storm. This storm followed a storm occurring about 2-weeks earlier that produced 10 cm of rainfall over three days. The total August 1928 rainfall was 3.4 standard deviations above the long-term 1890-1990 average. The floods of the 1892 and 1902 also are associated with extreme rainfalls. Monthly rainfalls for May and June 1892 were respectively 3.5 and 3.3 standard deviations above their long-term averages and monthly rainfalls for July of 1900 and 1902 were respectively 2.5 and 2.4 standard deviations above their long-term averages. For comparison, the July floods of 1940 and 1950 were associated with monthly rainfall totals that were respectively 2.4 and 3.4 standard deviations above their long-term average magnitudes. The 20 cm thick sandy unit that occurs immediately on top of the pre-settlement soil may be partially the result of extremely large floods that occurred in 1851 in southwestern Wisconsin watersheds [*Holford*, 1900]. On the other hand, the anomalous high sand percentage may also reflect the early lead miners' practice of in-stream washing of residuum from lead ores. Much of the early mining involved shallow extractions of lead ore from the red-clay residuum where sand and clay fractions were relatively high. In-stream washing of lead ore to remove residuum may have contributed significant amounts of sand and clay to the channel. This conclusion is supported by the observation that relatively low concentrations of coarse silt in the first 50 cm of overbank alluvium above the pre-settlement soil reflect high percentages of clay and fine silt there.

The stratigraphic section between 355 and 680 cm depth on Figure 11 represents late-Holocene floodplain alluvium that was deposited prior to significant human influence. Direct comparison of flood frequencies here with flood frequencies indicated in the overlying historical alluvium is difficult because the Holocene section was analyzed at intervals of 25 cm whereas the historical alluvium was analyzed at intervals of 2-3 cm. Furthermore, the Holocene vertical accretion of the floodplain occurred at an order of magnitude slower pace than the historical rate, allowing bioturbation processes to mix adjacent stratigraphic horizons. The average rate of vertical accretion since 1830 at the Stelpflug Site 2 has been about 2.2 cm yr^{-1}. Radiocarbon dating for a Holocene alluvial sequence in a comparable drainage area on the nearby Little Platte River floodplain indicated vertical accretion occurred at the rate of about 0.15 cm yr^{-1} between A.D. 735 and the time of Euro-American settlement in the region [*Knox*, 1999]. The slow rate of pre-settlement accretion in association with natural bioturbation processes means that recognition of small and even moderate magnitude floods is quite difficult in these settings. Large floods, however, leave a sandy imprint, provided sand is available in the system. Sandy horizons are normally well preserved in the stratigraphic record, but they may represent either singular or multiple floods.

The late-Holocene alluvium at the Stelpflug Site 2 involves sandy horizons in a striking example of reversal in the normal fining-upward tendencies that characterize floodplain overbank deposits. This reversal, which occurs between about 425 and 500 cm depth, is depicted by a relatively high concentration of sand (Figure 11). The sand concentration in the upper part of the section implies an episode when floods were generally larger than before or after. Similar sedimentary reversals have been observed for late-Holocene overbank alluvium at other southwestern Wisconsin sites which supports the idea that the reversal here is not site specific. Furthermore, the stratigraphic implication that floods increased in size in the late-Holocene is consistent with the proxy record of large overbank floods that have been estimated for flood stages competent to transport large cobbles and boulders found in floodplain alluvium elsewhere in southwestern Wisconsin [*Knox*, 1993].

The proxy record of large overbank floods that occurred during the Holocene in small tributary drainage basins of southwestern Wisconsin was developed by computing the depths of water competent to transport the largest cobbles and boulders found on the buried surfaces of Holocene point bars or in overbank Holocene alluvium (Figure 12). Former flood depths were computed with equation (1), presented earlier, which estimates the water depth competent to transport a clast equivalent to the average size of the 5 largest cobbles or boulders found in the paleoflood gravels. Sampling procedures and other methodology have been previously presented for Holocene overbank flood gravels, and only a brief review will be given here because the procedures remain the same for the new data related to point bar gravels [*Knox*, 1993]. The relict gravels were

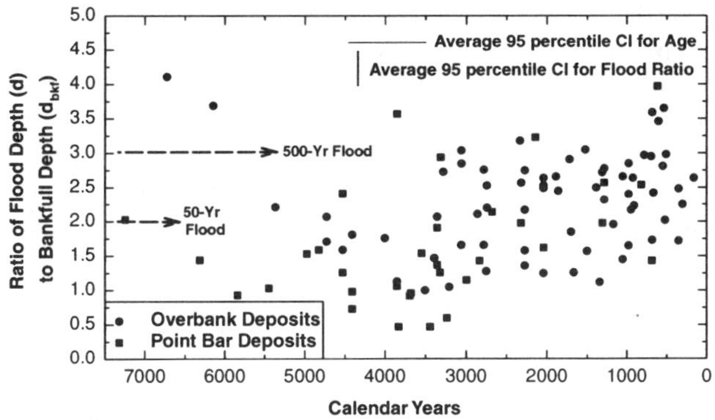

Figure 12. Depth magnitudes for southwestern Wisconsin overbank floods were computed by calculating the depth of water competent to transport a clast equivalent to the average size of the five largest cobbles or boulders in a relict flood gravel exposed in overbank alluvium or on the surface of buried point bars. The computed flood stages are expressed relative to their heights above the associated channel beds shown to have been stable throughout the Holocene. The thalweg flood depth is presented as a ratio to bankfull flood depth to minimize the effects of drainage area to facilitate comparisons between sites. A regression analysis showed that 5% of the variance in the flood ratio was explained by drainage area, but the relationship was not statistically significant at the 0.01 probability level. The data for overbank alluvium is modified from *Knox* [1993] and the data for point bars is new. Note that, following an episode of relatively small floods between about 5000 and 3300 years ago, floods abruptly increased. Floods appear to have experienced further change after about 1000 years ago, and four sites suggest that floods may have been especially large between about 700 and 500 years ago. The indicated levels for the 50-yr and 500-yr floods are based on present-day instrumental records. See text discussion and *Knox* [1993] for additional elaboration of methodology and explanation of the data.

sampled by excavation where they were exposed in stream cutbanks. Sampled gravels included 68 overbank floods and 33 point bar floods representing more than 100 sites in various southwestern Wisconsin watersheds. Efforts were undertaken to avoid sampling flood gravels that might represent the same event in different watersheds, but it is acknowledged that some large storms may possibly have been inadvertently represented more than once. However, such redundancy is thought to be minimal. The sampled watersheds varied in size from a few to several hundred square kilometers. The flood depth competent to transport the largest clasts in a deposit are expressed on Figure 12 as a ratio to the bankfull-depth dimension at the same site to facilitate plotting all sites on a common graph. Bankfull depth was approximated as the height of the modern point bar above the modern channel bed. Radiocarbon dating of Holocene alluvium in southwestern Wisconsin valleys has shown that the vertical position of the modern channel bed has remained nearly stable throughout the Holocene except in relatively steep tributaries where minor cutting and filling has occurred [*Knox*, 1999]. The vertical stability of the channel bed ensures that episodes of aggradation or degradation have not influenced relations between stage and discharge during the Holocene.

Age assignments for relict flood gravels were determined both directly and indirectly from radiocarbon dating. Organic carbon sufficient for radiocarbon analysis was rarely found directly in association with the flood gravels because of oxidation. However, it was possible to determine ages for many of the immediately underlying alluvial units either by direct radiocarbon dating of those units or by correlating those alluvial units with equivalent dated alluvium elsewhere in the region. Also, because of the long-term vertical stability of the channel bed during the Holocene due to armored boulder gravels preventing cutting and filling, vertical accumulations of Holocene alluvium are strongly time dependent with a long-term vertical accretion rate of about 0.02 cm yr^{-1} (Figure 13). The general relationship illustrated in Figure 13 was further analyzed in a second regression equation that predicted the age of alluvial units at the channel bed level:

$$\text{Age} = 43.4(X_{sed}) - 1207.7(A_{da}) - 367 \quad (4)$$

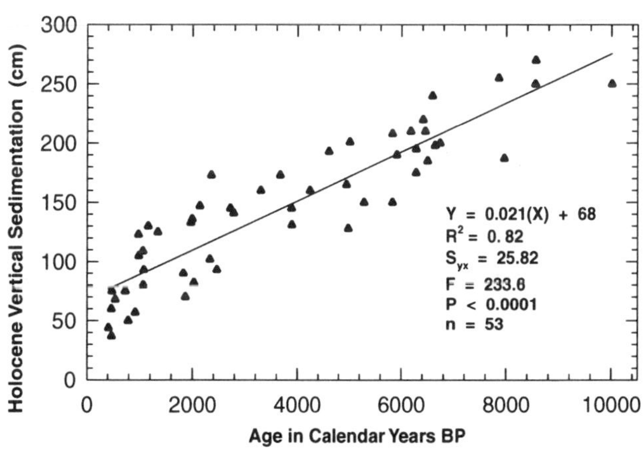

Figure 13. Southwestern Wisconsin alluvial channels with cobble and boulder armoring of channel beds have been characterized by long-term vertical stability of channel beds during the Holocene. Consequently, the amount of overbank vertical accretion on the area's floodplains is strongly time dependent and averages about 0.02 cm yr^{-1} for watersheds that range in size between about 5 and 675 km^2. The Y intercept at about 70 cm represents the component of alluvium that is point bar lateral accretion sediment typical of small tributaries in the region. (Data source: J. C. Knox).

where Age is time in calendar years before the radiocarbon base year 1950, X_{sed} is centimeters of vertical sedimentation between the channel bed and the top of the horizon representing the surface of the pre-agriculture soil, and A_{da} is the upstream contributing drainage area in square kilometers. Equation (4) explained 88 percent of the variance in the base data set of representing 34 sites. The equation is statistically significant at the 0.0001 level of probability and has a standard error of estimate of 900 years. While the standard error is large, the equation is sufficiently accurate for placement of alluvial units and affiliated flood gravels within sectors of Holocene time. Equation (4) was adapted for use in estimating ages of gravel units by determining the difference between two estimates of ages at each site. The first estimate involved inputting for X_{sed} the depth of the total pre-agriculture Holocene sedimentation between the surface of the pre-agriculture soil and the average channel bed base elevation. The second estimate involved inputting for X_{sed} only the depth of Holocene sedimentation between the base of the flood gravel in question and the average channel bed base. The difference between the two age estimates, in turn, provides an approximation of the number of years ago that the surface supporting the flood gravel in question was the floodplain surface.

The addition of 33 paleofloods estimated from relict point bar deposits to the original 68 paleofloods estimated from relict overbank gravels reinforces the temporal pattern previously estimated for late-Holocene large overbank floods (Figure 12). The most noteworthy patterns involve an episode characterized by small floods between about 5000 and 3300 years ago, followed by an abrupt increase in flood magnitudes after about 3300 years ago. The data also suggest that beginning about 1000 years ago even the smallest of the large floods increased in size relative to the period 3300 to 1000 years ago. Furthermore, four sites indicate exceptionally large floods occurred between about 700 and 500 years ago. These flood episodes have been shown to be strongly related to modest changes of climate inferred from independent fossil biotic and other evidence [*Knox*, 1993; 1999; *Baker et al.*, 1996]. The relatively strong evidence for larger floods in southwestern Wisconsin after about 1000 years ago, and especially between about 700 and 500 years ago, supports the idea that the sandy deposits between 425 and 500 cm depth at Stelpflug Site 2 were also a result of increased flood magnitudes then.

Depths of overbank sedimentation from individual floods in UMV watersheds since the beginning of agricultural influence occurred at either centimeter or millimeter scales. For comparable runoff events, depths of overbank sedimentation from individual floods in UMV watersheds prior to human disturbance were an order of magnitude smaller than those of the agricultural period [*Knox*, 1999]. Admittedly, the 25 cm sampling interval for the pre-agriculture Holocene sediment at Stelpflug Site 2 is too wide to differentiate contributions of singular floods. Nevertheless, the sandy zone between 425 and 500 cm depth at Stelpflug Site 2 is most likely the result of several to many large floods rather than a singular flood. Localized splays extending onto the floodplain from a present or former channel bank position normally produce thick sandy deposits from singular floods. There is no evidence at Stelpflug Site 2 that supports a splay interpretation. Most boundaries between depositional units of former Holocene floods were undoubtedly subtle, and bioturbation has subsequently further blurred these contact surfaces. Horizons of sandy silt or silty sand are common phenomena in the fine-grained overbank alluvium of UMV small and moderate scale watersheds. They become even more prominent as watershed scale increases to the size of the Mississippi River.

Mississippi River Scale Overbank Alluvium and Floods

The bedload sediment of the upper Mississippi River along western Wisconsin is dominated by sand [*Moody*, 1997]. Energy conditions during large floods are often sufficient to transport sand as suspended load and deposit it as overbank alluvium on islands and floodplain surfaces. The historical and Holocene overbank alluvium of the upper Mississippi River consists almost exclusively of fine sediment of sand, silt, and clay composition. Therefore, use of competent depth paleoflood methodologies to extend the instrumental record of flood history ordinarily is not possible. Neither is normal slackwater methodology because bioturbation and post-flood erosional processes have eliminated most slackwater evidence. Furthermore, the slackwater method is difficult to apply because the floodplain of the upper Mississippi River has been aggrading throughout the Holocene in response to large inputs from sediment stored in tributary valleys during the last glacial [*Knox*, 1996]. Since the beginning of Euro-American settlement in the early and middle 19th century, rates of overbank sedimentation on the Mississippi River islands and floodplain have also been accelerated by agricultural land use as noted above for the tributaries. Other influences on sedimentation rates relate to U. S. Corps of Engineers channel modifications since about 1878 and especially since the late 1930s when a system of locks and dams were built [*Anfinson*, 1993]. The alluvial stratigraphy of the upper Mississippi River floodplain and islands shows vertical variations in sedimentology that integrates all of these influences while at the same time providing a record of long-term changes in frequencies of large floods.

As is the situation for UMV tributary sites, the sedimentological record of large overbank floods on the upper Mississippi River is easiest to interpret where sediment of sandy texture has been transported to overbank sites. Backswamp environments, where sedimentation has been dominated by silt and clay, record passing floods, but individual depositional units are often difficult to differentiate without conducting detailed laboratory analyses. Furthermore, the grain size diameters of sand grains provide a more sensitive index of flood size than do silt and clay grain sizes. Nevertheless, bioturbation processes often make it difficult to differentiate singular deposits of large floods when several large floods occur in close temporal succession. An illustration is provided by Figure 14 which shows that sedimentological analyses at vertical intervals as small as 1 - 2 cm are inadequate to define the precise increments of large floods that have occurred since about 1930 on Island 146 in the upper Mississippi River at Lansing, Iowa. River stages at Lansing have been recorded since the late 1930s when the Lock and Dam No. 9 was put into operation. Large floods are relatively little affected by the dams on the Mississippi River because it is standard practice to fully open gates to allow flood passage during high stage flows. Only river stages equal or greater than 190.5 m above sea level have been plotted on Figure 14 because flow stages above this approximate magnitude are required to exceed the elevation of the present ground surface at Lansing Site 5. Annual maximum floods on the upper Mississippi River at Winona, Minnesota, located about 100 km upstream, confirm that zones of high sand percentages at Lansing Site 5 were each the result of multiple floods that occurred close together in time (Figure 14). Anomalously large increments of deposition from these floods have occurred at Lansing Site 5 because closure of Lock and Dam No. 9 in the late 1930s raised the normal low water stage about 2 m. After dam closure, the ground surface at Lansing Site 5 was initially only about 50 cm above normal low water stage of the river. The low relative elevation explains why relatively large increments of sedimentation occurred in response to relatively small flood stages of the mid-1940s and early 1950s while more modest sedimentation occurred in response to more extreme floods of the period 1965-1975. Decreasing sediment yields from tributary streams have also occurred since the 1930s in response to better land use conservation methods over time [*Knox*, 1987b; *Argabright et al.*, 1996]. This reduction probably also explains in part the smaller increments of deposition with large floods in the younger portion of the Lansing record. In spite of the high rates of historical sedimentation at Lansing Site 5, differentiating sediment contributions of singular large floods is difficult when large floods occur close together in time. The difficulty of differentiation increases by an order of magnitude for Holocene overbank alluvium because rates of floodplain vertical accretion were much smaller then in comparison to the historical period of landscape disturbance by human activities.

The overbank sedimentology at Lansing Site 5, calibrated against the record of water stage variations at the same locality demonstrates that vertical changes in grain size characteristics for upper Mississippi River floodplain and island sites provide useful information about long-term changes in frequencies of large overbank floods. Figure 15, for example, presents a >4000-year record of overbank sedimentation for a site on the Mississippi River near Bagley, Wisconsin. Note that a large increase in the sand

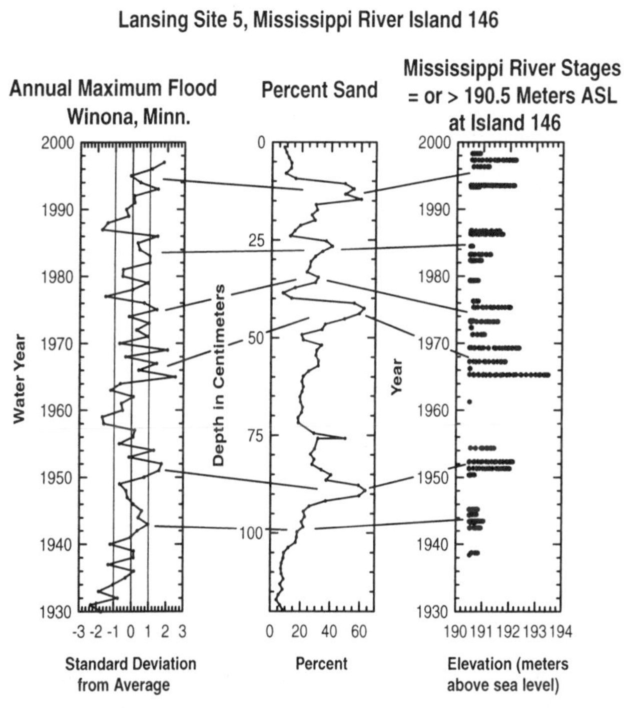

Figure 14. Differentiating the sedimentary contributions of singular floods in a sequence of large floods that have occurred close together in time is difficult when the overbank alluvium is very high in sand and sample analyses are spaced no closer than 1 - 3 cm intervals. Nevertheless, the comparison of the sedimentological data with the hydrological data of flood discharges and local flood stages shows that clusters of large floods are well represented in the stratigraphic record. The late 1930s closure of Lock and Dam No. 9 on the Mississippi River, a short distance downstream of Island 146, raised the water level about 2 m at this site and explains the initial high sedimentation rate for the floods of 1940s and 1950s when the island surface was little above the normal low water stage of the river. (Data sources: Discharge - U. S. Geological Survey; Island 146 water stages - U. S. Army Corps of Engineers).

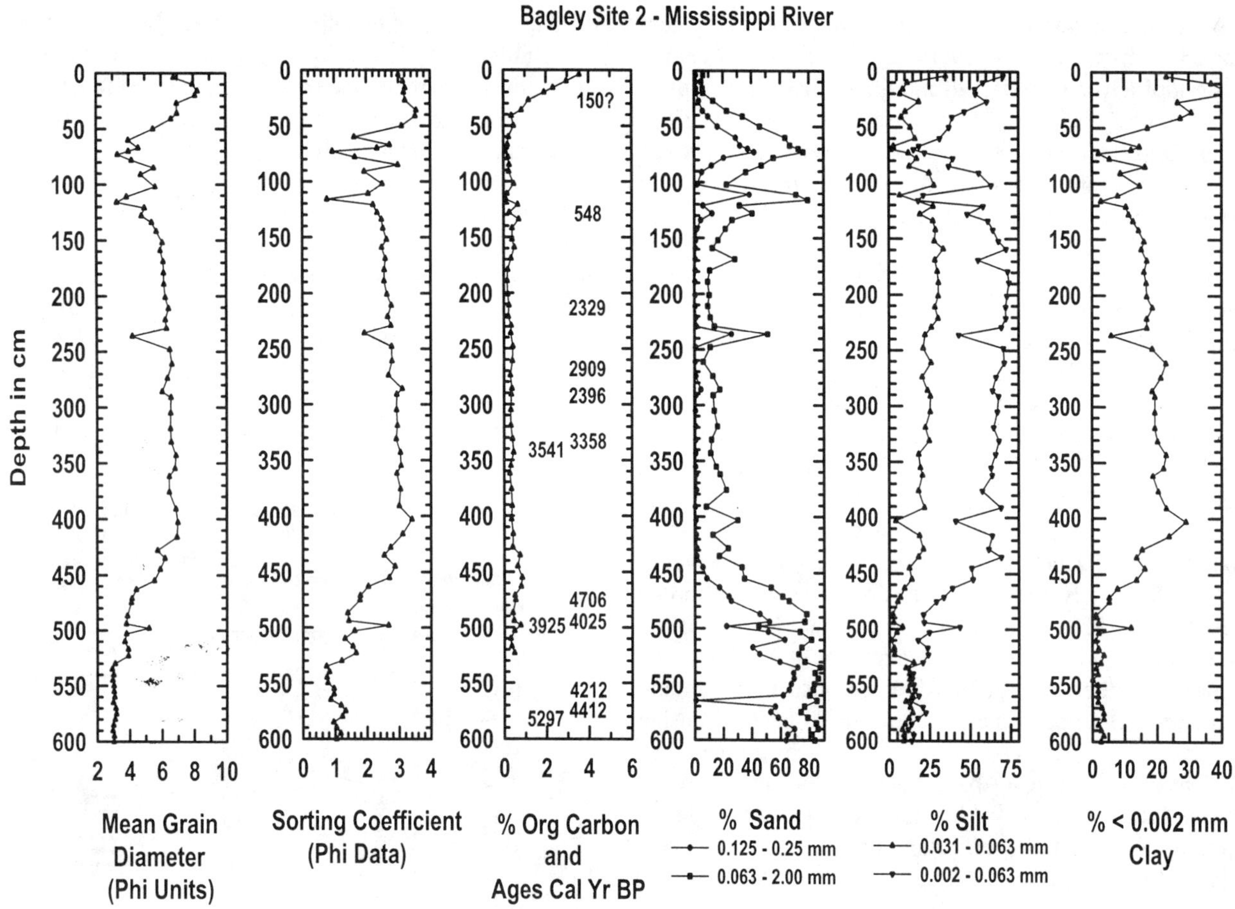

Figure 15. The horizons dominated by high sand components at Bagley Site 2 indicate that a pronounced episode of large floods on the upper Mississippi River probably began sometime shortly before about 550 years ago and ended about 100 to 150 years ago. Large floods also appear to have characterized the period prior to about 4000 years ago although some of the strong sandy character of the sediment below a depth of 450 cm may be related to changes in the morphology of the island and floodplain system then (see text). The high sand percentages extend to the core base at about 620 cm, but more work is needed to resolve their proper interpretation.

component of the overbank alluvium occurs near the top of the section beginning at a depth of about 135 cm below the present surface. Based on the causal relationships between sedimentology and flood stages shown in Figure 14, this sandy sequence implies an episode of overbank sedimentation associated with multiple large floods. Radiocarbon dating indicates an age of about 550 years (*c.* A. D. 1450) for the beginning of this episode. Observations by the senior author of sandy overbank depositional units in similar stratigraphic position at other floodplain and island sites along the upper Mississippi River indicate that the late-Holocene flood history observed at the Bagley Site 2 is characteristic of the region. It is noteworthy that more frequent occurrences of large floods on the upper Mississippi River after about A. D. 1450 agree in general with similar tendencies elsewhere. For example, large floods at about this time occurred at Stelpflug Site 2 and on other small and moderate size tributary systems of southwestern Wisconsin (Figure 12). The increased frequency of large floods about 550 years ago corresponds with the transition from the Medieval Warm Period to the cooler climate of the Little Ice Age [*Lamb*, 1977]. Times of rapid climate change and times of cooler climate generally favor more frequent occurrences of large floods in the UMV [*Knox*, 2000].

Variability in sand concentration and other sedimentological indices for overbank alluvium deposited from slightly before 550 years ago to about 4000 years ago at Bagley Site 2 is relatively conservative and suggests that large overbank floods were less frequent then. A prominent coarsening of the sediments at a depth of about 240 cm below the present floodplain surface suggests that a very large flood or floods occurred between about 2300 and 2900 years ago. A major change in sedimentology occurs at a depth of about 480 cm where an abrupt shift to higher sand content occurs. The increased importance of sand may be due either to more frequent large floods then or to a possible change in the proximity of the site to thalweg location of the Mississippi River or its secondary channels. The age of the sediment at this transition is about 3900-4000 years old at Bagley Site 2, but our work in progress representing other Mississippi River sites indicates a slightly older age. Because we have recognized this same abrupt change in sedimentology at comparable depth at several other sites along the upper Mississippi River we think it represents a major adjustment in the frequency and magnitude of overbank floods then. Alternatively, it may represent an indirect response to changes in flooding on the Mississippi River which then caused an abrupt shift in channel morphology and the behavior of overbank flooding. The occurrence of similar stratigraphic changes for the same general period at other upper Mississippi River sites suggests that the sedimentological change is not a result of local channel shifts. Furthermore, an interpretation of more frequent large floods on the Mississippi River at this time is consistent with the observation that small and moderate size tributaries of southwestern Wisconsin also display a modest trend for larger floods prior to about 4500 years ago (Figure 12).

It is significant that at least some of the late Holocene episodes of large floods on the upper Mississippi River occur at broadly similar times as episodes of large floods on the small and intermediate scale tributary watersheds. The similarity is surprising because the large overbank floods on the upper Mississippi River are mainly produced in association with spring snowmelt. The large overbank floods on the tributary watersheds, on the other hand, are mainly produced by heavy summer rainfalls. The broad similarities between the two flood records suggest that, over the long-term, changes in climate that are relevant to generation of large floods have influenced cold and warm seasons in comparable ways during the late-Holocene. Support of this idea is also indicated by the nearly identical Holocene records of high-frequency and low-frequency floods in the UMV tributaries of southwestern Wisconsin [*Knox*, 2000]. There, small magnitude floods of 1 - 2 years expected return frequency are mainly a product of spring snowmelt, whereas the large overbank floods result from heavy summer rain storms as previously noted. At present, the long-term paleoflood history of the upper Mississippi River is too poorly documented to provide firm correlation with the paleoflood history of the UMV tributaries. Nevertheless, the apparent associations suggest that better definition of these linkages could provide very useful information concerning how climate changes influence magnitudes and recurrence frequencies of floods in watersheds of different sizes.

SUMMARY AND CONCLUSIONS

Use of alluvial stratigraphic records for extending flooding histories beyond instrumental records is based on sedimentological characteristics of alluvial deposits and on the application of sedimentological evidence to derive flood-related high-water stages. This paper has focused on Holocene and historical paleofloods of large magnitude that have occurred in the humid-environment Upper Mississippi Valley where much of the paleoflood record has been blurred due to erosion by subsequent runoff and flooding and by mixing of alluvial sediments related to natural bioturbation processes. The results show that the type and quality of the stratigraphic record is strongly scale related, and this association limits the type of paleoflood methodologies that can be applied. Small (<25 km^2) and intermediate (25 to several hundred km^2) scale watersheds experience high-energy floods that provide stratigraphic records from which both frequency and magnitude information can be extracted. In contrast, most watersheds that are larger than several hundred to 1000's of km^2 experience large floods of relatively low energy. Consequently, large clasts suitable for application in a competent depth equation usually are not available in the stratigraphic record for these downstream locations. Furthermore, stratigraphic evidence of former maximum flood stages associated with slackwater deposition tends to be poorly preserved because of subsequent erosion and bioturbation. However, the fine-texture (sand, silt, and clay) alluvial stratigraphy at topographic positions on floodplains and low terraces below former high-water stages in the large watersheds commonly preserves relatively long and nearly continuous sedimentological records of floods.

The ubiquitous character of the fine-grained overbank flood sediment on floodplains and low terraces provides a detailed record from which recurrence frequencies of large overbank floods can be determined. The most prominent sedimentological indicator of a former large flood or series of large floods is a reversal in fining-upward sedimentology. The depositional unit representing a former large flood or floods is typically represented by a high sand fraction, provided sand is available in the drainage system. While reversals in fining-upward sedimentology and changes in grain size characteristics may occur in response to phenomena other than large flood occurrences, such effects can usually be identified by comparing alluvial stratigraphic records from multiple sites representing varying geomorphic settings. A 1-2 cm vertical sampling resolution in the UMV study watersheds generally is adequate to capture occurrences of large floods during both historical and Holocene times. However, higher resolution sampling is recommended for thick stratigraphic sequences characterized by high sand percentages where multiple large floods are a possibility. Under these conditions a sampling resolution of 1 - 2 cm may be too coarse to differentiate individual large floods because bioturbation processes tend to blur the boundaries between flood units when they involve temporal intervals of 1-3 years or less. Other factors being constant, the thickness of singular flood deposits usually increases with the magnitude of a flood.

The most frequent causes of large floods in the region were found to be highly scale related. Heavy summer rainfalls are the principal cause of the largest floods on small and intermediate size watersheds, whereas spring snowmelt is the most common cause of the largest floods on large watersheds. The difference in meteorological origin between rain floods and snowmelt floods provides a basis for evaluation of long-term changes in seasonal flooding. The results of this study disclose that late-Holocene times of frequent large snowmelt floods had a tendency to coincide with times of frequent large rainfall floods on the small and intermediate size tributaries in spite of their direct meteorological associations being different. Age control was too imprecise to determine exactly how closely in time the flood series were in agreement. Nevertheless, the data suggest that occurrences of large floods across all scales of watersheds were synchronous at time-interval grouping of at least as short as decadal to half-century lengths.

In conclusion, sedimentological properties of alluvial stratigraphy in UMV watersheds provide proxy records that characterize long-term changes in frequencies and magnitudes of large overbank floods. Because meteorological and physiographic factors that control flood generation and flood energy processes vary systematically with watershed size, it is possible to determine long-term behavior in seasonal occurrences of large overbank floods. As noted in the preceding paragraph, the present data indicate that synchronous behavior in seasonal floods probably has occurred for parts of the late-Holocene. These results suggest that more detailed chronologies of large overbank floods could be assembled by expanding the density of the sedimentological sampling network. Results from such an analysis would better define temporal patterns of large rainfall and snowmelt floods and better delineate times when they are synchronous and times when they are asynchronous across watersheds of different scales. In turn, these findings would provide important information that would be helpful for anticipating responses of large floods to possible future climate changes.

Acknowledgments. This research has been supported by the U. S. National Science Foundation, Grants EAR-8511280, EAR-9206854, EAR-9409778, and EAR-9807715. An Evjue-Bascom research award from the University of Wisconsin, Madison, provided additional support. William Becker, Peter Church, David Leigh, David May, and Frank Magilligan assisted with various components of field and laboratory work. We are also grateful to Kyle House, Gary Huckleberry, and Robert Brakenridge for contributing editorial and other suggestions related to the manuscript and figures.

REFERENCES

Anfinson, J.O., Commerce and conservation on the upper Mississippi River, *The Annals of Iowa*, *52*, 385-417, 1993.

Argabright, M.S., Cronshey, R.G., Helms, J.D., G.A. Pavelis, and H.R. Sinclair Jr., Historical changes in soil erosion, 1930-1992: The northern Mississippi Valley loess hills, USDA, Natural Resources Conservation Service, *Historical Notes, 5*, 92 pp., 1996.

Baker, R.G., E.A., Bettis, D.P. Schwert, D.G. Horton, C.A. Chumbley, L.A. Gonzàlez, and M.K. Reagan, Holocene paleoenvironments of northeast Iowa, *Ecol. Mon., 66*, 203-234, 1996.

Baker, V.R., Paleoflood hydrology and extraordinary flood events, *J. Hydrol., 96*, 79-99, 1987.

Baker, V.R., Geomorphological understanding of floods, *Geomorphology, 10*, 139-156, 1994.

Baker, V.R. and J.E. Costa, Flood power, in *Catastrophic Flooding*, edited by L. Mayer and D. Nash, pp. 1-21, Allen & Unwin, Boston, 1987.

Baker, V.R. and D.F. Ritter, Competence of rivers to transport coarse bedload material, *Geol. Soc. of Am. Bull., 86*, 975-978, 1975.

Colby, B.R., Fluvial sediments: a summary of source, transportation, deposition, and measurement of sediment discharge, U. S. Geol. Surv. Bull. 1181-A., 1963.

Costa, J.E., Response and recovery of a Piedmont watershed from Tropical Storm Agnes, June 1972, *Water Resour. Res., 10*, 106-112, 1974.

Costa, J.E., Paleohydraulic reconstruction of flash flood peaks from boulder deposits in the Colorado Front Range, *Geol. Soc. Am. Bull., 94*, 986-1004, 1983.

Costa, J.E., A history of paleoflood hydrology in the United States, 1800-1970, in *The History of Hydrology*, edited by E.R. Landa and S. Ince, pp. 49-53, AGU, Washington, D.C., 1987.

Ely, L.L., Y. Enzel, V.R. Baker, D.R. Cayan, A 5000-year record of extreme floods and climate change in the southwestern United States, *Science*, *262*, 410-412, 1993.

Freeman, J.E., Map of Grant County, Wisconsin Territory *1842*, Board of County Commissioners, Wisconsin Territory, 1842.

Graf, W.L., Downstream changes in stream power in the Henry Mountains, Utah, *Annals of the Assoc. of Amer. Geog., 73*, 373-387, 1983.

Heyl, A.V., A.F. Agnew, E.J. Lyons, and C.H. Behre, The geology of the Upper Mississippi Zinc-Lead District, *U.S. Geol. Surv. Prof. Pap. 309*, 1959.

Holford, C. N., History of Grant County, Wisconsin, The Teller Print, Lancaster, Wisconsin, 1900.

Hirschboeck, K. K., Flood hydroclimatology, in *Flood Geomorphology,* edited by V.R. Baker, R.C. Kochel, and P.C. Patton, pp. 27-49, John Wiley and Sons, New York, 1988.

Hirschboeck, K.K., L.L. Ely, and R.A. Maddox, Hydroclimatology of meteorologic floods, in *Inland Flood Hazards: Human, Riparian, and Aquatic Communities*, edited by E.E. Wohl, pp. 39-72, Cambridge University Press, 2000.

Hughes, P.E., J.S. Hannuksela, and W.J. Danchuk, Flood of July 1-5, 1978, on the Kickapoo River, southwestern Wisconsin U.S. Geol. Surv. Hydr. Inv. Atlas, HA-0653, 7 sheets, 1981.

Jarrett, R. D., Paleohydrology and its value in estimating floods and droughts, in *National Water Summary 1988-89— Hydrologic Events and Floods and Droughts*, edited by R.W. Paulson, E.B. Chase, R.S. Roberts, and D.W. Moody, pp. 105-116, U. S. Geol. Surv. Wat. Sup. Pap. 2375, 1991.

Jarrett, R.D. and E.M. Tomlinson, Regional interdisciplinary paleoflood approach to assess extreme flood potential, *Water Resour. Res.*, *36*, 2957-2984, 2000.

Knighton, A.D., Downstream variation in stream power, *Geomorphology*, *29*, 293-306, 1999.

Knox, J.C. Human impacts on Wisconsin stream channels, *Annals of the Assoc. of Amer. Geog.*, *67*, 323-342, 1977.

Knox, J.C., Responses of floods to Holocene climatic change in the upper Mississippi Valley, *Quat. Res.*, *23*, 287-300, 1985.

Knox, J.C. Stratigraphic evidence of large floods in the upper Mississippi Valley, in *Catastrophic Flooding*, edited by L. Mayer and D. Nash, pp. 155-180, Allen & Unwin, Boston, 1987a.

Knox, J. C., Historical valley floor sedimentation in the Upper Mississippi Valley, *Annals Assoc. Amer. Geog.*, *77*, 224-244, 1987b.

Knox, J. C., Climatic influence on Upper Mississippi Valley floods, in *Flood Geomorphology*, edited by V. R. Baker, R. C. Kochel, and P. C. Patton, pp. 279-300, John Wiley and Sons, New York, 1988.

Knox, J. C., Large increases in flood magnitude in response to modest changes in climate, *Nature*, *361*, 430-432, 1993.

Knox, J. C., Late Quaternary upper Mississippi River alluvial episodes and their significance to the Lower Mississippi River system, *Eng. Geol.*, *45*, 263-285, 1996.

Knox, J. C., Long-term episodic changes in magnitudes and frequencies of floods in the Upper Mississippi Valley, in *Fluvial Processes and Environmental Change*, edited by A. G. Brown and T. A. Quine, pp. 255-282, John Wiley & Sons Ltd., New York, 1999.

Knox, J. C., Sensitivity of modern and Holocene floods to climate change, *Quat. Sci. Rev.*, *19*, 439-457, 2000.

Kochel, R.C. and V.R. Baker, Paleoflood hydrology, *Science*, *215*, 353-361, 1982.

Kochel, R.C. and V.R. Baker, Paleoflood analysis using slackwater deposits, in *Flood Geomorphology,* edited by V. R. Baker, R. C. Kochel, and P. C. Patton, pp. 357-376, John Wiley and Sons, New York, 1988.

Komar, P.D., Selective grain entrainment and the empirical evaluation of flow competence, *Sedimentology*, *34*, 1165-1176, 1987.

Lamb, H.H., Climate: Present, Past and Future, Vol. 2, *Climatic History and the Future*, 835 pp., Methuen & Co. Ltd., London, 1977.

Lecce, S.A., Non-linear downstream changes in stream power on Wisconsin's Blue River, *Annals Assoc. Am. Geog.*, *87*, 471-486, 1997.

Magilligan, F.J., Historical floodplain sedimentation in the Galena River Basin, Wisconsin and Illinois, *Annals Assoc. Am. Geog.*, *75*, 583-594, 1985.

Magilligan, F.J., Thresholds and the spatial variability of flood power during extreme floods, *Geomorphology*, 5, 373-390, 1992.

Mandel, R.D., Geomorphic controls of the Archaic record in the Central Plains of the United States, in *Archaeological Geology of the Archaic Period in North America*, edited by E. A. Bettis III, pp. 37-66, Geol. Soc. of Amer. Special Paper 297, 1995.

Mason, J.A. and J.C. Knox, Age of colluvium indicates accelerated late Wisconsin hillslope erosion in the Upper Mississippi Valley, *Geology*, *25*, 267-270, 1997.

Miller, A.J., Valley morphology and boundary conditions influencing spatial patterns of flood flow, in *Natural and Anthropogenic Influences in Fluvial Geomorphology*, edited by J. E. Costa, A. J. Miller, K. W. Potter, and P. R. Wilcock, pp. 57-81, Amer. Geophys. Union Monograph 89, 1995.

Miller, A.J. and D.J. Parkinson, Flood hydrology and geomorphic effects on river channels and flood plains: the flood of November 4-5, 1985, in the South Branch Potomac River Basin of West Virginia, *U. S. Geol. Surv. Bull. 1981-E*, 96 pp., 1993.

Moody, J.A., Hydrologic, sedimentologic, and chemical data describing surficial bed sediments and water in the navigation pools of the upper Mississippi River, July 1991 – April 1992, *U. S. Geol. Surv. Open-File Rpt. 95-708*, 274 pp., Denver, CO, 1997.

Paramore, R., *Vertical accretion deposits and channel metamorphosis of the lower Grant River, southwestern Wisconsin*, M. S. Thesis, Dept. of Geog., University of Wisconsin, Madison, 167, 1992.

Parrett, C., N. B. Melcher, and R. W. James Jr., Flood discharges in the upper Mississippi River basin, 1993, *U. S. Geol. Surv. Water Sup. Pap. 1120-A*, 1-14, 1993.

Pitlick, J., A regional perspective of the hydrology of the 1993 Mississippi River Basin floods, *Annals of the Assoc. Am. Geog.*, *87*, 135-151, 1997.

U. S. Water Resources Council Hydrology Committee, *Guidelines for determining flood flow frequency, Bull. 17B*, Washington, D. C., 1981.

Williams, G. P., Paleohydrological methods and some examples from Swedish fluvial environments: I-cobble and boulder deposits, *Geografiska Annaler*, *65*, 227-243, 1983.

Wahl, K. L., K. C. Vining, and G. J. Wiche, Precipitation in the upper Mississippi River basin, January 1 through July 31, 1993, *U. S. Geol. Surv. Water Sup Pap. 1120-B*, 1-13, 1993.

James C. Knox and J. Michael Daniels, Department of Geography, 234 Science Hall, 550 North Park Street, University of Wisconsin, Madison, WI 53706-1491.

Slackwater Deposits As Paleostage Indicators In Canyon Reaches Of The Central Appalachians: Reevaluation After The 1996 Cheat River Flood

J. Steven Kite

Department of Geology & Geography, West Virginia University, Morgantown, West Virginia

Terry W. Gebhardt

ExxonMobil Production Company, USA, Houston, Texas

Gregory S. Springer

Department of Earth Resources, Colorado State University, Ft. Collins, Colorado

A 100-year flood on the Cheat River in January 1996 left sandy deposits less than 0.5 m below high-water marks indicated by flotsam and erosional trim lines. This relatively close match suggests slackwater deposits in Appalachian canyons have good potential as paleostage indicators that may allow regional flood records to be extended back several centuries. The accuracy of the stage record varies with landscape position; deposits in caves and on sheltered terrace surfaces are closer to high-water marks than deposits at tributary mouths, floodplain obstructions, or recesses along channel margins. The good match between slackwater deposits and high-water marks from the 3350 m^3/s 1996 flood contrasts markedly with the poor match found after an extraordinary (>500 year) 5380 m^3/s flood in November 1985. The 1985 flood stage was so much higher than floodplain and terrace surfaces that it lapped upon steep colluvial slopes, well above minor fluvial landforms favorable to deposition. Accordingly, depositional evidence of moderate paleofloods in Appalachian canyons may have greater preservation potential than evidence of extreme floods. All slackwater deposits in humid climates are vulnerable to post-flood modification, including those formed in caves. Those in canyon reaches are particularly vulnerable in the face of the construction and maintenance of transportation corridors. Reconstruction of long-term Appalachian flood chronologies based on slackwater deposits will require careful planning, detailed air-photo interpretation, and substantial fieldwork, including more studies documenting the relationships between high-water marks and stage indicators left by other floods on other rivers in the region.

INTRODUCTION

Despite remarkable successes in many regions of the world, slackwater deposits have not been used widely as paleostage indicators for flood studies in the Central Appalachians and the rest of the Eastern United States. This

Ancient Floods, Modern Hazards: Principles and Applications of Paleoflood Hydrology
Water Science and Application Volume 5, pages 257-266

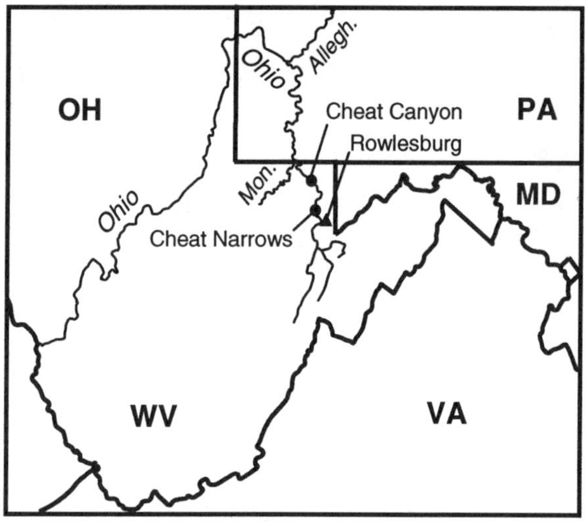

Figure 1. Location map showing two canyon study reaches: Cheat Narrows and Cheat Canyon. Rivers shown outside the Cheat River basin: Ohio, Mon = Monongahela, and Allegh = Allegheny.

shortcoming does not arise from an absence of slackwater deposits, as they have been well documented after major floods [*Holmes*, 1938; *Mansfield*, 1938; *Jahns*, 1947; *Moss and Kochel*, 1978]. However, this geographic region has drawn the attention of few workers and been the focus of few studies, in no small part because bioturbation, pedogenesis, and other post-depositional agents limit preservation of slackwater deposits under the region's humid climate. These natural limitations on preservation make it unlikely that subaerial Appalachian slackwater deposits will yield chronologies extending back several millennia, as has been the case on some streams in arid or semi-arid climates [e.g. *Kochel and Baker*, 1988; *Ely et al.*, 1993; *O'Connor et al.*, 1994]. Despite these limitations, widespread development of flood-prone properties for commercial, industrial and residential uses necessitates a better understanding of low-frequency, high-magnitude Appalachian flood events. Most historical flood records in the region only extend back a century or two [*Wiley et al.*, 2000]: an inadequate span of time to address the frequency of extraordinary floods. The widely used approach of studying floods in the region from tree-ring evidence [e.g. *Sigafoos, 1964, Yanosky, 1982, Osterkamp and Hupp, 1984, Hupp, 1988*] is temporally limited by relatively short-lived riparian tree species and the widespread deforestation of the 18th, 19th and early 20th centuries. Hence, flood chronologies obtained from slackwater deposits would be useful, even if these records are no more than a few centuries old.

The potential to use Appalachian slackwater sediments to augment historic flood series and aid better floodplain management remains largely untapped. The main purpose of the research discussed in this paper is to explore means of realizing this potential. An important step in assessing the utility of slackwater deposits is to determine their efficacy as paleostage indicators. This assessment is accomplished by comparison of fresh slackwater deposits to other high-water marks formed during recent floods. The research also seeks to advance slackwater methodology in the region by analyzing how the accuracy of slackwater deposits varies between different landforms.

Site Selection

Reach selection is critical to fruitful paleoflood studies. The studies discussed in this paper are focused on Appalachian canyons because these long, narrow, steep-walled valleys generally meet the criteria for good paleohydraulic reconstruction sites: fairly straight reaches with relatively uniform width, thin alluvial fill, and valley walls made of bedrock or thin colluvium over bedrock [*Costa*, 1983]. Other settings may contain slackwater deposits, but canyon reaches are more likely to meet the paleohydraulic criteria and yield meaningful flood reconstructions.

The Cheat River basin of north central West Virginia (Figure 1) provides several canyon reaches amenable to the analysis of slackwater deposits as stage indicators. The 3660 km^2 watershed is more than 80 percent forested, and although virtually all of it was once timbered [*Clarkson*, 1964], it is much more rural than most watersheds of comparable size in the Eastern United States. The Cheat River drains rugged uplands in the Allegheny Mountains section of the Appalachian Plateaus physiographic province. The highest summits in the watershed are over 1450 m in altitude, and 300 m or more local relief is common. Steep slopes throughout the entire basin promote rapid runoff. Narrow floodplains provide little over-bank storage so the system is very flashy. The Cheat River has a long history of destructive flooding [*Fansler*, 1962; *Kite and Linton*, 1993] and is a major source of flooding on the Monongahela River in Pennsylvania.

Two canyon reaches, Cheat Narrows and Cheat Canyon (Figure 1; Table 1), have provided sites for three studies of slackwater sedimentation [*Linton*, 1992; *Kite and Linton*, 1993; *Springer*, 1994; *Springer and Kite*, 1997; *Gebhardt*, 1998]. Both canyons occur where the Cheat River has incised through resistant quartz sandstones of the Pennsylvanian-age Pottsville Group and into relatively weak underlying Mississippian-age clastic and carbonate bedrock. Cheat Narrows includes several sites adequately suited for the study of recent slackwater sedimentation. However, the reach is less than ideal for the preservation or discovery of long-term slackwater sequences because it hosts State Highway 72 on one side of the canyon and a CSX Transportation rail line on the other. In contrast, Cheat Canyon is much better suited for slackwater analysis as it has seen little development, except for an abandoned

Table 1. Comparison of Cheat Narrows and Cheat Canyon. Approximate drainage-basin areas given for entrance to Cheat Narrows near Rowlesburg and at Jenkinsburg Bridge (a.k.a. Bull Run Bridge) in Cheat Canyon.

Attribute	Cheat Narrows	Cheat Canyon
Basin Area	~2550 km^2	~2930 km^2
Stream Gradient	0.0044 m/m	0.0062 m/m
Canyon Length	9 km	25 km
Local Relief	140-450 m	240-405 m
Mean Side Slope $\pm$ St. Dev.	0.52 $\pm$ 0.07 m/m	0.64 $\pm$ 0.16 m/m
Maximum Side Slope	0.98 m/m	1.21 m/m

logging railroad 50 m above the canyon floor, well above the upper limits of historic flooding. However, field access with survey equipment is difficult throughout much of Cheat Canyon because its many *Rhododendron* thickets are too dense to traverse and the channel thalweg is deep and swift, even during low flow. These obstacles limited the number of cross-sections completed in this study.

Historical Floods on the Cheat River

The most complete flood series on the Cheat River is from Rowlesburg, West Virginia (Figure 1), where USGS gages have been in nearly continuous operation since 1924 (Table 2). An older National Weather Bureau gage and historical accounts back to the middle 19th Century supplement the USGS record. The gage has been relocated several times, so drainage area has varied from 2360 km^2 to 2520 km^2 [*Ward et al.*, 1998; 2000]. Rowlesburg lies 2 km upstream from Cheat Narrows and 23 km upstream from Cheat Canyon. Drainage area increases to about 2930 km^2 at study sites in Cheat Canyon [*Ward et al.*, 1998], but nearby gage stations have been too short lived to capture most of the largest floods known from the Rowlesburg record.

The Rowlesburg flood series (Figure 2) includes large July floods in 1844 and 1888, followed by a relatively inactive period of 97 years during which only the Hurricane Hazel flood of October 1954 is a noteworthy event. Three of the six largest recorded floods have occurred since 1985, including the two largest. The disastrous November 1985 flood [*Bittinger*, 1985; *Teets and Young*, 1985] was caused by two low-pressure cells that migrated from the Gulf of Mexico into the region in the wake of Hurricane Juan [*Colucci et al.*, 1993; *Wiley et al.*, 2000]. In contrast, floods in February 1994 and January 1996 were spawned by rain-on-snow conditions. Although historical floods have had a variety of causes, there is little evidence that any historical flood-generating event caused widespread debris flows or debris slides in the Cheat River basin. Accordingly, floodwaters have not contained so much sediment as to significantly alter stage-discharge relationships.

Peak-flow data for large floods at Rowlesburg are imprecise approximations, as is normally the case when stage far exceeds data on the rating curve for a station. A preliminary discharge of 6500 m^3/s was published for November 1985 at Rowlesburg [*Lescinsky*, 1986], but that value was based on a slope-area measurement that was rated "poor" and may have been influenced by damming behind a railroad bridge [*Shearman*, 1988]. Subsequently, this estimate was reduced to 5380 m^3/s [*Ward et al.*, 1988], a value more consistent with slope-area discharge estimates elsewhere in the basin and with U.S. Army Corps of Engineers HEC-2 step-backwater and HEC-1 rainfall-runoff modeling [*Shearman*, 1988]. Despite a 45 percent increase in drainage area before reaching the mouth of the river, flood discharge did not increase significantly downstream from Rowlesburg during the 1985 event [*Kite and Linton*, 1993]. Precipitation was less in the lower basin [*Colucci et al.*, 1993] and rainfall had diminished before the flood crest arrival.

Discharge determinations for the 1994 and 1996 events have not been significantly revised since these floods, but recent revisions [*Ward et al.*, 1998] raised the peak discharge values for the 1844, 1888 and 1954 floods by 20 to 36 percent more than previously published [*Ward et al.*, 1997]. The lower values were cited in earlier works on slackwater deposits along the Cheat River [*Kite and Linton*, 1993; *Gebhardt*, 1998]. The revised discharge estimates indicate the 1996 flood was the second greatest flood of record, but it was not extraordinary relative to other large floods in the watershed. *Wiley et al.* [2000] determined it approximately equaled the discharge of a 100-year flood. The 1985 flood discharge estimate remains two-thirds greater that any other flood in the period of record, almost 20 percent greater than the calculated 500-year flood at Rowlesburg [*Wiley et al.*, 2000]. Accordingly, one of our

Table 2. Six largest recorded discharges on the Cheat River at Rowlesburg (USGS Station: 03070000: 2520 km^2) through the 1996 water year (from on-line USGS peak flow data).

Year, Date	Discharge m^3/s
1985, 5 Nov	5380
1996, 19 Jan	3230
1844, 6 Jul	3140
1888, 10 Jul	2860
1994, 9 Feb	2650
1954, 16 Oct	2560

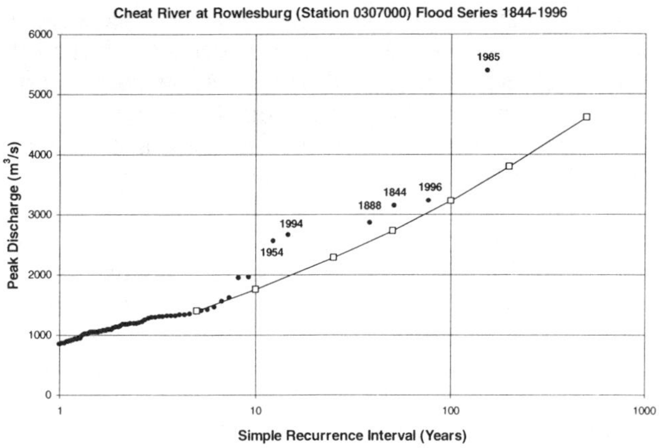

Figure 2. Flood series for the Cheat River at Rowlesburg through water-year 1996, based on revised U.S.G.S. peak-flow data. Solid circles denote individual events with recurrence interval determined from length of record and flood rank. Open squares denote recurrence-interval peak-discharge data [*Wiley et al.* 2000], calculated with guidelines established by Water Resources Council.

most basic research questions is to ask whether slackwater deposits left by the 1985 flood can demonstrate the extraordinary nature of this event.

Previous Investigations of Cheat River Deposits

Initial research on the Cheat River documented widespread slackwater deposition during the 1985 flood, particularly at the mouths of back-flooded tributaries where a consistent sequence of tributary-derived sandy loam capped by main-stem derived silt loam occurred [*Linton*, 1992; *Kite and Linton*, 1993]. This sequence is similar to that noted in Pennsylvania in the aftermath of Hurricane Agnes [*Moss and Kochel*, 1978]. Tributaries with gradients over 0.05 also contributed basal gravel and capping sandy loam deposits to the slackwater sequence [*Linton*, 1992].

Despite the consistency and widespread occurrence of these deposits at tributary mouths, the upper surface of significant (*i.e.* > 2 cm thick) slackwater deposits ranged from 2.9 to 5.6 m below other high-water marks from the 1985 flood, such as flotsam, eroded soils, and erosional trim lines in leaf litter. Slope-area discharge estimates based on the height of slackwater deposits under-represented actual flow by a substantial 51 to 67 percent [*Kite and Linton*, 1993]. Hence, the highest slackwater evidence of the exceptional 1985 flood was left on surfaces that would be inundated by a mere 10-year flood.

Isolated patches of sand, less than 2 cm thick, were identified at higher levels, but these thin deposits were rare and impossible to differentiate from similar tributary deposits left great distances from slackwater created by the main stem [*Kite and Linton*, 1993]. Bioturbation made these thin deposits indistinguishable from the pre-existing soil A horizon less than two years after the flood; consequently these thin deposits have almost no potential as long-term recorders of paleoflood stage. Inundated terrace surfaces near Rowlesburg were virtually devoid of sediments immediately after the 1985 flood, possibly because of the high velocity of flow in the "over-bank" environment. Flow depths over the terraces exceeded 2 m, and most of the Cheat River bottomland behaved as one large valley-wall-to-valley-wall channel. The rare, isolated 1 to 2 cm thick patches of sandy loam on these surfaces also were obliterated within two years.

Kite and Linton [1993] postulated several causes for the poor match between tributary-mouth slackwater deposits and high-water marks from the 1985 flood. These speculations included flooding on tributaries that prevented true slackwater conditions, density stratification of relatively clear tributary water masses over sediment-laden Cheat River water masses, and a mismatch in timing between maximum stage and maximum sediment transport. All three factors and relatively low sediment concentrations probably contributed to the disparity.

The 1985 flood also left slackwater deposits near floodplain obstructions and in sheltered recesses along channel margins. *Kite and Linton* [1993] examined many of these deposits in Cheat Narrows, but not in Cheat Canyon. These deposits were composed of thick (> 10 cm) laminated sands, but in every case occurred several meters below high-water marks. Collectively, the distribution of significant slackwater deposits in relation to other high-water marks led to the conclusion that "slackwater deposits in the study area are not good indicators of 1985 flood levels" [*Kite and Linton*, 1993]. This conclusion led Kite and Linton to question the accuracy and precision of slackwater deposits as paleostage indicators for older floods in the region. This question evokes an implication that slackwater deposits have poor potential as paleostage indicators to extend flood series for Appalachian rivers.

This implication provided a catalyst for a study of cave sediments by *Springer* [1994], completed before the February 1994 flood. Limestone occurs in Cheat Narrows, but limestone mining, filling, and road construction have obliterated natural caves in this reach. Hence, the search for caves with potential for slackwater sediments led into Cheat Canyon, where *Springer and Kite* [1997] found a variety of river-derived slackwater deposits, mostly thin veneers of laminar silt, sand, and organic matter.

The uppermost 1985 slackwater sediments are within 1 m of high-water marks indicated by flotsam in Coliseum Rapids Cave, whose mouth lies 14 m above summer low-flow stages [*Springer and Kite*, 1997], a much better match than occurred at Linton's subaerial sites in Cheat Narrows. Coliseum Rapids Cave occurs on the outside of a channel bend, just upstream from a constricted reach. The site's topography does not lend to either step-backwater modeling or indirect discharge estimates, and no discharge values have been calculated.

All of the near river-level caves in Cheat Canyon are very active hydrologically, with profuse infiltration through the ceiling and intermittent stream flow causing erosion of fine-grained slackwater deposits [*Springer and Kite*, 1997]. Moreover, the rate of organic decay in the years immediately following the 1985 flood was so rapid that it is unlikely organic matter from this flood would persist for more than 25 years. In short, Cheat Canyon suggests relatively precise paleostage reconstruction is possible from cave slackwater sediments, but the humid, active cave environment may not offer significantly better preservation potential than sheltered recesses or other canyon surfaces.

OBSERVATIONS AFTER THE 1996 FLOOD

The January 1996 flood, with a discharge of 3230 m^3/s at Rowlesburg, provided an opportunity to assess the relationship between stage and slackwater sedimentation for a large, but not extraordinary flood. Reconnaissance work was undertaken throughout the Cheat River basin during and immediately after the flood. Several slackwater settings were videotaped in Cheat Narrows on the rising limb of the flood; but diminishing daylight and concerns over safety precluded direct observation of peak stage or any observation of cave flooding. Two smaller 2-year recurrence flows of approximately 1100 m^3/s were observed in Cheat Canyon later in 1996. Field observations and videotape reveal that inundated slackwater sites along the margins of canyon reaches were repeatedly traversed by 10 to 50 cm amplitude waves generated during the collapse of standing waves in the river channel. The dominant processes in this environment were more typical of a low-energy beach than a still, ponded water column. Continuous wave action precluded deposition of silt or clay; hence, subaerial slackwater deposits in these reaches are composed primarily of sand.

Post-flood reconnaissance along Cheat Narrows and at tributary mouths elsewhere revealed that the 1996 flood left slackwater deposits in the same locations that experienced deposition during the much larger 1985 flood. Only 2 to 10 cm thick, deposits from the 1996 flood are up to an order of magnitude thinner than 1985 deposits, but this is the only sedimentological indication that the earlier flood's peak discharge was two-thirds greater that the 1996 flood's discharge. Indirect slope-area method discharge calculations for inundation of these depositional surfaces by *Linton* [1992] average 2730 m^3/s for the highest slackwater deposits in Cheat Narrows. This estimate is less than half of the 1985 discharge, but nearly 85 percent of the 3230 m^3/s peak flow in 1996.

Detailed field investigations after the 1996 flood focused in Cheat Canyon [*Gebhardt*, 1998]. This work included inventory and mapping of selected slackwater deposits, probing and digging test pits to determine stratigraphy, and channel profiling with a Leica TC400 electronic distance measurer for use in a model of flow through part of the canyon.

Slackwater deposits are more common in the relatively pristine Cheat Canyon than in the much-modified Cheat Narrows, in part because of a greater abundance and range of depositional surfaces. Small-scale landforms that can be identified on enlarged 1:20,000 scale aerial photos prove to be more important in controlling slackwater conditions than the channel and valley-width irregularities indicated on 1:24,000 scale topographic maps. Slackwater deposits formed in 1996 on low terraces, in sheltered channel-margin recesses, at flow obstructions, and at the mouth of one large tributary, Big Sandy Creek. Unfortunately, the tributary-mouth deposit occurred at a popular whitewater rafting take-out point and was obliterated before it could be studied.

The mouth of Coliseum Rapids Cave was not inundated during the 1996 flood, but smaller caves at lower levels received fresh slackwater deposits. The deposits are within 1 m of the height of high-water marks, but our survey transects and flow reconstructions did not extend far enough upstream to include this area; hence, we can not provide details on the spatial relationships between 1996 flood stage and slackwater deposits in caves.

Fieldwork elsewhere in the canyon shows that the match between slackwater deposits and high-water marks depends upon the setting. All slackwater deposits associated with flow obstructions and sheltered channel-margin recesses lie at least 1.4 m below other high-water marks (Table 3). In contrast, slackwater deposits on terraces extend within 0.5 m of a distinctive erosional trim line in leaf litter evident immediately after the flood. One deposit is within 0.2 m of the true high-water mark. Sand dominates the sedimentology of both settings; slackwater deposits in

Table 3. Slackwater deposits from the 1996 flood in Cheat Canyon. Many deposits in channel-margin recesses were identified but not mapped or surveyed. All of the unmapped deposits were >1.5 m below high-water marks. Sites 1-3 were used in HEC-RAS flow modeling (Figure 3).

Site	Distance from Jenkinsburg	Type of Deposit	Distance below High-Water Mark
1	0.45 km	terrace	0.5 m
2	0.7	terrace	0.5
3	1.2	channel-margin recess	1.4
4	2.7	terrace	0.2

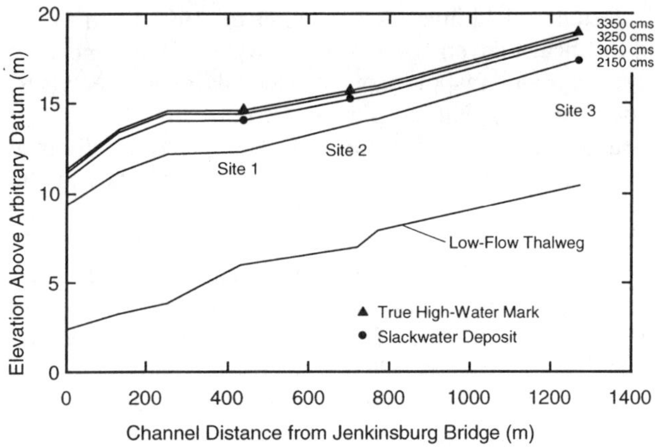

Figure 3. Longitudinal water-surface profiles for the 1996 flood computed with HEC-RAS models of a 1.27 km reach in Cheat Canyon. True high-water marks were determined from leaf-litter scour and flotsam. Slackwater deposits at sites 1 and 2 occur on terraces. Slackwater deposits at site 3 occur in a channel-margin recess.

channel-margin recesses may be 0.4 m thick, whereas those on terraces have maximum thickness of 0.1 to 0.15 m.

Step-backwater Modeling of the 1996 flood

In order to examine the relationships between true high-water marks and depositional stage indicators, longitudinal water-surface profiles of the 1996 flood were generated along a 1.27 km reach in Cheat Canyon (Figure 3) using Hydrologic Engineering Center River Analysis System (HEC-RAS) software. HEC-RAS is a step-backwater method of flow modeling *[U.S. Army Corps of Engineers Hydrologic Engineering Center,* 1998]. Step-backwater flow models are based on the continuity equation and conservation of energy; although not expressly designed this purpose, *O'Connor and Webb* [1988], *O'Connor* [1993], and *Enzel et al.* [1994] have demonstrated their utility in paleoflood reconstruction from slackwater deposits.

The 1.27 km reach in Cheat Canyon meets most criteria for use of step-backwater techniques [*Gebhardt*, 1998], but our modeling may have yielded better results if logistical concerns had not influenced our selection of a reach near the canyon's only vehicle access at Jenkinsburg Bridge (0.0 km on Figure 3). Modeling the lower segment of this reach with HEC-RAS is problematic because the bridge abutments constricted flow width at a point where the channel is exceptionally deep. Cross-sections were surveyed at the bridge and 4 upstream sites that control conveyance through the reach (Figure 3). Two cross-sections were added between site 1 and Jenkinsburg Bridge for the HEC-RAS model runs. These two cross-sections were based on topographic maps and were too close to the end of the reach for reliable output, given a basic model assumption of critical flow depth at the bridge. The additional cross-sections help obtain more realistic output further upstream, and the model runs show results consistent with high-water marks upstream from site 1.

The leaf-litter trim line and other true high-water marks best match a modeled flow of 3350 m^3/s. This value is slightly more than the 3230 m^3/s recorded at the USGS gauge at Rowlesburg, 23 km upstream of the canyon. If a liberal value of 0.5 m is used for the difference in height between slackwater deposits and high-water marks, a HEC-RAS model discharge of 3050 m^3/s best matches the deposits. This value is 91 percent of the discharge indicated by the leaf-litter trim line. If a more conservative value of 0.2 m is used, then the difference between discharges is ~ 100 m^3/s or $<$ 3 percent, which is less than the error expected in HEC-RAS modeling [*U.S. Army Corps of Engineers Hydrologic Engineering Center*, 1998]. The model results demonstrate that slackwater deposits can be used to reconstruct flood profiles and provide a basis for relatively accurate estimates of peak flow for floods of this magnitude.

New Observations on Stage Indicators from Earlier Floods

All observed slackwater deposits from the 1996 flood in Cheat Canyon occur in association with thick (up to 1 m) 1985 slackwater deposits. In fact, many 1996 deposits appear to have been derived almost entirely from reworking of pre-existing slackwater deposits, especially those occurring in channel-margin recesses. One exposure revealed a thin (0.1 m) slackwater deposit from the 1994 flood between those of the two larger flood. Both the 1994 and 1985 deposits supported very weak A horizons at this site. An unconformable contact lacking remnants of a pre-existing A horizon is more typical at the base of 1996 deposits. It appears that most 1994 deposits were reworked during the 1996 flood and that most slackwater sequences in Cheat Canyon are incomplete records of multiple flood events.

In contrast to the similar extent and elevations of deposits from the two floods farther upstream at Cheat Narrows, 1985 slackwater deposits in Cheat Canyon extend higher than 1996 deposits. This spatial relationship is significant because the 1996 deposits are very near the high-water marks of a flood that is approximately 60 percent of the 1985 flood at this point in the basin. This relationship implies that 1985 slackwater deposits in Cheat Canyon are better stage indicators than those *Kite and Linton* [1993] documented farther upstream.

Field reconnaissance identified one 1985 deposit only 1.5 m below a discontinuous trim line of wave-cut notches and truncated soils. This trim-line, indicative of erosion of up to 0.2 m depth, exists elsewhere in Cheat Canyon and lies at the upper limit of durable plastic or Styrofoam flotsam left by the 1985 flood; a limit not reached by any other historic flood. Unfortunately, this relationship was beyond the scope of *Gebhardt's* [1998] research and we can not make more specific statements on the accuracy of subaerial 1985 slackwater deposits as stage indicators in Cheat Canyon. Additional surveying and flow modeling are required to accurately estimate discharge, but it is possible that subaerial slackwater deposits in the Cheat Canyon record as much as 80 percent of the true 1985 discharge. This percentage might be sufficient to allow slackwater analysis to document that the 1985 flood was greater than all other floods of record.

DISCUSSION

The good match between slackwater deposits and other high-water marks in Cheat Canyon most likely relates to the position of the high-water marks with respect to depositional surfaces. The poor match in Cheat Narrows may stem in part from road and railroad construction eliminating potential depositional surfaces, but the disparity may also be related to the hydraulic geometry created by long-term flood frequency and magnitude. The relatively uniform 1985 peak discharge downstream from Rowlesburg translated into flood stages that were less extraordinary relative to other floods and approached the level of existing fluvial landforms as the flood progressed downstream. The less extraordinary a flood is relative to moderate-magnitude floods that control the geometry of depositional landforms, the more likely that high-water marks will coincide with relatively flat surfaces suited to slackwater deposition. The 1985 high-water marks documented in Cheat Narrows were almost entirely on steep colluvial slopes poorly suited to slackwater deposition, and were not sites of sedimentation.

The lack of large floods in the preceding 97 years may have contributed to low sediment concentrations during early stages of the extraordinary 1985 flood. Typical of Appalachians rivers, even this flood on the Cheat River was sediment starved compared to floods in arid or semiarid landscapes. Sediment concentrations remain relatively low, but many bars that were vegetated prior to 1985 have been bare since, suggesting that channel activity has increased. The abundance of unvegetated surfaces after the extraordinary 1985 flood and disturbance from subsequent floods, such as the 1994 flood, made more fine-grained sediment available on the Cheat River in 1996 than at the onset of the 1985 flood. Higher sediment concentrations in early stages of the 1996 flood may have promoted deposition in the low-energy wave environment near the upper limits of slackwater settings, thus promoting a better match between deposits and high-water marks.

Wave action along the flood margins is responsible for another type of flood-stage evidence that may extend into slackwater settings: a discontinuous trim line of wave-cut notches and truncated soils that is resilient enough to be readily identifiable over twelve years after the flood. This feature is unlikely to be incorporated in the alluvial stratigraphic record, but where it survives it may be a more precise stage indicator than slackwater deposits. Detailed investigation into subtle, small breaks in slope and soil-profile truncation may be able to determine if similar trim lines might persist on the landscape for many decades, if not centuries, and serve as a geomorphological tool for reconstructing large floods in humid climates.

Implications for Appalachian Paleoflood Studies

Slackwater deposits left by recent floods on the Cheat River indicate that carefully selected canyon reaches in the Appalachians have the potential to record paleofloods. However, this potential has been severely diminished in many canyon reaches because both slackwater deposits and suitable depositional surfaces have been obliterated by the development of transportation corridors. Many Appalachian canyon reaches, at one time or another, have been bottlenecks in extensive networks of canals, railroads, or highways. It was common for construction of transportation corridors to obliterate nearly all pre-existing landforms on both sides of a canyon reach. Subsequent up-grading and maintenance further diminished the likelihood of long-lived slackwater sequences. Hence, the greatest potential for preservation lies in canyons least developed for transportation. Conversely, the urbanized canyons in which flood studies are most needed may be the least likely to contain intact deposits. This problem calls for a approach in which trends and salient points regarding flood frequency deciphered in undisturbed canyons are extrapolated to more urbanized reaches with similar flood characteristics. Further problems stem from factors related to the climate and of the region: low sediment concentrations in most floods, nearly ubiquitous vegetative cover, and disruption of stratification because of rapid bioturbation and pedogenesis.

These problems do not eliminate the possibility of high-quality paleoflood records, but certainly make slackwater deposits difficult to locate, recognize, and document. Air-photo interpretation and fieldwork should focus on the best sites for capturing a high-resolution slackwater record. Sheltered terraces and caves should be given highest priority in field studies as these landscape elements yield more precise stage indicators than deposits in channel-margin recesses, tributary mouths, and flow obstructions. Two- or three-dimensional flow modeling may facilitate identification of specific sites favorable to deposition under flows of different magnitude.

Cave slackwater deposits may form near true high-water marks of extraordinary floods, but many caves near river level are nearly as active as subaerial surfaces, so not all cave flood records are long lived. In the right setting, an approach that integrates flood sedimentology and cave geomorphology may yield good paleoflood records [*Springer*, this volume]. Unfortunately, the ultimate limitation on using cave slackwater sediments is that many Appalachian canyon reaches, especially in the Blue Ridge and Piedmont provinces, do not have caves or other karst landforms.

The Cheat River record suggests that slackwater deposits in mountainous humid climates may be more precise recorders of moderate frequency floods than extraordinary low-frequency floods. Although landslides or bedrock structure and lithology may produce canyon constrictions that can only be modified by extreme floods, the relatively inconspicuous fluvial landforms scattered throughout canyon reaches coincide with more moderate stages. High-magnitude floods have not created depositional fluvial landforms near the upper limits of flooding, possibly because their low frequency makes them relative ineffective or because the floods are predominantly erosional in canyon reaches. Most canyon slopes are too steep for the accumulation of fine sediments under the wave action that sweeps slackwater settings during large floods. Sandy deposits that might be left are vulnerable to erosion by either ebbing floodwater or subsequent local surface runoff that inevitably follows any flood in this humid climate.

In contrast, high stages of moderate floods or diminished downstream phases of extraordinary floods are more likely to coincide with surfaces suitable for slackwater deposition. The deposits of extraordinary floods are likely to occur in the same settings as those of moderate floods; hence, deposit thickness, texture, sorting or other sedimentary attributes may be the only indicators of exceptional flood magnitude.

Colluvial slopes should not be totally ignored in paleoflood studies. Although too steep to allow long-lived slackwater deposition, these slopes are generally stable and may display wave-cut notches or truncated soils. Many steep hillslopes in the Appalachians, including canyon walls, are relatively stable surfaces underlain by relict Pleistocene colluvium [*Clark and Ciolkosz*, 1988; *Clark et al.*, 1989]. These colluvial slopes have been relatively stable throughout the Holocene, and support moderately well-developed soil profiles. Detailed transects may show distinct soil-profile truncations at the upper limit of paleofloods or, at least, reveal breaks between well-developed soils on stable surfaces from immature profiles on colluvium scoured by extraordinary floods. Age assignment for flood scour would be difficult, but if radiocarbon-dated soil chronosequences are established, it may be possible to make inferences about the age of past floods and speculate on the critical issue of flood recurrence.

One important caution must be extended: recent slackwater deposits on the Cheat River may not represent circumstances that develop under profoundly different types of floods. All large 20th century Cheat River floods had relatively low sediment concentrations compared to many floods during recent decades in nearby Appalachian river basins [*Williams and Guy*, 1973; *Jacobson et al.*, 1993; *Miller and Parkinson*, 1993; *Morgan et al.*, 1997; 1999; *Wieczorek et al.,* 2000]. High-magnitude floods may dominate the formation of slackwater deposits in reaches that experience floods laden with sediment introduced by debris flows or other slope failures. These slackwater deposits may be excellent long-lived paleostage indicators in the Appalachians, especially if they are preserved by rapid burial under debris-flow deposits. The November 1985 flood exposed many slackwater deposits in debris fans of the South Branch Potomac and Shenandoah river basins [*Kite*, 1987; *Tharp and Kite*, 1989]. These and similar deposits throughout the Appalachian region warrant further study when and where large floods create short-lived exposure.

In summation, a mixed prognosis exists in the Appalachians for paleoflood deposits as tools to extend flood series. Slackwater deposits have been well documented, but their efficacy as stage indicators varies from setting to setting. Additional studies are warranted on the basic relationships between flood variables, reach characteristics, and geomorphologic evidence such as stage indicators. These relationships are likely to vary greatly throughout the region and may require a basin-by-basin or even reach-by-reach analysis. The work may be never be as rewarding or visually stimulating as similar studies in arid and semi-arid landscapes, but the high population densities along constricted Appalachian valleys provide a pressing need for long-term flood records to allow better floodplain management.

Acknowledgments. This study was partially supported by the Geological Society of America, the Richmond Area Speleological Society, the Cave Conservancy of the Virginias, and Chevron, Inc. Fieldwork benefited form discussions with Andy Miller, Bob Jarrett, and all the participants in the 1997 Southeastern Friends of the Pleistocene Field Trip. Jeff Wiley was very helpful in verifying U.S.G.S. flood discharge data. The manuscript was improved by thoughtful reviews by Bob Jarrett, Kirk Vincent, Kyle House, Gary D'Urso, and Steven Taylor.

REFERENCES

Bittinger, W., The Flood of November 4-5, 1985 in Tucker, Preston, Grant and Hardy Counties, West Virginia, 96 pp., McClain Printing Co., Parsons, WV, 1985.

Clark, G. M., and E. J. Ciolkosz, Periglacial geomorphology of the Appalachian Highlands and Interior Highlands south of the glacial border - a review, *Geomorphology*, 1, 191-220, 1988.

Clark, G. M., E. J. Ciolkosz, J. S. Kite and D. A. Lietzke, Internat. Geol. Congr. Field Trip T150: Central and Southern Appalachians Geomorphology - Tennessee, Virginia, and West Virginia, 105 pp., 28th Internat. Geol. Congr, Washington, DC, 1988.

Clarkson, R. B., *Tumult on the Mountains: Lumbering in West*

Virginia 1770-1920, 385 pp., McClain Printing Co., Parsons, WV, 1964.

Colucci, S. J., R. B. Jacobson and Steven Greco, Meteorology of the storm of November 3-5, 1985, in Virginia and West Virginia, *U.S. Geol. Surv. Bull. 1981-B*, 31 pp., 1993.

Costa, J. E., Paleohydraulic reconstruction of flash-flood peaks from boulder deposits in the Colorado Front Range, *Geol. Soc. Amer. Bull.*, 94, 986-1004, 1983.

Ely, L. L., Yehouda Enzel, V. R. Baker and D. R. Cayan, A 5000 year record of extreme floods in climate change in Southwestern US, *Science*, 262, 410-412, 1993.

Enzel, Yehouda, L. L. Ely, J. Martinez-Goytre, R. G. Vivian, Paleofloods and a dam-failure flood on the Virgin River, Utah and Arizona, *J. Hydrology*, 153; 291-315, 1994.

Fansler, H. F., *History of Tucker County, West Virginia*, 702 pp., McClain Printing Co., Parsons, WV, 1962.

Gebhardt, T. W., Slackwater Deposits and Step-Backwater Modeling in the Cheat River Canyon, West Virginia, M.S. thesis, 90 pp., West Virginia University, 1998.

Holmes, R. S., Some observations on Ohio River alluvial deposits, *Soil Science Soc. of Amer. Proc.*, 3, 323-329, 1938.

Hupp, C. R., Plant ecological aspects of flood geomorphology and paleoflood history, in *Flood Geomorphology*, edited by V. R. Baker, R. C. Kochel and P. C. Patton, pp. 335-356, John Wiley and Sons, New York, 1988.

Jahns, R. H., Geological features of the Connecticut River Valley, Massachusetts, as related to recent floods, *U.S. Geol. Surv. Water-Supply Pap. 996*, 158 pp., 1947.

Jacobson, R. B., E. D. Cron, J. P. McGeehin, C. E. Carr, J. M. Harper and A. D. Howard, Landslides triggered by the storm of November 3-5, 1985, Wills Mountain anticline, West Virginia and Virginia, *U.S. Geol. Surv. Bull. 1981-C*, 33 pp., 1993.

Kite, J. S., Colluvial diamictons in the Valley and Ridge province, West Virginia and Virginia, *U.S. Geol. Surv. Circ. 1008*, 21-23, 1987.

Kite, J. S., and R. C. Linton, Depositional aspects of the November 1985 flood on Cheat River and Black Fork, West Virginia, *U.S. Geol. Surv. Bull. 1981-D*, 24 pp., 1993.

Kochel, R. C., and V. R. Baker, Paleoflood analysis using slackwater deposits, in *Flood Geomorphology*, edited by V. R. Baker, R. C. Kochel and P. C. Patton, pp. 357-376, John Wiley and Sons, New York, 1988.

Lescinsky, J. B., Flood of November 1985 in West Virginia, Pennsylvania, Maryland and Virginia: *U.S. Geol. Surv. Open-File Rep. 86-486*, 33 pp., 1986.

Linton, R. C., Slackwater Deposits Along the Cheat River in East-Central West Virginia, M.S. thesis, 173 pp., West Virginia University, 1992.

Mansfield, G. R., Flood deposits of the Ohio River January-February 1937, *U.S. Geol. Surv. Water-Supply Pap. 838*, pp. 693-732, 1938.

Miller, A. J., and D. J. Parkinson, Flood hydrology and geomorphic effects on river channels and flood plains: the flood of November 4-5, 1985, in the South Branch Potomac River Basin of West Virginia, *U.S. Geol. Surv. Bull. 1981-E*, 96 pp., 1993.

Morgan, B. A., G. F. Wieczorek and R. H. Campbell, Debris-flow hazards in area affected by the June 27, 1995, storm in Madison County, Virginia, *U.S. Geol. Surv. Open-File Rep. 97-438*, 15 pp., 1997.

Morgan, B. A., G. F. Wieczorek and R. H. Campbell, Historical and potential debris-flow and flood hazard map of the area affected by the June 27, 1995, storm in Madison County, Virginia, *U.S. Geol. Surv. Geologic Invest. Series Map I-2623-B*, 1 map (1:24,000 scale), 1999.

Moss, J. H., and R. C. Kochel, Unexpected geomorphic effects of the Hurricane Agnes storm and flood, Conestoga drainage basin, southeastern Pennsylvania, *J. Geol*, 86, 1-11, 1978.

O'Connor, J. E., Hydrology, hydraulics, and sediment transport of Pleistocene Lake Bonneville flooding on the Snake River, Idaho, *Geol. Soc. Amer. Special Pap. 274*, 83 pp., 1993.

O'Connor, J. E., L. L. Ely, E. E. Wohl, L. E. Stevens, T. S. Melis, M. S. Kale and V. R. Baker, A 4500-year record of large floods on the Colorado River in the Grand Canyon, *J. Geol.*, 102, 1-9, 1994.

O'Connor, J. E., and R. H. Webb, Hydraulic modeling for paleoflood analysis, in *Flood Geomorphology*, edited by V. R. Baker, R. C. Kochel and P. C. Patton, pp. 393-402, John Wiley and Sons, New York, 1988.

Osterkamp, W. R., and C. R. Hupp, Geomorphic and vegetative characteristics along three northern Virginia streams: *Geol. Soc. Amer. Bull.*, 95, 1093-1101, 1984.

Shearman, J. O., Floods--review of peak-discharge calculations for the Monongahela River basin flood of November 1985, unpublished memorandum to the U.S. Geol. Surv. record dated May 19, 1988, 11 pp., 1988.

Sigafoos, R. S., Botanical evidence of floods and flood-plain deposition, *U.S. Geol. Surv. Prof. Pap. 485-A*, 35 pp. 1964.

Springer, G. S., Geomorphic Implications of Cave Sediments in the Cheat River Canyon, North-Central West Virginia, M.S. thesis, 167 pp., West Virginia University, 1994.

Springer, G. S., and J. S. Kite, Cave sedimentation and erosional history in the Cheat River Canyon, north-central West Virginia, *Geomorphology*, 18, 91-100, 1997.

Teets, B., and S. Young, *Killing Waters: the Great West Virginia Flood of 1985*, 112 pp., Cheat River Publishing, Terra Alta, WV, 1985.

Tharp, S. J. and J. S. Kite, Stratigraphy of late Quaternary fans in the Ridge and Valley of Virginia, *Geol. Soc. Amer. Abstr. with Programs*, 21, 62, 1989.

U.S. Army Corps of Engineers Hydrologic Engineering Center, *HEC-RAS River Analysis System Users Manual Version 2.2*, U.S. Army Corps of Engineers Hydrologic Engineering Station CPD-68, Davis, CA, 1998.

Ward, S. M., E. A. Friel, F. M. Taylor and G. M. Ferrell, Water resources data West Virginia water year 1986, *U.S. Geol. Surv. Water-Data Rep. WV-86-1*, 196 pp., 1988.

Ward, S. M., B. C. Taylor and G. R. Crosby, Water resources data West Virginia water year 1996, *U.S. Geol. Surv. Water-Data Rep. WV-96-1*, 258 pp., 1997.

Ward, S. M., B. C. Taylor and G. R. Crosby, Water resources data West Virginia water year 1997, *U.S. Geol. Surv. Water-Data Rep. WV-97-1*, 474 pp., 1998.

Ward, S. M., B. C. Taylor and G. R. Crosby, Water resources data West Virginia water year 1999, *U.S. Geol. Surv. Water-Data Rep. WV-99-1*, 305 pp., 2000.

Wieczorek, G. F., B. A. Morgan and R. H. Campbell, Debris-flow hazards in the Blue Ridge of Central Virginia, *Environmental & Engineering Geoscience*, VI, 3-23, 2000.

Wiley, J. B., J. T. Atkins, Jr. and G. D. Tasker, Estimating magnitude and frequency of peak discharges for rural, unregulated, streams in West Virginia, *U.S. Geol. Surv. Water Resour. Invest. Rep. 00-4080*, 93 pp., 2000.

Williams, G. P., and H. P. Guy, Erosional and depositional aspects of Hurricane Camille in Virginia, 1969, *U.S. Geol. Surv. Prof. Pap. 804*, 80 pp., 1973.

Yanosky, T. M., Hydrologic inferences from ring widths of flood-damaged trees, Potomac River, Maryland, *Environmental Geology*, 4, 43-52, 1982.

Terry W. Gebhardt, ExxonMobil Production Company, USA, 800 Bell St., Houston, Texas, 77002-7426.

J. Steven Kite, Department of Geology & Geography, West Virginia University, Morgantown, West Virginia, 26506-6300.

Gregory S. Springer, Department of Earth Resources, Colorado State University, Ft. Collins, Colorado, 80523.

Historical Flood and Paleoflood Chronology of the Lower Verde River, Arizona: Stratigraphic Evidence and Related Uncertainties

P. Kyle House

Nevada Bureau of Mines and Geology, Reno, Nevada

Philip A. Pearthree

Arizona Geological Survey, Tucson, Arizona

Jeanne E. Klawon

U.S. Bureau of Reclamation, Denver, Colorado

A detailed analysis of slackwater flood deposits in the mouth of a tributary to the lower Verde River, Arizona, reconstructs the history of the largest floods on the river over approximately the last 1400-1600 years. The record provides abundant information about the magnitude and frequency of extreme floods in the late Holocene and places the river's short historical record into a realistic long-term context. Most importantly, however, is that the stratigraphic record and recent flood history of the river provide important insights into various uncertainties that combine to preclude confident compilation of complete paleoflood records in typical bedrock-canyon slackwater settings. Numerous processes act over time to compromise the integrity of paleoflood stratigraphy. In particular, vertical accretion in most slackwater sequences ensures that the stratigraphic record is progressively self-censoring and probably biased towards larger and younger floods. This process combines with other sources of uncertainty to hinder the compilation of complete records and the discrimination and accurate dating of individual events. These inherent uncertainties need to be explicitly addressed in slackwater paleoflood studies, particularly with respect to interpretation of climatic variability and flood-frequency analysis.

INTRODUCTION

The Verde River integrates runoff from slightly more than 14,000 km^2 of strikingly contrasting terrains in central Arizona (Figure 1, also on CD). It's historical flood regime is characterized by extreme flow events up to three orders of magnitude greater than average flow conditions. Spatial and temporal variations in the flood frequency in this region are strongly influenced by regional to global-scale climatic variability [*Hirschboeck*, 1985; *Webb and Betancourt*, 1992; *Ely et al.*, 1993; *House and Hirschboeck*, 1997]. Precipitation, runoff, and flood generation in the Verde watershed also vary significantly over space and time during and between individual events [e.g. *Chin et al.*, 1991; *House and Hirschboeck*, 1997]. The complex genesis

Ancient Floods, Modern Hazards: Principles
and Applications of Paleoflood Hydrology
Water Science and Application Volume 5, pages 267-293

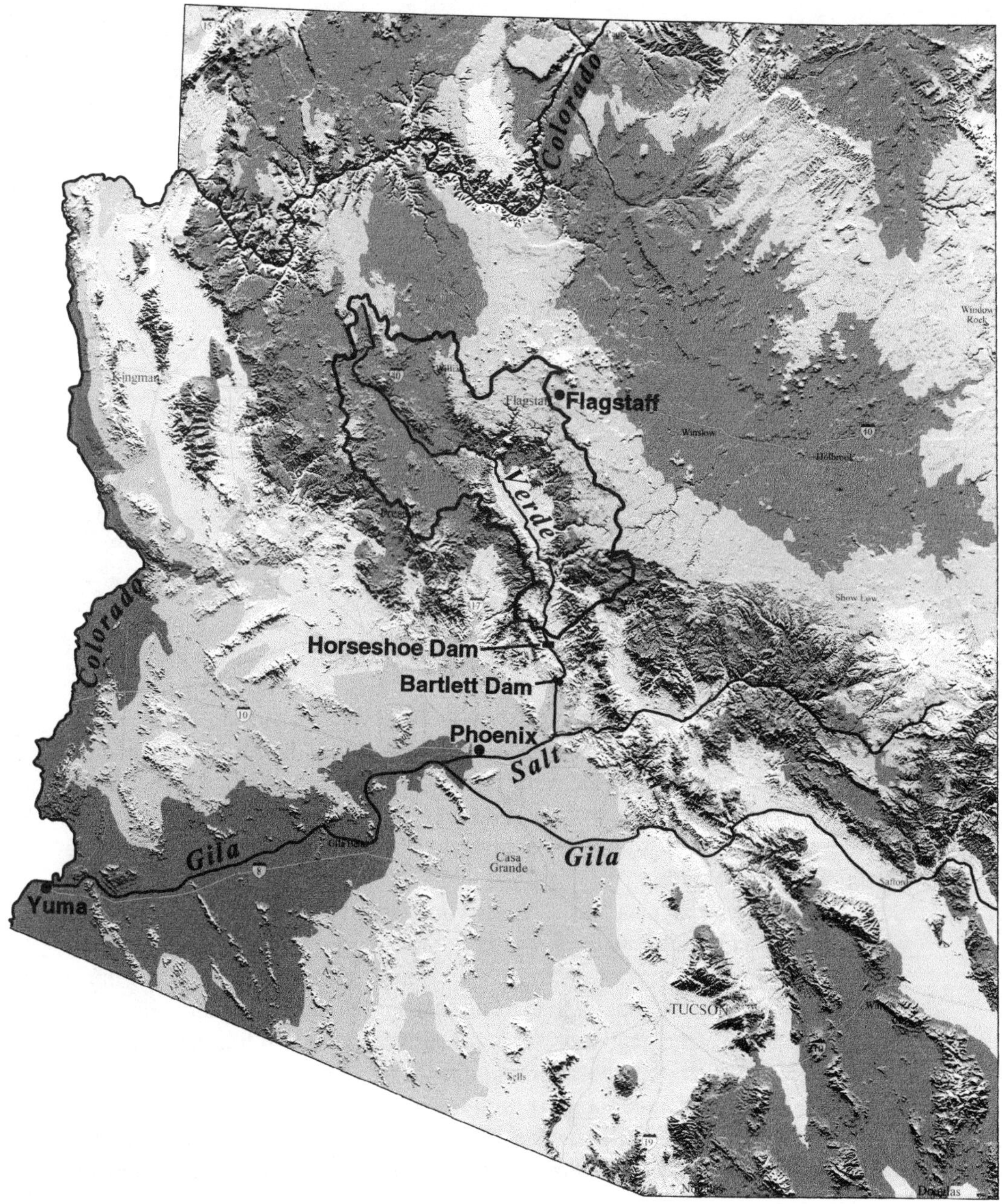

Figure 1. Shaded relief map of Arizona highlighting the location of the Verde River watershed and various locations mentioned in text.

of floods, their variability in occurrence over time, and a relatively short gage record confound attempts to adequately evaluate flood magnitude and frequency over time scales greater than about 50 to 100 years using the historical record alone. This presents a unique set of challenges for water resource management and flood control in the region.

Fluvial paleoflood data constitute a source of real information about discrete flood events that can augment the historical record. Here, we present the results of a detailed paleoflood study on the lower Verde River that documents the magnitude and timing of its floods over approximately the last 1500 years. The paleoflood chronology is based on detailed stratigraphic investigations of a complex slackwater flood deposit sequence in a backflooded tributary mouth. The sequence of slackwater deposits has a composite thickness of nearly 9 meters. The study site is 0.4 km upstream from a continuously recording U.S. Geological

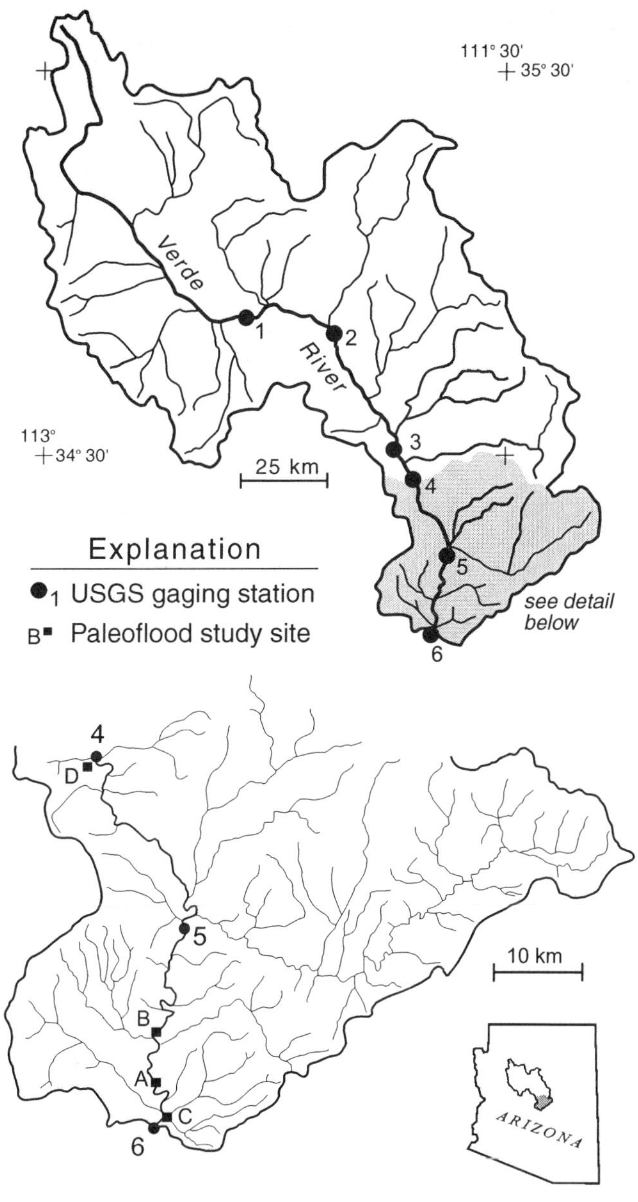

Figure 2. Maps of the Verde River watershed showing gages and paleoflood study sites. Inset map of lower Verde Basin shows location of this and previous paleoflood study sites in relation to the drainage net. See Table 1 for description of sites.

Survey stream gage (Figure 2, also on CD). This facilitates comparison of the paleoflood record with the systematic flood record. Moreover, we surveyed high-water marks at the study site following floods over the last seven years to compile a site-specific rating curve of stage vs. discharge. This allows for a direct comparison of the records.

In addition to extension of the record of floods in time, our analysis of slackwater flood deposits illustrates important uncertainties related to stratigraphic studies of late Holocene paleofloods. These uncertainties bear directly on related limitations in individual event discrimination and temporal resolution of typical slackwater paleoflood records. Analysis of historical flood deposits also indicates uncertainty in magnitudes of historical floods reported from the gage site.

This study was instigated by the occurrence of two extreme winter floods on the Verde River and its major tributaries in January and February 1993 [*House*, 1993; *Smith et al.*, 1994; *House et al.*, 1995; *House and Hirschboeck*, 1997]. These are the largest and second largest floods, respectively, of the systematic gage record on the lower Verde River (Tables 1 and 2). All other gages on the Verde River also recorded peaks of record in either January or February 1993 [*Smith et al.*, 1994]. Physical evidence from the 1993 floods provided an important baseline for interpreting paleoflood evidence.

Study Area

The Verde River watershed is located in central Arizona (Figure 1). It's headwaters drain semi-arid tablelands on the southwestern edge of the Colorado Plateau and alpine portions of the San Francisco Peaks near Flagstaff. The central portion of the watershed spans the Mogollon Rim, a major northwest trending, forested escarpment at the interface between the Colorado Plateau and the rugged mountains and basins of central Arizona. The Verde River and most of its tributaries draining the rim converge in Verde Valley, a broad basin prominently bounded on the east by the Mogollon Rim. The contributing drainage area to the river doubles along its course through the valley. At the southern end of the valley, the river enters a narrow bedrock canyon that transects rugged mountainous terrain in the Arizona central highlands. In the lower portion of this reach, the river enters the Sonoran Desert. Two major dams (Horseshoe and Bartlett) regulate flow on the lower river. The Verde joins the Salt River about 30 km east of Phoenix and ultimately joins the Gila River west of Phoenix, which continues to its confluence with the Colorado River near Yuma.

FLOOD HYDROLOGY AND HYDROCLIMATOLOGY

Four continuous recording gages are located along the Verde River (Figure 2 and Table 1). Their records indicate that the flood-producing portion of the Verde River basin is in the rugged central highlands and along the Mogollon Rim. This highland area is ideally situated to receive precipitation from orographic lifting of moist air derived from the Pacific Ocean and the Gulf of California. Nearly 50% of the watershed lies upstream of these runoff sources, and the upper portion of the basin contributes negligibly to large flood peak discharges downstream due to a long time of runoff concentration [see discussions in *Chin et al.*, 1991; *House and Hirschboeck*, 1997; and *Klawon*, 1998].

TABLE 1. Gaging stations on the Verde River and paleoflood study sites in the lower basin.

Label on Fig. 1	Station / Site Name	USGS #	Drainage Area (sq. km)	Record[1]	Max. Peak Q (cms)	Date
1	Verde River Near Paulden, AZ	09503700	5568	1963-date	657	2-20-93
2	Verde River Near Clarkdale, AZ	09504000	8091	1916, 1918, 1920; 1965-date	1510	2-20-93
3	Verde River below Camp Verde, AZ	09505550	11,105	1970-1980	1560	2-15-80
4	Verde River near Camp Verde, AZ	09506000	12,030	1934-1945; 1989-date	3370	2-20-93
5	Verde River below East Verde River near Pine, AZ	09508000	13,594	1934-1941[2]	3115	3-3-38
6	Verde River below Tangle Creek, AZ	09508500	14,229	1891, 1905, 1916, 1920; 1925-date	4110	1-8-93
PALEOFLOOD STUDY SITES						
A	Ely-Baker		13,932		*Ely, 1985* *Ely and Baker*, 1985	
B	Red Creek		13,683		*O'Connor et al.*, 1986; *House et al.* 1995 *House*, 1996	
C	Horse Creek		14,229		This study	
D	Chasm Creek		12,030		*House et al.* unpub. data	

1. Data from *Pope et al.* [1998]
2. Official dates published by USGS. Miscellaneous estimates have been made at this site through 1980 [*Aldridge et al.*, 1984; *Chin et al.*, 1991]

The largest historical floods on the Verde River have resulted from winter weather patterns involving multiple cyclonic storm fronts from the eastern Pacific Ocean (Figure 3, also on CD). Successions of these winter storms can culminate in heavy rain-on-snow in the upper watershed and heavy rain in lower elevation portions of the basin, potentially resulting in significant flooding. Historically, flood-producing winter storms have often been associated with El Niño conditions [*House and Hirschboeck*, 1997; *Ely et al.*, 1994]. At the Tangle Creek gage on the lower Verde River, only two floods in the 90th percentile of the partial duration series (> 1136 cms) did not occur in the winter (Figure 3, Table 2). Also, no flood in the record over 770 cms has been associated with summer conditions (i.e. convective thunderstorms). Events within the range 770 to 1795 cms represent the largest floods from dissipating tropical storms that occasionally affect Arizona in the fall or moderately large floods from winter frontal storms.

Flood runoff in the Verde River basin can vary considerably in space and time in response to similar hydroclimatic conditions. In recent floods, this has led to apparently significant differences in the size and timing of peak discharges estimated at various sites in the basin. This point was illustrated by the 1993 floods on the Verde River, when relatively small portions of the basin generated extremely large amounts of runoff because of heavy rain coupled with extremely wet antecedent conditions and a rapidly melting snow-pack [*House and Hirschboeck*, 1997]. *House et al.* [1995] used records from the Tangle

TABLE 2. Largest flood estimates reported from the Verde River gage below Tangle Creek (#9508500)

Date	Peak Discharge[1]	
	cms	cfs
24-Feb-1891	4248	150000
08-Jan-93	4106	145000
20-Feb-93	3511	124000
15-Feb-95	3058	108000
04-Mar-38	2832	100000
27-Nov-05	2718	96000
22-Feb-20	2690	95000
15-Feb-80	2684	94800
19-Dec-78	2662	94000
01-Mar-78	2588	91400
06-Mar-95	2543	89800
31-Dec-51	2311	81600
17-Feb-27	1982	70000
20-Jan-16	1951	68900
20-Feb-80	1863	65800
20-Oct-72	1795	63400
07-Feb-37	1784	63000
06-Sep-70	1753	61900
17-Jan-93	1509	53300
09-Feb-32	1501	53000
07-Dec-66	1501	53000
26-Jan-69	1297	45800
14-Mar-41	1240	43800
12-Mar-82	1192	42100

1. Values in the 90th percentile of the partial duration series

Creek and Camp Verde gages and supplemental discharge estimates from two sites between them to demonstrate that about 63% of an estimated 4140 cms peak discharge at the Tangle Creek gage [*Smith et al.*, 1994; *Pope et al.*, 1998] was supplied by runoff in the lower 25% of the basin during the flood of January 8, 1993. The discharge remained above approximately 2860 cms at Tangle Creek for 10 hours during the flood as runoff from the central part of the basin continued to arrive at the gage. The estimated peak discharge of the February 20, 1993 flood was nearly as large as the January peak at Tangle Creek (3540 cms) [*Smith et al.*, 1994]. It was also generated from about 25% of the whole basin, but in this case it was the central part of the basin between the Paulden and Camp Verde gages that produced the vast majority of the runoff (>95%). Relatively small variations in the timing of heavy rainfall in different portions of the basin during the 1993 storms could have led to a larger peak discharge on the lower Verde, especially during the early January event [*House et al.*, 1995; *House and Hirschboeck*, 1997].

Historical Flood Record

Archival and anecdotal accounts document large floods in central Arizona back to 1833 [*Dobyns*, 1981]. The earliest accounts from Native Americans and Anglo settlers describe floods on the middle and lower Gila River (Figure 1). Here, we invoke them only as a proxies for possible, similar flooding on the Verde River. In the systematic record between 1930 and 1999, only 33% of the annual floods on the Verde and the middle Gila occurred within 0 to 3 days of each other. In contrast, by using the partial duration series censored below the 10% annual exceedance probability level [*Pope et al.*, 1998] the correspondence is 44%; furthermore, only one of four floods near or above the 2% annual exceedance probability level on the Gila River [*Pope et al.*, 1998] did not have a corresponding flood on the Verde River. Thus, the comparison is reasonable, but certainly not definitive. The reliability of the anecdotal and archival accounts with respect to flood magnitude is also uncertain. Desert floods are still often perceived as exceptional occurrences, so it is possible that historical floods of moderate or even relatively low magnitude were remarked upon. Presumably, the largest floods received greater attention.

Accounts of the Pima Tribe and Anglo settlers indicate that notable floods on the lower Gila River (i.e. below its confluence with the Salt River) occurred in 1833, 1862, 1867, 1868, 1874, and 1891 [*Dobyns*, 1981]. The 1833 flood was apparently an extraordinary event based on its extent on the floodplain as reported by members of the Pima tribe [*Huckleberry*, 1995]. The calendar date of the flood is not precisely known, only that it occurred after November 13, 1833 (and presumably before January 1834). Significant flooding occurred across much of the western United States in the winter of 1862 [e.g. *Engstrom*, 1996], and the lower Gila evidently had a major winter flood that year. However, the Pima Indians living on the middle Gila River did not record it [*Dobyns*, 1981]. This implies that the flood could have been derived primarily from the Salt

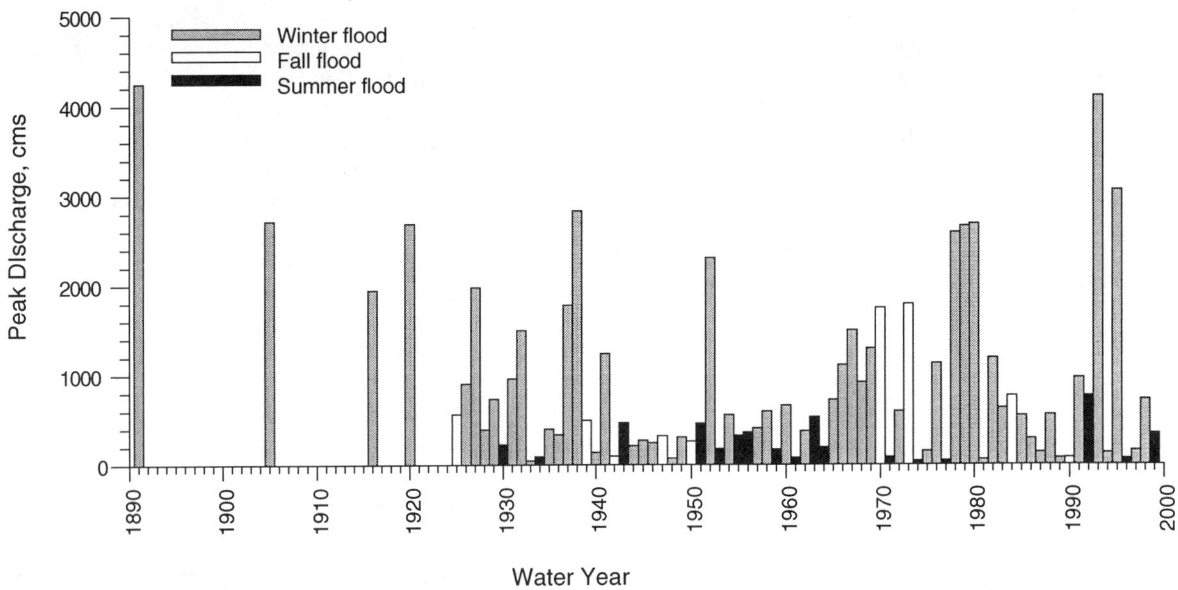

Figure 3. Annual flood series of the lower Verde River shown in relation to season of flood occurrence; data from *Pope et al.*, 1998.

and Verde rivers (downstream from the middle Gila). In September 1868, catastrophic flooding comparable to the scale of the 1833 flood occurred on the middle and lower Gila River [*Dobyns*, 1981]. The 1868 flood was regional in scale and presumably related to a dissipating tropical cyclone on the basis of its date. Historically, this type of storm has resulted in more significant flooding on the middle Gila River than on the Verde River.

A particularly significant episode of flooding affected most large rivers in Arizona in February 1891. Flooding on the Salt River below its confluence with the Verde in the winter 1891 was catastrophic and extensively documented because of its impact in the Phoenix area. This is the earliest historical flood in the Verde River record and the discharge "probably exceeded" 4248 cms [*Patterson and Somers*, 1966; *Pope et al.*, 1998]. This flood also caused dramatic channel changes in portions of Verde Valley [*Pearthree*, 1996]. Discharge estimates for this event on the Salt River in Tempe, Arizona, (below the confluence of the Salt and Verde rivers) range from 7362 to 8495 cms [*Fuller*, 1987]. This discharge range is similar to what would have transpired in the Phoenix area in January 1993 if not for the influence of multiple reservoirs on the rivers upstream.

Systematic Flood Record

Systematic documentation of large floods on the Verde River began with the flood of February, 1891 (Figure 3 and Table 2). Between 1891 and 1925, the largest floods on the river were measured indirectly at different sites above the Salt River confluence. In 1925, a formal gaging program was initiated on the lower Verde River. Large winter floods were documented in 1905, 1920, and 1938 (estimated discharges of 2718, 2690, and 2832 cms, respectively) [*Pope et al.*, 1998]. After 1938, there was a prolonged period of generally low to moderate magnitude flows through 1977. This changed dramatically in 1978, when large floods occurred in March 1978 (2588 cms), December 1978 (2662 cms); and February 1980 (2684). A 13-year period of low to moderate flows was followed by the extreme floods of January and February 1993 (estimates of 4106 and 3540 cms, respectively). This episode was soon followed by two large floods in February and March 1995 (estimates of 3058 and 2543 cms, respectively). Prior to the floods in 1993, the USGS estimate for the 1% annual probability flood on the lower Verde River was 4690 cms [*Garrett and Gellenbeck*, 1991]. Following the floods in 1993 and 1995, the estimate increased to 5720 cms [*Pope et al.*, 1998]. Our study of the river's flood history found no clear evidence substantiating the occurrence of a flood this large in at least the last 800 years and possibly as long as the last 1400-1600 years.

Dendrohydrologic Record

An annual dendrohydrologic (tree-ring-based) reconstruction of flow for the Verde River is available that spans the period 570 AD to 1984 AD [*Graybill*, 1989; *Van West and Altschul*, 1997] (Figure 4, also on CD). These data provide a reasonable record of annual flow volumes on the Verde River, but their relation to the occurrence of large

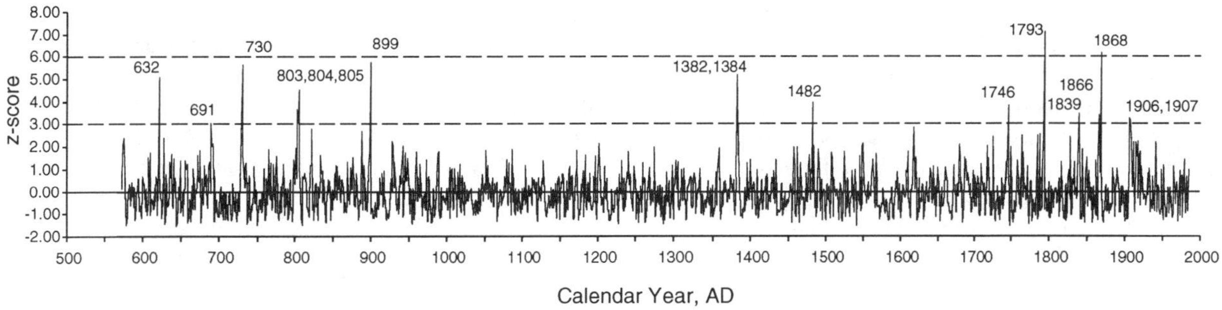

Figure 4. Dendrohydrologic reconstruction of annual flow on the Lower Verde River. Data shown as standardized departures from the overall mean value (z-scores) [data from *Graybill*, 1989 and *Van West and Altschul*, 1997]. Horizontal reference lines at z-scores of 0, 1, and 3 are shown for reference.

peak discharges is not particularly strong. Historically, most periods of particularly high flow volumes are associated with a higher probability of the occurrence of one or more large floods comprising a portion of that volume. Only one of three major peaks in the tree-ring record in 1839, 1866, 1868 corresponds to a year with a particularly notable flood; in this case a flood in the autumn of 1868. In fact, the reconstructed value for 1868 AD is the second largest departure from the mean in the 1414-year record (the largest is 1793 AD). Large floods in 1862 and 1891 are not associated with large departures in the tree-ring data, and floods in the 20th century are not particularly well represented. Thus, the linkage is indirect and variable, but the data provide an interesting point of comparison and a valuable record of hydrologic variability [*Redmond et al.*, this volume].

Flood-scarred trees are direct linkages to large floods because they reflect flood damage that can often be dated to the year of occurrence. *McCord* [1990] used flood-scarred trees to reconstruct the flood history of two tributaries to the Verde River (Rattlesnake Canyon and Oak Creek). Where records overlap, the flood-scar data are consistent with the historical record of floods. In Rattlesnake Canyon, floods recorded by tree-scars occurred in 1828, 1831, 1852, 1890, 1905, 1919, and 1970 (note that the events in 1919 and 1890 correspond to floods that occurred before the growing season in calendar years 1891 and 1920). This is a small tributary, however, and there is a high likelihood of it recording floods from isolated convective storms that did not significantly impact the Verde River. In Oak Creek, one of the larger Verde River tributaries, tree-scars record floods in 1885 and 1938, both of which were notable floods on Oak Creek, and one of which (1938) was a notable flood on the Verde [*McCord*, 1990].

THE LOWER VERDE PALEOFLOOD RECORD

We documented physical evidence of the Verde River's largest paleofloods and historical floods by analyzing slackwater flood deposits and fresh high-water marks in the mouth of Horse Creek, a small tributary that is extensively backflooded during high Verde River flows. The study site is in a bedrock-controlled reach of the Verde 0.4 km upstream from a stream gage (Verde River below Tangle Creek, USGS gage 09508500) and 14.5 km upstream from Horseshoe Dam, the first of two major dams on the river (Figures 1 and 2). The elevation of the study site is approximately 625 m asl (2050 ft). Average annual temperature and precipitation at Bartlett Dam, (approx. 25 km downstream) are 21.4 °C and 349 mm, respectively [*Western Regional Climate Center*, 2001]. Vegetation at the site is a typical Arizona Sonoran Desert assemblage with a rich diversity and relatively high density of cacti, trees, and shrubs [*Benson and Darrow*, 1981, pp. 10-11.].

Several large floods occurred on the Verde River and its tributaries immediately before and during our project. This resulted in increasingly improved exposures of a voluminous sequence of flood deposits and a unique opportunity to record the deposition and subsequent removal of the slackwater depositional record at the site.

Methods

Methods and concepts of paleoflood reconstruction using slackwater deposits have been described extensively in the literature [e.g. *Patton et al.*, 1979; *Baker*, 1987; *Kochel and Baker*, 1982, 1988; *Ely and Baker*, 1985]. Only some of the general concepts are briefly reiterated here. Compared to other methods for paleohydrologic and paleohydraulic reconstruction, slackwater paleoflood analysis in bedrock canyons is the most accurate technique for determining the magnitude and frequency of individual paleofloods [*O'Connor and Webb*, 1988].

Slackwater flood deposits. The most detailed flood records in bedrock canyons are afforded by slackwater flood stratigraphy. Slackwater flood deposits are fine-grained sediments deposited from suspension in areas of reduced flow velocities during large floods [*Kochel and Baker*, 1982,

1988]. Fine-grained slackwater deposits are emplaced under an unknown (but typically shallow) depth of water and represent only minimum flood stages; thus, corresponding discharges are typically minima when compared to flood stages predicted by hydraulic models [see discussions in *House et al.*, 1995; *Jarrett and England*, this volume; *Webb et al.*, this volume]. Peak paleoflood stages can be inferred from other types of paleostage indicators, including relict high-water marks such as flotsam lines, flood-scoured hillslopes, and flood-damaged vegetation. Recent studies have demonstrated that an overall limit can be established on paleoflood stage and discharge over time using landforms such as abandoned river terraces that show no evidence of having been significantly inundated for thousands of years [*Levish*, this volume].

Slackwater deposits in the canyon of the lower Verde River are preserved in extensive, high-standing terraces or "benches" along channel margins in wide, straight reaches, and thick slackwater terraces in tributary mouths [*Ely and Baker*, 1985; *O'Connor et al.*, 1986; *House et al.*, 1995]. We evaluated many of these types of sites at a reconnaissance level along much of the lower Verde River. We ultimately focused on a site in the mouth of Horse Creek, a small tributary of the Verde River, where there is a particularly voluminous accumulation of SWDs just upstream of a narrow bedrock constriction (Figure 5, also on CD). We described several exposures of slackwater flood deposits at this site in detail, emphasizing the discrimination of individual flood events with conventional stratigraphic and sedimentologic criteria [e.g. *Baker*, 1987, 1989; *Kochel and Baker*, 1982, 1988].

Geochronology. Analysis of datable materials is required to place flood stratigraphy into a numerical chronological context. Aside from relative dating techniques, radiocarbon dating is the most common method of geochronology used in paleoflood studies. Typical datable organic materials include charcoal, shells, and plant macrofossils. Archaeological materials, such as potsherds, are also common (particularly in the southwest US), and they can often be accurately dated to known cultural periods on the basis of diagnostic characteristics. Historical artifacts are common in recent deposits, and these can also be dated fairly accurately.

The stratigraphic context of organic materials must be carefully considered when interpreting radiocarbon dates [see discussion in *Trumbore*, 2000]. Many organic materials found in flood deposits, particularly charcoal, may be inherited detritus that do not accurately reflect flood timing. Their associated radiocarbon ages are maxima for the age of hosting deposits, possibly pre-dating the event by decades, or centuries in extreme cases [e.g. *Blong and Gillespie*, 1978]. Organic and cultural materials found *in situ* at boundaries between flood deposits typically provide a minimum age for the underlying deposit and a maximum age for the overlying deposit. Bioturbation of sandy flood deposits can transpose the two types of sample contexts described above and introduce younger organic materials into underlying flood deposits [e.g. *Johnson*, 1990]. This is an important source of uncertainty that is difficult to evaluate beyond the recognition that it may be a factor in strongly bioturbated, homogenized deposits.

We obtained 19 radiocarbon dates from samples collected from flood deposits at the Horse Creek site, 17 from charcoal and two from snail shells (Tables 3 and 4). We also encountered several prehistorical artifacts in clear stratigraphic contexts. In seemingly 'sterile' deposits bulk samples (often several kilograms) were mined for datable organic material by flotation and mechanical separation [e.g. *Pohl,* 1995]. This technique yielded sufficient quantities of datable material from numerous deposits, including a series of late Pleistocene flood deposits. Most organic samples were identified to the species or genus level [*Paleo Research Laboratories*, unpubl. data 1998, 1999, 2000] to evaluate their likely provenance in the watershed. This, in addition to stratigraphic context, can help determine if a sample is *in situ* or far-traveled, thus providing a general idea of the potential fidelity of the material's age to the date of the flood.

A systematic approach is required for interpreting radiocarbon dates to develop an accurate flood chronology. As noted in the preceding discussion, numerous sources of uncertainty combine to limit the resolution of radiocarbon dates associated with fluvial deposits. Resolution on the order of 200 years or less is nearly impossible to attain with much confidence. In interpreting radiometric dates, we adhered to the following guidelines:

1. Age estimates from clearly inherited detrital organic materials are maxima for host deposits and minima for underlying deposits. This assumes that hillslope and preexisting alluvial charcoal reservoirs are depleted during large floods, and subsequent floods are unlikely to carry charcoal older than a previous flood.

2. Age estimates from clearly *in situ* materials at bounding stratigraphic surfaces are assumed to be maxima for overlying deposits and minima for underlying deposits.

3. Discrepancies among ages of detrital materials in a single deposit indicate reworking of older alluvial deposits, biological reworking of younger organic materials [e.g. *Johnson*, 1990], or sampling problems. In these instances, interpretations are based on site-specific characteristics, ages from presumably correlative deposits in other exposures, and overall preponderance of evidence.

Previous Paleoflood Studies of the Verde River

Three paleoflood studies of the Verde River have been conducted just upstream of the Horse Creek site, two prior to the floods of 1993, and the third immediately after them

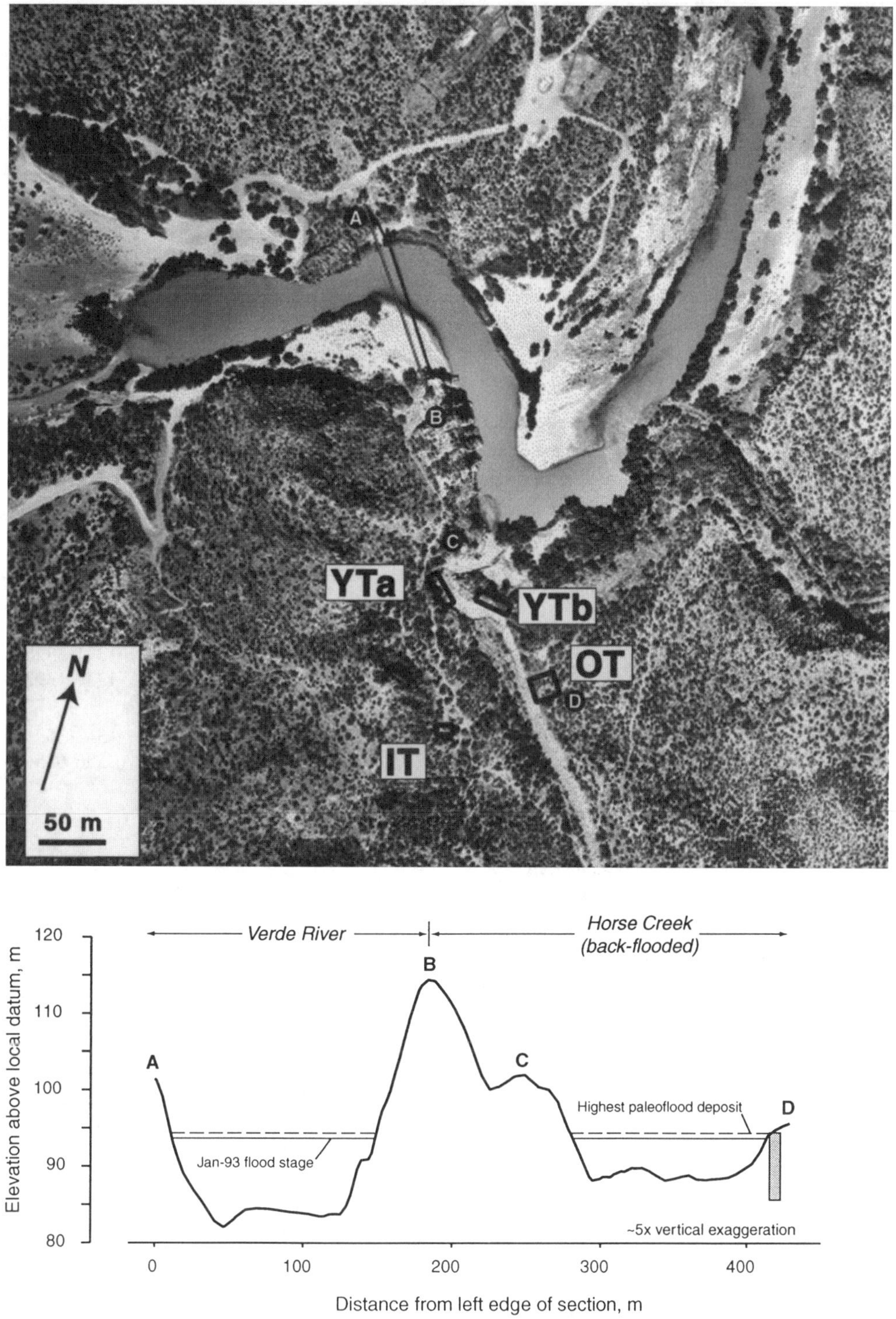

Figure 5. Location map and cross section of the Horse Creek study site and the principal stratigraphic exposures. Note that the box corresponding to the OT includes both the OTa and OTb. Prominent points on the cross section are indicated on the map. The stratigraphic section is generalized and shown schematically. Aerial photo from September 14, 1998.

TABLE 3. Late Holocene radiocarbon dates from flood deposits at the Horse Creek site.

Sample I.D.[1]	Context	Material	Type[2]	Conv. Date yr BP	Cal. Yr AD (1-sigma)[3]	Relative Prob.	Cal. yr AD (2-sigma)[3]	Relative Prob.
A	Detrital	Charcoal	*Pinus*	Modern	<1950	Na	na	na
B	Detrital	Charcoal	Unknown	280 ± 60	1494-1498 1514-1599 1615-1666 1783-1791	0.018 0.586 0.361 0.034	1456-1679 1739-1805 1935-1947	0.871 0.111 0.017
C	Detrital	Charcoal	*Juniperus*	360 ± 50	1472-1524 1560-1630	0.434 0.566	1447-1638	1
D	Detrital	Charcoal	Unknown	100 ± 50	1692-1727 1812-1919 1949-1950	0.246 0.741 0.013	1676-1764 1768-1775 1802-1939	0.333 0.014 0.624
E	Boundary	Charcoal	*Fabaceae*	320 ± 50	1495-1498 1511-1600 1614-1641	0.029 0.756 0.215	1460-1657	1
F	Detrital	Charcoal	*Fraxinus*	290 ± 50	1518-1596 1620-1657	0.675 0.325	1470-1672 1778-1799 1943-1945	0.962 0.036 0.002
G	Boundary	Charcoal	*Fabaceae*	380 ± 50	1446-1521 1586-1625	0.668 0.332	1439-1533 1539-1636	0.546 0.454
H	Detrital	Charcoal	*Fabaceae*	710 ± 50	1257-1305 1365-1386	0.771 0.229	1221-1326 1347-1392	0.745 0.255
I	Detrital	Charcoal	Conifer, hardwood, and *Pinus*	1430 ± 40	602-656	1	543-552 556-665	0.027 0.973
J	Boundary	Charcoal	Conifer, *Fabaceae*, *Quercus*, *Salicaceae*	1560 ± 50	430-542	1	408-604 607-616	0.989 0.011
K	Detrital	Charcoal	Unknown	1280 ± 40	688-732 736-774	0.536 0.464	661-783 789-828 840-862	0.891 0.072 0.037
L	Detrital	Charcoal	Unknown	130 ± 40	1679-1709 1718-1740 1753-1756 1804-1885 1912-1935 1947-1949	0.188 0.135 0.016 0.497 0.146 0.018	1673-1778 1800-1905 1905-1942 1946-1951	0.405 0.429 0.147 0.019
M	Boundary	Charcoal	Unknown	340 ± 50	1489-1528 1550-1663	0.318 0.682	1453-1644	1
N	Boundary	Charcoal	*Fraxinus*	840 ± 40	1163-1173 1180-1256	0.098 0.902	1045-1054 1056-1088 1121-1138 1156-1278	0.013 0.069 0.044 0.874

1. Sample I.D. is keyed to Figures 6, 7, and 8.
2. Plant species identified by Kathryn Puseman of Paleo Research Laboratories, Golden, CO.
3. Calibration and relative probability values from *Stuiver and Reimer* [1993, method b] and *Stuiver et al.* [1998].

TABLE 4. Early Holocene and Pleistocene radiocarbon dates from flood deposits at the Horse Creek site.

Sample ID[1]	Context	Material	Type[2]	Conv. Date yr BP	Cal. yr BC (1σ)[3]	Relative Prob.	Cal. yr BC (2σ)[3]	Relative Prob.
O	Unclear	Charcoal	Unknown	12320 ± 160	13962-12652 12485-12140	0.47 0.53	13493-12111 11973-11902	0.974 0.026
P	Detrital	Snail shells	Terrestrial	13600 ± 70	14620-14140	1.0	14889-13889	1.0
Q	Unclear	Mixed charcoal	Conifer, *Juniperus*, hardwood	13990 ± 50	15067-14590	1.0	15322-14368	1.0
R	Detrital	Mixed charcoal	Conifer, hardwood, unidentified	7000 ± 40	5970-5952 5910-5838 5823-5809	0.155 0.724 0.120	5985-5943 5925-5772 5759-5752	0.188 0.800 0.012
S	Detrital	Snail shells	Aquatic	14490 ± 50	15654-15155	1.0	15915-14928	1.0

1. Sample I.D. is keyed to Figures 7 and 8.
2. Plant species identified by Kathryn Puseman of Paleo Research Laboratories, Golden, CO.
3. Calibration and relative probability values from *Stuiver and Reimer* [1993, method b] and *Stuiver et al.* [1998].

(Figure 2). The first Verde River paleoflood study was performed approximately 6 km upstream of the Tangle Creek gage [*Ely and Baker*, 1985; *Ely*, 1985]. They reported stratigraphic evidence and related peak discharge estimates for several historical floods and concluded that the largest of these was the 1891 flood. *Ely and Baker* [1985] also found evidence for a substantially larger flood dated about 950 AD.

O'Connor et al. [1986] studied a reach approximately 6 km upstream of the Ely-Baker reach. They concluded that the paleoflood record at that site was less complete and peak discharge estimates for presumably correlative flood deposits were consistently lower than reported by *Ely* [1985] and *Ely and Baker* [1985].

House et al. [1995, also in *House*, 1996] performed a detailed follow-up study at the Red Creek site and a reconnaissance study at the Ely-Baker site following the floods in 1993. Similarly to *O'Connor et al.* [1986], they report modeling results for the peak discharge of the largest 1993 flood in the reach that are lower than that reported from the gage downstream as well as that interpolated from a rating curve derived from model results reported from the Ely-Baker site [*Ely*, 1985; *Ely and Baker*, 1985]. These differences were ascribed to underestimation of discharge on the basis of slackwater deposits and to augmentation of discharge by tributary input between the sites.

PALEOFLOOD STRATIGRAPHY

The floods of 1993 clearly demonstrated that floods on the lower Verde River deposit large amounts of slackwater sediment in the mouth and lower reaches of Horse Creek. The 1993 floods were large, amounting to the flood of record at every mainstem gage and most tributary gages in the basin (*Smith et al*, 1994; *Pope et al.*, 1998). These floods and less extreme events in 1995 and 1998, left distinct stratigraphic evidence that provides a useful framework for evaluating the context and integrity of the stratigraphic record and related peak discharges of past floods.

The Horse Creek site contains the most voluminous sequence of Holocene flood slackwater deposits that we have identified anywhere along the river. The Horse Creek-Verde River confluence is on the outside of a nearly 90° bend in the river that is immediately upstream of a narrow bedrock constriction (Figure 5). During large floods, this geometry induces a large eddy that submerges the mouth of Horse Creek. This results in the deposition of large volumes of suspended sediment and, in the case of the 1993 floods, tremendous amounts of flotsam in the lower reaches of Horse Creek [see, e.g., *Denlinger et al.*, this volume]. The site is less than 0.4 km upstream from the final stream gage on the unregulated portion of the Verde River, and no significant tributaries occur between the study site and the gage. A wide range of flood estimates have been reported from the gage over the course of our study—4106 cms (January 1993), 3510 cms (February 1993), 3058 cms (February 1995), 2543 cms (March 1995) and 731 cms (April 1998). We have surveyed high-water marks from most of these events to compile a stage–discharge rating curve at the site that allows a direct linkage of the paleoflood stage record with the systematic record of estimated flood discharges. Note, however, that there is some uncertainty in the gage record with respect to stage-discharge relations of the largest floods [*Denlinger et al.*, this volume; also described in subsequent section].

We evaluated the flood stratigraphy of the Horse Creek site in four exposures of three inset terraces, informally named the young (YT), intermediate (IT), and old (OT) terraces (Figure 5). Each terrace occupies a separate large niche along the lower channel of Horse Creek. The composite stratigraphic thickness of the sequence is nearly 9 meters. Taken together, these flood deposits provide a quasi-continuous record of the largest floods on the Verde River over nearly 1600 years. We qualify the record as 'quasi-continuous' because, as will be explained below, it is a continuous record only of those floods whose stages were sufficient to exceed local, rising censoring levels imposed by the surfaces of vertically accreting stacks of flood deposits.

Young terrace. We described extensive exposures of slackwater flood deposits in the young terrace (YT) on either side of Horse Creek (Figure 5). One exposure (YTa) measures approximately 9 m wide by 4 m high, is in a terrace that occupies a bedrock alcove on the left bank of Horse Creek, and is inset below the IT (Figure 6, also on CD). The other exposure (YTb) on the opposite bank of Horse Creek measures approximately 20 m wide by 5 m high and is in a terrace remnant farther upstream the channel (Figure 7, also on CD). Based on historical artifacts, radiocarbon dates, and physical characteristics of the deposits, it is possible to correlate with fair confidence most of the exposed units in the YTa and the YTb across the channel of Horse Creek. Historical artifacts and radiocarbon dates suggest that at least the six uppermost deposits in each exposure are from historical floods. The two uppermost deposits in each exposure are known to be from the 1993 floods on the basis of our post-flood observations in April, 1993.

The upper 0.5 m of the YTa consists of deposits from the January and February 1993. Their stratigraphic positions are inverted relative to their respective flood magnitudes because each flood overtopped the terrace (Units H1 and H2 in Figure 6; note that flood deposits are numbered sequentially with respect to their stratigraphic position). The January flood overtopped the YTa by more than 2 m but left a surprisingly thin deposit. The same flood left deposits >2 m thick in other places, including the YTb. The February 1993 flood overtopped the YTa surface and left a tabular deposit similar in thickness to the January deposit. The February flood left only a lenticular deposit in an erosional niche in the upper portion of the January deposit on the YTb (see Figure 7).

The February deposit (H1) is characterized by a predominantly flat-laminated bed of gray sand with alternating laminae of fine organic detritus. Locally, rip-up clasts from the January flood deposit are present in the lower 10-20 cm. The January deposit (H2) has multiple beds with colors ranging from orange to tan to gray, which probably reflect different subwatershed sources. Like the February deposit, the uppermost portion is a repeating sequence of horizontally-laminated fine sand with organic interlaminae deposited near the organic detritus-laden water surface in a swash environment [e.g. *Topping et al.*, 1999]. We observed this characteristic in fresh flood deposits throughout the basin following the 1993 events.

The deposit immediately below the January 1993 unit (H3) is discontinuously preserved along the extent of the YTa exposure as a couplet of gray sand overlain by orange sand. A similar and presumably correlative couplet is continuously exposed across the YTb exposure. This deposit is not significantly bioturbated and overlies historical flood deposits. We tentatively assign H3 to the February 1980 flood (Table 2). Unit H4 partially buries a grove of mesquite trees on the YTa surface and is penetrated by small roots. It is from a slightly older flood (December 1978 is likely). We recently collected a corroded 9-volt battery that had fallen out of the YTa exposure from either the H3/H4 or H4/H5 contact, indicating that at least one of them certainly post-dates the mid-20th century. A correlative unit to H4 in YTb is not particularly clear, but we tentatively assign it to a unit in the appropriate stratigraphic position. It is possible that H4 in the YTb is from the flood of March 1978 and that a correlative deposit is not present in the YTa.

Unit H5 is a distinctive bed in the YTa with clear upper and lower boundaries except portions that are extensively bioturbated by mesquite roots. It is separated from H4 by a thin bed of angular colluvium in YTa. The grove of mesquite trees on the YTa is rooted principally in unit H5, and several mesquites are also rooted in a presumably correlative level in the YTb. The YTb unit has a similar texture and color and is at a comparable elevation relative to the river. We extracted a radial cross-section from the main stalk of a relatively small mesquite tree that was rooted in H5. This species provides reasonably reliable annual ring counts [*Flinn et al.*, 1994] and 51±5 annual rings were discernible on the specimen that we collected [*F. Biondi*, written commun., 2000] indicating that it germinated between 1944 and 1954. A charcoal sample collected from H5 just beyond the left edge of the described exposure yielded a modern date (1950 AD). It was from a bioturbated zone, however, and may have been translocated downward. The tree-ring date represents a better minimum age of H5 (and maximum age for H4). The time between flood-deposition and germination of the tree is unknown. We conclude that H5 is the 1938 flood deposit on the basis of the tree-age and the reported magnitude of the flood. It is possible, however, that the deposit is from the 1905 or 1920 flood and the 1938 flood is not recorded as a stratum.

Historical artifacts collected from the base of Unit H5 indicate that it is certainly younger than the middle 19th Century. Small pieces of pure aluminum were collected from near the base of the unit in the YTa (Figure 6). These samples were collected in an intensely bioturbated portion

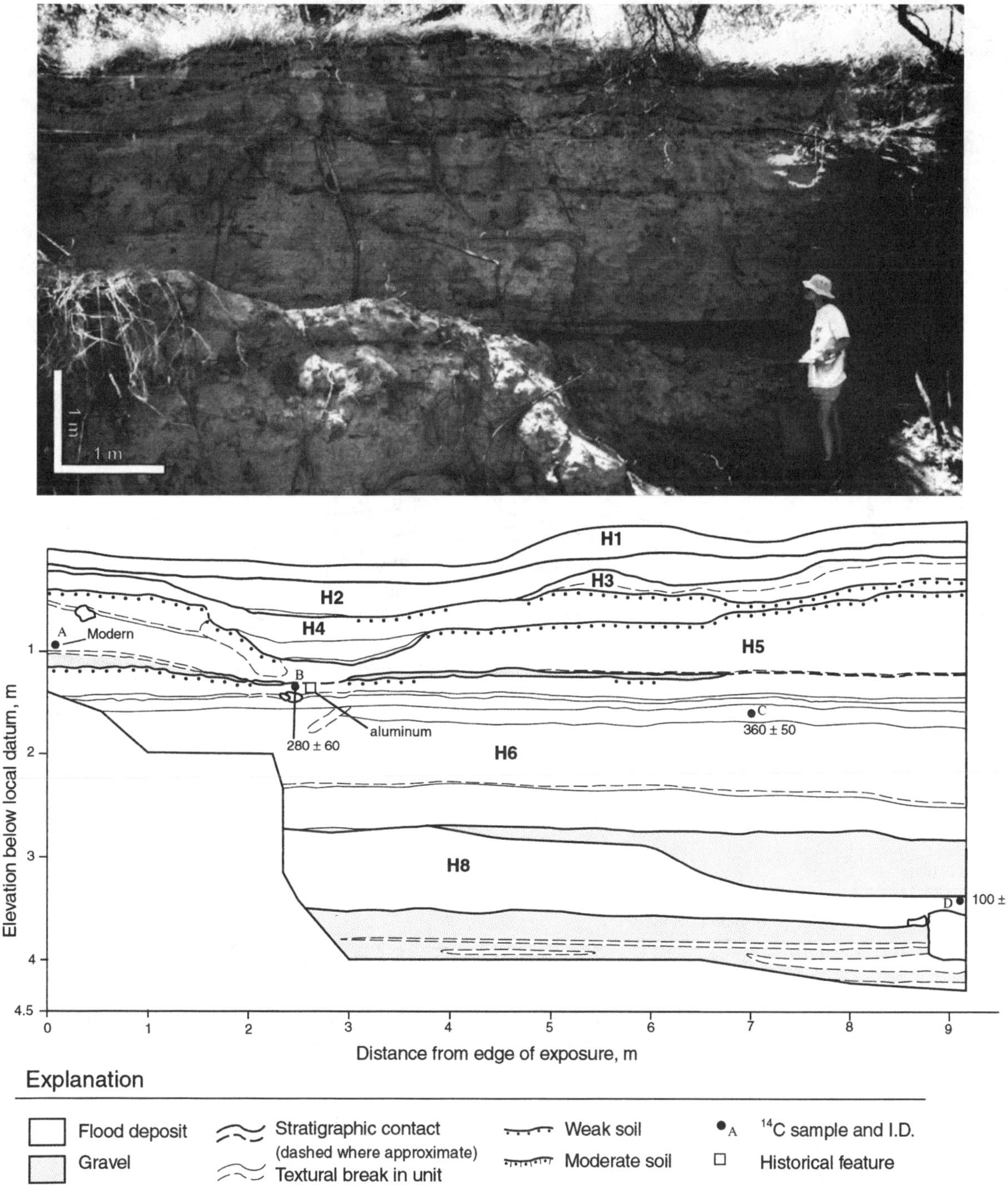

Figure 6. Photograph and diagram of the Young Terrace (YTa) stratigraphy. Radiometric dates in years BP and keyed to Table 3. Photograph of exposure taken August 1998. The photograph corresponds to the portion of the stratigraphic diagram from 2.5 to 9.25 m. The vertical extent is the same except that the portion of the photo that corresponds to the left one-half of the diagram is covered by a young, inset deposit. The inset was subsequently removed by a flash flood down Horse Creek (see text and Figure 11 for discussion and comparison). For reference, note the small, square unit delineated in the lower right-hand corner of the diagram. It is a rotated block of an old flood deposit and is immediately to the left of the person's left leg in the photograph.

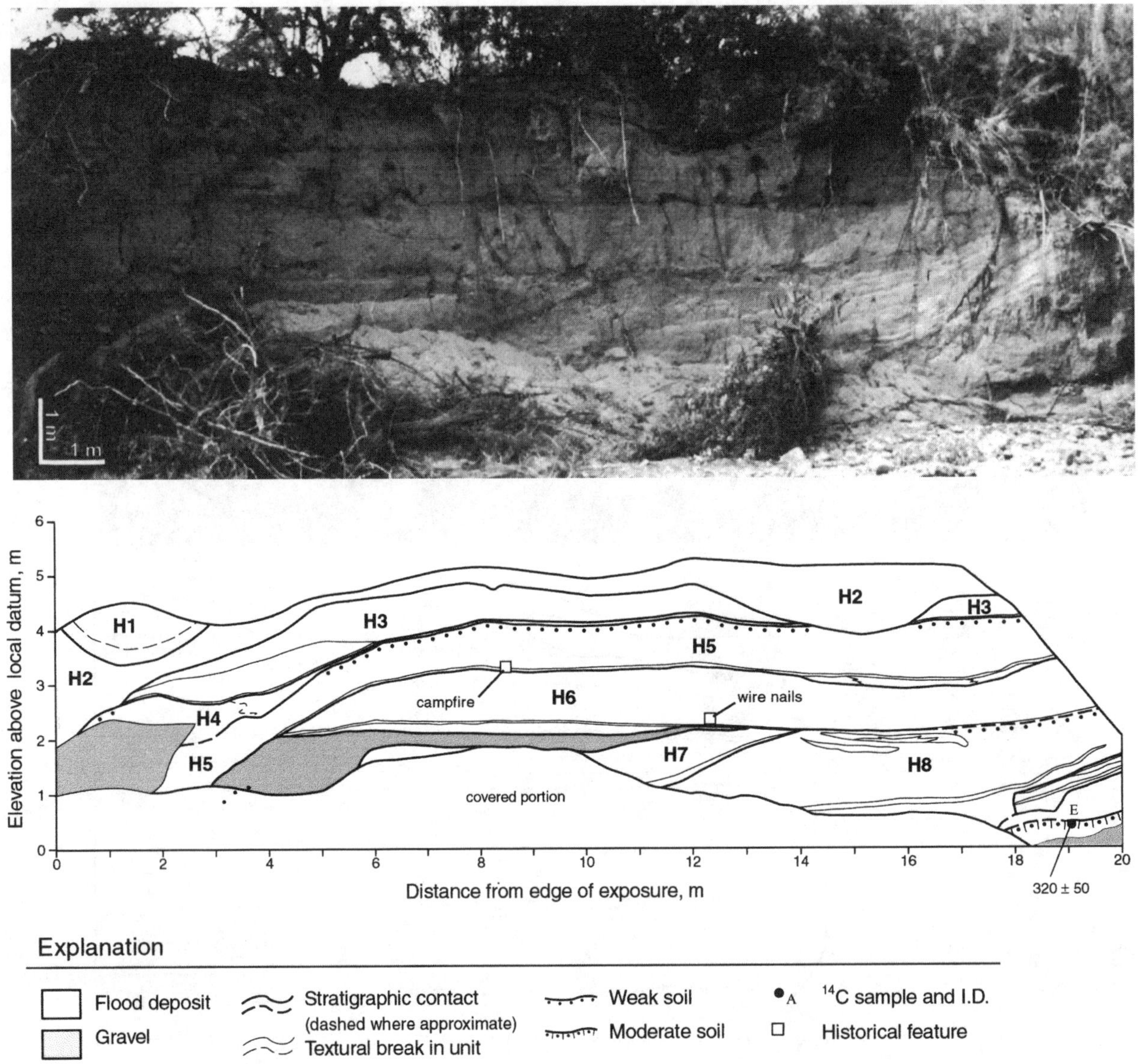

Figure 7. Photograph and diagram of the flood stratigraphy of the Young Terrace (YTb). Radiometric dates are shown in conventional years BP and are keyed to Table 3. The photograph was taken in August 1998. The exposure is nearly 5 m high and 20 m wide. The photograph corresponds to the portion of the stratigraphic diagram from approximately 4 m to 20 m. The entire vertical extent is the same in both. The top of the exposure in the photograph is mantled with a thick pile of organic flotsam from the floods in January and February 1993. For reference, note that the gravelly unit shown just to the right of the base of unit H5 in the diagram is immediately above the scale bar in the lower left-hand portion of the photograph.

of the unit, so their provenance is uncertain. Their position corresponds to the stratigraphic boundary between units H5 and H6 in undisturbed portions of the exposure. Mass production of aluminum was patented in 1880s. This provides a general constraint on the minimum age of H6 and the maximum age of H5. Charcoal was collected from a conspicuous campfire at the boundary between H5 and H6 in the YTb. It included numerous fragments of unidentified charred fruit [*Paleo Research Laboratories*, unpubl. data, 2000], suggesting that it is a historical feature..

In the YTa, Unit H6 is a thick deposit with multiple beds. In the YTb, an artifact containing wire nails was collected from the base of a similar and, presumably, correlative deposit (Figure 7). The nails date to no earlier than 1855-1891 [*Sutton and Arkush*, 1998, pp. 166-170], and we conclude that unit H6 in the YTb is probably from the 1891 flood. Floods in 1862 or 1868 are considered less likely possibilities based on the limited availability of wire nails at that time [*Sutton and Arkush*, 1998]. The 1891 designation is inconsistent with radiocarbon dates in the YTa, however. The oldest date from charcoal in H6 in the YTa is 1447-1638 AD, and we conclude that it is too old. The possibility remains, however, that the unit we have labeled H6 in the YTa may predate that in the YTb. The young date of 1676-1939 AD from near the base of YTa supports our preferred correlation.

In the YTb exposure, unit H6 has an erosional contact with two flood deposits (H7 and H8) that are also separated by an erosional unconformity. It is not clear which of these deposits underlies H6 in the YTa. The contact between H6 and the underlying deposit there varies from a flat, sharp contact between massive beds of silty sand to a sloping, erosional contact between gravel and massive silty sand. We assume that unit H7 is not present in the YTa, having been removed during the interval of erosion represented by the gravel layer between H6 and H8. Unconformities are also present between H6/H7 and H7/H8 contacts in the YTb. Charcoal just below the H6/H8 contact in the YTa yielded an age of 1676-1939 AD and a sample from a buried soil below H8 in the YTb yielded a date of 1495-1641 AD, indicating the likelihood that units H1 through at least H6 are historical. The date of Unit H8 is uncertain. It may be 1862, 1868, or possibly 1793, on the basis of the anecdotal and dendrohydrologic data described in previous sections.

The base of the YTa contains gravel deposits from Horse Creek. Their presence is indicative of erosion and removal of Verde River flood deposits. Locally, detached and rotated blocks of older flood deposits are intercalated with the tributary gravel. The lower 1 m of the YTb includes eroded remnants of three flood deposits that are not present in the YTa. These deposits are low in the sequence, so we do not include them in the tally of floods. The stratigraphic record of the YT indicates 3 to 4 m of net vertical accretion of slackwater deposits during the historical period. Any record of older floods at the sites of the YTa and YTb has been removed by erosion or is covered by the younger deposits.

Intermediate terrace. We described the flood stratigraphy of the intermediate terrace (IT) in an exposure approximately 2 m wide by 3 m high. It is located on the left (south) bank of Horse Creek upstream from both YT exposures (Figure 5). Evidence from the IT lengthens the flood record by approximately 1400 years (Figure 8, also on CD). We interpret the IT sequence as a continuous record of the largest floods over this time interval since no erosional unconformities were identified in the stratigraphy. The aggradation of the terrace to its present level began on a lower, and presumably erosional, surface prior to approximately 1600 years ago. Thus, as in the case of the YT, evidence of older floods has been removed or buried. The January 1993 flood locally lapped onto lower portions of the surface of the IT locally, but did not inundate the IT surface at the site where we examined the stratigraphy.

We have no direct constraints on the age of the uppermost deposit in the IT. An *in situ* charcoal sample from a buried soil surface at the base of the underlying deposit provides a maximum age for both units of 1439-1636 AD. The upper two units are also separated by a distinct paleo-

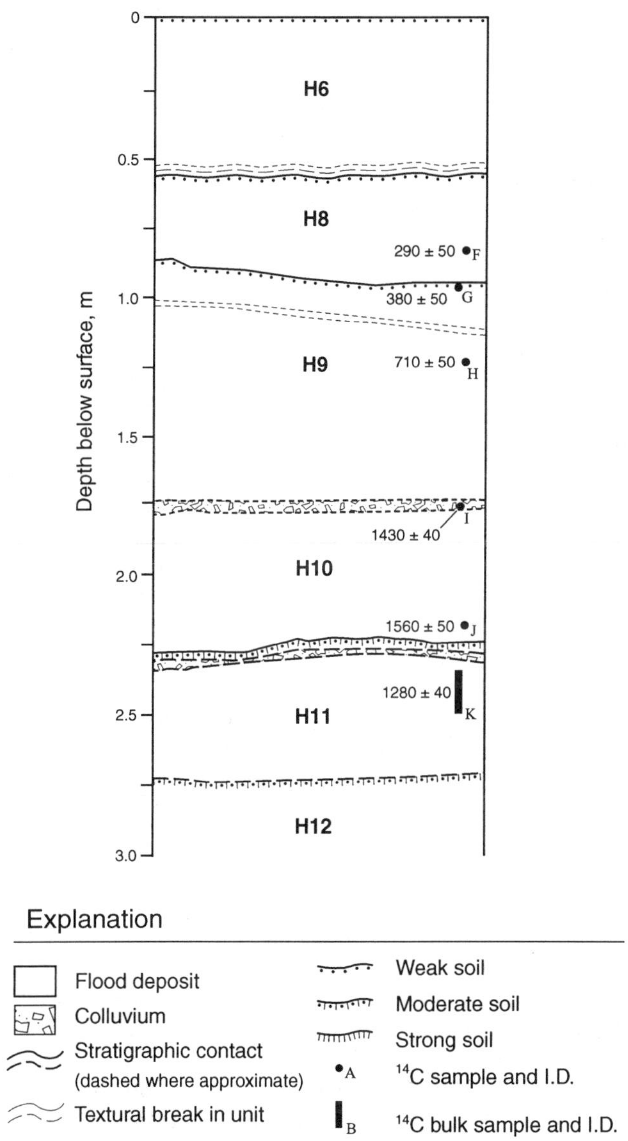

Figure 8. Flood stratigraphy of the Intermediate Terrace (IT) Radiometric dates in conventional years BP and are keyed to Table 3.

surface (buried soil A-horizon). We designate the second unit from the top as H8 on the basis of radiometric dates, stratigraphic position, and relation with units in the OT (discussed below). It may not correlate to H8 in the YT, but deposition would most certainly have occurred at that stratigraphic level during the flood that overtopped the IT. Direct correlation is not possible, however, but the potential discrepancy is largely inconsequential. We conclude that the overlying unit is H6, the presumed 1891 flood deposit. It is unlikely to be from any of the other historical floods because of the large flood magnitude implied by its position relative to the peak stage of the January 1993 flood, the largest in the systematic record.

As noted previously, H8 is separated from H9 by a paleosurface with a minimum age of 1439-1636 AD. A detrital charcoal sample from H9 yielded an age of 1221-1392 AD. This date represents a potential maximum for H9, although we cannot be certain that the charcoal was not introduced via bioturbation as it was collected below a slightly disturbed zone. The date is somewhat discordant with dates associated with a presumably correlative unit in the OT (described below).

The contact between H9 and H10 is characterized by a distinct textural change (increase in induration and carbonate accumulation), and we obtained a significantly older radiocarbon date from detrital charcoal (543-665 AD) in H10. The approximately 700-year difference in dates from within H9 and H10 and the lack of an erosional unconformity suggest the possibility of a multi-centennial hiatus in flood deposition above the level of H10, a conclusion that is corroborated somewhat by stratigraphic relations in the OT. The date from H10 is consistent with a date (408-616 AD) from a buried surface beneath a thin bed of colluvium separating units H11 and H12. The flood associated with H10 occurred between approximately 408 and 665 AD. The ages of units H11 and H12 are difficult to determine because of sparse datable material and a slight age-reversal. We conclude that the two units share a minimum date in the range of 408-616 AD. There is no basis for assigning a maximum date to either unit.

Old terrace. The old terrace (OT) is inset below a high Pleistocene river terrace on the north (right) bank of Horse Creek (Figure 5). The OT contains two late Holocene flood deposits overlying a sequence of approximately six indurated soils formed in late Pleistocene alluvial deposits. We described a section of the OT measuring 4 m wide by 3 m high that was exposed along the outside of an abandoned bend of the Horse Creek channel (OTa). We also excavated a shallow pit on the highest flat remnant of the terrace just upslope from the principal exposure (OTb) (Figure 9, also on CD).

We interpret the highest deposit in the OT sequence to be correlative with Unit H8 of the IT and YT sequences. Despite abundant lithic flakes and potsherds on the surface of the deposit, radiometric dates place it within the broad interval of approximately 1450-1950 AD which post-dates dense prehistoric occupation of the region. The artifacts are downslope from an archaeological site on the adjacent Pleistocene terrace. Abundant lithic flakes and potsherds on the slope between the OT and the terrace indicate translocation of the artifacts. The cultural litter on the OT surface therefore does not provide a good constraint on the surface age. Radiocarbon dates from H9 in the IT and the buried surface between H9 and H10 in the OT indicate that the H8 event occurred no earlier than 1435 AD. Data from the YT suggest that the event may be historic, but correlation of the OT deposit with the YT is uncertain.

The age-range of H9 is based on several radiometric dates and the age of a diagnostic potsherd found at its upper contact. In the OTb, unit H9 overlies an erosional unconformity on the uppermost Pleistocene soil in the OTa. An *in situ* charcoal sample from this contact yielded a date of 1045-1278 AD, a maximum date for unit H9. Pottery collected at the contact between H9 and H8 dates to approximately 1080-1200 AD [*C. Van West*, personal communication], thus potentially constraining the timing of the flood to a 200-year window (1045-1200 AD). However, the context of the pottery may be similar to that on the surface of the OT (top of H8). A more conservative time range can be established with the *in situ* date from between H8 and H9, (1453-1644 AD) as a minimum and the date from the base of H9 (1045-1278 AD) as a maximum.

Charcoal and gastropod shells from the deposits underlying unit H9 yielded late Pleistocene dates and a stratigraphically inconsistent early Holocene date (Table 4). The unconformity between latest Pleistocene deposits and late Holocene deposits raises the possibility that the two largest floods in the Holocene occurred within the past 1000 years. However, the irregular contact between the latest Holocene deposits and the Pleistocene section suggests an erosional unconformity, which implies that evidence of older large floods may have been removed from the OT prior to about 1000 years ago.

The face of OTa is mantled by at least two onlapping flood deposits, one that we know is from the January 1993 flood (unit H2) and one that we infer is correlative to unit H5, which we assigned to the 1938 flood. This correlation is uncertain, but it is not likely that H6 would be associated with such a thin, silty onlap at a level this low relative to the position of the sandy H6 deposit at the top of the IT. The two clear onlapping deposits are separated by a layer of fine-colluvium. Each is an upslope-thinning bed of laminated silt and silty sand extending from the floor of the alcove at the base of the OTa exposure. The alcove is an abandoned bend of Horse Creek that may have formed between the floods associated with H5 and H6, resulting in the removal of older flood deposits (including possibly

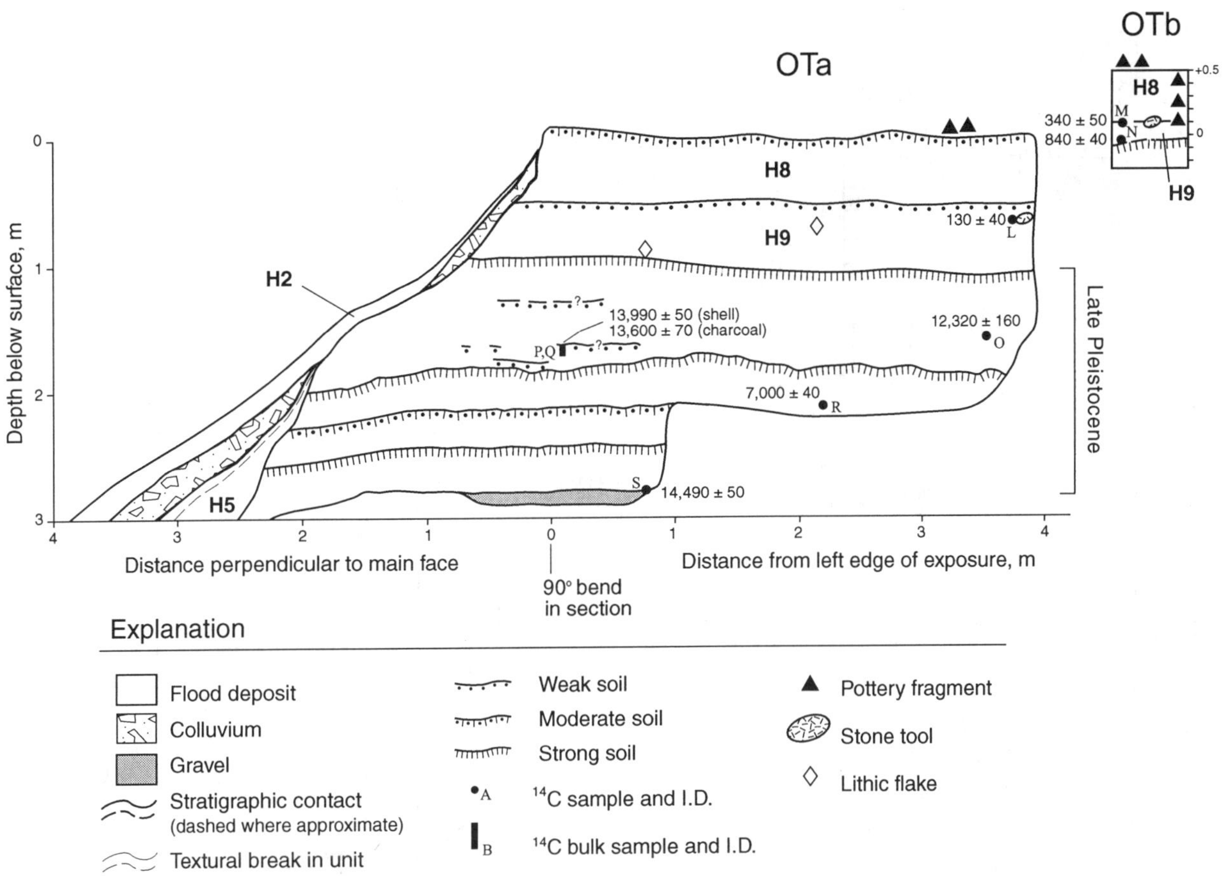

Figure 9. Stratigraphic diagram of the Old Terrace. Radiometric dates in conventional years BP and are keyed to Tables 3 and 4. The principal diagram corresponds to the OTa and the small diagram in the upper right corresponds to the OTb. Their relative elevations are indicated appropriately in the diagram, but the OTb is approximately 5 m from the main face of the OTa.

H6). In April 1993, we observed the edge of the 1993 onlap at the same elevation as the best high-water marks in the immediate area. If the underlying onlap is comparably faithful to the associated peak stage, the flood may have been comparable in magnitude to the February 1993 flood (reported estimate of 3511 cms; see Table 2).

Flood Discharge Estimates

The duration of our intermittent fieldwork at this site between 1993 and 2000 has allowed us to compile a rating curve based on reported peak discharges from the gage downstream (Tables 1 and 2; Figure 10, also on CD). We surveyed high-water marks from events in 1993 (two floods), 1995 (two floods), and 1998, with reported discharges of 4106 cms, 3511 cms, 3058 cms, 2543 cms, and 731 cms, respectively. Here we report our results in relation to the officially reported values for the sake of consistency and comparison with the systematic record of floods. However, as reported in *Denlinger et al.* [this volume], the resulting rating curve is at considerable variance with 2-D hydraulic modeling of a 6-km long reach of the river that contains the gage and the Horse Creek slackwater site. The 2-D model consistently predicts lower discharges than our rating curve indicates [see *Denlinger et al.*, this volume].

High-water marks from the largest floods in 1993, 1995, and 1998 were unambiguous in post-flood investigations at the Horse Creek site. The precise stage of the February 1993 and the March 1995 floods were more difficult to determine. In each case, the stage that we report is based on our best field estimate of the maximum stage of the flood. We fit a logarithmic curve to the points in the range of 2543-4107 cms and a linear curve between 731 and 2543 cms to estimate the minimum discharges corresponding to the various flood deposits in the study site (Table 5). We extrapolated the logarithmic curve beyond the peak stage of the January 1993 flood to estimate the minimum discharges associated with the highest flood deposits at Horse Creek. Based on this, the floods that

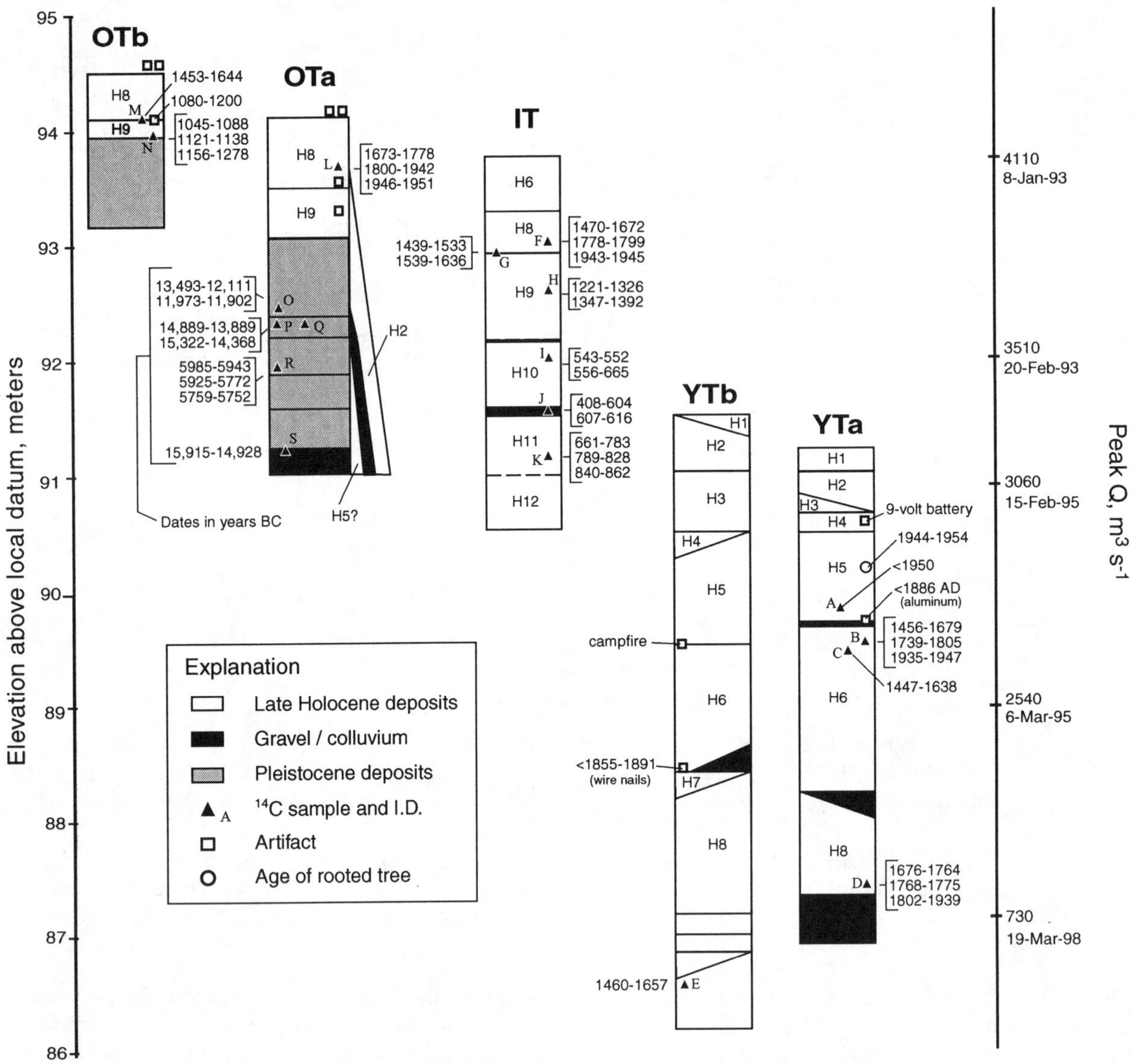

Figure 10. Composite stratigraphic column and stage-discharge relation for the Horse Creek paleoflood site. Each of the complete stratigraphic columns shown in Figures 6 through 9 has been simplified and generalized. Calibrated calendrical radiocarbon dates are shown with 2-sigma range and are keyed to Tables 3 and 4. Discharge ratings are based on high-water mark elevations we surveyed at the study site and with flood discharge estimates reported from the gage 0.4 km downstream.

emplaced units H8 and H9 must have had a peak discharges in excess of 4216 and 4326 cms, respectively.

Lower Verde River Paleoflood Chronology

The sequence of flood deposits that we studied at the Horse Creek site contains a record of at least 11 Verde River floods over the last 1400-1600 years, including the two large floods in 1993 and at least four other historical floods. The stratigraphic record of floods from the four exposures that we examined is generalized and compiled in Figure 10 (also on CD), the timing of each event is summarized in Table 6. Units H1 through H6 are historical flood deposits that correspond to winter floods, and on the basis of the flood hydroclimatology of the largest floods on the Verde River [e.g. *House and Hirschboeck*, 1997], it is likely that this is also the case for each of the prehistoric events (H8 through H12). It is possible, however, that some of the deposits may correspond to floods from intense dissipating tropical storms unprecedented in the historical record. The relatively short record of the YT is indicative

TABLE 5. Minimum discharge corresponding to flood units in each terrace.

Unit	YTa	YTb	IT	OTa	OTb
H1	3248	3085	—	—	—
H2	3242	3351	—	4067[1]	—
H3	3175	3192	—		—
H4	3121	2788	—	3540[1,2]	—
H5	3095	3004	—		—
H6	2893	2680	4083	—	—
H8	2321	2462	3927	4216	4326
H9	—	—	3810	3996	4216
H10	—	—	3563	—	—
H11	—	—	3345	—	—
H12	—	—	3178	—	—
Pleist.[3]	—	—	—	3891	4145

1. Onlapping deposit
2. Correlation uncertain
3. Base of Holocene deposits

of the transience of the lowest inset terraces in a typical tributary-mouth slackwater site. The stratigraphy of the YT is informative with respect to establishing age-control for relatively recent floods; however, due to the relative youth and low elevation of the two YT exposures, they are not particularly valuable sources of information about large paleofloods. The higher and more protected exposures in the IT and OT are most useful in this regard and, taken together, record the occurrence of six large floods over the past 1400-1600 years. The large floods of 1993 are not recorded in these SWD sequences aside from a thin onlap from the January flood on the steep face of the OT. The long-term preservation potential for this feature is extremely low.

Relation to Other Paleoflood Records in the Region

Ely and Baker [1985] described a series of three distinct slackwater deposit 'benches', or terraces, with inset relationships at several sites in a 3.5 km reach upstream from the Horse Creek site (Figure 2, site A). The middle bench included deposits from the most recent large floods on the river up to and including 1978 and 1980 floods. This sequence is probably correlative to the YT. The upper bench probably corresponds with the IT at Horse Creek. *Ely and Baker* [1985] concluded that it is composed of at least one deposit where preserved in their reach. The uppermost deposit in this bench was tentatively assigned to the 1891 flood on the basis of vegetation, height above the river and corresponding flood discharge, stratigraphic position, and detrital charcoal and wood from the deposit that yielded a radiocarbon date of 1510-1950 AD [*Ely*, 1985, p. 21]. This is consistent with our interpretation of unit H6.

A higher, isolated deposit was also described in their reach. It is probably correlative to the OT. A diagnostic potsherd on its surface dated to approximately 950-1000 AD, and disseminated charcoal 50 cm below the surface yielded a calibrated age of 900-1160 AD. It is likely that this deposit is correlative to unit H9, and that H8 was either not present or not recognized at their site. *Ely and Baker* [1985] also noted the presence of a higher surface underlain by fine-grained Pleistocene deposits that had no evidence of being overtopped by late Holocene floods.

O'Connor et al. [1986] identified a series of deposits at the Red Creek study site (Figure 2, site B) that is very likely correlative to the YT at Horse Creek. They found no evidence for any older, particularly large floods. In a follow-up study, *House et al.* [1995] largely confirmed the findings of the earlier study, but did find an isolated exposure of higher and older deposits that correlates to either the IT or the OT. The correlation is equivocal because the stratigraphy was not examined in detail.

Overall the paleoflood records from the Ely-Baker reach and the Horse Creek site correspond fairly well. However, the lack of a deposit clearly correlative with unit

TABLE 6. Age estimates for historical flood and paleoflood deposits

Unit	Minimum Age (yr AD)	Basis	Maximum Age (yr. AD)	Basis	Most Likely Date (preferred in boldface)	
H1	1993	1	1993	1	Feb-93	Historical Floods
H2	1993	1	1993	1	Jan-93	
H3	1980	5, 2	<1944	3	1980	
H4	1978	5, 2	<1944	3	Dec-78	
H5	1944	3	<1855-1891	4	1905,1920, **1938**	
H6	>1944	5, 3	<1855-1891	4	**1891**, 1905, 1920	
H8	>H6	5	1439-1636	6, 7		Paleofloods
H9	1435-1650	5, 6	1221-1392	6,7		
H10	1245-1390	8	543-665	6, 7		
H11	408-616	6,8	—	9		
H12	>H11	5	—	9		

1. Date of flood is known
2. On the basis of the gaging record
3. Age of rooted tree
4. Age of historic artifact
5. Stratigraphic position
6. In situ charcoal sample at stratigraphic boundary
7. Detrital charcoal sample in host deposit
8. Detrital charcoal sample in overlying deposit
9. No basis

H8 is a discrepancy. It is possible that the flood discharge increase between the sites associated with unit H9 was less than that of unit H8 such that their relative magnitudes were reversed upstream. Some significant tributaries enter the river between the two sites (see Figure 2). Historically, this small portion of the watershed may have contributed more than 425 cms to mainstem floods [*House et al.*, 1995]. It is also possible that they did not recognize a deposit overlying H9 because of extensive bioturbation and inadequate exposure (L.L. Ely, written communication, 2001). Stratigraphic evidence from the Chasm Creek site upstream [*House et al.*, unpublished data, 2000; location D on Figure 2] suggests that a fairly recent (within the past few hundred years) and large (greater than 1993) flood would be evident at a relatively high stratigraphic level at the Ely-Baker site. Moreover, *Fuller* [1987] reported that two exceptionally large floods occurred in the last 1000 years on the Salt River below its confluence with the Verde River. One of these occurred between approximately 1590 and 1890 AD. It is probably correlative with H8. The other occurred approximately 950-1150 AD and probably correlates to H9.

Relation to the Regional Paleoflood Record

Ely [1992] compiled paleoflood stratigraphic data from 19 rivers throughout Arizona and southern Utah (including the Verde River) to develop a regional flood chronology that spans the last 5000 years. The regional stratigraphic record is based on parsing an extensive radiocarbon chronology of paleofloods into 200-year increments. The data suggest that flood events are distributed non-randomly in time. *Ely* [1992, 1997] reported that late Holocene intervals associated with increased frequency of large floods include 3800-2200 BC, 400 BC-1100 AD, and 1400 AD to the present; and that periods with apparently low frequency of large floods include 2200-400 BC and 1200-1400 AD. The apparent variability in the regional paleoflood record corresponds with other types of paleoclimatic data consistent with hydrologic variability.

In this study, our approach to interpreting the radiocarbon chronology does not lend itself to subdividing the record into 200-year increments. However, it provides an important point of comparison with the regional record. In general the correspondence is marginal. For example, the possible 700-year hiatus in floods large enough to overtop the IT and OT between approximately 600 AD to 1055 AD is inconsistent with the regional chronology. Furthermore, the Verde River record does not show the apparent sharp decrease in flood frequency in the interval of 1200-1400 AD. It is possible that one particularly large flood (H9) actually occurred in that period given a broad interpretation of the radiocarbon dates. The YT record at Horse Creek does document the occurrence of numerous floods within the last 100-200 years, an observation that *Ely* [1992, 1997] reports from multiple sites in the Southwest. We postulate that this is due in part, at least, to the much higher preservation potential for younger flood deposits.

DISCUSSION

There is only marginal correspondence between the regional record and our record from the Lower Verde River. This may reflect the variable response of a single element of a larger composite. It may also reflect the influence of various sources of uncertainty in compiling fluvial paleoflood data, including limitations of geochronology, differing interpretation of the absence of evidence, and analysis of records that are often biased towards larger and younger floods. Our experience at the Horse Creek site suggests that it is most likely a combination of several sources of uncertainty that are largely inevitable in this type of analysis. These are outlined below.

Uncertainties Affecting Record Completeness

The principal goal of a typical slackwater paleoflood investigation is to enumerate floods represented in the stratigraphic record as accurately and completely as possible and to determine their timing as precisely as possible [*Kochel and Baker*, 1982]. In attempting to do this on the lower Verde River, we discovered that the ease of this task is influenced by the time between floods; deposit thickness, composition and texture; the adequacy of stratigraphic exposure; and the time available for the analysis. Our extensive investigation of the paleoflood history of the lower Verde River indicates that, aside from the vagaries of sedimentological and stratigraphic ambiguity, weathering, and basic logistical complications, it is essentially inevitable that the slackwater stratigraphic record is complex, incomplete, and often biased towards recent floods.

These observations are not new. They have been stated previously in the paleoflood literature with varying degrees of emphasis [e.g. *Baker*, 1989; *Kochel and Baker*, 1988; *Ely*, 1997] and have been explored recently in other slackwater settings [*Webb et al.*, this volume]. Our study at Horse Creek clearly illuminates several types of uncertainty and suggests several others. Specifically, these include the effects of stratigraphic ambiguity, erosion, internal stratigraphic complexity, incomplete exposure, pedogenesis, and stratigraphic record self-censoring. The effects of these alone, or in combination, have important implications for evaluating the information content of regional or site-specific fluvial paleoflood data.

Stratigraphic ambiguity. Individual flood deposits can be differentiated using stratigraphic, sedimentologic, and pedologic criteria, including changes in grain size, induration and color; the presence of intercalated deposits of different origin, organic horizons representing buried surfaces, and concentrations of archaeological materials [*Baker,* 1987, 1989]. Some of these criteria that are textural in nature may be present within a single flood deposit. It is critical and occasionally difficult to discriminate these features from actual stratigraphic breaks. For example, abrupt changes in grain size and sediment color can reflect variability in flow hydraulics and suspended sediment transport. Changes in color and induration can reflect development of soil horizons and the presence of buried surfaces, which are diagnostic indicators of a stratigraphic break. The key to interpreting stratigraphic breaks in flood deposit sequences and differentiating them from textural breaks within single deposits is to identify evidence of subaerial exposure at the contact between individual units [e.g. *Smith,* 1993; *Hattingh and Zawada*, 1996]. Thus, the best types of evidence for stratigraphic breaks include buried soils/paleosurfaces, erosional unconformities, and intercalated deposits that are not from mainstem river floods, such as colluvium, tributary sediments, or concentrations of archaeological materials [*Baker*, 1987]. In instances where separate flood deposits were emplaced in a very short interval of time (for example the January and February 1993 floods) stratigraphic breaks can be essentially indistinct from textural breaks within single flood units. This poses a problem for tallying individual floods.

Erosion. The extant stratigraphic record of floods in the mouth of Horse Creek was changed by the occurrence of three large winter floods down the Verde River and three to four floods down Horse Creek during the course of our field investigations. We compiled a photographic record of geomorphic and stratigraphic changes in the study area over the period 1992-2000 that illustrate the impact of these events at the mouth of Horse Creek and at the face of the YTa exposure (Figure 11, also on CD). Also, in our earliest investigations of the flood stratigraphy we observed and described deposits from Verde River floods that have since been completely eroded from the site.

The YTa is located along the outside of a tight bend in the Horse Creek channel. This geometry has contributed to the removal of a substantial amount of inset flood stratigraphy since the Verde River floods of 1993. We first visited the Horse Creek site in September 1992 and

Figure 11. Successive photos of the YTa exposure. White circles on each series of photos are matched points of reference. Series A: Looking up the mouth of Horse Creek to the face of the YTa exposure. The September 1992 photo shows a clear inset terrace below the main terrace. The inset terrace is assumed to be from floods in the 1970s and 1980s, and the person is standing on deposit inferred to be from the flood of 2-Mar-91 (971 cms). The April 1996 photo shows that the low inset terrace in previous photo has been removed and replaced by a flood deposit from the 15-Feb-95 flood. The photo from November 2000 shows the cumulative effect of at least 3 flash floods down Horse Creek between 1996 and 2000. The low inset terrace from 1995 has been completely removed and the bed elevation of Horse Creek has changed significantly. Series B: These photos indicate the change to the YTa exposure after one of the Horse Creek floods. The most significant change is the improved exposure of the base of the YTa section by complete removal of the low inset deposit.

observed two distinct terraces of slackwater flood deposits in the location of the YTa. The higher terrace was the YTa proper and the lower terrace was a large inset feature approximately 1-2 m below the (then) YT surface. We again visited the site two months after the large floods in 1993, and observed fresh deposits from the February flood at least 1 m thick overlying the inset terrace and onlapping the surface of the YT. We extracted an early to mid-1970s vintage beer can from the inset deposits, which we infer were laid down by one or more of the large floods in 1970, 1972, or 1978 (see Table 2). By 1996, most of the inset terrace and the underlying historical deposits had been

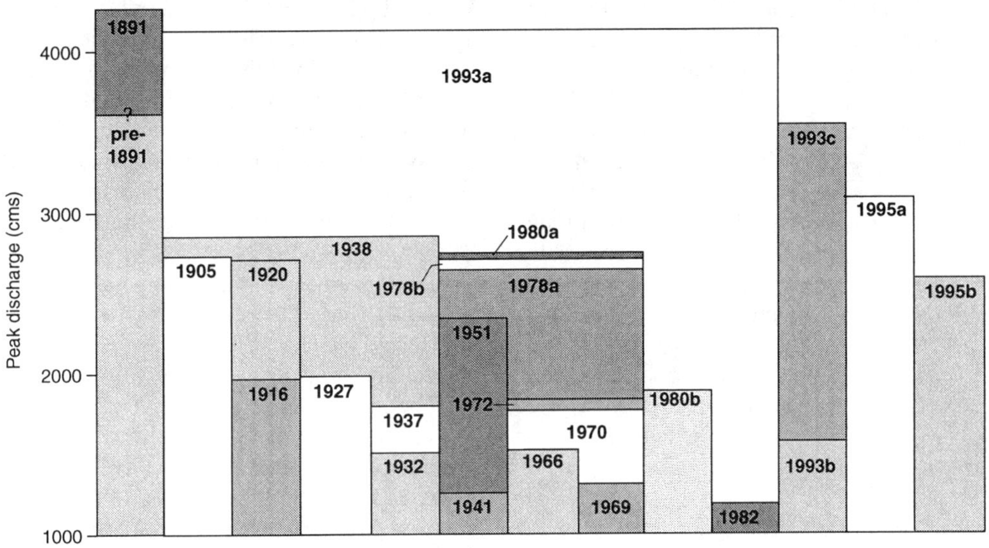

Figure 12. Idealized stratigraphy from historical floods on the lower Verde River. Only floods in the 90th percentile of the partial duration series are shown (see Table 2). The schematic assumes that all floods above 1000 cms are recorded, and no erosion occurs between floods. Comparison of this diagram with the paleoflood stratigraphy in Figures 6 through 9 demonstrates that, in reality, the record is significantly modified by various natural processes and logistical circumstances that influence the preservation and exposure of flood deposits.

removed or were buried by a Verde River flood in 1995. This event did not overtop the surface of the YT but instead left an inset terrace in essentially the same location as the previous inset. Between 1996 and 1998, the entire inset sequence (including the deposits from the 1995 flood) was completely removed by at least two substantial flows on Horse Creek. Horse Creek floods in the summers of 1999 and 2000 removed more of the YTa (Figure 11, series B).

The dynamics of this environment over this brief period indicate the great potential for changes over 10s to 100s of years. Evidence from the largest floods does seem to have a much higher likelihood of long-term preservation than does evidence from small and moderate magnitude floods because it is emplaced in locations farther from the main channel. However, gaps in that stratigraphic record and the overall geomorphic setting of the older deposits (the IT in particular) indicates that they too are subject to periodic removal.

Duration of subaerial exposure. The duration of subaerial exposure, deposit thickness, and elevation relative to the channel dictate the degree to which the integrity of the deposit and its differentiability from the underlying deposits may be compromised by bioturbation and pedogenesis. Bioturbation of flood deposits at the Horse Creek site was most pronounced where the grove of mesquites is rooted in the upper portions of the YT (in units H4 and H5), and in late Pleistocene deposits in the lower portion of the OT. In the OT, bulk samples taken from indurated, buried A-horizons in the late Pleistocene deposits were composed almost entirely of insect burrow casts [*K. Puseman*, written communication]. Pedogenesis in the Pleistocene deposits has advanced to the point of making stratigraphic analysis aside from the delineation of buried soil features very uncertain.

In the framework of vertically accreting flood deposits, the impact of pedogenesis is potentially enhanced as the threshold of inundation increases. This is true because a rising inundation threshold generally results in progressively longer periods of subaerial exposure as larger and rarer floods are required to exceed it. This concept is illustrated by the YT, where the lower portions of the exposure are characterized by relatively thick and minimally bioturbated deposits. Toward the top of the stack, the deposits are thinner, and bioturbation is more pronounced. Differentiation of individual flood deposits near the top of the YTa was not significantly compromised, however, because of the relative youth of the deposits and the large lateral extent of the exposure which allowed us to trace contacts over long distances through both disturbed and undisturbed zones.

Internal complexity and incomplete exposure. The overall scale and geometry of the depositional site can impart great internal complexity to paleoflood deposits. Potential stratigraphic complexity on the lower Verde River can be illustrated with a simple depositional model based on the gage record (Figure 12, also on CD). The resulting cartoon suggests that in some settings intricate stratigraphy should be expected. In cases where deposits are aggrading on multiple surfaces at different elevations, are being differentially trimmed back by tributary floods, and are subjected to differing degrees of bioturbation and soil development, more complex situations are feasible. A detailed picture of the internal three-dimensional structure of the flood stratigraphy at the Horse Creek site has not been fully realized, and such a task would be possible only through full-scale, piecemeal excavation of the entire site. Clearly, at this single site it is impossible to ensure the compilation of a complete flood record. However, it is likely that the floods large enough to exceed local censoring thresholds over approximately the last 1500 years are recorded. Overall, however, the flood record is incomplete because there is no evidence of floods that were not large enough to leave strata on the extant slackwater terraces at the time of the flood.

Progressive self-censoring. This phenomenon is the most important source of uncertainty that contributes to incomplete and biased stratigraphic records. It has been implied to some extent in each preceding discussion. Each stratigraphic section is an individual record of successive floods that constitute a rising threshold, or local censoring level, over time [*Baker*, 1989]. We call this the phenomenon of the *progressively self-censoring vertically accreting sequence*, and it is analogous to vertical accretion on the floodplain of an alluvial river. As the terrace height increases, it is inundated less frequently and for shorter durations, which results in progressively fewer and thinner deposits on average. Thus, fewer floods are recorded over time as the terrace level increases. Locally, lower inset deposits may record lower floods, but the fidelity of deposit height to flood magnitude is uncertain. Also, in this context, the superposition of individual deposits does not necessarily reflect differences in flood magnitude. It only indicates that each flood overtopped the concurrent terrace surface. The two 1993 flood deposits are inverted relative to their estimated magnitudes on the top of the YTa. Without additional information, it can only be concluded that the two events exceeded the threshold at the base of the January 1993 deposit, 3175 cms, approximately 75% of the reported discharge for the January flood. Only in rare cases where multiple exposures of correlative deposits exist at different elevations can relative flood magnitude be determined with more confidence [e.g., *Baker*, 1989].

The composite stratigraphic record from Horse Creek is an excellent example of progressive stratigraphic self-censoring at multiple levels. The diagram in Figure 13 (also on CD) illustrates this point. The stratigraphy associated with each of the principal terraces is shown as an incrementally increasing threshold over time. In this depiction the overall transience and low fidelity of the YT deposits to flood magnitude is apparent. The role of the IT as a progressively increasing censoring level for larger floods is also clear. For example, over an approximately 1000-year period (~400 AD through 1450 AD) there is no record of any flood that did not exceed the concurrent level of the IT. This generally corresponds to a discharge range of 3200-4100 cms. Thus, the steadily increasing discharge threshold imposed by the IT biases the record towards progressively larger floods, while simultaneously providing a record of non-occurrence of floods substantially in excess of that level. The OT serves a similar role that represents a higher threshold of flood non-occurrence over a similar period of time. This data characteristic is useful for flood frequency analysis, but it can complicate paleoenvironmental interpretations substantially.

The effect of the progressively increasing thresholds at the Horse Creek site has been to censor the content of the stratigraphic record such that there are no deposits from a range of moderate to large flood discharges that may have occurred within a period of approximately 1000 years. The floods that have occurred within approximately the last 200 years have been recorded in the YT sequences, but with potentially poor fidelity to the associated flood magnitude. It is possible that the former circumstance could be interpreted as evidence for no sizable floods as opposed to the equally likely situation of no potential for flood deposit preservation. The latter circumstance could result in significant underestimation of paleoflood discharges.

CONCLUSIONS

Our detailed paleoflood investigation on the lower Verde River demonstrates both the strengths and weaknesses of slackwater paleoflood hydrology as a technique for improving understanding of the frequency of large floods in bedrock channels. The greatest strength of this approach is that it can enumerate individual floods over time scales much longer than the systematic record. The greatest weakness arises from the collective effect of various physical processes, logistical circumstances, and methodological limitations that hamper the ability to confidently compile complete, accurate, high-resolution paleoflood chronologies. This is a good argument for compiling paleoflood data at a regional level to ensure the most representative record [e.g. the approach of *Ely*, 1992]. It is also a particularly strong argument for the combination of slackwater paleoflood methods with non-exceedance paleoflood methods [*Levish*, this volume] to most fully characterize the flood history of an individual river.

We compiled a quasi-continuous record of Verde River floods that have occurred over the last 1400-1600 years. However, there are a variety of uncertainties that combine to preclude confident accounting of all large floods that may have occurred during this period. Between ~1600 BP to 400 BP the record is incomplete for events with magnitudes less than about 3200 cms, based on the lowest

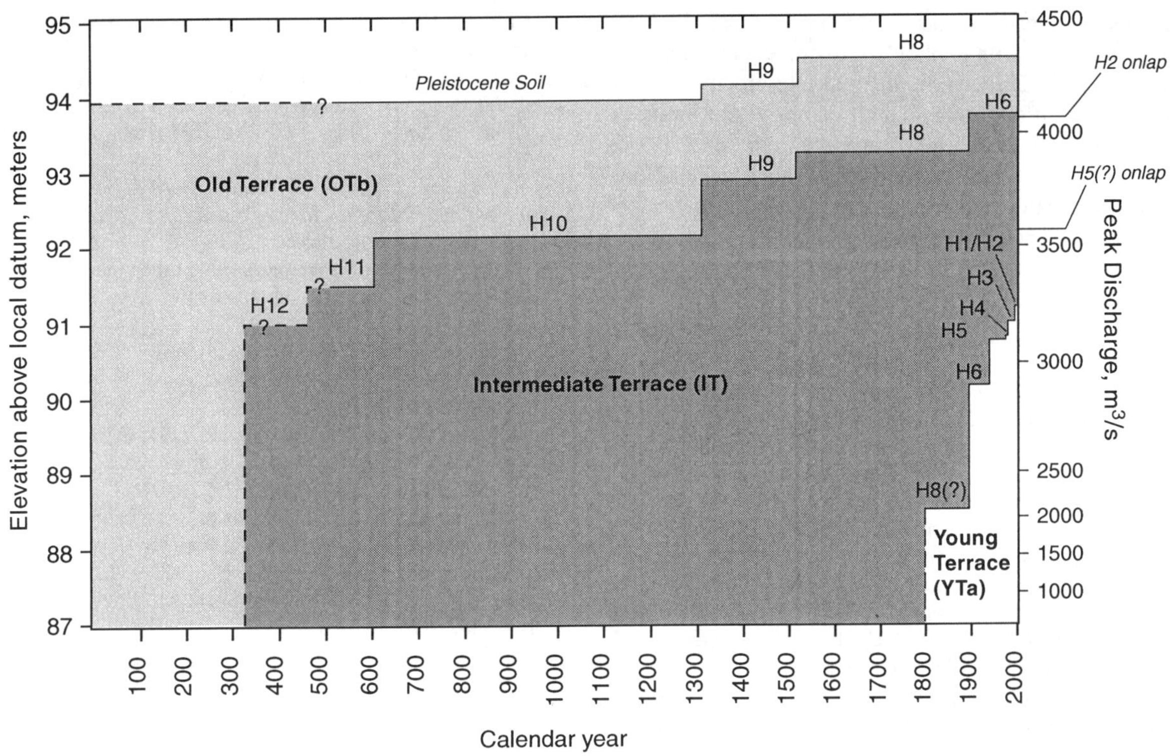

Figure 13. Schematic diagram of progressive self-censoring of flood stratigraphy at the Horse Creek site. Each flood event is expressed as a vertical increment along a horizontal continuum of time. Approximate dates of floods and lengths of interflood periods (terrace non-inundation) are based on information in Table 5. Related geochronological uncertainties are not indicated in order to simplify the diagram, but the time-ranges as shown are conservative. The YTa and OTb are chosen to represent the young and old terraces, respectively. Each line circumscribes a 'box' that corresponds to the extent of a local censoring level in space and time. Discharges on the right Y-axis are based on the rating curve derived from information shown in Figure 10. Deposits that onlap the face of the OTa are indicated but not used to develop the local censoring levels because of their geometry and ephemeral nature. This figure indicates that the IT and the OT contain the most information about the occurrence of the largest floods on the lower Verde River. There is no long record of floods that occurred but did not inundate the extant OT and IT surfaces. Thus, each of these terraces has become a censor of progressively larger floods over time, essentially guaranteeing that the stratigraphic record of floods is incomplete. The position of the YT as a low and transient inset relative to the higher terraces is clear. It represents an ephemeral record of floods unable to exceed the concurrent IT and OT levels. It is likely that one or more analogs to the YT have existed over the duration of the paleoflood records of the IT and OT but have been sequentially removed by erosion.

censoring level over that interval. Only the last 150-200 years of record is reasonably complete, but the overall fidelity of flood deposit elevation in the lower exposures to associated flood magnitude is low, and the vulnerability of the entire sequence to removal by erosion is very high.

The most important and largely intractable detriment to complete flood records in slackwater paleoflood hydrology is the phenomenon of progressive self-censoring. This is an internal filtering process that results in biased and incomplete stratigraphic records in a predominantly vertically accreting depositional environment. A variety of other, related processes can also compromise the completeness of the paleoflood record. At the Horse Creek site we are confident that floods able to exceed local, rising censoring levels over the last ~1500 years are represented; however, the occurrence of floods that were lower in stage than local censoring levels at any given point in time cannot be known. In this sense the flood record is quasi-continuous, or incomplete.

Fluvial paleoflood information is invaluable for extending flood records in real time. It can be a powerful component of flood-frequency analysis and can contribute to improving our understanding of the linkages between regional climatic variability and local extreme events [e.g., *Knox*, 2000; *Hirschboeck et al.*, 2000; *Redmond et al.*, this volume]. However, there are important and, arguably, inevitable sources of uncertainty that compromise record completeness. More effective incorporation of paleoflood information into water resource planning and applied hydrology is predicated on explicitly recognizing and addressing its limitations. Paleoenvironmental interpretations from the basis of paleoflood data should recognize the limits on the resolution of the size and timing of individual floods and problems of record completeness, and make allowances for alternative interpretations of the absence of stratigraphic evidence of floods.

Acknowledgements

Numerous colleagues contributed to our efforts on the lower Verde River. Thanks to Henry Allen, Diana Anderson, Roger Denlinger, Lisa Ely, Mark Fonstad, Jon Fuller, Matt Jones, Dan Levish, Eric McDonald, Matt Norocki, and Ann Youberg for assistance in the field. Interpretations in this report benefited from field discussions with Dan Levish, Dean Ostenaa, Eric McDonald, Lisa Ely, and participants in the Second International Paleoflood Conference (1999). An earlier version of the manuscript was significantly improved with critical reviews by Lisa Ely and Dean Ostenaa. Our multi-year research efforts were supported by The National Science Foundation (Grant EAR-9996036), the U.S. Geological Survey (WRRC AWARD#1434-HQ-96-GR-02684), the Salt River Project, the Arizona Geological Survey, the Nevada Bureau of Mines and Geology, the Desert Research Institute, and the Bureau of Reclamation. We thank Jeff Philips (USGS), Dallas Reigle, and Mike Charliebois (SRP) for providing data and logistical support. Susan Tingley and Kris Pizarro (NBMG) helped prepare several of the figures. Efforts of the Maricopa County Sheriff's Department helicopter rescue crew are greatly appreciated and will not be soon forgotten, nor will the comments, questions, and actions of various anonymous individuals who made our field experiences on the Verde River truly memorable.

REFERENCES

Aldridge, B.N., and J.H. Eychaner, *Floods of October 1977 in southern Arizona and March 1978 in central Arizona, U.S. Geol. Surv. Water-Supply Pap. 2223*, 143 pp., 1984

Baker, V.R., Paleoflood hydrology and extraordinary flood events: *J. Hydrol.*, v. 96, p. 79-99, 1987.

Baker, V.R., Magnitude and frequency of palaeofloods: in *Floods: Hydrological, Sedimentological, and Geomorphological Implications*, edited by K. Beven, and P. Carling, pp. 171-183, John Wiley, New York, 1989.

Benson, L., and R.A. Darrow, *Trees and Shrubs of the Southwestern Deserts*, 3rd ed., 416 pp., Univ. of Ariz. Press, Tucson, 1981.

Blong, R.J., and R. Gillespie, Fluvially transported charcoal gives erroneous ^{14}C ages for recent deposits, *Nature, 271*, 739-741, 1978.

Chin, E.H., B.N. Aldridge, and R.J. Longfield, *Floods of February 1980 in southern California and central Arizona, U.S. Geol. Surv. Prof. Pap. 1494*, 126 pp., 1991.

Dobyns, H.R., *From fire to flood, Historic human destruction of Sonoran desert riverine oases*, 222 pp., Ballena Press, Socorro, NM, 1981.

Ely, L.L., *Reconstructing paleoflood hydrology with slackwater deposits: Verde River, Arizona*, unpubl. M.S. Thesis, Univ. Ariz., Tucson, 48 pp, 1985.

Ely, L.L., J.E. O'Connor, and V.R. Baker, Paleoflood hydrology of the Salt and Verde rivers, central Arizona, in *U.S. Committee on Large Dams, 8th Annual Lecture Series*, Phoenix, Ariz., pp. 3.1-3.35, 1988.

Ely, L.L., *Large floods in the southwestern United States in relation to late Holocene climatic variations*, unpubl. Ph.D. thesis, Univ. Ariz., Tucson, 326 pp., 1992.

Ely, L.L., Response of extreme floods in the southwestern United States to climatic variations in the late Holocene, *Geomorphology, 19*, 175-201, 1997

Ely, L.L., and Baker, V.R., Reconstructing paleoflood hydrology with slackwater deposits: Verde River, Arizona, *Phys. Geog.*, 6, 103-126, 1985.

Ely, L.L., Y. Enzel, V.R. Baker, and D.R. Cayan, A 5000-year record of extreme floods and climate change in the southwestern United States: *Science*, 262, 410-412, 1993.

Ely, L.L., Y. Enzel, and D.R. Cayan, Anomalous north Pacific atmospheric circulation and large winter floods in the southwestern United States, *J. Climate, 7*, 977-987, 1994.

Engstrom, W.N., The California storm of 1862, *Quat. Res.*, 46, 141-148, 1996.

Flinn, R.C., S. Archer, T.W. Boutton, and T. Harlan, Identification of annual rings in an arid-land woody plant, *Prosopis Glandulosa*, *Ecology*, *75*, 850-853, 1994.

Fuller, J.E., *Paleoflood hydrology of the alluvial Salt River, Tempe, Arizona*, unpubl. MS Thesis, Univ. Ariz., Tucson, 70 pp., 1987.

Garrett, J.M., and D.J. Gellenbeck, *Basin characteristics and streamflow statistics in Arizona as of 1989, U.S. Geol. Surv. Water-Resour. Invest. Rept. 91-9041*, 612 pp., 1991.

Graybill, D.A., The reconstruction of prehistoric Salt River streamflow, in *The 1982-1984 Excavations at Las Colinas: Environment and Subsistence*, edited by D.A. Graybill, D.A. Gregory, F.L. Nials, S.K. Fish, R.E. Gasser, C.H., Miksicek, and C.R. Szuter, pp. 25-38, Archaeological Series 162, 5, Arizona State Museum, Univ. Ariz., Tucson., 1989.

Hattingh, J., and P.K. Zawada, Relief peels in the study of palaeoflood slackwater sediments, *Geomorphology, 16*, 121-126, 1996.

Hirschboeck, K.K., *Hydroclimatology of flow events in the Gila River Basin, central and southern Arizona*, unpubl. Ph.D. thesis, Univ. Ariz., Tucson, 335 pp., 1985.

Hirschboeck, K. K., L. L. Ely, and R. A. Maddox, Hydroclimatology of meteorologic floods, in *Inland Flood Hazards*, edited by E. E. Wohl, pp. 39-72, Cambridge University Press, 2000.

House, P.K., The Arizona floods of January and February 1993, *Ariz. Geol.*, 23, 1–9, 1993.

House, P.K., *Reports on applied paleoflood hydrological investigations in western and central Arizona*, unpubl. Ph.D. thesis, Univ. Ariz., Tucson, 356 pp., 1996.

House, P.K., P.A. Pearthree, and J.E. Fuller, *Hydrological and paleohydrological assessment of the 1993 floods on the Verde River, central Arizona, Ariz. Geol. Surv. Open File Rep. 95-20*, 38 pp., 1995.

House, P.K., and K.K. Hirschboeck, Hydroclimatological and paleohydrological context of extreme winter flooding in Arizona, 1993, in *Storm-induced Geologic Hazards: Case Histories from the 1992-1993 Winter in Southern California and Arizona*, edited by R.A. Larson and J.E. Slosson, pp. 1-24, Boulder, CO, *Geol. Soc. Am. Rev. Eng. Geol.*, 11, 1997.

Huckleberry, G.A., Archaeological Implications of Late-Holocene Channel Changes on the Middle Gila River, Arizona. *Geoarchaeology, 10*, 159-182, 1995.

Johnson, D.L., Biomantle evolution and the redistribution of earth materials and artifacts, *Soil Sci., 149*, 84-102, 1990.

Klawon, J.E., *Paleoflood hydrology and historic flood analysis in the upper Verde River basin, Central Arizona, Ariz. Geol. Surv. Open File Rep. 98-5*, 93 pp., 1998.

Knox, J. C., Sensitivity of Modern and Holocene floods to climate change, *Quat. Sci. Rev., 19*, 439-457, 2000.

Kochel, R.C., and V.R. Baker, Paleoflood hydrology, *Science, 215*, 353-361, 1982

Kochel, R.C., and V.R. Baker, Paleoflood analysis using slackwater deposits, in *Flood Geomorphology*, edited by V.R. Baker, R.C. Kochel, and P.C. Patton, pp. 357-376, John Wiley, New York, 1988.

McCord, V.A.S., *Augmenting flood frequency estimates using flood-scarred trees*, unpubl. Ph.D. thesis, Univ. Ariz., Tucson, 182 pp., 1990.

O'Connor, J.E. Fuller, and V.R. Baker, 1986, *Late Holocene flooding within the Salt River basin, central Arizona*, unpubl. report to the Salt River Project, Phoenix, Ariz., 84 pp., 1986.

O'Connor, J.E., and R. H. Webb, Hydraulic modeling for paleoflood analysis, in *Flood Geomorphology*, edited by V.R. Baker, R.C. Kochel, and P.C. Patton, pp. 393-402, John Wiley, New York, 1988.

Patterson, J.L., and W.P. Somers, *Magnitude and frequency of floods in the United States, Part 9, Colorado River Basin, U.S. Geol. Surv. Wat. Supply Pap.* 1683, 1966.

Patton, P.C., V.R. Baker, and R.C. Kochel, Slackwater deposits: A geomorphic technique for the interpretation of fluvial paleohydrology, in *Adjustments of the Fluvial System*, ed. by D.P. Rhodes and G.P. Williams, Kendall / Hunt Pub. Co., Dubuque, IA, 225-253, 1979.

Pearthree, P.A., *Historical geomorphology of the Verde River: Ariz. Geol. Surv. Open File Rep. 96-13*, 26 pp., 1996.

Pohl, M., Radiocarbon ages on organics from piedmont alluvium, Ajo Mountains, Arizona, *Phys. Geog., 16*, 339-353, 1995.

Pope, G.L., P.D. Rigas, and C.F. Smith, *Statistical summaries of streamflow data and characteristics of drainage basins for selected streamflow-gaging stations in Arizona through water year 1996, U.S. Geol. Surv. Wat. Res. Invest. Rept.* 98-4225, 907 pp., 1998.

Smith, C.F., P.D. Rigas, L.K. Ham, N.R. Duet, and D.W. Anning, *Water resources data, Arizona, Water Year 1993, U.S. Geol. Surv. Water-data Rep. AZ-93-1*, 360 pp., 1994.

Smith, G.A., Missoula flood dynamics and magnitudes inferred from sedimentology of slackwater deposits on the Columbia Plateau, Washington, *Geol. Soc. Am. Bull., 105*, 77-100, 1993.

Stuiver, M., and P.J. Reimer, Extended ^{14}C database and revised CALIB radiocarbon calibration program (ver. 4.1), *Radiocarbon, 35*, 215-230, 1993.

Stuiver, M., P.J. Reimer, E. Bard, J.W. Beck, G.S. Burr, K.A. Hughen, B. Kromer, F.G. McCormac, J. v. d. Plicht, and M. Spurk, INTCAL98 Radiocarbon age calibration 24,000-0 cal BP., *Radiocarbon, 40*, 1041-1083, 1998.

Sutton, M.Q. and Arkush, B.S., *Archaeological Laboratory Methods*, 381 pp., Kendall/Hunt Pub. Co., Dubuque, IA, 1998.

Topping, D.J., D.M. Rubin, J.M. Nelson, P.J. Kinzel III, and J.P. Bennett, Linkage between grain-size evolution and sediment depletion during Colorado River floods, in *The Controlled Flood in Grand Canyon*, edited by R.H. Webb, J.C. Schmidt, G.R. Marzolf, and R.A. Valdez, pp. 71-98, *Amer. Geophys. Union Geophys. Mono. Ser., 110*, Washington, D.C., 1999.

Trumbore, S.E., Radiocarbon geochronology, in *Quaternary Geochronology: Methods and Applications*, edited by J.S. Noller, J.M. Sowers, and W.R. Lettis, pp. 41-60, *Amer. Geophys. Union Ref. Shelf Ser. 4*, Washington, D.C., 2000.

Van West, C.R., and J.H. Altschul, Environmental variability and agricultural economics along the Lower Verde River, A.D. 750-1450, in *Vanishing River, Landscapes and Lives of the Lower Verde Valley: The Lower Verde Archaeological Project, Overview, Synthesis, and Conclusions*, edited by S.M. Whittlesey, R. Ciolek-Torrello, and J.H. Altschul, Chapter 9, pp. 337-392, SRI Press, Tucson, 1997.

Webb, R.H., and J.L. Betancourt, Climatic variability and flood frequency of the Santa Cruz River, Pima County, Arizona: *U.S Geol. Surv. Water-Supply Pap.* 2379, 40 p., 1992.

Western Regional Climate Center, Arizona Climate Summaries, *http://www.wrcc.dri.edu/summary/climsmaz.html*, 2001.

P. Kyle House, Nevada Bureau of Mines and Geology, University of Nevada, MS 178, Reno, NV 89557.
email: khouse@unr.edu

Philip A. Pearthree, Arizona Geological Survey, 416 W. Congress St. #100, Tucson, AZ 85701.
email: pearthree_phil@pop.state.az.us.

Jeanne E. Klawon, U.S. Bureau of Reclamation, P.O. Box 25007, Denver, CO 80225.
email: jklawon@do.usbr.gov

Paleoflood Hydrology of the Paria River, Southern Utah and Northern Arizona, USA

R.H. Webb

U.S. Geological Survey, Tucson, Arizona

J.B. Blainey

California State University Foundation, U.S. Geological Survey, Sacramento, California

D.W. Hyndman

Department of Geological Sciences, Michigan State University, Lansing, Michigan

Three critical assumptions in paleoflood studies are that (1) successively higher-stage floods leave depositional evidence in flood deposits, (2) that the height of the flood deposit can be used as evidence of the water-surface elevation, and (3) one-dimensional flow models accurately depict water movement in bedrock-constrained channels. Slackwater deposit/paleostage indicator (SWD/PSI) evidence preserved along the Paria River in southern Utah and northern Arizona can be used to challenge or verify these assumptions as well as improve flood-frequency estimates. The Paria River is a highly sediment-charged river that occasionally produces exceptionally large floods. Downstream from the Utah-Arizona border, the river flows through a narrow bedrock canyon in Navajo Sandstone where considerable SWD/PSI evidence is preserved under overhanging walls, particularly at one site, informally named Bonza Alcove. The paleoflood record, which begins at 4.2 ka with a massive clay, consists of 5 flood units with an inset deposit of an additional 6 flood units. The stratigraphy demonstrates that smaller-discharge floods may overtop the units of larger floods owing to the nature of flood deposition in three dimensions. Silt lines correlative with some of the flood deposits were preserved across the flood deposits and on the sandstone walls; these silt lines were 0.30 to 0.90 m above the corresponding flood deposit. Using one-dimensional step-backwater analysis, the discharge that best matches the silt-line evidence for the largest historical flood is 1,200 m^3/s, which is 2.6 times larger than the largest flood in the gaging record (1923-1998). Four different scenarios of varying amounts of paleoflood information are used in flood-frequency

Ancient Floods, Modern Hazards: Principles
and Applications of Paleoflood Hydrology
Water Science and Application Volume 5, pages 295-310

Published in 2002 by the American Geophysical Union

analysis, and the effect of the addition of paleoflood data in estimating flood recurrence quantiles is assessed.

INTRODUCTION

The Paria River is a major tributary of the Colorado River in southern Utah and northern Arizona (Fig. 1a). This river flows through ephemeral- and perennial-flow reaches from its headwaters, above 3,000 m elevation in southern Utah, to its southerly course near the Colorado, where it winds through spectacular scenery of bedrock cliffs, wide and fertile valleys, and badlands. Near the Colorado River, paleoflood deposits are preserved under overhanging sandstone cliffs, allowing dating of large paleofloods as well as reconstruction of peak discharges.

Paleoflood deposits and evidence on the Paria River are exceptionally well preserved for this region and thus can be used to illustrate many of the strengths and weaknesses inherent in paleoflood hydrology. In addition, the deposits can be used to test many of the assumptions inherent in reconstruction of paleoflood discharges. This paper presents an analysis of Paria River paleoflood deposits in the context of the general application of paleoflood data in flood frequency analysis. Because the Paria River paleoflood record contains floods that greatly exceed those recorded in the gaging record, we also discuss other rivers in the southwestern United States that have unusually large floods, either in the gaging record or the combined gaging and paleoflood record.

COMMON ASSUMPTIONS IN PALEOFLOOD STUDIES

Kochel and Baker [1982, 1988] discuss the common assumptions required to calculate paleoflood discharges from slackwater deposit/paleostage indicator (SWD/PSI) information. In most paleoflood hydrology studies, stable bedrock walls confine the channel, preventing lateral channel change, and the depth to bedrock in the channel is assumed to be low. Storage of alluvium in the river channel is assumed to be minimal, as is the potential for scour and fill. *Graf* [1987] addressed the former question and found that channels the size of the Paria River on the Colorado Plateau only stored moderate amounts of sediment in channel flood plains. *Patton and Boison* [1986] report two periods of late Holocene incision in Harris Wash, a tributary of the Escalante River to the east of the Paria River. Both of these studies involve channel systems that do not store flood deposits but may instead respond more like alluvial channels with some constraints by bedrock walls.

Slackwater deposits are fine-grained material, typically coarse silts and fine sands, deposited in low-energy environments that permit rapid settling of fine material. It is commonly assumed that a new sedimentary unit is deposited and preserved for each subsequent flood with a stage exceeding the pre-existing elevation of sediment accumulations [*Kochel and Baker*, 1988]. If sediments accumulate in sites protected from erosion by subsequent floods and runoff, these sites will continue to accumulate sediments from successive floods of greater magnitude [*Baker et al.*, 1983; *Baker*, 1984].

Paleoflood records begin for reasons that generally are unexplained because of lack of evidence. After deposition begins, slackwater deposits are stored in small eddy deposits. The size of flood necessary to begin slackwater deposition is commonly referred to as the preservation threshold. Radiocarbon analysis of carbon stored in the flood deposits

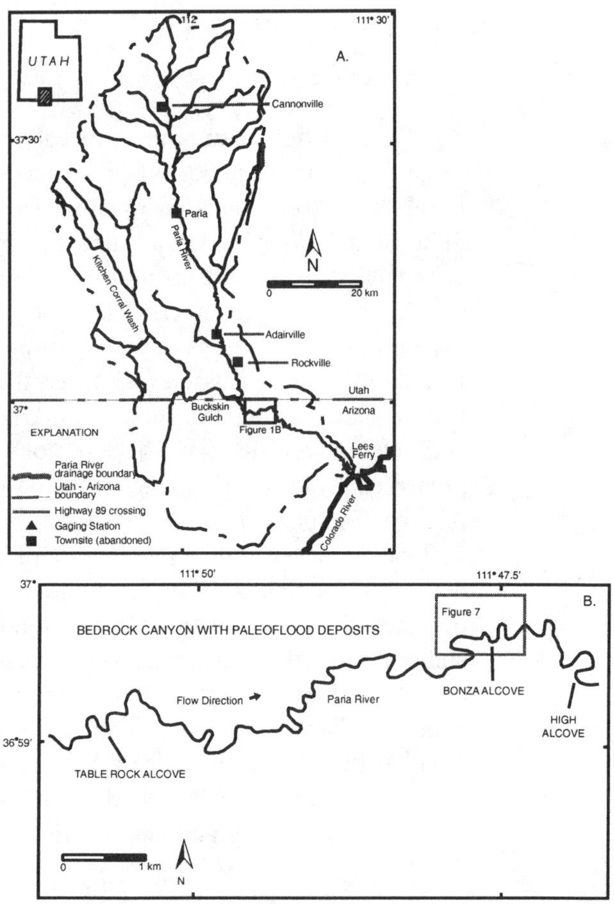

Figure 1. A. Map of the Paria River basin in southern Utah and northern Arizona showing the reach containing paleoflood study sites. B. Map of part of the bedrock canyon of the Paria River in northern Arizona showing the locations of paleoflood study sites.

provides accurate dates of the flood (minimal lag time within the error of radiocarbon analysis). Flow in the channel is assumed to be reasonably described using one-dimensional modeling using Manning's equation [see *Webb and Jarrett*, this volume]. The height of flood deposits are considered a minimum estimate of the water-surface elevation of the flood, and a flood deposit that overlies another unit is generally assumed to be deposited by a higher discharge; this is commonly assumed by is not always assured [*House et al.*, this volume]. Finally, all floods in the paleoflood record that are combined in flood-frequency analysis are assumed to occur in a climate similar to the modern hydrologic regime.

SETTING

Geologic and Hydrologic Setting

The Paria River drains 3,670 km^2 of Mesozoic sedimentary rocks in southern Utah and northern Arizona. Of particular note, much of the headwaters area of the Paria River consists of Tropic Shale; this unit is at least partially responsible for the high sediment concentrations that characterize storm runoff in the basin [*Beverage and Culbertson*, 1964; *Graf et al.*, 1991]. Upper Cretaceous rocks in the headwaters contain small coal seams [*Gregory and Moore*, 1931]. Downstream from the Arizona-Utah border, the river flows through a narrow bedrock canyon that is primarily in Navajo Sandstone (Fig. 1b).

The stream gage for the Paria River at Lees Ferry, Arizona (station 09382000) was established in 1923 (Fig. 2). This gage has recorded a mean annual discharge of 128 m^3/s with a maximum-recorded peak discharge of 456 m^3/s in 1925. Large floods typically occur during the late summer or early fall (Fig. 2), sometimes as a result of dissipating tropical cyclones [*Graf et al.*, 1991]. According to the historical record, floods larger than those recorded in the gaging record may have occurred in 1883 and 1909. Although *Topping* [1997] challenges the magnitudes of flood peaks for the Paria River at Lees Ferry, we use the flood peaks reported by the U.S. Geological Survey [*Pope et al.*, 1998]. The magnitude of floods in the gaging record appears to have decreased through the gage record (Fig. 2), probably as a result of decreased frequency of large storms in southern Utah in the 20th century [*Graf et al.*, 1991].

Pregaging History of Flooding in the Paria River

One of the largest floods in the history of southern Utah occurred in January and February, 1862, when the initial settlements near St. George were nearly destroyed after 44 days of precipitation [*Webb*, 1985]. The Colorado River at Topock had a large flood that reputedly reached 11,300 m^3/s during the winter of 1862 [*Anderson and White*, 1979]. Mormon settlers had not reached the Paria River by 1862, so there is no historical record of a flood occurring at this time. However, the Paria River is a potential flood-producing source contributing to the high discharge reported on the Colorado River in 1862, because this winter flood is likely to have originated from the lower-elevation tributaries of the Colorado River in southern Utah and northern Arizona and not from the higher-elevation tributaries in Colorado.

Caucasians first explored the Paria River basin in 1866 when a military contingent in pursuit of hostile Navajos passed through the basin [*Chidester and Bruhn*, 1949]. The Paria River Valley was permanently settled in 1871, and the towns of Cannonville, Paria, Rockhouse, and Adairville were built along the river in the following decade. John D. Lee, who is infamous in the history of southern Utah, established a ranch at the mouth of the Paria River in January 1872 [*Cleland and Brooks*, 1955]. The first written reports on the basin were made in 1872 by members of the Powell Expedition, who crossed through the basin en route to the mouth of the Dirty Devil River and described grass-filled valleys with shallowly entrenched channels [*Webb*, 1985].

Lee and other settlers had problems keeping irrigation dams across the Paria River because periodic floods washed the impoundments out. Rockhouse and Adairville were mostly abandoned by 1878 due to floods [*Gregory*, 1945] that in retrospect were relatively small. The next significant flood occurred on March 13, 1884, when a large rain-on-snow flood devastated towns, farmland, and irrigation dams in the basin following an unusually severe winter. A

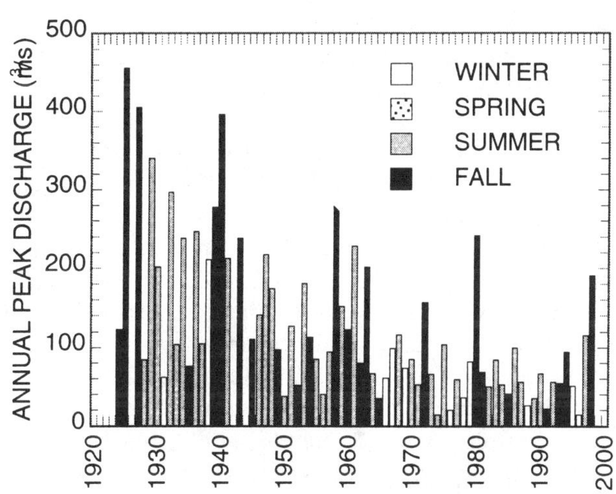

Figure 2. Annual flood series for the Paria River near Lee's Ferry, Arizona (09382000).

headcut in the channel at Rockhouse moved upstream, causing final abandonment of Rockhouse and Adairville [*Chidester and Bruhn*, 1949]. The townsite of Paria was damaged to the extent that half of its population left [*Gregory and Moore*, 1931], and the town was mostly abandoned by 1885 [*Gregory*, 1945]. The Paria River was transformed into a wide, braided stream with few alluvial terraces [*Gregory and Moore*, 1931]. Erosion brought the headcut upstream to Cannonville by 1886, where one third of the agricultural land was lost by 1890 [*Gregory*, 1945; *Chidester and Bruhn*, 1949].

A massive regional storm caused the largest historical flood on the Paria River in September 1909 [*Webb*, 1985]. The storm began in late August when a mid-latitude low-pressure system moving west to east stalled over south-central Utah [*Thiessen*, 1909]. Bridges were washed out in the upper Paria River basin, and channels downcut in the Escalante River drainage east of the Paria River basin [*Webb and Baker*, 1987]. The townsite of Paria, severely damaged by the 1909 flood, was finally abandoned in 1912 [*Carr*, 1972]. The 1909 flood is thought to be the largest historical flood on the Paria River [*Topping*, 1997] but could have been exceeded by floods in either 1862 or 1884.

Previous Studies of the Geomorphology of the Paria River

Because of its high sediment yield and primary source of fine-grained sediments to the upper reaches of Grand Canyon, the sediment transport and geomorphology of the Paria River has been extensively studied. Suspended-sediment concentrations were sampled daily from October 1947 to September 1976 [*Garrett et al.*, 1985]. For one sample collected in 1952, the average daily concentration of suspended sediment was 411,000 ppm, prompting *Beverage and Culbertson* [1964] to coin the term "hyperconcentrated flow" for flows with such high sediment loads. Sediment concentrations are highest during summer and fall floods [*Graf et al.*, 1991].

Hereford [1986] and *Graf et al.* [1991] report the development after 1940 of floodplain deposits adjacent to the low-flow channel of the Paria River. *Hereford* [1986] reports that channels deepened and widened in response to large floods between 1883 and 1890, and that development of floodplains was related to lower flood frequency. Despite the compelling physical evidence presented by *Hereford* [1986] and *Graf et al.* [1991], and despite a well-documented avulsion that moved the mouth of the Paria River downstream in 1912, *Topping* [1997] concluded that channel geometry near Lees Ferry had changed minimally between 1872 and 1996.

STRATIGRAPHY OF FLOOD DEPOSITS

Paleoflood Evidence

Paleoflood evidence is common in the bedrock canyon of the Paria River, particularly in a reach about 15-30 km downstream from the Utah-Arizona border (Fig. 1). In this reach, considerable SWD/PSI evidence is preserved under overhanging walls in the Navajo Sandstone. These overhanging walls are known as alcoves, and all well-preserved flood deposits along the Paria River occur in alcoves. In particular, massive slackwater deposits are present at Table Rock and Bonza Alcoves (Fig. 1b). In both of these cases, the alcoves are just upstream of a left-turning bend in the river channel, and both likely formed backwater (eddy) environments as a result of the combination of a wide channel and a downstream obstruction.

Silt lines are pervasive in the Paria River Canyon and can be visually correlated for long distances. Silt lines form and are preserved because the porous walls of Navajo Sandstone readily absorb water during floods, leaving behind a residue of the suspended sediment that in some cases contained datable organic material. This form of PSI evidence provides the best elevations for water-surface elevations for paleofloods on the Paria River, and correlation of silt lines and flood deposits was achieved with a combination of direct connection, correlation on the basis of particle size, color, and ^{14}C dating. In some cases, silt lines are directly associated with driftwood piles.

In certain alcoves high above the channel of the Paria River, finely bedded sedimentary deposits are inset around colluvium. In High Alcove (Fig. 1b), a deposit more than 20 m above the modern bed of the Paria River contained sedimentary structures (fine, laminated silts and clays) that indicate that they were not significantly above the active channel when they were deposited. These floodplain deposits suggest that the bed of the Paria River was higher at the time of deposition, either because bedrock downcutting to the modern level had not occurred yet or because a rockfall had blocked the channel downstream from the site.

Stratigraphy of Paleoflood Unit

Paleoflood units were described in detail at Bonza Alcove (Fig. 1b), where a large paleoflood deposit about 50 m long and 13 m high is preserved in an alcove on river right just before a left hand bend in the river (Fig. 3). The deposit consists of several levels corresponding to inset units of various ages (Figs. 4a, 4b). The sequence begins with a massive basal sand unit, which stratigraphically is the oldest

Figure 3. Photograph of the Paria River looking downstream from Bonza Alcove from the top of Unit 13. Near-vertical walls of Navajo Sandstone control channel width in the perennial channel. Riparian vegetation is mostly non-native and is not considered in assignment of roughness coefficients.

unit present. However, no organic material was present in this unit to allow radiocarbon dating.

Overlying the basal sand unit is a massive clay unit that is draped over the topography in the vicinity of Bonza Alcove. The clay unit is less than a meter thick but has been draped over more than 1 m of vertical range in Bonza Alcove, and thus has been interpreted to have been deposited during ponding due to a downstream obstruction. The probable location of this obstruction is downstream about 800 m, where the remnants of a rockfall remain in the channel of the Paria River. This rockfall likely dammed the Paria River for a short period of time until a flood breached it. The presence of this rockfall is highly significant because it forms the basis for the assumption of channel stability as well as providing a reason for the start of paleoflood deposition at Bonza Alcove.

Above the clay unit, six paleoflood units are preserved (Fig. 4b). Most of these units are faintly laminated or massive sands; particle-size distributions are shown in Figure 5 and sediment colors are given in Table 3. Most of the units are composed of fine to medium sand with less than 15% silt and clay (Fig. 5), and are well sorted. The lowermost flood units (Units 2.5-4) cannot be traced in three dimensions owing to the size of the deposit, but trenching through Unit 1 indicates that it is inset into Unit 2, which was a larger flood.

Three other units appear to be inset against the high sequence (Fig. 4b), although the appearance is deceptive. In the second highest sequence, Unit 9 overlies Unit 1, which then overlies Units 11 and 12. Although Units 9, 11, and 12 are inset against the higher and presumably older units, Unit 1 drapes over this layer, indicating the origin of this part of the sequence is complex. The stratigraphy and particle-size distribution of Unit 12 suggests that it was deposited by a hyperconcentrated flow, because gravel was transported in suspension and was deposited in a matrix of poorly sorted sediments (Fig. 5).

Inset against this sequence is a much lower sequence comprised of Units 13 and 14 (Fig. 4b), which forms a low and broad bench in Bonza Alcove. At the outer edge of this bench, another slightly lower level appears to be inset against Units 13 and 14 but instead consists of a still younger flood unit (Unit 15) on top of Units 13 and 14 (Fig. 4b).

The complex stratigraphy at Bonza Alcove indicates that careful stratigraphy is required to unravel the true relations between the discharges of successive floods. Three of the four stratigraphic columns consist of a stratigraphically younger flood unit overlying a higher discharge unit. This indicates that paleoflood studies that assume that each higher flood unit represents a higher discharge without careful stratigraphic analysis may be in error [see *House et al.*, this volume].

Dating of Paleoflood Units

Radiocarbon dating techniques provide an absolute date on organic material such as charcoal, fine-grained organics, and driftwood entrained in the flood units. All of the dates used in this study were standard liquid-scintillation analyses performed at the University of Arizona. Analyses of ^{137}Cs activities in 20th century units were performed by the U.S. Geological Survey laboratories in Menlo Park, California. We also used ring counting on non-native tamarisk trees to identify 1980 as the likely year of one flood. Despite intensive dating, we have no method for discriminating the dates of historical floods that occurred before 1952, and the dates of these floods must be inferred from other evidence, such as the historical record.

Association of radiocarbon dates with floods. Radiocarbon ages of organic material in a flood deposit may differ from the actual age of the slackwater deposit depending on the residence time of the organic material in the environment prior to deposition in a flood deposit. Erroneously old radiocarbon dates can occur when "dead" carbon, such as coal, is mixed with modern carbon. Young radiocarbon dates can occur when organic material is deposited after the flood, either by humans, insects, or plants. Seasonally pro-

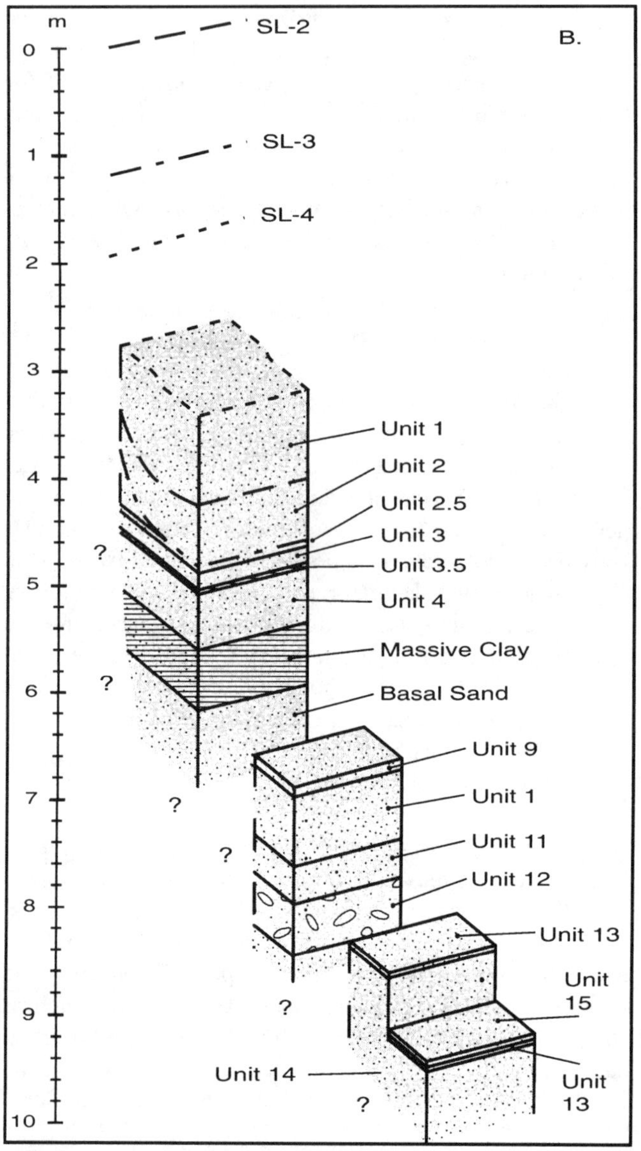

duced fine-grained organics such as leaves, twigs, and seeds generally provide the best material to estimate the age of a unit, as they readily decompose.

To evaluate the association of organic material with the age of paleofloods, we sampled two driftwood piles associated with flood units and separated various types of organic material for separate ^{14}C analysis. Table 2 shows the association between age of organic material from a flood unit and material in a driftwood pile. The data from the driftwood pile (Unit 15, Silt Line 7) show that the type of organic material dated is extremely important. Charcoal, which may be confused with coal, has a long residence time in the environment; in the case of Unit 15, the charcoal may pre-date the flood by as much as 300 years. Similarly, for Unit 1, we obtained ^{14}C dates that the flood could have occurred between AD 1472 and 1955; however, using the results from Unit 15, we accept a radiocarbon date of pre-AD 1955, indicating that the flood was historic and probably occurred in the 19th century.

Coal contamination. The clay unit and Units 4 and 3 at Bonza Alcove yielded organic material that resembled transported charcoal but was thought to be coal. These samples were submitted for ^{14}C analysis; the resulting radiocarbon dates indicate that the material was either pure coal or some combination with relatively modern carbon. One date was infinite age (>43,000 yrs BP), suggesting pure coal, and two radiocarbon dates are older than 30,000 yrs BP, suggesting mixing of dead and modern carbon (Table 1). The lack of accounting of coal contamination may be the reason why some floodplain deposits in the Escalante River basin lacked soil development yet were allegedly greater than 30,000 years old [*Boison*, 1983].

^{137}Cs dating. Nuclear testing between 1945 and 1963 introduced the non-natural radioactive isotope ^{137}Cs and increased the natural production of radioactive ^{14}C isotope [see references in Ely et al., 1992]. ^{137}Cs is strongly adsorbed to clay particles and transported with the suspended load and therefore can be used to as a reliable dating technique with direct association to floods. The date of deposition controls the absence or presence of ^{137}Cs in the Paria

Figure 4. Stratigraphy of flood deposits in Bonza Alcove. A. Photograph of the main flood deposit in Bonza Alcove. The photographer is standing on Unit 13; the highest flood layer is Unit 1, and the bench that the backpack is resting on is capped with Unit 9. B. Stratigraphy of major units of the large flood deposit in Bonza Alcove. Scale is above arbitrary datum in meters above the modern water surface.

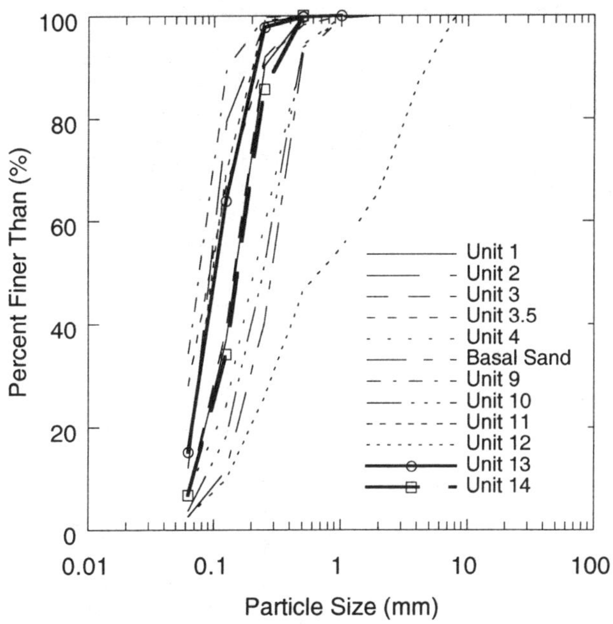

Figure 5. Particle size distributions for flood deposits at Bonza Alcove.

River flood deposits rather than the particle size, leaching, or direct fallout. ^{137}Cs analyses for flood deposits at Bonza Alcove on the Paria River, previously published in Ely et al. [1992], were used to constrain whether floods occurred before or after 1952. Because ^{137}Cs activities were not detectable in deposits known to be prehistoric, Ely et al. [1992] established that direct fallout of ^{137}Cs did not occur in alcoves such as Bonza Alcove. This indicates that any ^{137}Cs detected in those flood deposits was transported during the flood.

Chronology of Paleoflood Deposits

At High Alcove (Fig. 1b), sediment deposits were sampled in an alcove 24 m above the bed of the modern Paria River. The sediments appeared to be overbank floodplain deposits, suggesting that the bed of the Paria River was at least 20 m higher at the time of deposition. Radiocarbon dates of 20,110±240 and 23,660±430 yrs BP were obtained for fine-grained organic materials in these units (Table 1). The significance of these ages is that either the bedrock canyon of the Paria River has downcut at least 20 m in 20 kyr (1 m/kyr), or that some obstruction downstream (a rockfall?) caused temporary ponding to this height above the modern river.

At Table Rock Alcove (Figure 1b), the highest silt line (Silt Line 1; see section 3.4) corresponded to a flood unit in the back of the alcove. At the top of this flood unit, which is the only readily identifiable unit associated with the highest silt line, flood deposition had killed some herbaceous plants and driftwood was embedded in the sediments; one piece of wood was lying on the surface of the unit. The dead plants and embedded driftwood yielded dendro-corrected dates of BC 10,400 and 8,600, respectively (Table 1); the young age of the piece of wood (AD 300) suggests that it was moved to the site by humans. The best date for this flood is BC 10,400, owing to the types of material dated, and this early Holocene flood is the oldest flood yet documented on the Colorado Plateau.

At Bonza Alcove, we obtained 28 radiocarbon dates on 11 fluvial units (Tables 1 and 2). The chronology begins at BC 2890 with a rockfall and subsequent deposition of a massive clay unit that drapes the channel margin. The first flood unit with a radiocarbon date – BC 2850 or 2650 (Table 1) – is Unit 3.5, indicating that Unit 4, the oldest flood, occurred between BC 2890 and 2650. Unit 2.5 has a radiocarbon date of AD 1278; Unit 3 is undated and is bracketed by Units 3.5 and 2.5. Unit 2 is well dated using three ^{14}C dates between AD 1418 and AD 1437 (Table 1). As previously discussed, Unit 1 is pre-1955 and likely historic. Previously published ^{137}Cs analyses [*Ely et al.*, 1992] indicate none of this human-produced radioisotope is in Unit 1, which supports a pre-1955 age. The flood that deposited Unit 1 could have occurred in 1862, 1884, or 1909, years that are known to have large floods on the Paria River.

The second highest sequence of flood units (Units 9-12) are relatively poorly dated. Unit 9 is stratigraphically younger than Unit 1 and probably was deposited historically; no ^{137}Cs activity was detected in this unit [*Ely et al.*, 1992]. Units 11 and 12 underlie Unit 1 and probably are prehistoric. All of the lower flood units (Units 13-15; Fig. 4b) are younger than 1955. A post-bomb ^{14}C date of AD 1958 from the top of Unit 13, combined with detection of ^{137}Cs activity in the sediments [*Ely et al.*, 1992], indicates that Unit 13 was deposited by the 1958 flood. As described in *Ely et al.* [1992], Unit 15 had tamarisk trees that dated to germination in 1980 growing from the surface, post-bomb ^{14}C dates that indicate a 1980 deposition, and detection of ^{137}Cs activity in the sediments. Unit 14 is undated but is stratigraphically older than Unit 13 and probably is a historical flood.

Correlation of Silt Lines and Paleoflood Units

In addition to flood deposits, silt lines were preserved in Bonza Alcove, both across flood units and on sandstone walls (Fig. 6a). Preservation of the silt lines possibly is a result of the combination of porous walls of Navajo Sandstone and high sediment concentrations in floods. Correlation between the silt lines and corresponding units was

Table 1. Radiocarbon dates on paleofloods or related features in the Paria River, northern Arizona.

Alcove	Unit	Laboratory sample number	Material dated	Radiocarbon date (yrs BP ±2 σ)	Calendar[a] date (AD or BC)	Significance of date
High	na	5710	Leaves	20,110±240	na	High floodplain deposits
High	na	5711	Charcoal	23,660±430	na	High floodplain deposits
Table Rock	Silt Line 1	5146	Dead plants	10,440±35	BC 10,400	Highest silt line
Table Rock	Silt Line 1	5748	Driftwood	9,500±120	BC 8600	Highest silt line
Table Rock	Silt Line 1	5147	Wood	1760±30[b]	AD 300	Aboriginal wood?
Bonza	basal clay	5154	Plant fragments	4,280±80	BC 2890	Start of record
Bonza	basal clay	5155	Coal+charcoal	>43,000[b]	na	Coal contamination
Bonza	Unit 3.5	5157	Charcoal+wood	4,130±200	BC 2850, 2650	Brackets 2 paleofloods
Bonza	Unit 2.5	5150	Wood	770±140	AD 1278	Direct date on deposit
Bonza	Unit 2	5153	Plant fragments	520±60	AD 1421	Direct date on deposit
Bonza	Unit 2	5560	Plant fragments	525±80	AD 1418	Direct date on deposit
Bonza	Unit 2	5563	Plant fragments	475±60	AD 1437	Direct date on deposit
Bonza	Unit 1	5811	Conifer needles	0	pre-AD 1950	Highest flood deposit
Bonza	Unit 4	5158	Coal+charcoal	30,430±1990[b]	na	Coal contamination
Bonza	Unit 3	5156	Coal+charcoal	33,350±1580[b]	na	Coal contamination
Bonza	Unit 2.5	5149	Plant fragments	140±40[b]	AD 1678-1955	Too young stratigraphically
Bonza	Unit 1	5559	Plant fragments	180±55[c]	AD 1661-1954	Probable 1925 flood
Bonza	Unit 14	5813	Plant fragments	106.3±0.6 PMC	AD 1958	Probable 1958 flood
Bonza	Unit 13	5564	Plant fragments	129.5±0.9 PMC	AD 1980	1980 flood spillover?
Bonza	Unit 15	5818	Leaves	129.1±0.7 PMC	AD 1980	1980 flood

na, not applicable. PMC, percent modern carbon (ultramodern ^{14}C date). All radiocarbon dates are by the University of Arizona.

[a] Calendar dates were calculated using *Stuiver and Becker* [1993]. Date does not reflect 2σ uncertainty.

[b] Date rejected for reasons given in right column.

[c] This radiocarbon sample was collected from the remnant of Unit 1 that underlies Unit 9.

made on the basis of color, which generally appeared to be distinctive (Table 3), as well as physical connections to the units. Figure 6b shows the relation between the stages of the major flood units and the silt lines.

The silt lines show that flood units on the Paria River generally are less than 1 m below the water-surface elevation (Table 3). For example, Silt Line 4 is 0.45 to 0.70 m above Unit 1, Silt Line 7 is 0.38 to 0.56 m above Unit 15, and Silt Line 8, emplaced by a small flood (probably in 1986), was 0.25 m above a small unit. The elevation differential is fairly consistent, which indicates that the percentage of error is highest at low discharges and relatively low at the highest discharges. The impacts of this differential on discharge estimation will be discussed in the next section.

RECONSTRUCTION OF PALEOFLOOD DISCHARGES

Discharges were estimated using step-backwater analysis [*Webb and Jarrett*, this volume] incorporated in the U.S. Army Corp of Engineers computer program HEC-RAS version 2.0 [*U.S. Army Corps of Engineers*, 1998]. Step-backwater analysis is a one-dimensional fixed-bed hydraulic model that uses Manning's n and the Manning equation to describe energy losses due to roughness. We used modern channel geometry and roughness conditions as inputs to this model, and our analysis assumes that the present surveyed channel is adequate for hydraulic analysis for paleofloods over the previous hundreds to thousands of years. A total of 14 stages of Silt Line 4 (Unit 1) were available along the length of the reach (Table 4), and we used the flood represented by Unit 1 and Silt Line 4 to verify that our hydraulic model produced reasonably consistent results within the model reach.

Channel Geometry and Roughness

Surveys of the channel geometry and paleostage indicators were conducted at Bonza Alcove in 1989, 1990, and

Table 2. Radiocarbon dates on organic fractions segregated from a driftwood pile deposited during the 1980 flood (Unit 15, Silt Line 7) [from *Ely et al.*, 1992] and a driftwood pile on Unit 1 on the Paria River, northern Arizona.

Type of organic material	Radiocarbon date (PMC)	Calendar date[a]
	UNIT 15	
Pine cones	127.5±0.9	AD 1980
Leaves	129.1±0.7	AD 1980
Small twigs	130.5±0.6	AD 1980
Pine needles	132.0±1.0	AD 1979
Juniper needles	132.0±1.0	AD 1973
Juniper berries	146.8±0.8	AD 1970
Mixture	117.4±0.7	AD 1959-1961
Driftwood	110.2±0.8	AD 1958
Charcoal	130±55 yrs BP	AD 1669-1955
	UNIT 1	
Juniper and pine needles	0	pre-AD 1955
Unsorted fine organics	160±50	AD 1669-1955
Unsorted debris	240±50	AD 1644-1955
Twigs and driftwood	270±130	AD 1472-1955

[a]for methods used to determine calendar dates from ultramodern ^{14}C analyses, see *Ely et al.* [1992].

1998 (Fig. 7). An arbitrary elevation was set on river left opposite Bonza Alcove because absolute elevation could not be determined in the deep canyon of the Paria River. All channel cross sections are referenced to this arbitrary elevation, and known high-water marks are referred to as stages to correspond directly to model output. The downstream end of the reach that was used for hydraulic modeling is at the remains of a prehistoric rockfall (Fig. 7), which serves as a hydraulic control for critical depth. We assume that this rockfall occurred some time around 2890 BC (Table 1), was then breached by a flood before 2650 BC on the basis of the Flood Unit 3.5 radiocarbon date (Table 1), and then remained stable over the period of analysis because no contradictory information is available. An analysis of the influence of channel geometry uncertainty would be extremely complex, and thus was not attempted for this study. This surveyed and modeled reach continues 1.33 km upstream past Bonza Alcove.

Thirty-six cross sections were measured with an average spacing between sections of 42 m. In places where channel irregularities create non-uniform flow conditions, such as at constrictions, expansions, and bends, cross sections were more closely spaced, and the spacing increased through straight sections of channel (Fig. 7). The width of surveyed cross sections ranged from 1-2 times the distance between cross sections, suggesting that one-dimensional analysis may not be appropriate for modeling paleoflood hydraulics in this reach.

Manning-*n* values were assigned using average bed particle size and checked using photographic comparisons with reaches of known Manning-*n* values [*Barnes*, 1967; see discussion in *Webb and Jarrett*, this volume]. Uncertainties in Manning-*n* values and energy-loss coefficients generally have much less impact than uncertainties in cross-sectional data on discharge values estimated from hydraulic modeling [*O'Connor and Webb*, 1988]. Riparian vegetation was disregarded in the hydraulic analysis, because the vegetation is mostly non-native tamarisk that would not have been present at the time of the early paleofloods [see *Webb et al.*,

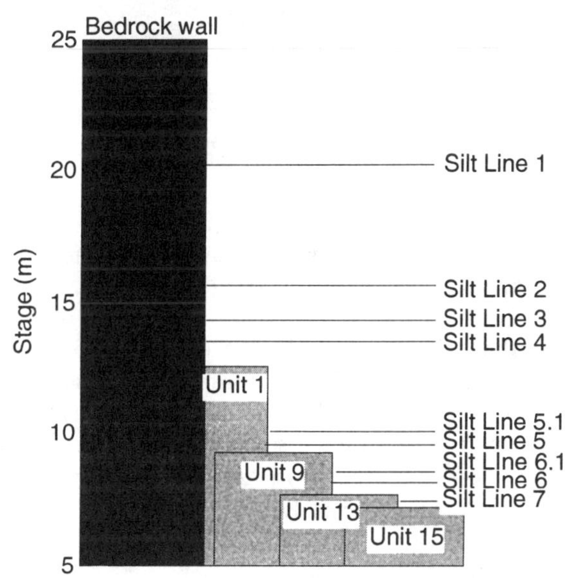

Figure 6. A. Photograph showing silt lines above flood deposits in Bonza Alcove. Silt lines were used for high-water marks, instead of paleoflood deposits, in discharge reconstruc-tions. The prominent silt line at left is Silt Line 4. B. Diagram showing the relation between stages of silt lines and the major deposits shown in Figure 4B.

Table 3. Correlations of silt lines and flood deposits at Bonza Alcove.

Unit	Silt line	Height of silt line above deposit (m)	Deposit[a] and silt line color	Deposit[b] textural class	Date of flood
n.d.	1	n.d.	n.m.	n.m.	10,400 BC
1	4	0.45-0.70	10YR 6/4	ms	Pre-historic
2	2	0.20-0.60	10YR 6/4	vfs	AD 1420
2.5	3	n.m.	n.m.	sivfs	AD 1278
3	n.s.l.	n.s.l.	10YR 6.5/3	fs	Brack-eted
3.5	n.s.l.	n.s.l.	10YR 6/4	sivfs	2650-2850 BC
4	n.s.l.	n.s.l.	10YR 6/3.5	ms	Brack-eted
9	5	0.60	7YR 5.9/4	vfs	AD 1909
n.d.	5.1	n.d.	10YR 6.3/4	ms	Un-known
11	n.s.l.	n.s.l.	10YR 6/3	vfs	Pre-historic
12	n.s.l.	n.s.l.	10YR 5.5/3	gsis	Pre-historic
13	6	0.38-0.50	9YR 6.5/3	sifs	AD 1958
14	6.1	0.50	8YR 6.3/3	ms	Historic
15	7	0.35-0.40	7.5 YR 5.5/4	vfs	AD 1980
n.d.	8	0.25	n.m.	n.m.	AD 1986?

n.d., no deposit; n.s.l., no silt line, n.m., not measured

[a]Colors are given in Munsell units.

[b]ms, medium sand; fs, fine sand, vfs, very fine sand; sivfs, silty very fine sand; gsis, gravelly silty sand (hyperconcentrated flow deposit). See Figure 5 for particle-size distributions.

1991]. Although it is possible that the post 1950 floods could have been affected by tamarisks, we did not adjust the roughness for these floods because they were relatively low stage events and thus only a small portion of the channel could have been influenced. Table 4 lists selected channel data used in the hydraulic analysis.

Hydraulic Modeling

The step-backwater model produced large drops, each on the order of 1 m, in the water-surface profile through bends, particularly upstream from cross section 27 (Table 4). This probably results from application of one-dimensional hydraulics to three-dimensional flow through a tightly meandering bedrock canyon (Fig. 7), which results in an inadequate characterization of energy losses in tight bends. Therefore, we considered the hydraulic analysis upstream from cross section 27 to be invalid and discarded those water-surface profiles from consideration in Figure 8. In addition, smaller drops occurred in cross sections 8, 10, 12, and 17, leading to instability in the calculation of water-surface profiles; because these cross sections were in close proximity to others nearby at a longitudinal spacing of less than the cross section width, we eliminated these cross sections from our model.

Table 4. Channel roughness and geometry data used in one-dimensional hydraulic analysis of paleofloods in the Paria River at Bonza Alcove. Locations of cross sections are shown on Figure 7. Sections 19-22 are at Bonza Alcove. Cross sections 8, 10, 12, and 17 were deleted for the final hydraulic analysis (see text for details).

Cross sect-ion	Distance upstream from previous section (m)	Channel center Manning's *n* value	Stage of Silt Line 4 (m)	Stage at 1,200 m^3/s (m)	Energy-grade line at 1,200 m^3/s (m)
1	0	0.077	n.m.	8.5	11.3
2	64	0.045	11.2	11.2	11.8
3	92	0.033	n.m.	11.2	12.2
4	47	0.035	12.0	11.7	12.3
5	50	0.033	n.m.	11.7	12.5
6	33	0.034	n.m.	11.9	12.5
7	23	0.036	n.m.	12.2	12.6
9	30	0.034	n.m.	12.3	12.6
11	32	0.034	n.m.	12.3	12.7
13	49	0.033	n.m.	12.3	12.8
14	41	0.035	n.m.	12.3	12.9
15	80	0.033	n.m.	12.8	12.9
16	53	0.034	n.m.	12.8	12.9
18	31	0.033	n.m.	12.8	13.0
19	57	0.032	n.m.	12.2	13.6
20	42	0.032	13.2	12.8	13.7
21	20	0.030	13.5	12.9	13.7
22	62	0.031	13.7	13.5	13.7
23	27	0.031	13.3	13.5	13.8
24	18	0.031	n.m.	13.5	13.8
25	47	0.033	13.6	13.3	14.1
26	35	0.037	n.m.	13.1	14.5
27	35	0.032	n.m.	14.0	14.6
28	39	0.032	13.6	14.2	14.6
29	31	0.040	13.7	14.2	14.5
30	37	0.033	n.m.	13.8	15.2
31	34	0.031	n.m.	14.9	15.3
32	29	0.036	13.6	15.0	15.3
33	59	0.032	14.0	15.0	15.4
34	42	0.033	13.9	15.0	15.4
35	38	0.035	14.2	15.1	15.5
36	56	0.036	14.5	15.2	15.5

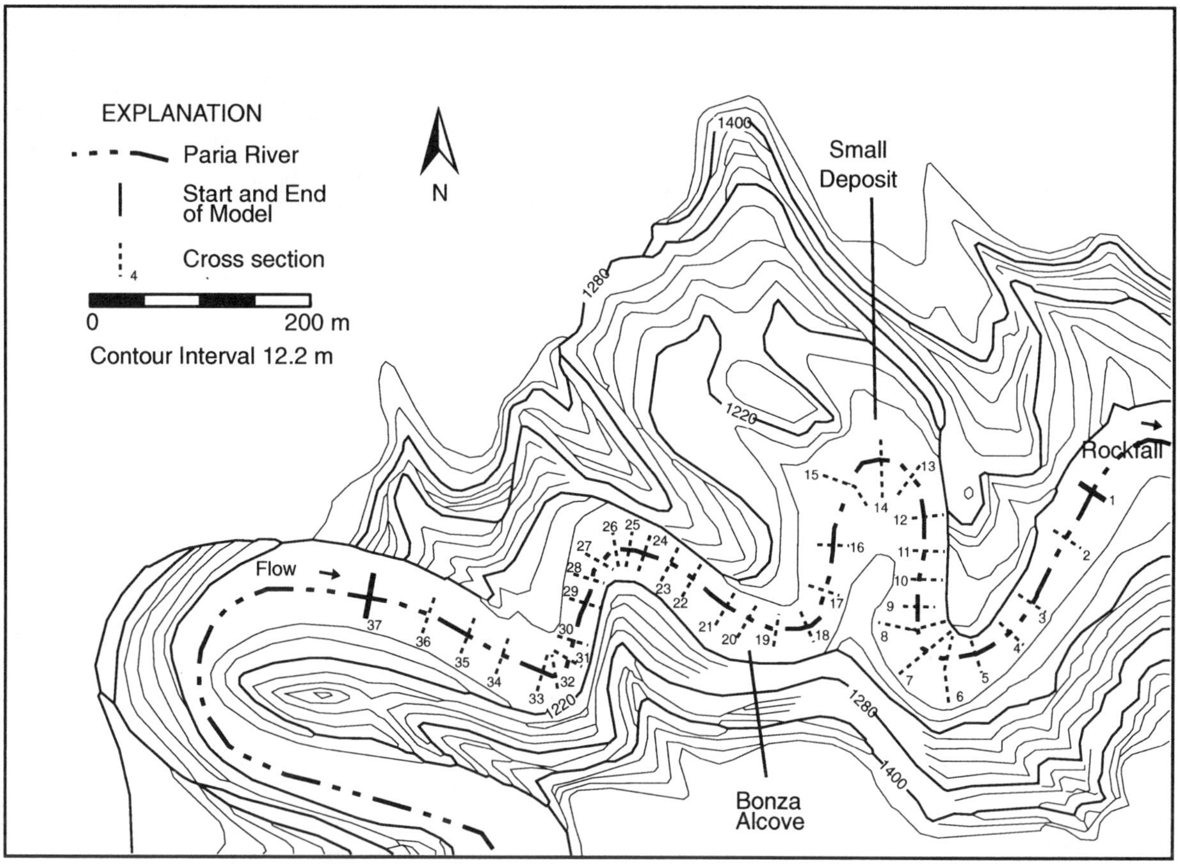

Figure 7. Map of the reach of the Paria River modeled using one-dimensional step-backwater analysis.

The step-backwater results indicate that a discharge of approximately 1,200 m^3/s best matches the known high-water evidence presented by Silt Line 4 downstream from cross section 27 (Fig. 8). At this discharge, all stages for Silt Line 4 are constrained between the calculated water-surface profile and the energy-grade line (Table 4). The range in discharges required to encompass all stages for Silt Line 4 is 1,150-1,300 m^3/s. This range does not represent a true uncertainty for the discharge estimate because it does not take into account potential changes in channel cross sections (although these are believed to have been slight), roughness, or inappropriate use of a one-dimensional model. The discharge that best matches the silt line evidence for the Silt Line 4 (1,200 m^3/s) is 2.6 times larger than the largest flood in the gaging record (Fig. 2). Similarly, 6 elevations for Silt Line 7 (1980 flood) were used to estimate a discharge of 210 m^3/s with a probable range of 200-300 m^3/s (Table 5). The discharge for this flood recorded at the downstream gaging station was 241 m^3/s.

High-water evidence (siltlines) suitable for estimating the discharges associated with flood units is only available at Bonza Alcove. We used the one-dimensional hydraulic model to develop a stage-discharge relation at cross section 18, which traverses the center of Bonza Alcove. Using this stage-discharge relation, the highest discharge (2,400 m^3/s) is for Silt Line 1, although the uncertainty in this discharge estimate cannot be evaluated (Table 5). Discharges for Silt Line 2 (Unit 2) and Silt Line 3 (Unit 2.5) were estimated to be 1,600 and 1,500 m^3/s, respectively (Table 5). Discharge estimates for other paleofloods appear in Table 5.

Topping [1997] attempted to revise the gaging record of the Paria River at Lee's Ferry according to his concepts of channel stability and flow in the vicinity of the gaging station. Comparison of the discharges estimated using one-dimensional modeling at Bonza Alcove, calculated at the gage station at Lee's Ferry (during the gaging record), and calculated according to the techniques presented by *Topping* [1997] appear in Table 5. Of the five floods compared among the three methods, all but one comparison (Silt Line 5.1) show reasonable agreement, and our assignment of a 1926 date to the Silt Line 5.1 flood may be wrong, as the flood may have occurred between 1909 and the start of the

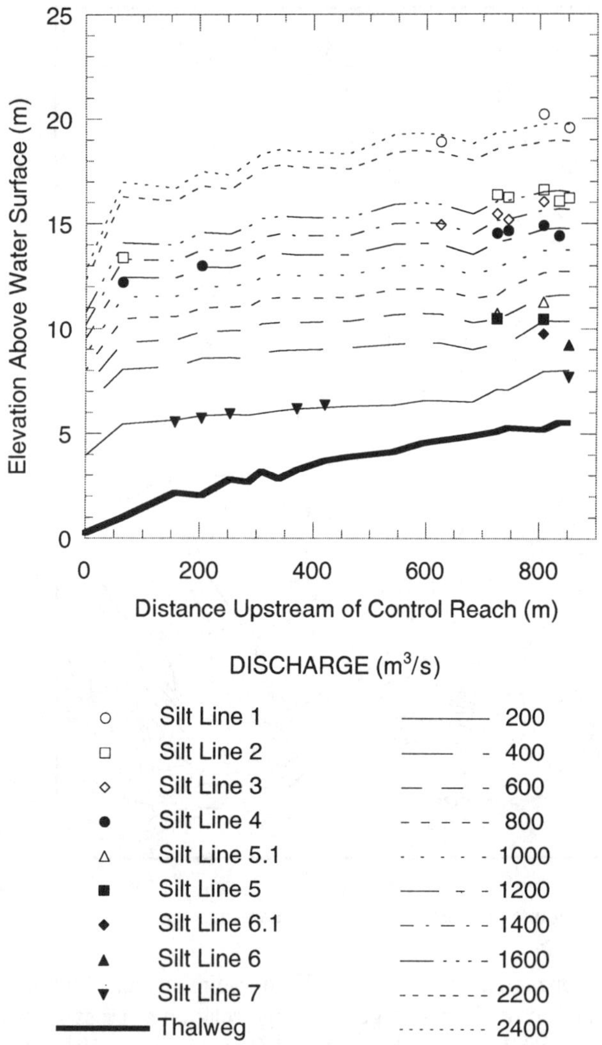

Figure 8. Water-surface profiles for the Paria River in the Bonza Alcove reach, calculated using one-dimensional step-backwater analysis.

gaging record in 1923. Although the gage is several miles below Bonza alcove, there are no major tributaries in this reach of the river, thus the flow at the gage should generally be slightly larger than the flow at Bonza Alcove unless the main rainfall is in the lower part of the drainage basin.

FLOOD-FREQUENCY ANALYSIS

We used the Expected Moments Algorithm (EMA) of *Cohn et al.* [1997] to fit the gaging record and paleoflood data to a log-Pearson type III distribution for flood-frequency analysis. We used Gringorten plotting positions [as suggested by *Hirsch and Stedinger,* 1987] to display the discharges for floods on the Paria River. Plotting positions for the largest floods vary considerable from the frequency curves because of sensitivity of the largest events to the type of plotting position used [*Hirsch and Stedinger*, 1986]. EMA is a method of moments procedure that adjusts the probability distribution estimated using gaging data with the expected moments of paleoflood data. In addition to the paleoflood information given in Table 5, we added a threshold record length of 4,888 years as determined by the radiocarbon dates on the massive clay unit (Table 1; Fig. 4b). To show the organization of the censored paleoflood data, we prepared a censoring threshold diagram [see *Blainey et al.*, this volume, for a discussion of these diagrams], which appears in Figure 9a.

We calculated four scenarios of flood frequency on the Paria River (Fig. 9b) by assuming the following:

• The flood frequency calculated from the gaging record alone.

• The discharge of 1,200 m^3/s for Unit 1 (Silt Line 4) represents a flood that is the largest discharge since 1870. In this case, we used a one-exceedence model with the exceedence threshold set to the discharge of 1,100 m^3/s represented by the top of Unit 2 (not the true discharge represented by Silt Line 2).

• The discharge of 1,600 m^3/s for Unit 2 (Silt Line 2) is the largest flood to occur after deposition of the massive clay unit with an exceedence threshold as the top of Unit 1.

• After deposition of the massive clay, we recognized four exceedence thresholds represented by the discharges for Units 4, 3.5, 3, and 2.5 (1,500 m^3/s). The three largest floods of the paleoflood record exceed the Unit 2.5 threshold.

The frequency curves for the gaging record alone, the gaging record with the largest historical flood, and the gaging record with the entire paleoflood record essentially overlie one another. The frequency curve based on Unit 2 (Silt Line 2) as the largest flood in 4,888 years causes a significant increase in quantile estimates for recurrence intervals greater than the 0.002 quantile (500-year flood). The largest historical flood plots above the frequency distribution based on that data, which is an artifact of plotting positions. It is important to note that although use of paleoflood data does not significantly change the estimates for extreme floods on the Paria River, the confidence limits for the extreme floods are closer to the frequency distribution, reducing the uncertainty in flood frequency estimates.

DISCUSSION AND CONCLUSIONS

Summary of Paleoflood Hydrology of the Paria River

The paleoflood record of the Paria River is one of the longest in the southwestern United States and provides considerable information for testing of assumptions made in

Table 5. Discharges for paleofloods in the Paria River estimated using one-dimensional hydraulic analysis and SWD/PSI evidence at Bonza Alcove

Unit	Silt line	Date of flood	Discharge (m^3/s)	Range in[a] discharge (m^3/s)	Discharge at[b] gaging station (m^3/s)	Other[c] discharge (m^3/s)
n.d.	1	BC 10,400	2,400	n.e.	n.m.	n.m.
1	4	prehistoric	1,200	1,000-1,400	n.m.	n.m.
2	2	AD 1420	1,600	1,400-2,000	n.m.	n.m.
2.5	3	AD 1278	1,500	1,400-1,600	n.m.	n.m.
3	none	bracketed	925	n.e.	n.m.	n.m.
3.5	none	2650-2850 BC	875	n.e.	n.m.	n.m.
4	none	Bracketed	850	n.e.	n.m.	n.m.
9	5	AD 1909	500	400-600	n.m.	460
n.d.	5.1	AD 1926?	590	400-600	456	270
n.d.	6.1	AD 1927, 1929?	380	n.e.	340-405	210-320
13	6	AD 1958	300	n.e.	326	n.m.
15	7	AD 1980	210	200-300	241	240

n.m., not measured; n.e., not estimated; n.d., no deposit

[a]Only represents the range in discharge calculated using one-dimensional step-backwater analysis where two or more high-water marks are available. The actual error in discharge estimate may be much higher.

[b]See Figure 2 and *Pope et al.* [1998].

[c]Reported by *Topping* [1997].

paleoflood studies. Radiocarbon dating of a deposit at High Alcove indicates that the bed of the Paria River was roughly 24 m higher about 20,000 years ago than it is presently. Assuming that the deposition of the Table Rock deposit was not related to a rockfall – a tenuous assumption at best – the overbank floodplain deposition in High Alcove suggests a downcutting rate through bedrock that if time invariant would represent 1 mm/yr. Therefore, at Bonza Alcove in the last 4,888 years of paleoflood record, the bed of the river would have lowered nearly 5 m. However, the rockfall downstream from Bonza Alcove, and deposition of the massive clay, shows that rockfalls do affect river base level in these bedrock canyons and that downcutting rates are not necessarily constant. Future paleoflood studies should attempt to account for these base-level changes that could affect the hydraulic interpretation of flood deposits.

At Bonza Alcove, the stratigraphic section begins with the massive clay, radiocarbon dated at 4.2 ka, that is likely the result of ponding behind a rockfall dam. This massive clay is overlain by 5 flood units that are primarily silty fine sand; six other historical flood units are inset against this primary sequence. Sixteen ^{14}C dates constrain the ages of this deposition (Table 1), and other ^{14}C dates show the problems of association of organic material with the transporting flood, including problems of coal contamination in rivers that drain sedimentary terranes.

In addition to the flood deposit, silt lines were preserved in Bonza Alcove, both across the flood units and on the sandstone walls. Correlation between silt lines and flood units was made on the basis of color, which appeared to be distinctive, and in some cases ^{14}C dating of entrained drift in the silt line. Silt lines ranged from 0.30 to 0.90 m above the corresponding unit. We used the silt lines to estimate paleoflood discharges from a one-dimensional step-backwater analysis, based on 32 surveyed cross sections over a 1.33 km reach. In the case of the most prominent silt line (SL4), which likely represents a historical flood, a total of 14 elevations of silt lines were available to compare with the step-backwater results.

The largest flood in the paleoflood record is 2,400 m^3/s and occurred about 10,000 years ago. The discharge that best matches the silt-line evidence for the highest historical flood is 1,200 m^3/s, or 2.6 times larger than the largest flood in the gaging record. However, there has clearly been channel change in this time period related to the rockfall, thus this discharge estimate is not used in the flood frequency analysis. Comparison of discharges for historical floods recorded at Bonza Alcove, in the gaging record, and in the revision by *Topping* [1997] is favorable, although considerable differences occur between several independent estimates of flood discharges for this river. Our discharge estimates are problematic because of inadequate characterization of energy losses in the tight bends of the Paria River canyon using one-dimensional step-backwater analysis. Our one-dimensional analysis indicated significant stage drops in these bends, which influenced the water-surface elevations upstream. Incorporation of energy losses owing to wall friction is an important facet of discharge reconstruction in narrow bedrock canyons.

Paleoflood deposits on the Paria River record floods of significantly higher discharges than those recorded at its long-term gaging station (1,200-2,400 m^3/s compared with

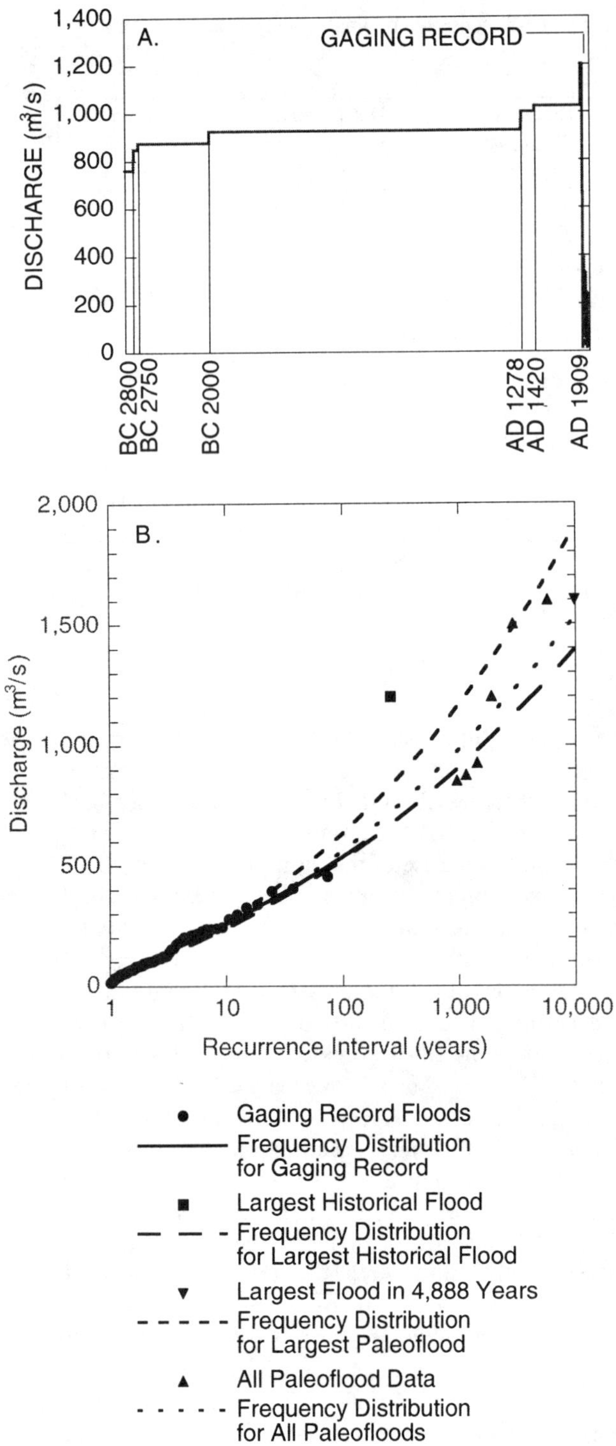

Figure 9. A. Censoring-threshold diagram for the Paria River at Bonza Alcove. B. Flood-frequency analysis for the Paria River at Lees Ferry, Arizona. Paleoflood data collected about 20 km upstream of Lees Ferry, with no significant intervening tributaries, was incorporated into the 75-year gaging record for the Paria River.

456 m^3/s). However, these floods appear to be of reasonable magnitude when compared with the regional envelope curve [*Enzel et al.*, 1993; see *Webb and Jarrett*, this volume], and flood-frequency analysis indicates that the paleoflood data do not significantly change the frequency curve for Paria River floods. Use of paleoflood data does, however, decreases the width of confidence intervals, making estimation of long-recurrence interval floods more certain.

Other Rivers in the Southwestern United States With Unusually Large Floods

The large paleoflood on the Paria River – with a discharge of about 1,200 m^3/s – is about 2.6 times larger than the largest flood in the gaging record at Lee's Ferry (Table 6). However, even the largest flood in 12,000 years at 2,400 m^3/s (Table 6) plots beneath the regional envelope curve for the southwestern United States [*Enzel et al.*, 1993]. The question remains whether any other rivers in the region have also experienced unusually large floods in the gaging and(or) historical record.

As given in Table 6, at least five other rivers in the region have ratios between the first and second largest floods of greater than 2.3. One of the reasons for these large floods is the hydroclimatology of the region. For example, the Paria River has floods that result from isolated summer thunderstorms, winter frontal systems, and tropical cyclones [*Graf et al.*, 1991]. This mixed population of causal mechanisms for floods can lead to unusually large floods in the region that appear to be statistical outliers owing to the short length of gaging records and the infrequency of large, regional storms. In this region, the presence of one unusually large discharge in an annual flood record should not be interpreted as a statistical outlier; instead, more research should be done on mixed populations of floods and the frequency of extreme events.

Significance of the Paria River Record for Other Paleohydrological Studies

The paleoflood record of the Paria River suggests several significant challenges to future work on paleoflood hydrology in bedrock canyons. As illustrated by the relations between silt lines and flood deposits, at least one silt line is present with no significant corresponding flood deposit. This suggests the possibility that floods may overtop existing flood deposits and not leave depositional evidence behind, except in fortuitous circumstances such as the Paria River. Recent work on the San Juan River (Lynn Orchard, University of Arizona, unpublished data, 1999) indicates that the largest flood left only small sedimentary deposits. If lack of depositional evidence can be shown in the general

Table 6. Rivers in Arizona and southern Utah with one unusually large flood in the historical record. Although the gaging records for these rivers are variable length, no attempt has been made to make the records more comparable.

River	Drainage area (km^2)	Date of flood	Discharge (m^3/s)	Second highest discharge (m^3/s)	Ratio of top two floods
Escalante River	832	1932[a]	600	129	4.65
Aravaipa Creek	1,400	1983	2,000	566	3.53
San Pedro River	3,210	1926	2,775	711	3.90
Paria River	3,664	unknown[b]	1,200	456	2.63
San Juan River	59,800	1911[c]	5,000	1,980	2.52
Little Colorado River	67,800	1923	3,400	1,420	2.39

a *Webb et al.* [1988].

b This paper. The flood is either prehistoric or at least before 1909.

c Orchard *et al.*, unpublished data, 1999.

case of slackwater deposits, flood frequency estimates using this type of paleoflood data may be biased too low.

Second, the assumption that the height of flood deposits equals the stage of paleofloods is not necessarily a good one, even in the case of relative magnitude of discharge. In the case of the Paria River, silt lines range from 0.30 to 0.90 m above the heights of flood units, indicating that flood units underestimate the true water-surface profile, even in this highly sediment-charged river. However, owing to undulations in the water-surface profile within a reach, discharge estimates based on the flood units may be within the uncertainty associated with reconstruction of floods using silt lines or drift lines. Moreover, the silt lines indicate that the second highest flood unit (Unit 2) was deposited by a larger flood than the one that deposited the highest flood unit (Unit 1).

Finally, one-dimensional hydraulic models, which are readily available and easily run, are not very appropriate to reconstructing discharges in bedrock canyons such as the Paria River. Multidimensional models, such as that of *Wiele et al.* [1996] and *Denlinger et al.* [this volume], and with the addition of accounting for wall roughness, are much more appropriate than other multidimensional models that are based on Manning's *n* [e.g., *Wohl et al.*, 1999].

Finally, radiocarbon dating of flood units offers a significant challenge because of the lag residence times of organic materials in watersheds and because of the potential for coal contamination in sedimentary terranes. Although in general these uncertainties are not expected to significantly alter flood-frequency analyses (the notable exception of the total record length), these errors are highly important if paleoclimatic reconstructions are attempted from paleoflood data. This is particularly true of historical floods that occurred before 1952, because without dendrochronological or other direct historical evidence, the dates for these floods must be inferred and are essentially guesses.

Acknowledgments. The authors thank many of their colleagues for their help in the field and conversations concerning the Paria River paleoflood sites. Lisa Ely, Ted Melis, Sara Rathburn, Glenn Rink, and Ellen Wohl helped with the field work; in particular, Wohl named Bonza Alcove. We also thank Lisa Ely, Dean Ostenaa, and Diana Anderson for their critical reviews and suggested changes to the manuscript.

REFERENCES

Anderson, T.W., and N.D. White, Statistical summaries of Arizona streamflow data, *U.S. Geol. Surv. Water-Resour. Invest. Rept. 79-5*, 416 pp., 1979.

Baker, V.R., Flood sedimentation in bedrock fluvial systems, *Can. Soc. Petrol. Geol., Mem. 10*, 87-98, 1984.

Baker, V.R., R.C. Kochel, P.C. Patton, and G. Pickup, Palaeohydrologic analysis of Holocene flood slack-water sediments, *Spec. Publs. Int. Ass. Sediment.*, 6, 229-239, 1983.

Beverage, J.P., and J.K. Culbertson, Hyperconcentrations of suspended sediment, *Amer. Soc. Civil Eng., J. Hydraul. Div.*, 90, 117-126, 1964.

Boison, P.J., *Late Pleistocene and Holocene alluvial stratigraphy of three tributaries in the Escalante River basin, Utah*, 118 pp., M.S. Thesis, Univ. Calif., Santa Cruz, 1983.

Boison, P.J., and P.C. Patton, Sediment storage and terrace formation in Coyote Gulch basin, south-central Utah, *Geology*, 13, 31-34, 1985.

Carr, S.L., *The historical guide to Utah ghost towns*, 166 pp., Western Epics, Salt Lake City, 1972.

Chidester, I., and E. Bruhn, *Golden nuggets of pioneer days, a history of Garfield County*, Garfield County News, Panguitch, Utah, 1949.

Cleland, R.G., and J. Brooks, *A Mormon chronicle: the diaries of John D. Lee 1848-1876*, Huntington Library, San Marino, Calif., 1955.

Cohn, T.A., W.L. Lane, and W.G. Baier, An algorithm for computing moments-based flood quantile estimates when historical information is available, *Water Resour. Res.*, 33(9), 2089-2096, 1997.

Ely, L.L, R.H. Webb, and Y. Enzel, Accuracy of post-bomb 137Cs and 14C in dating fluvial deposits, *Quat. Res.*, 38, 196-204, 1992.

Enzel, Yehouda, L.L. Ely, P.K. House, V.R. Baker, and R.H. Webb, Paleoflood evidence for a natural upper bound to flood magnitudes in the Colorado River basin, *Water Resour. Res.*, 29, 2287-2297, 1993.

Garrett, W.B., G.C. Lusby, and R.F. Hadley, Suspended sediment in the Paria River drainage basin at Lees Ferry, Arizona, and Sheep Creek drainage basin near Cannonville, Utah, *U.S. Geol. Surv. Water-Resour. Invest. Rept.*, 30 pp., 1985.

Graf, J.B., R.H. Webb, and R. Hereford, Relation of sediment load and flood-plain formation to climatic variability, Paria River drainage basin, Utah and Arizona, *Geol. Soc. Amer. Bull.*, 103, 1405-1415, 1991.

Graf, W.L., Late Holocene sediment storage in canyons of the Colorado Plateau, *Geol. Soc. Amer. Bull.*, 99, 261-271, 1987.

Gregory, H.E., Population of southern Utah, *Econ. Geogr.*, 21, 29-57, 1945.

Gregory, H.E., and R.C. Moore, The Kaiparowits Region, a geographic and geologic reconnaissance of parts of Utah and Arizona, *U.S. Geol. Surv. Prof. Paper 164*, 161 pp., 1931.

Hereford, R., Modern alluvial history of the Paria River drainage basin, southern Utah, *Quat. Res.*, 25, 293-311, 1986.

Kochel, R.C., and V.R. Baker, Paleoflood analysis using slackwater deposits, pp. 357-376 in *Flood Geomorphology*, ed. V.R. Baker, R.C. Kochel, and P.C. Patton, John Wiley and Sons, New York, 1988.

Kochel, R.C., and V.R. Baker, Paleoflood hydrology, *Science*, 215, 353-361, 1982.

Kochel, R.C., V.R. Baker, and P.C. Patton, Paleohydrology of southwestern Texas, *Water Resour. Res.*, 18, 1165-1183, 1982.

O'Connor, J.E., and R.H. Webb, Hydraulic modeling for paleoflood analysis, pp. 393-402 in *Flood Geomorphology*, ed. V.R. Baker, R.C. Kochel, and P.C. Patton, John Wiley and Sons, New York, 1988.

Patton, P.C., and P.J. Boison, Processes and rates of formation of Holocene alluvial terraces in Harris Wash, Escalante River basin, south-central Utah, *Geol. Soc. Amer. Bull.*, 97, 369-378, 1986.

Pope, G.L., P.D. Rigas, and C.F. Smith, Statistical summaries of streamflow data and characteristics of drainage basins for selected streamflow-gaging stations in Arizona through Water Year 1996, *U.S. Geol. Surv. Water-Resour. Invest. Rept. 98-4225*, 907 pp., 1998.

Stuiver, M., and B. Becker, High-precision calibration of the radiocarbon time scale AD 1950 - 6000 BC, *Radiocarbon*, 35, 35-65, 1993.

Thiessen, A.H., Climatological data for September, 1909, District No. 10, Great Basin, *Monthly Weather Rev.*, 37, 657-658, 1909.

Topping, D.J., *Physics of flow, sediment transport, hydraulic geometry, and channel geomorphic adjustment during flash floods in an ephemeral river, the Paria River, Utah and Arizona*, Ph.D dissertation, Univ. Wash., Seattle, 1997.

U.S. Army Corps of Engineers, *HEC-RAS river analysis system, version 2.2, Hydraulic Reference Manual CPD-69*, U.S. Army Corps of Engineers, Hydrol. Engin. Center, Davis, Calif., variable pp., 1998.

Webb, R.H., *Late Holocene flooding on the Escalante River, south-central Utah*, Ph.D dissertation, Univ. Ariz., Tucson, 1985.

Webb, R.H., and V.R. Baker, Changes in hydrologic conditions related to large floods on the Escalante River, south-central Utah, pp. 306-320 in *Regional Flood-Frequency Analysis*, ed. V. Singh, V., Dordrect, The Netherlands, D. Reidel Publ., 1987.

Webb, R.H., J.E. O'Connor, and V.R. Baker, Paleohydrologic reconstruction of flood frequency on the Escalante River, south-central Utah, pp. 403-418 in *Flood Geomorphology*, ed. V.R. Baker, R.C. Kochel, and P.C. Patton, John Wiley and Sons, New York, 1988.

Webb, R.H., and S.L. Rathburn, Paleoflood hydrologic research in the southwestern United States, *Trans. Res. Board Rec.*, 1201, 9-21, 1989.

Webb, R.H., S.S Smith, and V.A.S. McCord, Historic channel change of Kanab Creek, southern Utah and northern Arizona, 91 pp., *Grand Canyon Natur. Hist. Assoc. Mono. No. 9*, 1991.

Wiele, S.M., J.B. Graf, and J.D. Smith, Sand deposition in the Colorado River in the Grand Canyon from flooding of the Little Colorado River, *Water Resour. Res.*, 32(12), 3579-3596, 1996.

Wohl, E.E., D.M. Thompson, and A.J. Miller, Canyons with undulating walls, *Geol. Soc. Amer. Bull.*, 111(7), 949-959, 1999.

Robert H. Webb, U.S. Geological Survey, 1675 W. Anklam Road, Tucson, AZ 85745; email: rhwebb@usgs.gov; 520-670-6671 ext 238

Joan B. Blainey, U.S. Geological Survey, 6000 J Street Placer Hall, Sacramento, CA 95819; email: jblainey@usgs.gov

David W. Hyndman, Department of Geological Sciences, Michigan State University, East Lansing, MI S 48824; email: hyndman@msu.edu

Paleofloods and the Estimation of Long Term Transmission Losses and Recharge to the Lower Nahal Zin Alluvial Aquifer, Negev Desert, Israel

Noam Greenbaum

Departments of Geography and Natural Resources & Environmental Management, University of Haifa, Haifa, Israel

Uri Schwartz, Asher P. Schick, and Yehouda Enzel[1]

Department of Geography, Hebrew University of Jerusalem, Jerusalem, Israel.

The long term, event-based recharge of the alluvial fan aquifer of Nahal Zin, a 1400 km^2 catchment in the hyperarid Negev Desert, is derived from a combination of: (a) measurements of transmission losses and recharge in an intensively instrumented alluvial reach and alluvial fan of the lower Nahal Zin, (b) a paleoflood study of Nahal Zin that estimates the long-term frequency of the largest floods, and (c) statistically significant relationships between the measured flood parameters—peak discharge, volume, and duration over the last 65 years. These relationships are integrated here with the paleoflood record to estimate long-term transmission losses. Transmission losses were found to be highly correlated (R^2=0.95; significant at 95%) with inflow volume to the reach. The relations between peak discharge and flood volume and flow duration to flood volume were R^2= 0.81 and R^2=0.55 respectively (at 95% significance). Transmission losses are substantially reduced when the time interval between two successive floods is less than one week. The results show that transmission losses and recharge are related to the flood volume by a power decay function. Flood volume increases exponentially with flood peak. The cumulative transmission losses during the combined (1935-1998) measured and historical records were 25 x 10^6 m^3, 52% of which were contributed by medium size floods (30-200 m^3s^{-1}) and the remaining 48% by large floods (>200 m^3s^{-1}).

[1] Also at: Institute of Earth Sciences, Hebrew University of Jerusalem, Jerusalem, Israel.

Ancient Floods, Modern Hazards: Principles and Applications of Paleoflood Hydrology
Water Science and Application Volume 5, pages 311-328

INTRODUCTION

Floods are the main contributors to the recharge of the shallow alluvial aquifers as well as deeper aquifers in the hyperarid Negev Desert [*Schwartz*, 1986; *Schwartz et al*, 1995; *Shentsis et al.*, 1997; 1998; 1999; and *Lane et al*, 1985]. While in temperate climates the main cause for downstream changes in the flood hydrograph is flood attenuation without any volume loss, in desert alluvial channels losses by infiltration into the alluvium are the prime factor of flood hydrograph attenuation [*Dubief*, 1953; *Vanney*, 1960; *Williams*, 1970 *Schick*, 1971: *Buono and Lang*, 1980; *Enzel et al*, 1989; *Enzel*, 1992; *Murphey et al.*, 1972; *Schick*, 1980; *Schwartz*, 1986; *Walters*, 1990; *Renard et al.*, 1993; *Osterkamp*, 1995; and others]. These losses may be so large that floods disappear on their way downstream [*Vanney*, 1960; *Schick*, 1988; and *Enzel*, 1992]. Water losses by infiltration into the alluvial channel during desert floods initiate the recharge process into shallow and deep groundwater [*Abdulrazzak and Morel-Seytoux*, 1983; *Stephens and Knowlton*, 1986; and *Abdulrazzak et al.*, 1987]. Infiltration in the channel, or anywhere else in the arid drainage basin, is a dynamic and complex process. It differs in magnitude, volume, and duration from place to place. It is a function of several factors, the important ones being the hydro-morphological characteristics of the watershed, the hydraulic parameters of the medium between the surface and the water table, and the balance of moisture in the environment. Although this process may control the hydrology of arid streams, there are very little data on surface flows and much less on infiltration in the alluvial channels of the arid environment. Therefore, infiltration and recharge are very difficult to quantify. Even when some data exist, the record is short and incomplete and therefore estimations, especially of long-term recharge, are inaccurate. Combining short-term measurements on the infiltration and recharge with long-term paleoflood data provides a new way of estimating the long-term recharge.

There are number of approaches to estimate transmission losses. All of them are limited by the available data. The first one is calculating the water-balance for a given reach [*Lane* et al. 1971;1983; Lane, 1985; 1990; *Walters,* 1990; *Osterkamp* et al., 1994; and others]. This approach demands high quality flood measurements at the gauging stations, a reach without tributaries that complicate the calculations [e.g., *Shentsis et al.*, 1999], and a sufficiently long flood record to estimate frequency. In alluvial arid streams it is difficult to satisfy all these demands.

Another approach uses an infiltrometer for the simulation of infiltration into valley floors and alluvial channel beds [*Talsma*, 1969; 1970; 1973; *Kuells et al.*, 1995; and others]. The main advantage of this method is that it provides a relatively good estimate of the rate of infiltration because the water input is controlled. The problem lies in the variability of channel characteristics and therefore extrapolation of point measurements to the entire alluvial reach is difficult. A further approach to study infiltration is based on monitoring the profile response to water supply using a neutron moisture meter. Such information is processed for temporal changes in cumulative infiltration, which in turn may be interpreted for the bulk hydraulic properties of a given profile [*Rose et al.*, 1965; and *Hillel et al.*, 1972]. These measurements suffer from the same problems as infiltration experiments. An additional approach is to use tracers. However, the arid environment is very problematic for most existing tracers because of their varying susceptibility to heterogeneous media. A feasibility study of dye tracers, conducted in the study reach of Nahal Zin [*Kuells et al.*, 1995], tested the applicability of this technique to surface flow measurements. The separation of the flow in braided desert channels complicates the dilution-discharge equation. The results of this experiment have shown variable flow conditions that require a further modification of the method.

Shentsis et al. (1999) demonstrated that in the hyperarid Negev Desert the recharge consists almost entirely of the larger events; therefore, paleoflood records, which are especially informative as to the largest floods, provide the data needed for estimating long-term recharge. The systematic record, if it exists, contains only few measurements of large floods because they are rare and tend to destroy the gauging stations. Hence, flood frequency analyses of the systematic record suffer from a low statistical confidence because the large floods are unreliably represented.

Paleoflood hydrology reconstructs the magnitude and frequency of past floods [*Baker*, 1987], and can be very useful in ungaged drainage systems or where the systematic records are short. The method uses slackwater deposits (SWD) and other paleostage indicators (PSI) as evidence of the high stage of past flows [*Baker*, 1987]. Most of the paleoflood studies have been conducted in the semi-arid regions of the United States [e.g., *Patton et al.*, 1979; *Kochel et al.*, 1982; *Baker,* 1987; *Ely and Baker*, 1985; *Partridge and Baker*, 1987; *Webb et al.*, 1988; *Enzel et al.*, 1994; and *O'Connor et al.*, 1994], arid Australia [*Baker et al.*, 1983; *Baker et al.*, 1987; and *Wohl*, 1988], monsoonal central India [*Ely et al.*, 1996], and semi-arid Spain [*Benito et al.*, 1998].

The first study of paleofloods in a hyperarid region was conducted in the Negev Desert, southern Israel, where all catchments receive < 100 mm/y of rainfall [*Greenbaum*, 1996; *Greenbaum et al.*, 1998; 2000]. The paleoflood study of Nahal Zin (Figure 3) yielded a high quality record of 28 paleofloods, with peak discharges > 200 m^3 s^{-1} during the last 2000 years [*Greenbaum et al.*, 2000].

At appropriate sites, paleoflood hydrology can reconstruct two parameters of past floods: the peak discharges and the temporal distribution of the largest floods. This can serve various applications such as risk analysis [*Ely et al.*, 1991; *Enzel et al.*, 1991] and hydrological and hydro-

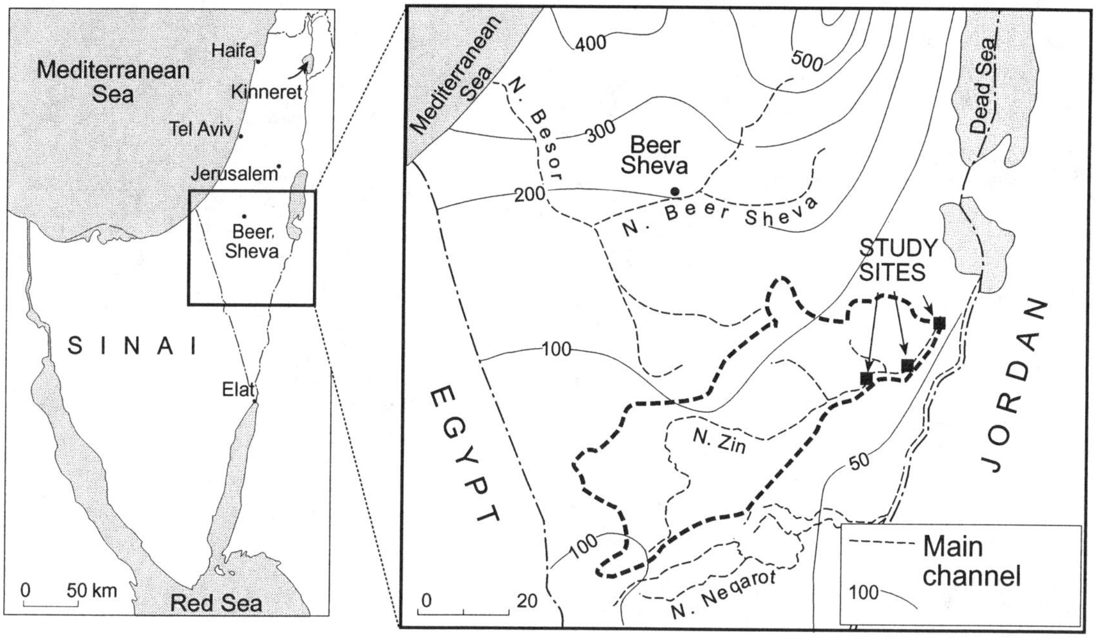

Figure 1. The location map of Nahal Zin within the Negev Desert, and the study areas. The heavy dashed line is the catchment boundary. The isohyets are average annual rainfall [after Greenbaum et al., 2000]

climatological interpretations connected with climatic change [e.g., *Ely et al.*, 1993; *Greenbaum et al.*, 2000; and *Frumkin et al.,* 1998].

In addition to the assumptions associated with paleoflood hydrology, i.e. a relatively stable bedrock reach over the last few thousand years and preservation of evidence of the largest floods during that time interval, we assume that the hydrological characteristics of the flows, i.e. the interrelations between the flood peak discharge, volume, and duration, did not change much during the last 2000 years.

OBJECTIVES

The aim of this study is to estimate event-based long-term transmission losses in the alluvial channel of the lower Nahal Zin (Figure 1). These losses are the maximal potential recharge into the alluvial aquifer of the lower Nahal Zin. Our approach combines results of three types of data: (a) measurements of transmission losses and recharge in an intensively instrumented alluvial reach and alluvial fan of the lower Nahal Zin (Figure 2), (b) paleoflood study of Nahal Zin that estimates the long-term frequency of the largest floods, and (c) statistically significant relationships between the measured flood parameters—peak discharge, volume, and duration over the last 65 years. These relationships are coupled here with the paleoflood record to estimate long-term transmission losses.

STUDY AREA

Nahal Zin is a 1400 km^2 watershed in the hyperarid central Negev Desert, southern Israel (Figure 1). Mean annual rainfall of the Nahal Zin catchment varies from 90 mm in its upper parts to 50 mm and less in the lower reaches (Figure 1). It drains the eastern flanks of the Negev Mountains (elevation 1000 m above sea level) to the Dead Sea (415m below sea level). Its upper basin (Figure 2) drains and incises hard limestone and chalk of moderate permeability and forms a large canyon. Downstream, the channel turns eastward (Figure 2) and forms a 25 km long, wide, gravelly alluvial channel formed in erodible clays, marls and chalks. At the lower end of this section a large tributary—Nahal Hava (120 km^2) enters the main channel. Between the stations Massos and Aqrabim (Figure 2), the channel crosses a body of Pleistocene and Holocene alluvial fill up to 20 m thick, which forms an aquifer that feeds two springs located at the downstream end of this reach. This thick alluvial reach serves as a sink for floods from the upper reaches that lose a large part of their volume and stage by rapid infiltration into the permeable bed [*Polak*, 1980]. The largest tributary, Nahal Hatira (266 km^2), enters the main channel of Nahal Zin close to the lower end of this reach (Figure 2). Downstream of station Aqrabim, Nahal Zin incises hard carbonate rocks and forms a 20 km long bedrock canyon, 150-250 m wide. The canyon is

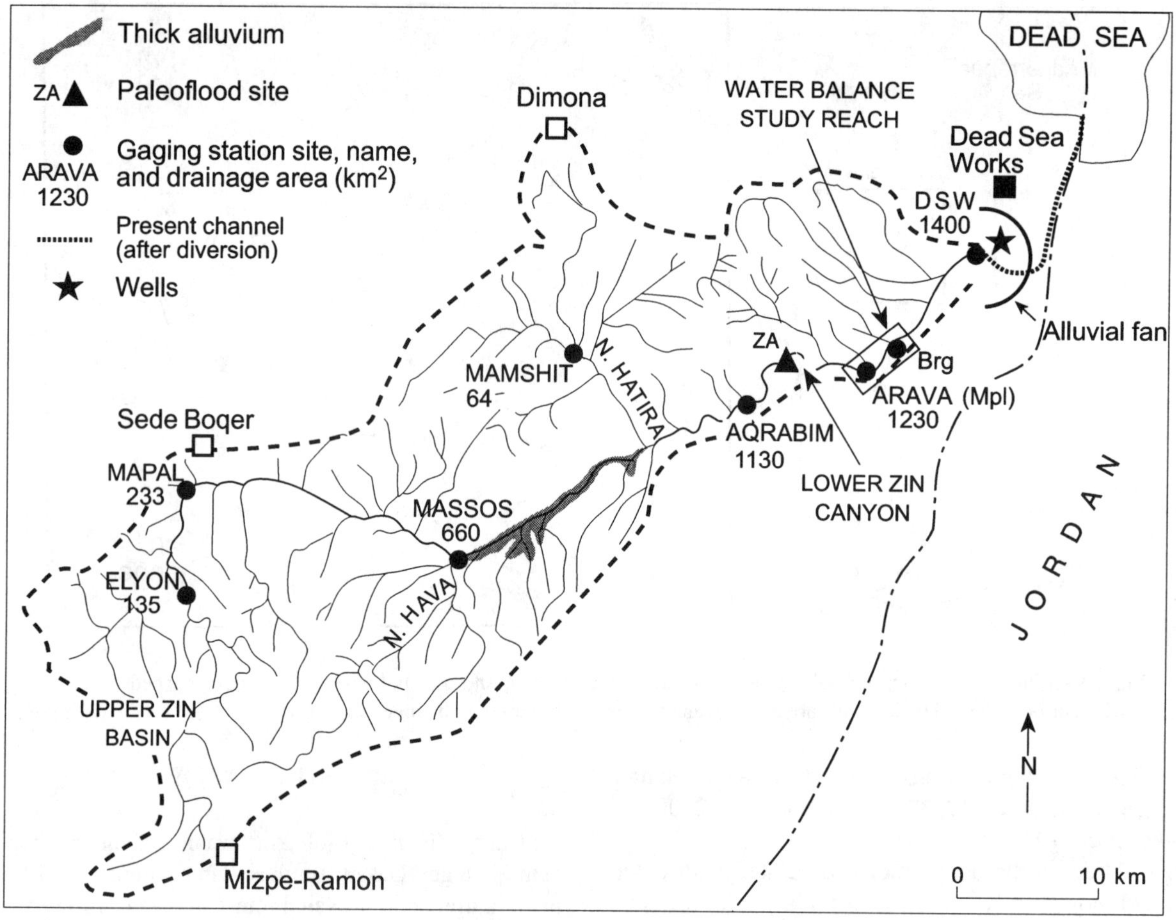

Figure 2. Nahal Zin catchment showing the main tributaries, gauging stations, their drainage areas, and the study areas [after Greenbaum et al., 2000]

partly filled with highly cemented Pleistocene conglomerates into which the Holocene channel is incised, forming an inner canyon 8-10 m deep. From station Arava (Mpl), Nahal Zin, over its 15 km final reach, is incised in the impermeable lacustrine sediments of late Pleistocene Lake Lisan the predecessor of the Dead Sea [*Bentor and Vroman*, 1960; *Begin et al.*, 1974]. The natural mouth of Nahal Zin was located in the southwestern edge of the Dead Sea where it formed a 9 km^2 alluvial fan (Figure 2). In 1985 the channel was diverted from the fan apex southeastward into Nahal Arava (Figure 5) in order to protect an industrial complex, the Dead Sea Works, from floods [*TAHAL*, 1961; 1981].

The Paleoflood Sites

The paleoflood studies in Nahal Zin have been described by detail by *Greenbaum et al.* [2000] and are summarized below. The three paleoflood sites are located in tributary mouths in a 800 m long reach within the bedrock channel of the lower Nahal Zin canyon. The SWD are fine sand and silt in texture. The 25-30 m wide channel at the studied reach is almost straight, and the average channel slope is 0.010. The site is located 8 km downstream of the Aqrabim station (Figure 2) which operated from 1951 to 1981, and 6 km upstream of the Arava gauging station operating since 1989. Discrimination of individual flood deposits in the Nahal Zin SWD sequences was based on sedimentological markers such as: abrupt vertical grain size change; colluvial deposits between flood sediment; fine silt, clay and organic layers; and buried incipient soils. The soils also offer an estimate as to the duration of exposure, indicating a hiatus in sedimentation by large floods for a period represented by the age and degree of soil development (Figures 3 and 4). The dateable material in the Nahal Zin sites includes fine organic debris (e.g., wood and charcoal) which are found on top of the flood sediments in the form of organic beds, or may form driftwood

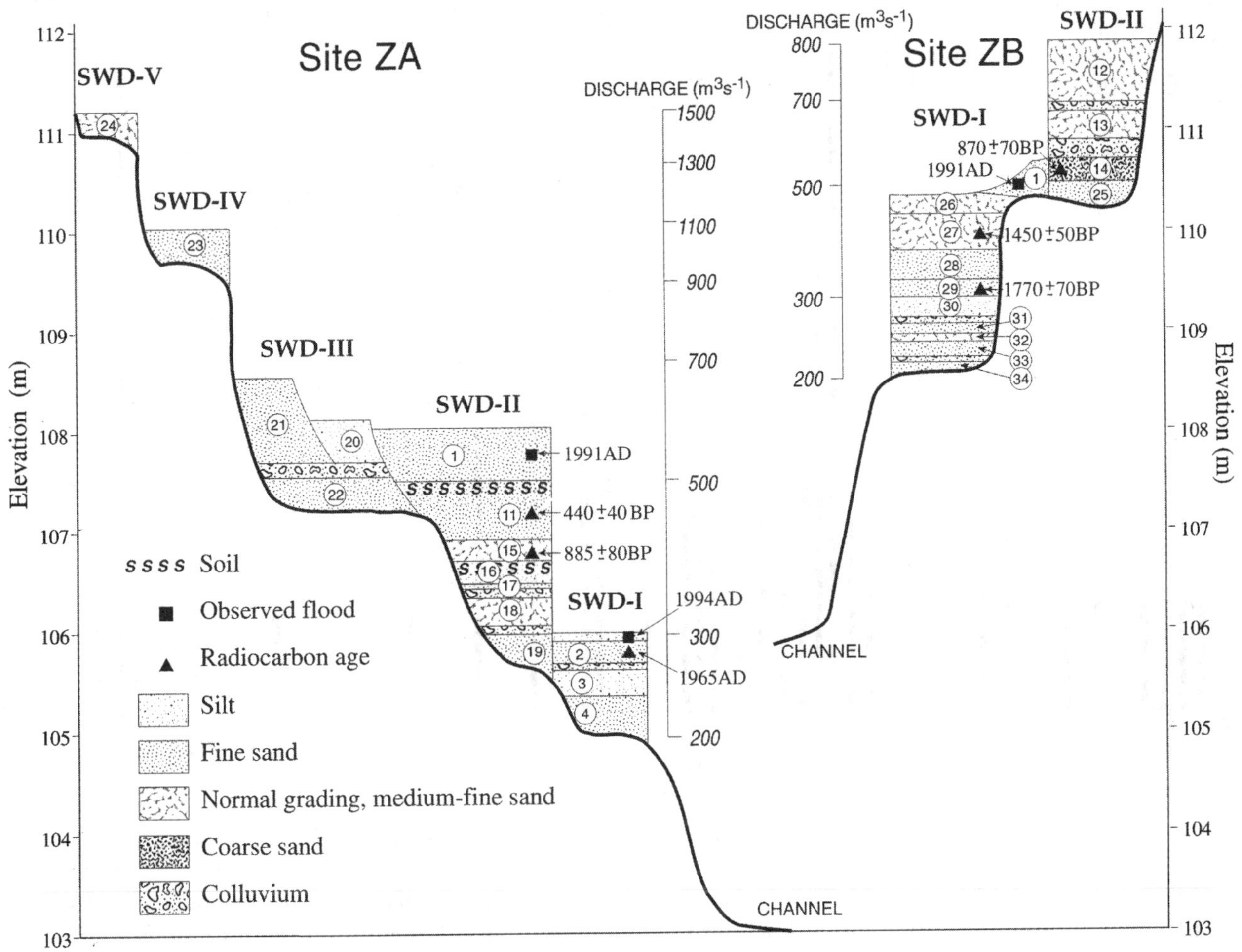

Figure 3. Paleoflood stratigraphy of Nahal Zin at sites ZA and ZB, and the associated peak discharges (see Table 1) [after Greenbaum et al., 2000]

lines. Single pieces of wood and charcoal are also found within the flood sediments.

Study Reach

The channel reach for the study of infiltration and water-balance is 5.5 km long and is a part of the last 15 km long channel in which Nahal Zin flows along the Arava Rift Valley toward the Dead Sea (Figure 2). It starts at the lower waterfall, where gauging station Arava (Mpl) was located, and ends at the new bridge of the Arava road. The channel is incised into the impervious lacustrine deposits of Lake Lisan. The channel slope is 0.010 and the width is 70-100 m. The channel bed is covered by alluvium 0.1-4.0 m thick. The Lisan deposits isolate (hydraulically) the alluvium from the deep geologic formations that constitute the regional aquifers.

The Alluvial Fan Aquifer

The Nahal Zin alluvial aquifer is a local, shallow aquifer within the modern and Holocene alluvial fan of Nahal Zin (Figure 5). Its total area is 9 km². It is bounded by the Amatsia fault in the southwest, and it extends north to the southern end of the Dead Sea Works. The eastern fringes of the aquifer lie adjacent to the Dead Sea evaporation pans. The maximum thickness of the aquifer is approximately 100 m in the center of the paleochannel of Nahal Zin [*Katz*, 1976; *Naor et al.*, 1986; 1991] and it thins to 40 m at the northern and eastern margins of the alluvial fan.

The main water sources contributing to the aquifer in the order of their relative importance are: (a) Infiltration into the distributary channel bed in the alluvial fan and the channel of Nahal Zin [*Schwartz*, 2001] (b) Water from the near-shore clastic beds of the Lisan and pre-Lisan forma-

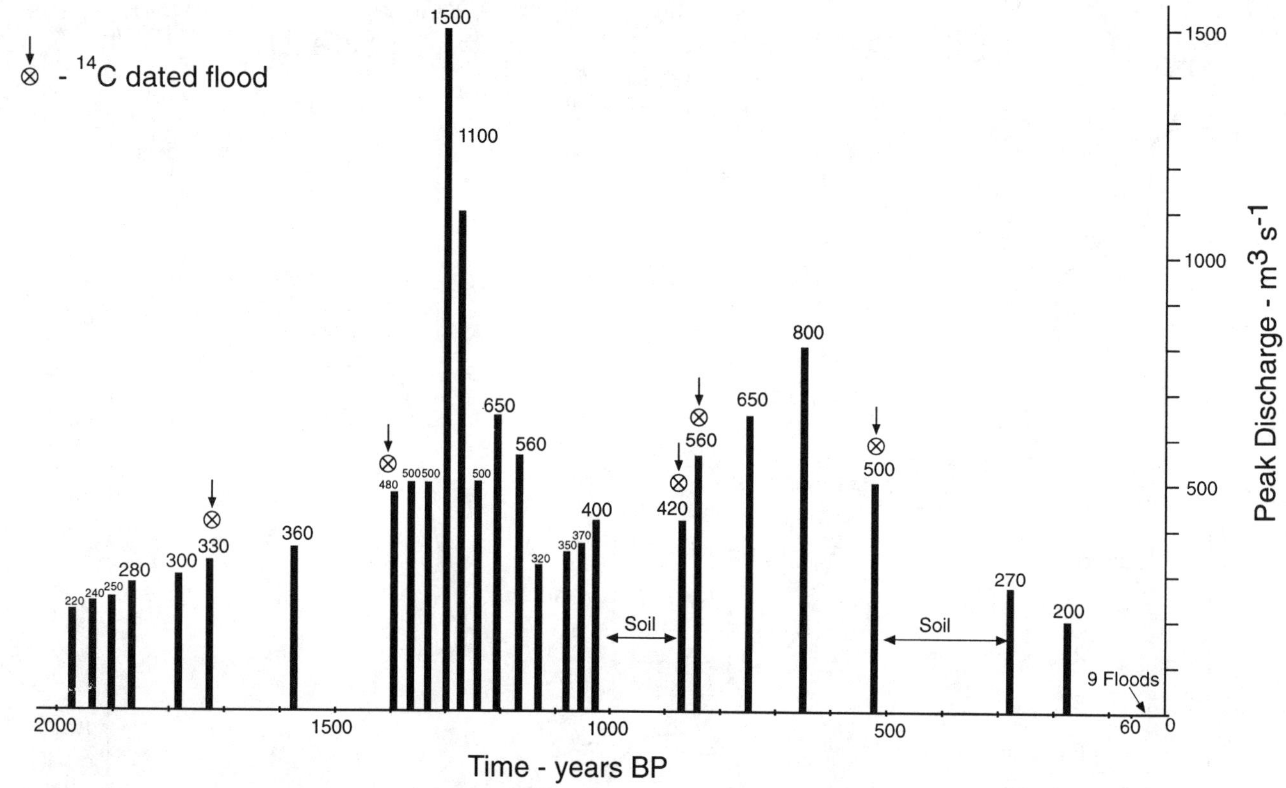

Figure 4. Temporal distribution and magnitude of paleofloods in Nahal Zin over the past 2000 years.

tions (Lisan Aquifer), and (c) water rising along the rift's margin faults from lower bedrock aquifers [*Naor*, 1994]. The transmissivity values are very high, reaching thousands of meters per day [*Naor et al.*, 1991].

HYDROLOGICAL RECORD OF NAHAL ZIN

Floods

The first gauging station in Nahal Zin, established at the outlet by the Dead Sea Works, operated between 1935 and 1946 [Palestine Potash Ltd; *DeLeeuw*, 1945; DSW in Figure 2]. It recorded 38 floods, ten of which exceeded 100 m^3s^{-1}. This record is considered as historical data [*Greenbaum*, 1996] because it is not continuous with the gauged record and all flood magnitudes are based on indirect measurements.

The 50-year systematic record of Nahal Zin is derived from seven gauging stations installed by the Israel Hydrological Service (IHS) in 1951. Five stations are along the main channel and two in tributaries (Figure 2). Only three stations have a continuous record since 1951. The IHS Arava Station (Mpl in Figure 2) is located at the lower waterfall, which is the inlet to the channel reach in which the detailed water balance and transmission losses study [*Schwartz*, 2001] was carried out. The largest number of floods recorded at the Massos station is 98. Only ten floods in that station had a discharge exceeding 100 m^3s^{-1}.

In 10% of the hydrological years no flows were observed in the lower stations. Nahal Zin, in its lower section, had an average of 2.4 flows per year during the instrumental record.

The largest floods in the historical and gauged records are: 550 m^3s^{-1} (Jan. 1, 1945); 600 m^3s^{-1} (May 13, 1945); 550 m^3s^{-1} (Oct. 13, 1991). All three flood magnitudes were calculated using peak stage watermarks [*DeLeeuw*, 1945; *Greenbaum*, 1996].

Ground Water

The systematic measurements of water elevation in Zin Wells located on the alluvial fan (Figures 2 and 5), started in the early 1960s, and include 2–3 readings per year. The measurements are conducted by the Dead Sea Works.

During the course of our studies, substantial efforts were made to locate records of water levels in wells downstream of flow gauging stations. These efforts failed to reveal, for the entire Israeli side of the Arava valley, even a

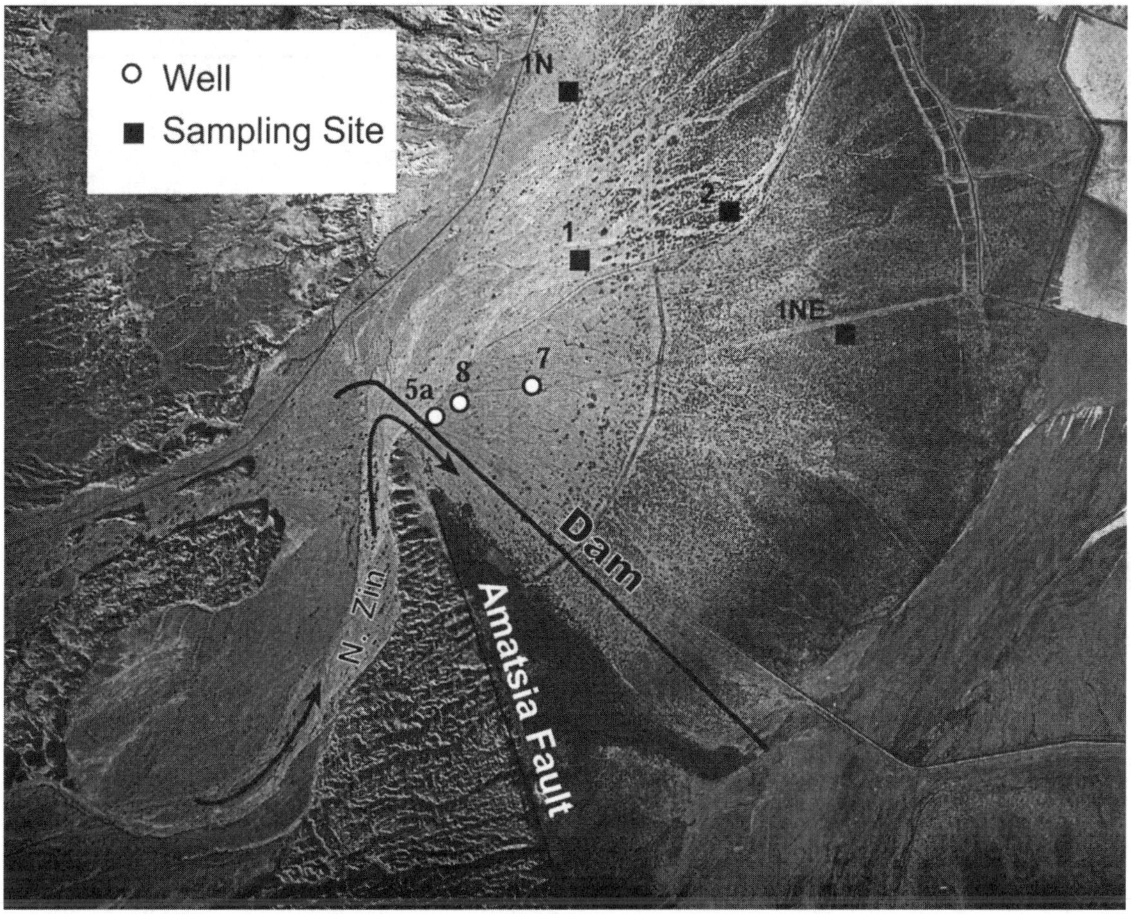

Figure 5. Nahal Zin alluvial fan, dam and wells.

single case in which both data sets exist to satisfy the test of floodwater control over water table variations. The reasons were: (a) the meager number and poor quality of the stream gauging records and (b) the long intervals between water level recordings.

The record presented here [*Schwartz*, 2001] is the first one made in Israel and perhaps one of the first in the world. It records both floods and water levels in a shallow alluvial aquifer at the same temporal resolution and quality aimed at resolving the flood influence on water table variations. The high resolution measurements started in November 1994 and lasted five years.

Paleofloods

The geometry and sedimentology of the three sites—ZA, ZB and ZC were studied in detail (Figures 2 and 3) [*Greenbaum et al.*, 2000]. Twenty-eight paleofloods during the last 2000 years were reconstructed in Nahal Zin (Table 1). The peak discharges range from 200 m^3s^{-1} to 1500 m^3s^{-1} (Figure 4) and extend the range of measured peak discharges from 600 m^3s^{-1} to 1500 m^3s^{-1} [*Greenbaum et al.*, 2000].

The distribution of floods through the late Holocene indicates that paleofloods are clustered in certain periods in which they are relatively abundant (Figure 4). These periods alternate with periods of few floods or with gaps in the occurrence of floods as indicated by the formation of soils [*Greenbaum et al.*, 2000]. In addition, larger floods also characterize periods of numerous floods, while periods of few floods tend to have lower magnitudes (Figure 4). The results show two periods with numerous high magnitude floods—1380-880 cal. yr B.P. and during the last 65 years. These periods follow periods with few and smaller floods—1730-1380 cal. yr B.P., and 880-65 cal. yr B.P. A transitional period, with a small number of very large floods, occurs during the period 880-530 cal. yr B.P. Prior to 1730 cal. yr B.P., 5 small floods are recorded at the sites but they contain no datable material.

Table 1: Systematic, historical and paleofloods (>200 m^3/s^{-1}) hydrological parameters in Nahal Zin

Flood No.	Peak [1] Discharge m^3s^{-1}	Age [1] years	Duration [2] hours	Volume [2] 10^6 m^3	Transmission [2] Losses (10^6 m^3)
0	300 [3]	1994 AD [3]	58 [3]	4.8 [3]	0.8-0.9
1	550 [3]	1991 AD [3]	70 [3]	7.0 [3]	1.0-1.2
2	300 [3]	1965 AD [3]	24 [3]	5-6	0.8-0.9
3	200 [3]	1946 AD [3]	24 [3]	3-4	0.7-0.8
4	600 [3]	1945 AD [3]	75-80	11-12	1.3-1.5
5	550 [3]	1945 AD [3]	52 [3]	24.0 [3]	2.8-3.0
6	200 [3]	1941 AD [3]	31 [3]	6.1 [3]	1.0-1.1
7	450 [3]	1940 AD [3]	30 [3]	8-9	1.2-1.3
8	335 [3]	1937 AD [3]	55-60	6-7	1.0-1.1
9	270		50-55	5-6	0.8-0.9
10	200		40-45	3-4	0.7-0.8
11	500	440±40 BP	70-75	10-11	1.2-1.3
12	800		90-95	17-18	2.0-2.2
13	650		85-90	12-13	1.5-1.7
14	560	870±70 BP	75-80	11-12	1.3-1.5
15	420	885±80 BP	65-70	8-9	1.2-1.3
16	400		65-70	8-9	1.2-1.3
17	370		60-65	7-8	1.0-1.2
18	350		60-65	7-8	1.0-1.2
19	320		55-60	6-7	1.0-1.1
20	560		75-80	11-12	1.3-1.5
21	650		85-90	12-13	1.5-1.7
22	500		70-75	10-11	1.2-1.3
23	1100		100-110	24-26	2.8-3.0
24	1500		140-150	38-40	3.8-4.0
25	500		70-75	10-11	1.2-1.3
26	500		70-75	10-11	1.2-1.3
27	480	1450±50 BP	65-70	9-10	1.2-1.3
28	360		60-65	7-8	1.0-1.2
29	330	1770±70 BP	60-65	6-7	1.0-1.1
30	200		40-45	3-4	0.7-0.8
31	280		50-55	5-6	0.8-0.9
32	250		50-55	5-6	0.8-0.9
33	240		40-45	4-5	0.8-0.9
34	220		40-45	4-5	0.7-0.8

1 after Greenbaum et al. (2000)
2 calculated values
3 measured values

The oscillations in frequency and magnitude of paleofloods indicates changes in the hydrological regime in the central Negev during those periods [*Greenbaum et al.*, 2000] and are remarkably well correlated with the Dead Sea lake level variations during the last 2000 years [*Greenbaum et al.*, 2000; *Frumkin et al.*, 1998].

METHODS

Paleoflood Hydrology

We used radiocarbon dated fine-grained slackwater flood deposits (SWD) located mainly in tributary mouths, and other paleostage indicators (PSI) such as driftwood lines along the bedrock canyon which represent high stages of large floods [*Patton et al.*,1979; *Kochel et al.*, 1982; *Baker*, 1987; *Baker and Kochel*, 1988; and *Greenbaum et al.*, 2000].

Paleoflood discharge estimates associated with these deposits and high water marks were calculated using the HEC-2 model (*Hydrologic Engineering Center*, 1982), which generates water surface profiles using step-backwater calculations for various discharge values [*O'-Connor and Webb*, 1988]. In Nahal Zin only floods larger than 200 m^3s^{-1} backflood the tributaries in the studied sites and thus only their deposits are preserved [*Greenbaum et al.*, 2000].

Observations of modern floods at the sites and at the Aqrabim and Arava gauging stations (Figure 2) enabled the calibration of the HEC-2 water surface profiles. Three measured floods could be accurately compared: (a) The flood of October 1965 (300 m^3s^{-1}) was recorded at the Aqrabim station and identified clearly in the paleoflood record at site ZA (Figure 3). (b) The floods of October 1991 (550 m^3s^{-1}) and November 1994 (300 m^3s^{-1}) were recorded at the Arava station and their deposits were identified at the paleoflood sites within hours after the flood recession.

Flood Frequency Analysis

For flood frequency analysis we used the MAX program [*Stedinger et al.*, 1988], which computes the maximum likelihood estimates of the parameters of selected probability distributions. The program combines and integrates flow frequency and magnitude information from various types and sources [*Stedinger and Cohn*, 1986; *Stedinger and Baker*, 1987; and *Stedinger et al.*, 1988]. MAX uses maximal flows of the various types of records, assuming that all the flows that exceed a certain threshold within a known time period are recorded. The paleoflood record of Nahal Zin shows clustering, and a combined analysis for the whole period would lead to average probability values which ignore the fluctuations of flood frequency during different hydrologic periods. Therefore, a separate analysis was conducted for each of the combined periods of a similar hydrologic regime (Figure 6), using the log-normal distribution which shows the best fit to the variously combined sets of records [*Greenbaum et al.*, 2000]. The systematic record only (1951-1998) provided a calculated 100-year flood of 850 m^3s^{-1} (curve B, Figure 6). By adding the historical record (1935-1946) the 100-year flood increases to 970 m^3s^{-1} (curve A, Figure 6). Combining these records with the paleoflood data of the period 1380-880 cal. yr B.P. that is characterized by abundant large floods, the 100-yr flood decreases to 810 m^3s^{-1} (curve C, Figure 6), probably because during that period the frequency of floods was lower than during the last 65 year period. The period of few floods and low magnitudes (1730-1380 cal. yr B.P.) was combined with the similar period 880-65 cal. yr B.P. and provided a calculated 100-year flood of only 180 m^3s^{-1} [*Greenbaum et al.*, 2000]. The present study uses curve C for the long-term flood frequency.

Transmission Losses and Recharge

Schwartz [2001] studied the process of infiltration and transmission losses in the alluvial channel of Nahal Zin. Direct measurements of moisture distribution in the active channel during flood were carried out, using Time Domain Reflectometry (TDR) backed up by electric resistivity plates (leaf wetness). These measurements provided evidence for the pattern of water movement in the alluvium by demonstrating the effect of antecedent moisture on infiltration rate and transmission losses. Fourteen flow events occurred during the period 1994-1998, 11 of which were recorded by two gauging stations. These measurements enabled a detailed water balance for the calculations of transmission losses and effective hydraulic conductivity [*Lane et al.*, 1983].

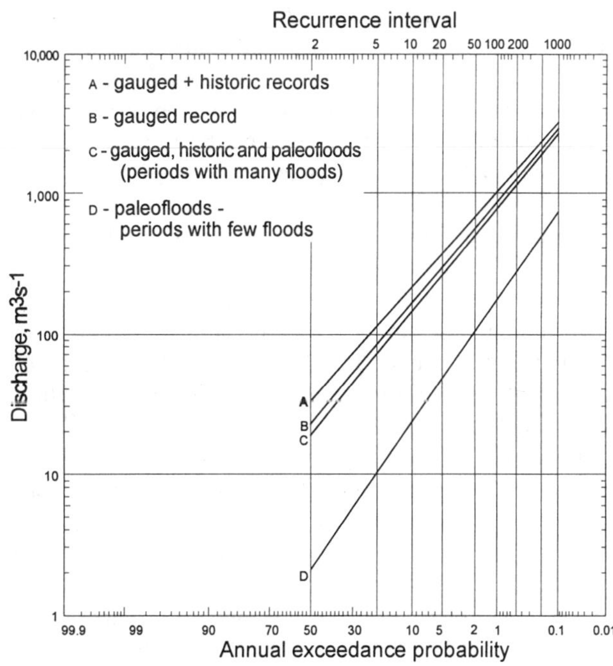

Figure 6. Results of log-normal flood frequency analysis, based on the MAX program (Stedinger *et al.* 1988): **A,** for the gauged and historical record of Nahal Zin; **B,** for the gauged record; **C,** for the gauged, historical and palaeofloods of the periods with many floods; **D,** for paleofloods during periods with few floods [after Greenbaum et al., 2000]

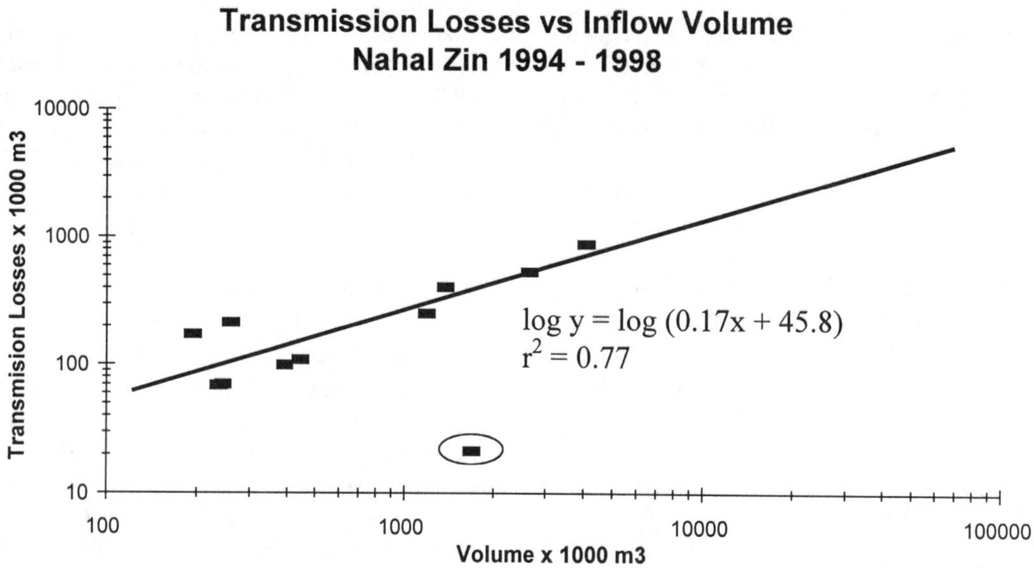

Figure 7. Relations between transmission losses volume and inflow volume for Mpl–Brg reach, Nahal Zin.

We used pressure transducers (PT, type Druck, 3 min resolution), for the measurement of water stage. These measurements were supported by manually taken, direct measurements of water stage during most of the floods. Slope-area rating curve and HEC-2 computations provided a 3 min resolution hydrograph. This setup is considered accurate for hydrological measurements in deserts. We consider that gauging errors based on the methods used in the present study have a probable error of $< \pm$ 15%. This is markedly less than the difference between inflow and outflow at most periods of time considered. At high flows, the accuracy of discharge determination increases because of the nearly static width of the floods at the study sites. Transmission losses in the study reach vary from 20% for the large floods to over 85% for the small floods, i.e., the accuracy of the transmission loss estimation is greater than the potential gauging error.

In-channel measurements were coupled with measurement of ground water level in the alluvial aquifer of Nahal Zin. We used the effective hydraulic conductivity of the channel, flood duration and recharge area to calculate the recharge to the alluvial aquifer.

The measurements of channel flow, sub-channel moisture and alluvial fan groundwater level were aimed to study the relations between channel flow and groundwater recharge. The unique natural setup of the alluvial channel reach, lack of tributaries and a terminal alluvial fan that forms a shallow aquifer overlying extremely impermeable bedrock enabled a deterministic study of the distribution process between the surface and subsurface domains and of the aquifer recharge.

RESULTS

Transmission Losses in the Alluvial Reach

In the lower Nahal Zin transmission losses increase with increasing flood volume [*Schwartz*, 2001] (Figure 7). The relative transmission losses, in percent, decrease with increasing flood volume. When these relationships are extended to the paleoflood record, the calculated transmission loss for the largest flood (1500 m^3s^{-1}) exceeds 4 x 10^6 m^3 (Table 1).

Figure 8 shows the water losses between stations Mpl and Brg during the first 14 to 30 hours of three selected floods. It describes an hourly water-balance for the reach based on 3-minute resolution hydrographs of the stations Mpl and Brg and indicates the rate of transmission loss during every stage of the flood. The data presented in Figure 8 indicate that the transmission losses increase steadily during the first hours. At point S in Figure 8 (a) and (b), a sharp change takes place and the losses decrease at a steady rate to point G where the losses become negative. The difference among the three floods is in the accumulated volume of infiltrated water and the time needed to reach point S. The curve between the origin and point S represents a phase in the flood during which the alluvial reach absorbs water and approaches saturation. Point G, where the losses turn negative, reflects, both the flood recession at the upper station (Mpl), and the continuation of drainage, by gravitation, from the reach under nearly saturated conditions.

Transmission losses are linearly correlated (R^2=0.77; significant at 95%) with inflow volume into the reach, shown on a log-log scale (Figure 7). One particular flood stands out by its very small transmission losses (black circle, Figure 7). This flood occurred only a few days after a previous one. *Schwartz* [2001] showed that absorption of water by the alluvium is significantly inhibited for at least 5 days after the occurrence of a preceding flood. This effect disturbs the relationship between transmission losses and inflow volume. Their correlation increases to over R^2=0.97 if floods occurring on wet surface are excluded.

Estimation of the effective hydraulic conductivity was based on the flood parameters and water balance for the study reach. We used the procedure developed by *Lane et al.* [1983] to estimate the effective hydraulic conductivity

The total volume of losses for a channel reach is KD, where K is the effective hydraulic conductivity and D is the duration of flow. Also, the total losses are [P – Q(x,w)], where P = inflow volume (m^3), Q = outflow volume (m^3), x = Length of reach (km), and w = average width of flow so that,

$$KD = [P - Q(x,w)] \quad (1)$$

and therefore

$$K = [P - Q(x,w)] \backslash D \quad (2)$$

The average width of each flow event has been derived from the water stage record. The range of effective hydraulic conductivities [*Schwartz*, 2001] reflects the antecedent conditions of the reach and the flood magnitude in terms of discharge and water coverage.

Flood Peak Discharge, Flood Volume and Flood Duration

The 65 years of measurements in the lower Nahal Zin are derived from three stations: Aqrabim, Arava and Dead Sea Works (Figure 2). Only minor tributaries enter the main channel between these stations, therefore additional inflow is negligible and flows in and out of the reaches between these stations are known. The hydrological data from these stations were combined and analyzed for the relations between peak discharge, flood volume and flow duration (Figures 9 and 10). The non-linear relationships between the above-mentioned flood parameters are shown on a log-log scale with a linear regression line of the logarithmic values of the flood parameters. The relations indicate that the volume of the flood increases with increasing peak discharge and duration, and the increase of volume is exponentially related to peak discharge or duration. The relationship between flood volume and flood duration is less significant (R^2=0.55) than between peak discharge and flood volume (R^2= 0.81). The relationship R^2=0.55 in-

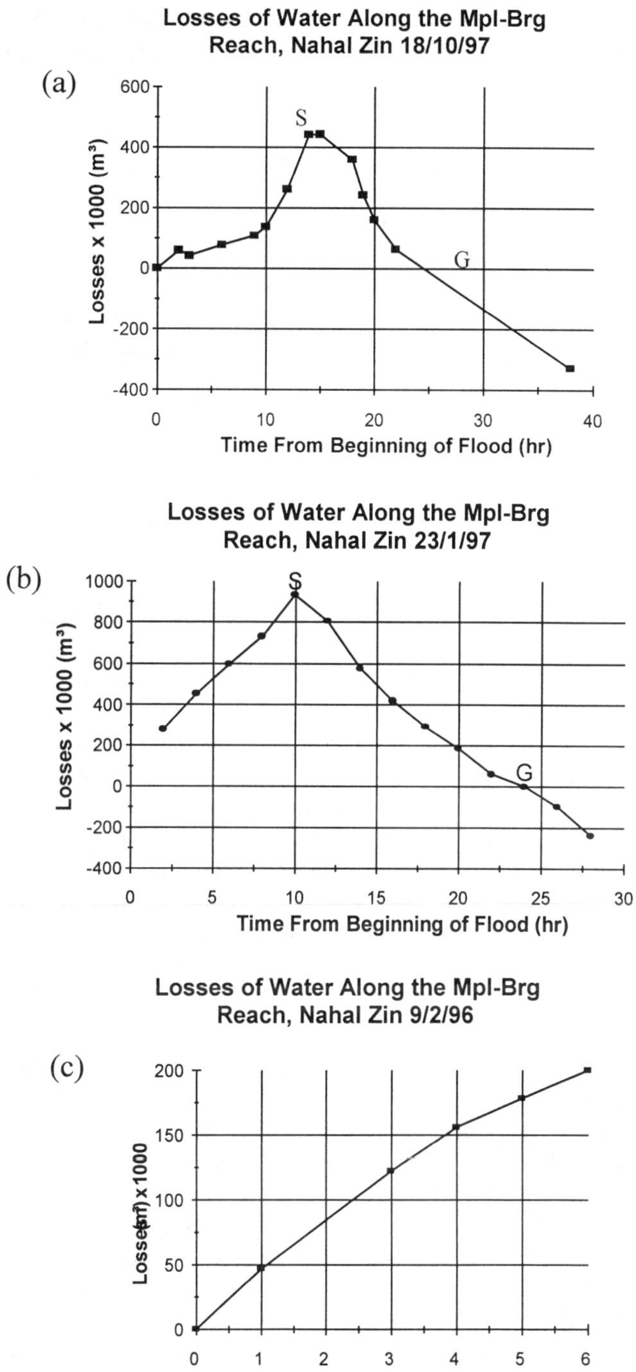

Figure 8. An hourly water balance for Mpl–Brg reach, Nahal Zin. (a) Medium flood, water covers the whole width of the channel when in high stage. Time to point S is 15 hours. (b) Large flood, water covers the whole width of the channel already at the beginning of the flood. Time to point S is 10 hours. (c) Small flood, the surface of the channel is partially covered with water during the entire flood. The balance is for 6 hours because a failure had occurred in the Brg gauging station.

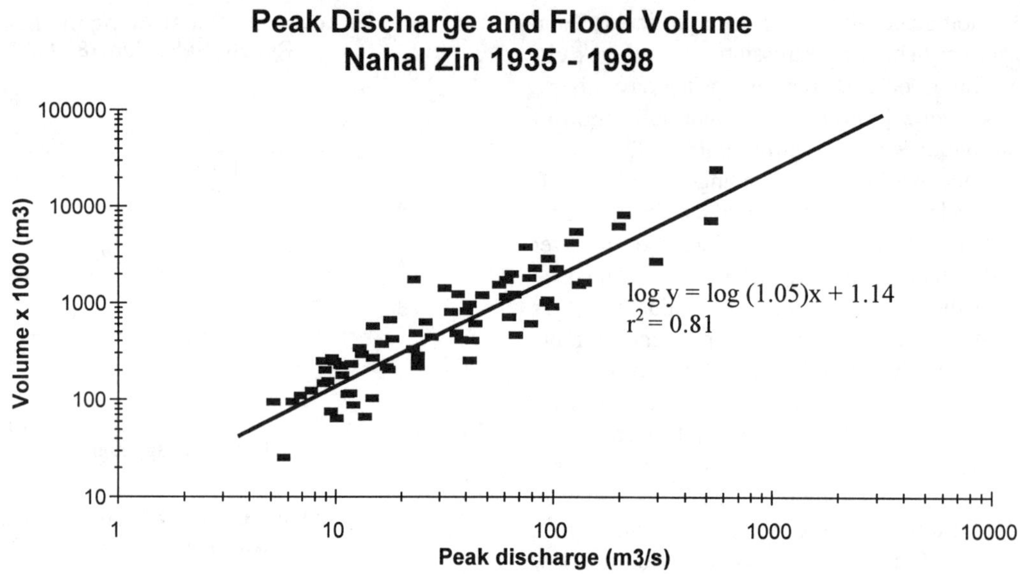

Figure 9. Relations between peak discharge and flood volume for the 65 years record of Nahal Zin.

creases if floods occurring on a wet surface are excluded (Figure 10).

The established relations for the measured record are then extrapolated to the paleoflood record. The presence of floods of up to 600 m^3s^{-1} in the modern record allows a more reliable extrapolation to the range of the largest floods in the paleoflood record. The estimated flow volumes for the larger floods in Nahal Zin exceed 40 x 10^6 m^3 (Table 1). The estimated duration of these floods is up to 150 hours.

Alluvial Fan Recharge

As evident in Figure 11, the water level in the Well Zin 8 began to rise 34 hours after the February 14 flood which pushed the water level by 0.9 m. Recharge to Nahal Zin alluvial aquifer is considered here by two methods: (1) based on statistical relationships between water losses during floods to flow parameter, i.e. flood peak and flow volume; and (2) the estimation of effective hydraulic conductivity of the alluvium in the channel, for various flood magnitudes.

The volume of water contributed to the alluvial fan aquifer during floods in Nahal Zin is calculated by equation 3, based on equations 1 and 2, which constitute the procedure to estimate the effective hydraulic conductivity. The value 0.3 m/hr/m^2 that was chosen for the calculations of the total volume of recharge will be discussed below. The total volume Q (m^3) of recharge (by floods) into the alluvial fan aquifer is:

$$Q = KDA \qquad (3)$$

where K = effective hydraulic conductivity, D = duration of flow, and A = total recharge area.

Substituting the parameters in Equation 3 with the actual constants of the study reach, i.e. KDA, yields a recharge rate of 12,000 m^3/hr. The recharge area is the channel reach between Amatzia Fault and the point where Nahal Zin turns southeastward (Figure 5). A primary assumption in the model is that the effective hydraulic conductivity of the recharge area is equal to that of the Mpl – Brg reach.

The calculation of annual recharge by floods is based on the annual effective duration of flow, as derived from the historical and modern records. The calculation included an identification of flood producing recharge, i.e. peak flow > 30 m^3s^{-1}, and the summation of the total duration of these floods is divided by the 65 y of record. The result, 50 hours, is the annual effective duration of flow, which yields a mean annual recharge of 0.6 x 10^6 m^3.

DISCUSSION AND CONCLUSIONS

It is customary to predict transmission losses using a "black-box" water balance in which the infiltration process and quantities in the reach are unknown. A further complication is the scarcity of inflow data from tributaries. *Shentsis et al.* [1999] estimated the annual recharge by performing flood-volume frequency analysis. This procedure can be applied only when a full set of hydrographs at the various stations exists, and it depends on the duration of measurements, which, as usual, suffers from the problem of extrapolation to an extended temporal scale. The approach in our study is to deal with the recharge estimation

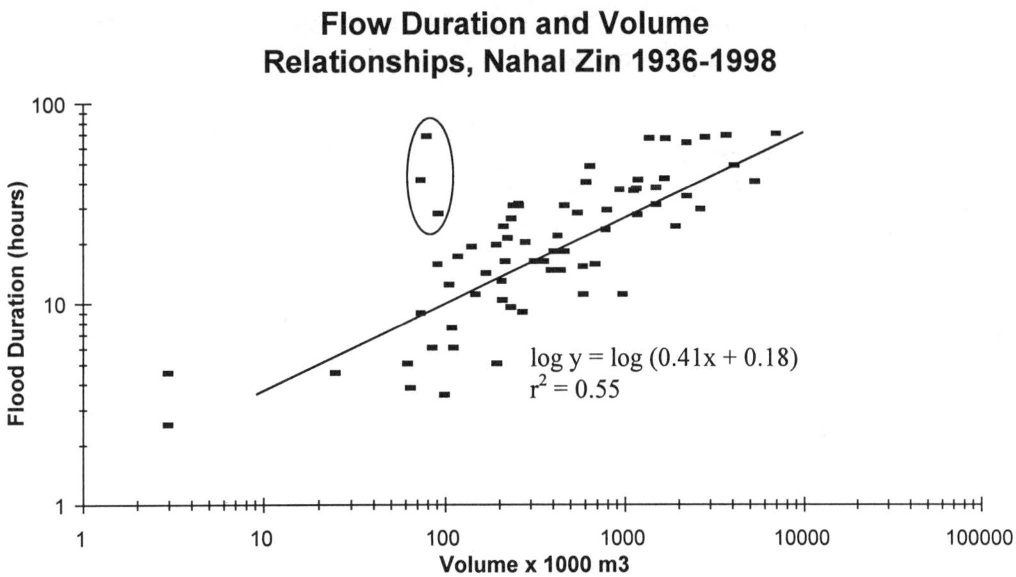

Figure 10. Relations between flood volume and flood duration for the 65 years record of Nahal Zin. The encircled dots represent floods occurring on wet surface.

as an event-based procedure and to address (a) the temporal scale issue by extending the record of the significant recharge-contributing floods, (b) actual process of transmission losses and recharge by measuring them during individual events, (c) the flood volume-transmission losses relationships by analyzing the intensively instrumented record, and (d) the statistical relationships between flow parameters, i.e. peak discharge, duration and volume based on the historical and modern records. Such an approach enables the calculation of long-term transmission losses. The existence of a clear connection between hydrological flow parameters and groundwater response in the alluvial aquifer enables an additional verification of transmission losses by way of calculating the resultant penetrating flow from the effective hydraulic conductivity multiplied by flow duration.

The procedure in this study is based on a water balance for a channel reach free of tributaries. It is further based on direct measurements of infiltration in the reach and parallel measurements of the water levels in the recharged alluvial aquifer downstream.

The alluvium in this reach is bounded by impervious bedrock, which, on the one hand, limits the total water storage of the alluvium, but on the other raises the water balance calculations to a more accurate level [*Schwartz*, 2001].

The rate of infiltration is controlled by alluvial texture, structure and antecedent moisture. Transmission losses are determined by the rate of infiltration and the area covered by water. Higher discharge may cause larger transmission losses but not necessarily higher rates of water losses. Increase of discharge covers dry parts of the channel and floodplain surfaces with high rates of infiltration that increase transmission losses. The limited alluvium in the study reach and the rectangular shape of the cross-section emphasize the effect of antecedent moisture. In the study area, peak discharges greater than 30 m^3s^{-1} (recurrence interval 3 years) covers the full width of the channel. Therefore, this discharge is regarded as the threshold for a full potential of infiltration. Lower discharges cover only a partial width and thus transmission losses are reduced. Peak discharges larger than 30 m^3s^{-1} occupy the same width with higher stage and therefore have a low effect on transmission losses, which are compensated by the increased flow duration and increased head. In the Massos–Aqrabim reach (Figure 2), where the alluvium is thick and the width is essentially unlimited relative to flood magnitudes, transmission losses are highly related to the discharge, which is directly related to the area covered by the floodwater.

Results of direct measurements [*Schwartz*, 2001] in the Mpl-Brg water balance study reach (Figure 2) indicate that transmission losses are significantly reduced when the time interval between two successive floods is less than one week. The influence of antecedent moisture from an earlier flood is still noticeable even after one month, which is the time needed for the moisture content in the alluvium to be reduced by gravitational drainage and evaporation down to values of 3-4% [*Schwartz*, 2001]. Measured transmission losses over dry alluvium for floods with peak discharges of

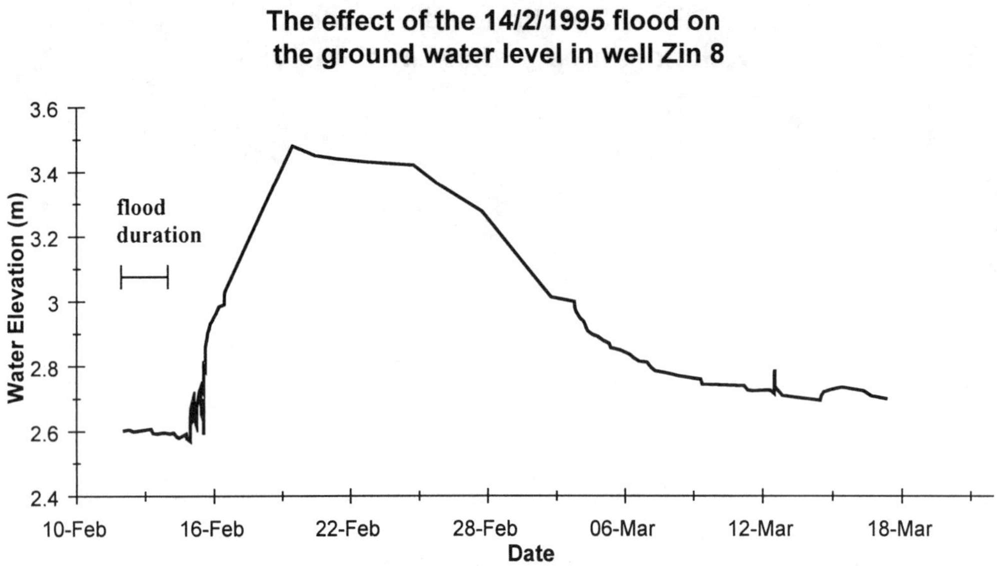

Figure 11. Alluvial fan water table response to floods in Nahal Zin.

30-300 m^3s^{-1} decrease from 35% to 17%, respectively. Based on Figure 11, the expected specific transmission losses of the Jan. 23, 1997 flood (122 m^3s^{-1}) are 26%, whereas the directly measured transmission losses were 21%. The minor overestimation of the losses by using the empirical model (Figures 7, 9 and 10) compared to the measured losses is probably within the accepted error of reach hydrology, but can be attributed to the antecedent moisture caused by a small (1.5 m^3s^{-1}) flood that occurred a week earlier. Floods with peak discharges < 30 m^3s^{-1} lose a higher percentage of their volumes (86% for a flood with a peak discharge of 9 m^3s^{-1}), and even disappear completely in the alluvium if they arrive on an initially dry channel. This sharp decline in individual floods in downstream reaches was observed in other arid parts of the world in streams of various scales [*Dubief*, 1953; *Vanney*, 1960; *Schick*, 1980; 1988; *Williams*, 1970; *Buono and Lang*, 1980; *Enzel et al.*, 1989; *Enzel*, 1992; *Murphey et al.*, 1972; Schwartz, 1986; *Walters*, 1990; *Renard et al.*, 1993; and *Osterkamp*, 1995].

The correlation between volumes of floods and transmission losses (Figure 7) is used to estimate the transmission losses of the large modern floods and of the paleofloods. Flow volumes are derived from the relationship of peak discharge and volume (Figure 9). These significant correlations enable a reasonable extrapolation to the high magnitude floods which comprise the paleoflood record. The general trend of the extrapolated relations between paleoflood volumes and their transmission losses is clear; however, the confidence of the graphical solution would benefit from a further statistical study. Low-frequency high-magnitude floods have a large volume (Figure 9) and therefore large transmission losses per event (Table 2). For example, the calculated transmission losses for the above threshold discharge (30 m^3s^{-1}) are 0.1-0.2 x 10^6 m^3 whereas for a flood with a peak discharge of 810 m^3s^{-1} it is 2.0-2.2 x 10^6. The paleoflood record is composed of peak discharges from 200 m^3s^{-1} (recurrence interval of 16 years) to 1500 m^3s^{-1} (recurrence interval of 300 years), which yields flow volumes of 3-4 x 10^6 m^3 to 40-42 x 10^6 m^3 (Table 2) and transmission losses volumes of 0.7-0.8 x 10^6 m^3 to 3.7-4.0 x 10^6 m^3, respectively. The calculated transmission loss volumes for the extrapolated 500 and 1000 year floods (if such floods ever occurred) are 4.0-5.5 x 10^6 m^3. It should be emphasized that the transmission losses of individual events increase at a markedly lesser rate than the flood volume.

Measurements of groundwater level show a rise of 0.4 m and 0.9 m during the floods of Dec. 5, 1994 and Feb. 14, 1995 (Figure 11). The measured flow volumes were 1.2 x 10^6 m^3 and 0.25 x 10^6 m^3, respectively, and the calculated transmission losses were 0.33 and 0.2 x 10^6 m^3, respectively. The difference in the groundwater rise between the two events may reflect the form and dimensions of the alluvial fan aquifer, i.e. a wide base and a narrow top, typical of alluvial fan morphology.

Transmission losses for various reaches of Nahal Zin were evaluated for measured floods using water balances [*Polak*, 1980; *Shentsis et al.*, 1999]. The upstream channel reach of Nahal Zin between gauging stations Massos and Aqrabim (Figure 2) is characterized by up to 20 m alluvium —much thicker than in the studied reach in the lower

Table 2: Calculated volumes of floods and transmission losses for various peak discharges

Recurrence Interval (years)	Peak Discharge [1] (m^3s^{-1})	Volume (10^6 m^3)	Transmission Losses (10^6 m^3)
3	30	0.4-0.5	0.1-0.2
4	50	0.7-0.8	0.2-0.3
5	70	1.0-1.2	0.3-0.4
8	100	1.6-1.8	0.4-0.5
10	150	2.4-2.6	0.5-0.6
16	200	3-4	0.7-0.8
20	270	5-6	0.8-0.9
30	300	6-7	0.9-1.0
50	520	10-11	1.2-1.3
100	810	17-18	2.0-2.2
200	1220	32-34	3.3-3.6
300	1500	40-42	3.7-4.0
500	2000	50	4.0-4.5
1000	2830	70	5.0-5.5

[1] after Greenbaum et al. (2000)

Nahal Zin. The width of the channel ranges between 300 and 500 m. The decrease in peak discharge and volume along this section are dramatic. For example, during the Oct. 22, 1979 flood, the peak discharge and flow volumes were reduced from 210 m^3s^{-1} to 42 m^3s^{-1} and from 1.6 x 10^6 m^3 to 0.96 x 10^6 m^3, respectively, with no inflow from tributaries. The calculated transmission losses during that flood along this section were 0.64 x 10^6 m^3 (40 %). *Polak* [1980] estimated the maximum transmission losses per event for the Massos - Aqrabim reach as 1.0 x 10^6 m^3 (0.04 x 10^6 m^3 per 1 km length of channel reach), while *Shentsis et al.* [1999] estimated the mean annual transmission losses for the same reach as 1.6 x 10^6 m^3 (0.064 x 10^6 m^3/km), based on statistical analyses.

Shentsis et al. [1999] estimated mean annual transmission losses of 0.7 x 10^6 m^3 for the Aqrabim - Arava reach. They hypothesized that most of the losses occur along "hydrogeological windows" in the limestone bedrock (i.e. wide openings in the bedrock such as widened fractures, caves, etc.) of the Lower Nahal Zin canyon where the alluvial cover is thin. They estimated that these transmission losses feed directly into deeper and regional bedrock aquifers. However, our measurements at the upper and lower ends of the canyon indicate that there was no significant or measurable decrease in the peak discharge during the floods of Oct. 13-14, 1991 and Nov. 5-7, 1994 [*Greenbaum,* 1996]. This indicates that even minor flood attenuation did not occur and therefore, we conclude that no significant losses occur in the bedrock canyon of the lower Nahal Zin. We propose that an alluvial cover on the bedrock is a prerequisite for an efficient loss of discharge by infiltration. This volume of infiltrating floodwaters is the maximum potential of floodwaters to reach deeper aquifers.

Transmission loss estimates in arid environments throughout the world are scarce. *Schwartz and Schick* [1990] carried out a water balance in Nahal Yael (0.5 km^2) in the Negev Desert. They estimated transmission losses between 357 - 2020 m^3/km into the 3 - 28 m wide channel reach with an alluvial body of 16,000 m^3. At the Habawna-1-Saudi Arabia, transmission losses ranged from 46 to 98.5 % [*Walters*, 1990]. The alluvium properties of the studied reach were classified as group 2 (defined by *Lane,* 1983). Percent of water losses over a wider range (17 – 91%) were calculated for Walnut Gulch [*Lane et al.*, 1971; 1980]. Transmission losses at the Guir-Saoura-Messaoud drainage system (44,000 km^2) in the Sahara Desert (Algeria and Morocco) were estimated by *Dubief* [1953] and *Vanney* [1960] between 1.0–2.5 x 10^6 m^3/km. Transmission losses along the Mojave River for the 1969 floods were estimated as 50 – 60 x 10^6 m^3 (0.5 x 10^6 m^3/km) [*Buono and Lang*, 1980; *Enzel*, 1990]. These floods (RI>18-yr) caused a water table rise of ~10 m.

Estimating the long-term cumulative transmission loss and recharge is problematic. Medium size floods in Nahal Zin—peak discharges of 30 to 200 m^3s^{-1} (not included in the paleoflood record)—may reach volumes of between 0.4-0.5 x 10^6 m^3 and 3-4 x 10^6 m^3, respectively. The estimated transmission losses for these floods range between 0.1-0.2 x 10^6 m^3 and 0.7-0.8 x 10^6 m^3, respectively (Table

2). These amounts of transmission losses equal to between 25-40% and 20-23% of the flood volumes, respectively. The large floods (200-1500 m^3s^{-1}) lose only between 16% and 9% of their volumes. For example, hereby we compare the hydrological characteristics of the 8-yr flood with the 300-yr flood (Table 2)- peak discharges of 100 m^3s^{-1} and 1500 m^3s^{-1} respectively (15 fold), volume of 1.6-1.8 x 10^6 m^3 and 40-42 x 10^6 m^3 (24 fold). The estimated transmission losses are 0.4-0.5 x 10^6 m^3 for the 8 yr flood and 3.7-4.0 x 10^6 m^3 for the 300 yr flood (9 fold). This decline of the relative transmission losses emphasizes the importance of the medium-size floods as contributors to the long-term recharge, although they are not included in the paleoflood record.

One of the solutions for the lack of data on medium-size floods is to remove the large floods (>200 m^3s^{-1}) from the measured record (65 years), then calculate the total transmission losses, and apply the result on the 2000 year long record, and add the paleoflood data. This procedure assumes no major change in the frequency of the medium-size floods. In the case of Nahal Zin, the calculation of the transmission losses of all floods between 30-200 m^3s^{-1}, within the 65 years measured and historical records, yielded a total of 13 x 10^6 m^3, whereas the total transmission losses of the large (>200 m^3s^{-1}) floods is 12 x 10^6 m^3.

Understanding the relations between recharge from large and medium floods enables an estimation of long-term transmission losses based solely on paleoflood records, assuming no major change between the frequencies of large and medium size floods in periods with abundant large floods as well as in periods with few large floods.

SUMMARY

This paper introduces an event-based approach to the estimation of transmission losses and recharge in Nahal Zin and its alluvial fan. Relations between peak discharge, flood volume, and flood duration were established for the 65 years of measured and historical records. We further established high-confidence statistical relations, based on direct measurements, between transmission loss volumes and flood volumes. Transmission losses are significantly reduced when the time interval between two successive floods is less than one week. The antecedent moisture caused by an earlier flood affects transmission losses for up to one month. The relations between transmission losses and flood volume were extrapolated over the 2000-year paleoflood record and provided estimates of transmission losses for particularly large floods. We found that transmission losses of the largest flood in the paleoflood record (peak discharge 1500 m^3s^{-1}) were 4 x 10^6 m^3. These estimates comprise the long-term record of transmission losses of the large (> 200 m^3s^{-1}) floods. The flood duration was equated with the hydraulic and physical characteristics of the alluvial fan aquifer for recharge calculations and compared with the transmission losses. The results of these procedures show that transmission losses and therefore recharge are related by a power decay function to the flood volume, which increases exponentially with flood peak discharge. The cumulative transmission losses during the measured and historical records were 25 x 10^6 m^3, 52% of which were contributed by medium size floods (30-200 m^3s^{-1}) and the remaining 48% by large floods (>200 m^3s^{-1}).

Acknowledgments. Reviews by R. H. French, D. P. Boyle and R. Naranjo have contributed to the improvement of this paper. Michal Kidron from the cartography laboratory of the Hebrew University drew the figures. Figures 1, 2, 3, and 6 are derived from "The Paleoflood Record of a Hyperarid Catchment, Nahal Zin, Negev Desert, Israel," by Noam Greenbaum, Asher P. Schick, and Victor R. Baker, copyright 2000 John Wiley and Sons, Ltd. This study was funded by the Dead Sea Works. This paper forms a part of Uri Schwartz's PhD dissertation at the Hebrew University.

REFERENCES

Abdulrazzak, J.M., and H.J. Morel-Seytoux, Recharge from an ephemeral stream following wetting front arrival to water table, *Water Resour. Res.*, 19, 1, 194-200, 1983.

Abdulrazzak, J.M., A.U. Sorman, and A.O. Rizaiza, Estimation of natural groundwater recharge under Saudi Arabian arid climatic conditions, NATO Advanced Study Institute Series, Series-C, *Mathematical and Physical Sciences*, 222, 125-138, 1987.

Baker, V. R., Paleoflood hydrology and extraordinary flood events, *J. Hydrol.*, 96, 79-99, 1987.

Baker, V.R., and R.C. Kochel, Flood sedimentation in bedrock fluvial systems, pp. 123-128, in *Flood Geomorphology*, edited by V.R. Baker, R.C. Kochel, and P.C. Patton, John Wiley and Sons, New York, 1988.

Baker, V.R., R.C., Kochel, P.C. Patton, and G., Pickup, Paleohydrologic analysis of Holocene flood slack-water sediments, Special Publication, *International Association Sedimentologists*, 6, 229-239, 1983.

Baker, V.R., G. Pickup, and R.H. Webb, Paleoflood hydrologic analysis at ungauged sites, central and northern Australia, pp. 325-328, in *Regional Flood Frequency Analysis*, edited by V.P. Singh, Reidel, Boston, 1987.

Begin, Z.B., A. Ehrlich, and Y. Nathan, Lake Lisan: the Pleistocene precursor of the Dead Sea, 30 pp., *Israel Geol. Surv. Bull.*, 63, 1974.

Benito, G., M.J. Machado, A. Pérez-González, and A. Sopeña, Paleoflood hydrology of the Tagus River, central Spain, pp. 317-333, in *Paleohydrology and Environmental Change*, edited by G. Benito, V.R. Baker, and K.J. Gregory, John Wiley and Sons, 1998.

Bentor, Y.K., and A. Vroman, The geological map of the Negev, 1:100,000, sheet 16, Mt. Sedom, with explanatory text, 117 pp., *Israel Geological Survey*, 1960.

Bueno, A., and D.J. Lang, Aquifer discharge from the 1969 and 1978 floods in the Mojave River basin, California. Denver.

Colorado. *U.S. Geol. Surv. Water Resour. Invest. Open-File Rep.* 80-207, 1980.

DeLeeuw, A., Protection against floods, Rep. 8 and 11, *Palestine Potash Ltd.*, 1945.

Dubief, J., Ruissellement superficiel au Sahara, *Colloq. Int. C.N.R.S.*, 35, 303-314, 1953.

Ely, L.L. and V.R. Baker, Reconstructing paleoflood hydrology with slackwater deposits, Verde River, Arizona, *Phys. Geog.*, 6, 103-126, 1985.

Ely, L.L., E. Enzel, V.R. Baker, and R.C. Cayan, A 5000 years record of extreme floods and climate change in the southwestern United States, *Science*, 262, 410-412, 1993.

Ely, L.L., Y. Enzel, V.R. Baker, V.S. Kale, and S. Mishra, Changes in the magnitude and frequency of Holocene monsoon floods on the Narmada River, Central India, *Geol. Soc. Am. Bull.*, 108, 1134-1148, 1996.

Ely, L.L., Y. Enzel, J.E. O'Connor, and V.R. Baker, Paleoflood records and risk assessment: examples from the Colorado River basin, pp. 105-112, in *Water Resources Engineering Risk Assessment, NATO ASI Series*, G29 edited by J. Ganoulis, Springer-Verlag, Berlin, Heidelberg, 1991.

Enzel, Y., Hydrology of a large, closed arid watershed as a basis for paleohydrological and paleoclimatological studies in the Mojave River drainage system, southern California, 315 pp., PhD. Dissertation, University of New Mexico, 1990.

Enzel, Y., Flood frequency of the Mojave River and the formation of late Holocene playa lakes, southern California, *The Holocene*, 2, 11-18, 1992.

Enzel, Y., D.R. Cayan, R.Y. Anderson, and S.G. Wells, Atmospheric circulation during Holocene lake stands in the Mojave Desert: evidence of regional climate change, *Nature*, 341, 44-47, 1989.

Enzel, Y., E.E. Ely, P.K. House, and V.R. Baker, The contribution of long-term records of hydrologic extremes to risk analysis, pp. 97-104, in *Water Resources Engineering Risk Assessment, NATO ASI Series*, G29, edited by J. Ganoulis, Springer-Verlag, Berlin, Heidelberg, 1991.

Enzel, Y., L.L. Ely, J. Martinez-Goytre, and R. Gwinn-Vivian, Paleofloods and dam-failure flood on the Virgin River, Utah and Arizona, *J. Hydrol.*, 153, 291-315, 1994.

Frumkin, A., N. Greenbaum, and A.P. Schick, Paleohydrology of the Northern Negev: comparative evaluation of two catchments, in *Water, Environment and Society in Times of Climatic Change*, pp. 97-111, edited by A.S. Issar, and N. Brown, Kluwer, the Netherlands, 1998.

Greenbaum, N., Paleofloods in the large ephemeral stream systems of the Central Negev, Israel, 154 pp., PhD. Dissertation (in Hebrew with abstract in English), Hebrew University of Jerusalem, Jerusalem, Israel, 1996.

Greenbaum, N., A. Margalit, A.P. Schick, D. Sharon, and V.R. Baker, A high magnitude storm and flood in a hyperarid catchment, Nahal Zin, Negev Desert, Israel, *Hydrol. Proc.*, 12, 1-23, 1998.

Greenbaum, N., A.P. Schick, and V.R. Baker, The paleoflood record of a hyperarid catchment, Nahal Zin, Negev Desert, Israel, *Earth Surf. Proc. Landf.*, 25, 951-971, 2000.

Hillel, D., V.D. Drentos, and Y. Stylianou, Procedure and test of an interval drainage method for measuring soil hydraulic characteristics in situ, *Soil Sci.*, 114, 395-400, 1972.

Hydrologic Engineering Center, *HEC-2 Water Surface Profiles: Program User's Manual, U.S.* Army Corps of Engineers, Davis, CA., 1982.

Katz, D., Kikar Sedom area-geohydrological investigations, *TAHAL*, 01/75/36, (in Hebrew), Tel Aviv, 1976.

Kochel, R.C., V.R. Baker, and P.C. Patton, Paleohydrology of southwestern Texas, *Water Resour. Res.*, 18, 1165-1183, 1982.

Kuells, C., C. Leibundgut, U. Schwartz, and A.P. Schick, Channel infiltration study using dye tracers. *Proceedings of the IAHS Symposium on Application of Tracers in Arid Hydrology, Vienna, August 1994*, 232, pp. 429-436, 1995.

Lane, L.J., M.H. Diskin, and K.G. Renard, Input-output relationships for an ephemeral stream channel system, *J. Hydrol.*, 13, 22-40, 1971.

Lane, L.J., V.A. Ferreira, and E.D. Shirley, Estimating transmission losses in ephemeral stream channel. *Proceedings of the 1980 Meeting of the Arizona Section, American Water Resources Association and the Hydrology Section on Hydrology and Water Resources in Arizona and the Southwest Meeting 1980,* Arizona-Nevada Academy of Science, Las Vegas, NV, 1980.

Lane, L.J., Estimating transmission losses, *Proceedings of the ASCE, Irrigation Division, Specialty Conference on the Development and Management Aspects of Irrigation and Drainage Systems, San Antonio, TX, July 1985*, pp. 106-113, 1985.

Lane, L.J., Transmission losses, flood peaks, and groundwater recharge, *Proceedings of the ASCE, Hydrology & Irrigation Division, International Symposium on Hydraulics/Hydrology of Arid Lands, San Diego, CA, July 30-August 2, 1990*, pp. 343-348, 1990.

Naor, H., I. Granit, and Y. Harash, Nahal Zin-Nahal Ye'elim area, inspection of the hydrological and operative situation of the aquifers and the wells, *TAHAL*, 04/86/8, (in Hebrew), Tel Aviv, 1986.

Naor, H., M. Fink, and I. Granit, Nahal Zin-Nahal Ye'elim area, inspection of the hydrological and operative situation of the aquifers and the wells, *TAHAL Rep.*, 04/91/2377, (in Hebrew), Tel Aviv, 1991.

Naor, H., Proposals for hydrogeological studies in the area of Sedom – Nahal Zin, *TAHAL Rep*, 03/94/1, (in Hebrew), Tel Aviv, 1994.

O'Connor, J.E., L.L. Ely, E.E. Wohl, L.E. Stevens, T.S. Melis, V.S. Kale, and V.R. Baker, A 4500-years record of large floods on the Colorado River in the Grand Canyon, Arizona, *J. Geol.*, 102, 1-9, 1994.

O'Connor, J. E. and R.H. Webb, Hydraulic modeling for paleoflood analysis, pp. 383-402, in *Flood Geomorphology,* edited by V.R. Baker, R.C. Kochel, and P.C. Patton, John Wiley and Sons, New York, 1988.

Osterkamp, W.R., L.J. Lane, and C.S. Savard, Recharge estimates using a geomorphic/distributed-parameter simulation approach, Amargosa River Basin, *Water Resour. Bull.*, 30, 3, 493-507, 1994.

Osterkamp, W.R., L.J. Lane, and C.M. Menges, Techniques of ground-water recharge estimates in arid/semi-arid areas, with examples from Abu Dhabi, *J. Arid Environ.*, 31, 349-369, 1995.

Partridge, J.B. and V.R. Baker, Paleoflood hydrology of the Salt River, central Arizona, *Earth Surf. Proc. Landf.*, 12, 109-125, 1987.

Patton, P.C., V.R. Baker, and R.C. Kochel, Slack-water deposits: A geomorphic technique for the interpretation of fluvial paleohydrology, pp. 225-253, in *Adjustment of the Fluvial System*, edited by D.D. Rhodes, and G.P. Williams, Kendall/Hunt Publication Corporation, Dubuque, Iowa, 1979.

Polak, S., Nahal Zin catchment; hydrological analysis for the period 1951-1978 (in Hebrew), *Israel Hydrol. Serv. Rep.* Hydro./5/1980, 39, 1980.

Renard, K. G., L.J. Lane, J.R. Simanton, W.E. Emmerich, J.J. Stone, M.A. Weltz, D.C. Goodrich, and D.S. Yakowitz, Agricultural impacts in an arid environment: Walnut Gulch case study, *Hydro. Sci. Techno.*, 9, 1-4, 145-190, 1993.

Rose, C.W., W.R. Stern, and J.E. Drummond, Determination of hydraulic conductivity as a function of depth and water content for soil in situ, *Aust. J. Soil Res.*, 3, 1-9, 1965.

Schick, A.P., Fluvial processes and settlement in arid environments, *GeoJournal*, 3, 351-360, 1977.

Schick, A.P., Arid Zone Geosystems, 170 pp., Division of Physical Geography, The Hebrew University of Jerusalem, 1980.

Schick, A. P., Hydrologic aspects of floods in extreme arid environments, pp. 189-203, in *Flood Geomorphology*, edited by V.R. Baker, R.C. Kochel, and P.C. Patton, John Wiley and Sons, New York, 1988.

Schwartz, U., C. Kuells, A.P. Schick, and C. Leibundgut, Floods in Nahal Zin: infiltration of floodwater into the alluvium and the recharge to the Zin Aquifer, Rep. 1, *The Flood Research Group*, Hebrew University, Jerusalem, 16, 1995.

Schwartz, U., Water in the alluvial fill in an area of extreme aridity, 100 pp., M.Sc. Thesis, Department of Physical Geography, The Hebrew University of Jerusalem, Jerusalem, (in Hebrew), 1986.

Schwartz, U., Surface and near-surface responses to floods in a large channel (Nahal Zin) in the context of an alluvial aquifer in a hyper-arid environment, PhD. Dissertation, Hebrew University of Jerusalem, Jerusalem, Israel, in preparation, 2001.

Schwartz, U., and A.P. Schick, Water in an alluvial fill in an area of extreme aridity, *Ofaqim Be'Geographia (Horizons in Geography)*, 31, 109-122 (in Hebrew), 1990. .

Shentsis, I., L. Meirovich, A. Ben-Zvi, and E. Rosenthal, Assessment of transmission losses in Nahal Neqarot and its tributaries and their contribution to the regional aquifer, 62 pp., *Israel Hydrol. Serv. Rep.* Hyd/7/98, (in Hebrew), 1998.

Shentsis, I., L. Meirovich, A. Ben-Zvi, and E. Rosenthal, Assessment of transmission losses in the Tsin water course and its contribution to the regional aquifer, *Israel Hydrol. Serv. Rep.* Hyd/1/97, 88, (in Hebrew), 1997.

Shentsis, I., L. Meirovich, A. Ben-Zvi, and E. Rosenthal, Assessment of transmission losses in Nahal Paran water course (Halamish-Bottle Neck reach) and monitoring its contribution to the regional aquifer, 37 pp., *Israel Hydrol. Serv. Rep.* Hyd/10/98, (in Hebrew), 1998.

Shentsis, I., L. Meirovich, A. Ben-Zvi, and E. Rosenthal, Assessment of transmission losses and groundwater recharge from runoff events in a wadi under shortage of data on lateral inflow, Negev, Israel, *Hydrol. Proc.*, 13, 1649-1663, 1999.

Stedinger, J. R., and V.R. Baker, Surface water hydrology: historical and paleoflood information, *Rev. Geophys.*, 25, 119-124, 1987.

Stedinger, J. R., and T.A. Cohn, Flood frequency analysis with historical and paleoflood information, *Water Resour. Res.*, 22, 785-793, 1986.

Stedinger, J. R., R. Surani, and R. Therivel, *MAX Flood Frequency Analysis Package, Version 3.0 and MAX User's Guide*, 51 pp., Department of Environmental Engineering, Cornell University, Ithaca, NY, 1988.

Stephens, D.B., and R.K. Knowlton, Soil water movement and recharge through sand at a semiarid site in New Mexico, *Water Resour. Res.*, 22, 6, 881-889, 1986.

TAHAL, Protection of the Dead Sea Works from floods, 28 pp., Tahal Consulting Engineers Ltd, (in Hebrew), Tel Aviv, 1961.

TAHAL, The diversion of Nahal Zin, Preliminary program, 10 pp., Tahal Consulting Engineers Ltd, (in Hebrew), Rep. 04/81/40, (in Hebrew), Tel Aviv, 1981.

Talsma, T., In situ measurement of sorptivity, *Aust. J. Soil Res.*, 7, 277-284, 1969.

Talsma, T., Some aspects of three dimensional infiltration, *Aust. J. Soil Res.*, 8, 179-184, 1970.

Talsma, T., Infiltration in field soils, *10th International Congress on Soil Science*, Moscow., 1973.

Vanney, J.R., 1960. Pluie et crue dans le Sahara nord-occidental, 4, Memoirs regionaux de l'Institute des Recherches Sahariens, Algiers, 1960.

Walters, M.O., Transmission losses in arid region, *J. Hydrau. Engin.*, 116, 1, ASCE, 129-138, 1990.

Webb, R.H., J.E. O'Connor, and V.R. Baker, Paleohydrologic reconstruction of flood frequency on the Escalante River, south central Utah, pp. 403-418, in *Flood Geomorphology*, edited by V.R. Baker, R.C. Kochel, and P.C. Patton, John Wiley and Sons, New York, 1988.

Williams, G.E., The Central Australian stream floods of February-March 1967, *J. Hydrol.*, 11, 185-200, 1970.

Wohl, E.E., Northern Australian paleofloods as paleoclimatic indicators, 285 pp., Unpublished PhD dissertation, University of Arizona, Tucson, 1988.

Noam Greenbaum, Departments of Geography and Natural Resources & Environmental Management, University of Haifa, Mount Carmel, Haifa, 31905, Israel, E mail nryg@zahav.net.il

Uri Schwartz, Department of Geography, Hebrew University of Jerusalem, Mount Scopus, Jerusalem 91905, Israel, E mail msuris@mscc.huji.ac.il

Asher P. Schick, Department of Geography, Hebrew University of Jerusalem, Mount Scopus, Jerusalem 91905, Israel, E mail schick@vms.huji.ac.il

Yehouda Enzel, Institute of Earth Sciences, Hebrew University of Jerusalem, Givat Ram, Jerusalem 91904, Israel, E mail yenzel@vms.huji.ac.il

Caves and Their Potential Use in Paleoflood Studies

Gregory S. Springer

Department of Earth Resources, Colorado State University, Colorado

Caves may act as repositories for geomorphic information. However, few studies have explicitly addressed subterranean paleoflood deposits and their relationships to adjacent surface streams. True caves, as opposed to shelters and alcoves, may possess complex relationships with overlying valley walls, base level streams, and adjacent catchments. The resulting sedimentary record may co-mingle sediments from many inputs. Discerning the origin and meaning of the sediments is possible through analysis and mapping of sediment facies, textures, and stratigraphic discontinuities. Observations made in caves along the Cheat and Greenbrier rivers, West Virginia show that historical flood deposits form allostratigraphic packages bounded by scour surfaces, fungal-infested organic mats, and mud-cracks. Deposits consist primarily of loamy silts, fine sands, and abundant woody debris. Preservation of the deposits is only assured where influx of water from valley walls is minimal. Locally, a humid, temperate climate yields high in-cave humidities and a nearly constant year round temperature of ~12° C. These conditions favor bioturbation and decay of paleoflood deposits. True caves are usually integrated components of fluvial systems and much is to be gained by integrating fluvial and karst studies, particularly in climatic regions unfavorable to preservation of surficial paleoflood deposits.

INTRODUCTION

Modern studies of paleoflood hydrology rest upon the pioneering work of *J Harlan Bretz* [1923a, 1923b, 1929, 1969]. His name is indelibly associated with catastrophic paleofloods of the American northwest, but among his many works are at least seven papers concerning caves. Caves were of special interest to Bretz and the most influential of his speleological papers was, "Vadose and phreatic features of limestone caverns." The central theme of the paper was that caves are formed by groundwater flow independent of surficial processes [*Bretz*, 1942]. This hypothesis was adopted from *Davis* [1930] and presumed that observable cave streams are underfit, recent invasions of caves that have emerged from beneath the water table. This hypothesis came to be known as the "American school" of thought. The school's tenets were abandoned in the 1960s when caves were shown to be components of overlying landscapes [see *Watson and White*, 1985]. As a result, while we acclaim Bretz for his prescient work on paleofloods his contributions to karst studies are largely forgotten. More importantly, Bretz's favorite disciplines, karst and fluvial studies, continue to be pursued independently of one another because of the mistaken perception that karst is essentially a groundwater phenomenon.

The estrangement of karst and fluvial studies has hindered the use of caves in geomorphic studies and only a handful of detailed studies have explicitly addressed subsurface flood deposits in relation to the flood histories of surface streams [e.g. *Kochel*, 1982; *Patton and Dibble*, 1982; *Gillieson et al.*, 1991; *Springer and Kite*, 1997]. This partly reflects the fact that the vast majority of paleoflood studies have been pursued in arid and semi-arid regions where exposed deposits may survive for lengthy periods of time and subsurface deposits

Ancient Floods, Modern Hazards: Principles and Applications of Paleoflood Hydrology
Water Science and Application Volume 5, pages 329-343

may be superfluous [see *Gillieson et al.*, 1991]. In contrast, studies in humid regions have found that surface deposits are not readily preserved and that caves may record information absent on the surface [*Kite and Linton*, 1993; *Springer and Kite*, 1997].

The ability to recognize paleofloods in humid climates will be an important avenue of research as the field of paleoflood hydrology expands to cover all climatic regimes. Although ancient floods may be recorded on alluvial floodplains in humid regions, work along a bedrock river has shown that preservation of paleostage indicators (PSIs) is inhibited by biogenic activity, scour, and lack of available deposition surfaces [*Kite et al.*, this volume]. As a result, extending studies of paleoflood hydrology to humid regions may require the use of new sources of information. The subterranean world, a realm of particular interest to the father of paleoflood hydrology, is an underutilized tool for extending flood records in such regions. Shelters, alcoves, and true caves (voids with lengths exceeding entrance widths) are found along rivers in virtually all climatic settings. True caves are the focus of this paper and consist principally of solution-formed voids; the word 'cave' should not be applied to shelters or alcoves.

This article provides a general review of existing studies of cave-borne paleoflood deposits and presents an overview of the relationships between caves, surface streams, and subterranean deposits. The purpose of the article is to encourage and aid the use of subterranean paleoflood deposits by fluvial geomorphologists.

CAVES AND STUDIES OF SURFACE STREAMS

An erosional void, regardless of whether it is sculpted above or below ground, may serve as a repository of geomorphic information (e.g. PSIs). Thus, caves may preserve evidence of past geomorphic events whose effects extended into the subsurface. Karst geomorphologists have exploited this fact to quantify long-term river incision rates, reconstruct the timing of continental-scale changes in drainage basin organization, and to reconstruct climate fluctuations [e.g. *Burns et al.*, 1998; *Granger and Fabel*, 1999; *Granger et al.*, 1997; *Springer et al.*, 1997]. Despite the relevance of such studies to fluvial geomorphology, the over-lapping disciplines of karst and fluvial geomorphology are rarely treated simultaneously. Some of the estrangement is inherited from those who thought that accessible caves were formed by deeply-seated processes whose geomorphic records could tell us little about surface processes [e.g. *Davis*, 1930]. Advances in karst geomorphology have since produced a general understanding of the relationship between caves, surface processes, and associated deposits while showing that most subterranean streams behave similarly to surface streams [*Ford and Williams*, 1989].

Subterranean paleoflood deposits were used to extend the flood record within the ~500 km^2 Lennard River catchment in the Kimberley region of northwestern Australia [*Gillieson et al.*, 1991]. However, high temperatures and a long dry season favor preservation of surface deposits and were the primary sources of paleoflood information used by *Gillieson et al.* [1991]. The recognized deposits consisted mostly of fine sands, driftwood deposits, and trim lines on lichen-encrusted gorge walls. Four of 25 recognized PSIs were found in caves near the downstream terminus of a limestone gorge. Correlation of PSIs, based upon elevations and a few analytical ages, suggests that the cave deposits record minimum peak stages similar to those of PSIs found further upstream. At the very least, the study confirms that subterranean paleoflood de-posits can be useful. However, the usefulness of subterranean PSIs will be greatest where surface deposits are much less likely to be preserved.

Surface deposits are least likely to be preserved in temperate regions that receive ample rainfall, but do not possess high evapotranspiration rates. In one such area, the usefulness of combining speleologic and surficial studies can be seen in the complementary research of *Kite and Linton* [1993] and *Springer and Kite* [1997]. The two papers examine slackwater sediments deposited in caves and tributary mouths along the Cheat River, West Virginia during a catastrophic, November 1985 flood (Figure 1). The 1985 flood was generated by the combined forces of a decaying tropical storm and a frontal system and remains twice as large as any other historic flood on the Cheat River, with a peak discharge of ~5400 m^3 s^{-1} [*Kite and Linton*, 1993]. Simultaneous flooding along the Greenbrier River in southern West Virginia produced extensive damage and the event was the flood of record until being slightly exceeded by a rain on snow event in January 1996. The 1985 flood caused extensive flooding and loss of lives throughout Maryland, West Virginia, and Virginia while the 1996 flood was associated with extensive property damage along the Cheat and Greenbrier rivers.

Several small towns were largely destroyed by the Cheat River in 1985. The magnitude of the flood led *Kite and Linton* [1993] to examine flood-plain and tributary mouth stratigraphy associated with the flood while simultaneously seeking paleoflood deposits. *Kite and Linton* [1993] failed to find any deposits that could conclusively be labeled as paleoflood deposits and found that slackwater sediments lay far below high water marks. High water marks are unlikely to be preserved in a humid climate and the comparatively low-lying slackwater sediments are poor estimates of peak stage [*Kite and Linton*, 1993]. Therefore, the record of the 1985 flood was not representative of the magnitude of the event.

Flotsam and slackwater sediments were deposited in caves up to 14 m above the banks of the Cheat River and these

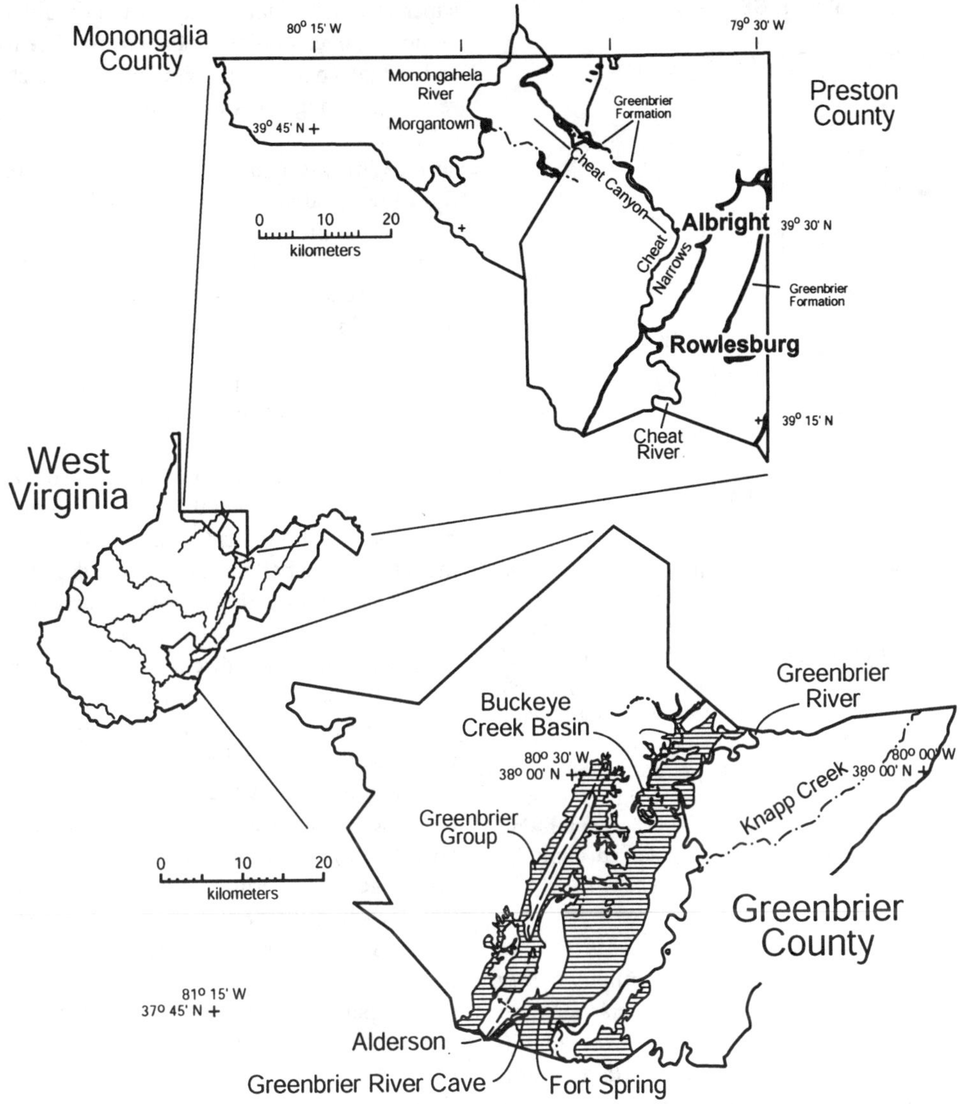

Figure 1. Locations of the Cheat and Greenbrier rivers of West Virginia. The subterranean paleoflood studies discussed in this chapter were conducted in the Cheat River Canyon (labeled) and between Fort Spring and Alderson on the Greenbrier River. Limestone outcrops are shown in black (Greenbrier Formation) or stripping (Greenbrier Group). Both basins possess extensive limestone outcrops, but the rivers only flow across limestone along a few short segments.

deposits represent the most definitive estimates of peak stage in a cavernous reach of the river [*Springer and Kite*, 1997]. Following the 1996 flood, additional work on the Cheat River suggested that the lack of surficial paleoflood deposits may be related to the absence of low-lying deposition surfaces capable of pre-serving deposits along this high gradient, bedrock river [*Kite et al.*, this volume]. *Springer and Kite* [1997] found some low-lying paleoflood deposits as well as some high above the modern river and unrelated to the modern flood regime (Figure 2). They did not find subsurface paleoflood deposits at comparable elevations to the highest 1985 flood deposits. As an indirect outgrowth of the work by *Springer and Kite* [1997], a subterranean paleoflood study has been initiated along the Greenbrier River in southeastern West Virginia and the earliest observations of this project are presented below.

SUBTERRANEAN CHANNELIZED FLOW

In general, cave passages act as extensions of or substitutes for surface channels. Holokarst, also known as sinkhole karst, comprises those terrains wherein the drainage net lies almost entirely under-ground. In such terrains, relief is usually less than limestone thicknesses, surface streams are short,

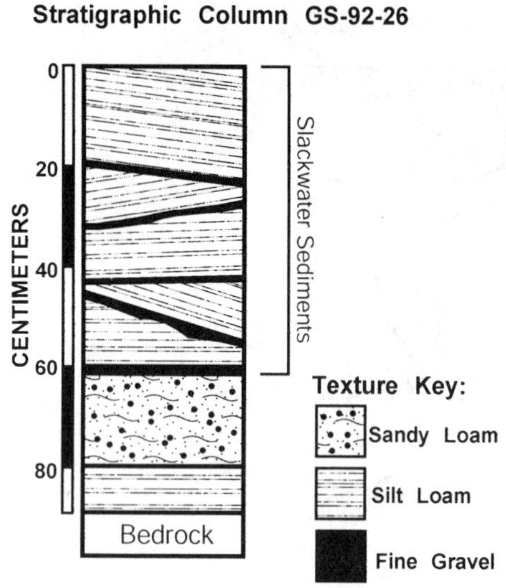

Figure 2. Paleoflood deposits derived from the Cheat River, West Virginia and preserved as allostratigraphic packages in Cornwell Cave. Individual paleoflood deposits, silt loams, are bounded by scour surfaces and gravelly infiltrates. The varying inclinations of individual beds is attributable to parallel accretion along scour surfaces with varying orientations. Parallel accretion is a process whereby fine-grained sediments accumulate parallel to inclined surfaces in slowly moving floodwaters [*Reams*, 1968]. The interbedded gravels are largely composed of red, argillaceous shale chips derived from overlying strata. Identical deposits are forming in the cave today where vertical domes and enlarged joints lie shallowly beneath the valley wall. Modified from *Springer and Kite* [1997].

and only the largest streams penetrate or pass entirely across extensive expanses of soluble outcrops without disappearing [*Miller*, 1996]. This discuss-ion need not explore the intricacies of holokarst except as they are related to large, undiverted streams. Fluviokarst, as the name implies, includes those terrains wherein the surface drainage network is well-developed, but caves divert flow under-ground (Figure 3). Geomorphologic textbooks often place considerable emphasis on holokarst and little if any mention is made of fluviokarst [e.g. *Easter-brook*, 1993]. Nonetheless, fluviokarst comprises a significant fraction of the world's karst terrains. Fluviokarst comprises the bulk of all karst in eastern North America and Europe and limestones are present along many rivers of the world.

Relative to studies of paleofloods, the importance of fluviokarst lies in the interaction between surface and subsurface processes. As in all karst, fluviokarstic caves act as water and sediment conduits. Caves are discrete pathways and may transmit more than 80% of all water passing through fluviokarst terrains [*Atkinson*, 1977]. Subsurface conduits are not indicative of water table position and are often independent of it [*White*, 1988; *Ford and Williams*, 1989]. In spite of superficial appearances, most caves should not be regarded as being in the realm of traditional subsurface hydrology studies. Individual conduits are fed by surface streams and point recharge through sinkholes, display the morphological traits of surface streams, transmit water at velocities typical of surface streams, and are capable of transmitting large quantities of bedload and suspended load [*White*, 1988]. Ignoring intermittent pressure flow, the equations that describe subterranean stream flow are the same as those that describe flow in surface channels. However, not all caves are likely to prove useful in paleoflood studies.

UTILITY OF CAVES IN PALEOFLOOD STUDIES

The usefulness of subterranean paleoflood information will be dependent upon the goals of an investigation. Cave-borne paleoflood information should be most valuable when the information pertains to a stream that flows entirely underground or almost entirely above ground. Surface streams with complex interactions between subsurface conduits may not have consistent stage-discharge relationships over centennial time scales because flow paths, bank storage, and the fraction of subsurface stream flow may change as a result of sediment occlusion or removal. Similarly, if sub-surface flow paths and fractions are unknown, evidence of paleofloods may not translate to useful interpretations of paleofloods, because it will be impossible to reliably determine past stages and discharges. Furthermore, hydraulic modeling will be extremely difficult for even the simplest mixed surface-subsurface systems when open and closed conduit flow occur in parallel. Therefore, investigators interested in obtaining reliable paleostage and discharge information should avoid all systems wherein both surface and subsurface flow paths convey or store significant or unknown fractions of flood flow.

Studies of streams diverted entirely underground may be useful, but typical paleoflood investigations are concerned with the flood histories of surface streams and their import for human lives and property. Therefore, we shall not concern ourselves with paleofloods in predominately subterranean fluvial systems. This begets the question, "What surface-subsurface systems are suitable for reconstructing the flood histories of surface streams and how can we recognize them?" As suggested above, the criteria that a stream must flow almost entirely above ground seems an appropriate answer to the first half of the question. However, discerning whether sub-surface flow is significant can be extremely difficult. In general, the challenge is to determine the relative volumes of caves to surface channels and whether caves transmit significant amounts of stream flow in parallel with the surface channel. Such determinations are made on the basis of cave surveys and studies of subsurface pathways. Fortunately, amateur surveys of caves are available for many karst regions in developed nations [e.g. *Garton and Garton*, 1976; *Dasher and Balfour*, 1994].

Assuming subsurface flood storage and conveyance are negligible, subterranean paleoflood deposits should, ideally, be interpreted the same as surface deposits and this is an assumption of *Gillieson et al.* [1991]. The fact that river-associated caves must necessarily lie in bedrock banks eases one principal concern of paleoflood hydrologists. Namely, whether the geometry of the surface channel has remained fixed. Discerning whether observed deposits are the product of flooding with-in the surface channel is a far greater problem. The recognition and interpretation of flood-related sediments in valley wall caves is complicated by the diversity and abundance of subterranean, fine-grained deposits. To better understand subsurface deposits, it is necessary to first have a general understanding of the relationships between caves and surficial processes.

CAVES AS COMPONENT LANDFORMS

Intimate relationships exists between surface landforms and the vast majority of caves. Solutional caves that form independent of surface processes are comparatively few [see *Palmer*, 1991]. This article focuses entirely upon caves formed in conjunction with surface processes, particularly large rivers, although much of the discussion is relative to all caves adjacent to surface streams. The tendency of large streams to pass entirely across soluble outcrops reflects the inability of subsurface conduits to grow large or fast enough to completely capture streams with large catchment areas [*Smart*, 1988; *Miller*, 1996]. However, during base flow even large streams may flow entirely underground and surface channels may only be utilized during floods.

Solutional caves display a wide variety of relationships to surface streams. While caves are commonly classified according to plan geometries [*White*, 1988], process-based classifications will be preferred when caves and interior deposits are used to interpret geomorphic histories; plan geometries are associated with particular processes, but not exclusively [see *Palmer*, 1991]. For the discussion at hand, we may categorize caves relative to master surface streams as either tributaries, diversions, or subterranean floodplains [e.g. *Jones*, 1971].

Tributary caves may reach surface streams through subsurface connections, surface flow from springs, or both. As an example, Buckeye Creek Cave drains a 14 km^2, topographically enclosed basin in southeastern West Virginia and contributes to Spring Creek (Figure 3). Buckeye Creek reaches the bed of Spring Creek via subsurface pathways during low stages, but can overflow to create a large spring 5 m above the banks of Spring Creek. Backflooding of the cave is produced by flooding in the Buckeye basin or along Spring Creek. As such, slackwater sediments in the cave can not be uniquely associated with flood events on Spring Creek. For this same reason, many tributary caves will prove unsuitable for reconstructing flood histories on adjacent surface streams unless there is unambiguous evidence of the floodwater source.

Diversion caves pirate water from one channel segment to another within the same stream. Diversion of surface waters is not necessarily accomplished by a single cave passage; commonly multiple surface-subsurface connections divert flow into a master passage. Subsurface water may follow conduits parallel to the valley wall, through meander necks, or beneath surface drainage divides. It is common for valley-parallel caves to transmit base flow and small magnitude floods while surface channels remain dry. In such cases, overflow during large floods produces a complex mixture of

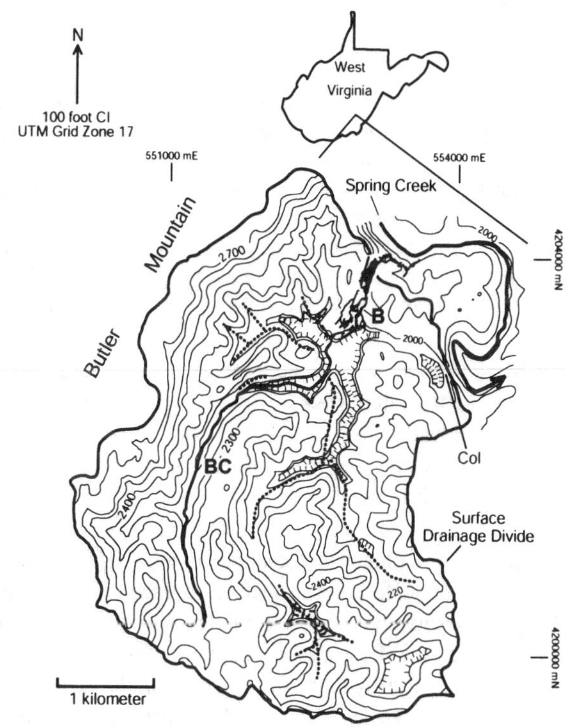

Figure 3. A classic fluviokarst basin, the Buckeye Creek (BC; solid line) catchment in southeastern West Virginia. Dotted lines denote primary subsurface pathways within the catchment. Sediment and recharge are constrained to exit the basin by subsurface pathways, principally the southeastern-most passage of Buckeye Creek Cave (B). Note that flow is largely parallel to existing valley axes and that Buckeye Creek once reached Spring Creek via the designated col. At least four relict passages (tiers) lie to the northwest of the present stream passage in Buckeye Creek Cave. Sediment occlusions near the downstream terminus of Buckeye Creek Cave cause backflooding and sedimentation that precludes recognition of paleoflood deposits derived from Spring Creek. Flow paths and cave plan from *Dasher and Balfour* [1994].

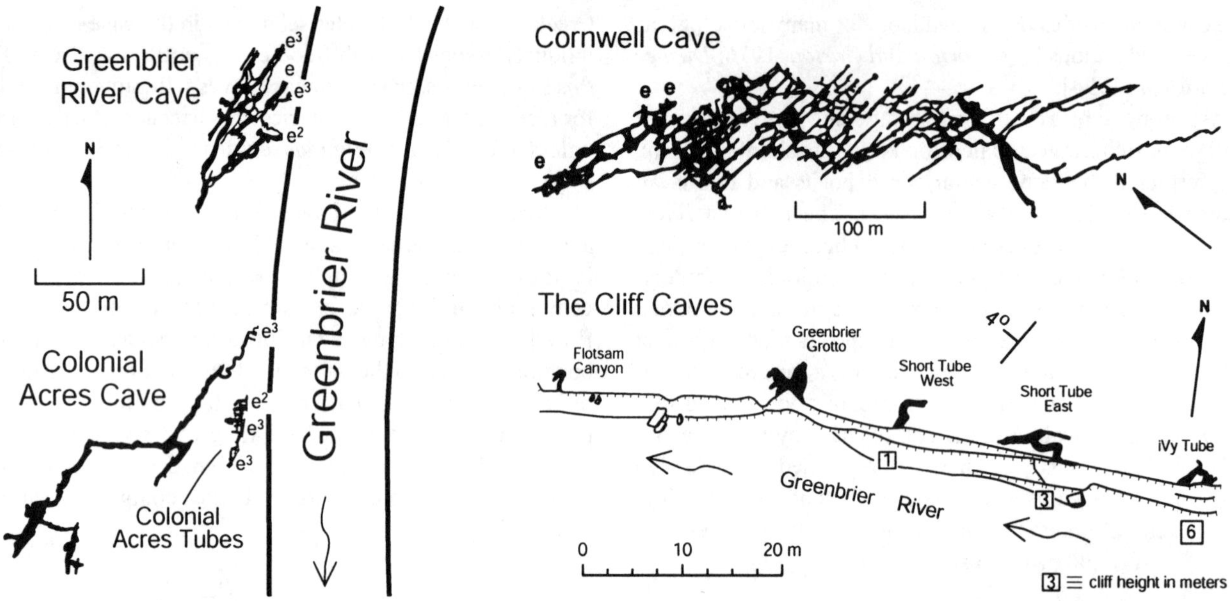

Figure 4. Plan views of injection caves along Cheat and Greenbrier rivers. Lower case e's denote entrances, powers denote number of multiple entrances in a cluster. Anastomotic mazes, such as Greenbrier River Cave and Colonial Acres Tubes, are developed along bedding planes. Network mazes, such as Cornwell Cave along the Cheat River, form along rectangular joint systems and intersections of joints and bedding planes. Only maze remnants are observed where maze caves have been largely filled or removed by lateral erosion. The Cliff Caves exemplify such a fate, but are useful for paleoflood studies despite their small sizes. The caves span a vertical range of 9 m, which is believed to be greater than the historic flood range at this site, and can potentially record a wide range of flows. Equally important, their small sizes and dependance upon the river for sediment input simplifies interpretation of their deposits. Cornwell Cave plan from *Garton and Garton* [1976].

subsurface and surface flow that may extend for kilometers. Because the associated caves often lie near or below the channel bed elevation, they can offer little information about paleostages or discharges. Cave passages in relict or intermittently active diversion cave segments that lie above the channel bed may yield paleoflood deposits [*Traugh et al.*, 1999], but the usefulness of the deposits will be determined by knowledge of surface/subsurface flow fractions and stage-discharge relationships. Such knowledge will be least attainable where karst is best developed and where large cave systems parallel significant portions of channels. Because of their size, such cave systems are the most likely to have undergone exploration and study, but the co-mingling of sediments derived from diverted flood-waters, backflooding, valley wall sources, and re-working may yield extremely complex paleoflood records.

Channel and bank injection caves are analogous to subterranean floodplains. Such caves are small-scale diversions that receive recharge from a surf-ace channel [*White*, 1969]. As a sub-class of diversions, floodwater injection caves are here de-fined as those with shallow flow paths that transmit or store floodwaters along short channel segments (≤300 m; Figure 4). Referring to them as 'phreatic caves', *Springer et al.* [1997] use analyses of cave passage morphologies, sediment facies, and geomorphic relationships to show that floodwater inject-ion caves form near or below the bed of Cheat River. At present, it is sufficient to note that floodwater injection caves are almost entirely dependent upon the adjacent surface channel for sediment and recharge. Sediment received from overlying valley walls is often easily separable from river-associated deposits, hence floodwater injection caves may act as ideal repositories of information about mainstem flooding and erosion histories (discussed below). The small size of most floodwater injection caves enhances their appeal as paleoflood recorders, because they often lack the complex, multiple generation sediment deposits of large cave systems.

CAVE SEDIMENTS IN PALEOFLOOD STUDIES

Origins and Classification of Sediment Facies

In a very real sense, caves may be regarded as the ultimate bedrock channels. However, the majority of cave streams flow atop detrital sediments that are intermittently transported as bedload or suspended load. As a result, stream-borne deposits predominate in caves although most surficial deposits are either present or have subterranean analogs [see *White*,

Table 1: Subterranean sediment facies as they relate to adjacent surface streams. Adapted from White (1988).

Class and Facies	Surface Equivalents
Water-Borne Facies Class: Detrital sediments transported by flowing water.	
Bedload: Comprised of grains mobilized by tractive forces at the base of stream flows.	Channel Deposits
Slackwater: Comprised of grains carried in suspension and deposited during discrete flood events.	Overbank Deposits
Quietwater: Comprised of grains carried in suspension and deposited within standing bodies of water.	Lacustrine Deposits
En Mass: Comprised of grains transported as semi-coherent units by floodwaters (sliding beds) or debris flows.	Debris Flows Deposits
Gravity Facies Class: Sediments or blocks transported by gravity, but independent of fluid flow.	
Collapse: Comprised of blocks detached from cave ceilings, walls, and entrance areas.	Talus and Scree
Infiltrates: Comprised of detrital grains transported by gravity through discrete, sub-vertical pathways.	Creep soils?
Chemical Facies Class: Precipitates, evaporites, and other deposits created by chemical means.	
Travertine: Carbonate precipitates, including speleothems. Principally aragonite and calcite.	Tufa Dams
Residuum Facies Class: Bedrock weathering products preserved in situ with primary grains or sedimentary structures observable.	Saprolites

1988]; the principal task of a conduit is to transport or store sediment, dissolved load, and water. An abbreviated table of subterranean sedimentary facies is presented in Table 1. Sediments of the water-borne facies class include those of direct interest to paleoflood hydrologists. The en mass facies will be of little interest to hydrologists seeking information about past discharges and stages, but *Gillieson* [1986] has recognized multiple paleoflood deposits created by debris flows that enter caves of the New Guinea highlands and travel as much as 3 km underground before depositing sediment in backflooded passages as 10-m-thick foresets and diamictons. Such deposits attest to the ability of caves to convey sediment and record catastrophic events.

Fine-grained sediments pervade many caves. These sediments typically belong to the slackwater, quietwater, or residuum facies, although sand de-posits may reflect transport as bedload or suspended load. In caves with multiple tiers, large floods can produce intermittent flooding of upper level passages and slowly fill passages to their ceilings with silt or clay. A mantle of fine sediments often caps the bedload facies to produce what some authors have called an abandonment suite [*Ford and Williams*, 1989]. The fining-upward sequences are ubiquitous in many fluviokarst caves of the eastern United States. However, recent work by *Granger and Fabel* [1999] can be interpreted as evidence that some thick deposits actually represent slackwater or quietwater conditions created in response to local base level rises during Pleistocene climatic oscillations. Adding some credence to this possibility, the caps of many abandonment suites are finely laminated and lack evidence of repeated wetting and drying from ephemeral flooding (e.g. multiple mud-cracked beds or scour surfaces). In either event, abandonment suites are paleoflood deposits.

Quietwater conditions may be produced by ponding behind passages that are constricted or occluded by collapse, log jams, or sediment. Quietwater sediments observed along the Cheat River are typified by fine, gleyed sandy clay loams inter-bedded with laminated, brown silt loams [*Springer et al.*, 1997] (Table 2). Sand grains of the quietwater facies were derived from slow dissolution of overlying, sandy bedrock in flooded passages with simultaneous, slow deposition of allogenic detrital sediments. The resulting particle size distributions are paradoxically rich in sand and clay, but depleted in silt (Table 2). Other limestones may likewise yield comparatively large calcite grains, sand, chert, or dolomite [*Reams*, 1968]. The incorporation of such autogenic grains makes it possible to differentiate quietwater and slackwater sediments, so long as we assume that slackwater sediments accumulate comparatively fast relative to the rain of autogenic sediments. Within river-associated caves, quietwater conditions may occur when floodwaters are left standing in caves. However, unless passages are flooded there should not be a significant accumulation of autogenic detritus because overlying ceilings will not be sub-merged.

Sediments of the water-borne facies are easily distinguishable from those of the gravity and travertine facies (Table 2). Gravitational sediments are the subsurface equivalent of colluvium [*White*, 1988]. Collapse deposits may consist of single blocks to massive rock piles spanning hundreds of square meters. Individual blocks are angular and chaotically arranged. Coarse infiltrates are often poorly sorted, possess diamicton fabrics, and are intimately associated with valley wall inputs. Fine infiltrates can form cones beneath enlarged wall or ceiling joints, but may be disseminated through large portions of flooded passages to form laminated deposits almost identical to more traditional water-borne sediments [*Bull*, 1981]. Travertine is the principal chemical precipitate found in carbon-ate caves. Travertine includes stalactites,

Table 2: Sediment Characteristics of Selected Samples

		Percent by Class[a]				
Facies Class	Sample	Sand	Silt	Clay	Texture	Munsell Color
Slackwater	GS-92-6C[b]	69.5	22.6	7.9	sandy loam	10YR 5/4
Slackwater	GS-92-9A	17.5	66.5	16.0	silt loam	5YR 3/2
Quietwater	GS-92-21C	38.5	17.0	44.5	clay	5YR 5/8
Quietwater	GS-92-29A	59.7	13.6	26.7	sandy clay loam	2.5YR 6/2
Residuum	GS-92-27D	68.7	20.1	11.2	sandy loam	5Y 7/2

[a] Data and sample numbers from Springer (1994).

stalagmites, and flowstone which are collectively known as speleothems. The crystalline nature of speleothems allows ready discernment of their origin and an opportunity to constrain the age of associated sediments using U-Th dating [*Ford and Williams*, 1989].

Allostratigraphic Packages

Texturally, water-borne detrital sediments in caves range from fine clays to boulders. Individual beds may be remarkably homogeneous and it is tempting to classify individual units on the basis of lithostratigraphy. *Springer et al.* [1997] identified and classified sedimentary packages within floodwater injection caves of the Cheat River on the basis of lithostratigraphy. However, such a classification scheme is only applicable within the context of a single study area wherein grain size and sedimentary structures may be uniquely associated with singular depositional processes. Many cave deposits are comprised of discrete event- or process-based packages that can be classified as allostratigraphic units. Bounding discontinuities include scour surfaces, planes of mud-cracks, organic horizons, and bedrock. Though rarely stated explicitly, many surficial paleoflood studies have divided stratigraphic columns into allostratigraphic packages [e.g. *Gillieson et al.*, 1991]. Indeed, the complexity of fluvial environments almost requires the use of allostratigraphy because multiple beds of varying textures, structures, and depositional processes can be present within a single genetically-related package. As an example from the speleological literature, *Wolfe* [1973] reports that greater than 3.5 m of sediment were deposited during a single subterranean flood, but that an individual package graded upward from cobbles to silts.

Surficial paleoflood deposits may be bounded by soils, organic horizons, and scour surfaces [e.g. *Gillieson et al.*, 1991]. Soils are absent within caves, though weathering is not. *Springer and Kite* [1997] report that mats of fungal Mycelia lay beneath silt loams and sands deposited in caves along the Cheat River by the 1985 flood. In a cave far above the present flood range, *Springer and Kite* [1997] also found ancient paleoflood deposits comprised of up to five laminated silt loams divided into individual packages by thin, gravel-rich infiltrates lying atop scour surfaces (Figure 2). Recently deposited infiltrates observed elsewhere in the relict cave were identical to those in the paleoflood deposit, but the interbedded silt loams possessed sedimentary structures attributable to pipe-full flow. The juxtaposition of sedimentary structures and deposits attributable to both open and closed-conduit flows offers a decisive means for recognizing subterranean paleoflood deposits and recorded repeated deposition and erosion of flood-borne sediments.

Traugh et al. [1999] report that a 3 m thick deposit of subterranean sediments lying 30 m above the Obey River, Tennessee contains individual clastic beds bounded by travertine. The clastic beds are interpreted as paleoflood deposits. Travertine increases in abundance toward the top of the deposit and the researchers suggest that the fine-grained clastic units are paleoflood deposits. Intriguingly, *Traugh et al.* [1999] attribute the gradual increase in travertine content to recession of the Obey River as it incised, climate change, and censoring due to sediment accumulation. If their various assumptions are borne out, the cave records paleofloods spanning as much as 600 ka. This raises the possibility of extended regional paleoclimatic studies, using paleofloods as proxies for climate, far beyond the range currently being explored. In any event, the presence of clastic sediments interbedded with travertine offers un-ambiguous evidence of intermittent inundation and the hope of dating paleoflood deposits independent of the presence of organic carbon; speleothems may be dated using radionuclides [see *Ford and Williams*, 1989].

RECENT FLOOD DEPOSITS IN WEST VIRGINIA CAVES

Study Settings

The work of *Springer and Kite* [1997] and an in-progress study along Greenbrier River, West Virginia shed some light on what researchers might expect of subterranean paleoflood sediments along rivers. Both rivers drain mountainous por-

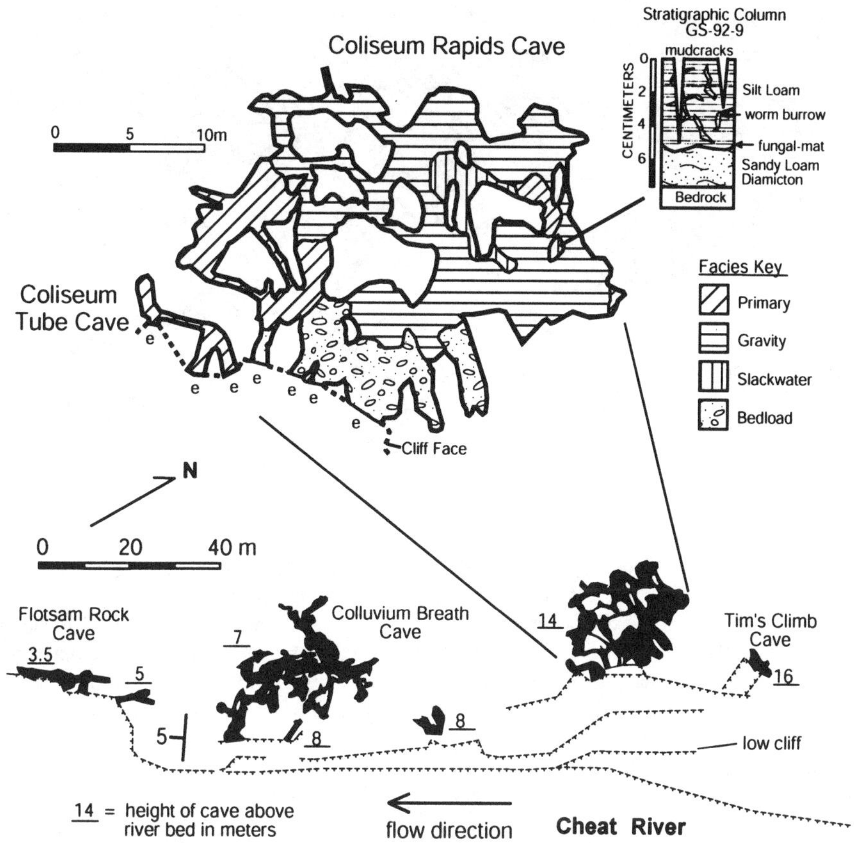

Figure 5. The caves of Coliseum Rapids on the Cheat River. The largest caves are anastomotic mazes. Smaller caves are erosional remnants of mazes and younger caves developed along joints. The highest 1985 flood deposits are found in Coliseum Rapids Cave. As indicated by the facies map, deposits are concentrated in a small portion of the cave. Scour by infiltrating waters is actively removing sediments deposited by the flood, which lies shallowly beneath a narrow strath and valley wall stream. The gravity and bedload facies dominate where vertical infiltration is greatest. Vertical infiltration has produced abundant collapse and erosion, but adjacent passages preserve undisturbed elliptical passages nearly filled to their ceilings with sediments of the primary facies. Lower case e's denote entrances.

tions of the Allegheny Plateau, flow parallel to strike for long distances, are underlain by Paleozoic sedimentary rocks, and cross thin belts of limestone where they turn across strike (Figure 1). The Cheat River is a northward flowing tributary of the Monongahela River whereas the Greenbrier River is a southward flowing tributary of the New River. Throughout their lengths, the Cheat and Greenbrier vary from rock to boulder bed rivers and their average gradients in the study reaches are 0.006 and 0.002, respectively.

The Cheat and Greenbrier rivers cross Mississippian-age carbonates of the Greenbrier Group or Formation at similar basin areas (see Figure 1). Along the Cheat River, the Greenbrier Formation is thin (57 m) and sandy to the point that some beds can be classified as calcareous sandstones. Along the Greenbrier River, the limestones thicken to ~370 m and achieve group status [*Carney and Smosna*, 1989].

The largest historic floods on the Cheat and Greenbrier rivers occurred in 1985 and 1996. On the Cheat River, the peak 1985 discharge was ~5400 $m^3 s^{-1}$ which is 66% larger than the next largest flood, the 1996 flood (~3200 $m^3 s^{-1}$). Respectively, discharges of the 1985 and 1996 floods on the Greenbrier River were ~2550 and ~2650 $m^3 s^{-1}$. As such, the highest historic deposits along both rivers are likely to be from either the 1985 or 1996 floods. The term historic is meant to imply those deposits containing trash or that show relatively little weathering. Evidence of the 1985 and 1996 floods are preserved in at least 24 caves along the two rivers. The highest and most extensive deposits are found in Coliseum Rapids Cave along the Cheat River and Greenbrier River Cave (Figures 5 and 6).

Coliseum Rapids and Greenbrier River caves are relatively large anastomotic mazes with aggregate passage lengths in excess of 150 m, most other caves are very small remnants (<20 m long) of once larger mazes that have either been plugged with sediment or eroded away. Typically, many small caves are found in association with one another (Fig-

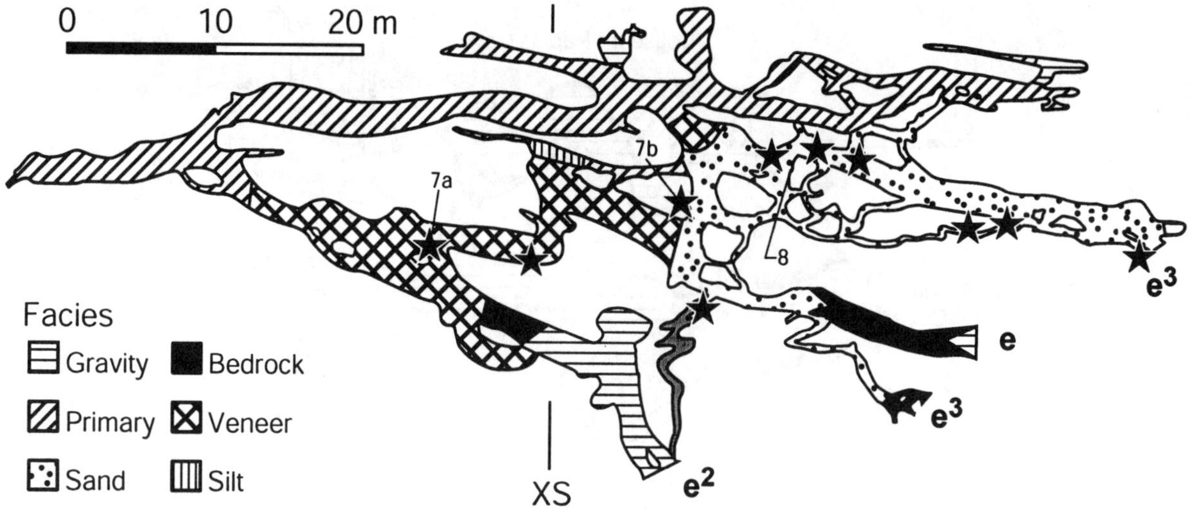

Figure 6. Facies map of deposits associated with the 1985 and 1996 floods in Greenbrier River Cave. The silts and sands are slackwater deposits of the floods. Veneer facies refers to areas with millimeter thick veneers of silt. Other facies are described in Table 2. Stars represent rafts or jams of woody debris. Jams are found in vertical or horizontal passage constrictions whereas the rafts lie in sheltered alcoves. Generally, sands deposits are thin (<2 cm) and the silt deposits do not exceed 3 cm in thickness. Numbers are the figure numbers of pictures taken in the cave. Numbered locations refer to location of figure pictures. The cross sections in Figure 9 pass through the cave along the line XS.

ures 4 and 5). Such collections offer limited, but valuable information about flood stages. Where the vertical extent of caves do not overlap, a non-recording gap exists, but the highest inundated caves record evidence of minimum peak stage.

Characteristics of Host Caves

The vast majority of caves observed along the two rivers are floodwater injection caves. Steep bed gradients, large discharges, and narrow channels combine to create steep hydraulic gradients perpendicular and parallel to channel banks during floods. The most significant solution features are those where injection occurs along prominent, sub-horizontal bedding planes and vertical joints. Anastomose tubes, miniature versions of the anastomotic mazes seen in Figure 4, are well-developed along these discontinuities and their presence is intimately associated with maze caves and injection of floodwaters [*Palmer*, 1975]. Hence, maze caves alongside the two rivers are merely large injection features that are extensions of the rivers.

The plans of the floodwater injection caves can be classified as either anastomotic or network mazes [*Springer et al.*, 1997]. Anastomotic mazes typically form where many flow paths are simultaneously enlarged along sub-horizontal bedding planes in the absence of strong joint control [*Palmer*, 1975] (Figures 4 and 5). In contrast, network mazes preferentially exploit intersecting joints. Passage cross sections are commonly smooth-walled tubes that often enlarge faster along the guiding pathway (e.g. a bedding plane) to form flattened ellipses. As mazes, both network and anastomotic caves typically have multiple entrances along their river-side margins. Maze caves may transmit floodwaters parallel to adjacent channel margins. Others backflood for lack of widely-spaced connections with a river and behave as complex stilling basins.

Characteristics of Deposits

Springer et al. [1997] report that cave sediments become finer away from the mouths of floodwater injection caves. Cobble and boulder diamictons are found in the mouths of low-lying caves alongside the Cheat River. Sands and silts predominate further inside the caves. Clays are found in the furthest recesses of the largest floodwater injection cave examined (Cornwell Cave, Figure 4). Basal sediments in these caves are often massive and lack scour surfaces and multiple generations of mud-cracks. Some sequences display soft-sediment deformation and others contain bedrock pendants delicately detached by solution from passage walls and syndepositionally entombed upright in sediments. These pieces of evidence led *Springer et al.* [1997] to conclude that such sediments reflect deposition while the caves were still submerged. As such, these sediments can be considered as primary deposits associated with initial cave development. An alternative explanation might be that the fine-grained sediments represent aggradation of the river and caves during Pleistocene climatic oscillations [e.g. *Granger and Fabel*,

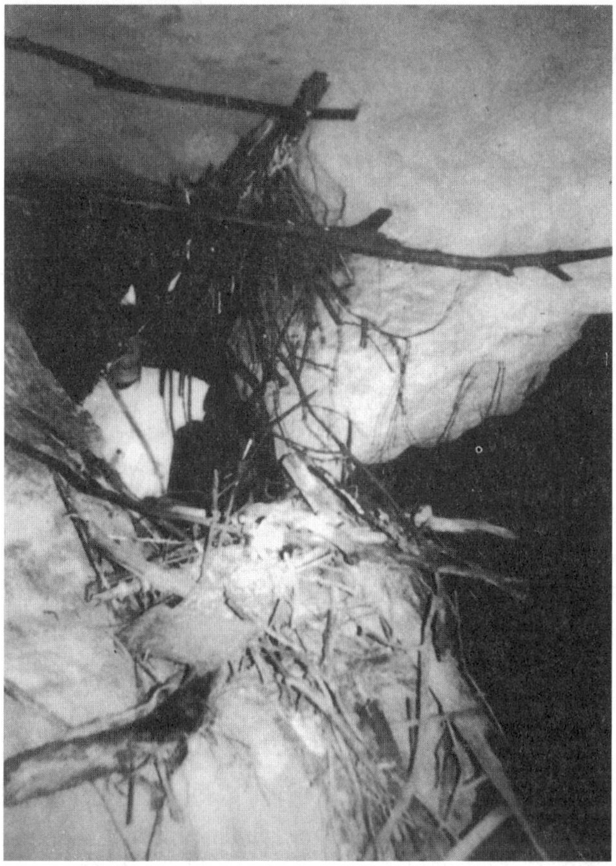

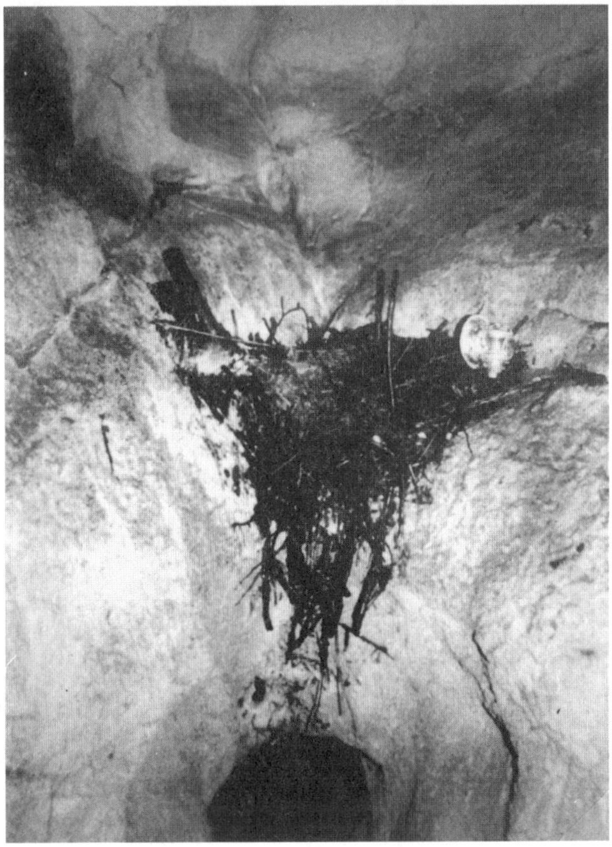

Figure 7. Log jam (a) and stranded raft (b) deposited by the 1996 flood in Greenbrier River Cave (locations shown in Figure 6). The raft lies in an alcove of a ~3.5 m wide passage and the log jam lies on the inside of a tight passage turn. As of July 1999, the woody debris is moist and infested with fungal Mycelia. Spider webs seen at the bottom and top left of the raft (b) reflect the presence of invertebrates that are actively processing these fresh deposits and attracting a thriving arachnid population. The brass carbide lamp at right is 10 cm high.

1999]. Along the Greenbrier River, fine-grained deposits also fill many passages and differ little from those of the Cheat River except in color. Where present, the 'primary' deposits are laterally extensive and fairly homogeneous. Regardless of their true origin, it is important that primary sediments be differentiated from slackwater sediments deposited in caves by occasional floods after they became emergent. Both record paleofloods, but floods of very different meanings.

Within caves of the Greenbrier and Cheat rivers, the slackwater facies can be sub-divided into flotsam and sediment sub-facies. Volumetrically, the flotsam sub-facies is the most important of the two. Large woody debris, logs and sticks, are found as stranded rafts and jams (Figure 7). The largest subterranean rafts observed are ~4 m long and 0.5 m thick. Along the Cheat River, the rafts mantle breakdown surfaces in large rooms at the back of Coliseum Rapids Cave and were absent from Colluvium Breath Cave where a log jam occludes the entrance (Figure 5). In Greenbrier River Cave, rafts and log jams are found in passage constrictions (Figure 6). Among the three caves, the distribution of woody debris is not random. Flood-waters are largely constrained to enter Colluvium Breath Cave through a single entrance, 2 m wide and 0.8 m high, that quickly chokes down to a height of only 0.3 m. Large woody debris readily clogs the entrance and prevents flotsam from washing further into the cave. Coliseum Rapids Cave has four entrances along a 20 m long portion of the bank. The largest entrance is 3 m wide and 1 m high, but because the entrances lie close together there is no through current; Coliseum Rapids Cave forms a backwater that efficiently traps woody debris. Greenbrier River Cave possesses at least nine entrances and four of these are ~4 m high and 2 m wide. The entrances lie along a 30 m long cliff face with the upstream entrances pointing directly into the approaching channel. Flow entered the cave through the upstream entrances and exited through the downstream-most entrances. Log jams formed on the upstream side of passage constrictions and all but one of the jams and rafts lie along the principal throughflow path which averages 2.5 m wide and

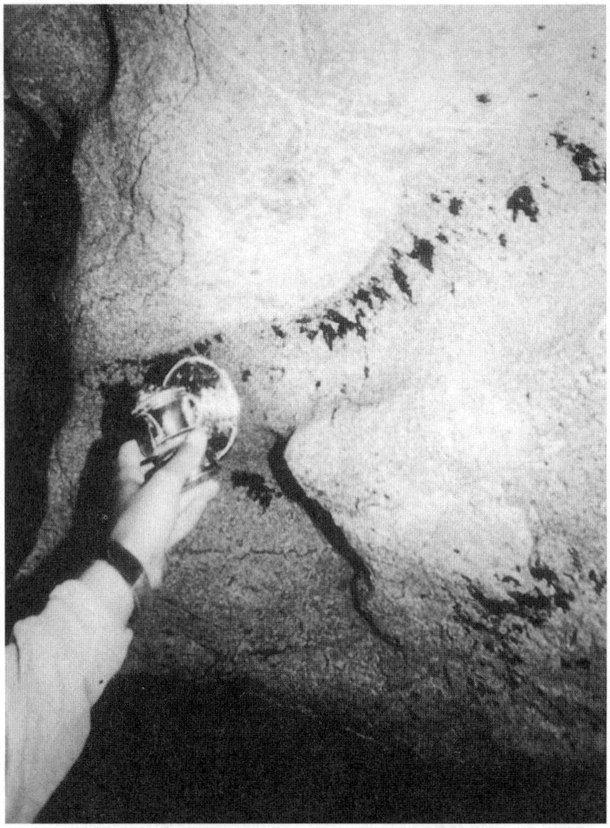

Figure 8. A silt line deposited by the 1996 flood on the walls of an alcove in Greenbrier River Cave. Note the discoloration of walls below the silt line and the presence of macerated organics clinging to walls immediately below the top of the silt line. Styrofoam balls are found within the strand line, which lies 55 cm above the highest observed silt loams. The vertical proximity of the strand line and sediments bodes well for future paleoflood studies in caves. Location of the picture is indicated in Figure 6.

high (Figure 6). Flow also entered nearby Colonial Acres Cave and exited by a downstream path (Figure 4), but flow exits Colonial Acres Cave through a perennial sump that hinders buoyant material from easily exiting the cave. As such, woody material accumulates throughout the cave and covers passage walls nearest the downstream sump. The resulting 'sump facies' intermingles woody material from multiple floods and prevents ready paleoflood interpretation. The sump facies can be readily identified by the presence of mud-caked walls, abundant organics, and trash on all passage surfaces (including ceilings).

Anthropomorphic trash is a common constituent of wood rafts. In the Cheat River caves, the trash includes a wide range of materials such as sporting balls, toys, household items, and fragments of houses. Along the Greenbrier River, the material largely consists of plastic drink containers, Styrofoam, and aerosol cans. The contrast in contents reflects the fate of towns upstream of the two caves. Two small towns were virtually destroyed by the 1985 Cheat River flood with entire blocks of houses completely obliterated. Concurrent flooding on the Greenbrier River produced extensive inundation, but not catastrophic destruction [*Teets and Young*, 1985]. Thus, the relative magnitudes of the two floods is recorded by the types of trash deposited in the caves.

The sediment sub-facies largely consists of loamy silts and fine sands. Silt can be found throughout the entire flood range, which is ~11 m on the Greenbrier River and 14 m on the Cheat River. Sands are found as much as 10 m above the Greenbrier River. In Greenbrier River Cave, slack-water sediments are virtually absent from conveying passages, but occupy wall niches and protected ledges that were probably zones of flow separation. In some wall recesses, the 1996 flood deposited up to 3 cm of silt loam and sand while the adjacent throughflow passage received only a dusting of sediment. Even small caves may not act as passive backwater features, but as active participants in channel processes.

Evidence of Peak Stages

Peak stage of the 1996 flood is recorded in Greenbrier River Cave as 'bathtub rings' or 'silt lines' of silt, macerated organics, seed husks, and Styrofoam balls that cling to passage walls (Figure 8). The highest sediments lie within 55 cm of the silt lines (Figure 9). In Coliseum Rapids Cave, macerated organics infiltrate from an overlying strath through ceiling cracks and render organics as poor indicators of peak stage. However, Styrofoam balls cling to some passage walls and offer conclusive evidence of peak stage for the 1985 flood; there is no hillside source for Styrofoam. Silt loams lie within 1 m of the highest Styrofoam balls in Coliseum Rapids Cave. In both cases, the highest surface deposits lie at least 3 m lower than the cave deposits and are therefore considerably less accur-te indicators of peak stage [*Springer and Kite*, 1997].

Both the 1985 and 1996 floods are recorded in Greenbrier River Cave. However, the only recognized deposit of the 1985 flood is a deposit of silt and macerated organics in a wall niche. The deposit is heavily decomposed and lacks sedimentary structures. The deposit lies at an elevation similar to that of comparatively unweathered deposits attributed to the 1996 flood, but is considerably more weathered. Given the lack of comparable, recent floods and the presence of still undigested organic material in the deposit it is believed to be from the 1985 flood. Hence, the comparable discharges of the 1996 flood on the Greenbrier River produced recognizable deposits at similar elevations above the channel bed.

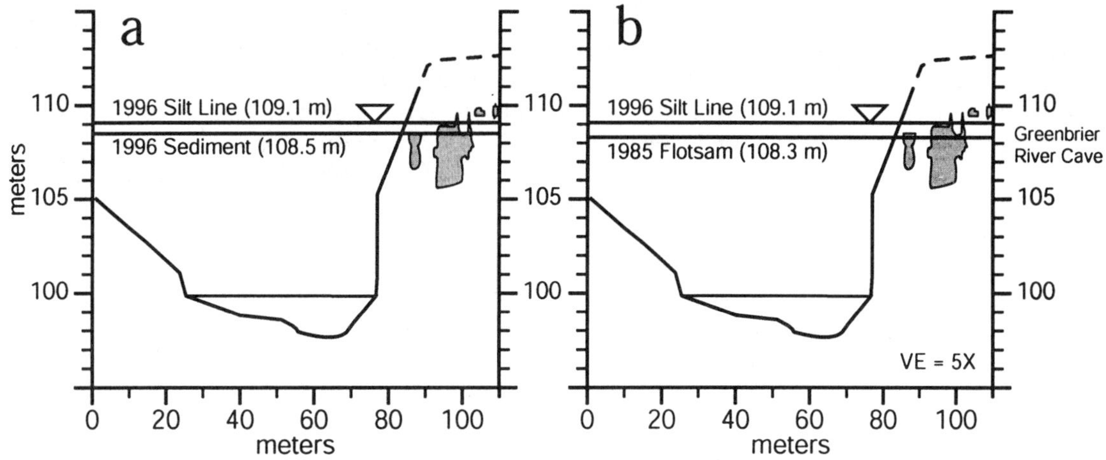

Figure 9. Cross section of the Greenbrier River and Greenbrier River Cave with heights of stage indicators shown. The cave lies within a ~14 m high cliff face on river right. Some passages were above peak stage of the 1996 flood. Note the close agreement between the 1996 silt line and cave sediments (a). Discharges of the 1985 and 1996 floods on the Greenbrier River were ~2550 and ~2650 m^3 s^{-1}, respectively. The similarity in discharges appears to recorded by the relative stages of the highest PSIs assigned to each flood (b). The highest surface deposits lie between 105 and 106 m. Elevations are relative to an arbitrary datum. Location of cross section is indicated in Figure 6.

Preservation Potential

Although both the 1985 and 1996 floods are recorded in Greenbrier River Cave, long-term preservation of the deposits in useful forms is by no means assured. *Springer and Kite* [1997] sought evidence of paleofloods along Cheat River, but only found evidence of recent floods in low-lying caves. By chance, most caves in the modern flood zone of the Cheat River lie shallowly beneath hillslopes and capture surface flow. During an intense thunder-storm, *Springer and Kite* [1997] report observing water droplets and small streams scouring sediment deposited by the 1985 flood. Water clinging to ceilings reached sediments beneath overhangs and virtually no portion of the cave was unaffected. Only in relict caves that lay beneath thick overburdens did the authors find ancient paleofloods. Of the 14 caves examined on the Greenbrier River, none appear to receive significant surface runoff and scour appears to be associated with floodwaters of the river itself.

In the absence of scour, bioturbation and decay offer the greatest threats to preservation. Cave deposits examined by *Springer and Kite* [1997] had undergone extensive bioturbation within 7-to-9 years after the 1985 flood. Woody debris deposited in caves is fungal covered, moist and easily disaggregated (Figure 7). In contrast, surface rafts are decaying from their undersides upward and the upper few decimeters of woody debris is well-preserved and sunbleached.

Within the Greenbrier River caves, the relatively fresh silt deposits of the 1996 flood are mud-cracked, bioturbated, and capped with earthworm mounds while bioturbation has destroyed all sedimentary structures in loamy sands. The sole 1985 deposit in Greenbrier River Cave has been reduced to a compact, homogenized deposit wherein bioturbation has co-mingled sediments and organics. It should be evident that in a humid environment, where caves typically possess relative humidities of ~99% and year-round temperatures of ~12° C, decay and bioturbation may render caves as less than perfect repositories of paleoflood information, although still of great potential use.

EXPANDING SUBTERRANEAN PALEOFLOOD STUDIES

Caves have only marginally aided the study of paleofloods. Recent work along the Cheat and Greenbrier rivers has, thus far, failed to extend flood records. Excellent subterranean records of recent floods does generate considerable optimism that caves will be useful in future paleoflood studies [e.g. *Gillieson et al.*, 1991], but fundamental work remains to be done. Computational models of floods have yet to be made using stage information gleaned only from caves. The most crucial test will be whether models using only subterranean paleostage indicators yield results comparable to known flood properties obtained from gages.

The purpose of this article is to encourage and aid additional studies along other rivers. The first and foremost hurdle to be crossed by such studies will be locating suitable caves along target rivers. Although geologic maps are useful for finding river segments that lie atop soluble bedrock, such maps almost never offer information as to the presence, absence, or exact locations of caves. Within the United States, such information is accumulated by amateur local and state cave surveys. The majority of surveys are allied with the National Speleological Society (NSS; http://www.caves.org), the largest cave organization in the country and interested geomorphologists should contact the NSS for information regarding surveys.

In any given region, the existence of a cave database does not guarantee that river-associated caves will actually appear in the database. Small caves are systematically under-reported and some areas have yet to be fully canvassed for caves. For instance, in excess of 900 caves have been reported to the West Virginia Speleological Survey in Greenbrier County, West Virginia [*Jones*, 1997]. The county is home to numerous world-class cave systems and has undergone intense cave exploration since the late 1940's. Until 1998, only one river-level cave had been reported from nearly 20 km of limestone outcrop in the vicinity of Greenbrier River Cave. Two days of searching uncovered an additional 14 caves in two small reaches. Inquiries revealed that these caves were known to other cave explorers, but had not been deemed significant because they were 'too small.' All but one of the caves contained deposits from the 1985 or 1996 floods. Paleoflood researchers should be prepared to search for, explore, and map many of the caves they will use.

CONCLUSIONS

Caves are unique depositories of geomorphic information that display a variety of relationships to base level streams, drainage basins, and overlying hillsides. As such, each cave possesses a unique geomorphic history and depositional environment. No assumptions about the utility of subterranean deposits should be made *a priori*. Just as with surface landforms, each cave and its deposits will require careful study to assess whether information contained therein is useful. More importantly, such studies will reveal just how geomorphic information should be interpreted. Interpretation of sub-terranean information is dependent upon know-ledge of the cave's geomorphic setting and relation-ships to surficial landforms. In caves, such information is recorded as sedimentary facies, cave plans, and passage cross sections. Equally valuable information can be gleaned from the spatial relationships of caves and their sediments to surface landforms.

Conceptually, the most useful information will probably be yielded by those caves that possess unique relationships to individual surface streams; a greater number of possible sources increases the difficulty of assigning sediments to a specific origin while simultaneously increasing the likelihood that deposits will be scoured or reworked. Protected floodwater injection caves offer great opportunities for paleoflood studies. However, even where sediments are protected from scour by secondary streams or infiltrating recharge, information may be lost to bioturbation and decay. The small size of many river-associated caves may mean that their existence will only be discovered by concerted searches. The preliminary work done to date clearly shows that such caves stand to offer valuable in-formation to fluvial geomorphologists.

Acknowledgments. This research was funded by the Cave Conservancy Foundation, the Cave Conservancy of the Virginias, National Science Foundation (Grant EAR-0000725), and the Richmond Area Speleological Society, whom I thank. I thank Ellen Wohl for providing the opportunity to attend the Second Inter-national Paleoflood Conference in Prescott, Arizona and The Department of Earth Resources at Colorado State University for graciously providing a travel grant that facilitated study of the Greenbrier River caves. I thank Mike Blum, Ted Melis, and Ira Sasowsky for their reviews of this article and Alison Krauss for her assistance.

REFERENCES

Atkinson, T.C., Diffuse flow and conduit flow in limestone terrain in the Mendip Hills, Somerset (Great Britain), *Journal of Hydrology*, 35, 93-110, 1977.

Bretz, J H., Glacial drainage on the Colombia Plateau, *Geological Society of America Bulletin*, 34, 573-608, 1923a.

Bretz, J H., The Channeled Scablands of the Colombia Plateau, *Journal of Geology*, 31, 617-649, 1923b.

Bretz, J H., Valley deposits immediately east of the Channeled Scablands of Washington, *Journal of Geology*, 31, 617-649, 1929.

Bretz, J H., Vadose and phreatic features of limestone caverns, *Journal of Geology*, 50, 675-811, 1942.

Bretz, J H., The Lake Missoula floods and the Channeled Scabland, *Journal of Geology*, 77, 505-543, 1969.

Bull, P., Some fine-grained sedimentation phenomena in caves, *Earth Surface Processes and Landforms*, 6, 11-12, 1981.

Burns, S.J., Matter, A., Frank, N., and Mangini, A., Speleothem-based paleoclimate record from northern Oman, *Geology*, 26, 499-502, 1998.

Carney, C. and Smosna, R., Carbonate deposition in a shallow marine gulf, the Mississippian Greenbrier Limestone of the central Appalachian Basin, *Southeastern Geology*, 30, p. 25-48, 1989.

Dasher, G.R., and Balfour, W.M., The caves and karst of the Buckeye Creek Basin, West Virginia Speleological Society Bulletin 12, West Virginia Speleological Society, Barracksville, West Virginia, 238 p., 1994.

Davis, W.M., Origin of limestone caverns, *Bulletin of the Geological Society of America*, 41, 475-628, 1930.

Easterbrook, D.J, Surface processes and landforms, Macmillan Publishing Company, New York, 520 p., 1993.

Ford, D.C. and Williams, P., Karst Geology and Hydrology. Unwin Hyman Limited, London, Great Britain, 582 p., 1989.

Garton, R. and Garton, M.E., Caves of northcentral West Virginia, Bulletin 5, West Virginia Speleological Survey: Barrackville, West Virginia, 109 p., 1976.

Gillieson, D., Cave sedimentation in the New Guinea Highlands, *Earth Surface Processes and Landforms*, 11, 533-543, 1986.

Gillieson, D., Smith, D.I., Greenaway, M., and Ellaway, M., Flood history of the limestone ranges in the Kimberley region, Western Australia, *Applied Geography*, 11, 105-123, 1991.

Granger, D.E., Kirchner, J.W., and Finkel, R.C., Quat. downcutting rate of the New River, Virginia measured from differential decay of cosmogenic ^{26}Al and ^{10}Be in cave-deposited alluvium, *Geology*, 25, 107-110, 1997.

Granger, D.E. and Fabel, D., Comparison of cosmogenic AL-26/BE-10 burial dating and paleomagnetic results from Mammoth Cave, Kentucky (abstract). In: *Abstracts with Programs of the 1999 Annual Meeting & Exposition of the Geological Society of America*, Denver, Colorado, A-90, 1999.

Jones, W.K., Characteristics of the underground floodplain, *National Speleological Society Bulletin*, 33, 105-114, 1971.

Jones, W.K., Karst Hydrology Atlas of West Virginia, Karst Waters Institute, Martinsburg, WV,111 p., 1997.

Kite, J.S. and Linton, R.C., Depositional aspects of the November 1985 flood on Cheat River and Black Fork, West Virginia', Geomorphic studies of the storm and flood of November 3-5, 1985 in the upper Potomac and Cheat River basins in West Virginia and Virginia: U.S. Geological Survey Bulletin 1981, edited by Jacobson, R., D1-D24, 1993.

Kochel, R.C., Quaternary geomorphology of Seminole Canyon State Historical Park, Val Verde County, Texas, *In*: Seminole Canyon: The Art and the Archeology, Turpin, S.A. (ed.), Archeological Survey Resource Report No. 83, 227-276, 1982.

Miller, T.E., Geologic and hydrologic controls on karst and cave development in Belize, *Journal of Cave and Karst Studies*, 58, 100-120, 1996.

Palmer, A.N., The origin of maze caves, *National Speleological Society Bulletin*, 37, 56-76, 1975.

Palmer, A.N., Origin and morphology of limestone caves, *Geological Society of America Bulletin*, 103, 1-21, 1991.

Patton, P.C. and Dibble, D.S., Archeologic and geomorphic evidence for the paleohydrologic record of the Pecos River in west Texas, *American J. of Science*, 282, 97-121, 1982.

Reams, M.W., Cave sediments and the geomorphic history of the Ozarks, unpublished PhD thesis, Washington University, Missouri, 1968.

Smart, C.C., A deductive model of karst evolution based on hydrological probability, *Earth Surface Processes and Landforms*, 13, 271-288, 1988.

Springer, G. S., Geomorphic implications of cave sediments in the Cheat River canyon, North-central West Virginia. MS thesis, West Virginia University, Morgantown, WV, 167 p., 1994.

Springer, G.S., and Kite, J.S., River-derived slackwater sediments in caves along Cheat River, West Virginia: *Geomorphology*, 18, p. 91-100, 1997.

Springer, G.S., Kite, J.S., and Schmidt, V.A., Cave sedimentation, genesis, and erosional history in the Cheat River Canyon, West Virginia. *Geological Society of America Bulletin*, 109, p. 524-532, 1997.

Teets, R., and Young, S., Killing waters, the great West Virginia flood of 1985, Cheat River Publishing, Terra Alta, West Virginia, 122 p., 1985.

Traugh, K.L., Sasowsky, I.D., and Barrera, E., A record of prehistoric extreme flood events preserved in clastic and calcitic cave sediments (abstract). In: *Abstracts with Programs of the 1999 Annual Meeting & Exposition of the Geological Society of America*, Denver, Colorado, A-152, 1999.

Watson, R.A. and White, W.B., The history of American theories of cave origin, *Geol. Soc. Amer. Centennial* Special Volume 1, 109-123, 1985.

White, W.B., Conceptual models for limestone aquifers, *Ground Water*, 7, 15-21, 1969.

White, W. B., Geomorphology and Hydrology of Karst Terrains, Oxford University Press, New York, 464 p., 1988.

Wolfe, T.E., Sedimentation in karst drainage basins along the Allegheny Escarpment in southeastern West Virginia, unpublished PhD thesis, McMaster University, Canada, 1973.

Gregory S. Springer, Department of Earth Resources, Colorado State University, Fort Collins, CO, 80523. E-mail: laramide@bedrockstreams.org.

Modeled Paleoflood Hydraulics as a Tool for Interpreting Bedrock Channel Morphology

Ellen Wohl

Dept. of Earth Resources, Colorado State University, Ft. Collins, Colorado

Reach-scale bedrock channel morphology is the product of adjustments between hydraulic driving forces and substrate resisting forces. Because the formative discharges along bedrock channels are likely to be infrequent, unpredictable, and difficult to measure directly, much of the information on hydraulic driving forces must be obtained indirectly using paleoflood records and flow modeling. The spatial distributions of hydraulic variables can then be compared to substrate characteristics and channel morphologies in order to detect consistent patterns that may provide insights into the controls on bedrock channel morphology. Four case studies from Japan, the United States, Israel and Australia illustrate the research approach. Field methods include channel surveys of geometry, PSIs, and sediment deposits, grain-size characterization, and use of the Selby rock-mass strength and a substrate heterogeneity index to characterize bedrock channel substrate. Modeling utilizes 1d and 2d flow simulations and flume experiments, and analyses utilize statistical methods to identify correlations between channel morphology and potential control variables. Stepwise discriminant analysis was employed to detect significant correlations between channel morphology and other variables for a dataset of forty bedrock channel reaches. The control variables substrate heterogeneity, channel gradient, and Selby rock-mass strength correlated significantly with channel morphology at p-values of less than 0.05. This suggests that bedrock channel morphology, like alluvial channel morphology, results from an adjustment between the controlling variables of substrate and hydraulics, and that the manner of this adjustment is predictable if the controlling variables can be quantified.

INTRODUCTION

Data on flood magnitude and frequency obtained from paleohydrologic indicators have been applied to the study of flood hydroclimatology and paleoclimatic patterns [*Ely et al.*, 1993], to flood-frequency analyses [*Baker et al.*, 1979, 1987; *Enzel et al.*, 1993; *Salas et al.*, 1994], and to understanding channel morphology [*O'Connor et al.*, 1986; *Baker and Pickup*, 1987; *Wohl*, 1992a,b, 1993; *Wohl et al.*, 1994]. Paleoflood data are particularly important in the examination of potential controls on the morphology of resistant-boundary channels, because these channels may only change substantially at timescales which are long relative to timespans of direct observation or systematic gage measurements. Because only infrequent, large floods may be capable of exceeding boundary resistance in many bedrock channels, paleoflood information on the magnitude, frequency, and hydraulics of large floods is vital to an understanding of rates and processes of bedrock channel change, and hence an understanding of bedrock channel morphology.

Bedrock channels have recently received increasing attention because of the growing recognition that these channels differ from alluvial channels in some fundamental ways. Past

Ancient Floods, Modern Hazards: Principles
and Applications of Paleoflood Hydrology
Water Science and Application Volume 5, pages 345-358

inattention to bedrock channels [e.g. *Leopold et al.*, 1964; *Knighton*, 1998] may be attributed to the perceptions that these channels were uncommon relative to other channel types, that they were substrate-dominated with little mutual adjustment between hydraulics and substrate, and that bedrock channels were therefore not important. The latter view in particular is overly simplistic. Like alluvial channels, bedrock channels may also largely be the authors of their own geometries [*Leopold and Langbein*, 1962], although the scale of formative events is likely to be different. Fortunately for paleohydrologic studies, bedrock channels are likely to be the chroniclers of their own cataclysms [*Baker and Pickup*, 1987], thus facilitating inferences about the scale of formative events. Bedrock channels differ from alluvial channels in that the standard alluvial-channel conceptual models of hydraulics (log-velocity profile), flow regime (bankfull discharge; definition of flood discharge), sediment transport (suspended sediment of sand size or finer, with most transport during bankfull discharge), channel morphology (the continuum of straight, meandering, and braided channel planforms), and channel evolution (incision rates directly proportional to stream power) often do not adequately describe bedrock channel processes. Bedrock channels have also received increasing attention in connection with studies of landscape evolution. The rates of bedrock channel incision may govern hillslope stability [*Burbank et al.*, 1996], the transmittal of baselevel change along a drainage basin, and rates of landscape evolution. Understanding and being able to parameterize the processes acting along bedrock channels is key to modeling long-term landscape evolution [*Howard et al.*, 1994; *Montgomery,* 1994; *Rosenbloom and Anderson,* 1994].

Contemporary research on bedrock channels tends to focus on either the drainage-basin scale, as when modeling longitudinal profile evolution using rate laws, or on the reach scale, as when the emphasis is on channel geometry, hydraulics and sediment transport, or erosional and depositional processes. The research approach outlined in this chapter focuses on the reach scale, where a reach is defined as a segment of channel with consistent geometry, which may be tens of meters to tens of kilometers in length.

Reach-scale channel morphology is of interest for at least two reasons. First, numerous studies have demonstrated that the erosion rate law for bedrock channels is not the same among basins or throughout a basin [*Seidl and Dietrich,* 1992; *Stock and Montgomery,* 1999]. This law commonly takes the form

$$-(\partial z/\partial t) = k\, A^m S^n \qquad (1)$$

where A is drainage basin area, S is bed gradient, k is an erodibility constant that depends on bedrock properties and the magnitude and frequency characteristics of the flow regime, and m and n are exponents that depend on the choice of shear strength or stream power as the erosion surrogate [*Howard*, 1980]. Changes in the (m/n) ratio have been explained in terms of different erosive processes, such as debris-flow erosion in headwater reaches, and knickpoint retreat in downstream reaches [*Seidl et al.*, 1994]. These differences will be a function of the interactions among lithology, uplift style and rate, sediment supply, and discharge. This leads to the possibility of predicting erosive processes or channel morphologies likely to be present along different portions of a drainage basin if the potential control variables can be adequately characterized.

Secondly, reach-scale channel morphology may provide insight into adjustments between hydraulics and substrate. *Grant* [1997] proposed that steep, competent alluvial channels adjust to maintain flow conditions close to critical, with interactions between hydraulics and channel geometry preventing supercritical flow for more than limited spatial and temporal (during high discharges) scales. For example, any channel irregularity that constricts flow and creates Froude numbers close to one is likely to be maintained because sites of supercritical flow are subject to relatively rapid erosion of the constriction until flow returns to critical or subcritical, whereas sites with low Froude numbers and subcritical flow experience little erosion, as demonstrated by *Kieffer's* [1985, 1987, 1989] work on constrictions along the Colorado River in Grand Canyon. *Tinkler* [1997] proposed that sites of critical flow in bedrock systems dictate the hydraulics and dependent channel geometry elsewhere in the system. For example, *Thompson et al.'s* [1998] work on laterally constricted pools demonstrates how the strength of the jet of (super)critical flow which forms at the constriction during high discharges controls the strength and location of marginal eddies and associated deposition, as well as deposition on the pool exit-slope. *Carling* [1987, 1989, 1995] demonstrated that boulder berms deposited during supercritical flows minimized energy losses and ensured that high flows were transmitted efficiently through channel transitions.

Reach-scale bedrock channel morphology is the product of adjustments between hydraulic driving forces and substrate resisting forces [*Wohl*, 1998a]. Because the formative discharges along bedrock channels are likely to be infrequent, unpredictable, and difficult to measure directly, much of the information on hydraulic driving forces must be obtained indirectly using paleoflood records and flow modeling. The spatial distributions of hydraulic variables can then be compared to substrate characteristics and channel morphologies in order to detect consistent patterns that may provide insights into the controls on bedrock channel morphology. This may eventually lead to the empirical definition of "stability fields" for bedrock channel morphologies, analogous to the empirically-defined gradients, bank erodibility, and sediment transport for straight, meandering, and braided alluvial channels.

Empirical definition of such stability fields for bedrock channels will be limited by the ability to separate and quantify

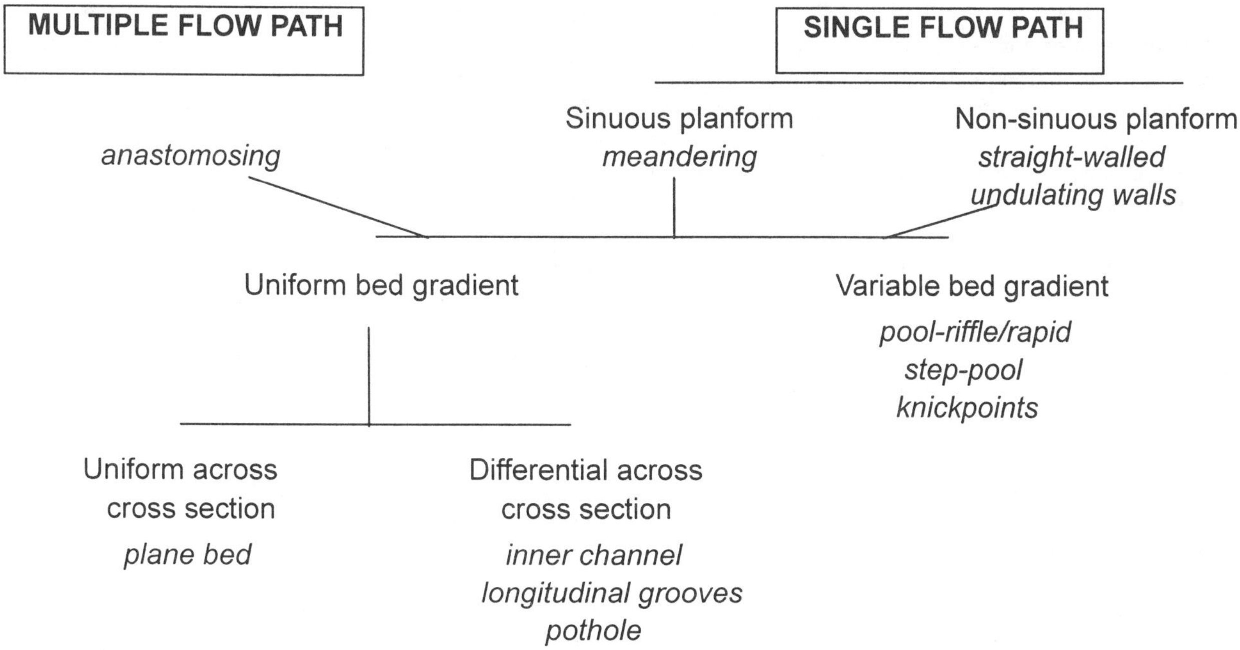

DESCRIPTIONS
anastomosing: multiple low sinuosity channels that branch and rejoin
meandering: single channel with sinuosity > 1.5 and walls in-phase
straight-walled: single channel with sinuosity < 1.5
undulating walls: single channel along which walls have regular concavities and convexities that approximate a sine-generated curve, with undulations on each wall being out-of-phase relative to the opposite channel wall
pool-riffle/rapid: periodic bed undulations between topographic highs (riffles) and lows (pools)
step-pool: downstream bed undulations in form of vertical steps with pools between them
knickpoints: aperiodic vertical steps along channel
plane bed: uniform bed gradient and rectangular to trapezoidal cross section form
inner channel: continuous low-flow channel incised below high-flow strath
longitudinal grooves: straight or sinuous grooves incised parallel to each other and to flow; no single dominant flow path
pothole: irregularly spaced vertical scour holes

Figure 1. Bedrock channel morphologic classification (after *Wohl*, 1998a, Figure 1).

specific driving and resisting forces, and the resulting erosional processes of abrasion, corrosion, cavitation, and quarrying and plucking, as well as by the ability to infer process from form.

The remainder of this chapter outlines field methods used to characterize bedrock channels, and the modeling and statistical analyses used to detect significant correlations between channel morphology and control variables. A series of four case studies are briefly summarized, followed by a discussion of the patterns now emerging from a dataset large enough to support generalizations and statistical analyses. The chapter concludes with a discussion of the implications of the assumption behind these analyses, which is that different erosive processes reflect different balances between driving and resisting forces, as expressed through different bedrock channel morphologies.

METHODS

Field Methods

Ideally, field methods would characterize all potential aspects of the hydraulic driving forces and the substrate resisting forces, as well as channel geometry. In practice, I have used a few relatively simple techniques to characterize the most obvious variables. Channel surveys are designed to include the components of channel geometry necessary for classifying channel morphology (Figure 1), and for modeling paleoflood discharges. The spatial resolution of the survey is usually a function of site accessibility and the complexity of channel geometry. Site accessibility will control the type of equipment used (laser theodolite; theodolite; automatic level; hand-held level, clinometer and rangefinder), and the ability of the operators to actually reach all points in the channel. The complexity of channel geometry dictates the spacing of cross sections or mesh nodes to be used in flow modeling [*O'Connor and Webb*, 1988; *Miller and Cluer*, 1998]. Survey spatial resolution may thus vary widely between channel reaches (Table 1).

Channel surveys also establish the height of paleostage indicators (PSIs) such as tree impact scars, scour lines, silt lines, and slackwater deposits. These indicators can be used to define water-surface profiles for use in flow modeling. In addition to providing information for use in modeling flood magnitude, PSIs may provide a flood chronology that can be used to establish flood magnitude-frequency relations [*Baker,* 1987; *Ely et al.,* 1992].

Although channel surveys indicate the areal extent and possibly the volume of sediment deposition along the channel, grain-size distribution is determined via random-walk clast counts [*Wolman*, 1954] for gravel- to boulder-size sediments, and sieve analysis of sediments finer than gravel. Data on clast lithologies may also prove useful in assessing the abrasive potential of sediment in transport. The thickness of discontinuous or irregular channel-bed sediments may in some cases be determined by pounding in rebar, or by augering through the sediments.

The bedrock channel substrate can be characterized using the Selby rock-mass strength index [*Selby*, 1980], and a substrate heterogeneity index. The Selby index was developed for bedrock hillslopes, but it may also be usefully applied to bedrock channels. The index has five numerical categories for each of six variables (Table 2). The numerical ratings for these variables are summed to provide a single numerical value for a site. Although the numerical value for some of the categories (e.g. weathering) is slightly subjective, tests among different operators at the same site produce identical ratings [*Wohl*, unpublished data]. The number of sites along a channel used to produce a reach-averaged Selby value will depend on the variability of the substrate and the channel morphology (Table 1). The substrate heterogeneity index is a simple rating of 1-4 which characterizes within-reach variability in lithology, Selby index, jointing, and bedding (Figure 2). These indices may be supplemented by thin sections from the channel boundaries to estimate porosity and permeability, by detailed measurements of bedrock bedding characteristics (thickness, dip), or by other measurements deemed relevant to a particular channel reach.

Modeling and Analyses

Surveyed channel geometry and paleostage indicators provide the input for flow models used to estimate flood magnitude and hydraulics. The one-dimensional step-backwater computer model HEC-2, known as HEC-RAS after 1996, has been most widely used for bedrock channel studies [*Hydrologic Engineering Center*, 1997]. *O'Connor and Webb* [1988] provide an excellent review of this modeling approach. HEC-RAS is a user-friendly model which uses surveyed cross sections and user-defined Manning *n* values, starting water-surface elevation, and discharge to model water-surface elevations for subsequent cross sections. Sensitivity analyses of the effects of uncertainty in estimating *n* values for HEC-RAS modeling indicate that the percentage change in discharge associated with varying *n*-values is proportional to roughness, but inversely proportional to channel gradient and to width/depth ratio [*Wohl*, 1998b]. For channels with a gradient ≤ 0.01, varying *n* by $\pm 25\%$ produces a maximum change of 20% in discharge. The HEC-RAS user may specify an output table including such hydraulic variables as velocity, shear stress, stream power per unit area, and energy gradient. These variables may be cross-sectional averages, or they may be averaged across some portion of the cross section, such as main channel or right overbank. With sufficiently closely spaced cross sections, spatial variability among hydraulic variables can thus be related to changes in channel morphology and/or substrate. HEC-RAS may be run for supercritical or subcritical flow, but the model does not approximate flow transitions well. As a one-dimensional (1d) model, HEC-RAS also does not adequately simulate transverse flow, or flow separation and associated turbulence. Because of these limitations, two-dimensional flow models may be more appropriate for some analyses.

Two-dimensional (2d) models such as RMA2 [*Donnell et al.*, 1997] and HIVEL2D [*Berger and Stockstill*, 1995] are much more time-consuming and difficult to use than HEC-RAS. Two-dimensional models are based on equations for conservation of mass and conservation of momentum [*Miller and Cluer*, 1998]. In RMA2 and HIVEL2D, the user inputs a survey grid representing channel geometry, boundary roughness coefficients, eddy viscosity, and discharge. The model output consists of water depth and velocity at each node in the finite-element mesh used to represent the channel reach. Spatial resolution of model results is thus a function of the density of nodes derived from channel geometry surveys. RMA2 models only subcritical flow, but HIVEL2D can model both subcritical and supercritical flow. *Miller and Cluer* [1998] provide a comprehensive overview of modeling considerations for simulation of flow in bedrock channels. Other models, such as TELEMAC2D, have received much more calibration and testing for floods, but these have not been applied to bedrock channels [*Bates et al.*, 1997; *Siggers et al.*, 1999; *Stewart et al.*, 1999].

Simulations of flow hydraulics may also be supplemented by direct measurements of hydraulic variables, usually during lower discharges, and by flume experiments. Direct measurements of velocity may be at-a-point measurements obtained with a 1d or 2d flow meter, measurements of the velocity field obtained with a broad-band acoustic Doppler current profiler [*Cluer*, 1997], reach averages obtained using conductivity meters and a salt tracer [*Curran*, 1999], or videography of surface flow fields using visual tracers [*Meselhe et al.*, 1998].

These data may be used to calibrate and verify flow model parameterizations. Flume experiments with a cohesive substrate may use either a pre-formed channel configuration [*Wohl et al.*, 1999], or a channel geometry formed in response to a specific flow regime [*Wohl and Ikeda*, 1997]. In either case, measurements of water-surface elevations and velocity distributions may be used to verify the ability of flow models

Table 1. Sample bedrock channel reach-scale studies.

River	Reach length (m)	Channel morphology	No. of cross sections	Type of PSI	No. of Selby[a]	Drainage area (km^2)	Avg. reach gradient (m/m)	Avg. channel width (m)	Avg. Selby rating	Substrate hetero-geneity	Reference
Shichiri-gawa (Japan)	1130	plane bed	28	scour lines veg. limits	2	10	0.009	16.5	50	3	Wohl and Ikeda, 1998
Futama-gawa (Japan)	190	step-pool	34	scour lines veg. limits	3	3.8	0.082	5	57	1	Wohl and Ikeda, 1998
Torii-zawa (Japan)	450	plane bed/ step-pool	82	scour lines veg. limits	8	0.3	0.051	4	57	3	Wohl and Ikeda, 1998
Coyote Narrows (Utah)	1334	undulating walls	252	silt lines wood	15	65	0.021	1.4	56	1	Wohl et al., 1999
Piccaninny Creek (Australia)	4700	inner channel long. grooves	13	silt lines scour lines debris lines slackwater deposits	23	61	0.004	34	60	1	Wohl, 1993
Nahal Paran (Israel)	5600	knickpoints	44	debris lines slackwater deposits	20	2950	0.012	92	65	4	Wohl et al., 1994

[a]Number of sites along channel where Selby rock-mass strength was measured

Table 2. SELBY ROCK-MASS STRENGTH

Intact rock strength	Weathering	Joint spacing	Joint orientation	Width of joints	Continuity of joints	Outflow of groundwater	TOTAL RATING
100-60 r: 20	unweathered r: 10	> 3m r: 30	steep dips into slope, cross joints interlock r: 20	< 0.1 mm r: 7	none continuous r: 7	none r: 6	100-91
60-50 r: 18	slightly weath. r: 9	3-1 m r: 28	mod. dips into slope r: 18	0.1-1 mm r: 6	few continuous r: 6	trace r: 5	90-71
50-40 r: 14	mod. weath. r: 7	1-0.3 m r: 21	horiz. dips, or nearly vertical (hard rocks only) r: 14	1-5 mm r: 5	continuous, no infill r: 5	slight (25 l/min/10m^2) r: 4	70-51
40-35 r: 10	highly weath. r: 5	300-50 mm r: 15	mod. dips out of slope r: 9	5-20 mm r: 4	continuous, thin infill r: 4	moderate (25-125 l/min) r: 3	50-26
35-10 r: 5	completely weathered r: 3	< 50 mm r: 8	steep dips out of slope r: 5	> 20 mm r: 1	continuous, thick infill r: 1	great (> 125 l/min) r: 1	< 26

Intact rock strength determined with Schmidt hammer

Weathering: *unweathered* -- parent rock showing no discoloration, loss of strength, etc

slightly weathered -- slightly discolored, especially adjacent to discontinuities; intact rock not noticeably weaker than fresh rock

moderately weathered -- rock is discolored throughout more of its mass, but less than half the rock mass is decomposed and disintegrated; alteration has penetrated discontinuities which may be zones of weakly cemented alteration products or soil

highly weathered -- rock is discolored throughout; discontinuities may be open and have discolored surfaces; up to one half of rock mass is decomposed and disintegrated so that it can be excavated with rock hammer

completely weathered -- rock is discolored and changed to soil, but some original rock fabric and texture preserved

1. Homogeneous
single lithologic unit
and
1-2 Selby categories per unit
and
bedrock dips in a consistent direction
(< 20° variability)
and
joint orientation and spacing consistent
(≤ 2 Selby categories)

2. Moderately homogeneous
single lithologic unit
and
> 2 Selby categories within the unit
or
bedrock dip is variable (channel trends
updip and downdip,
or > 20° variability)
or
joint characteristics variable
(> 2 Selby categories)

3. Moderately heterogeneous
more than 1 lithologic unit with
1-2 Selby categories of difference

4. Heterogeneous
more than 1 lithologic unit with
> 2 Selby categories of difference
or
bedrock dip is variable
or
joint characteristics are variable

Figure 2. Substrate heterogeneity index.

to adequately simulate the hydraulics present along a given channel geometry under varying discharges [*Thompson and Wohl*, 1998].

The next step in investigating bedrock channel morphology involves identifying correlations between channel morphology and potential control variables. This type of analysis may be performed for specific features within a reach. For example, stream power minima may be correlated with changes in channel geometry at channel bends or sites of coarse-clast deposition [*O'Connor et al.*, 1986; *Baker and Pickup*, 1987; *Wohl*, 1992a,b]. Magnitude of erosional bed features may correlate with unit stream power and energy gradient, or with substrate resistance [*Wohl and Ikeda*, 1998]. Correlations between morphology and potential control variables may also be drawn for datasets with multiple reaches. Using data from 40 bedrock channel reaches representing a range of lithologies, channel morphologies, and hydroclimatic regimes, *Wohl and Merritt* [in review] employed discriminant analysis to demonstrate that channel morphology is significantly correlated with substrate heterogeneity, channel gradient, and Selby rock-mass strength. In each of these examples, a correlation between channel morphology and a potential control variable does not necessarily imply a causal relation. However, this approach does provide insight into bedrock channel dynamics, in the absence of an ability to directly manipulate bedrock channel control variables and observe the resulting adjustments.

CASE STUDIES OF MODELED BEDROCK CHANNEL PALEOFLOOD HYDRAULICS

Channels of the Boso Peninsula, Japan [*Wohl and Ikeda*, 1998]

The Boso Peninsula of east-central Japan has numerous channels formed in interbedded Miocene sandstone and mudstone units. Three of these channels were surveyed and modeled in 1996. Shichiri-gawa has a consistent plane-bed morphology, Futama-gawa is characterized by step-pool reaches, and Torii-zawa has alternating plane-bed and step-pool reaches (Table 1). The channels are ungaged. In this wet climate, persistent PSIs take the form of scour lines and vegetation limits. We also used stage indicators produced by a September 1996 flood associated with typhoon rainfall. This flood is estimated to have a 100-year recurrence interval. Channel geometries were surveyed using a theodolite; the progressively greater density of cross sections from Shichiri-gawa through Futama-gawa to Torii-zawa reflects increasing variability in channel morphology and gradient (Table 1).

Hydraulic modeling used HEC-2 with Manning n values of 0.045 for all sections except the boulder-mantled portions of Futama-gawa, for which we used a Manning value of 0.07. Varying n by ± 25% caused flow stage to change by less than ± 5% in these channel reaches. We used the HEC-2 default expansion and contraction coefficients of 0.1 and 0.3, respectively. The modeled September 1996 flood flow along Shichiri-gawa was subcritical. Modeled maximum discharges were at or close to critical conditions along most of the Futama-gawa and Torii-zawa reaches. Flow stage as modeled with HEC-2 was consistently within ±5% of stage indicators surveyed in the field.

Channel substrates were characterized using coarse-clast counts, Selby indices, and substrate heterogeneity indices, as described in the previous section. For each of the three study channels, we subdivided the channel longitudinally into subreaches of consistent channel morphology and bed gradient. Subreach average values of unit stream power and energy gradient obtained from HEC-2 runs were regressed against subreach averages of Selby rock-mass strength, and depth and volume of erosional features. Magnitude of erosion at the subreach scale correlates with stream power and energy gradient along Futama-gawa, which has a homogeneous mudstone substrate. Where sandstone and mudstone are interbedded along Shichiri-gawa and Torii-zawa, magnitude of erosion correlates with Selby rock-mass strength. The resistance,

Figure 3. Looking upstream along Wire Pass Canyon, a tributary of the Paria River in southern Utah. This channel, incised into the Navajo Sandstone, has highly undulating walls and a deep, narrow cross section (stadia rod is extended to 7 meters).

thickness, and orientation of the bedrock strata also correlate with the dimensions and type of erosional features. Elongate pools develop where strata strike parallel to the channel, whereas chute-and-pool knickpoints develop where resistant strata cross the channel. In general, where the softer sandstone predominates, the channel has a plane bed with uniform gradient. Where soft sandstone and resistant mudstone are evenly interbedded, steps and pools or potholes commonly form, presumably in response to localized erosion associated with differential substrate resistance.

If the potential controls on bedrock channel morphology are conceptualized as a continuum from hydraulics-dominated morphologies to substrate-dominated morphologies, the channels of the Boso Peninsula represent a case where substrate dominates. Erosional and depositional features correlate directly with changes in the strike, thickness, and relative resistance of sandstone and mudstone strata exposed along the channel. Substrate variability is thus inferred to be controlling both channel morphology and the associated spatial variability in hydraulic forces.

Channels with Undulating Walls, Colorado Plateau, USA [*Wohl et al.*, 1999]

The Navajo Sandstone of the Colorado Plateau in northern Arizona and southern Utah has numerous deep, narrow slot canyons. The walls of these canyons are often out-of-phase with one another, creating alternating expansions and constrictions (Figure 3). The Navajo Sandstone tends to be relatively soft (Selby ratings average 56) and homogeneous (substrate heterogeneity index of 1). A series of 8 study reaches surveyed in these channels during 1993-94 was used to examine potential controls on channel morphology. Longitudinally continuous silt lines are well-preserved in this dry climate and provide high-quality PSIs against which to compare water-surface profiles generated by HEC-2 and HIVEL2D. Maximum PSIs take the form of occasional pieces of wood wedged between the channel walls. Channel geometries were surveyed using a theodolite; closely spaced cross sections were necessary to adequately characterize the variations in channel width associated with wall undulations. Manning n values of 0.03 were assigned to the wetted perimeter, and we used the maximum expansion and contraction coefficients recommended for HEC-2, 1.0 and 0.6, respectively, after conducting sensitivity analyses on the effect of varying these coefficients. The only change that is hydraulically present in the HEC-2 model with the presence of the undulating walls is a change in the wetted perimeter. The undulating walls cause an increase in the wetted perimeter, which in turn reduces the hydraulic radius and the velocity. Hydraulic changes caused by flow separation and turbulence are not explicitly accounted for in the model, although they are accounted for implicitly through n values and expansion and contraction coefficients. The HEC-2 results thus likely represent a minimum estimate of hydraulic changes resulting from wall undulations. Because of these limitations we also used the two-dimensional numerical model HIVEL2D to estimate hydraulics along the Coyote Narrows reach. The highest silt line preserved along the study reach was best matched by a discharge of 16 m^3s^{-1} in the HEC-2 program, with a maximum deviation of 27% of the flow depth. The HIVEL2D model produced a best fit at 13 m^3s^{-1}, with a maximum deviation of 3% of the flow depth. It is not clear from these results that the 2d model conferred a substantial advantage in matching silt lines, although the 2d model does provide hydraulic data on flow separation zones which are not available in the 1d model.

Channel substrates were characterized as described previously. Our initial assumption was that the spacing and magnitude of the wall undulations would correlate with joint characteristics, differential porosity and permeability, or some other source of substrate variability. This proved not to be the case, however. Based on flume experiments and upstream channel geometries, the wall undulations appear to originate as the remnants of breached, longitudinally offset potholes along the channel center line, or from sinuous longitudinal grooves formed during incision of a knickzone. As the channel widens with time, the undulations are maintained as velocity maxima immediately downstream from the constrictions dissipate in turbulent eddies and upwellings which erode the side walls of the expansions (Figure 4a). As this effect relaxes downstream the channel narrows, causing a backwater, which in turn causes longitudinal acceleration of flow entering the next constriction.

At some point in time, the undulations stop enlarging because the zone of lateral flow separation has become sufficiently wide that much of the flow energy is being dissipated as shearing between the rapid flow along the channel center and the secondary flow along the channel margins (Figure 4b). The lower velocity flow along the outer channel margins no longer effectively erodes the side walls of the expansions. Thus, the wall undulations are preserved not because they are associated with differential substrate resistance, but because the undulations and flow hydraulics create a feedback that stabilizes channel morphology. These channels with undulating walls represent a situation where hydraulics dominate channel morphology. Although very subtle differences in substrate characteristics may initiate the flow perturbations which create offset potholes or sinuous longitudinal grooves, differential substrate resistance does not appear to play a substantial role in controlling the subsequent development of channel morphology.

Nahal Paran, Israel [*Wohl et al.*, 1994]

Nahal Paran flows north from Egypt's Sinai peninsula through Israel's Negev Desert. Much of the channel is alluvial, but a 10.5-km-long canyon has been incised into interbedded late Cretaceous chert and dolomite in the lower portion of the basin. A shallow, discontinuous layer of sand to boulder size alluvium is present along the channel bed in the canyon. The upper canyon has a series of bedrock knickpoints mostly buried in alluvium, whereas the lower canyon has exposed knickpoints and an inner channel. Slackwater deposits and debris lines record approximately 10 floods of 200 to 2500 m^3s^{-1} over a period of at least 350 years. A 5.6-km length of the Paran was surveyed in 1991-92 using a theodolite. HEC-2 modeling for this reach used a Manning n value of 0.045, and the default expansion and contraction coefficients. The modeled flows were mostly subcritical, although supercritical flow occurred at the three largest knickpoints and along portions of the inner channel. Surveyed channel cross sections were very closely spaced at these locations in anticipation of such changes in flow. Well-preserved driftwood lines continuous through these reaches indicated that the knickpoints (which have up to 5.6 m of vertical drop) and inner channel are submerged at high flows.

Most of the sites of boulder deposition within the low-flow channel are associated with tributary junctions, where shear stress and unit stream power decrease as a result of increasing flow width. Bed gradient in the lower canyon is three times that in the upper canyon, and the lower canyon has no substantial boulder deposits. The active knickpoints and steeper gradient of the lower canyon correlate with greater exposures of resistant chert units in the lower canyon. The inner channel at the downstream end of the study reach is associated with a series of chert layers with a cumulative thickness of at least 5 m, which is the depth of the inner channel. The length of the inner channel corresponds to the length along which these layers are exposed at river level. Channel incision has not yet penetrated to any less resistant units that may underlie the chert units, so that the inner channel has not widened following knickpoint migration. In contrast, the active knickpoints

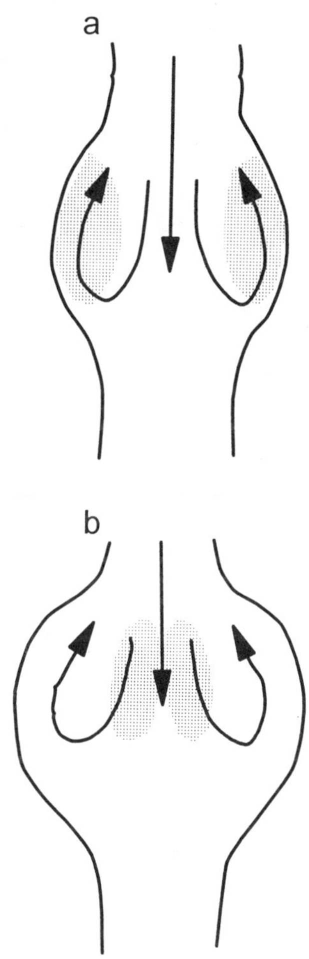

Figure 4. Conceptual model of undulating wall formation. In this plan view diagram, the early phase of undulating wall formation (a) involves substantial energy dissipation in upwelling and transverse flow along the channel margins in the expansions (shaded areas). During later phases of channel adjustment (b), as expansion width increases, areas of substantial flow energy dissipation shift to the zone between the central jet of rapid flow and the marginal eddies (shaded areas).

Figure 5. Looking upstream along Piccaninny Creek, Australia. This channel, incised into sandstone, has a sinuous inner channel formed from incision through longitudinally offset potholes.

present in the canyon have a chert layer 1-2 m thick forming a caprock or lip which is underlain by interbedded chert layers and much less resistant dolomite units. Once the upper layer of chert is breached by channel incision, the underlying weaker units are rapidly eroded, undermining the upper layer and permitting the formation of an undercut knickpoint and the subsequent widening of the channel as the knickpoint retreats upstream.

As for the channels of the Boso Peninsula, changes in channel morphology along Nahal Paran correlate strongly with changes in substrate characteristics. Although downstream hydraulic variability is substantial and also correlated with changes in channel morphology, substrate controls are probably dominating channel morphology in the canyon of Nahal Paran.

Piccaninny Creek, Australia [*Wohl*, 1993]

Piccaninny Creek has incised a spectacular gorge into interbedded Devonian conglomerate and sandstone units in northwestern Australia. The ephemeral channel has downstream alternating reaches of sand and gravel deposition and exposed bedrock, and the bedrock reaches include plane-bed, longitudinal grooves, potholes, and inner channel morphologies (Figure 5). Slackwater deposits, silt lines, scour lines, and driftwood lines indicate flood flows of 50 to 100 m^3 s^{-1}. We surveyed a 4700-m-long reach in the canyon using a handlevel, stadia rod, and tape in 1991. Flows were modeled as subcritical using the HEC-2 program, with Manning *n* values of 0.03 and 0.045 for the main channel and overflow zones, respectively, and the default expansion and contraction coefficients.

As in the case of the channels with undulating walls on the Colorado Plateau, lithologic variability and rock-mass strength showed no significant correlation with channel morphology. The substrate was remarkably uniform throughout the study reach, although channel geometry varied dramatically. Long subreaches of exposed bedrock commonly have a plane-bed upstream, which progresses to parallel longitudinal grooves and finally an inner channel in the downstream direction. The lower half of the canyon has a much higher average bed gradient, more exposed bedrock, and longer inner channel segments. The lower canyon is probably a knickzone in which channel incision is occurring via the erosion of inner channels. As with the Colorado Plateau channels, the absence of detectable substrate variability is interpreted to indicate that hydraulic processes dominate channel morphology along Piccaninny Creek. The downstream erosional sequence from plane beds to inner channel is probably related to the shedding of turbulent vortices off irregularities in the channel bed. These vortices could create structure in the flow which lasts long enough to initiate erosional bedforms which become sinuous with continued erosion. Once these features are initiated, they further perturb the flow, causing a self-enhancing feedback. If these inferences are correct, the length of any inner channel segment (40-114 m along Piccaninny Creek) would be controlled by the downstream distance over which structured flow could persist, with the inner channel ending where the individual paired vortices disintegrate to generalized turbulence.

DISCUSSION

The case studies summarized above represent fairly crude characterizations of the driving and resisting forces operating to shape bedrock channel morphology. The use of averaged hydraulic variables, for example, does not indicate the high spatial variability in velocity, shear stress, or unit stream power commonly present along bedrock channels [*Tinkler*, 1997; *Tinkler and Wohl*, 1998](Figure 6). Similarly, the use of a single maximum discharge for flow modeling and for statistical analyses does not take into account the possibility of flow overprinting, such as adjustments in channel morphology in response to a sequence of flows of differing magnitudes. The substrate heterogeneity index was developed in recognition of the inability of a single reach- or subreach-averaged Selby rock-mass strength index to adequately represent substrate resistance along a reach with multiple lithologies or with variable bedrock bedding and jointing. However, even with the caveat that the method outlined in this chapter represents only an approximation of the forces at work in bedrock channels, consistent patterns of bedrock channel morphology do appear.

As discussed by *Wohl and Merritt* [in press], initial hypotheses in relating bedrock channel morphology to potential control variables were that (1) some measure of hydraulic driving force and some measure of substrate resisting force would correlate significantly with reach-scale channel morphology, and (2) channel-bed gradient would correlate with channel morphology. If driving forces greatly exceed resisting forces, we expect to see a graded stream with a concave-upward longitudinal profile and a lower gradient. If driving forces are much lower than resisting forces, we expect to see a straight or convex longitudinal profile, and some form of channel adjustment that has the effect of maximizing erosive force, such as the presence of an inner channel. The final data matrix included Selby rock-mass strength, substrate heterogeneity index, total stream power, unit stream power, channel-bed gradient, drainage area, maximum discharge, the ratios of unit stream power to Selby rock-mass strength and total stream power to Selby rock-mass strength, and channel morphology.

Stepwise discriminant analysis was employed to detect significant correlations between channel morphology and other variables. Discriminant analysis is designed specifically for datasets with one or more quantitative variables and a classification variable defining groups of observations (here, channel morphology). The analysis develops a discriminant criterion to classify each observation into one of the groups. Stepwise discriminant analysis was used to determine how many of the bedrock channels studied would be correctly classified with respect to channel morphology using the discriminant criterion.

The control variables substrate heterogeneity, channel gradient, and Selby rock-mass strength correlated significantly with channel morphology at p-values of less than 0.05. Of the 40 data points, 28 (70%) were correctly classified in stepwise analysis as the channel morphologic type that we defined in the field. Step-pool channels form a cluster on the high end of axis 1 (high stream gradient), whereas pool-riffle channels fall out on the low end of the axis (low gradient), and channels with undulating walls, an inner channel, meandering, or a plane-bed tend to have intermediate gradients (Figure 7). Step-pool and plane-bed channels tend to form in more heterogeneous substrates, pool-riffle, undulating wall, meandering and inner channels are associated with homogeneous substrates. Channels with meanders or with undulating walls are associated with lower values of Selby rock-mass strength, pool-riffle channels tend to have higher Selby values, and the other channel types span the range of Selby values. The ratio of hydraulic driving to substrate resisting forces, as measured by the ratio of total stream power to Selby rock-mass strength, is lowest for step-pool channels, and increases through meandering, pool-riffle, plane-bed, and inner channels, reaching a high value for channels with undulating walls.

Alluvial bedforms and variations in channel geometry have been interpreted as roughness elements that increase the expenditure of flow energy by generating flow separation and turbulence [*Knighton*, 1998]. Changes in alluvial bedform configuration in response to water and sediment discharge reflect changes in energy availability. Similarly, channel geometries in bedrock channels may reflect energy availability and the competence of the flow to erode the channel boundaries. Our results suggest that channel morphologies which have the effect of increasing bed erosion for a given discharge (step-pool, undulating wall, inner channel) tend to be associated with higher gradient reaches, whereas morphologies that evenly distribute flow energy (plane-bed) or that dissipate a greater proportion of flow energy internally rather than against the channel boundaries (pool-riffle) tend to be associated with lower gradient reaches. The specific type of channel morphology present along a higher or lower gradient channel appears to be a function of substrate resistance and heterogeneity.

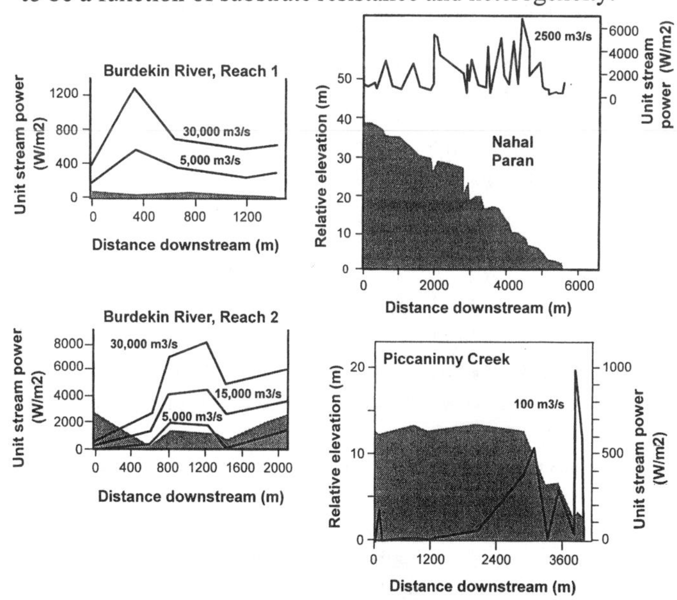

Figure 6. Examples of the downstream variability in stream power per unit area along bedrock channels. The shaded area at the lower portion of each graph represents the channel thalweg. Solid lines above represent the average value of unit stream power at each surveyed cross section, for the discharge listed next to the line. The Burdekin River in northeastern Australia is incised into welded rhyolitic tuffs [*Wohl*, 1992a]. Reach 1 is a fairly straight, uniform reach, whereas reach 2 has a more complex channel geometry. Piccaninny Creek, Australia and Nahal Paran, Israel are described in the text.

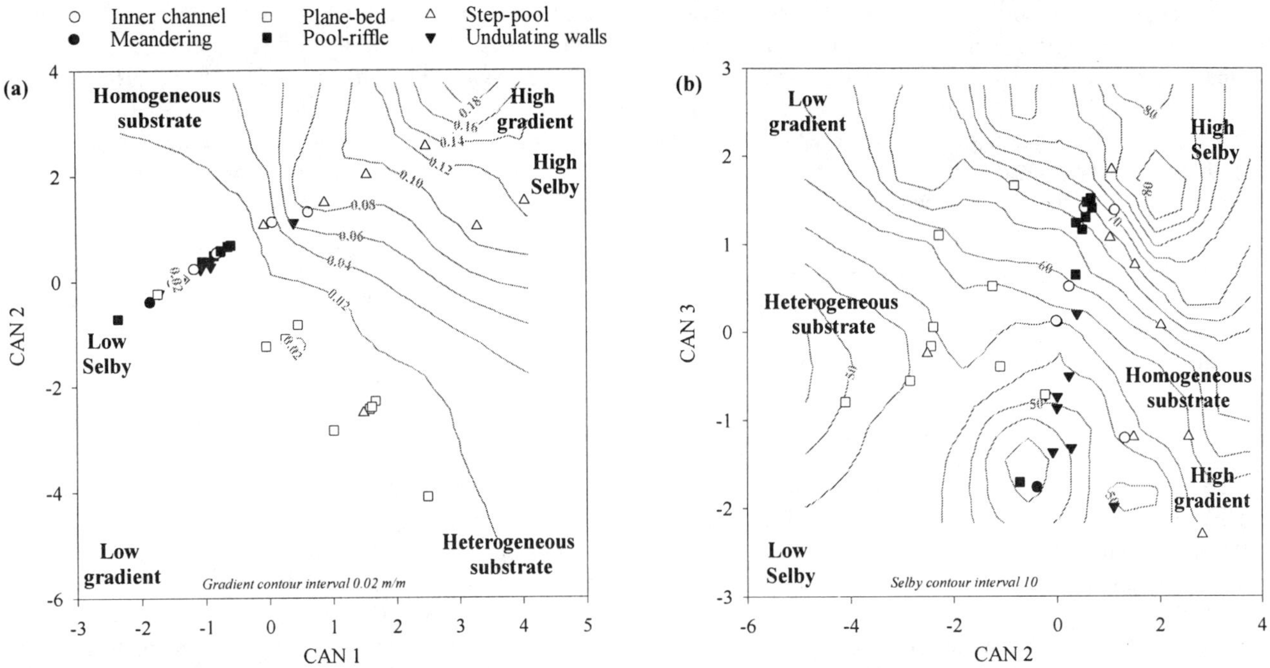

Figure 7. (a) Plot of first two axes from canonical discriminant analysis (CDA) of bedrock channel data from 40 stream reaches. Symbols represent the six channel types from the classification scheme in Wohl [1998a]. Weightings of the independent variables Selby rock-mass strength, gradient, and substrate heterogeneity, the variables used to classify channel type, were 0.12, 0.68, and 0.62 on canonical axis 1 and 0.40, 0.66, and -0.78 on canonical axis 2, respectively. Canonical axes 2 and 3 are plotted in (b). Canonical weightings for axis 3 were 0.91, -0.33, 0.06 for Selby rock-mass strength, gradient, and substrate heterogeneity, respectively. Wilks' Lambda for the discriminant function analysis using within class covariance matrices was $F = 9.66$, $p < 0.0001$ and the classification error rate was 0.32. Contour lines of reach gradient in frame a and Selby rock-mass strength in frame b were generated and overlayed onto the CDA plots to illustrate the distributions of reach types in ordination space relative to each of these individual factors. Contour intervals are indicated on each of the plots.

IMPLICATIONS OF THESE ANALYSES

Discriminant analyses indicate that independent variables representing the hydraulic driving forces and substrate resisting forces acting along bedrock channels classify more than two-thirds of the bedrock channel reach morphologies correctly. These results suggest that bedrock channel morphology, like alluvial channel morphology, results from an adjustment between the controlling variables of substrate and hydraulics, and that the manner of this adjustment is predictable if the controlling variables can be quantified. If we can predict likely channel morphology at the reach scale, we can better understand reach-scale segmentation and inter-reach adjustment along bedrock channels. Investigations of the morphology of coarse-grained alluvial channels have also identified substrate resistance, channel gradient, and coarse sediment supply as controls on channel morphology [*Grant et al.*, 1990; *Montgomery and Buffington*, 1997]. A primary difference between bedrock and most alluvial channels is that bedrock channels are more likely to be shaped by the infrequent, high-magnitude flows that are capable of exceeding boundary resistance. Because these flows are unlikely to be adequately captured by gaging records or direct measurements, flow modeling based on paleostage indicators and surveyed channel geometry represents the most effective means of estimating flood hydrology and hydraulics. The interpretation of bedrock channel morphology thus relies heavily on paleohydrologic information and techniques, and provides a good example of applied paleoflood hydrology.

Acknowledgements. Gregory Springer, Sara Rathburn, Paul Carling and Andrew Miller provided very useful reviews of this chapter. The original field work described in the referenced case studies was supported by the National Geographic Society, the Japan Society for the Promotion of Science, the US-Israel Educational Foundation, and the Geological Society of America. David Merritt performed the statistical analyses summarized in the discussion section of this paper.

REFERENCES

Baker, V. R., Paleoflood hydrology and extraordinary flood events, *J. Hydrol.*, 96, 79-99, 1987.

Baker, V. R., L. L. Ely, J. E. O'Connor, and J. B. Partridge, Paleoflood hydrology and design applications, in *Regional Flood Frequency Analysis*, edited by V. P. Singh, pp. 339-353, D. Reidel Publ. Co., Dordrecht, 1987.

Baker, V. R., R. C. Kochel and P. C. Patton, Long-term flood frequency analysis using geological data, in *The Hydrology of Areas of Low Precipitation*, pp. 3-9, IAHS Publ. No. 128, 1979.

Baker, V. R. and G. Pickup, Flood geomorphology of the Katherine Gorge, Northern Territory, Australia, *Geol. Soc. Am. Bull.*, 98, 635-646, 1987.

Bates, P. D., M. G. Anderson, J. M. Hervouet, and J. C. Hawkes, Investigating the behaviour of two-dimensional finite element models of compound channel flow, *Earth Surf. Proc. Land.*, 22, 3-17, 1997.

Berger, R. C. and R. L. Stockstill, Finite-element model for high-velocity channels, *J. Hydr. Eng.*, 121(10), 710-716, 1995.

Burbank, D. W., J. Leland, E. Fielding, R. S. Anderson, N. Brozovic, M. R. Reid and C. Duncan, Bedrock incision, rock uplift and threshold hillslopes in the northwestern Himalayas, *Nature*, 379, 505-510, 1996.

Carling, P. A., Hydrodynamic interpretation of a boulder berm and associated debris-torrent deposits, *Geomorph.*, 1, 53-67, 1987.

Carling, P. A., Hydrodynamic models of boulder berm deposition,*Geomorph.*, 2, 319-340, 1989.

Carling, P. A., Flow-separation berms downstream of a hydraulic jump in a bedrock channel, *Geomorph.*, 11, 245-253, 1995.

Cluer, B. L., *Eddy Bar Responses to the Sediment Dynamics of Pool-Riffle Environments*, Ph.D. dissertation, Colorado State University, Ft. Collins, 1997.

Curran, J. H., *Hydraulics of Large Woody Debris in Step-Pool Channels, Cascade Range, Washington*, M.S. thesis, Colorado State University, Ft. Collins, 1999.

Donnell, B. D., J. L. Finnie, J. V. Letter, Jr., W. H. McAnally, Jr., L. C. Roig and W. A. Thomas, *Users Guide to RMA2 WES Version 4.3*, U.S. Army Corps of Engineers, Waterways Experiment Station Hydraulics Laboratory, 1997.

Ely, L. L., Y. Enzel, V. R. Baker and D. R. Cayan, A 5000-year record of extreme floods and climate change in the southwestern United States, *Science*, 262, 410-412, 1993.

Ely, L. L., R. H. Webb and Y. Enzel, Accuracy of post-bomb ^{137}Cs and ^{14}C in dating fluvial deposits, Quaternary Research, 38, 196-204, 1992.

Enzel, Y., L. L. Ely, P. K. House, V. R. Baker and R. H. Webb, Paleoflood evidence for a natural upper bound to flood magnitudes in the Colorado River basin, *Water Resour. Res.*, 29, 2287-2297, 1993.

Grant, G. E., Critical flow constrains flow hydraulics in mobile-bed streams: a new hypothesis, *Water Resour. Res.*, 33, 349-358, 1997.

Grant, G. E., F. J. Swanson, and M. G. Wolman, Pattern and origin of stepped-bed morphology in high-gradient streams, Western Cascades, Oregon, *Geol. Soc. Am. Bull.*, 102, 340-352, 1990.

Howard, A. D., Thresholds in river regimes, in *Thresholds in Geomorphology*, edited by D. R. Coates and J. D. Vitek, pp. 227- 258, George Allen and Unwin, London, 1980.

Howard, A. D., W. E. Dietrich and M. A. Seidl, Modeling fluvial erosion on regional to continental scales, *J. Geophys. Res.*, 99 (B7), 13,971-13,986, 1994.

Hydrologic Engineering Center, *HEC-RAS River Analysis System v.2.0 User's Manual*, U.S. Army Corps of Engineers, Davis, California, 1997.

Kieffer, S. W., The 1983 hydraulic jump in Crystal Rapid: implications for river-running and geomorphic evolution in the Grand Canyon, *J. Geol.*, 93, 385-406, 1985.

Kieffer, S. W., The rapids and waves of the Colorado River, Grand Canyon, Arizona, *U.S. Geol. Survey Open-File Report* 87-096, 1987.

Kieffer, S. W., Geologic nozzles, *Rev. Geophys.*, 27, 3-38, 1989.

Knighton, A. D., *Fluvial forms and processes*, Arnold, London, 1998.

Leopold, L. B. and Langbein, W. B., The concept of entropy in landscape evolution, *U.S. Geol. Survey Prof. Paper* 500-A, 1962.

Leopold, L. B., Wolman, M. G. and Miller, J. P., *Fluvial Processes in Geomorphology*, W. H. Freeman and Company, San Francisco, 1964.

Meselhe, E. A., A. A. Bradley, A. Kruger, and M. V. I. Muste, PIV and numerical modeling for flow analysis, *Water Resources Engineering 98*, edited by S. R. Abt, J. Young-Pezeshk, and C. C. Watson, p. 7, Am. Soc. Civ. Eng., Reston, VA, 1998.

Miller, A. J. and B. L. Cluer, Modeling considerations for simulation of flow in bedrock channels, in *Rivers Over Rock: Fluvial Processes in Bedrock Channels*, edited by K. J. Tinkler and E. E. Wohl, pp. 61-104, Am. Geophys. Union Geophys. Mon. 107, Washington, D.C., 1998.

Montgomery, D. R., Valley incision and the uplift of mountain peaks, *J. Geophys. Res.*, 99 (B7), 13,913-13,921, 1994.

Montgomery, D. R. and J. M Buffington, Channel-reach morphology in mountain drainage basins, *Geol. Soc. Am. Bull.*, 109, 596-611, 1997.

O'Connor, J. E. and R. H. Webb, Hydraulic modeling for paleoflood analysis, in *Flood Geomorphology*, edited by V. R. Baker, R. C. Kochel, and P. C. Patton, pp. 393-418, Wiley, New York, 1988.

O'Connor, J. E., R. H. Webb and V. R. Baker, Paleohydrology of pool-and-riffle pattern development, Boulder Creek, Utah, *Geol. Soc. Am. Bull.*, 97, 410-420, 1986.

Rosenbloom, N. A. and R. S. Anderson, Hillslope and channel evolution in a marine terraced landscape, Santa Cruz, California, *J. Geophys. Res.*, 99 (B7), 14,013-14,029, 1994.

Salas, J. D., E. E. Wohl and R. D. Jarrett, Determination of flood characteristics using systematic, historical and paleoflood data, in *Coping With Floods*, edited by G. Rossi, N. Harmancioglu, and V. Yevjevich, pp. 111-134, Kluwer Academic, Dordrecht, 1994.

Seidl, M. A. and W. E. Dietrich, The problem of channel erosion into bedrock, *Catena Supple.* 23, 101-124, 1992.

Seidl, M. A., W. E. Dietrich and J. W. Kirchner, Longitudinal profile development into bedrock: an analysis of Hawaiian channels, *J. Geol.*, 102, 457-474, 1994.

Selby, M. J., A rock mass strength classification for geomorphic purposes: with tests from Antarctica and New Zealand, *Zeit. Geomorph.*, 24, 31-51, 1980.

Siggers, G. B., P. D. Bates, M. G. Anderson, D. E. Walling and Q. He, A preliminary investigation of the integration of modelled floodplain hydraulics with estimates of overbank floodplain sedimentation derived from Pb-210 and CS-137 measurements, *Earth Surf. Proc. Land.*, 24, 211-231, 1999.

Stewart, M. D., P. D. Bates, M. G. Anderson, D. A. Price and T. P. Burt, Modelling floods in hydrologically complex lowland river reaches, *J. Hydrol.*, 223, 85-106, 1999.

Stock, J. D. and D. R. Montgomery, Geologic constraints on bedrock river incision using the stream power law, *J. Geophys. Res.*, 104

(B3), 4983-4993, 1999.

Thompson, D. M., J. M. Nelson and E. E. Wohl, Interactions between pool geometry and hydraulics, *Water Resour. Res.*, 34, 3673-3681, 1998.

Thompson, D. and E. Wohl, Flume experimentation and simulation of bedrock channel processes, in *Rivers Over Rock: Fluvial Processes in Bedrock Channels*, edited by K. J. Tinkler and E. E. Wohl, pp. 279-296, Am. Geophys. Union Geophys. Mon. 107, Washington, D.C., 1998.

Tinkler, K. J., Critical flow in rockbed streams with estimated values for Manning's *n*, *Geomorph.*, 20, 147-164, 1997.

Tinkler, K. and Wohl, E., A primer on bedrock channels, in *Rivers Over Rock: Fluvial Processes in Bedrock Channels*, edited by K. J. Tinkler and E. E. Wohl, pp. 1-18, Am. Geophys. Union Geophys. Mon. 107, Washington, D.C., 1998.

Wohl, E. E., Bedrock benches and boulder bars: Floods in the Burdekin Gorge of Australia, *Geol. Soc. Am. Bull.*, 104, 770-778, 1992a.

Wohl, E. E., Gradient irregularity in the Herbert Gorge of northeastern Australia, *Earth Surf. Proc. Land.*, 17, 69-84, 1992b.

Wohl, E. E., Bedrock channel incision along Piccaninny Creek, Australia, *J. Geol.*, 101, 749-761, 1993.

Wohl, E. E., Bedrock channel morphology in relation to erosional processes, *Rivers Over Rock: Fluvial Processes in Bedrock Channels*, edited by K. J. Tinklcr and E. E. Wohl, pp. 133-151, Am. Geophys. Union Geophys. Mon. 107, Washington, D.C., 1998a.

Wohl, E. E., Uncertainty in flood estimates associated with roughness coefficients, *J. Hydraul. Eng.*, 124, 219-223, 1998b.

Wohl, E. E., N. Greenbaum, A. P. Schick and V. R. Baker, Controls on bedrock channel incision along Nahal Paran, Israel, *Earth Surf. Proc. Land.*, 19, 1-13, 1994.

Wohl, E. E. and H. Ikeda, Experimental simulation of channel incision into a cohesive substrate at varying gradients, *Geol.*, 25, 295-298, 1997.

Wohl, E. E. and H. Ikeda, Patterns of bedrock channel erosion on the Boso Peninsula, Japan, *J. Geol.*, 106, 331-345, 1998.

Wohl, E. E. and Merritt, D. M., Bedrock channel morphology, in press, *Geol. Soc. Am. Bull.*

Wohl, E. E., D. M. Thompson and A. J. Miller, Canyons with undulating walls, *Geol. Soc. Amer. Bull.*, 111, 949-959, 1999.

Wolman, M. G., A method of sampling coarse river-bed material, *EOS, Trans. Am. Geophys. Union*, 35, 951-956, 1954.

Ellen Wohl, Dept of Earth Resources, Colorado State University, Ft. Collins, CO 80523

The Geology and Geography of Floods

Jim E. O'Connor

U.S. Geological Survey, Portland, Oregon

Gordon E. Grant

U.S. Forest Service, Corvallis, Oregon

John E. Costa

U.S. Geological Survey, Portland, Oregon

From a global perspective, the magnitude, duration, and volume of riverine floods vary widely depending on source mechanism, global location, geology, and physiography. But despite this variance, there are real physical limits as to how large a flood can be. Meteorological floods result from individual storms of various types that deliver water volumes as great as 10^{11} m^3 over a few days or weeks as well as from seasonally persistent climate patterns that deliver upwards of 10^{12} m^3 over several weeks or months and cause peak discharges of as much as 10^5 m^3/s within large, continental-scale basins. The fundamental limits to meteorologic flooding depend primarily on precipitation rates and volumes, and the efficiency with which distributed rainfall concentrates into channel networks. Global and U.S. flood records show that for basins primarily affected by individual storms, the largest flows are in areas where atmospheric moisture or storm tracks are intercepted by local topographic relief. For larger basins, the largest flows result from seasonably persistent climate patterns and, on a global basis, are generally in the tropical regions where there is more precipitable moisture. Especially notable is the contribution of high-relief topography to meteorologic flooding because of its double effect of (1) lifting of atmospheric moisture, thus increasing local precipitation, and (2) promoting faster concentration of flow into channels. Floods from dam failures and terrestrial freshwater sources such as lakes and ice caps involve similar total volumes as meteorological floods, up to 10^{13} m^3, but with much larger discharges—as great as $2{\cdot}10^7$ m^3/s. The largest well-documented dam-failure floods have been during Pleistocene ice ages when glaciers and changing hydrologic conditions deranged drainage systems, causing large lakes to breach ice dams and basin divides. The

Ancient Floods, Modern Hazards: Principles
and Applications of Paleoflood Hydrology
Water Science and Application Volume 5, pages 359-385

Published in 2002 by the American Geophysical Union

largest historic floods from terrestrial water sources have had discharges as great as $1.5 \cdot 10^6$ m^3/s and include floods from volcanic-induced melting of ice caps and failures of ice and landslide dams. The fundamental limits to dam-failure floods primarily arise from the size of the outlet channel and the volume of the impounded waterbody.

INTRODUCTION

Riverine floods come in a wide range of frequencies and magnitudes from various geophysical mechanisms. Viewed over geologic time and space scales, however, fundamental climatologic, topographic, and geologic factors substantially influence and limit local flood volumes and magnitudes in predictable ways. Many of these influences are obvious, such as the relationship of certain types of floods to local topographic and climate patterns. But some are less obvious, including geologic controls on the timing, location, and magnitude of natural dam failures. In this paper, we outline some of the primary factors that influence the magnitude of floods and how these factors vary temporally and spatially over geologic time and space scales.

Our motivation is the perception that the human perspective of floods is often narrow, usually focusing on experiences or hazards of floods on particular rivers. Even seasoned hydrologists and geologists treat floods outside of modern experience, such as the great Pleistocene Missoula Floods, as scientific curiosities. Such great floods are intrinsically interesting because of their perceived uniqueness and dramatic role in shaping the landscape, and have served as lightning rods for inquiry and controversy, but they are not usually considered as a component of normal geophysical behavior. But indeed floods of all sizes and sources can be viewed as a geologic process on par with other Earth-system processes, such as tectonics and atmospheric circulation, with attributes, causes, effects, and even interrelationships that span similar ranges of scale.

Practitioners of paleoflood hydrology [*Kochel and Baker*, 1982], who study floods of all scales and from many different geologic eras and environments, are well situated to appreciate, describe, and even quantify these interrelationships. Indeed, many of the papers in this volume provide examples of floods resulting from these geologic, physiographic, and climatologic interactions in specific settings. The purpose is this paper to provide an overall context for these individual paleoflood studies by systematically describing spatial and temporal distribution of large floods and flood-generating mechanisms, thereby establishing a framework for the environmental setting of floods over Earth history. These results are consistent with physical limits for terrestrial floods determined from general consideration of flood generation processes.

What is a flood? Simply stated by a prominent geomorphologist and hydrologist: "Floods come from too much water" [*M.G. Wolman*, pers. comm., 1983]. For the purposes of this paper, we consider floods as notable flows of channelized water. In other words, there are no specific criteria that we are using to define a flood. Many of the floods that we describe or include in our analyses were damaging and geomorphically effective, but our discussion is not limited to such events. Nor do we rely on other measures of flow magnitude such as recurrence interval or the relation of flow stage to channel banks or floodplain elevations. Such measures are either difficult to characterize or irrelevant for floods of diverse origins and scales. Our discussion is restricted to riverine floods, so we are excluding flooding of shorelines associated with rising lake or sea levels.

This paper is organized by the following five premises, each of which builds upon the previous one and is elaborated in the following sections with data compilations and specific examples.

1. Floods are the product of the release of stored water into channels and floodplains.
2. The magnitude of floods depends on storage, release, and transfer processes and rates.
3. Certain geologic, climatologic, and physiographic settings result in storage volumes and release rates that produce larger floods than in other settings.
4. These large-flood settings are generally predictable within an overall geologic and geographic framework.
5. Observations of flood magnitude, duration, and volume are consistent with global physical limits to flooding that result from factors such as topography, rainfall rates, and basic hydraulic principles.

STORAGE RESERVOIRS AND RELEASE MECHANISMS

All floods result from the rapid release of water onto the landscape. The geography of floods is therefore closely tied to those parts of the Earth where large quantities of water can both accumulate and be released quickly to channels and floodplains. Water can be stored both in the atmosphere as vapor, or terrestrially as liquid water or solid ice and snow. Both atmospheric and terrestrial sources of floods have characteristic volumes and release rates that affect

Table 1. Global freshwater budget (from *Barry and Chorley*, 1992).

Location	Source	Percent
Terrestrial	Ice Sheets and Glaciers	74.6
	Ground water	25.0
	Lakes	0.3
	Soil Moisture	0.06
	Rivers	0.03
Atmosphere	Precipitable Moisture	0.035

their potential to cause large flows. The non-uniform and non-random geographic distribution of these flood-generating sources and processes means that the potential for large floods is concentrated in specific arrangements of physiographic and atmospheric features. In this section we review atmospheric and terrestrial water sources, including their areal extents and volumes, geographic distribution, and rates and mechanisms of release.

Global Water Storage

Most of the Earth's water (97%) is stored in the oceans [*More*, 1967]. Of the world's fresh water (3% of total), almost 75% is frozen for centennial to millennial timescales in ice sheets and glaciers (Table 1). Virtually all remaining fresh water is stored as groundwater. All water readily available to cause floods—that is water stored as lakes, rivers, soil moisture and the atmosphere—accounts for only 0.4% of total fresh water and 0.01% of global water volume. The amount of water vapor stored in the atmosphere, expressed as average depth over the Earth's surface, is only 2.5 cm [*Barry and Chorley*, 1992].

Atmospheric Storage and Delivery

Despite this almost minuscule amount of atmospheric water storage relative to the total volume of water on Earth, under the proper conditions huge fluxes of moisture can be concentrated and delivered to the landscape, either in individual storms or from persistent patterns of weather or climate (Figure 1, also on CD). For individual storms, these conditions depend primarily on micro- to macro-scale vertical and, to lesser extent, horizontal motions of the atmosphere that cause moisture-laden air to converge and accumulate. At time scales greater than a few days or weeks, synoptic and global patterns of horizontal atmospheric motion control rates and volumes of precipitation delivery [*Hayden*, 1988].

Individual storms. As summarized by *Hirschboeck* [1991], individual storms precipitate moisture because of air-mass uplift and consequent cooling and release of water as rain or snow. Such air-mass uplift generally results from: (1) thermal convection, (2) large-scale frontal convergence of warm and cold air masses, (3) upper air disturbances, and (4) orographic lifting. Convectional processes include thunderstorms, tropical and extratropical cyclones, and mesoscale convective complexes or systems. These span a wide range of spatial scales, from 10^{-1} to 10^{2} km^2 for individual thunderstorms, to thunderstorm systems (mesoscale convective complexes) spanning 10^3 to 10^5 km^2, to hurricanes that can cover 10^6 km^2 or more. Such convective systems can produce extremely intense precipitation, leading to flash floods in small watersheds or regional floods in larger basins. Frontal convergence occurs over large (10^5 to 10^6 km^2) areas, leading to large basin floods. Upper air disturbances and orographic lifting can occur over small or large areas and have similarly local or widespread effects on flooding.

Storm systems can be thought of as moisture reservoirs, with permeable and often ill-defined boundaries, but distinctive spatial and temporal scales. Of specific importance to flooding is the volume of water stored in and delivered by these types of storm systems and delivery processes. There have been few systematic studies to compare storms of different types in terms of their total moisture volumes. Storm systems are typically described in terms of total precipitable water [e.g. *Bradley and Smith*, 1994]: the liquid equivalent of water vapor in an atmospheric column or from the surface to a specified barometric surface (i.e., 300 mb). For example, *Bradley and Smith* reported that total precipitable water from 33 extreme rainstorms in the southern U.S. ranged from 26.4 to 56.1 mm. For storms where the precipitable water content is known, the entire storm system can be considered an atmospheric source, with the total volume of water available for runoff being the product of the precipitable water and the areal extent of the storm system. For the more typical case in which areal extent and total precipitable water are not reported, we can approximately calculate the total storage volume as the product of storm area, average rainfall intensity, and storm duration. Representative values of these parameters for various storm types as reported in the meteorological literature are shown in Table 2. Total storm volumes range from $1.8 \cdot 10^5$ to $2.1 \cdot 10^7$ m^3 of water for small and intense convective cells to $6.1 \cdot 10^9$ m^3 for mesoscale convective complexes to $4.7 \cdot 10^{10}$ m^3 for hurricanes.

Rainfall volumes from individual storms can also be estimated from measurements of precipitation depth, area, and duration for individual storms (Figure 2, also on CD). For example, *Leopold* [1942] summarized such data for six convective summer thunderstorms in Arizona and New

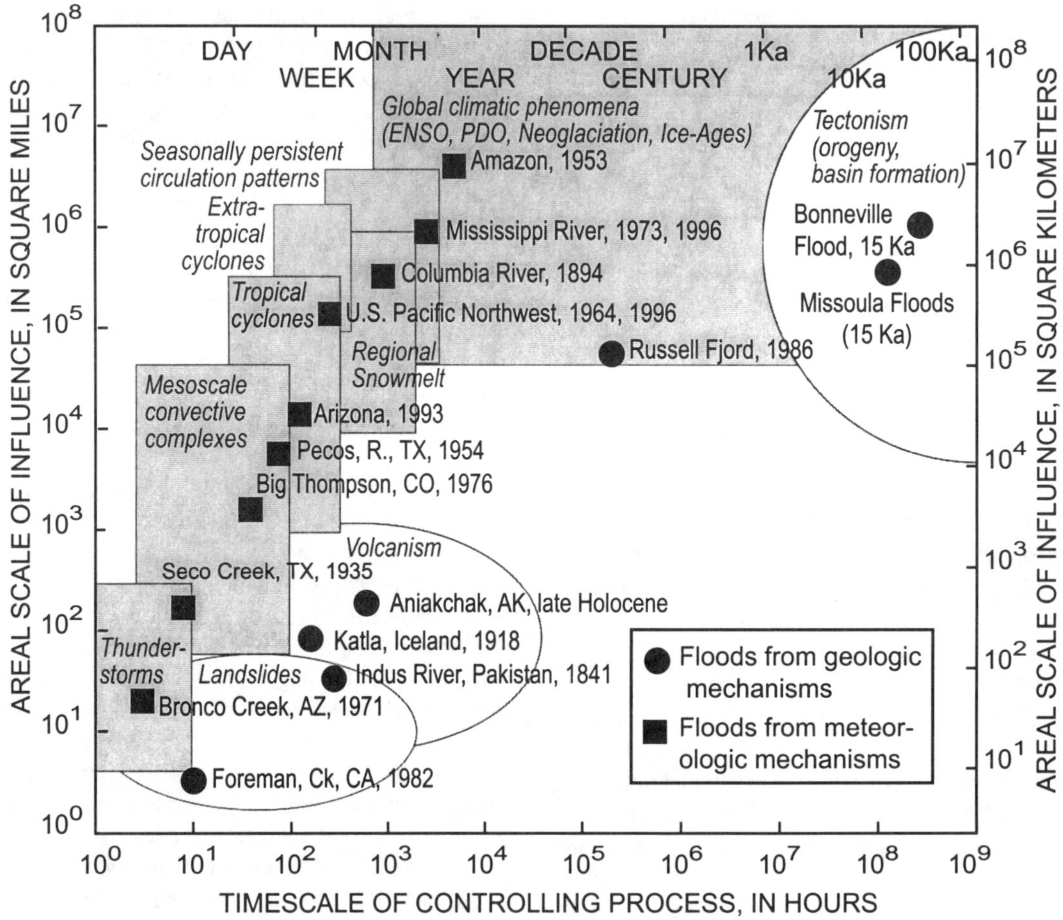

Figure 1. Space-time domain for selected meteorologic and geologic processes that can cause floods. Selected floods also shown. After Hirschboeck (1988).

Mexico, for which total rainfall exceeded volumes ranging between $6.5 \cdot 10^5$ and $4 \cdot 10^7$ m^3. The *World Meteorological Organization* [1986, p. 261-263] has compiled observations of maximum depth-area-duration for individual storms (Figure 2b), for which the largest measured rainfall volume resulted from a 1916 hurricane centered in Florida that delivered an average of 226 mm of rainfall over 260,000 km^2, for a total of $6 \cdot 10^{10}$ m^3 of water. Similarly, $4.3 \cdot 10^{10}$ m^3 of water was precipitated by a 1967 hurricane in Texas. The thunderstorm that produced the record 4.5-hour rainfall at Smethport, Pennsylvania, during July 17-18, 1942 (Figure 2a) had a total storm volume in excess of $2.7 \cdot 10^9$ m^3. Exceptionally large and wet hurricanes not included in the World Meteorological Organization database likely produced even more precipitation. For example, maps of total precipitation resulting from the Oct.-Nov. 1998 Hurricane Mitch in Central America (http://cindi.usgs.gov/hazard/event/mitch/mitch/atlas/hmprecip.html; May 25, 2001) indicate a total precipitation volume of about $2 \cdot 10^{11}$ m^3.

Persistent patterns of moisture delivery. Floods in larger basins more typically result from persistent precipitation-enhancing synoptic weather or climate. Precipitation is still delivered by individual storms or convective cells as described in the previous section, but the frequency, intensity, and duration of such storms can be affected by regional or global atmospheric patterns that can persist for months to decades. Such patterns include quasi-stationary trough-and-ridge patterns of upper atmospheric circulation that can promote continued atmospheric convergence over areas of 10^5 to 10^7 km^2 for several days or weeks. Such a condition over the north-central United States was the cause of the Mississippi River floods of 1973 and 1993 [*Hirschboeck*, 1991; 2000]. Climatic conditions of annual to decadal duration that can cause regionally enhanced precipitation include coupled atmospheric and oceanic phenomena of hemispheric or global scales such as the El Nino/Southern oscillation (ENSO) and the Pacific Decadal Oscillation (PDO; e.g. *Cayan and Peterson*, 1989; *Dettinger and Cayan*, 2000). Even longer-term climate fluctuations,

Table 2. Selected estimates of precipitation volume for representative storm types.

Storm type	Area (km^2)	Duration (hrs)	Peak intensity (mm/hr)	Total volume precipitated ($m^3 \cdot 10^6$)	Reference
Small convective cell (intense phase)[a]	25	0.5		21.6	Ludlam, 1980 (p. 271)
Large cumulonimbus convection cell	32,400	20	90.0	450[b]	Ludlam, 1980 (p. 298-299)
Mesoscale convective complex	82,000	24	6.8	6,100[c]	Cotton and Anthes, 1989
Hurricane (avg)	31,000	72[d]	21.0	47,000	

[a] Average precipitation for storms with and without hail
[b] Total storm volume quoted as $4.5 \cdot 10^{12}$ kg over an area of 32,400 km^2
[c] From time-averaged integral of 82 storms
[d] Assumes a 200-km inner-area diameter
[e] Assumes 3-day duration

such as century- and millennial-scale periods of global cooling and warming (e.g. the Neoglacial Period, Medieval Optimum, and Pleistocene ice ages), may be relevant to flood frequency and magnitude [e.g. *Ely et al.*, 1993].

It is more difficult to determine directly the volumes of water delivered by persistent climate and weather regimes. Indirect minimum estimates can be obtained by considering the size of floods that have resulted from periods of enhanced precipitation, although this substantially underestimates total volumes by not accounting for water lost to evapotranspiration or long-term ground-water storage. Nevertheless, some values that may be representative include $4.9 \cdot 10^{11}$ m^3 of water for the 1973 flood on the Mississippi River and $4 \cdot 10^{12}$ m^3 for the 1953 flood on the Amazon River in Brazil (Table 3). These values suggest that persistent weather and climate patterns over large basins can produce water volumes several times greater than the wettest individual storms (Table 2).

Rates of atmospheric moisture delivery. A primary factor controlling the discharge of meteorological floods regardless of atmospheric source is the rate of transfer of water from atmospheric storage to the landscape by precipitation. Record rainfall amounts vary from 38 mm in 1 minute at Barot, Guadeloupe (West Indies), to over 40 m during 2 years of strong monsoon circulation at Cherranpunji, India (Figure 2a). In spite of the broad range of volumes of water delivered by various types of storm systems and persistent weather patterns, maximum rainfall intensities follow a well-defined relationship,

$$\mathbf{P = 350D^{0.49}}\ (r^2=0.99), \quad (1)$$

over six orders of magnitude, where P equals maximum observed precipitation depths in millimeters and D is duration in hours (Figure 2a). Although there are no

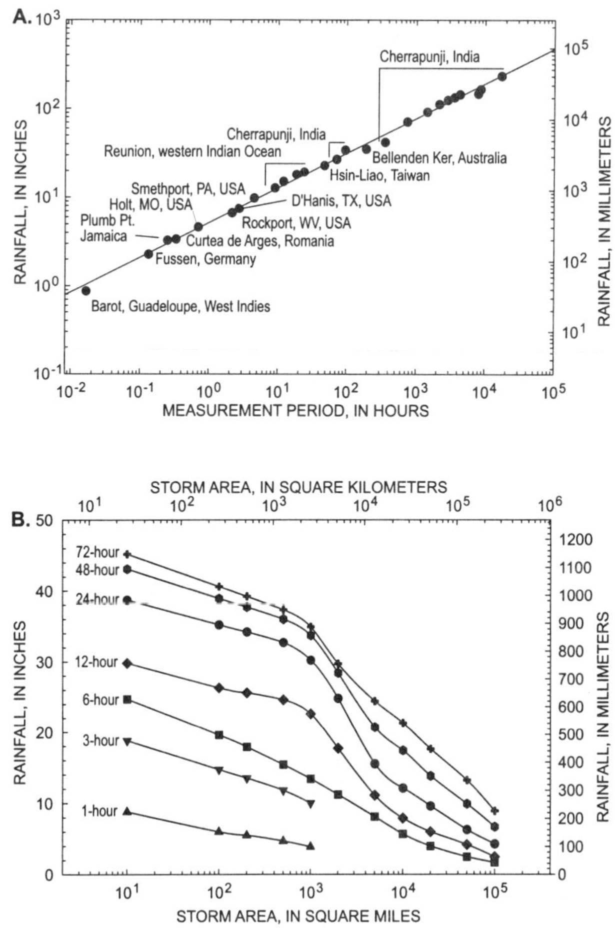

Figure 2. Maximum rainfall observations. A) Global maximum point rainfall measurements for various time intervals. Data from *World Meteorological Organization*, 1986, p. 258. B) Maximum depth-area-duration precipitation measurements for U.S. storms. Data from *World Meteorological Organization*, 1986, p. 261.

Table 3. Selected floods from natural dam failures, constructed dam failures, and meteorological events ("–" indicates no data; "n.a." indicates "not applicable.")

River/dam-/lake	Location	Date of Flood	Type of Dam or Flood	Dam Height (m)	Breach Depth (m)	Volume (m^3)	Peak Discharge (m^3/s)	Duration[a] (hours)	Reference
Natural dams failures									
Kuray	Altai, Russia	Pleistocene	Glacier	--	900	$1.0\cdot10^{12}$	$1.8\cdot10^{7}$	$3.1\cdot10^{1}$	Baker et al., 1993
Lake Missoula	Montana, US	Pleistocene	Glacier	--	525	$2.2\cdot10^{12}$	$1.7\cdot10^{7}$	$7.2\cdot10^{1}$	O'Connor and Baker, 1992
Lake Bonneville	Idaho, US	Pleistocene	Basin sill	280	108	$4.8\cdot10^{12}$	$1.0\cdot10^{6}$	$2.6\cdot10^{3}$	O'Connor, 1993
Aniakchak River	Alaska, US	Holocene	Caldera	183	183	$3.7\cdot10^{9}$	$1.0\cdot10^{6}$	$2.1\cdot10^{0}$	Waythomas et al., 1996
Columbia River	Oregon, US	Holocene	Landslide	75	60	$2.1\cdot10^{10}$	$2.2\cdot10^{5}$	$5.3\cdot10^{1}$	O'Connor et al., 1996
Russell Fiord	Alaska, US	Oct. 8, 1986	Glacier	25.5	25.5	$5.4\cdot10^{9}$	$1.1\cdot10^{5}$	$2.9\cdot10^{1}$	Mayo, 1989
Indus River	Pakistan	June, 1841	Landslide	150	150	$1.2\cdot10^{9}$	$5.7\cdot10^{4}$	$1.2\cdot10^{1}$	Hewitt, 1968; Shroder et al., 1991
Rio Mantaro	Peru	Oct. 28, 1945	Landslide	96.3	55.8	$3.0\cdot10^{8}$	$1.2\cdot10^{4}$	$1.4\cdot10^{1}$	Snow, 1964
Rio San Pedro	Chile	July 25, 1960	Landslide	--	43	$2.0\cdot10^{9}$	$7.5\cdot10^{3}$	$1.5\cdot10^{2}$	Pena and Klohn, 1989
Lag. Jancaruish	Peru	Oct. 20, 1960	Moraine	21	21	$8.0\cdot10^{6}$	$7.5\cdot10^{3}$	$5.9\cdot10^{-1}$	Lliboutry et al., 1977
Lag. Josefina	Ecuador	May 10, 1993	Landslide	80	40	$1.2\cdot10^{8}$	$5.0\cdot10^{3}$	$1.3\cdot10^{1}$	Zevallos-Moreno, writ. comm., 1993
Isfayramsay R.	Russia	June 18, 1966	Landslide	90	90	$6.5\cdot10^{6}$	$5.0\cdot10^{3}$	$7.3\cdot10^{-1}$	Glazyrin and Reyzvikh, 1968
Lake Nostetuko	Canada	July 19, 1983	Moraine	--	38.4	$6.5\cdot10^{6}$	$5.7\cdot10^{2}$	$6.3\cdot10^{0}$	Blown and Church, 1985; Clague and Evans, 1994
Rio Pisque	Ecuador	Jan. 26, 1990	Landslide	45	30	$2.5\cdot10^{6}$	$4.8\cdot10^{2}$	$2.9\cdot10^{0}$	Plaza-Nieto et al., 1990; Zev.-Mor., w. comm., 1993
N. Fork Toutle R.	Wash., US	Aug. 27, 1980	Landslide	9	9	$3.1\cdot10^{5}$	$4.5\cdot10^{2}$	$3.8\cdot10^{-1}$	Jennings et al., 1981; Costa and Schuster, 1991
Rio Toro	Costa Rica	July 13, 1992	Landslide	52	50	$5.0\cdot10^{5}$	$4.0\cdot10^{2}$	$6.9\cdot10^{-1}$	Mora et al., 1993
Squaw Creek	Oregon, US	Sept 7, 1970	Moraine	22.4	22.4	$3.2\cdot10^{5}$	$2.8\cdot10^{2}$	$6.3\cdot10^{-1}$	O'Connor et al., in press
Tunawaea R.iver	New Zealand	July 22, 1992	Landslide	50	20	$6.3\cdot10^{5}$	$2.5\cdot10^{2}$	$1.4\cdot10^{0}$	Jennings et al., 1993
Constructed dam failures									
Teton	Idaho, US	June 5, 1976	Earthen	93	67	$3.1\cdot10^{8}$	$6.5\cdot10^{4}$	$2.6\cdot10^{0}$	Costa, 1988; MacDonald and Langridge-Monopolis 1984;
Malpasset	France	Dec. 2, 1959	Masonry	61	61	$2.2\cdot10^{7}$	$2.8\cdot10^{4}$	$4.3\cdot10^{-1}$	Costa, 1988
St. Francis	Calif., US	Mar. 22, 1928	Masonry	--	56.4	$4.7\cdot10^{7}$	$2.0\cdot10^{4}$	$1.3\cdot10^{0}$	Costa, 1988
Oros	Brazil	Mar. 26, 1960	Earthen	35.4	35.4	$6.5\cdot10^{8}$	$1.4\cdot10^{4}$	$2.7\cdot10^{1}$	Costa, 1988; MacDonald and Langridge-Monopolis 1984;
Hell Hole	Calif., US	Dec. 23, 1964	Earthen	67	35.1	$3.1\cdot10^{7}$	$7.4\cdot10^{3}$	$2.3\cdot10^{0}$	Costa, 1988; MacDonald and Langridge-Monopolis 1984;
Lawn Lake	Colo., US	July 15, 1982	Earthen	--	7.9	$8.4\cdot10^{5}$	$5.1\cdot10^{2}$	$9.2\cdot10^{-1}$	Jarrett and Costa, 1986
Porter Hill	Oregon, US	Feb. 27, 1993	Earthen	2.5	2.5	$1.5\cdot10^{4}$	$3.1\cdot10^{1}$	$2.7\cdot10^{-1}$	Costa and O'Connor, 1995
Meteorologic floods									
Amazon River	Obidos, Brazil	Mar.-Sept. 1953	Rainfall	n.a.	n.a.	$4.0\cdot10^{12}$	$3.7\cdot10^{5}$	$4.3\cdot10^{3}$	Oltman et al., 1964
Mississippi River	Vicksburg, Miss., US	March-July, 1973	Rainfall	n.a.	n.a.	$4.9\cdot10^{11}$	$5.6\cdot10^{4}$	$3.4\cdot10^{3}$	USGS NWIS (http://water.usgs.gov/nwis)
Columbia River	The Dalles, Oreg., US	May-July, 1894	Rain/ Snowmelt	n.a.	n.a.	$1.3\cdot10^{11}$	$3.5\cdot10^{04}$	$1.4\cdot10^{3}$	U.S. Army Corp of Engineers, 1958, plate 13
Willamette River	Portland, Oreg., US	Dec. 1964	Rain/ Snowmelt	n.a.	n.a.	$8.6\cdot10^{9}$	$1.3\cdot10^{4}$	$3.4\cdot10^{2}$	Waananen et al., 1970
Plum Creek	Colo., US	June 15, 1965	Rainfall	n.a.	n.a.	$3.9\cdot10^{7}$	$4.4\cdot10^{3}$	$1.0\cdot10^{1}$	Osterkamp and Costa, 1987
Rapid Creek	S.Dak.., US	June 9, 1972	Rainfall	n.a.	n.a.	$2.5\cdot10^{7}$	$1.4\cdot10^{3}$	$1.0\cdot10^{1}$	Larimer, 1973
Big Thompson R.	Colo., US	July 31, 1976	Rainfall	n.a.	n.a.	$4.7\cdot10^{6}$	$8.8\cdot10^{2}$	$3.0\cdot10^{0}$	McCain et al., 1979
McKenzie Creek	Oregon, US	Feb. 6, 1996	Rain/ Snowmelt	n.a.	n.a.	$1.3\cdot10^{8}$	$8.7\cdot10^{2}$	$6.6\cdot10^{1}$	G.E. Grant, unpublished data
Lookout Creek	Oregon, US	Feb. 6, 1996	Rain/ Snowmelt	n.a.	n.a.	$1.9\cdot10^{7}$	$1.9\cdot10^{2}$	$6.6\cdot10^{1}$	G.E. Grant, unpublished data
H.J. Andrews Exp. Forest (W1)	Oregon, US	Feb. 6, 1996	Rainfall/ Snowmelt	n.a.	n.a.	$2.9\cdot10^{5}$	$2.4\cdot10^{0}$	$6.7\cdot10^{1}$	G.E. Grant, unpublished data

[a]Durations for natural and constructed dam failures calculated from reported values of discharge and volume by assuming a triangular hydrograph. Durations for meteorological floods estimated from published hydrographs or discharge records.

apparent *a priori* reasons why maximum rainfall intensities over timescales ranging from 1 minute to years and delivered by a wide assortment of atmospheric processes should scale so closely with measurement interval, the explanation must lie with the water-holding capacity of the atmosphere, which constrains the amount of rainfall that can occur in time.

These record precipitation values are point measurements. Meteorological floods, however, develop from rainfall over finite areas for which these point measurements are not necessarily representative. While the duration of local maximum precipitation follows a simple scaling relation with measurement interval, depth/area/duration relations, which define the total volume of water delivered to a drainage basin, do not scale so simply (Figure 2b). For all storm durations greater than about 6 hours in the United States, maximum precipitation amounts diminish rapidly for areas greater than about 2,000 km^2, resulting in total precipitation inputs that are less on a unit area basis for larger basins. This result, which may reflect the maximum size of convection cells within large storms in the U.S. (Figure 1), closely corresponds with patterns of peak discharge per unit area for large floods in the U.S., as will be discussed in following sections.

From these relations (Figure 2), it is apparent that maximum precipitation rate decreases with both duration and the area over which it is measured. But because areally averaged maximum precipitation rates diminish much faster with increasing duration (by about $D^{-0.5}$; Figure 2a) than with increasing area (by $A^{-0.13}$ for storm durations less than 6 hours, and $A^{-0.05}$ for longer duration storms, where A is basin or storm area; Figure 2b), it is smaller basins with faster times of concentration (proportional to $A^{1/2}$) that generate by far the largest discharges per unit area.

Because of seasonal cycles that exist almost everywhere on Earth, a single year is probably the longest timeframe to consider maximum precipitation values with respect to individual flood events. Notable large floods from enhanced precipitation at such long timescales primarily affect very large basins, such as the 1953 flood on the Amazon River (Table 3). Climatic patterns with durations greater than a year, such as ENSO, PDO, and ice ages, are more likely to affect the frequency of flooding rather than being directly responsible for individual flood volumes and magnitudes [e.g. *Knox*, 1993; *Ely et al.*, 1993], although systems subject to groundwater flooding may be primarily affected by longer duration climatic patterns [e.g. *Jones et al.*, 2000].

The translation of precipitation volumes to flood volumes involves consideration of many other factors that will be described in more detail in following sections. Most of these, such as evapotranspiration during and after precipitation, and losses to snowpack, canopy vegetation, subsurface, and groundwater storage, reduce the volume of flooding. Additionally, the structure of the surface and subsurface drainage network, which in turn relates to basin geomorphology and geology, controls the delivery of water to channels. Especially for large basins, these factors have substantial influence on flood peak discharges and volumes.

Terrestrial reservoirs

Most floods in human experience are meteorological in origin and derived from atmospheric sources. Indeed, meteorological floods motivate most flood hazard assessment. But the largest floods in Earth history, at least in terms of peak discharge, resulted from rapid release from terrestrial water storage. There are two primary classes of terrestrial storage that can produce large floods: (1) frozen water in glaciers and snowpacks; and (2) waterbodies impounded by natural or constructed dams.

Ice and snow. Although most of the fresh water on Earth is stored within icecaps and glaciers (Table 1), little of this water is available for runoff under the current climate regime. Of the approximately $1.5 \cdot 10^7$ km^2 of the Earth's surface that is covered by ice, over 86% is in the South Polar regions, with an additional 12% in the North Polar regions [*van der Leeden et al.*, 1990]. The water within these polar ice bodies is in long-term storage with little opportunity to generate runoff in such concentrations as to produce large floods. Much of the remaining 2% of the Earth's surface covered by ice consists of high-elevation glaciers in mid-latitude orogenic belts, and the water within these glaciers is mostly unavailable for flood generation. On timescales relevant for flood generation, rapid melting of ice from glaciers or icecaps probably occurs only during volcanic eruptions, such as Icelandic jőkulhlaups and the 1985 eruption of Nevada del Ruiz and consequent lahars in Columbia.

Seasonal snowpacks also store water, with total water storage varying throughout the snow accumulation and melt cycle. Depending on time and location, annual snow water storage can range up to $4 \cdot 10^6$ m^3/km^2, based on maximum-recorded snowfall values [*van der Leeden et al.*, 1990]. Average maximum monthly values for mid-latitude mountain ranges in the U.S. are on the order of $1.7 \cdot 10^6$ m^3/km^2 for the northern Cascade Range, $0.4 \cdot 10^6$ m^3/km^2 for the central Sierra Nevada, and $0.5 \cdot 10^6$ m^3/km^2 for the Rocky Mountains in Colorado, based on SNOTEL data (available at http:// www.wrcc.dri.edu/snotel/snoareas.html; May 25, 2001). Although volumes of water of this magnitude are potentially available for conversion to streamflow, melt-season release of water from snowpack storage due to sensible heat transfer is generally slower than maximum precipitation rates. Therefore, snow accumulation generally acts to reduce local peak discharges. For example, melt rates for the 1999 snowpack on Mt. Rainier averaged 8 mm/day, which, if converted directly into runoff, would produce $8 \cdot 10^3$ m^3/km^2/day—a value several orders of magnitude lower than rainfall volumes produced by typical storm precipitation intensities (Table 2). Snowmelt floods are generally only notable in very large basins, such as the 1884 flood on the Columbia River (Table 3), where the integrated effects of regional melting can produce larger volumes of water for durations and areas longer and larger than most flood-producing storms.

Snowmelt can substantially contribute to flood flows in basins of all sizes when there is significant snowmelt in conjunction with storm runoff. During rain-on-snow events, rapid melting of snowpacks and release of stored snow water results from both latent and sensible heat fluxes from condensation and warm rain, respectively. In severe rain-on-snow events, such as the December and January 1964-1965 floods in the western United States and the February 1996 floods in the Pacific Northwest, snowmelt may augment total precipitation by 20 to 30% [*Risley et al.*, 1997]. Such events are the dominant flood-producing mechanism for most medium-to-large basins in the Cascade Range of Oregon and Washington. Predicting the magnitude of rain-on-snow events is much more complex than typical watershed routing to predict storm runoff because the amount of local snowmelt and runoff depends on a variety of factors, including elevation, temperature, local wind conditions, snowpack thickness and moisture content, all of which vary tremendously over short distances in the mountainous settings for which rain-on-snow floods are most common. During the February 1996 flood in the Pacific Northwest, some areas of snowpack actually accumulated moisture, thus reducing total storm runoff (unpublished data for the 1996 flood at the H.J. Andrews Long Term Experimental Research Site, Oregon, available at http://sequoia. fsl.orst.edu/lter/pubs/spclrpfr.html)

Lakes and Reservoirs. Because floods result from rapid release of large volumes of stored water, lakes and reservoirs are sources of terrestrial water commonly involved in large floods. Such impounded water bodies are an effective cause of large floods because they (1) commonly store large volumes and (2) release water instantaneously to the stream network. Unlike floods generated from atmospheric moisture, floods from rapidly evacuated terrestrial waterbodies can immediately affect a channel without the attenuation due to the time and space scales associated with precipitation rates and integrating flow through a drainage network.

Terrestrial waterbodies commonly involved in floods include ice-, moraine-, and landslide-dammed lakes, tectonic basins, caldera lakes, and artificial reservoirs impounded by constructed dams. Volumes of stored water in modern terrestrial waterbodies range up to $2.3 \cdot 10^{13}$ m^3 contained in Lake Baikal, Russia—some three orders of magnitude larger than the volume of water contained in a typical hurricane (Table 2). Nevertheless, the volumes of water involved in floods from historic releases from terrestrial waterbodies are much smaller, the largest being $5.4 \cdot 10^9$ m^3 released from a 1986 breach of an ice-dammed lake at Russell Fjord, Alaska (Table 3; Mayo, 1989). This release is approximately equivalent to the volume of water precipitated by large mesoscale convective systems, but smaller than the largest tropical cyclones (Table 2). The largest dam-break floods in the geological record, however, such as the Pleistocene Missoula and Bonneville Floods, had volumes of 10^{12} to 10^{13} m^3, eclipsing flood volumes generated by historic lake releases and individual storms by two orders of magnitude, but perhaps equivalent in terms of total volume to the largest meteorological floods in large basins from persistent periods of enhanced rainfall (Table 3).

For dam-break floods, breach shape (especially depth) and erosion rate are primary factors controlling peak discharge at the point of failure. Unlike meteorological floods, which develop from spatially distributed inputs that are concentrated through a drainage network, geologic and geomorphic factors do not affect the translation of stored water volumes to the channel during dam breaks, aside from their influence on the nature of the impoundment. For impounded waterbodies that are relatively large with respect to the breach depth (h), peak discharge is proportional to $h^{2.5}$ (also assuming that breach width scales with breach depth) because of the limiting condition of critical flow through the breach [*Walder and O'Connor*, 1997]. Peak discharge is less strongly correlated with impounded volume because incomplete failure may result in only partial evacuation of storage reservoirs [*Webby and Jennings*, 1994], and because of the imperfect correlation between breach height and storage volume [*Walder and O'Connor*, 1997].

To summarize, floods are generated from both atmospheric and terrestrial sources of water, both of which can be defined by both their storage volumes and timescales of release. Individual storms can produce up to $2 \cdot 10^{11}$ m^3 of water over areas of $1 \cdot 10^6$ km^2 for periods of up to several days and are the primary cause of meteorological floods in small basins. Climate and weather patterns that persist for weeks to years generate even larger volumes over larger areas and are the primary cause of large meteorological floods in large basins, with total flood volumes as great as 10^{10} to 10^{12} m^3. The most important terrestrial source of water for floods is impounded waterbodies. Floods from breached impoundments directly enter the channel, and thus

Table 4. Quaternary floods with discharges greater than 100,000 m^3/s.

Flood/River	Location	Date[a]	Peak Discharge (10^6 m^3/s)	Mechanism	Reference
Kuray	Altai, Russia	Late Pleistocene	18	Ice-dam failure	Baker et al., 1993
Missoula	Northwestern USA	Late Pleistocene	17	Ice-dam failure	O'Connor and Baker, 1992
Darkhat Lakes	Mongolia	Late Pleistocene	4	Ice-dam failure	Grosswald, 1987 (in Rudoy, 1998)
Jassater Lakes	Altai, Russia	Late Pleistocene	2	Ice-dam failure	Grosswald, 1987 (in Rudoy, 1998)
Yaloman lakes	Altai, Russia	Late Pleistocene	2	Ice-dam failure	Grosswald, 1987 (in Rudoy, 1998)
Ulymon Lakes	Altai, Russia	Late Pleistocene	1.9	Ice-dam failure	Grosswald, 1987 (in Rudoy, 1998)
Katla	Iceland	AD1918	1.5	Sub-glac. vol. eruption	Jonsson, 1982
Lake Agassiz	Alberta, Canada	9900 yr BP	1.2	Proglacial-lake overflow	Smith and Fisher, 1993
Aniakchak	Alaska, USA	Late Holocene	1.0	Caldera-lake breach	Waythomas et al., 1996
Lake Bonneville	Northwestern USA	Late Pleistocene	1.0	Lake-basin overflow	O'Connor, 1993
Lake Regina	Canada/USA	Late Pleistocene	0.8	Ice-dam failure	Lord and Kehew, 1987
Jökulsá á Fjöllum	Iceland	Early Holocene	0.7	Sub-glac. vol. eruption	Waitt, in press
Indus River	Pakistan	AD1841	0.54	Landslide-dam falure	Shroder et al., 1991
Amazon River	Obidos, Brazil	AD 1953	0.37	Rainfall	Rodier and Roche, 1984
Wabash River	Indiana, USA	Late Pleistocene	0.27	Ice-dam failure	Vaughn and Ash, 1983
Toutle River	Northwestern USA	2,500 yr BP	0.26	Landslide-dam falure	Scott, 1989
Amazon River	Obidos, Brazil	AD 1963	0.25	Rainfall	Rodier and Roche, 1984
Amazon River	Obidos, Brazil	AD 1976	0.24	Rainfall	Rodier and Roche, 1984
Columbia River	Northwestern USA	500 yr BP	0.22	Landslide-dam falure	O'Connor et al., 1996
Lake Agassiz	Canada/USA	9500 yr BP	0.20	Proglacial-lake overflow	Teller and Thorliefson, 1987
Lena River	Kasur, Russia	AD 1967	0.19	Ice-jam and snowmelt	Rodier and Roche, 1984
Lena River	Kasur, Russia	AD 1962	0.17	Ice-jam and snowmelt	Rodier and Roche, 1984
Lena River	Kasur, Russia	AD 1948	0.17	Ice-jam and snowmelt	Rodier and Roche, 1984
Lake Agassiz	Northcentral USA	Late Pleistocene	0.13	Ice-dam failure	Matsch, 1983
Porcupine River	Alaska, USA	Late Pleistocene	0.13	Ice-dam failure	Thorson, 1989
Russell Fjord	Alaska, USA	AD 1986	0.11	Ice-dam failure	Mayo, 1989
Yangtze River	China	AD 1870	0.11	Rainfall	Rodier and Roche, 1984

[a]Dates reported as "yrs BP" are on basis of radiocarbon dating and have not been calibrated to calendar years.

are capable of producing tremendous peak discharges. Maximum volumes from historic breached impoundments are about 10^9 m^3, but Pleistocene ice-dam failures released volumes of 10^{12}-10^{13} m^3—volumes similar to the largest meteorological floods.

THE WORLD'S LARGEST FLOODS.

Specific consideration of the Earth's largest known floods provides perspective on the causes of the most extreme floods and patterns of their geographic and geologic settings. We use two complementary data compilations: (1) geologic and historic records of the world's largest known Quaternary floods, of cause and (2) the largest gaged meteorological floods from the world's largest basins. The first compilation emphasizes natural dam failures and provides insights on the great diversity and scales of flood-generating mechanisms and their particular settings in space and time. The second provides a basis for discussing the geologic, climatologic, and physiographic settings of large meteorological floods from large basins on a global basis. This latter discussion in turn leads to the next section, where we describe the distribution of relatively large meteorological floods from basins of all sizes in the U.S.

The Largest Floods in the Quaternary

We have compiled from various sources a listing of 27 freshwater floods known to have had peak discharges greater than $1 \cdot 10^5$ m^3/s during Quaternary time (Table 4, Figure 3, also on CD). Undoubtedly, there have been many more for which there are no systematic or paleohydrologic

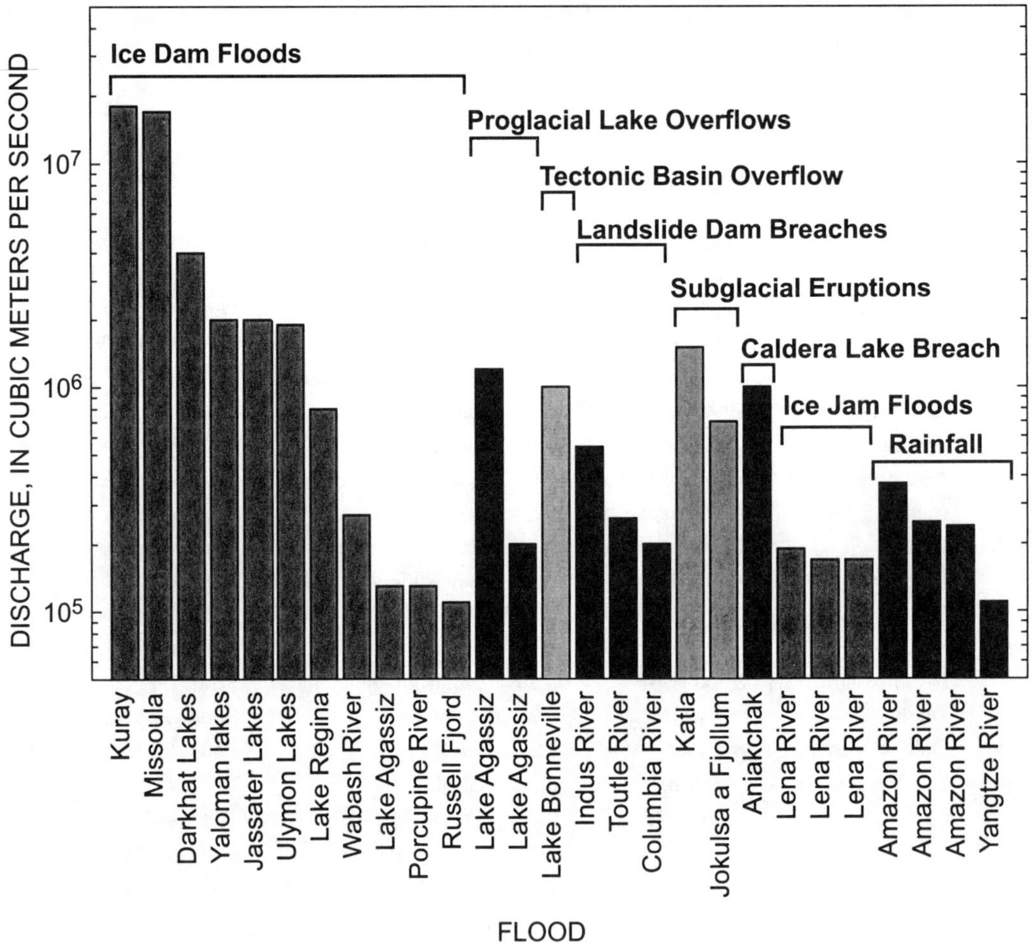

Figure 3. Quaternary floods with discharges greater than 100,000 m^3/s, arranged by causative process. From data of table 4.

measurements. Nevertheless, we can draw some fundamental inferences from this partial list as well as from other documented large floods and paleofloods.

All known terrestrial floods with discharges greater than $5 \cdot 10^5$ m^3/s resulted from rapid release of water stored behind natural dams or within glaciers (Table 4, Figure 3). The largest known floods of the Quaternary had peak discharges of about $2 \cdot 10^7$ m^3/s and resulted from breaches of glacial-age ice dams that blocked large mid-continent drainage systems. Most of the other largest floods ever documented resulted from breaches of natural dams, including landslide dams, ice dams, releases from caldera lakes, and ice-jam floods. Only 4 of the 27 largest documented floods were primarily the result of meteorological conditions and atmospheric water sources. However, if only historic events are considered, the proportion of large meteorological floods rises to 4 of 10.

Floods from Ice-Dammed Lakes

The largest floods in Earth history have resulted from derangement of drainage networks. During the Quaternary, interactions of continental ice sheets with marginal drainage systems have been a primary source of landscape-shaping floods. These floods include the $2 \cdot 10^7$ m^3/s Missoula Floods along the Cordilleran Ice Sheet margin [e.g. *Bretz*, 1923, 1969; *Baker*, 1973; *O'Connor and Baker*, 1992], giant floods along the margins of the Laurentide Ice Sheet [e.g. *Kehew and Lord*, 1987; *Lord and Kehew*, 1987; *Thorson*, 1989; *Teller and Kehew*, 1994], and more recently discovered giant Pleistocene floods in central Asia [e.g. *Baker et al.*, 1993; *Rudoy and Baker*, 1993, *Rudoy*, 1998]. These floods all resulted from rapid release of water impounded either in (1) preexisting river valleys dammed by ice or (2) proglacial lakes formed along isostatically depressed glacial margins. Similar but much smaller floods have occurred historically [*Walder and Costa*, 1996], including the 1986 flood of 105,000 m^3/s resulting from the failure of the dam formed by Hubbard Glacier across Russell Fjord (Table 4; *Mayo*, 1988).

The large ice-sheet margin floods occurred during glacial ages when, ice sheets had advanced across preexisting drainage networks. Known cases are primarily from the northern mid-latitudes where the present geometry of landmasses supported large continental ice sheets that advanced southward and disrupted large drainage systems. The largest floods (in terms of peak discharge) have been in areas of significant relief at icesheet margins, causing deep valleys to be filled by relatively tall ice dams that impounded very large lakes. When tall dams cataclysmically fail, such as the circa 600-1000 m high dams for the Kuray and Missoula Floods in Asia and North America (Tables 3 and 4), very large peak discharges result because of the exponential dependence of discharge to breach depth [*Walder and Costa*, 1996].

Basin Breach Floods

Filling and spilling of tectonic basins can also cause large floods. A common scenario is when a hydrologically closed tectonic basin fills with water, resulting in erosion of the basin divide and release of a large volume of water into adjacent drainages. Such was the case for the Bonneville Flood of North America [*Gilbert*, 1890; *Malde*, 1968; *O'Connor*, 1993], which resulted from Pleistocene Lake Bonneville filling the Bonneville Basin, overtopping and eroding alluvial fan sediment that formed the drainage divide at Red Rock Pass, Idaho, and then spilling northward into the Snake River basin. Nearly $5.0 \cdot 10^{12}$ m^3 of water was released at a peak discharge of about $1.0 \cdot 10^6$ m^3/s during incision of about 100 m of the alluvial barrier. There have likely been similar floods from the many other closed basins in the North American Basin and Range province, where there is evidence of incision of surface overflow channels at basin rims [e.g. *Currey*, 1990] or where there are downstream bouldery deposits [e.g. *Anderson*, 1998].

Geologic and archeologic records also document tremendous marine floods into tectonically closed basins. The largest known example is the flooding of the Mediterranean basin through a breach developed at the Straits of Gibraltar [*Hsü*, 1983]—an event now recognized to have caused the faunal upheaval used by Charles Lyell to divide the Miocene and Pliocene epochs [*Hsü*, 1972]. More recently, *Ryan and Pitman* [1999] proposed flooding of the Black Sea basin about 7,500 years ago by water from the rising Mediterranean Sea when it overtopped and eroded a divide at the present-day Bosporus Strait. *Ryan and Pitman* [1999] further speculated that this event was the biblical Noah's Flood.

A common element of these floods from and into hydrographically closed basins is that they all occurred in semi-arid environments during times of changing hydrologic conditions. Very large terrestrial freshwater floods are generally associated with sustained periods of positive water balances accompanying changing climatic conditions. Likewise, marine floods are most likely during times of rising sea level due to warming and deglaciation. Large closed basins are preferentially located in tectonically active regions within the arid subtropical belts. In these regions, long periods of negative water balances allow basin development to proceed without continuous overflow and incision of the flanking topographic divides, thus creating the potential for large water volumes to descend over steep sills during rare times of basin filling. Likewise, episodes of marine floods into closed basins are mainly from semi-arid areas where high evaporation rates kept water levels well below sea level, thus providing the hydraulic head necessary for cataclysmic marine water inflow.

Floods Related to Volcanism

Volcanism has been directly or indirectly responsible for 5 of the 27 largest global floods (Figure 3). Some of the largest volcanic floods resulted from eruptions melting substantial snow and ice, including the largest known historic flood of $1.5 \cdot 10^6$ m^3/s that accompanied a subglacial volcanic eruption at Katla, Iceland, in 1918 (Table 4; Jonnson, 1982). Similarly, a $0.7\text{-}1.0 \cdot 10^6$ m^3/s flood down Jökulsá á Fjöllum, Iceland, was caused by a subglacial eruption about 7,000 years ago [*Waitt*, in press].

Volcanic eruptions have also caused floods by means other than direct melting. Debris dams formed during volcanic eruptions by lava flows, lahars, pyroclastic flows, and landslides have blocked drainages, allowing lakes to form and breach. A flow of about $2.6 \cdot 10^5$ m^3/s coursed down the Toutle River, Washington, about 2,500 years ago when a landslide associated with an eruption of Mount St. Helens blocked a river valley, and the resulting ancestral Spirit Lake breached cataclysmically [*Scott*, 1989]. Similar floods would have been likely after the 1980 Mount St. Helens eruptions if mitigation measures had not been undertaken [*Sager and Chambers*, 1986]. Large floods resulting from breached volcanogenic blockages have also been reported for the circa 700 yr BP eruption of Tarawera Volcano, New Zealand (pers. commun. with Katy Hodgson, Lincoln University, New Zealand, 2000) and for the circa 7,700 yr BP Mt. Mazama eruption in the Oregon Cascades [*Conaway*, 1999]. Large historic floods in conjunction with volcanism have been reported in southern Mexico, Japan, and Alaska [*Costa*, 1988]. There have been at least two large Quaternary floods on the Colorado River in the Grand Canyon, Arizona, from breaches of lava dams [*Hamblin*, 1994; *Fenton et al.*, this volume].

Floods from breaches of water-filled calderas have also caused large flows, including a late Holocene flood of about $1 \cdot 10^6$ m^3/s from Aniakchak Volcano in Alaska [*Waythomas et al.*, 1996], a flood 1800 years ago from Lake Taupo, New Zealand [*Manville et al.*, 1999], and a late Holocene flood from Newberry Caldera in Oregon [*Chitwood and Jensen*, 2000]. *Waythomas et al.* [1996] suggested that such floods are a common and important geomorphic process in volcanic terrains—an observation corroborated by conditions in the central volcanic zone of New Zealand's North Island, where much of the present drainage system has been formed or affected by large floods from breached caldera lakes (pers. commun. with Vernon Manville and Colin Thorne, Institute of Geological and Nuclear Science, Taupo, New Zealand, 2000).

With the exception of Iceland, which is the only oceanic hot spot or mid-island ridge volcanic center with substantial glacial ice, most documented volcano-related floods are within the Pacific Rim. This could be primarily a function of the detailed hazard analyses that have been done for many of these volcanoes, but it may also reflect the abundant volcanoes in locations of high moisture regimes, especially snow- and ice-covered continental margin volcanoes at temperate latitudes.

Floods From Breached Landslide Dams

Failures of landslide dams not directly related to volcanic activity were the cause of 2 of the 27 largest floods (Table 4). Dam-breach modeling indicates that a flood with a peak discharge of about $5.4 \cdot 10^5$ m^3/s resulted from breaching of an 1841 rockslide that temporarily blocked the Indus River within the western Himalayas, Pakistan [*Shroder et al.*, 1991]. Similarly, there is stratigraphic evidence of a flood of about $2.2 \cdot 10^5$ m^3/s from breaching of a circa 500 yr BP landslide dam across the Columbia River in western Washington [*O'Connor et al.*, 1996].

Landslide dams can form in a wide range of physiographic settings, from high alpine debris avalanches to quick-clay failures in wide valley floors [*Costa and Schuster*, 1988]. The most common mechanisms triggering dam-forming landslides are rainstorms, rapid snowmelt, and earthquakes. Landslide dams can be unstable and subject to failure because they have no controlled outlet. The vast majority of dam failures and ensuing floods result from overtopping and incision of the blockage, generally beginning soon after impounded water first reaches the low point of the blockage [*Costa and Schuster*, 1988].

Despite representing only a small percentage of the world's largest floods, breaches of landslide dams have been recognized as a significant hazard [e.g. *Evans*, 1986; *Schuster and Costa*, 1986; *Costa*, 1988; *Costa and Schuster*, 1988, 1991; *Clague and Evans*, 1994; *Walder and O'Connor*, 1997; *Hewitt*, 1998] and have been the focus of several paleoflood studies [e.g. *Shroder et al.*, 1991; *O'Connor et al.*, 1996, *O'Connor and Grant*, 1999]. There have been at least six historic landslide dam floods with peak discharges greater than $1 \cdot 10^4$ m^3/s (see compilation of *Walder and O'Connor* [1997]). Most of these large floods were associated with tall blockages of large rivers, including cases where breach depths ranged up to 150 m through landslide dams that were as high as 250 m. Similar to the case of glacial dams, the potential peak discharge through a landslide dam scales with the 1.5-to-2.5 power of blockage height [*Walder and O'Connor*, 1997]. Consequently, landscapes that generate large landslides forming tall blockages in confined valleys have the greatest potential for extreme floods. Larger floods are typically on larger rivers because there are generally larger volumes of water impounded behind dams that form in large river valleys, which in turn results in more erosion and larger breach depths at the dam [*Webby and Jennings*, 1994].

Landslide dams can dwarf human-constructed dams and therefore produce much larger floods. The largest landslide dam on Earth is the 550 m high Usoy landslide dam in Tajikistan, which formed following a large earthquake in 1911 [Schuster, 2000]. This dam is nearly twice the height of the largest constructed dam in the world today, the 300-m-high Nurek rockfill dam, also in Tajikistan. The largest flood documented from failure of a constructed dam is the Teton Dam, Idaho, which failed in 1976 by piping (Table 3) and for which the peak discharge was about $6.5 \cdot 10^4$ m^3/s, less than half the peak discharge of floods resulting from the landslide dam failures listed in Table 4.

Ice-Jam Floods

Three of the world's largest 27 recorded flows were ice-jam floods on the Lena River, (Table 4) Russia—and this does not include the "unprecedented flooding" of May 1998 (http://www.ncdc.noaa.gov/ol/climate/research/1998/topB99.html) for which we have been unable to obtain discharge information. Floods on large rivers from ice jams result from "breakup jams" [*Hoyt and Langbein*, 1955, p. 31-34; *Church*, 1988] in which dislodged river ice accumulates at constrictions or river bends, forcing ponding upstream and rapid release of water if the ice dams fail rapidly. Such was the case in April 1952 on the Missouri River, North Dakota, where a failing ice dam resulted in discharge rising from about 2,100 m^3/s to more than 14,000 m^3/s in less than 24 hours [*Hoyt and Langbein*, 1955, p. 32-33].

Extreme ice-jam floods are most common in high-latitude north-flowing continental river systems like the Lena River and adjacent large river systems of northern

Eurasia and the MacKenzie River of North America. Large poleward-flowing rivers are especially susceptible to large breakup floods because headwater areas may melt before downstream areas, increasing the potential for large blockages.

Large Meteorological Floods

We have found records of four floods that had discharges greater than $1 \cdot 10^5$ m^3/s and were due primarily to meteorological causes (Table 4)—the floods of 1953, 1963, and 1976 on the Amazon River at Obidos, Brazil, and the 1870 flood of 110,000 m^3/s on the Yangtze River, China. The Amazon River, by far the world's largest river with a drainage basin of $7.05 \cdot 10^6$ km^2 that encompasses a vast portion of the equatorial South American Cordillera, is also Earth's most favorably sized and positioned basin to generate very large floods. The Yangtze (Chang Jiang) River flood of 1870 was the largest of a historic record that goes back to AD 1153 [*Baker and Kochel*, 1988].

Inspection of records of the largest floods from the Earth's largest drainage basins (Table 5, Figures 4 and 5, also on CD) shows that, in general, larger basins produce larger floods. But scatter about this relation is at least partly produced by a pronounced geographic pattern of larger unit discharges (defined as peak discharge per unit area) in the tropics, primarily between 10°S and 30°N (Figure 6, also on CD). The largest floods in large basins within the tropics are primarily derived from rainfall within areas affected by tropical cyclones or strong monsoonal airflow (including the Brahmaputra, Ganges, Yangtze, Mekong, and Huangue River basins) or eastward-draining continental basins such as the Amazon and Orinoco River basins that intercept easterly flows of tropical moisture. The distribution of relatively large floods is skewed northward of the equator by the preponderance of land in the northern hemisphere, promoting northward migration of monsoonal moisture flow driven by orographic lifting over large mountain belts such as the Himalayas [*Gupta*, 1988].

There are several large basins in the tropics that do not have records of relatively large peak discharges on a global basis. These include the Congo, Niger, Chari, and Sao Francisco River basins that drain large areas of low relief or are isolated from zones of major precipitation. Likewise, many of the horse latitude (20-40°) and mid-continental drainage basins outside areas of seasonal tropical moisture influxes do not produce large floods. Examples include the Murray and Darling River basins in Australia, the Nile River and Zambezi Rivers in Africa, and the Colorado and Mississippi Rivers of North America.

Northward of 40°N, snowmelt and ice jams are important contributors to peak discharges from large basins, forming a group of rivers with flood discharges greater than the apparent latitudinal limits of flood flows derived primarily from rainfall (Figure 6). Exceptional discharges on the Lena, Yenisey, and Yukon Rivers were augmented by ice jams, but relatively large flows on rivers such as the Columbia and Dnieper Rivers indicate the importance of melting snow to peak flows, especially in mid-latitude basins with substantial relief. With the present configuration of continents, there are no southern hemisphere river basins greater than 500,000 km^2 with peak discharges substantially affected by snowmelt.

The distribution of large drainage basins is itself influenced by geologic environment [*Potter*, 1978]. The vast majority of river basins greater than 500,000 km^2 drain toward passive continental margins, and many have headwaters in active orogenic belts. Examples include the Amazon, Parana, Congo, and Mississippi Rivers—four of the five largest basins in the world. Fewer basins, such as the Colorado and Nile Rivers, drain towards zones of regional extension although many large basins may have originally formed in zones of continental rifting and extension [*Potter*, 1978]. Of global basins larger than 500,000 km^2, only the Columbia River in western North America drains toward a plate boundary of active convergence.

Relevance to Landscapes and Paleoflood Studies.

The incidence of floods caused by different processes changes through time. Most of the largest documented floods in the past 100,000 years resulted from the failures of natural dams. But conditions that caused some of these natural dams, such as the large ice-dam failures at the margins of continental ice sheets, are now absent. Consequently, Earth history contains "flood epochs"—times when climate and topography collude to produce higher-than-typical frequencies of large floods. For floods with discharges on the order of 10^6 m^3/s, times of advanced continental ice sheets and rapidly changing global water balances are required. In many locations, these outsized floods from previous flood epochs are dominant forces in establishing regional drainage networks and landscape patterns that persist into times of fewer extreme floods [*Baker*, 1983]. Even during the Holocene, there is evidence of changing flood frequency that can be attributed to subtle climate change [e.g. *Ely et al.*, 1993; *Knox*, 1993].

Other geologically controlled cataclysmic floods are probably more evenly distributed in time, at least at timescales of 10^2 to 10^5 years, and the future likelihood and magnitude of such events can be more directly guided by analysis of past events. Within active volcanic provinces, the likelihood of large volcanic eruptions and the consequent flood-generation processes probably does not change significantly over hundreds or thousands of years.

Table 5. Largest meteorologic floods from river basins greater than about 500,000 km^2 (Fig. 4). Data from Rodier and Roche (1984) except as noted. River and station locations shown on Figure 4.

Basin No.	River Basin[a]	Country	Basin Area (10^3 km^2)[b]	Station	Station Area (10^3 km^2)	Station Lat. (deg.)[c]	Station Long. (deg.)[d]	Peak Discharge (m^3/s)	Date	Flood Type
1	Amazon	Brazil	5,854	Obidos	4,640	-1.9	-55.5	370,000	June, 1953	Rainfall
2	Nile	Egypt	3,826	Aswan	1,500	24.1	32.9	13,200	Sept. 25, 1878	Rainfall
3	Congo	Zaire	3,699	Brazzaville B.	3,475	-4.3	15.4	76,900	Dec. 27, 1961	Rainfall
4	Mississippi[e]	USA	3,203	Arkansas City	2,928	33.6	-91.2	70,000	May, 1927	Rainfall
5	Amur	Russia	2,903	Komsomolsk	1,730	50.6	138.1	38,900	Sept. 20, 1959	Rainfall
6	Parana	Argentina	2,661	Corrientes	1,950	-27.5	-58.9	43,070	June 5, 1905	Rainfall
7	Yenisey	Russia	2,582	Yeniseisk	1,400	58.5	92.1	57,400	May 18, 1937	Snowmelt
8	Ob-Irtysh	Russia	2,570	Salekhard	2,430	66.6	66.5	44,800	Aug. 10, 1979	Snowmelt
9	Lena	Russia	2,418	Kasur	2,430	70.7	127.7	189,000	June 8, 1967	Snowmelt/Ice Jam
10	Niger	Niger	2,240	Lokoja	1,080	7.8	6.8	27,140	Feb. 1, 1970	Rainfall
11	Zambezi	Mozambique	1,989	Tete	940	-16.2	33.6	17,000	May 11, 1905	Rainfall
12	Yangtze	China	1,794	Yichang	1,010	30.7	111.2	110,000	July 20, 1870	Rainfall
13	Mackenzie	Canada	1,713	Norman Wells	1,570	65.3	-126.9	30,300	May 25, 1975	Snowmelt
14	Chari	Chad	1,572	N'Djamena	600	12.1	15.0	5,160	Nov. 9, 1961	Rainfall
15	Volga	Russia	1,463	Volgograd	1,350	48.5	44.7	51,900	May 27, 1926	Snowmelt
16	St. Lawrence	Canada	1,267	La Salle	960	45.4	-73.6	14,870	May 13, 1943	Snowmelt
17	Indus	Pakistan	1,143	Kotri	945	25.3	68.3	33,280	1976	Rain/Snowmelt
18	Syr Darya	Kazakhstan	1,070	Tiumen-Aryk	219	44.1	67.0	2,730	June 30, 1934	Rain/Snowmelt
19	Orinoco	Venezuela	1,039	P. Angostura	836	8.1	-64.4	98,120	Mar. 6, 1905	Rainfall
20	Murray	Australia	1,032	Morgan	1,000	-34.0	139.7	3,940	Sept. 5, 1956	Rainfall
21	Ganges	Bangladesh	976	Hardings Br.	950	23.1	89.0	74,060	Aug. 21, 1973	Rain/Snowmelt
22	Shatt al Arab	Iraq	967	Hit/Euphrates	264	34.0	42.8	7,366	May 13, 1969	Rain/Snowmelt
23	Orange	South Africa	944	Buchuberg	343	-29.0	22.2	16,230	1843	Rainfall
24	Huanghe	China	894	Shanxian	688	34.8	111.2	36,000	Jan. 17, 1905	Rainfall
25	Yukon	USA	852	Pilot Station	831	61.9	-162.9	30,300	May 27, 1991	Snowmelt
26	Senegal	Senegal	847	Bakel	218	14.9	-12.5	9,340	Sept. 15, 1906	Rainfall
27	Colorado[e]	USA	808	Yuma	629	32.7	-114.6	7,080	Jan. 22, 1916	Rainfall
28	Rio Grande[e]	USA	805	Roma	431	26.4	-99.0	17,850	1865	Rain/Snowmelt
29	Danube	Romanis	788	Orsova	575	44.7	22.4	15,900	April 17, 1895	Snowmelt
30	Mekong	Vietnam	774	Kratie	646	12.5	106.0	66,700	Sept. 3, 1939	Rainfall
31	Tocantins	Brazil	769	Itupiranga	728	-5.1	-49.4	38,780	April 2, 1974	Rainfall
32	Columbia[e]	USA	724	The Dalles	614	45.6	-121.2	35,100	June 6, 1894	Snowmelt
33	Darling	Australia	650	Menindee	570	-32.4	142.5	2,840	June, 1890	Rainfall
34	Brahmaputra[f]	Bangladesh	650	Bahadurabad	636	25.2	89.7	81,000	Aug. 6, 1974	Rain/Snowmelt
35	Sao Francisco	Brazil	615	Traipu	623	-9.6	-37.0	15,890	April 1, 1960	Rainfall
36	Amu Darya	Kazakhstan	612	Chatly	450	42.3	59.7	6,900	July 27, 1958	Rain/Snowmelt
37	Dnieper	Ukraine	509	Kiev	328	50.5	30.5	23,100	May 2, 1931	Snowmelt

[a]Basins larger than 500,000 km^2 for which we were unable to find reliable data include the Nelson River in North America; the Jubba, Irharhar, Araye, Tafassasset and Qattar Rivers in Africa, and the Kolyma and Tarim Rivers in Asia.
[b]Basin areas from Vörösmarty et al. (2000).
[c]Decimal degrees; positive indicates east of the Greenwich meridian.
[d]Decimal degrees; positive indicates north of the equator.
[e]Station and discharge data from U.S. Geological Survey National Water Information System (http://water.usgs.gov/nwis).
[f]Station area and drainage basin data from Global Runoff Data Centre in the Federal Institute of Hydrology, Germany (http://www.bafg.de/grdc.htm)

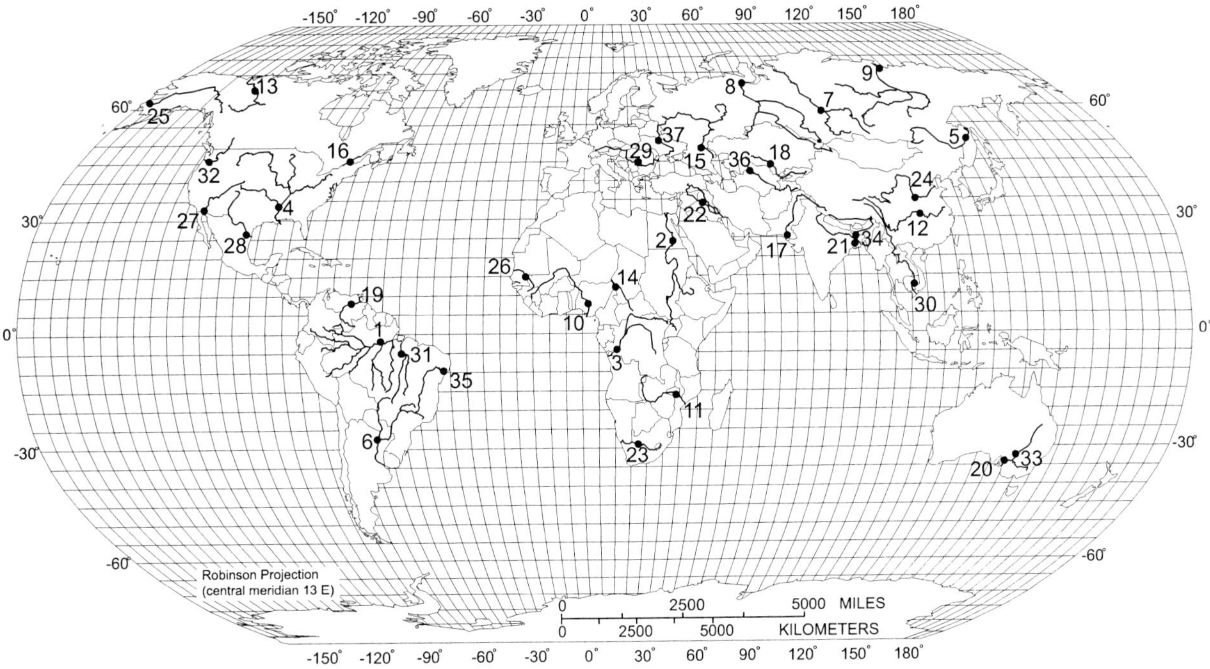

Figure 4. Rivers with drainage areas larger than 500,000 km^2 listed in Table 5.

Similarly, the frequency of large valley-blocking landslides that accompany large subduction earthquakes probably remains relatively constant over such timescales.

SPATIAL DISTRIBUTION OF THE LARGEST GAGED FLOODS IN THE UNITED STATES.

The types of broadscale relations between physiography and climatology evident in the global distribution of meteorological floods in large basins also appear at continental scales, although the controlling factors involve correspondingly finer scales of physiographic and climatologic interactions. Such continental-scale relations can perhaps be most clearly portrayed by the spatial distribution of large floods in the United States, where the U.S. Geological Survey has a stream-gaging and data-archival program that measures and records flows at several thousand locations throughout the country.

U.S. Geological Survey Annual Flood Data

The basic source of data for this analysis was the "peak flow files" maintained as part of the U.S. Geological Survey National Water Information System [*Lepkin and DeLapp*, 1979]. These peak flow files are separately maintained for each stream-gaging station and include values for the largest instantaneous discharge for each water year (Oct. 1 to Sept. 30). Data are available for current as well as discontinued streamgaging stations. In June of 2000, there were records for 23,216 stations that together have more than a million annual peak discharge measurements and estimates. All data were originally measured, reported, and analyzed in English units. Consequently, converting to metric units results in some awkward values in this report and accompanying figures.

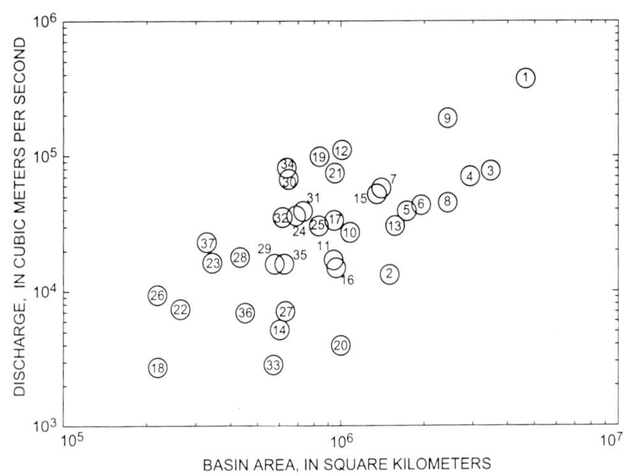

Figure 5. Largest recorded floods from basins larger than 500,000 km^2 (although many measurements made at locations where only part of the basin contributes). Data keyed to "Basin No." identified in Table 5.

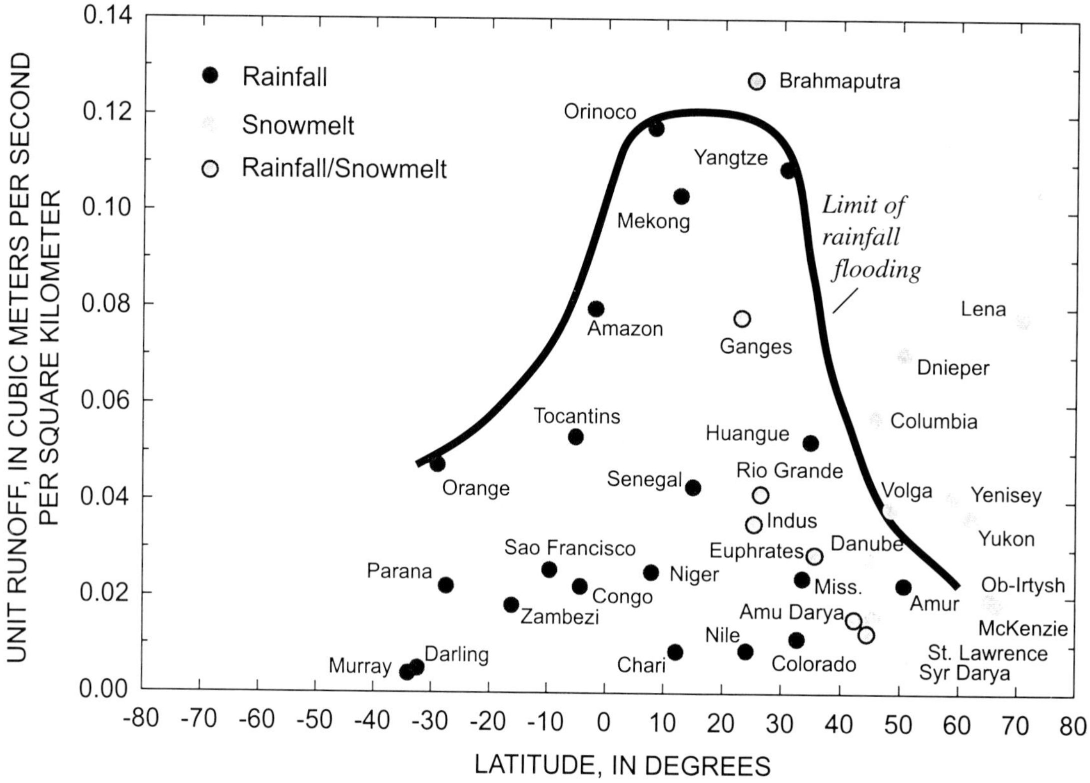

Figure 6. Largest recorded floods from basins larger than 500,000 km^2 plotted as unit discharge (peak discharge per square kilometer), showing strong latitudinal control on flood size and mechanism. From data of Table 5.

From these peak flow files, we extracted the single largest non-null peak flow (excluding dam failure floods) from all stations with longitude and latitude information and drainage areas greater than 0.26 km^2 (0.01 mi^2), resulting in records of maximum annual peak discharge from each of 22,063 stations (Plate 1, also on CD). Discharge values ranged from 0.03 m^3/s to 70,000 m^3/s from drainage areas up to the $2.93 \cdot 10^6$ km^2 drained by the Mississippi River at Arkansas City, Arkansas (Figure 7, also on CD). These records include estimates for historic peaks at gaging stations dating back to the late 18th century as well as all systematic gaged records through water year 1997 (Sept. 30, 1997). The vast majority of these floods were meteorologic. Some peak discharges have been affected by regulation and diversion, and many are estimates from indirect procedures or from extrapolation of rating curves well beyond measured flows.

To further distinguish the largest of these maximum annual flows at gaging stations relative to drainage area, we selected the flows that exceeded the criteria

$$\mathbf{Q > 15A^{2/3} \text{ for } A < 2{,}590,}$$
$$\text{and } \mathbf{Q > 206A^{1/3} \text{ for } A \geq 2{,}590,} \qquad (2)$$

where Q is the peak discharge in cubic meters per second and A is the drainage area in square kilometers. These criteria were formulated to yield a subset of stations with exceptionally large maximum flows (Figure 7), but with an overall distribution of basin areas similar to that of all the stations (Figure 8, also on CD). The maximum discharge at each station exceeds this large-flow threshold for about 13% of the 22,063 stations. The locations of these stations are plotted on Plate 1b, which forms the basis of our conclusions of the physiographic and climatologic controls on large flows within the United States.

The slope decrease for the criterion for drainage areas greater than 2,590 km^2 (1000 mi^2) is necessary to produce a subset of large-flow stations with a distribution of drainage areas similar to that of all stations, thus ensuring an even selection of large-flow stations relative to the distribution of drainage basin areas of gaged streams. Similar slope decreases in unit area peak discharges for larger drainage basins appear in envelope curves bounding the maximum global and U.S. rainfall-runoff floods [*Crippen and Bue*, 1977; *Costa*, 1987], as well as for regional compilations [e.g. *Enzel et al.*, 1993]. This decrease in slope for the large-flow criterion separating the largest

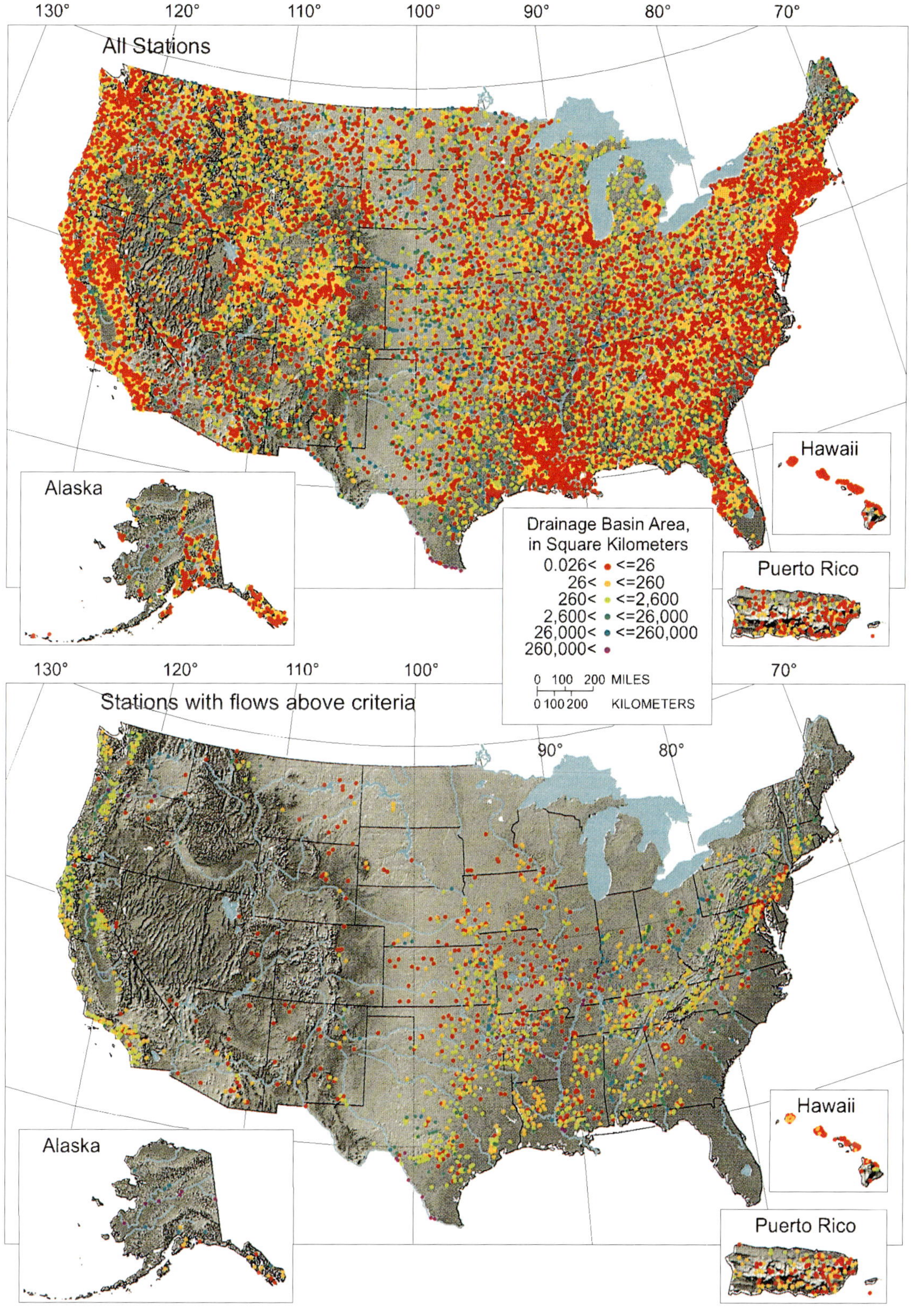

Plate 1. Former and active U.S. Geological Survey stream gages with peak discharge records used in analysis of spatial distribution of large floods, color-coded by basin area. A) All 22,063 former and active sites. B) Large flow sites (2,929) with peak discharges greater than the large-flow criterion described in text.

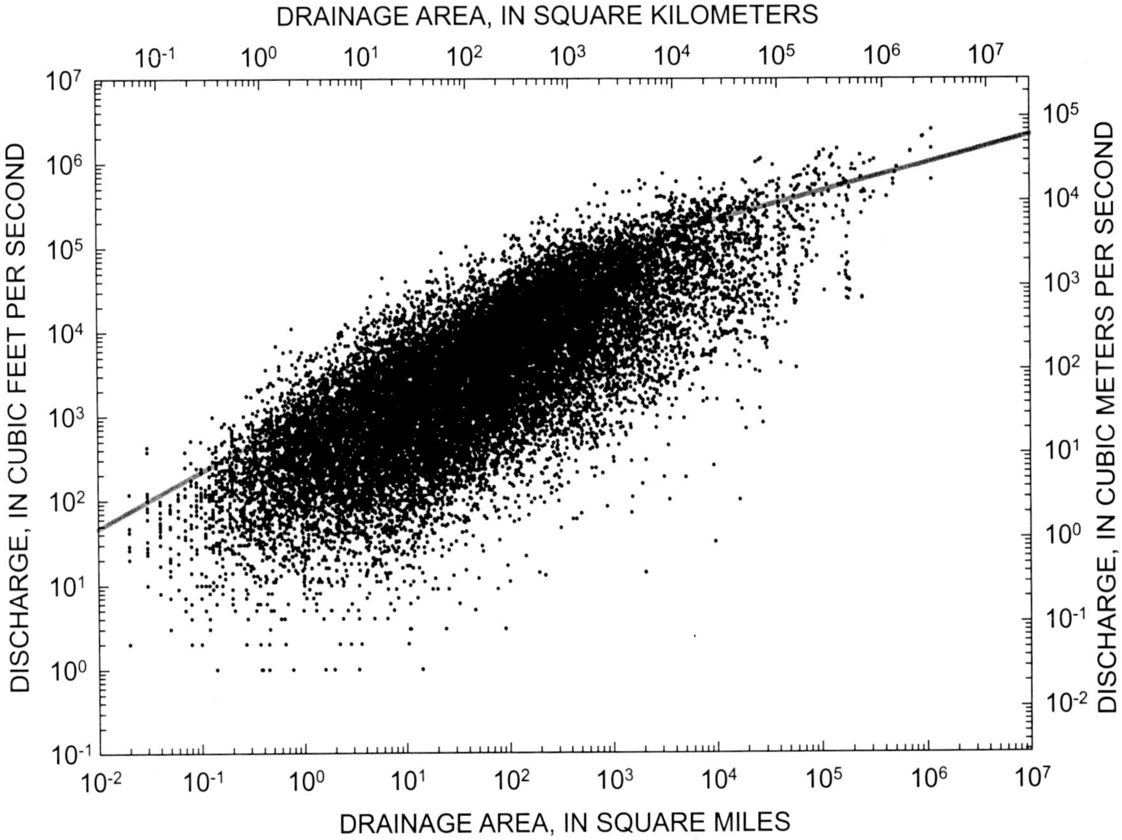

Figure 7. Single maximum flow of record for each of 22,063 former and active U.S. Geological Survey stations with peak discharge records, plotted by station drainage area. Includes records through September 1997.

U.S. gaged flows corresponds with the marked decrease in areally averaged rainfall depth for U.S. storms with areas larger than 2,000 km^2 and durations greater than 6 hours (Figure 2b) and must reflect physical aspects of moisture delivery [*Wolman and Costa*, 1984], such as limits on storm size and precipitation rates—perhaps corresponding to the spatial extent of intense convective precipitation within tropical storms. Another compounding factor may be that large flows in large basins are probably preferentially attenuated by valley storage [*Miller*, 1990].

The Spatial Distribution of Relatively Large Flows

The spatial distribution of stations with the largest flow magnitudes relative to drainage area (Plate 1b) is substantially different from the distribution of all stations (Plate 1a), indicating that large flows are concentrated in particular areas. Puerto Rico and Hawaii, where 78% of the 441 stations have maximum station discharges above the delineating threshold, are most subject to relatively large flows. Within the conterminous United States, Plate 1b echos figure 24 of *Hoyt and Langbein* [1955, p. 75] in which they depicted relative "flood-discharge

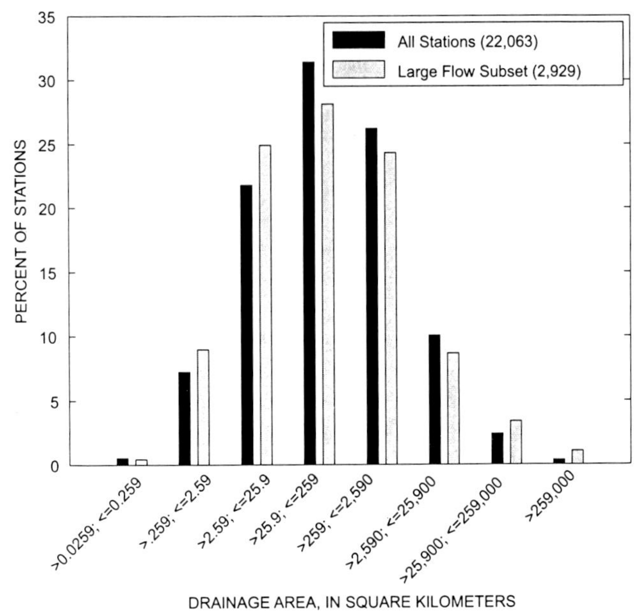

Figure 8. Distribution of all stations and large-flow stations by drainage area.

potentialities." Especially prominent in *Hoyt and Langbein's* map and in Plate 1b are the concentrations of relatively large flows in central Texas and in the central and northern Appalachians. Also indicated on *Hoyt and Langbein's* map and in Plate 1b are relatively high flow magnitudes along the U.S. west coast. One difference that emerges between Plate 1b and *Hoyt and Langbein's* map is in the southeastern Great Plains, an area where the gage network shows many relatively large flows (see also *Michaud et al.* [2001]), but one not noted by *Hoyt and Langbein* for large floods.

There are also broad areas where few measured discharges exceed our large-flow threshold. Only a few stations with small drainage basins in the interior west, especially within the Basin and Range province, have had exceptionally large flows, although this might partly reflect the relative sparseness of stations in this region (Plate 1b). Likewise, the northern tier states of the eastern and midwestern parts of the Unites States have had few large flows. There have been very few relatively large flows along the Coastal Plain and Florida.

Relations to Regional Climatology and Topography

The overall gradient of decreasing numbers of large floods to the north in general reflects the northward-decreasing gradient of atmospheric water content. But in detail, the flood potential of specific areas is determined by combined roles of geographic setting, regional climatology and topography, which in turn relate to geologic setting. Many of the climatologic and topographic factors controlling flooding in the United States were described in detail by *Hirschboeck* [1991] and are corroborated by the distribution of large floods plotted in Plate 1b.

Within this overall pattern are finer scale relations between regional climatology, topography, and the likelihood of large flows. Puerto Rico and Hawaii both have high frequencies of large floods because they are areas of local orographic lifting in the midst of the tradewind belt, where moisture-laden convective systems and tropical storms are intercepted and convectively lifted by the island mountain ranges. It follows that a greater proportion of stations with the largest flows are on the windward (east) side of these island complexes.

Within the eastern U. S., the pattern of large flows closely corresponds to the gradients of subtropical moisture derived from the Gulf of Mexico and the Atlantic Ocean (compare with figure 23 in *Hirschboeck* [1991]). Finer-scale patterns within this region, however, are clearly linked to topography and geology. Orographic lifting of Gulf of Mexico moisture has caused concentrations of large floods at topographic features such as the Balcones Escarpment in south-central Texas, the Sand Hills of western Louisiana, the Ozark Mountains of western Missouri, the Flint Hills of eastern Kansas, and along the eastern edge of the Appalachians on the eastern seaboard. *Baker* [1977] described in detail the role of climatic and physiographic controls on flood runoff for streams draining the Balcones Escarpment. The important role of orographic lifting is also illustrated by the lack of exceptional discharges in Florida and along the Coastal Plain of the southern and eastern tier states. Despite being closest to the moisture sources and subject to frequent landings of major hurricanes, these relatively flat areas lack the relief necessary to drive orographic precipitation. Once, however, moisture from hurricanes and their remnants intercepts eastern Appalachia, orographic lifting and efficient runoff concentration can produce exceptional flows [*Miller*, 1990]. The high incidence of relatively large flows in the central and southern midwest and upper Mississippi Valley is partly due to mesoscale convective complexes—large, multiple-celled, and persistent thundercell systems fed largely by moisture from the Gulf of Mexico [*Hayden*, 1988; *Hirschboeck*, 1991]. The Rapid City, South Dakota, flood of 1972 and the Big Thompson River, Colorado, flood of 1976 were caused by such systems. Within the zone of influence of Gulf of Mexico moisture, there are a variety of moisture delivery processes that have caused large flows, including individual thunderstorm cells, mesoscale convective complexes, Atlantic hurricanes, and quasi-stationary upper atmosphere circulation patterns producing several weeks or months of persistent precipitation. These processes are of such diverse temporal and spatial scales that there does not seem to be any systematic pattern of drainage basin size associated with exceptionally large flows (Plate 1b). The only exception is along the Coastal Plain of the southeastern U.S., where only larger rivers draining upland areas have produced large flows, primarily from hurricanes.

West of about longitude 105° west, approximately corresponding to the eastern limit of the Rocky Mountains, moisture from the Pacific Ocean chiefly influences flooding within the conterminous United States and Alaska. Consequently, the stations with largest flows are concentrated along the states contiguous to the Pacific Ocean. The interior western U.S. states of Idaho, Nevada, Utah, Wyoming, and Colorado have very few stations with large flows, and, furthermore, the interior west stations that have had relatively large flows generally represent small basins (less than 26 km^2) influenced by isolated thunderstorms. The distribution of stations with the largest flows within the interior West clearly reflects orographic lifting associated with local topographic features such as (1) the eastern limits of the Cordilleran orogenic belt in eastern Wyoming, central Colorado, and eastern New Mexico, (2) the Wasatch Front in Utah, and (3) the Central Highlands and Mogollon Rim in Arizona.

Orographic effects are even more pronounced, however, along the western continental margin of North America. The mountains of southeastern Alaska, the Cascade Range of Oregon and Washington, the Coast Ranges of Washington, Oregon, and California, and the Sierra Nevada and San Gabriel Mountains of California all strongly influence the distribution of stations with large flows. In Oregon, Washington, and Northern California, north-south trending mountain ranges intercept the westerly flow of extratropical cyclones from the Pacific Ocean. The largest floods for most basins are winter floods caused by intense cyclonic storms or the combined effects of cyclonic storms and coincident snowmelt. The extensive winter floods of 1964 and 1996 in the Pacific Northwest were the product of substantial rain-on-snow events. Because these types of events persist for several days, it is the larger basins—generally those with areas of 10^2 to 10^5 km^2—that are preferentially affected by these types of events (Plate 1b).

The big basins of the western U.S. that drain extensive areas of the North American Cordillera affected by winter snowfall, such as the Yukon and Columbia Rivers, also produce relatively large flows, invariably in late spring and summer. These large northern basins are the only ones for which snowmelt alone appears to be capable to produce flows greater than our large flow criteria.

Large flows in southern California and Arizona are from mixed sources [*Hirschboeck*, 1985; 1991], including extratropical cyclonic systems, dissipating tropical storms moving northward from the Pacific, mesoscale convective complexes, and local convective thunderstorms. Consequently, there is substantial variation in the size of basins and seasonality of the largest flows [*Hirschboeck*, 1985]. The east-west orientation of the Transverse Ranges of southern California and the Mogollon Rim of central Arizona promotes interception of moisture from northward incursions of Pacific tropical storms; consequently, there are distinct concentrations of stations with large floods aligned along these features.

Geologic Controls and Paleohydrologic Implications

Continental position and topography have great influence on the distribution of large floods. In the western United States, the combination of an extensive cordillera associated with a leading continental edge oriented transverse to the mid-latitude belts of westerly cyclonic moisture flow causes relatively large flows to concentrate closely along the continental margin. Similar situations also exist in southern South America and New Zealand [*Hayden*, 1988]. Smaller areas of uplifted terrain dating to older orogenies, such as the Rocky Mountain Front Range and the Central Highlands and Mogollon Rim of Arizona also produce small areas of exceptional flows because of their orientation relative to particular tracks of moisture flow. The distribution of large floods in the central and eastern parts of the U.S. corresponds more closely to the broad regional climatology of the overall decreasing moisture gradient away from the Gulf of Mexico. But the specific locations of relatively large floods still clearly relate to distinct topographic features produced by ancient orogenies, such as the Balcones Escarpment, Ozark Mountains, and the margins of the Appalachians.

The spatial distribution of large gaged floods throughout the United States shows that the locations of most of the largest flows can be related to specific combinations of regional climatology and topography. Besides corroborating earlier studies of flood-climate-topography relationships [e.g. *Hayden*, 1988, and *Hirschboeck*, 1991], this result leads to more general conclusions about the relation of meteorological floods and geologic environment. In the broadest context, the regional climatology of the conterminous United States is due to the present mid-latitude setting of the continent. At finer regional scales, climatology is influenced by topography, which in turn relates to the geologic history of the continent. Major factors controlling North American topography include the Paleozoic continental collisions that created the Appalachians during amalgamation of supercontinent Pangaea, and subsequent Mesozoic and Cenozoic rifting that allowed weathering and erosion of the Appalachians to form a broad piedmont and coastal plain on the now passive margin of the North American plate. Also during the Mesozoic and Cenozoic, volcanism and deformation along the now active western continental margin have formed the present high-relief topography of the western North American Cordillera. These types of combinations of continental location and geologic history are mirrored on most continents, as well as for past geologic epochs. Thus, understanding the intimate relation between flooding and physiography allows for generalized predictions of regions of relatively large flows elsewhere in the world and back in time.

This perspective can be meaningful for Holocene paleohydrologic research in that it provides context for identifying regions susceptible to exceptionally large flows. Moreover, it shows that regional climate is important in controlling flood magnitude, thus providing rationale for conducting paleoflood studies at sites where millennial scale climate change may be likely to affect flood magnitude and frequency. Further back in time, understanding the relations between climate, continental position and physiography and their effects on flow generation provide grist for understanding paleoenvironments and landscape evolution.

Additionally, paleohydrologic studies can be a useful complement to the gaged records of large floods such as those depicted on Plate 1b. Numerous studies have shown the utility of combining results from paleohydrologic investigations with gaged records to produce better estimates of flood frequency [*Kochel et al.*, 1982; *Stedinger and Cohn*, 1986; *Webb et al.*, 1988; *O'Connor et al.*, 1994]. Moreover, paleohydrologic methods and studies can perform important checks on gaged records of extreme flows [*Enzel et al.*, 1993; *House and Pearthree*, 1995], especially for small basins in western North America and other sparsely gaged regions worldwide where critical facilities are commonly designed on the basis of procedures that rely on a few extreme flood measurements.

SUMMARY—LIMITS ON FLOODS.

The preceding sections have described in varying levels of detail the controls on the magnitude of floods from both terrestrial and atmospheric sources as well as the geographic, geologic, and geomorphic environments that foster large floods. At fine scales of space and time, such as the incidence of historic U.S. flooding, important factors are the volumes of water available, and the rate in which it is delivered to the landscape and enters the channel. These factors are dominantly influenced by proximity to moisture sources and local physiography. At broader temporal and spatial scales, such as the global distribution of flooding through the Quaternary epoch, exceptional floods result from particular environmental settings. The Earth's largest floods have resulted from natural-dam failures associated with deranged drainage systems, such as floods resulting from failed ice dams at the margins of continental ice sheets and from filling and spilling of tectonic basins. The largest meteorological floods are primarily associated with the world's largest drainage basins, but are also influenced by topography and position with respect to global patterns of atmospheric circulation. Topography, drainage basin size, and position are the result of geologic processes operating over millions of years and can be linked to particular geologic environments. For example, almost all large basins drain toward passive or extensional continental margins.

Some of these concepts can be further generalized by considering the physical limits of flooding regardless of physiographic or geologic setting. These limits can be formulated and visualized to a first order by considering floods within a Cartesian coordinate space defined by axes of flood volume, peak discharge, and duration (Figure 9, also on CD). These axes are not independent, so such plots cannot be used to derive empirical relations, but they are helpful to show limits to floods posed by physical properties. A fundamental limit that affects both meteorological and dam-failure floods is continuity: for a triangular hydrograph, continuity requires that,

$$\mathbf{V = 0.5Q \cdot t}, \tag{3}$$

where V is flood volume, Q is peak discharge, and t is flood duration, in any consistent set of units. For alternatively shaped hydrographs, the relation is slightly different but of the same order.

For meteorological floods, another fundamental limit arises from interactions between the total volume of water delivered to the basin (a product of the area and precipitation) and the time required to concentrate rainfall distributed across a drainage basin into a channel. This constraint on flood magnitude can be approximately formulated by considering the maximum precipitation rates and volumes summarized previously in conjunction with plausible rates of flow concentration. The maximum discharge at a point of concentration is limited by the product of the contributing area, A, and the precipitation rate measured over some time interval, t:

$$\mathbf{Q_{max} = A(P/t)}, \tag{4}$$

where P is precipitation depth. If the maximum velocity of water moving across a drainage basin towards a point of concentration is on the order of 1 m/s, then the contributing area at a given time will be on the order of t^2; thus:

$$\mathbf{Q_{max} = t^2(P/t) = t \cdot P}. \tag{5}$$

From figure 2a, the maximum point rainfall data define the empirical relation:

$$\mathbf{P_{max(point)} = 0.0063t^{0.49}}, \tag{6}$$

which is equivalent to (1) but in units of meters and seconds. In the U.S., and perhaps worldwide, however, such point maximum values are applicable for areas on the order of $2.5 \cdot 10^7$ m^2, but diminish when measured over large areas [*World Meteorological Organization*, 1986, p. 103]. For longer-duration storms, areally averaged rainfall decreases by a factor of approximately $A^{-0.05}$ for areas up to $2 \cdot 10^{10}$ m^2 and by a substantially larger factor for larger areas (Figure 2b). For the purposes here, we will assume that for a given area,

$$\mathbf{P_{max(areal\ average)} = 0.0063t^{0.49}\,(A/(2.5 \cdot 10^7))^{-0.05}}, \tag{7}$$

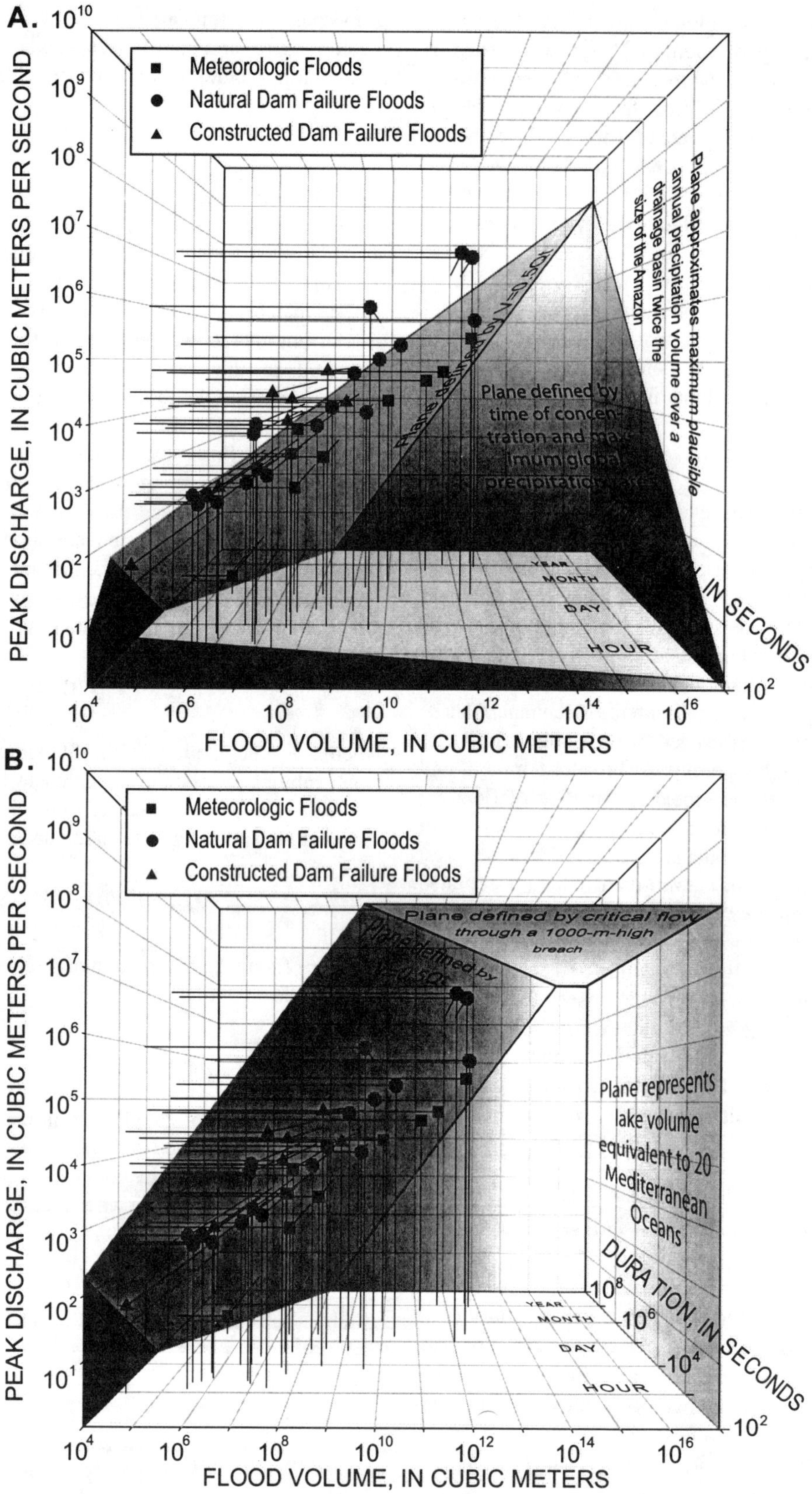
A.
Meteorologic Floods
Natural Dam Failure Floods
Constructed Dam Failure Floods
Plane approximates maximum plausible annual precipitation volume over a drainage basin twice the size of the Amazon
Plane defined by time of concentration and maximum global precipitation rate
YEAR
MONTH
DAY
HOUR
IN SECONDS
PEAK DISCHARGE, IN CUBIC METERS PER SECOND
FLOOD VOLUME, IN CUBIC METERS
B.
Meteorologic Floods
Natural Dam Failure Floods
Constructed Dam Failure Floods
Plane defined by critical flow through a 1000-m-high breach
Plane represents lake volume equivalent to 20 Mediterranean Oceans
DURATION, IN SECONDS
YEAR
MONTH
DAY
HOUR
PEAK DISCHARGE, IN CUBIC METERS PER SECOND
FLOOD VOLUME, IN CUBIC METERS

which reduces to:

$$P_{max(areal\ average)} = 0.015t^{0.49}A^{-0.05}. \quad (8)$$

This relation is likely to overestimate maximum plausible precipitation for larger areas, but because the area dependence is small relative to the duration, this approximation is not likely to substantially affect conclusions.

Substituting this relation into (5) yields

$$Q_{max} = 0.015t^{1.49}A^{-0.05}, \quad (9)$$

which can be expressed in terms of t and V by combining the continuity relation,

$$V = P_{max(areal\ average)} \cdot A, \quad (10)$$

and (8), which then results, after algebraic manipulation,

$$Q_{max} = 0.01t^{1.52}V^{-0.056}. \quad (11)$$

This result may seem somewhat counterintuitive, but recall that both the total flood volume (calculated from maximum global precipitation rates) and the area contributing discharge are highly dependent on the duration.

The plane defined by (11) intersects the plane defined by (3) to enclose a tent-shaped space of meteorological floods with a 'ridgeline' that climbs from low values of flood volume, duration, and peak discharge in the near left corner to large values of volume, duration, and discharge in the far right corner (Figure 9a). Representative data for some exceptional historical meteorological floods (Table 3) plot near this ridgeline for flood volumes ranging between 10^7 to 10^{13} m^3.

These relations also allow for speculation on the largest plausible meteorological floods on Earth. For example, a basin twice the size of the Amazon subject to a year's duration of the maximum global annual precipitation (approximately 26.5 m) (without evapotranspiration or loss to long-term groundwater storage) could hypothetically generate a flood of 10^{17} m^3 with a peak discharge of 10^9 m^3/s (Figure 9a). These values are unrealistic, however, for on a vegetated Earth, evapotranspiration reduces total annual runoff volume—generally by more than 50% for latitudes less than 60°, and by as much as 80 to 100% for latitudes less than 30° and for closed basins [*Baumgartner and Reichel*, 1975]. Consequently the volume of floods from several months of persistent precipitation over large drainage basins will be substantially less than the total precipitation volume [*Langbein et al.*, 1949]. But prior to the Devonian establishment of land plants and during times when different land mass configurations hosted exceptionally large and high-relief drainage basins, meteorological floods were likely much larger than at present [*Schumm*, 1968], perhaps with volumes of 10^{13} to 10^{17} m^3 with discharges of 10^6 to 10^9 m^3/s.

Flood volumes and peak discharges resulting from dam failures are not limited by precipitation rates or the time required for flow to be concentrated into a channel, but primarily by volume of water stored behind the impoundment and the condition of critical flow through the breach [*Walder and O'Connor*, 1997]. Volumes of existing natural impoundments range up to the $2.3 \cdot 10^{13}$ of water in Lake Baikal and similarly sized freshwater lakes have been cataclysmically released during the Pleistocene (Table 3). Even larger flood volumes have likely been released in more ancient geologic settings, including the circa 10^{16} m^3 of water that spilled into the Mediterranean basin at the end of the Miocene Period. Peak discharge for most breached dams is regulated by the condition of critical flow at the breach. For critical flow through a rectangular cross section,

$$Q = 0.54g^{1/2}B \cdot H^{3/2}, \quad (12)$$

where g is the acceleration of gravity (9.8 m/s^2), B is the channel width in meters, and H is the difference in elevation in meters between the impounded water surface (away from the zone of drawdown influence) and the bottom of the outlet channel (after *King* [1954, p. 8.8-8.11]). Other cross-section shapes yield slightly different coefficients but overall results of the same magnitude.

The largest known natural-dam failures have been the 500 to 900 m breaches of Pleistocene ice dams (Table 3). Thus, a blockage 1,000 m high might be an approximate upper limit for terrestrial impoundments subject to failure. If such a blockage were to fail completely with an outlet channel width five times its depth, the maximum discharge

Figure 9. Selected dam-failure and meteorologic floods (from Table 3) in discharge-duration-volume space, showing physical limits to flood magnitude. A) Shaded planes represent physical limits to meteorologic floods, resulting from continuity, time of concentration, and global maximum precipitation rates. All meteorological floods fall within these limits, but many dam-failure floods do not. B) Shaded planes represent physical limits for dam-failure floods, resulting from continuity and critical flow through a maximum plausible terrestrial dam breach. These limits enclose all types of floods.

would be about 10^8 m^3/s, approximately five times greater than the largest documented terrestrial floods (Figure 9b). This limit, combined with the limit forced by continuity [expressed by equation (3)], defines a much larger permissible space (within axes defined by discharge, volume, and duration) for dam-failure floods than exists for meteorological floods (Figure 9b). This result is borne out by the observation that several dam-failure floods plot within this space, but outside the limits of potential meteorological floods.

From empirical and theoretical considerations, both meteorological and dam-failure floods can potentially involve volumes of 10^{15} to 10^{17} m^3 of water with discharges of 10^6 to 10^8 m^3/s. However, meteorological floods with volumes greater than 10^{13} m^3 and discharges greater than 10^6 m^3/s probably require landscape conditions much different from those that have existed throughout most of the Phanerozoic. In contrast, floods from dam failures have achieved discharges of 10^6 to 10^8 m^3/s during the Quaternary, although the largest of these dam failure floods have been associated with altered hydrologic systems during Pleistocene ice ages.

Paleohydrologic research, aided by the tools of stratigraphy, geochronology, and hydrologic and hydraulic analysis, is uniquely poised to investigate floods that span these ranges of sizes and timescales. Nevertheless, the primary concepts brought forward by this paper—that there are likely to be quantifiable limits to floods, that specific arrangements of physiography and climatology can produce especially large floods from a variety of mechanisms, and that certain geologic conditions can foster more and larger floods than others—is not particularly new. Indeed, subplots embedded within these overarching precepts have motivated most of the paleohydrologic research of the last three decades, including much of the research reported in this volume. The framework established here, however, should serve as a rational basis for defining the context of specific floods or flood environments within the changing constellations of geology, physiography, and climatology that produce large riverine floods over the course of Earth history.

Acknowledgments. Discussions with P.K. House, V.R. Baker, and J.A. Smith aided in development of the ideas presented here. T. Haluska, L. Orzol, and R. Viger assisted with the compilation and analysis of the U.S. Geological Survey flow data. Reviews by E.E. Wohl, P.K. House, A. Gupta, and J. Williams improved the final manuscript.

REFERENCES

Anderson, D.E., Late Quaternary paleohydrology, lacustrine stratigraphy, fluvial geomorphology and modern hydroclimatology of the Amargosa River / Death Valley hydrologic system, California and Nevada, Ph.D. Dissertation, 521 pp., University of California, Riverside, 1998.

Baker, V.R., Paleohydrology and sedimentology of Lake Missoula flooding in eastern Washington, *Geol. Soc. Amer. Spec. Pap. 144*, 79 pp., 1973.

Baker, V.R. Stream-channel response to floods, with examples from central Texas, *Geol. Soc. Amer. Bull., 88*, 1057-1071, 1977.

Baker, V.R., Late Pleistocene fluvial systems, in Volume 1, The Late Pleistocene, edited by S.C. Porter, in *Late-Quaternary Environments of the United States*, edited by H.E. Wright, Jr., pp. 115-129, University of Minnesota Press, Minneapolis, Minnesota, 1983.

Baker, V.R., G. Benito and A.N. Rudoy, Paleohydrology of late Pleistocene superflooding, Altay Mountains, Siberia, *Science*, 259, 348-350, 1993.

Baker, V.R., and R.C. Kochel, Flood sedimentation in bedrock fluvial systems, in *Flood Geomorphology*, edited by V.R. Baker, R.C. Kochel and P.C. Patton, pp. 123-138, John Wiley and Sons, New York, 1988.

Barry, R.G. and R.J. Chorley, *Atmosphere, Weather and Climate*, 392 pp., Routledge, New York, 1992.

Baumgartner, A. and E. Reichel, *The World Water Balance; Mean Annual Global, Continental and Marine Precipitation, Evaporation and Runoff*, 179 pp., Elsevier, New York, 1975.

Blown, I.G. and M. Church, Catastrophic lake drainage within the Homathko River basin, British Columbia, *Can. Geotech. Jour., 22*, 551-563, 1985.

Bretz, J H., The Channeled Scabland of the Columbia Plateau, *J. Geol., 31*, 617-649, 1923.

Bretz, J H., The Lake Missoula floods and the Channeled Scabland, *J. Geol., 77*, 505-543, 1969.

Bradley, A.A. and J.A. Smith, The hydrometeorological environment of extreme rainstorms in the Southern Plains of the United States, *J. Applied Met., 33*, 14-18, 1994.

Cayan, D.R. and D.H. Peterson, The influence of North Pacific atmospheric circulation on streamflow in the west, in *Aspects of Climate Variability in the Pacific and the Western Americas, Amer. Geophys. Union Mono. 55*, edited by D.H. Peterson, pp. 375-398, 1989.

Chitwood, L.A. and R.A. Jensen, Large prehistoric flood along Paulina Creek, in *What's New at Newberry Volcano, Oregon*, edited by R.A. Jensen and L.A. Chitwood, pp. 31-40, U.S.D.A. Forest Service, Bend, Oregon, 2000.

Church, M., Floods in cold climates, in *Flood Geomorphology*, edited by V.R. Baker, R.C. Kochel and P.C. Patton, pp. 205-230, John Wiley and Sons, New York, 1988.

Clague, J.J. and S.G. Evans, Formation and failure of natural dams in the Canadian Cordillera, *Geol. Surv. Can. Bull. 464*, 35 pp., 1994.

Conaway, J.S., Hydrogeology and paleohydrology of the Williamson River basin, Klamath County, Oregon, M.S. thesis, 119 pp., Portland State University, Portland, Oregon, 1999.

Costa, J.E., A comparison of the largest rainfall-runoff floods in the United States with those of the People's Republic of China and the world, *J. Hydrol., 96*, 101-115, 1987.

Costa, J.E., Floods from dam failures, in *Flood Geomorphology*, edited by V.R. Baker, R.C. Kochel and P.C. Patton, pp. 439-464, John Wiley and Sons, New York, 1988.

Costa, J.E. and J.E. O'Connor, Geomorphically effective floods, in *Natural and Anthropogenic Influences in Fluvial Geomorphology, Amer. Geophys. Union Mono. 89,* edited by J.E. Costa, A.J. Miller, K.W. Potter and P.R. Wilcock, pp. 45-56, 1995.

Costa, J.E. and R.L. Schuster, The formation and failure of natural dams, *Geol. Soc. Amer. Bull.*, *100*, 1054-1068, 1988.

Costa, J.E. and R.L. Schuster, Documented historical landslide dams from around the world, *U.S. Geol. Surv. Open-File Rep. 91-239*, 485 pp., 1991.

Cotton, W.R. and R.A. Anthes, *Storm and Cloud Dynamics*, 883 pp., Academic Press, San Diego, California, 1989.

Crippen, J.R. and C.D. Bue, Maximum floodflows in the conterminous United States, *U.S. Geol. Surv. Water-Supply Pap. 1887*, 52 pp., 1977.

Currey, D.R., Quaternary palaeolakes in the evolution of semidesert basins, with special emphasis on Lake Bonneville and the Great Basin, U.S.A., *Palaeogeography, Palaeoclimatology, and Palaeoecology*, *76*, 189-214, 1990.

Dettinger, M. and D. Cayan, The Pacific Decadal Oscillation and flood frequencies in the United States, *Geol. Soc. Amer. Abstr. with Prog.*, *32*(7), 460, 2000.

Ely, L.L., Y. Enzel, V.R. Baker and D.R. Cayan, A 5000-year record of extreme floods and climate change in the southwestern United State, *Science*, *262*, 410-412, 1993.

Enzel, Y., L.L. Ely, P.K. House, V.R. Baker, and R.H. Webb, Paleoflood evidence for a natural upper bound to flood magnitudes in the Colorado River basin, *Water Resour. Res.*, *29*, 2287-2297, 1993.

Evans, S.G., The maximum discharge of outburst floods caused by the breaching of man-made and natural dams, *Can. Geotech. Jour.*, *23*, 385-387, 1986.

Evans, S.G., The breaching of moraine-dammed lakes in the southern Canadian Cordillera, in *Proceedings of the International Symposium on Engineering Geological Environment in Mountainous Areas*, pp. 141-150, Beijing, China, v. 2, 1987.

Gilbert, G.K., Lake Bonneville, *U.S. Geol. Surv. Mono. 1*, 438 pp., 1890.

Glazyrin, G.Y. and V.N. Reyzvikh, Computation of the flow hydrograph for the breach of landslide lakes, *Soviet Hydrology*, Selected Papers (published by the American Geophysical Union), Issue no. 5, 492-496, 1968

Gupta, A., Large floods as geomorphic events in the humid tropics, in *Flood Geomorphology*, edited by V.R. Baker, R.C. Kochel and P.C. Patton, pp. 301-315, John Wiley and Sons, New York, 1988.

Hamblin, W.K., Late Cenozoic lava dams in the western Grand Canyon, *Geol. Soc. Amer. Mem. 183*, 139 pp., 1994

Hayden, B.P., Flood climates, in *Flood Geomorphology*, edited by V.R. Baker, R.C. Kochel and P.C. Patton, pp. 13-26, John Wiley and Sons, New York, 1988.

Hewitt, K., Record of natural damming and related floods in the Upper Indus Basin, *Indus*, *10*(3), 11-19, 1968.

Hewitt, K., 1998, Catastrophic landslides and their effects on the Upper Indus streams, Karakoram Himalaya, northern Pakistan, *Geomorphology*, *26*, 47-80, 1998.

Hirschboeck, K.K., Hydroclimatology of flow events in the Gila river basin, central and southern Arizona, Ph.D. Dissertation, 354 pp., University of Arizona, Tucson, Arizona, 1985.

Hirschboeck, K.K., Flood hydroclimatology, in *Flood Geomorphology*, edited by V.R. Baker, R.C. Kochel and P.C. Patton, pp. 27-49, John Wiley and Sons, New York, 1988.

Hirschboeck, K.K., Climate and floods, in *National Water Summary 1988-89, Floods and Droughts, U.S. Geol. Surv. Water-Supply Paper 2375*, R.W. Paulsen, E.B. Chase, R.S. Roberts and D.W. Moody, compilers., pp. 67-88, 1991.

Hirschboeck, K.K, Overlapping scales and the atmospheric causes of floods, *Geol. Soc. Amer. Abst. with Prog.*, *32*(7), 460, 2000.

House, P.K., and P.A. Pearthree, A geomorphologic and hydrologic evaluation of an extraordinary flood discharge estimate: Bronco Creek, Arizona, *Water Resour. Res.*, *31*, 3059-3073, 1995.

Hoyt, W.G. and W.B. Langbein, *Floods*, 469 pp., Princeton University Press, Princeton, New Jersey, 1955.

Hsü, K.J., Origin of saline giants: a critical review after the discovery of the Mediterranean evaporite, *Earth-Sci. Rev.*, *8*, 371-396, 1972.

Hsü, K.J., *The Mediterranean was a Desert; a Voyage of the Glomar Challenger*, 197 pp., Princeton University Press, Princeton, New Jersey, 1983.

Jarrett, R.D. and J.E. Costa, Hydrology, geomorphology, and dam-break modeling of the July 15, 1982, Lawn Lake Dam and Cascade Dam failures, Larimer County, Colorado, *U.S. Geol. Surv. Prof. Pap. 1369*, 78 pp., 1986.

Jarrett, R.D. and H.E. Malde, Paleodischarge of the late Pleistocene Bonneville Flood, Snake River, Idaho, computed from new evidence, *Geol. Soc. Amer. Bull.*, *99*, 127-134, 1987.

Jennings, M.E., V.R. Schneider and P.E. Smith, Computer assessments of potential flood hazards from breaching of two debris dams, Toutle River and Cowlitz River systems, in *U.S. Geol. Surv. Prof. Pap. 1250*, pp. 829-836, 1981.

Jennings, D.N., M.G. Webby and D.T. Parkin, Tunawaea landslide dam, King Country, New Zealand, *Landslide News*, *7*, 25-27, 1993.

Jones, M.A., J.L. Jones and T.D. Olsen, Ground-water flooding in glacial terrain of southern Puget Sound, Washington, *U.S. Geol. Survey Fact Sheet FS-111-00*, 4 pp., 2000.

Jonnson, J., Notes on the Katla volcanoglacial debris flows, *Jökull*, *32*, 61-68, 1982.

Kehew, A.E. and M.L. Lord, Glacial-lake outburst along the mid-continent margins of the Laurentide ice-sheet, in *Catastrophic Flooding*, edited by L. Mayer and D. Nash, pp. 95-120, Allen and Unwin, Boston, Massachusetts, 1987.

King, H.W., *Handbook of Hydraulics (fourth edition)*, uneven pagination, McGraw-Hill, New York, 1954.

Knox, J.C., Large increases in flood magnitude in response to modest changes in climate, *Nature*, *361*, 430-432, 1993.

Kochel, R.C., and V.R. Baker, Paleoflood hydrology, *Science*, *215*, 353-361, 1982.

Kochel, R.C., V. R. Baker, and P.C. Patton, Paleohydrology of southwestern Texas, *Water Resour. Res.*, *18*, 1165-1183, 1982.

Langbein, W.B. and others (unidentified), Annual runoff in the United States, *U.S. Geol. Surv. Circ. 52*, 14 pp., 1949.

Larimer, O.J., Flood of June 9-10, 1972, at Rapid City, South Dakota, *U.S. Geol. Surv. Hydro. Inv. Atlas HA-511*, 1 sheet, 1973.

Leopold, L.B., Areal extent of intense rainfalls, New Mexico and Arizona, *Amer. Geophys. Union Trans.*, *32*, 347-357, 1942.

Lepkin, W.D. and M.M. DeLapp, Peak-flow file retrieval (program 1980), *U.S. Geol. Surv. Open-File Rep. 79-1336-I*, 64 pp. 1979.

Lliboutry, L., V.M. Arnao, A. Pautre and B. Schneider, Glaciological problems set by the control of dangerous lakes in Cordillera Blanca, Peru: I. Historical failures of morainic dams, their causes and preventation, *J. Glaciology*, *18*, 239-254, 1977.

Lord, M.L., and A.E. Kehew, Sedimentology and paleohydrology of glacial-lake outburst deposits in southeastern Saskatchewan and northwestern North Dakota, *Geol. Soc. Amer. Bull.*, *99*, 663-673, 1987.

Ludlam, F.H., Clouds and storms: *The Behavior and Effect of Water in the Atmosphere*, 405 pp., The Pennsylvania State University Press, Philadelphia, Pennsylvania, 1980.

MacDonald, T.C. and J. Langridge-Monopolis, Breaching characteristics of dam failures, *J. Hydraul. Eng.*, *110*, 567-586, 1984.

Malde, H.E., The catastrophic late Pleistocene Bonneville Flood in the Snake River Plain, Idaho, *U.S. Geol. Surv. Prof. Pap. 596*, 52 pp., 1968.

Manville, V., J.D.L. White, B.F. Houghton and C.J.N. Wilson, Paleohydrology and sedimentology of a post-1.8 ka breakout flood from intracaldera Lake Taupo, North Island, New Zealand, *Geol. Soc. Amer. Bull.*, *111*, 1435-1447, 1999.

Matsch, C.L., River Warren, the southern outlet of glacial Lake Agassiz, in *Geol. Assoc. Can. Spec. Pap. 26*, pp. 231-244, 1983.

Mayo, L.R., Advance of Hubbard Glacier and closure of Russell Fiord, Alaska—Environmental effects and hazards in the Yakutat area, in *U.S. Geol. Surv. Circ. 1016*, pp. 4-16, 1988.

Mayo, L.R., Advance of Hubbard Glacier and 1986 outburst of Russell Fiord, Alaska, U.S.A., *J. Glaciology*, *13*, 189-194, 1989.

McCain, J.F., L.R. Hoxit, R.A. Maddox, C.F. Chappell and F. Caracena, Storm and flood of July 31-August 1, 1976, in the Big Thompson River and Cache la Poudre River Basins, Larimer and Weld Counties, Colorado: Part A. Meteorology and hydrology in Big Thompson River and Cache la Poudre River Basins, in *U.S. Geol. Surv. Prof. Pap. 1115*, pp. 1-85, 1979.

Michaud, J.D., K.K. Hirschboeck, and M. Winchell, Regional variations in small-basin floods in the United States, *Water Resour. Res.*, *37*, 1405-1416, 2001.

Miller, A.J., Flood hydrology and geomorphic effectiveness in the central Appalachians, *Earth Surf. Proc. Landf.*, *15*, 119-134, 1990.

Mora, S., C. Madrigal, J. Estrada and R.L. Schuster, The 1992 Rio Toro Landslide Dam, Costa Rica, *Landslide News*, *7*, 19-22, 1993.

More, R.J., Hydrological models and geography, in *Models in Geography*, edited by R.J. Chorley and P. Haggett, pp. 145-185, Methuen, London, 1967.

O'Connor, J.E. and V.R. Baker, Magnitudes and implications of peak discharges from glacial Lake Missoula, *Geol. Soc. Amer. Bull.*, *104*, 267-279, 1992.

O'Connor, J.E., Hydrology, hydraulics, and geomorphology of the Bonneville Flood, *Geol. Soc. Amer. Spec. Pap. 274*, 83 pp., 1993.

O'Connor, J.E., L.L. Ely, E.E, Wohl, L.E. Stevens, T.S. Melis, V.S. Kale, and V.R. Baker, A 4,500-year record of large floods on the Colorado River in the Grand Canyon, Arizona, *J. Geol.*, *102*, 1-9, 1994.

O'Connor, J. E., T.C. Pierson, D Turner, B.F. Atwater, and P.T. Pringle, An exceptionally large Columbia River flood between 500 and 600 years ago–Breaching of the Bridge-of-the-Gods Landslide?, *Geol. Soc. Amer. Abstr. Prog.*, *28*(7), 1996.

O'Connor, J.E. and G.E. Grant, Floods, floods, and floods on the Deschutes River, Oregon, in *Proc. of the Second Intern. Paleoflood Conf.*, edited by P.K. House, Prescott, Arizona, p. 35, 1999.

O'Connor, J.E., J.H. Hardison, III, and J.E. Costa, Debris flows from failures of Neoglacial moraine dams in the Three Sisters and Mt. Jefferson Wilderness areas, Oregon, *U.S. Geol. Surv. Prof. Pap. 1608*, 2001 (in press).

Oltman, R.E., H.O. Sternberg, F.C. Ames and L.C. Davis, Jr., Amazon River investigations reconnaissance measurements of July 1963, *U.S. Geol. Surv. Circ. 486*, 15 p., 1964.

Osterkamp, W.R. and J.E. Costa, Changes accompanying an extraordinary flood on a sand-bed stream, in *Catastrophic Flooding*, edited by L. Mayer and D. Nash, pp. 201-224, Allen and Unwin, Boston, Massachusetts, 1987.

Pena, H. and W. Klohn, Non-meteorological flood disasters in Chile, in *Hydrology of Disasters*, edited by Ö. Starosolszky and O.M. Melder, pp. 243-258, James and James, London, 1989.

Plaza-Nieto, G., Y. Yepes and R.L. Schuster, Landslide dam on the Pisque River, northern Ecuador, *Landslide News*, *4*, 2-4, 1990.

Potter, P.E., Significance and origin of big rivers, *J. Geol.*, *86*, 13-33, 1978.

Risley, J., D. Marks and T. Link, Application of a quasi-energy balance model to simulate snowmelt under variable canopies during a major rain-on-snow event, *EOS, Trans. Am. Geophys. Union (1997 Fall Meeting Supplement)*, *78*(46), 320, 1997.

Rodier, J.A. and M. Roche, World catalogue of maximum observed floods, *Intern. Assoc. Hydrol. Sci. Publ. No. 143*, 354 p., 1984.

Rudoy, A., Mountain ice-dammed lakes of southern Siberia and their influence on the development and regime of the intracontinental runoff systems of North Asia in the late Pleistocene, in *Palaeohydrology and Environmental Change*, edited by G. Benito, V.R. Baker and K.J. Gregory, pp. 215-234, John Wiley and Sons, Chichester, 1998.

Rudoy, A.N. and V.R. Baker, Sedimentary effects of cataclysmic Late Pleistocene glacial outburst flooding, Altay Mountains, Siberia, *Sed. Geol.*, *85*, 53-62, 1993.

Ryan, W and W. Pitman, Noah's flood: *The New Scientific*

Discoveries about the Event that Changed History, 319 pp., Simon and Schuster, New York, 1999.

Sager, J.W. and D.R. Chambers, Design and construction of the Spirit Lake outlet tunnel, Mount St. Helens, Washington, in *Landslide Dams—Processes, Risk, and Mitigation, Amer. Soc. of Civil Eng. Spec. Publ. No. 3*, edited by R.L. Schuster, pp. 42-58, 1986.

Schumm, S.A., Speculations concerning paleohydrologic controls of terrestrial sedimentation, *Geol. Soc. Amer. Bull.*, *79*, 1573-1588, 1968.

Schuster, R.L., A worldwide perspective on landslide dams, in *Usoi Landslide Dam and Lake Sarez–An Assessment of Hazard and Risk in the Pamir Mountains, Tajikistan, International Strategy for Disaster Reduction Series No. 1*, edited by D. Alford and R. Schuster, pp. 19-22, United Nations, New York, 2000.

Schuster, R.L. and J.E. Costa, A perspective on landslide dams, in *Landslide Dams—Processes, Risk, and Mitigation, Amer. Soc. of Civil Eng. Spec. Publ. No. 3*, edited by R.L. Schuster, pp. 1-20, 1986.

Scott, K.M., Magnitude and frequency of lahars and lahar-runout flows in the Toutle-Cowlitz River system, *U.S. Geol. Surv. Prof. Pap. 1447-B*, 33 pp. 1989.

Shroder, J.F. Jr., K. Cornwell and M.S. Khan, Catastrophic breakout floods in the western Himalaya, Pakistan, *Geol. Soc. Amer. Abstr. Prog.*, *23*(5), 87, 1991.

Smith, D.G. and T.G. Fisher, Glacial Lake Agassiz: the northwestern outlet and paleoflood, *Geology*, 21, 9-12, 1993.

Snow, D.T., Landslide of Cerro Condor-Seneca, Department of Ayacucho, Peru, in *Geol. Soc. Amer. Eng. Geol. Case Histories, no. 5*, pp. 1-6, 1964

Stedinger, J.R., and T.A. Cohn, Flood frequency analysis with historical and paleoflood information, *Water Resour. Res.*, *22*, 785-793, 1986.

Teller, J.T., and L.H. Thorleifson, The Lake Agassiz-Lake Superior connection, in *Catastrophic Flooding*, edited by L. Mayer and D. Nash, pp. 121-138, Allen and Unwin, Boston, Massachusetts, 1987.

Teller, J.T., and A.E. Kehew, Introduction to the late glacial history of large proglacial lakes and meltwater runoff along the Laurentide Ice Sheet, *Quat. Sci. Rev.*, *13*, 795-799, 1994.

Thorson, R.M., Late Quaternary paleofloods along the Porcupine River, Alaska: Implications for regional correlation, in *U.S. Geol. Surv. Circ. 1026*, pp. 51-54, 1989.

U.S. Army Corps of Engineers, *Water Resource Development, Columbia River Basin*, 396 pp., U.S. Army Engineer Division, North Pacific, 1958.

van der Leeden, F., F.L. Troise and D.K. Todd, *The Water Encyclopedia*, 808 pp., Lewis Publishers, Chesea, Minnesota, 1990.

Vaughn, D., and D.W. Ash, Paleohydrology and geomorphology of selected reaches of the upper Wabash River, Indiana, *Geol. Soc. Amer. Abstr. Prog.*, *15*(6), 711, 1983.

Vörösmarty, C.J., B.M. Fekete, M. Meybeck and R.B. Lammers, Geomorphometric attributes of the global system of rivers at 30-minute spatial resolution, *J. Hydrol.*, *237*, 17-39, 2000.

Waananen, A.O., D.D. Harris and R.C. Williams, Floods of December 1964 and January 1965 in the far western states—Part 2. Streamflow and sediment data, *U.S. Geol. Surv. Water-Supply Pap. 1866-B*, 861 pp., 1970.

Waitt, R.B., Great Holocene floods along Jökulsá á Fjöllum, North Iceland, in *Flood and Megaflood Processes, Recent and Ancient Examples, Intern. Assoc. Sedimentologists Spec. Publ.*, edited by I. P. Martini, V.R. Baker and G. Garzón, 2001 (in press).

Walder, J.S. and J.E. Costa, Outburst floods from glacier-dammed lakes: The effect of mode of lake drainage on flood magnitude, *Earth Surf. Proc. Landf.*, *21*, 701-723, 1996.

Walder, J.S. and J.E. O'Connor, Methods for predicting peak discharge of floods caused by failure of natural and constructed earthen dams, *Water Resour. Res.*, *33*, 2337-2348, 1997.

Waythomas, C.R., J.S. Walder, R.G. McGimsey and C.A. Neal, A catastrophic flood caused by drainage of a caldera lake at Aniakchak volcano, Alaska, and implications for volcanic-hazards assessment, *Geol. Soc. Amer. Bull.*, *108*, 861-871, 1996.

Webb, R.H., J.E. O'Connor, and V.R. Baker, Paleohydrologic reconstruction of flood frequency on the Escalante River, south-central Utah, in *Flood Geomorphology*, edited by V.R. Baker, R.C. Kochel and P.C. Patton, pp. 403-418, John Wiley and Sons, New York, 1988.

Webby, M.G. and D.N. Jennings, Analysis of dam break flood caused by failure of Tunawaea landslide dam, in *Proc. Intern. Conf. Hydraulics in Civil Eng.*, University of Queensland, Brisbane, Australia, pp. 163-168, 1994.

Wolman, M.G., and J.E. Costa, Envelope curves for extreme flood events: Discussion, *J. Hydraul. Eng.*, *110*, 77-78, 1984.

World Meteorological Organization, *Manual for Estimation of Probable Maximum Precipitation, World Met. Org. Operational Hydrology Report No. 1 (second edition)*, 269 pp., Secretariat of the World Meteorological Organization, Geneva, Switzerland, 1986.

J.E. Costa, U.S. Geological Survey, 10615 S.E. Cherry Blossom Dr., Portland OR, 97216.

G.E. Grant, USDA Forest Service, Pacific Northwest Research Station, 3200 Jefferson Way, Corvallis, OR 97331.

J.E. O'Connor, U.S. Geological Survey, 10615 S.E. Cherry Blossom Dr., Portland OR, 97216. (oconnor@usgs.gov)